EPS Grand Challenges

Physics for Society in the Horizon 2050

Online at: https://doi.org/10.1088/978-0-7503-6342-6

EPS Grand Challenges

Physics for Society in the Horizon 2050

Edited by
Carlos Hidalgo
CIEMAT, Laboratorio Nacional de Fusión, Spain

Coordinators
Ralph Assmann
Deutsches Elektronen-Synchrotron (DESY), Germany
and
GSI Helmholtzzentrum für Schwerionenforschung, Darmstadt, Germany

Felicia Barbato
Gran Sasso Science Institute, Italy

Christian Beck
Queen Mary University of London, London, UK

Giulio Cerullo
Politecnico di Milano, Milan, Italy

Luisa Cifarelli
University of Bologna, Bologna, Italy

Felix Ritort
University of Barcelona, Barcelona, Spain

Christophe Rossel
IBM Research Europe - Zurich, Switzerland

Mairi Sakellariadou
King's College London, London, UK

Kees van Der Beek
Institut Polytechnique de Paris, Paris, France

Luc van Dyck
Euro Argo ERIC, Plouzané, France

Bart van Tiggelen
*Laboratoire de Physique et Modélisation des Milieux Condensés,
University Grenoble Alpes/CNRS, Grenoble, France*

Claudia-Elisabeth Wulz
Institute of High Energy Physics, Austrian Academy of Sciences, Vienna, Austria

IOP Publishing, Bristol, UK

ISBN   978-0-7503-6342-6 (ebook)
ISBN   978-0-7503-6338-9 (print)
ISBN   978-0-7503-6339-6 (myPrint)
ISBN   978-0-7503-6341-9 (mobi)

DOI   10.1088/978-0-7503-6342-6

Version: 20240301

IOP ebooks

British Library Cataloguing-in-Publication Data: A catalogue record for this book is available from the British Library.

Published by IOP Publishing, wholly owned by The Institute of Physics, London

IOP Publishing, No.2 The Distillery, Glassfields, Avon Street, Bristol, BS2 0GR, UK

US Office: IOP Publishing, Inc., 190 North Independence Mall West, Suite 601, Philadelphia, PA 19106, USA

Contents

4   Physics for understanding life   **4-1**

*Patricia Bassereau, Angelo Cangelosi, Jitka Čejková, Carlos Gershenson,
Raymond Goldstein, Zita Martins, Sarah Matthews, Felix Ritort, Sara Seager,
Bart Van Tiggelen, David Vernon and Frances Westall*

6 Physics for the environment and sustainable development

Ankit Agarwal, Philippe Azais, Luisa Cifarelli, Jacob de Boer, Deniz Eroglu,
Gérard Gebel, Carlos Hidalgo, Didier Jamet, Juergen Kurths, Florence
Lefebvre-Joud, Søren Linderoth, Alberto Loarte, Norbert Marwan,
Natalio Mingo, Ugur Ozturk, Simon Perraud, Robert Pitz-Paal,
Stefaan Poedts, Thierry Priem, Bernd Rech, Marco Ripani,
Shubham Sharma, Tuan Quoc Tran and Hermann-Josef Wagner

6-1

8   Science for society   8-1

Tobias Beuchert, Alan Cayless, Frédéric Darbellay, Richard Dawid,
Bengt Gustafsson, Sally Jordan, Philip Macnaghten, Eilish McLoughlin,
Christophe Rossel, Pedro Russo, Luc van Dyck, François Weiss
and Ulrich von Weizsäcker

Preface

Over the last decades, the European Physical Society (EPS) has raised concerns on some of the main problems humankind faced up to now and examined the role of physics to address them. But what are the urgent societal challenges in the future and what holds the world's physics agenda to solve them? One objective of the *EPS Grand Challenges project* is indeed to explore our ability in imagining and shaping the development of physics at the Horizon 2050.

The scientific committee of this *Grand Challenges* publication chaired by Carlos Hidalgo, includes the following chapter coordinators: Ralph Assmann, Felicia Barbato, Christian Beck, Kees van der Beek, Giulio Cerullo, Luisa Cifarelli, Luc van Dyck, Felix Ritort, Christophe Rossel, Mairi Sakellariadou, Bart van Tiggelen, Claudia-Elisabeth Wulz. They played the leading role in the development of the project that addresses two pillars: *Physics as global human enterprise for understanding nature* and *Physics developments to tackle major issues affecting the lives of citizens*. The essays, written by more than 70 leading scientists, outline the in-depth analysis of the strong links between basic research, its applications and their impact on a sustainable society. The interplay of natural sciences with social and human sciences is also discussed together with the role of open science, education, ethics, and responsible citizens in an interdisciplinarity environment.

The scientific committee would like to express its gratitude to the Editorial Board members[1] for their contribution in defining the structure and the topical content of this project. The constant support of the successive EPS Presidents (Luc Bergé, Petra Rudolf, Rüdiger Voss, Christophe Rossel), the EPS Secretaries General (Anne C Pawsey, David Lee) and the EPS secretariat during the whole process is deeply acknowledged. Finally, the excellent work of the editorial team of *Europhysics News* [EPN] in preparing the special issue on the *EPS Grand Challenges* [1] is also well recognised.

Reference

[1] EUROPHYSICS NEWS 2023 Special issue the EPS Grand Challenges for physics *EUROPHYSICS NEWS* **53** 5

---

[1] Members Editorial Board: Tariq Ali, Hans-Peter Beck, Els de Wolf, Karl Gradin, Giuseppe Grosso, Eduard Kontar, Eva Kovacevic, Marko Kralj, Douglas Mac Gregor, Eugenio Nappi, JoeNiemela, José Antonio Paixao, Kaido Reivelt, Sylvie Rousset, Petra Rudolf, Enrique Sánchez, Miguel Angel Sanchis, David Sands, Rüdiger Voss, and Victor Zamfir.

Editor

Carlos Hidalgo

Carlos Hidalgo received his PhD degree from Madrid Complutense University with his work on structural defects in solids and positron annihilation spectroscopy. His next area of research was related to plasma turbulence, transport and plasma diagnostics at CIEMAT where he is currently leading the Spanish National Fusion Laboratory.

Carlos has worked in different international laboratories, initially as a PhD student [Technical University of Denmark, Nuclear Research Centre of Grenoble, Technical University of Helsinki] and later as visiting scientist [Fusion Centre at the University of Austin (US), Oak Ridge National Laboratory (US), Joint European Torus (UK), Max Planck Institute (Germany), National Institute Fusion Studies (Japan), Southwestern Institute of Physics (SWIP, China)].

He has led different research teams in the framework of the International Fusion Programme and participated in European and International Advisory Committees on Fusion Science and Technology. He has chaired the Division of Plasma Physics and the Forum on Physics and Society of the European Physical Society.

Coordinators

Ralph Assmann

Ralph Assmann has obtained his doctorate in physics from the Ludwig-Maximilians-University in Munich. His PhD research was performed at the Max Planck Institute for Physics in Munich and at CERN in the ALEPH experiment on the mass of the Z boson, spin polarized particle beams and precise energy calibration. He then spent almost four years as research associate and staff at Stanford University and SLAC, where he worked on operation, modelling and design of colliders. For the next 15 years he worked at CERN in leading roles on the LEP and LHC colliders. He was LHC machine coordinator in run I of the LHC operation, during which the Higgs boson was discovered in 2012. From 2012 to 2023 he worked as Leading Scientist for Accelerator R&D at DESY, where he researched new, compact accelerators. He was awarded an ERC synergy grant in 2014. Over 10 years, until 2024, Dr Assmann has been the proposer and founding coordinator of the EuPRAXIA ESFRI project, a 569 M€ project on building the world-wide first user facility based on plasma-based accelerators, today supported by more than 50 institutes. He has been the Chair of the Accelerator Group in the European Physical Society from 2020–2023, the leader of several large Helmholtz and European funding grants and the coordinator of the European Network for Novel Accelerators. Presently, Dr Assmann is the Head of Accelerator Operation and Development at GSI in Germany, responsible for GSI's heavy ion beams and the future beam commissioning of the FAIR accelerators.

Felicia Barbato

Felicia Barbato is proposer and principal investigator of the project Crystal Eye: a wide sight to the Universe looking for the electromagnetic counterpart of gravitational waves, a satellite detector for gamma ray astronomy. Thanks to her interest in cosmic rays physics and space experiment she recently joined the HERD and DAMPE collaborations. She is now detector responsible for the Zirè payload of the NUSES space mission.

Her missions as researcher are innovation and education. With these aims in 2015 she joined the Physics and Optics Naples Young Students, a group of students and young researchers dedicated to the diffusion of the scientific culture, and became member of the European Physical Society (EPS) joining the Young Minds Project, giving many contributions for the diffusion of astroparticle physics and technology. This passion for the diffusion of the scientific culture together with the passion for technology, led her to be part of the management board of the EPS Technology and Innovation Group for some years.

Christian Beck

Christian Beck is a Professor at Queen Mary University of London, UK. He is Head of the Dynamical Systems and Statistical Physics Group and a Fellow of the Alan Turing Institute (the UK's national institute for data science and artificial intelligence). Currently he is the Chairman of the EPS Statistical and Nonlinear Physics Division.

Christian Beck got his PhD in Theoretical Physics at RWTH Aachen in Germany and spent some time as a postdoc in Warwick, Copenhagen, Budapest, and Maryland before joining the University of London. He has more than 150 journal publications and is the author of two books. His research interests are in the general area of statistical physics, dynamical systems, data-driven analysis and stochastic modelling of complex systems. Together with E G D Cohen from Rockefeller University, New York, he has developed so-called superstatistical techniques, which have become a well-known method to understand heterogeneous systems in non-equilibrium statistical physics. More recently he has also dealt with the physics of sustainable energy systems, frequency fluctuations in power grid networks, spatio-temporal patterns of air pollution concentrations and other environmental problems where methods from statistical physics can be applied.

Giulio Cerullo

Giulio Cerullo is a Full Professor with the Physics Department, where he leads the Ultrafast Optical Spectroscopy laboratory. He is a Fellow of the Optical Society of America, the European Physical Society, the Accademia dei Lincei and past Chair of the Quantum Electronics and Optics Division of the European Physical Society. He has participated in numerous European projects, including two that received ERC grants. He has co-founded three spin-offs, among which CRI and NIREOS. He is on the Editorial Advisory Board of the journals *Optica, Laser & Photonics Reviews, Scientific Reports* and *Journal of Raman Spectroscopy*. He has been General Chair of the conferences CLEO/Europe 2017, Ultrafast Phenomena 2018 and International Conference on Raman Spectroscopy 2022.

Luisa Cifarelli

Luisa Cifarelli was until 2022 Full Professor of Experimental Physics at the University of Bologna (Italy) where she is now Emeritus. Her research interests have mostly been in very high energy subnuclear physics in major European laboratories such as CERN, Geneva (Switzerland) and DESY, Hamburg (Germany). She has been a member of the CERN Council, of the INFN Board of Directors (Italy), of the Scientific Council of CNRS (France) and of JINR (Russia). She has been President of the Italian national research institute Centro Fermi, of the European Physical Society (EPS) and of the Italian Physical Society (SIF). She has been Chair of the Forum on

International Physics of the American Physical Society (APS), a member of the APS Committee on International Scientific Affairs and she is now APS International Councillor. She is member of the ALICE Collaboration at the CERN LHC and of the Scientific Advisory Committee of DIPC (Spain). She is a member of the Academia Europaea and of the Accademia delle Scienze dell'Istituto di Bologna. She has been a member of the Governing Board of the Bologna Academy and is currently a member of its Scientific Committee. She is the founder and director of the Joint EPS-SIF International School on Energy, which is held in Varenna (Italy) on a biennial basis.

Felix Ritort

Felix Ritort is full professor in condensed matter physics. He got his PhD in statistical physics in 1991 under the supervision of G Parisi and M Rubi. Until 2002 he made contributions to the field of disordered and nonequilibrium physics. Afterward, he started an experimental career in single-molecule biophysics to investigate energy processes at the molecular level. Ritort's group is recognized worldwide as a leader in applying the finest methods to extract quantitative information about thermodynamics and kinetics of molecular interactions. He has been awarded several prizes including the ICREA Academia Award 2008, 2013, 2018 and the Bruker Prize 2013 from Spanish Biophysical Society. He chairs the Division of Physics for Life Sciences of the European Physical Society.

His scientific research is highly multidisciplinary at the frontiers of physics, chemistry and biology. His lab is referenced worldwide on merging theory and experiments to investigate the thermodynamics and nonequilibrium behavior of small systems using single-molecule methods. Dr Ritort applies the finest concepts and tools from statistical physics to extract valuable information about a wide range of molecular processes: from the energetics of nucleic acids and proteins to the intermolecular binding kinetics in proteins, peptides and other macromolecular structures. A recurring theme in his research is the understanding of how molecular systems embedded in noisy thermal environments outperform the efficiency of macroscopic systems: being small has advantages that nature has exploited. Recently he has directed his interest to the study of energy and information and the search for principles that govern the emergent complexity of evolutionary ensembles in the molecular and cellular world.

Christophe Rossel

Christophe Rossel is a condensed matter physicist with education and academic professional experience in Switzerland (University of Neuchâtel and Geneva) and in the United States (Temple University, Philadelphia and University of California, San Diego). In 1987 he joined the IBM Research-Zurich Laboratory pursuing a scientific career focused on the physics of superconductors and later on nanoscience and the integration of advanced functional materials for semiconductor technology. As a member of various panels and president of the Swiss (SPS, 2008–12) and European Physical Societies (EPS, 2015–17) he has engaged in science policy issues, representing the community of physicists. A fellow of the EPS and of the Institute of Physics (IOP, UK) he was a member of the Open Science Policy Platform (OSPP, 2016–20) in Brussels and has been an executive board member of the Swiss Academy of Sciences (SCNAT) since 2018. He also chairs the EPS technology and Innovation group (TIG) as well as the working group on Physics and Industry of the International Union of Pure and Applied physics (IUPAP). He is emeritus senior researcher still affiliated with IBM.

Mairi Sakellariadou

Mairi Sakellariadou is a professor of Theoretical Physics at King's College London (University of London). She has studied Mathematics at the National and Kapodistrian University of Athens, Astrophysics at the University of Cambridge and obtained her Doctor of Philosophy in Physics at Tufts University (USA). She has worked at the Universities of Brussels, Tours, Pierre and Marie Curie (Sorbonne University), Zürich, Geneva and the theory division of CERN. She was also professor of General Relativity at the National and Kapodistrian University of Athens. She is a member of the LIGO Scientific Collaboration (LSC), the LISA Consortium, the Einstein Telescope Consortium, as well as the MoEDAL experiment at LHC (CERN). She chairs the Gravitational Physics Division (GPD) of the European Physical Society (EPS), and is a member of the Executive Committee of EPS and Editor-in-Chief of the International journal 'General Relativity and Gravitation' (Springer Nature). Her research covers various aspects of theoretical physics, early universe cosmology, classical and quantum gravity, particle physics, noncommutative geometry, as well as astrophysics. She has co-authored more than 300 papers published in international specialised journals and has given hundreds of invited talks in international conferences.

Kees van Der Beek

A condensed matter physicist and citizen of the world formerly working for the Centre National de la Recherche Scientifique (CNRS) in Palaiseau, **Kees van der Beek** is now Vice President for Research at the Institut Polytechnique de Paris and Vice-Provost for Research at Ecole polytechnique in Palaiseau, near Paris in France. His scientific interests cover superconductivity and its mechanisms, the ensemble of quantized flux vortices in type II superconductors, the physics of elastic manifolds in a random potential and other soft matter systems and more generally materials science. He has worked on irradiation damage on materials, and used large-scale equipment such as particle accelerators and synchrotrons. More recently, he has been interested in magnetic systems and in semiconductor physics, mainly in the light of topological insulators.

Before moving to France, he completed his secondary education in Singapore as well as in the Netherlands. He obtained his PhD from Leiden university in the Netherlands, graduating on the exotic properties of the vortex lattice in the then recently discovered cuprate high-temperature superconductors, after which he held post-doctoral fellowships at Argonne National Laboratory in the United States and at EPFL in Switzerland. He continued work on superconductivity at the Laboratory of Irradiated Solids of the Ecole polytechnique in France, of which he later became the head. Kees has had a long-term engagement in the French and European Physical Societies, and had the pleasure of successively chairing the condensed matter divisions of both. He was the head of the Physics of Light and matter department of the Paris Saclay University south of Paris, and one of the architects of the new Graduate School of Physics at Paris-Saclay. He is currently scientific delegate at the National Institute of Physics at CNRS.

Luc van Dyck

Luc van Dyck holds a degree in biochemical engineering and a PhD in the life sciences. After twelve years of research in Belgium and Germany, he moved to the private sector where he managed public–private collaborations and public funding in the research department of a global animal health company. From 2001 to 2011 he worked at the European Molecular Biology Laboratory (EMBL), where he served as executive coordinator of a platform of scientific organizations and learned societies, the Initiative for Science in Europe (ISE), involved in policy and advocacy at the European level. ISE is widely recognized for having been instrumental in the creation of the European Research Council (ERC). As a free-lance consultant, he worked for various organizations such as the OECD/Global Science Forum, EuroScience, the European Physical Society (EPS) and the AXA Research Fund. He now serves as Senior Advisor for Policy and Partnership Relations at the research infrastructure Euro-Argo ERIC.

Bart Van Tiggelen

Bart van Tiggelen got his master's degree in astronomy at the Leyden University and his PhD degree in physics and astrophysics at the University of Amsterdam. After some postdocs he settled in at Grenoble as a full research professor with CNRS and studied the propagation of all kinds of waves (seismic waves, acoustic and elastic waves, electromagnetic and matter waves) and is presently interested in magneto-optics and QED. He has been the scientific coordinator of many interdisciplinary programs in France, especially on the interface of physics and mathematics or biology. He is an engaged member of the French Physical Society and was the former Editor-in-Chief of EPL. Today he defends the new challenges in Open Science, especially towards younger generations.

Claudia-Elisabeth Wulz

 **Claudia-Elisabeth Wulz** studied Technical Physics at TU Vienna, where she obtained her PhD 'sub auspiciis', a special form of graduation under the auspices of the Federal President, who awards this highest possible distinction for academic achievements for a doctoral degree in Austria. During her studies she took up research at CERN, the world's largest laboratory for particle physics in Geneva. First as a summer student, and later also as a fellow, she worked in the group of Carlo Rubbia at the UA1 experiment, where she took part in the discovery of the W and Z bosons. As a member of the Institute of High Energy Physics of the Austrian Academy of Sciences she then became a founding member of the CMS experiment, one of the two large multi-purpose detectors at the Large Hadron Collider where the Higgs boson was discovered in 2012. She has held several leading roles, including the Chair of the CMS Collaboration Board, the highest decision-making body of the experiment. She was a member of the High-Energy and Particle Physics Board of the European Physical Society. She is also a member of the Board of Trustees, the Scientific Advisory Board, at Carinthia University of Applied Sciences and an adjunct professor at TU Vienna, lecturing on particle and astro-particle physics.

List of contributors

Ankit Agarwal
Potsdam Institute for Climate Impact Research, Potsdam, Germany
and
Indian Institute of Technology Roorkee, Roorkee, India

Janos K Asboth
Wigner Research Centre for Physics, Budapest, Hungary

Ralph Assmann
Deutsches Elektronen-Synchrotron (DESY), Hamburg, Germany
and
GSI Helmholtzzentrum für Schwerionenforschung, Darmstadt, Germany

Philippe Azais
Université Grenoble-Alpes, CEA, DPE, Grenoble, France

Marco Baldovin
Université Paris-Saclay, Orsay, France
and
CNR - Instute for Complex Systems, Rome, Italy

Felicia Barbato
Gran Sasso Science Institute, Italy

Marc Barthelemy
Université Paris Saclay, Gif-sur-Yvette, France

Patricia Bassereau
Institut Curie, Paris, France

Christian Beck
Queen Mary University of London, London, UK

Tobias Beuchert
German Aerospace Center (DLR), Earth Observation Center (IMF-DAS), Oberpfaffenhofen, Weßling, Germany

Jacob D Biamonte
NASA Ames Research Center, California, United State

Freya Blekman
Deutsches Elektronen-Synchrotron (DESY), Hamburg, Germany
and
Institut für Experimentalphysik, Universität Hamburg, Hamburg, Germany

Angela Bracco
University of Milano, Milano, Italy

Antoine Browaeys
Institut d'Optique, Université Paris-Saclay, France

Darwin Caldwell
Instituto Italiano di Tecnologia, Genoa, Italy

Angelo Cangelosi
University of Manchester, Manchester, UK

Alan Cayless
School of Physical Sciences, The Open University, Milton Keynes, UK

Jitka Čejková
University of Chemistry and Technology, Prague, Czechia

Giulio Cerullo
Politecnico di Milano, Milan, Italy

Henry Chapman
Center for Free-Electron Laser Science (CFEL), Deutsches Elektronen-Synchrotron (DESY), Hamburg, Germany
and
Department of Physics, Universität Hamburg, Hamburg, Germany
and
Centre for Ultrafast Imaging, Hamburg, Germany

Nelson Christensen
Observatoire de la Côte d'Azur, Nice, France

Dana Cialla-May
Leibniz Institute of Photonic Technology, Germany

Luisa Cifarelli
University of Bologna, Bologna, Italy

J Ignacio Cirac
Max Planck Institute of Quantum Optics, Garching, Germany

Frédéric Darbellay
Université de Genève, Geneva, Switzerland

Richard Darwin
Stockholm University, Stockholm, Sweden

Seamus Davis
Oxford University, Oxford, UK

Richard Dawid
Stockholm University, Stockholm, Sweden

Jacob de Boer
Vrije Universiteit Amsterdam, Amsterdam, Netherlands

José María de Teresa
Instituto de Nanociencia y Materiales de Aragón (INMA, CSIC-Universidad de Zaragoza), Zaragoza, Spain

Emmanuel Dormy
Ecole Normale Supérieure, Paris, France

Claudia Draxl
Humboldt-Universität zu Berlin, Berlin, Germany

Frédéric Druon
Laboratoire Charles Fabry, France

Marco Durante
GSI Helmholtzzentrum für Schwerionenforschung, Darmstadt, Germany

Patrick Eggenberger
University of Geneva, Geneva, Switzerland

Deniz Eroglu
Kadir Has University, Istanbul, Turkey

Angeles Faus-Golfe
IN2P3-CNRS—Université Paris-Saclay, Orsay, France

Luigi Fortuna
University of Catania, Catania, Italy

Gérard Gebel
Université Grenoble-Alpes, CEA, Liten, Grenoble, France

Carlos Gershenson
Universidad Autónoma de México, Mexico City, Mexico

Raymond Goldstein
University of Cambridge, Cambridge, UK

Giacomo Gradenigo
Gran Sasso Science Institute, L'Aquila, Italy

Hayit Greenspan
Tel Aviv University, Tel Aviv, Israel

Jean-Jacques Greffet
Institut d'Optique, Université Paris-Saclay, France

Bengt Gustafsson
Department of Physics and Astronomy, Uppsala University, Uppsala, Sweden
and
Nordita, Hannes Alfvéns väg 12, S-10691, Stockholm, Sweden

Susanne Hellwage
Leibniz Institute of Photonic Technology, Germany

Jens Hellwage
Leibniz Institute of Photonic Technology, Germany

Carlos Hidalgo
CIEMAT, Laboratorio Nacional de Fusión, Spain

Didier Jamet
Université Grenoble-Alpes, CEA, Liten, Grenoble, France

Sally Jordan
School of Physical Sciences, The Open University, Milton Keynes, UK

Clause Kiefer
University of Cologne, Germany

Christoph Krafft
Leibniz Institute of Photonic Technology, Germany

Jürgen Kurths
Humboldt-Universität zu Berlin, Berlin, Germany and Potsdam Institute for Climate Impact Research, Potsdam, Germany

Florence Lefebvre-Joud
Université Grenoble-Alpes, CEA, Liten, Grenoble, France

Franck Lépine
CNRS-ILM-Lyon, Lyon, France

Søren Linderoth
Technical University of Denmark, Lyngby, Denmark

Alberto Loarte
ITER Organization, Saint-Paul-lez-Durance, France

Jan Lüning
Helmholtz Institute, Berlin, Germany

Philip Macnaghten
Knowledge, Technology and Innovation group, Department of Social Science, Wageningen University, Wageningen, Netherlands

Daniel Malz
Department of Mathematical Sciences, University of Copenhagen, Copenhagen, Denmark

Timo Mappes
Friedrich Schiller University Jena, Jena, Germany

Antigone Marino
Institute of Applied Sciences and Intelligent Systems, National Research Council (CNR), Naples, Italy

Zita Martins
Centro de Química Estrutural, Institute of Molecular Sciences and Department of Chemical Engineering, Instituto Superior Técnico, Universidade de Lisboa, Lisbon, Portugal

Norbert Marwan
Potsdam Institute for Climate Impact Research, Potsdam, Germany
and
University of Potsdam, Potsdam, Germany

Sarah Matthews
University College London, London, UK

Thomas G Mayerhöfer
Leibniz Institute of Photonic Technology, Jena, Germany

Eilish McLoughlin
Dublin City University, Dublin, Ireland

Natalio Mingo
Université Grenoble-Alpes, CEA, Liten, Grenoble, France

Thierry Mora
Ecole Normale Supérieure, Paris, France

Ugur Ozturk
University of Potsdam, Potsdam, Germany

Simon Perraud
Université Grenoble-Alpes, CEA, Liten, Grenoble, France

Anreas Peters
HIT GmbH at University Hospital Heidelberg, Heidelberg, Germany

Robert Pitz-Paal
German Aerospace Center (DLR), Cologne, Germany

Stefaan Poedts
KU Leuven, Leuven, Belgium
and
Institute of Physics, University of Maria Curie-Skłodowska, Lublin, Poland

Chiara Poletto
INSERM and Sorbonne Université, Paris, France

Juergen Popp
Friedrich Schiller University Jena, Jena, Germany

Thierry Priem
Université Grenoble-Alpes, CEA, DPE, Grenoble, France

Bernd Rech
Institut für Silizium-Photovoltaik, Berlin, Germany

Lucia Reining
Laboratoire des Solides Irradiés, CNRS, École Polytechnique, Palaiseau, France

Marco Ripani
National Institute for Nuclear Physics, Genova, Italy

Felix Ritort
University of Barcelona, Barcelona, Spain

Petra Rösch
Friedrich Schiller University Jena, Jena, Germany

Christophe Rossel
IBM Research Europe - Zurich, Switzerland

Lucio Rossi
Università degli Studi di Milano, Milano, Italy

Pedro Russo
Department of Science Communication & Society and Leiden Observatory, Leiden University, Leiden, Netherlands
and
Ciência Viva, National Agency for Scientific and Tecnological Culture, Portugal

Mairi Sakellaridadou
King's College London, London, UK

Marta Sales
Universitat Rovira I Virgili, Tarragona, Spain

Pascal Salières
Université Paris-Saclay, CEA, LIDYL, Gif-sur-Yvette, France

Manos Saridakis
University of Athens, Athens, Greece

Eirini Sarigiannidou
LMGP Grenoble INP, Grenoble, France

Iwan Schie
Leibniz Institute of Photonic Technology, Germany

Jochen Schieck
Institut für Hochenergiephysik der Österreichischen Akademie der Wissenschaften, Wien, Austria
and
Technische Universität Wien, Austria

Michael Schmitt
Friedrich Schiller University Jena, Germany

Sara Seager
Massachusetts Institute of Technology, Cambridge, MA, USA

Pierre Seneor
Université Paris-Saclay, France

Shubham Sharma
Helmholtz Centre Potsdam–GFZ German Research Centre for Geosciences, Potsdam, Germany

Luis Silva
Instituto Superior Técnico, Universidade de Lisboa, Lisbon, Portugal

Friedrich Simmel
Technische Universität München, Garching, Germany

Thomas Tchentscher
European XFEL GmbH, Hamburg, Germany

Daniela Thrän
Deutsches Biomasseforschungszentrum—DBFZ, Leipzig, Germany

Tuan Quoc Tran
Université Grenoble-Alpes, CEA, Liten, INES campus, Le Bourget du Lac, France

Kees van Der Beek
Institut Polytechnique de Paris, Paris, France

Luc van Dyck
Euro Argo ERIC, Plouzané, France

Bart Van Tiggelen
Laboratoire de Physique et de Modélisation des Milieux Condensés, University Grenoble Alpes/CNRS, Grenoble, France

Javier Ventura-Traveset
European Space Agency, Toulouse, France

David Vernon
Carnegie Mellon University Africa, Kigali, Rwanda

Antje Vollmer
Helmholtz Institute, Berlin, Germany

Ernst Ulrich von Weizsäcker
Professor, Freiburg, Germany

Angelo Vulpiani
Università 'Sapienza', Rome, Italy

Hermann-Josef Wagner
Ruhr-University Bochum, Bochum, Germany

Aleksandra Walczak
Ecole Normale Supérieure, Paris, France

François Weiss
LMGP Grenoble INP, Grenoble, France

Frances Westall
CNRS Orleans Campus, Orleans, France

Claudia-Elisabeth Wulz
Institute of High Energy Physics, Austrian Academy of Sciences, Vienna, Austria

IOP Publishing

EPS Grand Challenges
Physics for Society in the Horizon 2050

Mairi Sakellariadou, Claudia-Elisabeth Wulz, Kees van Der Beek, Felix Ritort, Bart van Tiggelen, Ralph Assmann, Giulio Cerullo, Luisa Cifarelli, Carlos Hidalgo, Felicia Barbato, Christian Beck, Christophe Rossel and Luc van Dyck

Chapter 1

Introduction

Science begins when someone raises a general question and sets about answering it by methodical investigation, including and combining experimentation and logical argumentation. Such scientific action spawns understanding of our world in its broadest sense and therefore the power of predicting and describing the behaviour of different bodies and objects. The corollaries are significant practical advantages—ranging from agriculture to medical applications. The dawn of science is therefore as old as the dawn of man as we know it—the mastery of fire, tools, agriculture and, later, the isolation of alloys and pure metals were, as such, great scientific advances. These were to be followed by the inevitable questions, borne out of sole curiosity, regarding consciousness, the place of man in the Universe and the workings of the cosmos. Thus, one of the first problems to be tackled truly scientifically by ancient historical cultures was to conceive explanations of the seasons and of how heavenly bodies move. Without this initial curiosity, on which our scientific and technological knowledge are based, humanity would be radically different. The build-up of knowledge concerning the workings of the distant Universe and the world at hand on one side, and the properties and behaviour of materials on the other have led, in the second half of the 18th century, to the advent of the industrial revolution. The harnessing of electromagnetism and its phenomenal stream of applications followed during the 19th century. The huge social impact of both events is unrivalled—it is difficult to think of any political, religious or economical doctrine that has brought about such radical and robust changes in society.

Creativity plays a vital role in the development of science, insofar as one of its main objectives is to imagine and shape the future. Scientific knowledge, innovation, progress and even new paradigmes arise mostly from curiosity-driven research but also often from serendipitous discovery. Moreover, one should not overlook the cultural impact of scientific research, education and training. It is the principal method by which citizens can mature to critical, rational and independent thinking. A modern developed society must therefore nurture a strong scientific sector, both in education and research, in order to address its technological and societal challenges.

Although the quest for knowledge is not necessarily susceptible to ethical evaluation, science abandons its ethical neutrality when it crititically checks how knowledge is generated, and how its technological applications impact individuals' lives and society. This is particularly clear in health research, where commercial or financial interests should never prevail over individual liberties and well-being. The opposite is true in research for sustainable development. The desires of individuals should not prevail over the common benefits for society and humanity—a topic worth scientific analysis and debate. Furthermore, the recent progress in artificial intelligence will lead us through a fascinating landscape of novel applications linked to ethical considerations. In fact, science as a whole is not always ethically neutral, i.e., impartial and fair . Quoting Berthold Brecht in his play 'Life of Galileo Galilei', *should people dedicated to science develop something like a Hippocratic Oath with the promise of using Science solely for the benefit of mankind?* Science should aim to raise global life standards, requiring long-term perspectives on international cooperation with investment and cooperation in research, education and sustained development in global challenges such as energy management or climate change.

One of the most dazzling realizations of physics is that of scale and the place of mankind in the Universe. From its smallest constituent parts to its largest structures, the description of the Universe spans an improbable 45 orders of magnitude in length scale[1]. It describes the most fascinating of journeys, from the smallest things that we have ever explored—quark particles that are less than 10^{-18} m across—to the scale of the nucleus of an atom made up of protons and neutrons—10^{-15} m—or to the atoms dreamed by the ancient Greeks, with diameters of about 10^{-10} m. The journey continues in the living world from the size of a living cell that is about 10^{-5} m, to the human scale of 1 m in our natural environment, up to the Earth's diameter of 10^7 m. Stepping into space one evaluates the size of the Solar System to some 10^{11} m, the distance to the nearest stars outside the Solar System to 10^{16} m, the diameter of our Galaxy to 10^{21} m, until reaching the largest things we have ever measured, the greater breadth of the Universe with 10^{27} m.

What is also remarkable is the amazing effectiveness of mathematics in describing the most fundamental laws of physics. The list of achievements is impressive, ranging from Maxwell electrodynamics that holds at the scale of particles to that of distant galaxies, to Einstein's relativity theory that describes classical newtonian mechanisms as well as quantum mechanisms dealing with the interaction of matter and radiation on the atomic and subatomic scales. In particular, quantum

---

[1] Scientific notation is a way of writing very large or very small numbers. A number is written in scientific notation when a number between 1 and 10 is multiplied by a power of 10. For example, the Universe is about 10^{27} m across, that is, 1 followed by 27 zeros: 1 000 000 000 000 000 000 000 000 000 m.

electrodynamics, which combines quantum mechanics with Einstein special relativity, is known to be accurate in about one part in 10^{11}.[2]

We have to appreciate that one challenge is to know the laws of Nature, which are few and amazingly accurate, but another one is to predict the outcomes of these laws, which are numerous and quite often complex. In a complex system it is not so much the size of the components that is of primary importance, but the number of interconnections between them. This separation of the scientific perspective into laws and outcomes would help to understand why some disciplines of physics are so different in outlook.

Although important progress has been made and is further expected in specific areas of knowledge, the interlinking between separate areas or topics of science is crucial to addressing some of the grand scientific and societal challenges such as climate change or understanding life. Interdisciplinary allows interconnections to be made between many fields like physics, mathematics, biology, or chemistry in such a way that the whole body of connected individual ideas merge and expand into a successful global output.

There are many different images of science and of the activities of scientists in the public. Some people imply that science may eventually reach the limits of knowledge while others believe in endless horizons. Some think that science has or will provide the answers to key open questions, while others mistrust its development. The COVID-19 epidemic spreading has been more than a health and economic crisis. It illustrates our vulnerability but also the importance of interdisciplinary and multilateral science in addressing such a global challenge that affects societies at their core. Numerous multidisciplinary actions have been developed to tackle the spread of coronavirus and react against other desease outbreaks.[3]

In this book, we look at all these remarkable aspects, going from elementary particles, atoms, living cells, to stars, galaxies, and asking about our place in the Universe. We explore also what makes us, human beings, really unique in nature: our self-consciousness and our ability to imagine the world around us and shape the future by making use of the scientific method.

The book is an EPS enterprise designed to address the social dimension of science and the grand challenges in physics. Hopefully it will succeed to convince the reader that science and physics might help bringing positive changes and solutions to our societies, raising standards of living worldwide and providing further fundamental comprehension of Nature and the Universe on the Horizon 2050.

---

[2] This is equivalent to measuring the distance from Madrid to Berlin with an accuracy better than the width of a human hair.

[3] https://ec.europa.eu/info/research-and-innovation/research-area/health-research-and-innovation/coronavirus-research_en

Part I

Physics as global human enterprise for
understanding Nature

IOP Publishing

EPS Grand Challenges
Physics for Society in the Horizon 2050

Mairi Sakellariadou, Claudia-Elisabeth Wulz, Kees van Der Beek, Felix Ritort, Bart van Tiggelen, Ralph Assmann, Giulio Cerullo, Luisa Cifarelli, Carlos Hidalgo, Felicia Barbato, Christian Beck, Christophe Rossel and Luc van Dyck

Chapter 2

Physics bridging the infinities

Freya Blekman, Angela Bracco, Nelson Christensen, Emmanuel Dormy, Patrick Eggenberger, Claus Kiefer, Franck Lépine, Mairi Sakellariadou, Emmanuel N Saridakis, Jochen Schieck and Claudia-Elisabeth Wulz

2.1 Introduction

Mairi Sakellariadou[1] and Claudia-Elisabeth Wulz[2]
[1]King's College London, London, UK
[2]Institute of High Energy Physics, Austrian Academy of Sciences, Vienna, Austria

At the horizon 2050, our physics textbooks will have to be rewritten.

When the Higgs boson was found in 2012 at CERN's Large Hadron Collider (LHC) in Geneva, history was made. This particle, and its associated field, is the reason why atoms, stars, galaxies, and you, reading this book, are tangible entities. In addition, a tiny asymmetry between matter and antimatter that developed soon after the Big Bang made it possible for us to exist at all. Without the Higgs boson and this asymmetry, only radiation would permeate the Universe.

Detailed studies of the Higgs boson, at current or future colliders, as well as precision measurements of the properties of matter and antimatter at a multitude of different experiments will reveal how the Standard Model of particle physics has to be amended. That it needs to be extended is evident. It does not contain dark matter, whose existence was already manifest decades ago. Another phenomenon, only discovered in 1998 through the study of the brightness of supernovae as a function of their distance, is dark energy, which makes our Universe expand in an accelerated fashion. Known or 'visible' matter, the so-called baryonic matter, only accounts for 5% of the Universe, and is well described by the Standard Model of particle physics. The rest are dark matter (27%) and dark energy (68%). We have hardly any clues, but many ideas of what they could be.

© 2024 The Authors.

2-1 Published under license by IOP Publishing Ltd

Although postulated already in 1930, neutrinos are another category of particles that are still mysterious. It was only ascertained in the 1990s that they have mass, in contrast to the assumption in the Standard Model of particle physics, and that they come in different flavours that can transform into each other. There might even be more varieties—sterile neutrinos—which do not interact through the forces described by the Standard Model, but only through gravity.

Although the latter is so present in our everyday life and the movements of objects in the cosmos, it is the least understood force, and is not part of the Standard Model of particle physics. We do not even have a quantum-mechanical formulation of the theory of gravity yet, which would allow us to describe this force down to the smallest scales of the Universe. Our current understanding is based on Einstein's theory of general relativity, which however breaks down at the centre of black holes, for example. Everywhere else, it has so far been proven to be perfectly descriptive and accurate. The spectacular direct discovery of gravitational waves—ripples in space time predicted to arise from violent events in the cosmos such as mergers of black holes or neutron stars—in 2015 confirmed it once more. This discovery has further opened up a new field called multimessenger astronomy. We are no longer limited to observing the sky with our eyes or with telescopes detecting light or other electromagnetic waves, but we now also have gravitational waves, and neutrinos, at our disposal as messengers from cosmic sources. We can study all kinds of signals in a coordinated fashion in experimental facilities around the globe and even in space.

There are many bridges between the smallest and the largest scales. Nuclear physics, with its quest to understand the origins of known matter, from the primordial soup made of quarks and gluons, the protons and neutrons, the atomic nuclei, to the formation of the heavy chemical elements in explosions of stars, connects these infinities. It also has a large potential for technological spin-offs such as nuclear fusion to ensure the supply of electric power and medical applications such as cancer therapy, as well as efficient and affordable isotope production for diagnostic purposes. For the latter, imaging techniques using artificial intelligence and other means are drivers for improving diagnostic accuracy, rapidity, and the comfort of patients. Astrophysics and high-energy particle physics are also connecting the scales, and have given rise to the new field of astroparticle physics.

Cosmology, with its quest to understand the largest scales and nothing less than the fate of our Universe, needs information about its smallest components. Amongst others, measurements by space observatories such as Planck operated by the European Space Agency have helped establish the now widely accepted Standard Model of cosmology. For the time being, we are at a turning point in our knowledge of the future of our Universe. Soon we should know more about its evolution, and in particular, whether it will be confirmed to expand forever, or to rip apart, or even contract again.

The stars, the Sun, and the planets, including our own, still have secrets themselves. Their formation and evolution are vibrant research areas, tackled through computations and observations, and exploiting a multi-disciplinary approach. The study of exoplanets has also become a central subject in astrophysics. More down-to-earth, geophysics addresses topics that can affect us all, such as volcanic eruptions, earthquakes, or even changes in the Earth's magnetic field. The understanding of these phenomena can help predict their occurrence, for the benefit of mankind.

Will all or many of the open questions be answered at the horizon 2050? There is justified hope, supported by a plethora of theoretical developments and experimental facilities on Earth and in space. Will new questions arise? You bet.

2.2 Particle physics: physics beyond the Standard Model

Freya Blekman[1,2]
[1]Deutsches Elektronen-Synchrotron (DESY), Hamburg, Germany
[2]Institut für Experimentalphysik, Universität Hamburg, Hamburg, Germany

2.2.1 Particle physics: the extremely small connects to the big scientific questions

Particle physics explores Nature at the tiniest scales by studying the properties and interactions of elementary particles. Many of these properties and interactions between elementary particles are predicted in a theoretical framework known as the Standard Model (SM) (figure 2.1).

The predictions of the SM have been experimentally confirmed to an extraordinary degree of precision by a series of historic particle accelerators and most recently using the collisions at the LHC at CERN. Understanding the behaviour of these elementary particles has consequences to other fields of physics and beyond. For example, knowledge of the SM is necessary to obtain insight into the Universe just after its creation during the Big Bang. Another of these questions is that only about 5% of all matter and energy that constitutes our environment that we experience can be described with the SM, meaning that the remaining almost 95% is presently unaccounted for by our current knowledge. The remaining 'dark' energy in the Universe is referred to as Dark Matter and Dark Energy.

Particle physics does not only deal with the fundamental constituents of matter but also with the processes of how particles interact. While in our macroscopic world, an almost infinite number of forces and interactions can be experienced, on the microscopic scale of particles, these can be reduced to only four known types of interactions:

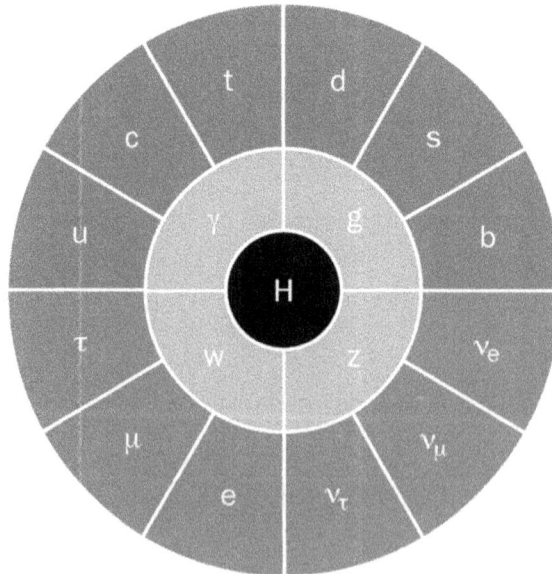

Figure 2.1. The Standard Model, as represented in [1].

gravitational, weak, electromagnetic, and strong. Gravitation is the force that people may be most familiar with, but it is counter-intuitively the least understood when dealing with particles. The weak interaction is responsible for, for example, the radioactive interactions in stars like our Sun. Electromagnetism, responsible for most of our daily life including light, magnetism, and the stability of molecules and the orbits of electrons in atoms, is deeply connected to the weak interaction and shares many of its properties. Finally, the strong interaction is the mechanism that ensures that the particles inside the atomic nucleus remain stable. The weak, electromagnetic, and strong forces are all included in the SM. The weak, electromagnetic, and strong interactions in the SM are mediated through particle interchanges between the quarks and leptons, the constituents of matter. The most commonly known quantum mediator is the one associated with electromagnetism, the photon. For the strong and weak forces, other equivalent particles exist that are called the gluon and the $W^{\pm}$ and Z bosons, respectively. Substantial work both on the theoretical side and extraordinary experimental proof to confirm any such new theory will be necessary to achieve the inclusion of gravity in the SM or in a Grand Unification Theory, one of the noble goals of physics. For example, particle physics uses conservation laws equivalent to the well-known rule of conservation of energy in the form of symmetries that a quantum field theory can describe. This framework is the SM.

In total, the SM describes the behaviour of all known elementary particles. The quarks can be used to create composite particles such as protons and neutrons. There are also three charged leptons such as electrons, and three neutrinos. These fermions are grouped into three so-called 'generations' of increasing mass. The origin of the generational structure is currently not understood. The forces are represented by the photon, $W^{\pm}$ and Z massive gauge bosons, and the gluon. The Higgs boson completes the present picture of the SM by inducing mass to the elementary particles, which would otherwise be mathematically massless in the SM.

It is essential to be aware that there are critical weak points in the SM as presented in the previous paragraphs. Additional unexplained arguments that the SM is not the complete picture of particles at the smallest scale include the apparent conflict in the equal treatment of matter and antimatter in the SM. After all, we live in a world dominated entirely by matter, a fact that is largely unexplained by the SM. Also, the SM does not explain why particles have different masses; mathematically, it allows all quarks and leptons to have equal mass. The proton would not be stable at that point, and atoms and molecules would not be stable as we know them to be.

The SM importantly also relies on neutrinos being massless. However, it was proven in the early 2000s [2] that neutrinos have a tiny but non-zero mass. Not only does the SM not need the neutrinos to have mass, but the neutrino mass also creates problems. The mass of neutrinos cannot be easily explained the same way as other particles and naturally suggest that the SM needs to be extended with more symmetries, forces, or interactions that solve the neutrino conundrum.

This means that physicists look for more explanations to create a better description of Nature than the SM. The diverse Beyond-the-Standard-Model (BSM) physics theories typically involve a new mathematical formalism that controls the known types of interactions and particles and almost always predict the existence of additional undiscovered particles. The search for evidence of the (in) direct production and existence of these particles and mechanisms is an important research challenge in fundamental physics.

A large number of theoretical extensions to the SM fix one or many of the previously mentioned problems. Unfortunately, it is currently impossible to identify which of these extensions is most likely to solve the problems of the SM. However, certain physicists indeed have a preference. Particle physics is at the moment very much a data-driven science, meaning that without further empirical input from experiments, it seems to be impossible to solve these questions. The goal of the programme of the LHC now that the Higgs particle has been discovered is to probe the particle world at the smallest possible scale to find answers to these urgent fundamental questions. While the LHC runs at energies around 13–14 TeV, LHC measurements access higher scales accessible in specific cases where the particle production involves intermediate undiscovered particles. The search for new particles is an important aspect of particle physics.

The link between particle physics and gravitation is one of the large open questions in physics in general, as it, to the best of our knowledge, would imply combining Einstein's theory of relativity with the quantum physics of the elementary particle world.

2.2.2 The LHC and the discovery of the Higgs boson

The LHC started its first preparation and feasibility studies in the 1980s. Still, construction began in strides when the previous accelerator in the same tunnel, the Large Electron–Positron collider (LEP), ceased operation in the early 2000s. In 2008 the LHC had its first collisions, and in 2010 the physics quality data started to be collected. Accelerators such as the LHC shoot bunches of tens to hundreds of billions of protons on each other. When bunches are shot at each other, the chance that a proton collision occurs is still relatively small. The number of these bunches inside the accelerator, how many protons each are filled with and how closely they can be packed, and how the beams are aimed at each other determine how many collisions occur. Improving each of these is extremely challenging and requires substantial and detailed work by the accelerator physicist teams at CERN. To implement substantial improvements in beam intensity work is necessary on the cutting-edge CERN magnet and accelerator system, which can take years. To facilitate data taking with similar conditions the LHC operation is organised in Runs. Run 1 took place between 2009 and 2012, Run 2 started in 2015 and lasted until 2018, and Run 3 is has started in 2022. For some reference, Run 2 provided approximately four times more data than Run 1, and Run 3 is expected to produce yet again double the datasets compared to Run 2. The substantial data increase is expected to occur in Runs 4 and later, at which point the so-called *High-Luminosity LHC* will produce datasets that are a factor ten larger than what was collected in the 15 years of Runs 1 through 3.

The Higgs particle was discovered by the ATLAS and CMS Collaborations in 2012. This elusive particle was proposed initially in 1964 to solve some of the inconsistencies in the theory that occurred when combining the weak and electro-magnetic interaction (which is why particle physicists talk about the electroweak interaction). Generalising, the Higgs boson is the result of the same mechanism that gives particles mass. While the discovery of the Higgs boson solved the general question of the origin of mass, it raised more questions. For example, it does not explain why different particles have different masses or why the range of masses in the SM is as extreme as it is, ranging from feather-light neutrinos to enormously massive top quarks and anything in between (figure 2.2).

Fortunately, the mathematical machinery of the SM does predict very accurately how the Higgs particle will interact with particles with different masses, so an essential part of the study of the Higgs boson has moved from discovery to important consistency checks of the SM. This is particularly relevant as the SM predicts that the frequency that the Higgs boson decays to certain particles is directly connected to the masses of those particles. As the range of particle masses in the SM is so extensive and unexplained, these studies offer an important avenue into probing any consistencies in our understanding of the origin of mass. In addition, these decay probabilities are difficult to measure for lighter particles just because these particles will be produced less by the Higgs boson. At the Higgs boson discovery in 2012, the decay to Z and W bosons and to two photons[1] were the only sensitive channels. Since then, the Higgs boson decays to two tau leptons, two bottom quarks, and two muons has been observed. The associated production of the Higgs boson together with two top quarks allows studies of the interaction between top quarks and Higgs boson.

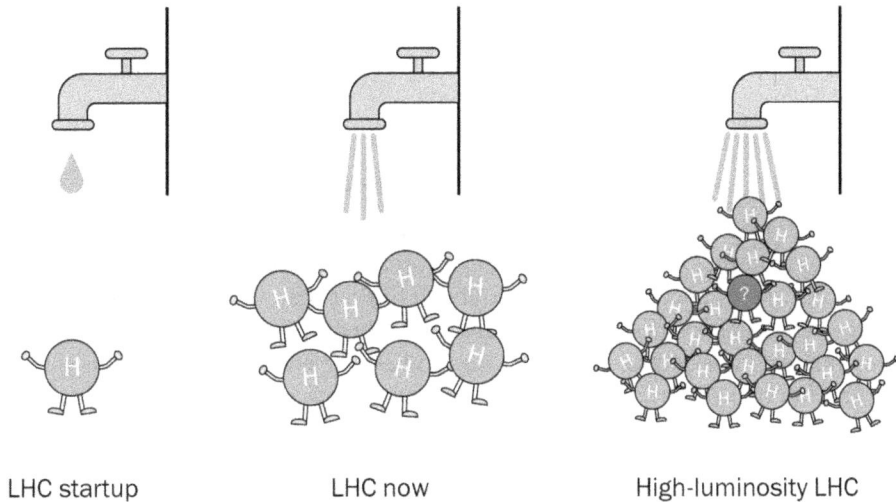

LHC startup   LHC now   High-luminosity LHC

Figure 2.2. The number of Higgs bosons produced at the LHC has already increased by a factor of about ten since the discovery in 2012. But to get very large samples of Higgs bosons and examine if they are truly consistent with the Standard Model, the High-Luminosity LHC or even future colliders will be necessary.

---

[1] Hey, but the photon is massless, and the Higgs boson decays to mass, that's not right? Indeed, the Higgs boson decay to photons is a complicated exception, where the Higgs boson first decays to heavy particles such as top quarks and W/Z bosons. After that, those *virtual* particles create the two photons. These virtual particles become important in the next section.

After the discovery in 2012, it also became possible to measure the mass of the Higgs boson. At that point, it became increasingly clear that the value of the Higgs boson mass, about 125 GeV, or the approximate equivalence of the mass of an Iodine atom (that obviously contains many protons and neutrons), is not light enough to immediately confirm that new particles need to be present to make the SM work, but *also* not heavy enough to decidedly indicate that the SM is complete. It is in a metastable *grey zone* where the only way is to confirm which of those two is true. This makes the study for the study of the Higgs boson particularly relevant also in the context of, for example, understanding the evolution of the Universe and explanation for the age and evolution of the Universe [3].

Each of these observations at discovery has been refined over time. With more data, the connections between the Higgs boson and the heavier particles in the SM have now been confirmed to be consistent with the Standard Model's predictions to within, on average, about 10%. The majority of the lighter SM particles has not been confirmed to interact with the Higgs boson yet. To further improve and confirm if the Higgs mechanism is truly consistent with great accuracy, the ten-times larger datasets of the High-Luminosity LHC or even future colliders will be necessary.

Studying new rare decays of the Higgs boson (to the lighter quarks and leptons and invisible particles) is the highest priority for the LHC and one of the driving topics for future colliders. Another important topic is examining the already observed interactions of the Higgs boson as precisely as possible to identify if they also are consistent with the SM at the sub-percent level. Both these endeavours require a vast number of Higgs bosons. Producing these large samples of Higgs bosons is the goal of the High-Luminosity LHC and one of the metrics used to evaluate future colliders' scientific potential.

2.2.3 The challenge of breaking the Standard Model

One of the reasons why the SM is one of the most successful scientific theories is because it can make accurate predictions. The Higgs boson is only one of the SM particles, and internal consistency checks of the SM go much further than the precise examination of the Higgs boson properties and interactions described in the previous paragraphs.

One interesting aspect of the quantum physics of elementary particles is that it is possible to produce *any* particle for a very short time as long as that particle can interact with other particles. Due to the Heisenberg uncertainty principle, it is even possible to produce very heavy particles for an infinitesimally short time. These virtual particles cannot be directly observed but affect the behaviour of the other, observable, SM particles. This means that the effect of undiscovered particles can already be observed indirectly in the behaviour of well-understood particles. Figure 2.3 represents this as the known particles behaving as if they are shielding the undiscovered world. Still, it is a shield that can be understood by measuring smart and precisely. This specific quality of indirect measurements of particle physics was why there were already indirect constraints on the Higgs boson and top quark properties, many years before they were discovered by direct production.

These subtle changes in behaviour can be observed if the properties, kinematic behaviour, and relative production rates are measured accurately. A massive

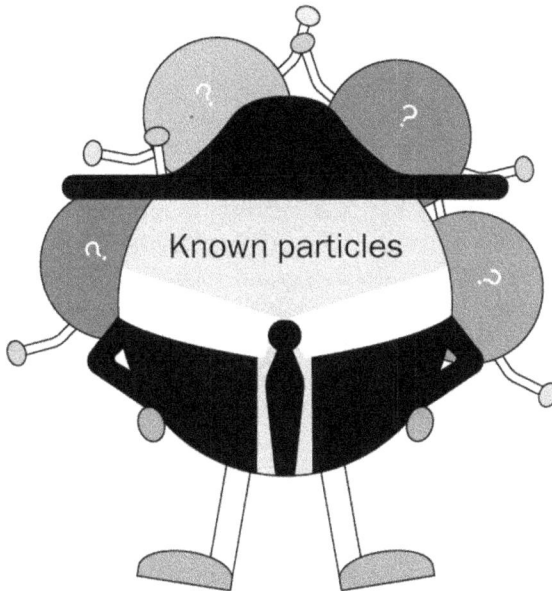

Figure 2.3. The detailed behaviour of known particles can be used to indirectly study undiscovered particles.

challenge for these precision measurements is that when the SM is challenged at such precisions, the uncertainties due to the experimental knowledge, the simulations, and the theoretical calculations that are providing the prediction become relevant. Testing and identifying the subtle changes that virtual particles introduce is another one of the main challenges in particle physics, as these precise measurements require combinations of extremely detailed understanding of the detectors, backgrounds, and SM predictions, together with the humongous datasets produced by extremely high-intensity colliders.

A great example of the power and challenges of these precision measurements is in the extremely challenging measurement of the μ g-2 experiment at Fermilab [4]. This experiment measures the muon's magnetic moment, a property that can be extremely precisely predicted by the SM. The magnetic moment is the strength of a muon's interaction with a surrounding magnetic field, where virtual particles (so that means at least all the other SM particles, and if other particles exist, those as well) modify the behaviour of the muon. This makes this experiment a great test for telltale signs of new particles and forces. To do so, the knowledge of the detectors and particularly an unprecedented knowledge of the exact value of the magnetic field that the muon is moving in is essential. In the spring of 2021, the first result from this experiment confirmed previous measurements of the magnetic moment are off from the SM prediction by a precision of about one part per million. The eventual goal of the experiment is to test the Standard Model's predictions of this value by measuring the precession rate experimentally to a precision of one-tenth part per million, and new data is currently being collected and analysed to do so. However, it is crucial to be aware that the measured numbers have to be compared to a SM prediction.

Predicting something that accurately is not easy, and the theoretical calculations by hundreds of theoretical physicists are another aspect of this test that receive serious scrutiny before it is definitive that there is an inconsistency and the SM is incomplete and in need of revision.

There are similar examples from the LHC; recently, the LHCb experiment measured unexpected discrepancies between the decays of beauty and charm quarks to the different leptons. In the SM, the electron, muon, and tau particles all behave identically except for their mass. Only virtual particles that interact differently with the different leptons can easily be used to explain such discrepancies. The ATLAS and CMS collaborations, and in Japan the Belle2 collaboration, are now studying similar decays to shed light on whether these flavour anomalies are reproducible. Results are expected in the coming years, and the study of these flavour anomalies is a very actively developing field of study within experimental particle physics. It is important to note that several boundary conditions need to be achieved to make progress in these precise measurements. For example, to understand the difference in behaviour between two similar particles, such as the muon and electron, the experimental measurement of both particles needs to be exquisitely understood. In addition, not all experiments were specifically designed to measure these signatures. Still, with the fast improvements in both detector knowledge and statistical analysis techniques such as deep learning, it turns out even *general-purpose detectors* such as ATLAS and CMS can be expected to make contributions to the study of these flavour anomalies.

To achieve these precision tests of the SM, two other requirements are essential that will provide challenges in the near future: the simulation that in such analyses is used needs to be accurate enough to represent the detectors and SM to the best degree possible, and the theoretical predictions that these precise measurements will be compared to will need to be of similar precision as the experimental results. For both of these, substantial difficulties will need to be overcome. In the case of simulation, this is partially a scientific computing challenge, as the number of simulated collisions will need to increase substantially, and this will put a substantial strain on the software and storage solutions used by the collaborations. In addition, such detailed simulation requires that no corners be cut as far as the modelling of tiny differences in different detector components and regions, making the simulation substantially slower. There currently are multiple innovative efforts ongoing that try to address these challenges, both as far as using different hardware, using smarter software solutions, including modern computer science and data science techniques that will be tested to their limits on data that may reach the size of up to exabytes. The simulation also needs to be sufficiently realistic as far as the computation of the SM, implying that substantial improvements will need to be made in the precision of the physics in the simulation, which is now typically performed at leading order or maybe next-to-leading order precision in perturbation theory. This challenge to increase in accuracy even more affects the SM calculations, which will need to be performed as a reference that measurements use to compare in hypothesis tests. In recent times, multiple precision measurements that implied discrepancies with respect to the SM were proven to be correct but still consistent with the SM when it was calculated to higher precision. To reliably and effectively improve the higher-order calculations, substantial strides in theoretical physics, both as far as computational techniques and mathematical methods, will be necessary.

2.2.4 The challenge of searching for new particles

There is obviously a more direct way to prove that the SM is incomplete, and this is to identify new particles in the collisions at the LHC. There are also a substantial number of smaller, dedicated experiments that may not even need a collider, and that aim to identify such signatures; the direct and indirect experiments looking for dark matter and searches for axions and sterile neutrinos are good examples.

Before the LHC startup, there was overwhelming positivity that new particles would appear as soon as the accelerator was turned on. By now history (and the Standard Model) has taught physicists some more humble attitudes. Twelve years after the first collisions of the LHC there are no significant signs of new particles, and this is driven partially by the fact that about 5% of the lifetime data has been collected, but more importantly by the fact that the focus up to now was mostly on the easy, *low-hanging fruit* new physics scenarios. It is, however, useful to consider how the searches programme, which is responsible for about half of the research output of the ATLAS and CMS collaborations, is performed and what the current and future challenges are for the direct search for new particles.

Many of the new physics predictions for the LHC are relying on assumptions that are inspired by the general idea of supersymmetry. Supersymmetry relies on the introduction of a new symmetry to the SM, which implies that every known particle has a partner particle. The idea of supersymmetry is definitely not a new one, but the Higgs boson mass of 125 GeV is very consistent with the preference for low mass Higgs bosons in even the most general versions of supersymmetric models. As of today, no signs of supersymmetry have been spotted in the LHC data, and the experimental searches are now focusing on the more challenging scenarios where supersymmetric particles may still survive undetected.

It is important to realise that the number of possible signatures that supersymmetric models is so large that after initial data taking, the LHC experiments focused on a strategy where signatures were investigated instead of specific models. In general, the number of varied parameters is substantially reduced and only very simple *Minimal Supersymmetric Models* are considered. The *simplified model spectra* of these models that are being used tend to assume that there is only one supersymmetric particle that can be produced at the LHC (instead of the full spectrum of a doubling of the number of particles). In addition this single supersymmetric particle then only decays only in one (or few) ways, creating a very specific signature that can easily be searched for. This well-defined signature does come at the sacrifice that the connection to realistic collections of supersymmetric particles are produced together. This means that when statements are made regarding 95% confidence level excluded mass ranges of supersymmetric particles (e.g., for the gluino, the supersymmetric partner of the gluon), the constraints will be given for a 100% production of gluinos with no other particles being accessible, and a 100% assumption on the gluino decaying to one specific particle (e.g., a gluon and an undetectable neutral particle). If these assumptions are loosened, the excluded mass range of gluinos may be reduced substantially and many scenarios where the particle decays to multiple different particles at some probability are still possible and not considered. The interpretation challenge of simplified models is important to be considered when drawing broad conclusions as far as whether supersymmetry is still accessible at the LHC; particularly in summaries this fact is regularly not neglected in the discussion of supersymmetry as still being a viable model to use for inspiration of searches for new physics.

Even in the rudimentary world of simplified model spectra there are still substantial fractions of supersymmetry scenarios for which the LHC does not yet have sensitivity. These tend to come in two general categories: scenarios where the particles for some reason are extremely rarely produced (these would mostly be the supersymmetric partners of the W, Z, and Higgs bosons and leptons), or the considered masses are such that the supersymmetric particles are kinematically suppressed (this is commonly referred to as *compressed spectrum* supersymmetry).

Beyond the simple supersymmetry predictions and searches a whole diverse world of *exotic* other models opens. There are many possible ways to combine multiple supersymmetric extensions, also with extra dimensions, with gravity and so forth. The famous representation in figure 2.4 is a very good way to represent the huge number of different scenarios that non-minimal supersymmetry models predict. There are many different scenarios that predict new particles that would create resonances in di- and tri-particle invariant mass in the SM continuum; for example, these tend to be related to various models predicting new bosons, extra dimensions, and string theory-inspired new symmetries that introduce one or more new mediator particles.

There are as many extensions to the SM as there are enthusiastic theoretical physicists, so at an experimental level the challenge is typically to identify a signature

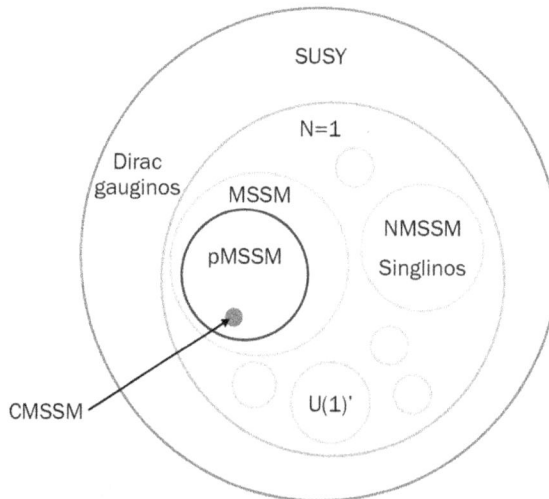

Figure 2.4. The green dot represents what the typical LHC experiments consider when searching for supersymmetry. Obviously many other scenarios are possible, and many of these are considered at the LHC as well but are grouped under the exotic nomenclature [5].

that does have either a very low or very well-understood background from the SM. However, one of the weaknesses of searches for such resonances is that they are driven predominantly by the collider energy, so additional data due to increased accelerator intensity do not necessarily increase the mass sensitivity of searches for resonances but do continue to provide sensitivity to smaller and smaller cross-sections. The topological choices tend to focus on signatures such as:

- New resonances, where practically every combination of jets, including flavour tagged jets, are considered, as are all charged leptons. Such resonances on backgrounds that are typically following simple power laws are extremely striking.
- Other striking signatures, such as deviations in event mass and energy distributions. These are predicted by a wide range of new physics models, such as microscopic black holes, and undiscovered new particles at energies not directly accessible by the LHC energy that do modify the behaviour or the amount of SM particles produced in general.
- Searches for invisible objects, such as direct production of dark matter particles, tend to focus on the production of an additional particle together with the invisible particles. The visible, from SM processes such as initial state radiation, in that case recoil against invisible momentum. Again, such mono-jet signatures are by now also examined with jet flavour identification tools and for mono-W/Z/Higgs bosons with jet substructure techniques.
- New fermions that decay in some way to known SM particles. Examples are new quarks (e.g., vector-like quarks), but also leptoquarks, particles that have both leptonic and strong interacting properties and that can be used to explain some tantalising deviations in SM precision measurements in the flavour sector, by, for example, the LHCb experiment.

As the LHC datasets increase, new signatures continue to become accessible. This is particularly relevant in searches for new physics theories that might have suppressed couplings in the resonance or missing energy signatures, but would create deviations from the SM in other signatures that are so rare that at LHC startup there was not enough data to examine them. With the ever-increasing improvements in the understanding of the detector, some other searches are also being developed, using techniques that were not even predicted when the LHC was designed.

2.2.5 Leaving no stone unturned

In recent years the searches at the LHC have become more creative. This development is partly driven by the lack of observed excesses, innovation on the experimental side, and the next generation of reconstruction algorithms becoming available. Improved understanding of the detectors has led to opportunities to reconstruct jets and their structure better. It is now possible to identify hadronic decays of Higgs bosons, W and Z bosons, and top quarks inside jets. These substructure tools have made enormous leaps in the last years, are now widely available, and perform at similar usability levels as, for example, b quark jet flavour tagging. This innovation at the reconstruction level has created a whole new class of possible analyses. ATLAS and CMS have developed effective research programs using substructure tools to study the SM precisely and search for new particles that decay to Lorentz-boosted objects.

Another avenue that has recently become very fruitful is searching and identifying long-lived signatures for physics beyond the SM. The LHC experiments were designed to identify particles that are produced close to the collision point. Still, over the last years, it has become clear that many new physics models have a parameter space where particles are limited in their decay somehow (usually either through phase space or coupling considerations) which creates displaced signatures. Examining these non-prompt signatures is possible as the experiments can detect such decays but previously rejected them. Experimental challenges tend to be driven by the fact that the backgrounds can be much more challenging to study or identify potential background sources. The methods to compare simulation to data become more complex. However, the diversity of physics models and signatures that can be probed by studying long-lived signatures is so bountiful that this has by now been established as a very effective LHC program, including links beyond the main LHC experiments. Long-lived particles have also inspired an LHC-wide collaborative community that includes a very effective collaboration between theoretical and experimental physicists [6].

The understanding of the LHC detectors is continuously improving still, and this also continually opens new potential signatures. Currently, the experimental innovation beyond substructure and long-lived signatures focuses on improving background rejection for lower-energy lepton signatures and improved performance in high jet multiplicity regions. Both are only marginally explored up to now; the existing methods focused on the low-hanging fruit of leptons coming from electro-weak boson decays for which the detectors were designed. Still, the previously mentioned flavour anomalies have motivated strong drives for the experiments to go as low as possible in transverse momentum and examine more and more rare signatures at high multiplicity that were not sufficiently populated in smaller datasets to study in more detail than would be possible in simple counting experiments.

At hadron colliders, traditionally, only a small fraction of the data is analysed. At the same time, the rest of the data is rejected to save disk space. Strategies to collect more difficult-to-access data are also being implemented. The LHCb experiment plans to implement an online selection strategy where almost all of the data will be selected by reducing the individual event size so that a larger fraction of events are saved and analysed [7]. Similar strategies are also being implemented in CMS and ATLAS, such as the *scouting* and *parking* efforts by CMS [8].

2.2.6 Beyond the general-purpose experiments

The two giant experiments at the LHC, ATLAS and CMS, are by no means the only endeavours to break the SM. Even the collisions at these so-called general-purpose experiments are examined by smaller experimental collaborations that aim to detect dedicated signatures, such as potential detection of very long-lived particles and detection of anomalous behaviour of neutrinos at tens or even hundreds of metres from the collision point [9–11]. In addition there are multiple very forward detectors that study LHC collisions where the protons only barely collide, which provide important measurements and potential tests of particularly the strong force [12–14]. At the LHC ring, there are two more collision points which host dedicated experiments, such as the already mentioned LHCb experiment that was originally predominantly designed to study the properties of beauty and charm quarks, including any potential differences between quarks and antiquarks and other properties of bound state of these quarks. LHCb has since changed its development into a broad and internationally leading collaboration that also studies many of the same questions addressed by ATLAS and CMS, often using complementary techniques and kinematic properties. At other colliders, similar experiments are coming online, such as the Belle2 experiment that is ramping up to perform an essential precision physics programme in the heavy flavour sector using electron–positron collisions at the SuperKEKB accelerator complex in Japan. The expected precision of measurements that are expected to be performed soon by Belle2 will create a very dynamic period in the study of heavy flavour quarks, particularly as many of the current results by LHCb that show some tension with respect to SM predictions can be performed independently at this facility.

2.2.7 Beyond the LHC

Beyond collider physics, there is a wealth of other particle physics experiments that also examine the SM. The study of neutrinos, the most abundant particles in the Universe, has a vibrant program that complements the study of the SM at colliders. Some may argue that neutrino physics has already moved to physics beyond the SM, as it is clear the particles have a mass that is not the default in the SM. Open questions such as the number of neutrinos, their mass hierarchy, and the possibility of violation of the tenets of the SM such as CP violation and the potential unification of forces can all be examined by studying the physics of neutrinos.

There are experiments where neutrinos are studied coming from particle accelerators, from reactors, and from the cosmos. Neutrinos are notoriously difficult to detect, so most of these experiments look for very few interactions and use detectors and experimental underground locations where almost no background is present. Producing such sensitive detectors creates unprecedented experimental challenges. In the future, new experiments will come online that will be able to answer many of the open questions. For example, the Deep Underground Neutrino Experiment (DUNE), a leading experiment for long-baseline neutrino physics and proton decay studies that uses a neutrino beam between Fermilab, Illinois and Sanford lab, South Dakota in the USA. In recent years many European particle physicists have joined these efforts. The goals are to use neutrinos to study similar underlying questions as accelerators, such as the origin of matter, unification of forces, and searches for new physics. Some of the challenges in neutrino physics are different than at colliders. Still, many of the final physics goals are shared. In addition, many large neutrino detectors have the sensitivity to study particles coming from the cosmos, such as the potential observation of thousands of neutrinos from supernovas and other transient astronomical phenomena.

The hunt for a solution to the dark matter puzzle is also becoming competitive, with the Xenon-based experiments XENONnT and LUX-ZEPLIN both starting their science runs recently [15, 16]. Both experiments rely on extremely pure large quantities of liquid Xenon atoms that should create very challenging and tiny signals when dark matter particles interact with them. These experiments rely on extremely low background environments, so are housed deep underground and aim to have very pure conditions so only a few dark matter interactions could be enough to establish a potentially paradigm-shifting discovery. A large number of smaller experiments that aim at specific dark matter signatures are also running, and more are in development. In combination with the Xenon-based experiments, dark matter programmes at neutrino experiments, and the dark matter signatures that could potentially be visible at the LHC experiments, it is likely that if dark matter is a particle, it will be cornered in the coming years.

2.2.8 Challenges

There are also technical and theoretical physics challenges that are vital to make progress beyond the challenge of identifying potential new physics that were described in all its facets previously. These are partially technical, such as the genuinely overwhelming data volumes that the LHC will produce in the future and the improvement of detector technologies to robustly detect these large volumes, but also address the more detailed understanding of the SM in enough detail to continue testing it with better accuracy.

2.2.9 Challenges: the data challenge

In the coming years the amount of data produced by particle physics experiments will increase exponentially. In addition there are challenges in producing the simulated samples that are necessary to analyse this data. To successfully analyse the data produced by the high luminosity LHC, the computational tools available will have to improve substantially, both as far as being able to process data volume, data processing speed, and simulation methods.

Each of those poses unique challenges that also have links to other big challenges in broader scientific context and in society. After all, the big techniques from particle physics have by now made large changes in other scientific fields possible, and with the large majority of the world's population using social media, the data volumes produced by humanity are, just like the data produced in particle physics, only expected to continue growing. It is also clear that the evolution of computing technology will alone not cover the increase in data resource needs. Improvements in computational techniques including modern artificial intelligence, but also potential new hardware developments and new computational methods like quantum computing and quantum sensing are expected to open opportunities to reduce the expanding cost of collecting, processing, and storing the data.

2.2.10 Challenges: the accelerators of the future

Since the dawn of particle physics in the early 1950s, particle physicists have relied on colliders that collide either (anti)protons, or electrons and positrons. Historically, these machines have played different roles; due to lack of synchrotron radiation, hadron collider machines can reach typically higher energies but with substantial backgrounds that create experimental challenges. In contrast, the lepton machines are extremely powerful at performing precise measurements with substantially reduced interference from other SM processes. One crucial piece of knowledge that the LHC has taught us is that with modern analysis techniques, hadron machines can also be partly used as precision machines. This precision functionality is limited chiefly to measurements that test the SM through ratios, where many uncertainties are reduced and the SM tends to be well understood, and the case where the backgrounds are understood to such a level that SM processes can be very precisely measured.

The particle physics community is currently performing an international procedure examining the potential, feasibility, and support of future colliders. Broadly, multiple machines are under consideration, divided into two categories: Higgs boson factories and search machines. The Higgs boson is still hiding many of its properties and has only been tested up to about 10% in its most sensitive scenarios and will be tested up to a few percent at the end of LHC running. Such a machine has already been deemed the highest priority by the European physics community [17]. Questions, however, remain as to how such a machine will be realised. It is important to have an international consensus and unanimous support on what machine be built and potentially where. The huge number of different options, including even a muon-antimuon collider, makes this decision making a daunting task. One of the considerable organisational challenges of the field will be to achieve the completion of such a machine and a collider physics program that will reach well into the 21st century.

2.2.11 Challenges: timing detectors and detector innovation

As the data volumes increase and the collisions become more busy, it becomes also more and more challenging to precisely detect interesting collisions. A vibrant and innovative instrumentation effort is part of the particle physics community, and creative new ideas that will allow high fidelity detectors that will perform well under an ever-increasing number of particles in a single collision.

Detector techniques that use not only spatial or energy measurements are of particular promise, as with accurate time detection it is easier to disentangle otherwise overlapping particles that did arrive tens of nanoseconds apart. In the future, such timing technology is expected to play an important role for detecting the collisions at future colliders, and the collaborations at the LHC are already developing first versions of these detector technologies that will be available for high-luminosity LHC.

There are of course many challenges in the developments of all these detectors; particle physics experiments have unique needs as far as the uniformity of the individual detection elements but also as far as radiation hardness and material budget (so indirectly power-usage). Many of these aspects are still actively being researched and are expected to continue to improve in the years to come. Innovation and strengthening of existing instrumentation efforts are critical to maximize the scientific outcomes of future accelerators; after all, the investment of these machines makes it paramount that powerful detectors are present in these precious collisions, and to push forward the innovation in non-accelerator-based experiments as well. In addition, a strengthening of the link between detector innovation and collaboration with industry is one that is expected to become more and more important in the future.

2.2.12 Challenges: the theory challenge

With ever-increasing experimental data, challenging the SM more and more precisely, it becomes also ever-more important that the SM can be calculated more accurately so that the hypothesis testing as the classical experimental method described can continue to flourish. Theoretical physics calculations are an essential part of this endeavour, where potential new lines of research can be identified and the precise tools are made that allow to fully exploit experimental data.

A broad programme of theoretical research that covers the full spectrum of particle physics is essential to achieve these goals. This includes both exploratory new ideas, foundations in mathematical physics and links to cosmology, astroparticle physics, and nuclear physics. As the calculations become more and more complex, the tools to do so also become an important part of the research. For example, to fully exploit the study of the Higgs boson at the high-luminosity LHC or at future colliders, the experimental precision is expected to very quickly not be the most important uncertainty, and uncertainties due to theoretical assumptions that are in principle reducible through challenging intellectual innovation will become dominant for many scenarios.

2.2.13 What is the ultimate status of the constants of Nature?

Particle physics is a field where that has the sensitivity for probing the most basic concepts of Nature. To do so, solutions to complex challenges both on the detector and on the accelerator side are necessary. Further challenges in computing, simulation, and data processing are also creating challenges to make full use of the available data. The field benefits from a fertile cross-pollination between theoretical physics and experiment. The apparent weaknesses of the SM, in combination with a wealth of potential extensions of the SM, put the fields in an exciting but very precarious position. Inconsistencies of the SM in the neutrino sector, the dark matter puzzle, the fact that the SM may very well be incomplete at higher energies or is at least not able to describe all processes reliably. These and many other arguments all point to the fact that the SM is incomplete and creates the clear potential for new particles to be discovered at the LHC in direct production or indirect measurements of SM properties.

On the other hand, the Standard Model's predictions tend to agree to very high accuracy in most measurements, and direct searches for low-hanging fruit at the start of the LHC data taking have not produced any convincing sign for a new particle beyond the discovery of the Higgs boson in 2012. This creates a potentially exciting but extremely challenging dichotomy where the focus has moved from predictions to searching for deviations for unpredicted (or at least, not overly model-dependent) signatures. The transition from a more theory-prediction-oriented field into an experimentally driven research program is currently in full swing in this field. Only time will tell what the future has in store for particle physics, but considering the wealth of experiments, the large number of extremely fundamental questions that are corroborated with data and that the SM cannot answer, and the fact that the flagship LHC has only collected one tenth of its data, the future is very bright.

2.3 The origin of visible matter

Angela Bracco[1]

[1]Università degli Studi di Milano, Milano, Italy and INFN

Nuclear physics is the science of the atomic nucleus and of nuclear matter. The atomic nucleus is the dense core of the atom and is the entity that carries essentially all the mass of the familiar objects that we encounter in Nature, including the stars, the Earth, and indeed human beings themselves. The quest for the origin of visible matter requires very good knowledge of nuclear matter and reactions under a large variety of conditions up to the most extreme ones. The physics of nuclei, starting from the hot dense soup of quarks and gluons in the first microseconds after the Big Bang, up to the formation of the chemical elements drives intense and innovative research in both experiment and theory.

The experimental investigations of nuclear properties require the use, and the continuous development, of an arsenal of experimental techniques and require different types of research infrastructures and theoretical approaches. The technical developments for the nuclear research are in large part also applied to the benefit of our society.

The distinctive feature of nuclear physics research is the breadth of the topics it covers. Here these topics are grouped under two titles, 'The fundamental description of the heart of matter' and 'Nuclear structure and reactions for the origin elements'.

The progress made in these research areas and the future perspectives are briefly presented and discussed. It is clear from the short narrative in the following pages that the physics of atomic nuclei and their constituent components is rich, varied, and extremely complex at many levels. Moreover, research in the next decades will be challenging and hopefully full of surprises.

2.3.1 Introduction

Understanding the origin of visible matter is a very complex but fascinating problem that drives experimental and theoretical research involving strongly the physics of nuclei.

From the hot dense soup of quarks and gluons (the latter carrying the interaction among quarks) in the first microseconds after the Big Bang, through the formation of protons and neutrons to the evolution of the chemical elements, the physics of nuclei is fundamental to our understanding of the Universe and, at the same time, is intertwined in the fabric of our lives. Nuclei also constitute a unique test bench for a variety of investigations of fundamental physics, which in several cases are complementary to elementary particle physics, but use different approaches. Nuclear physicists and chemists are creating totally new elements in the laboratory and producing isotopes of elements that have previously only existed in stellar explosions or in the mergers of neutron stars. For nuclear physics research new tools like accelerators and detectors are developed that often find broad applications in industry, medicine, or national security.

The overarching goal of present-day nuclear physics research is to explain the origin of visible matter and for this one needs to study in-depth several different physical phenomena involving nuclei and their constituents (see references [18–20]). Some central questions in this connection are:

- How is the mass generated and what are the static and dynamical properties of nuclei and of their constituents?
- How does the strong force in nuclei emerge from the internal structure of its constituents, the protons, and neutrons?
- What are the properties of nuclei and strongly interacting matter as encountered shortly after the Big Bang, in catastrophic cosmic events, and in compact stellar objects?
- How does the complexity of nuclear structure arise from the interactions of protons and neutrons in the nuclei?
- What are the limits of nuclear stability?
- How and where in the Universe are the chemical elements produced?

The experimental investigations addressing these questions require the development and skillful use of a variety of experimental techniques and theoretical approaches. Consequently, there are different types of research infrastructures where nuclear research is carried out. The distinctive feature of the research in nuclear physics is the presence of different research areas which can be grouped together under two titles: 'The fundamental description of the heart of matter' and 'Nuclear structure and reactions for the origin of elements'.

In Europe there are several laboratories (see, e.g., references [21–34] and references therein) that are either mainly devoted to nuclear physics or are carrying out a rich scientific program in this area. In figure 2.5 'Open access' laboratories are indicated in the map of Europe while other smaller sized, often university-based, laboratories are not shown. These smaller laboratories are good niches for specific timely activities and contribute very much to R&D and training. The scientific activities in nuclear research are distributed over different facilities which are not necessarily placed at the same location (see also references [35–40] and references therein). These facilities conduct research at the forefront, have minimal overlap among themselves, and are well coordinated via European projects. In addition, in Europe, for more than 30 years, an expert committee, NuPECC, including a representative of the Nuclear Physics division of EPS, has overseen research programs and prepares strategies for the field every 5–7 years, which are reported in published Long Range Plans.

2.3.2 The fundamental description of the heart of matter

It was just about hundred years ago that Lord Rutherford uncovered the existence of the proton as one of the basic building blocks of the atomic nucleus, and a decade later the other major building block, the neutron, was discovered. After that, many features of the force that binds protons and neutrons (the nucleons) within an atomic nucleus were established, and because of its strength this force is called 'strong

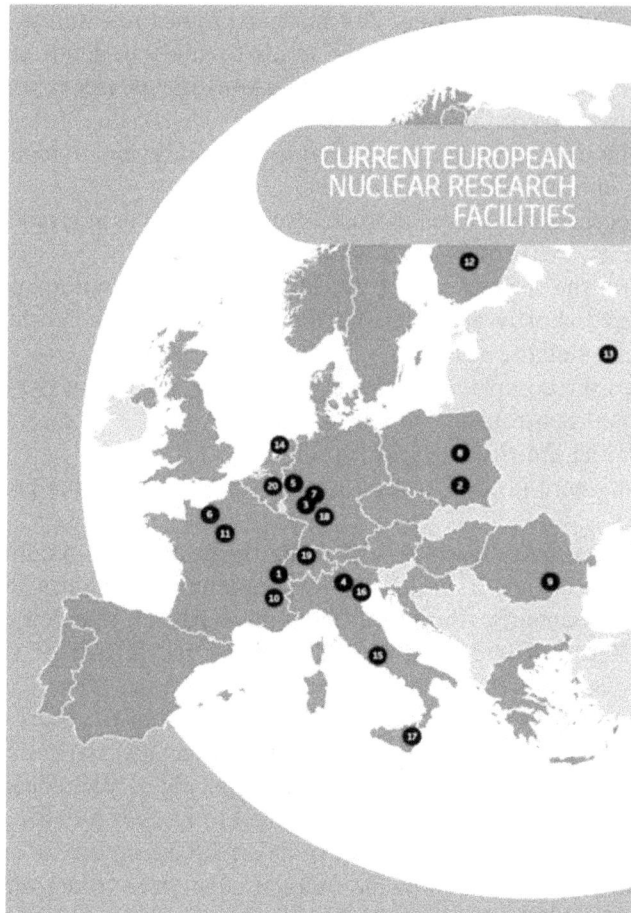

Figure 2.5. The current 'open access' nuclear research facilities in Europe. Other existing smaller size facilities, which are mainly university based, are not indicated in this map, although they contribute to the European research in this field, particularly in the area of applications. The facilities in this figure are: (**1**) CERN—European Organisation for Nuclear Research (ALICE, AD, COMPASS, and ISOLDE), Genève, Switzerland; (**2**) CCB—Cyclotron Centre Bronowice (IFJ PAN) Kraków, Poland; (**3**) ELSA—Electron accelerator, Bonn, Germany; (**4**) ECT*—European Centre for Theoretical Studies in Nuclear Physics and Related Areas, Trento, Italy; (**5**) FZJ—Forschungszentrum Jülich (COSY and HPC), Jülich, Germany; (**6**) GANIL—Grand Accélérateur National d'Ions Lourds (SPIRAL and SPIRAL2), Caen, France. (**7**) FAIR and GSI—Helmholtzzentrum für Schwerionenforschung GmbH, Darmstadt, Germany; (**8**) HIL—Heavy Ion Laboratory, Warsaw, Poland; (**9**) IFIN-HH Horia Hulubei—National Institute of Physics and Nuclear Engineering, Bucharest, Romania; (**10**) ILL—Institut Laue–Langevin, Grenoble France; (**11**) IPN—Institut de Physique Nucléaire, Orsay, France; (**12**) JYFL—Accelerator Laboratory, University of Jyväskylä, Finland; (**13**) JINR—Joint Institute for Nuclear Research, Dubna, Russia; (**14**) KVI-CART—Kernfysisch Versneller Instituut, Groningen, The Netherlands; (**15**) LNF-INFN—Laboratori Nazionali di Frascati of INFN, Frascati, Italy; (**16**) LNL-INFN—Laboratori Nazionali di Legnaro of INFN, Legnaro, Italy; (**17**) LNS-INFN—Laboratori Nazionali del Sud of INFN, Catania, Italy; (**18**) MAMI—Mainzer Microtron, Institute for Nuclear Physics, Mainz, Germany; (**19**) PSI—Paul Scherrer Institute, Villigen, Switzerland; (**20**) SCK*CEN—Mol, Belgium. These are the facilities with scientific programs dealing with the topics addressed in this chapter 'Origin of visible matter'. Adapted from http://www.nupecc.org/pub/lrp17/nupecc_lrp_bro-chure_2017.pdf [19] courtesy of NuPECC (nupecc.org).

2-22

force'. Of particular interest is the strong interaction of nucleons that are relatively close to each other, so that their internal structure might be expected to come into play. This 'short-range interaction' is responsible for the fact that most nuclei have roughly the same density, which ultimately determines the stability and size of neutron stars. While nucleons mostly move at moderate speeds inside nuclei (up to 30% of the speed of light), their short-range encounters can impart significantly higher momentum to each of them and make them move swiftly in opposite directions. Such high-momentum correlations, long predicted by nuclear theory, provide an excellent way to study the short-range part of the nucleon–nucleon interaction and have now been observed.

In spite of the progress made to establish the features of the nucleon–nucleon interaction and of the multitude of other strongly interacting particles (named hadrons) a fundamental understanding of the underlying laws of physics was missing until quantum chromodynamics (QCD) was recognized as the fundamental theory governing nuclear matter.

2.3.2.1 The structure of nucleons and other hadrons and the theory of quantum chromodynamics

According to the theory of QCD protons, neutrons, and all other hadrons are made from quarks, their antimatter siblings (antiquarks), and particles called gluons, which carry the force that binds quarks to each other (see, e.g., references [41–45]). Protons and neutrons can be thought of as containing three so-called valence quarks, immersed in a shimmering cloud of quarks, antiquarks, and gluons, all continually winking into and out of existence according to the laws of quantum mechanics. Quarks and gluons are the building blocks of all hadrons. No single quark or gluon, however, can be observed in isolation and thus one says that they are confined within a hadron. This implies that any process by which one tries to rip a quark out of a proton or neutron makes new hadrons, without ever isolating a single quark. Therefore, the strong interaction between quarks decreases at short distances and increases at large distances (in contrast to quantum electrodynamics where the interaction has an opposite behavior), and at very short distances the quarks are free (this property is named 'asymptotic freedom'). Gluons carry the force between quarks in much the same way that electromagnetic forces are carried by the photon, but in contrast with photons that do not interact between themselves, gluons do interact with each other.

It turns out that these unusual features of QCD imply that the intrinsic mass of the three valence quarks in the nucleons gives rise only to a small fraction of the proton and neutron masses and hence gluons also contribute to it. Since protons and neutrons account for nearly all the mass of atoms, almost all of the mass of the visible matter in the Universe is due to these seemingly exotic QCD effects. And while these general features of nuclear matter are well established, a detailed understanding of how this originates from QCD is only now emerging. Understanding the structural complexity of protons and neutrons in terms of quarks governed by the laws underlying the QCD theory is one of the most important challenges facing physics today, in spite of the impressive progress made so far.

Recent advances in computational power now allow for precise calculations of the masses which at the same time predict novel exotic particles to be identified experimentally.

The significant increase in the capabilities of large-scale supercomputers in conjunction with the enormous progress made in developing efficient simulation algorithms have allowed huge increases in the accuracy of first principles predictions (e.g., of the mass difference between the proton and neutron). Nevertheless outstanding challenges remain, among which is the understanding of how the proton gets its magnetic property, namely its spin (due to a rotation of a particle on an intrinsic axis), which is measured to have the value 1/2 (in units of $\hbar$, which is the reduced Planck constant of fundamental importance in quantum mechanics). Figure 2.6 is a pictorial representation of possible mechanisms that produce the spin of the proton. The individual spin of the valence quarks inside the proton accounts for only 1/3 of the spin of the protons. Theorists predict that gluon and a 'sea' of transiently existing quarks also contribute together with magnetic effects due to the orbital motion of the quarks.

Only with new measurements and new calculations of how quarks and gluons can combine into new particles with quarks of different types one can learn more regarding these fundamental questions. Because there are six quarks, the number and variety of hadron permutations possible is large, and laboratory experiments are devoted to creating and studying these combinations. In addition, one needs to study the spectra of hadrons, their internal structure and interactions—particularly those composed of the heavier quarks. Such hadrons are rare and exotic, yet tell us a lot about how the strong force works.

Similarly to atomic spectroscopy, which has been a crucial tool for studying the electromagnetic interactions that bind electrons to the nucleus, hadron spectroscopy experiments are carried out to illuminate the interaction that binds quarks. While the proton and neutron contain the two lightest valence quarks (up and down), other hadrons composed of these light quarks or of more massive quarks (strange, charm, bottom, and top) and their corresponding antiquarks can be created in energetic collisions produced by particle accelerators. Once produced, these hadrons decay promptly, allowing one to measure only a few of their properties, such as their mass, charge, and angular momentum.

Studying patterns of hadrons classified by their properties provides insight into the theory of QCD. The observed patterns of states suggest that almost all hadrons

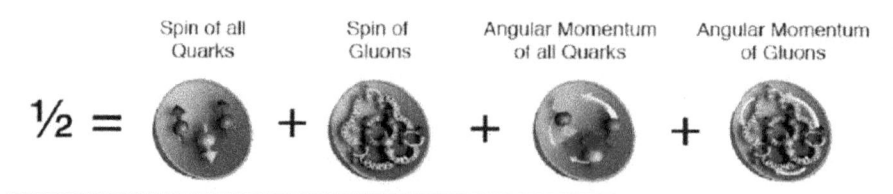

Figure 2.6. Schematic view of the proton and the potential contributions to its spin. Adapted from https://science.osti.gov/~/media/np/nsac/pdf/2015LRP/2015_LRPNS_0918155.pdf [20] courtesy of U.S. Department of Energy (energy.gov).

fall into two classes: baryons that contain three valence quarks, like the proton and neutron, and mesons that contain a valence quark and a valence antiquark. In principle, the QCD theory allows also hadrons made of two quarks and two antiquarks (tetraquarks), four quarks, and an antiquark (pentaquarks), and infinitely many other more complex configurations. Recently, physicists studying the spectrum of heavy mesons formed with charm and bottom quarks have uncovered evidence that supports the existence of tetraquark and pentaquark hadrons. Understanding the properties of these new states of QCD may provide insight on why nature prefers hadrons with relatively few quarks.

It is clear that the theory of QCD has to be tested at different energies so that experiments addressing the present open problems in this research area are conducted not only at the LHC accelerator at CERN but a large fraction of experiments are performed using accelerators providing particle beams with energy from few GeV to few hundreds of GeV. These particle accelerators are located at CERN and in other different places in Europe (see figure 2.5). In addition, several European research groups, of quite a large size, collaborate and give important contributions to experiments carried out in laboratories outside Europe, in particular at the Thomas Jefferson Laboratory in Virginia (USA).

The type of beams used for experiments addressing the problem of the strong force and of the structure of hadrons are hadron particles (protons, pions, and kaons), but also electrons and muons that interact via the electromagnetic force.

2.3.2.2 Symmetries and fundamental interactions

The presently known fundamental interactions governing Nature and the Universe from the largest to the smallest distances display symmetries and symmetry-breaking effects. High-precision studies allow tests of our understanding of Nature that are complementary to experiments at the highest energies and sometimes offer higher sensitivities to new effects beyond the SM of particle physics.

Nuclear Physics has played a major role in finding and establishing the laws that govern physics at the most fundamental level, with the nucleus being a very useful test bench for the weak, electromagnetic, and strong interactions. One of the most notable examples is the maximal violation of spatial inversion symmetry (named parity) in the weak interaction that was measured in the beta decay of unstable nuclei. Shortly after the discovery of parity violation other symmetries were found to be broken, and this triggered intense research on symmetry violations. Experimental activities include parity violation studies on atoms, ions, and molecules and searches for time reversal violating electric dipole moments of particles, including neutrons.

High-precision tests of fundamental symmetries and the SM with low-energy antiprotons are conducted at the facility AD/ELENA at CERN. This research has progressed considerably, and in figure 2.7 two examples are illustrated schematically, one showing the goal of comparing the gravitational mass of proton and antiproton and the other had the aim of comparing the mass of the antiproton with that of the electron (see also references [46, 47]).

Antiprotonic helium

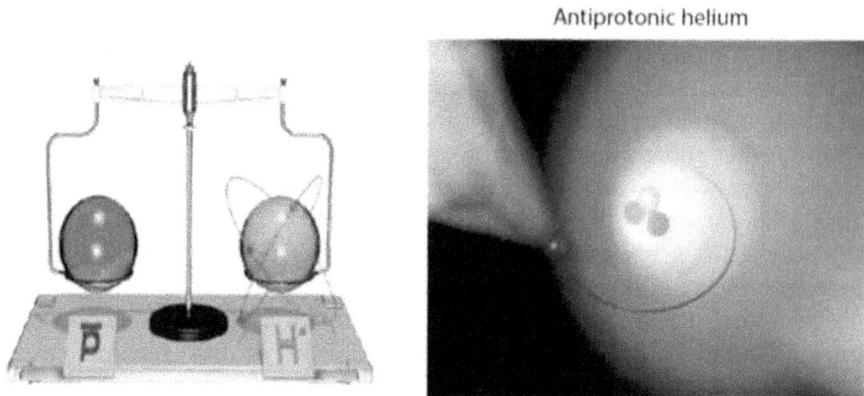

Figure 2.7. Left: Artistic views of the comparison of the gravitational mass of an antiproton and of a proton. Right: Artistic view of the antiprotonic helium. These researches are performed at CERN. Adapted from http://www.nupecc.org/pub/lrp17/lrp2017.pdf [18] courtesy of NuPECC (nupecc.org).

Today, many experiments on fundamental interactions bridge the areas of nuclear, atomic, particle, and astrophysics, and major advances in technology have made novel approaches feasible.

At the same time the most advanced theoretical and computational techniques are developed and applied by nuclear, atomic, and particle theory from low-energy particles, ultracold atoms, ions, and molecules. A number of these precision experiments use dedicated tabletop setups in small-size laboratories at universities where innovative techniques are tested before they are installed at large-scale infrastructures.

Another major class of experiments requiring high precision and the capability of detecting rare events concerns the search for the double beta decay without neutrino emission, performed worldwide, and in Europe in the underground laboratories LNGS, Modane, and Canfranc. These experiments are intended to determine the nature of neutrinos and to investigate the possible violation of conservation laws.

2.3.2.3 The phases of strongly interacting matter and the quark-gluon plasma
Nuclear collisions of heavy nuclei such as lead and gold at energies in the interval 100 GeV—few TeV per nucleon produce nuclear matter with temperatures in the trillions of degrees. The main motivation to measure these collisions is to recreate the matter in the same conditions taking place in the microseconds-old Universe after the Big Bang. It was understood in the 1970s that ordinary protons and neutrons could not exist at temperatures above two trillion degrees Celsius (see also references [48–50]).

The predicted new form of matter, which can be recreated by heating protons and neutrons until they 'melt,' was named quark-gluon plasma (QGP). In the hot soup of quarks and gluons produced by the Big Bang process the quarks are free to move within the region of the plasma over relatively large distances. A few fractions of a second after the Big Bang the temperature of the Universe dropped and the quarks became confined inside complex particles such as protons and neutrons that later formed the nuclei of the atoms. The transition between the primordial QGP and the hadron formation has, as far as we know, not left any imprint that is visible in present-day astronomical observations. Nevertheless, QGP matter is believed to exist inside the core of neutron stars, at baryon densities much higher than normal nuclear densities.

The energy or baryon densities necessary to form the QGP may be recreated in the laboratory via heavy ion collisions at sufficiently high energies, within volumes of the order of the nuclear size. The idea to collide heavy ions accelerated at ultrarelativistic energies for bringing nuclear matter into the QGP phase and studying its properties in the laboratory dates back to the early 1980s. Pioneering studies promptly demonstrated that the energy deposit and the nuclear stopping in the central part of the collision were quite large. The acceleration of heavy ions at the SPS accelerator at CERN, since 1994, and later the experiments at the BNL Relativistic Heavy Ion Collider (RHIC) energy for nucleon collisions of 200 GeV, created a system with energy density above the critical value and allowed the observation of the predicted signatures of the QGP. At higher energy, the colliding system was found to enter a new regime with an increase in the lifetime of the quark-gluon plasma.

The feature of the strength of the strong interaction decreasing at short distances has a very profound implication for nuclear matter under extreme conditions because at sufficiently high nuclear density or temperature, the average inter-parton distance becomes small, and therefore their interaction strength weakens. Above a critical energy density, of the order of 0.3 GeV fm^{-3}, a gas of hadrons undergoes a deconfinement transition and becomes a system of unbounded quarks and gluons. Predictions for this transition were obtained with simulations based on the QCD theory, in the form of a rapid increase of the entropy density around the critical energy density. The formation of QGP is also accompanied by a restoration of a particular symmetry, the chiral symmetry (a system is chiral if it is distinguishable from its mirror image).

The so-called phase diagram for nuclear matter is illustrated in a schematic way in figure 2.8, where the laboratories involved in experimental programs concerning the physics of the quark-gluon plasma are indicated by stars. The goal of the high-energy heavy-ion programs at these laboratories is to identify and characterize the properties of the QGP. These programs naturally have two steps: understanding the dynamics of heavy-ion collisions (e.g., via comparison to phenomenological models) and the extraction of fundamental properties to be compared to QCD predictions.

The heavy-ion collisions can be described as being characterized by three main stages: (i) an early non-equilibrium stage, (ii) an expansion stage, and (iii) a final freeze-out stage. An advantage of this modular structure is that it allows for the use of more or less advanced theoretical tools in each stage. In this way the modeling of heavy-ion collisions can gradually be improved and used to constrain further the properties of strongly interacting matter. This picture, and the associated phenomenology, has evolved over the last years as observables have been identified that are sensitive to specific processes in each phase.

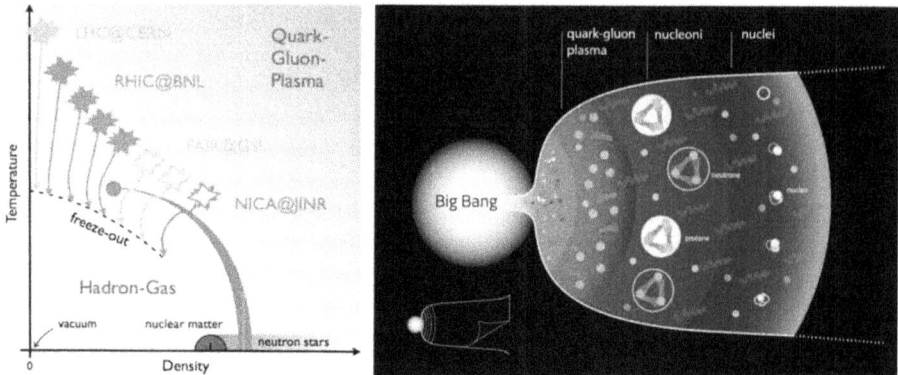

Figure 2.8. Left: Illustration of the phase diagram of quantum chromodynamics. The facilities probing the different sectors of this diagram are indicated. Right: Pictorial illustration of the Big Bang indicating the evolution from quark-gluon plasma up to the formation of nuclei. Adapted from http://www.nupecc.org/pub/lrp17/lrp2017.pdf [18] courtesy of NuPECC (nupecc.org).

The first stage, which also provides initial conditions (spatial distribution of the deposited energy and pressure, initial flow velocity) for the subsequent hydro-dynamical stage, is the least known and is often described by simple geometrical models using the underlying strong nucleon–nucleon interactions. More in-depth descriptions treating the collision at quark and gluon level are being developed today. Although some productions of particular particles via electromagnetic processes that were measured by LHC experiments in lead-lead collisions (e.g., J/ψ photo-production) provide evidence for shadowing due to nuclear gluons, further efforts are required to extract its amount.

During the second stage—the fireball expansion—the bulk evolution is described by relativistic viscous hydrodynamics. Due to the near-perfect fluid nature of the QGP, the initial geometrical anisotropy is efficiently converted into momentum anisotropy of the final particles. The systematic measurement of the flow of particles at different directions (as for example the elliptic flow) provides an avenue for constraining initial-state model calculations and transport properties of the QGP. The emergent hydrodynamic properties of the QGP are not apparent from the underlying QCD theory and thus they have been quantified with increasing precision via experiments at both RHIC and the LHC over the last several years. New theoretical tools have allowed the degree of fluidity to be characterised. In the temperature regime created at RHIC, QGP is the most liquid-like liquid known, while the hotter QGP created at the LHC has a somewhat larger viscosity. This temperature dependence will be more tightly constrained by the ongoing and upcoming measurements. Moreover, this bulk evolution of the collision of ions at relativistic energy provides the substrate for the medium modifications of hard probes, although a better integration of these two aspects of the description is certainly needed.

The formation of hadron particles takes place when the system reaches the pseudo-critical temperature, and after that the scattering rate decreases quickly and a kinetic description becomes more appropriate than hydrodynamics. This kinetic description can in principle account for the decoupling of the hadrons from the fireball produced by the collision. The measured relative abundances of hadrons indicate that chemical freeze-out happens at a temperature which is very close to the hadronisation temperature and at nearly zero chemical potential. Subsequently, the hadrons continue to rescatter elastically until they reach the kinetic freeze-out temperature where they decouple and freely stream to the detectors.

The complex analysis of data collected so far has produced many interesting results addressing the bulk properties of the collisions and the details of particle production whose explanation relies to large extent on the comparison with data from proton collisions. Among the results from the data from ALICE at LHC there is the copious production of antinuclei whose investigation opens interesting perspectives in understanding the role of antimatter in the Universe, the latter based on astronomical observations.

2.3.3 Nuclear structure and reactions for the origin of elements

The subfield of nuclear structure and reactions focuses on measuring, explaining, and using nuclei to improve scientific understanding as well as for scientific applications. This research addresses the underlying nature of atomic nuclei and the limits to their existence. The complex physics of the atomic nucleus has not only shaped our Universe but also ourselves (see, e.g., references [51–55]).

Nuclear astrophysics aims to understand the evolution of matter in all its complexity across cosmological times, just after the Big Bang, during the life-cycles of stars when the primary elements needed for life were created (see schematic picture in the top-left part of figure 2.9) and in the violent cosmic events that delivered the heaviest elements (top-right part of figure 2.9). The lightest elements, hydrogen, helium, and lithium, were made from primordial Big Bang matter, before the first stars were formed. The rest of the elements are thought to be built up in stars via processes occurring in their plasma involving complex chains of nuclear reactions and radioactive decays. In the bottom part of figure 2.9 different explosive nucleosynthesis processes responsible for the creation of the heaviest elements are shown by the coloured arrows (see, e.g., references [56–59]).

During a typical lifetime of a star, its supply of hydrogen fuel is 'burnt' to helium, followed by burning to carbon, nitrogen, and oxygen. Eventually the heavier elements such as silicon and finally iron and nickel are reached. The reaction chains can be mapped on a landscape charting the nuclear species that form with increasing numbers of protons and neutrons. It climbs slowly up along the central 'valley' of stable nuclei, continuing until all the nuclear fuel of the star is entirely consumed and thus the star eventually collapses.

We know little about how the heavier elements beyond iron in the Periodic table are created. Theory suggests that many of them are generated in extremely violent environments such as a supernova explosion that ends the life of a very massive star, followed by its core collapse into a neutron star or black hole. Other stellar processes of interest are those in which a white dwarf in a binary star system is

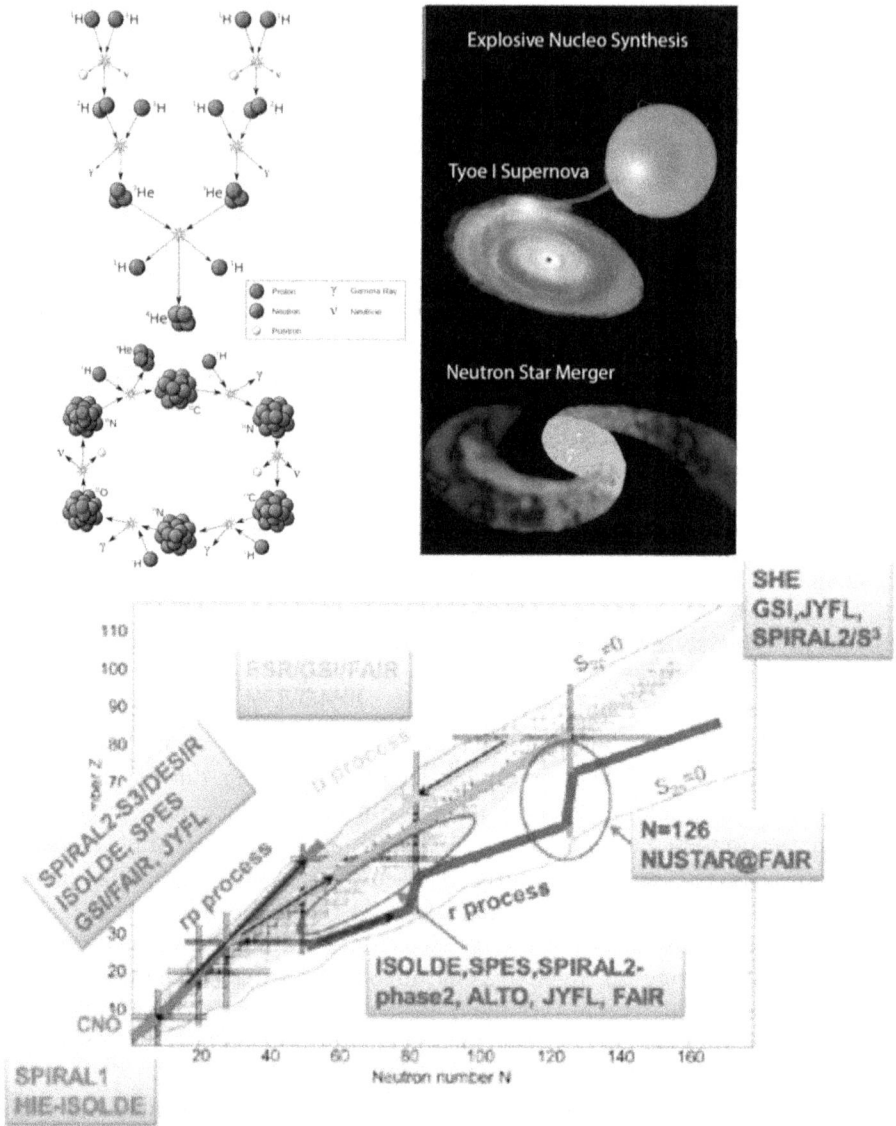

Figure 2.9. Top left: Artistic pictures of the nucleosyntheses of light nuclei. The p–p and the CNO cycles are shown. Some of the reactions involved in these cycles were measured at stellar energy in the laboratory LNGS-INFN with the experiment LUNA and the neutrinos with the experiment BOREXINO. Top right: Two pictures illustrating schematically two types of explosive nucleosynthesis (supernovae and neutron star mergers). Bottom: Schematic illustration of the nuclear chart, with the number of neutrons on the horizontal axis and the number of protons on the vertical one. The different regions relevant for the various nucleosynthesis processes in stellar explosions are indicated together with the European facilities involved in the study of these nuclei. The major facilities outside Europe addressing these issues are TRIUMF in Canada, FRIB in USA, and RIBF-RIKEN in Japan. Adapted from http://www.nupecc.org/pub/lrp17/lrp2017.pdf [18] courtesy of NuPECC (nupecc.org).

explosively resurrected by drawing off gas from its companion. Even more violent events like the merger of two neutron stars or black holes may also trigger these rapid nucleosynthetic processes. They are thought to follow paths up the nuclear landscape involving thousands of unstable nuclear species on both sides of the valley of stability, around the central ensemble of stable nuclei.

The challenge of understanding the origin and evolution of the chemical elements, and the role of nuclear physics in the lives and deaths of stars, requires state-of-the-art experimental, theoretical, and observational capabilities. Nature has supplied the Earth with about 300 stable as well as long-lived radioactive isotopes that we can study in the laboratory. However, a much greater variety of unstable isotopes, those created during stellar explosions, can be produced as radioactive beams in the laboratories using different systems of accelerator facilities. Developments during the years of these facilities have provided access to an increasing number of these exotic nuclei. However, only approximately 3000, of the possibly more than 8000 different nuclei that should exist, have been produced and characterized. A large 'terra incognita' dominated by the very neutron-rich nuclei and superheavy elements has still to be uncovered.

The nuclear properties relevant for the description of astrophysical processes depend on the environmental conditions. Nuclear theory is fundamental to connect experimental data with the finite temperature and high-density conditions in the stellar plasma. Advances in the description of nuclear interactions and of nuclear structure are crucial for our understanding of the evolution and explosion of stars, the chemical evolution of the Galaxy and its assembly history. In this context, the physics of neutron stars is of particular interest to the nuclear physics community. Indeed, the structure and composition of neutron stars are uniquely determined by features of the neutron-rich matter, and moreover by the relation between the pressure and energy density of this matter. Measurements of neutron star masses and radii place significant constraints on this pressure and density relation. Conversely, the measurement of both masses and the pressure of neutrons at the nuclear surface of exotic nuclei will provide critical insights into the composition of the neutron star crust. Neutron star crust models will have to rely on a combination of experimental and theoretical data, especially modifications to masses and effects such as super-fluidity and neutrino emissivity.

The features of nuclei are probed via specific nuclear reactions and the key questions are related to the evolution of nuclear structure, how the structure and the nuclear shapes change with temperature and angular momentum, and the complexity of nuclear excitations. Complex systems often display surprising simplicities and symmetries and nuclei are no exception.

The resulting emergent phenomena are the appearance of nucleonic shells, saturation of nuclear matter, rotation, superfluidity, and phase transitions. In addition, it is remarkable that a system with a hundred nucleons or more exhibits collective properties of many nucleons operating together. The detailed investigations of these varieties of nuclear features have progressed considerably, also entering into a high precision regime, thanks to powerful instrumentation, such as that for gamma-ray detection. For gamma spectroscopy the AGATA and

GRETINA arrays (in Europe and USA, respectively, see, e.g., references [60–62]) are the best-performing detector systems currently available, and there are plans to make them more powerful in the future, also in view of their great potential for applications in imaging techniques.

Extensive research is addressing the study of the neutron-rich matter which is less well known and for which very long isotopic chains are expected to exist. In heavier neutron-rich nuclei, the excess of neutrons collects at the nuclear surface creating a 'skin,' a region of weakly-bound neutron matter that is our best laboratory access to the diluted matter existing in the crusts of neutron stars. In spite of the progress made there are still mass and excitation energy regions that need to be explored to understand nuclear interactions, configurations, and nucleosynthesis processes. To access the unexplored mass regions one needs the beams that will be developed at the facilities undergoing major upgrades or under construction.

2.3.4 Challenges and opportunities up to 2050

For the next decades and in particular up to 2050 very appealing scientific challenges and opportunities exist in Nuclear Physics for both of the major research areas: 'The fundamental description of the heart of matter' and 'Nuclear structure and reactions for the origin of elements'.

At the FAIR facility, presently under construction, activities in both these research areas, together with many types of applications, will be carried out for several decades. This worldwide unique accelerator and experimental facility is composed of several research infrastructures that will allow for a large variety of unprecedented leading-edge research.

A schematic illustration of the FAIR facility, showing its four pillars, is shown in figure 2.10. These pillars are:

- The Super-FRS (super fragment separator) together with storage cooler rings and the versatile instrumentation (denominated as NUSTAR). These will allow decisive breakthroughs in the understanding of nuclear structure and nuclear astrophysics.
- The ultrarelativistic heavy-ion collision experiment CBM with its high rate capabilities. This will permit the measurement of extremely rare probes that are essential for the understanding of strongly interacting matter at high densities.
- The PANDA detector system at the antiproton storage cooler ring HESR. This facility will provide a unique research environment for an extensive program in hadron spectroscopy, hadron structure, and hadronic interactions.
- The APPA experiments will exploit the large variety of ion beam species, together with the storage rings and precision ion traps, for a rich program in fundamental interaction physics and applied sciences.

2.3.5 The fundamental description of the heart of matter: future perspectives

Our view of the structure of the nucleons has undergone a huge transformation in the last few decades. The most common picture is that the inside of the nucleon is rather a complex many-body system with a large number of gluons and sea quarks.

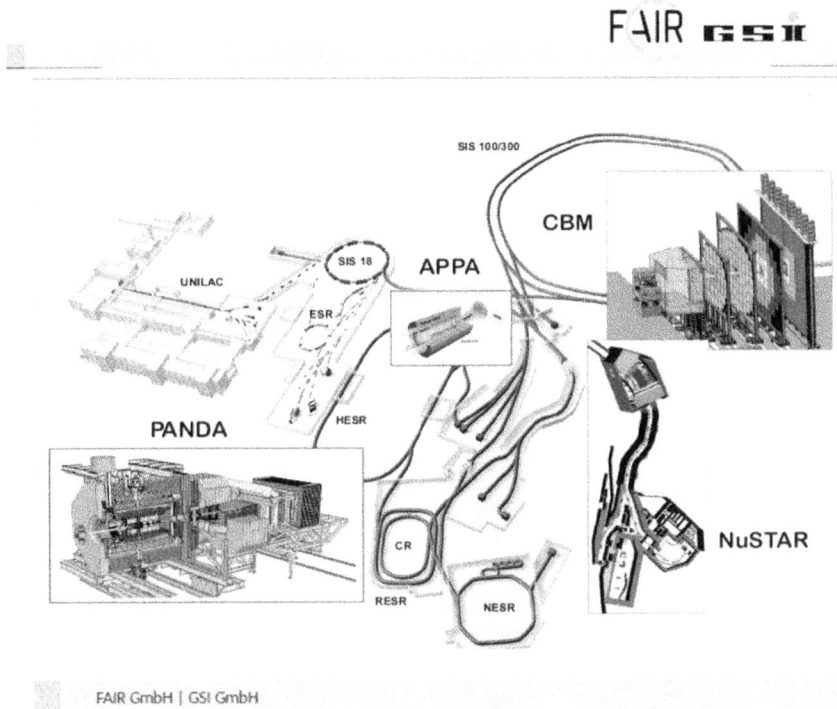

FAIR GmbH | GSI GmbH

Figure 2.10. The FAIR accelerator complex. The picture illustrates schematically the accelerators and the areas in which the detector systems devoted to the experiments concerning different areas of research. PANDA and CBM are devoted to the study of the fundamental constituents of matter, with CBM specifically designed for the hot highly compressed matter; NUSTAR is for investigations of nuclear structure, reactions, and nuclear astrophysics; APPA is for applications involving also atomic and plasma physics. Adapted from [21], copyright IOP Publishing Ltd. All rights reserved.

There is unambiguous evidence that they both play surprisingly important roles for defining the structure of nuclear matter around us. Their quantitative study requires a novel sophisticated tool such as the collision of high-energy electrons and ions.

In the USA a new facility providing Electron–Ion Collisions (EIC) is under construction at the Brookhaven National Laboratory, and several groups in Europe are involved primarily in the preparation of the experiments to be made there (see, e.g., references [63, 64]). The most precise information physicists have about the internal structure comes from scattering electrons, which interact with the nucleus and transfer momentum and energy. The layers of structure exposed change as one varies the value of the transferred momentum and energy.

The electron–ion collisions at the EIC facility will allow scientists to address immediate and profound questions about neutrons and protons and how they form the nuclei of atoms. The first question is how the mass of the nucleon arises. Protons and neutrons are bound systems of very light quarks so that their mass is generated dynamically through interactions carried by gluons. These collisions will map the gluon distribution in the proton, both in space and in momentum, with unprecedented precision, using the new technique of parton tomography. These images can be used to analyse the coupling between spin and orbital angular momentum. It would be possible not only determine the distribution of gluons but also measure the distribution of gluonic energy density and pressure in the proton. Other questions to be answered concern the origin of the nucleon spin and the emergent properties of dense systems of gluons.

The force mediated by gluons is fundamentally different from the electromagnetic force, since it strengthens as the quarks get farther apart so that quarks are permanently confined in neutrons and protons. Two questions concerning the gluons arise when nucleons are combined into nuclei: how are they modified as compared with a proton and is it possible that the whole nucleus becomes a dense gluon system? This abundance of gluons provides the opportunity to address fundamental questions about nucleons and nuclei. The number of gluons grows significantly in the high-energy limit. This means that gluons must overlap in the plane transverse to the electron–ion collision. The most interesting case is when this limit can be achieved at high resolution so that the number of gluons that can be packed into the transverse area of a proton or nucleus is large. Under such conditions, a quantum state of cold dense gluonic matter may exist. Such a state is possibly analogous to Bose–Einstein condensates of clouds of cold atoms created in atomic physics laboratories.

The EIC will be able to reach unprecedented gluon densities because it can accelerate ions with high atomic weight. Figure 2.11 shows how the scientific areas of investigation map onto the required collider luminosity and center-of-mass energy. The machine must collide electrons with protons and other atomic nuclei (ions) over a range of energies. There must be enough collisions for the experiment to gather adequate data to elucidate or settle the known physics questions, and other questions that may emerge, in a reasonable time. A collider's ability to squeeze many particles of two beams into a tiny volume where they collide defines its luminosity. The luminosity ultimately required of an EIC is comparable to those of the highest performing colliders built to date, such as the LHC at CERN.

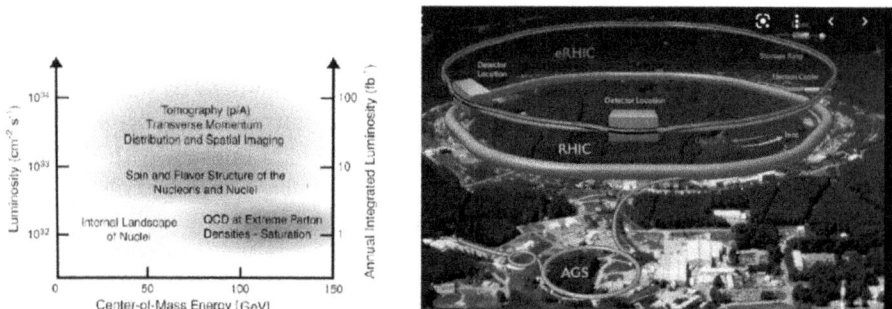

Figure 2.11. Left: The picture highlights the different scientific areas that can be addressed depending on the luminosity of the Electron Ion Collider (EIC) and the centre-of-mass energy. Right: Schematic layout of the planned EIC accelerator based on the existing RHIC complex at Brookhaven National Laboratory. Adapted from [34] courtesy of Sissy Körner.

Furthermore, given the crucial role of spin, both beams must be polarized. At present, it is anticipated that the EIC facility will begin operation in the early 2030s.

Beyond its importance for the understanding of fundamental properties of quarks and their interaction when they are in extreme conditions, the study of highly compressed nuclear matter is of great interest in the context of astrophysics, and the planned future measurements based on fixed target experiments (such as those at NICA and at FAIR) will provide highly selective data which are relevant for the physics of neutron stars.

The discovery of neutron stars with masses of about two solar masses and the observation of gravitational waves from neutron stars merging impose severe constraints on the nuclear matter in these stars. For example, it seems that the appearance of particles containing strange quarks could indicate that heavy neutron stars may collapse into black holes. While the structure of neutron stars is governed by the features at small temperature, present simulations of neutron star mergers indicate that the shocks which eject the matter relevant for formation of heavy elements (with approximately 200 nucleons) could reach temperatures of 100 MeV and densities of several times nuclear matter density. These conditions are very similar to those that are responsible for the collective expansion of the matter in heavy ion collisions that will be measured with very high rates at FAIR. High rates will allow the collision process to be probed in a selective way and thus permit a comprehensive analysis of collective flow and the particle production—quantities that are important inputs for simulations of neutron star mergers.

2.3.6 Nuclear structure and reactions and the origin of elements: future perspectives

Up to now many rare and unusual nuclei have been created and studied using accelerator facilities. The result has been the discovery of a wonderful variety of complex structures and shapes, from lightweight 'halo' nuclei in which an outer neutron orbits far from the central structure, through nuclei arranged in smaller clusters of nucleons, or in concentric shell-like structures, which when fully occupied by nucleons are very stable ('magic' nuclei), to dense, heavyweight nuclei with unusual shapes that behave more like wobbling liquid droplets. Experiments probing these nuclei attempt to understand how these various structures emerge as the numbers and proportions of protons and neutrons change across the nuclear landscape (see figure 2.12), and what controls their stability and behaviour when their energy content is varied. Of particular interest are nuclei with neutron–proton proportions that are only just bound, and are on the edge of shedding neutrons or protons (the neutron or proton 'driplines'), and 'superheavy' nuclei containing up to 300 nucleons.

The territory of neutron-rich nuclei is the most fertile ground for research in nuclear structure and further progress in this area requires measurements of key isotopic chains that encompass multiple magic numbers. With future experiments with radioactive beams, in particular those at FAIR with its suite of unique instrumentation, these chains will be accessible from the proton drip line to the neutron drip line (namely the boundary delimiting the zone beyond which atomic

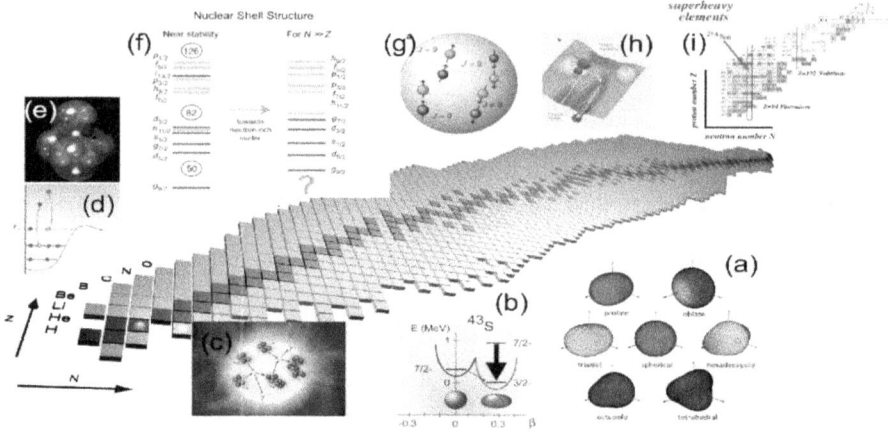

Figure 2.12. Artist's view of the nuclear landscape illustrating some of the key properties currently being studied using high-resolution spectroscopy: (a) the variety of nuclear shapes, (b) shape coexistence and isomerism, (c) reactions of astrophysical interest, (d) coupling to the continuum of unbound states, I cluster structure in nuclei, (f) evolution of the shell structure, (g) nuclear superconductivity, (h) understanding nuclear fusion and fission reactions, and (i) the journey towards the heaviest elements. Adapted from [60] CC BY 4.0.

nuclei decay by the direct emission of a proton or a neutron), permitting the study of the N/Z dependence of the nuclear force and continuum effects over very broad ranges. Such investigations will allow us to explore new paradigms of nuclear structure in the domain where many-body correlations, rather than the nuclear mean field, dominate. Single- and even multiple-neutron emissions are expected to characterise nuclei at the neutron drip line, while beta-delayed-neutron decay is prevalent among neutron-excess nuclei before the drip line is reached. Both forms of radioactivity only occur among nuclei far from stability.

To explain the nature of neutron-rich matter across a range of densities, as seen in the crust of neutron stars, an interdisciplinary approach is essential in order to integrate low-energy nuclear experiments and theory with knowledge from astrophysics, atomic physics, computational science, and electromagnetic and gravitational-wave astronomy. The nuclear input to this mix is essential. It includes the studies of nuclear matter at around both supranuclear and subnuclear densities, the analysis of high-frequency nuclear oscillations, and ab-initio approaches to the equation of state of nuclear matter.

Laboratory measurements using specific nuclear reactions induced by light and heavy ions are necessary to constrain nuclear matter compressibility and the symmetry energy. The symmetry energy is the energy difference between nuclear matter with protons and neutrons, and the pure neutron matter properties determine a range of neutron star features such as cooling rates, the thickness of the crust, the mass-radius relationship, and the moment of inertia. The precise characterisation of the nuclear symmetry energy through future experiments will be, therefore, a crucial step towards our capability of interpreting neutron star matter and its properties. Likewise, studies of masses, giant resonances, dipole polarisabilities, and neutron skin thicknesses of neutron-rich nuclei will provide key insights for astrophysics. Extending such measurements to more neutron-rich nuclei and increasing the precision, especially of neutron skin thickness measurements, will be an important component of the future studies.

The territory at and beyond the proton drip line offers instead unique opportunities to study other exotic nuclear decays and correlations, such as ground-state one- and two-proton decay, a class of radioactivity that exists nowhere else but that provides unique insight.

Superheavy nuclei will be the object of intensive research with new dedicated facilities, like at the new one at JINR which has just started its operation. It was recognized long ago that, in spite of the huge electric repulsion between all the protons in the nucleus, the binding that comes from the strong force could tip the balance in favour of the existence of superheavy nuclei. Precise calculations at the limits of mass and charge are difficult and thus the progress in this field has come largely from experimental efforts. The important steps forward in heavy-element discovery have to rely on the decay chains of alpha particles, as demonstrated by pioneering measurements in the recent past. To find the most favourable conditions for producing even heavier elements, extensive measurements with fusion reactions of different types of beams will be employed. The future on this topic seems to be bright, in particular because the new experiments will inform us on how to reach the expected region of long-lived superheavy nuclei. These nuclei will have to be identified and investigated to determine their properties, and x-ray detection will help to pin down their new high values of protons. One-atom-at-a-time chemistry studies will also expand into this region of superheavy elements.

2.3.6.1 Nuclear reactions and decays in the plasma

The availability of very intense ion beams will open new research directions concerning the formation and characterization of hot plasma at very high density, a region that needs to be explored. In such very compressed hot plasmas it is very interesting to study the acceleration acquired by nuclear particles (proton, heavy ions), the occurrence of nuclear reactions, and how nuclear decay properties change. The attraction by these issues is connected to the fact that when such plasmas are created in the laboratories, one can reproduce conditions on Earth similar to those occurring in some stars and planets.

Another efficient way to produce high-density plasmas is to use lasers (free-electron lasers and high-power lasers) which, depending on their power, can create different conditions of density and temperature. The combined and coherent efforts at the facilities either in operation or under construction or planned worldwide will allow the temperature–density plane to be mapped, where one can locate different elements of the cosmos, when characterised in relation of the features of their plasma (figure 2.13).

At FAIR research using the worldwide unique combination of intense laser radiation with beams, started at the existing laboratory GSI, will be boosted in the coming years. This combination opens the door for interesting experiments in the field of plasma physics, nuclear physics, and atomic physics. In addition, standalone

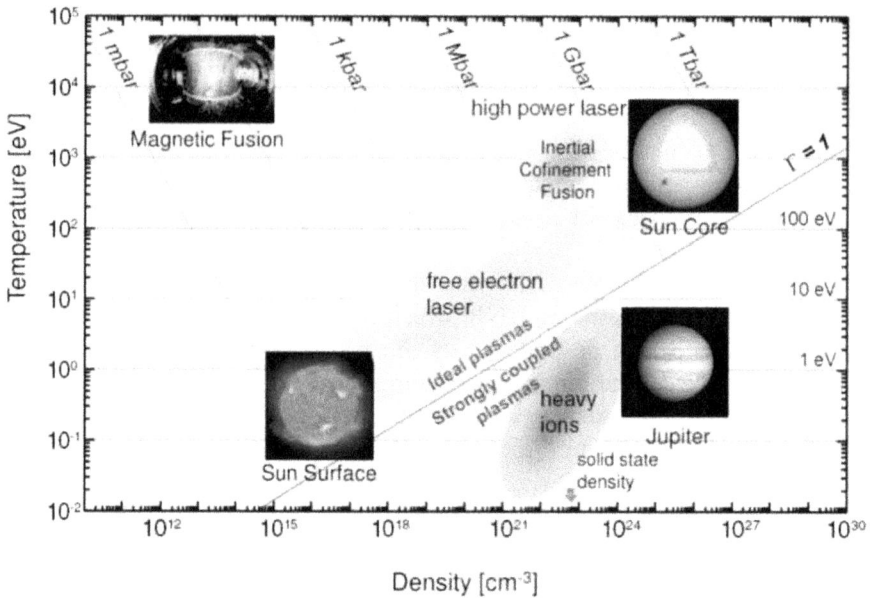

Figure 2.13. The plasma in plane of temperature and density with indicated some typical processes in the cosmos. The techniques used in laboratories to create plasma in different regions of temperature and density are also schematically indicated. Adapted from [21], copyright IOP Publishing Ltd. All rights reserved.

experiments with intense laser beams can be carried out to study proton acceleration or x-ray laser generation. These experiments complement other ones at facilities mostly devoted to develop new laser-based acceleration techniques.

A very interesting new direction in the field of nuclear physics is represented by the measurements of nuclear reactions induced by ions of different types produced in the plasma which can be created at high density by very powerful lasers. Although the concept of producing fusion reactions with the lightest nuclei (H, D, T, and He) in the plasmas is at the basis of reactors designed mainly for energy-generation purposes (as in the case of the very large infrastructure ITER), the measurements in laboratories of stellar reactions involving heavier nuclei, as in several types of stars, are very few. New efforts are made exploiting the ignition process that is used within the inertial confinement fusion research. In that context, the initiation of nuclear fusion reactions is produced by heating and compressing a fuel contained in solid targets made of different types of materials. The energy to compress and heat the fuel is delivered to the outer layer of the target using laser light beams, electrons, or ions.

It is important to point out that the high-power lasers at the new ELI International Facility (with three pillars, in Hungary, the Czech Republic, and Romania) that have recently become operational will offer the possibility to carry out interesting programs on the investigation of several different nuclear processes in the plasmas. The high-power lasers of ELI will be capable of producing high-energy

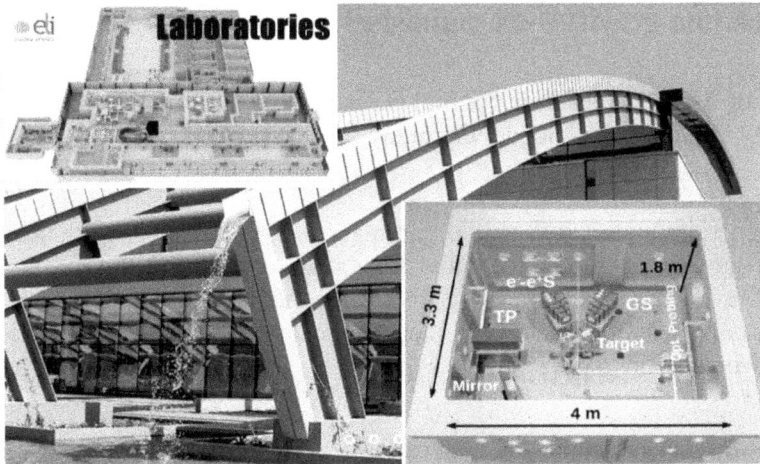

Figure 2.14. Schematic illustration of the ELI facility in Bucharest. Adapted from https://www.eli-np.ro courtesy of Sissy Körner.

charged particles, gamma rays, and neutrons, with a peak flux orders of magnitude higher than achievable with conventional accelerators.

In particular, at the ELI-NP pillar in Romania (see figure 2.14), which is more focused on nuclear physics, the short-duration pulses of its lasers will produce high fluxes of nuclear particles. These nuclear particles will be studied together with the new kinds of nuclear physics phenomena, such as:

- Exotic, heavy neutron-rich nuclei produced using new methods involving sequential reactions in plasmas
- The stopping power of charged particles bunches in dense plasmas
- Nuclear reactions in hot and dense plasmas simulating in the laboratory the astrophysical phenomena
- Nuclear excitations and de-excitations in plasmas leading to changes in (apparent) nuclear lifetimes

The investigations of exotic nuclear reactions aim primarily to explore the production of neutron-rich nuclei at around the $N = 126$ waiting point, which is relevant for the astrophysical r-process of nucleosynthesis. For this purpose, bunches of heavy ions generated and accelerated to around 10 MeV/nucleon by the laser through the radiation pressure acceleration mechanism will be employed. In this way a sequential fission–fusion will follow. This planned research will help to elucidate the mystery of high-Z element formation in the Universe. The proposed scheme is complementary to other approaches existing or planned for the production of radioactive nuclei and if successful, beyond demonstration, it will open interesting opportunities for the future.

2.4 Quantum gravity—an unfinished revolution

Claus Kiefer[1]

[1]Faculty of Mathematics and Natural Sciences, Institute for Theoretical Physics, University of Cologne, Cologne, Germany

It is generally assumed that the search for a consistent and testable theory of quantum gravity is among the most important open problems of fundamental physics. I review the motivations for this search, the main problems on the way, and the status of present approaches and their physical relevance. I speculate on what the situation could be in 2050.

2.4.1 Present understanding and applications

2.4.1.1 The mystery of gravity

Already one year after the completion of his theory of general relativity, Albert Einstein predicted the existence of gravitational waves from his new theory. At the end of his paper, he wrote[2]:

> ...the atoms would have to emit, because of the inner atomic electronic motion, not only electromagnetic, but also gravitational energy, although in tiny amounts. Since this hardly holds true in nature, it seems that quantum theory will have to modify not only Maxwell's electrodynamics, but also the new theory of gravitation.

Thus already in 1916 Einstein envisaged that quantum theory, which at that time was still in its infancy, would have to modify his newly developed theory of relativity. More than hundred years later, we do not have a complete quantum theory of gravity. Why is that and what are the prospects for the future?

Gravitation (or simply gravity) is the oldest of the known interactions, but still the most mysterious one. It was Isaac Newton's great insight to recognize that gravity is responsible for the fall of an apple as well as for the motion of the Moon and the planets. In this way, he could unify astronomy (hitherto relevant for the region of the Moon and beyond) and physics (hitherto relevant for the sublunar region) into one framework. In the Newtonian picture as presented in his *Principia* from 1687, gravity is understood as action at a distance: any two bodies in the Universe attract each other by a force which is inversely proportional to the square of their distance (see appendix). For this, he had to introduce the so far unknown concepts of *absolute space* (which has three dimensions) and *absolute time* (which has one dimension). These entities exist independent of any matter, for which they act like a fixed arena that cannot be reacted upon by the dynamics of matter. Newton's discovery marked

---

[2] This is my translation from the German. The original reference can be found in Kiefer [65], p 26.

the beginning of modern celestial mechanics, which allowed the study of the motion of planets and other astronomical bodies with unprecedented accuracy.

The strength of the gravitational force between two bodies is proportional to their *masses*. Masses can only be positive, in constrast to electric charges, which can be both positive and negative. This difference is the reason why charges can attract each other (if they, unlike the masses in gravity, differ in sign) as well as repel each other (if they have the same sign). For elementary particles, mass, by which we mean rest mass, is an intrinsic property (the same holds for charge). There can also exist particles with zero mass, of which the only observed one is the photon; such particles must propagate with the speed of light c. Elementary particles are also distinguished by their intrinsic angular momentum (spin), by which they can be divided into bosons (having integer spin) and fermions (having half-integer spin).

Newton's theory of gravity was superseded only with the advent of general relativity in 1915. It was Einstein's great insight to recognize that gravity can be understood as representing the *geometry* of space and time as unified to a four-dimensional entity called spacetime. In this way, gravity acquires its own dynamical local degrees of freedom. Spacetime then no longer plays the role of a fixed background acting on matter, but takes itself part in the game and can be reacted upon—both by matter and by itself. Gravity itself creates a gravitational field as is reflected by the non-linear nature of Einstein's field equations (see appendix). That gravity possesses its own degrees of freedom can best be seen by the existence of gravitational waves, which propagate with the speed of light and which were detected directly for the first time by the laser interferometers of the Laser Interferometer Gravitational-wave Observatory (LIGO) collaboration in 2015. That gravity (and thus spacetime) is fully dynamical is also called *background independence*.

Gravity is very weak. The gravitational attraction between, say, electron and proton in a hydrogen atom is about 10^{40} times smaller than their electric attraction. A metallic body can be prevented from falling to the massive Earth by holding it with a small magnet. Still, because masses are only positive, it is the dominating force for the Universe at large scales, because positive and negative electric charges, being present in roughly equal amounts, average to zero at those scales.

Newton had carefully distinguished between gravity (interaction between bodies) and inertia (resistance of bodies to changes in their momenta). These two concepts are unified in Einstein's theory as expressed by the equivalence principle. The geometry of spacetime thus leads in appropriate limits to the traditional gravitational interaction as well as to inertial forces such as centrifugal or Coriolis forces.

Gravity is of a universal nature. Everything in the world is in spacetime and is thus subject to its geometry, that is, to gravity. So far, Einstein's theory successfully explains all observed gravitational effects from everyday life (e.g., the working of the GPS) to the Universe as a whole. Figure 2.15 presents a famous photograph showing galaxies at distances that cover cosmic scales in space and in time—because light propagates at the finite speed c we see these galaxies in a very early state of their evolution, billions of years ago. Astronomers measure cosmic scales in Megaparsec (Mpc) and Gigaparsec (Gpc). In conventional units, 1 Mpc $\approx 3.09 \times 10^{22}$ m, and

Figure 2.15. A glimpse into the macroscopic world: the *Hubble Ultra Deep Field*, a photograph taken from September 2003 to January 2004 in a small celestial region in the constellation Fornax. Figure credit: NASA and the European Space Agency.

1 Gpc is thousand Mpc. The size of the observable Universe is estimated to be about 14 Gpc³.

Strictly speaking, there are two features for which it is presently open whether they can be fully accommodated into Einstein's theory or not: Dark Matter and Dark Energy. The two can only be observed by their gravitational influence; Dark Matter exhibits the same clumpiness as visible matter (and exhibits itself, for example, in the rotation curves of galaxies), but Dark Energy is of a homogeneous and repulsive nature and is responsible for the present accelerated expansion of our Universe (as measured by observing supernovae at increasing distances). Some scientists speculate that new physics is needed to account for Dark Matter and Dark Energy, but at present this is far from clear.

General relativity is what one calls a classical, that is, non-quantum, theory. Our current theories for the other interactions are all *quantum* theories or, more precisely, these interactions are described within a quantum framework, which uses concepts drastically different from classical physics. For example, whereas classical mechanics makes essential use of *trajectories* for bodies, the equations of which are determined by their initial positions and momenta, quantum mechanics no longer

---

³ This is the so-called particle horizon: the distance in today's Universe up to which we can see objects, that is, the distance over which information (basically in the form of electrodynamic or gravitational waves) had enough time since the Big Bang to reach us. The age of our Universe is estimated to be about 13.8 billion years.

contains such trajectories in its mathematical description[4]. It instead features wave functions Ψ from which observable quantities such as energy values for spectra and interference patterns of particles can be obtained. The relation to positions, momenta (and other classical concepts) proceeds via the probability interpretation, and the limits can be expressed by the indeterminacy (or uncertainty) relations. The quantum-to-classical transition can be understood and experimentally studied using the concept of decoherence [66, 67]. The quantum framework and formalism seems to be of universal nature.

Quantum theory is usually applied in the realm of microphysics. This is the world of molecules, atoms, nuclei, and elementary particles. Quantum theory thus lies at the basis not only of physics, but also of chemistry and biology. The smallest scales investigated experimentally so far are the scales explored by particle accelerators such as the LHC at CERN. Figure 2.16 shows a glimpse into these smallest scales—the decay of the Higgs particle into other particles. Such 'microscopic' pictures are far more abstract than photos of the kind shown in figure 2.15; a great amount of theoretical insight is involved to construct them.

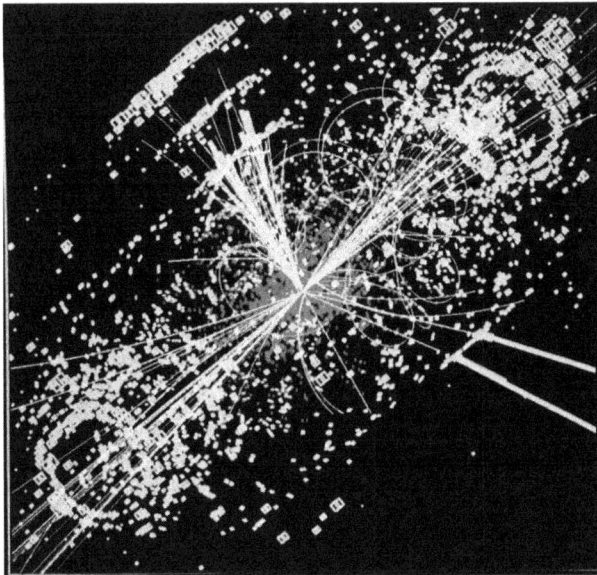

Figure 2.16. A glimpse into the microscopic world: simulation of the hypothetical decay of a Higgs particle into other particles at the detector CMS at CERN. Figure credit: Lucas Taylor/CERN—http://cdsweb.cern.ch/record/6284699 (Creative Commons License).

---

[4] The trajectories that appear in the so-called de Broglie–Bohm interpretation of quantum theory are of a non-classical nature.

These smallest explored scales are of the order of 10^{-18} m. Comparing it to the above cosmic scale of 14 Gpc, which is about 4×10^{26} m, we see that this corresponds to a difference of about 44 orders of magnitude[5].

The non-gravitational degrees of freedom are described by the Standard Model (SM) of particle physics. It provides a partial unification (within the framework of gauge theories) of strong, weak, and electromagnetic interaction. The SM is a quantum *field* theory, that is, a quantum theory with infinitely many degrees of freedom. So far, there are no clear hints for physics beyond the SM. For theoretical reasons, one expects a unification of interactions at high energies. Some approaches to unification make use of supersymmetry (SUSY) in which fermions and bosons are fundamentally connected. Despite intensive search at the LHC, no evidence for SUSY was found.

Particle physicists measure energies in electron volts (eV). For high energies, one uses Megaelectronvolts (MeV), 1 MeV $= 10^6$ eV, Gigaelectronvolts (GeV), 1 GeV $= 10^3$ MeV, and Terraelectronvolts (TeV), 1 TeV $= 10^3$ GeV. The LHC reaches a collision energy of 13 TeV. Because of Einstein's famous relation $E = mc^2$, masses can be measured in eV over c^2. The proton mass is about 938 MeV c^{-2}, and the mass of the famous Higgs particle discovered at the LHC in 2012 is about 125 GeV c^{-2}.

The fields in the Standard Model all carry energies and thus generate a gravitational field. Because they are quantum fields, they cannot be inserted directly into the classical Einstein field equations. Only a consistent unification of gravity with quantum theory can describe the interaction of all fields at the fundamental level.

2.4.1.2 What are the main problems?

What do we mean when we talk about quantum gravity? Unfortunately, this term is not used in a consistent way. Here, we call quantum gravity any theory (or approach) in which the superposition principle is applied to the gravitational field.

The superposition principle is at the heart of quantum theory: for any physical states of a system (described, e.g., by wave functions Ψ and ϕ), any linear combination $\alpha\Psi + \beta\phi$, where α and β are complex numbers, is again a physical state. This principle is confirmed by an uncountable number of experiments. For more than one system it leads to entanglement between systems, which is relevant for atoms (e.g., the qubits used in quantum information), for particles (e.g., neutrino oscillations), and many other cases.

Now, because gravity couples universally to all degrees of freedom, this should entail also a superposition of different gravitational fields, for which a quantum theory of gravity is needed. At a famous conference at Chapel Hill (USA) in 1957, Richard Feynman explained this by a gedanken experiment (see figure 2.17).

In this, the superposition of microscopic states (e.g., electron spins) is transferred to the spatial superposition of a macroscopic ball, for which the gravitational field is measurable. But how do we describe the gravitational field of an object which is in a spatial superposition at different locations? Only a theory of quantum gravity can

---

[5] It is interesting to see that the geometric mean of the largest and the smallest explored distance corresponds to about 10 km, which is an everyday scale.

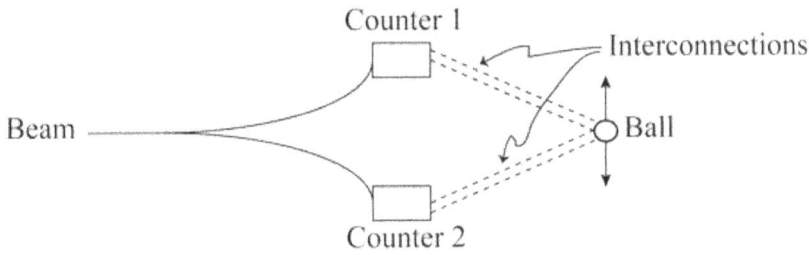

Figure 2.17. Stern–Gerlach type of gedanken experiment, in which the detectors for spin up respective spin down are coupled to a macroscopic ball. If the particle has spin right, which corresponds to a superposition of spin up and down, the coupling leads to a superposition of the ball being moved up and down, leading to a superposition of the corresponding gravitational fields. Figure adapted from DeWitt and Rickles, p 251, see DeWitt [68] CC BY 3.0.

achieve this. There exist attempts to realize superpositions à la Feynman in the laboratory; see, for example, Bose *et al* [69] and Marletto and Vedral [70]. Whether this is possible and whether one can draw conclusions on quantum gravity from this is currently subject of discussion.

There are other reasons in favour of the search for quantum gravity. As already mentioned above, if one aims at a unification of all interactions in Nature (a 'theory of everything' or TOE), one has to accommodate gravity into the quantum framework, since the quantum fields of the non-gravitational interactions act as sources for gravitational fields. One may, of course, envisage in principle a unified theory in which gravity stays classical. But there are at least two reasons that speak against this possibility. First, it is not very satisfactory to have such a fundamental hybrid theory. Second, there are various counter-arguments from the observational point of view against some hybrid theories (see, e.g., Kiefer [65] for details). But there exist no logical arguments that would force the quantization of gravity, and hybrid theories can indeed be constructed [71].

Einstein's theory, by itself, is incomplete. One can prove singularity theorems which state that, given some assumptions, there are regions in spacetime where the theory breaks down (Hawking and Penrose 1996). Concrete examples include the regions inside black holes and the origin of our Universe ('time zero'). Only a more general theory, such as a quantum theory of gravity, may be able to resolve these singularities and thereby allow a full description of black holes and the Universe.

There is also another kind of singularity. Quantum field theories are plagued by divergences which arise from probing space time at arbitrarily small scales, leading (by the indeterminacy relations) to momenta and energies of arbitrarily high values. On paper, these 'infinities' can be handled by regularization and renormalization. Regularization means that divergent expressions can be made finite by a mathematical procedure of 'isolating' the divergences (infinities). Renormalization means that the isolated divergences can be absorbed in physical parameters of the theory. These parameters cannot, of course, be calculated from the theory, but can only be determined empirically. The paradigmatic example is quantum electrodynamics (QED) and the parameters swallowing the infinities are the electric charge and the mass of the electron. Once this is done for finitely many parameters (typically a small number), the theory becomes predictive. While this procedure is consistent and can be successfully applied to the SM, the question arises whether a fundamental theory including gravity is finite by construction, that is, whether no divergences occur in the first place. Perhaps the root for both types of singularities (gravitational and quantum field theoretical) lies in the assumed continuum nature of space time.

Before we embark on a brief discussion of the main approaches, let us address the physical scales where we definitely expect quantum effects of gravity to become relevant (due to the universality of the superposition principle, such effects can, in principle, become relevant at any scale).

In the most recent version of the *Système International d'unités* (SI), which is valid since 2019, physical units are based as much as possible on fundamental constants[6]. In this, Planck's constant h, the speed of light (c), and the electric charge (e) are attributed fixed values. The units metre (m) and kilogram (kg) can then be inferred from h and c, while the second (s) is determined from atomic spectra. For us, h and c are relevant:

$$c = 299\ 792\ 458\ \mathrm{ms}^{-1}, \tag{2.1}$$

$$h = 6.626\ 070\ 040 \times 10^{-34}\ \mathrm{J \cdot s}, \tag{2.2}$$

The gravitational constant G is known with much lower accuracy. On the *NIST Reference on Constants, Units, and Uncertainty*, one finds the following 2018 value for G:

$$G = 6.674\ 30(15) \times 10^{-11} \frac{\mathrm{m}^3}{\mathrm{kg \cdot s}^2}. \tag{2.3}$$

It thus cannot serve the same purpose as h and c (otherwise, we could base our time unit on G). Einstein's theory also contains the cosmological constant Λ, which has dimension of an inverse squared length. From current observations one finds the value

$$\Lambda \approx 1.2 \times 10^{-52}\ \mathrm{m}^{-2} \approx (0.35\ \mathrm{Gpc})^{-2}, \tag{2.4}$$

which, however, is not precise enough for using Λ as a standard of units.

The three constants G, h (resp. $\hbar = h/2\pi$), and c provide the relevant scales for quantum gravity, because one can construct from them (apart from numerical factors) unique expressions for a fundamental length, time, and mass (or energy). Because Max Planck had formulated them already in 1899, they are called Planck units in his honour. The Planck length reads:

$$l_\mathrm{P} := \sqrt{\frac{\hbar G}{c^3}} \approx 1.616 \times 10^{-35}\ \mathrm{m}, \tag{2.5}$$

---

[6] See, e.g. Hehl and Lämmerzahl [72] for a thorough discussion.

the Planck time is:

$$t_{\mathrm{P}}: = \frac{l_{\mathrm{P}}}{c} = \sqrt{\frac{\hbar G}{c^5}} \approx 5.391 \times 10^{-44} \text{ s}, \qquad (2.6)$$

and the Planck mass is:

$$m_{\mathrm{P}}: = \frac{\hbar}{l_{\mathrm{P}} c} = \sqrt{\frac{\hbar c}{G}} \approx 2.176 \times 10^{-8} \text{ kg} \approx 1.22 \times 10^{19} \text{ GeV } c^{-2}, \qquad (2.7)$$

from which one can derive the Planck energy

$$E_{\mathrm{P}}: = m_{\mathrm{P}} c^2 \approx 1.22 \times 10^{19} \text{ GeV} \approx 1.96 \times 10^9 \text{ J} \approx 545 \text{ kWh}. \qquad (2.8)$$

Whereas Planck length and Planck time are far remote from everyday (and experimentally accessible) scales, Planck mass (energy) seems to be of a more everyday nature. The point, however, is that the Planck mass is more than 10^{19} times the proton mass m_{pr} and more than 10^{15} times the maximal collision energy attainable at the LHC. This means that to generate particles with masses of order the Planck mass or higher, one needs to construct an accelerator with galactic dimensions. This is one of the most important problems in the search of quantum gravity: we cannot probe the Planck scale directly by experimental means.

The size of structures in the Universe is determined by the squared ratio of proton mass and Planck mass, sometimes called the 'finestructure constant of gravity':

$$\alpha_{\mathrm{g}}: = \frac{G m_{\mathrm{pr}}^2}{\hbar c} = \left(\frac{m_{\mathrm{pr}}}{m_{\mathrm{P}}} \right)^2 \approx 5.91 \times 10^{-39}. \qquad (2.9)$$

It is the smallness of this ratio that is responsible for the usual smallness of quantum-gravitational effects in astrophysics. It is an open question whether this number can be calculated from a fundamental theory or whether it remains unexplained as a phenomenological parameter that can only be determined from observations.

2.4.1.3 What are the main approaches and applications?

Before addressing the full quantization of gravity, it is appropriate to have a brief look at what is known about the relation between quantum theory and classical gravity[7].

The relation between quantum *mechanics* (quantum theory with finitely many degrees of freedom) and gravity is studied by using the Schrödinger (or Dirac) equation in a Newtonian gravitational field. This is the regime where experiments are available, such as by observing interference fringes of neutrons or atoms. The combination of quantum *field* theory (QFT) with general relativity ('QFT in curved spacetime') is much more subtle. The perhaps most famous prediction there is that black holes are, in fact, not black but radiate with a thermal spectrum. This effect

---

[7] References on this and the following sections can be found e.g. in Kiefer [65]. See also Carlip [73] and Woodard [74] for general accounts of quantum gravity.

was derived from Stephen Hawking in 1974 and is called Hawking radiation. The temperature of a black hole is given by

$$T_{\mathrm{BH}} = \frac{\hbar\kappa}{2\pi k_{\mathrm{B}}c}, \tag{2.10}$$

where κ is the surface gravity characterizing a stationary black hole. Within Einstein–Maxwell theory (coupled gravitational and electrodynamical fields), one can prove the *no-hair theorem* for stationary black holes: they are uniquely characterized by the three parameters mass (M), electric charge (Q), and angular momentum (J). Astrophysical black holes are described by the two parameters M and J (Kerr solution).

For a spherically symmetric (Schwarzschild) black hole with mass M, the Hawking temperature is

$$T_{\mathrm{BH}} = \frac{\hbar c^{3}}{8\pi k_{\mathrm{B}}GM} \approx 6.17 \times 10^{-8}(\frac{M_{\odot}}{M})\ \mathrm{K}. \tag{2.11}$$

The smallness of this value means that this effect cannot be observed for astrophysical black holes, which have a mass of at least three solar masses ($3M_{\odot}$).

Figure 2.18 shows an example of an observed black hole—a supermassive black hole with $M \approx 6.5 \times 10^{9}M_{\odot}$ in the centre of the galaxy M87. For such black holes, the Hawking effect is utterly negligible.

There is, in fact, an analogue of the Hawking effect in flat (Minkowski) spacetime. An observer moving with constant acceleration a through the standard vacuum state of flat space time experiences a bath of thermally distributed particles with 'Unruh temperature':

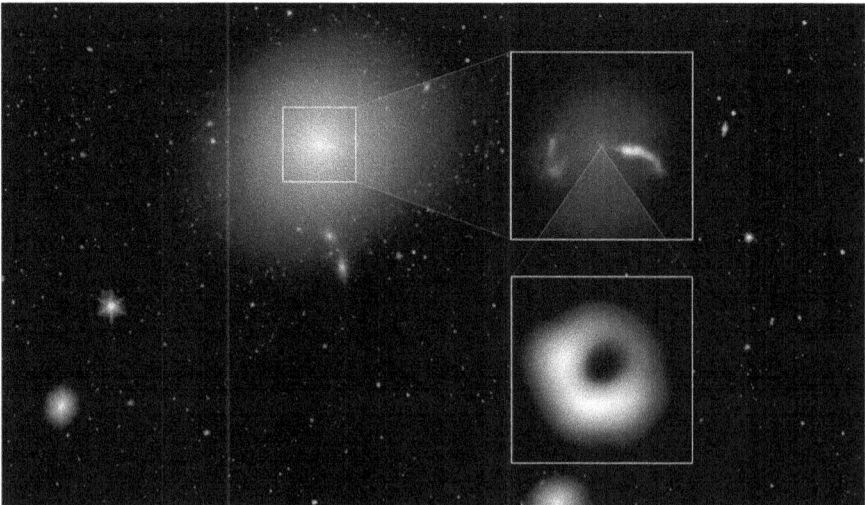

Figure 2.18. Shadow of the supermassive black hole in the centre of the bright elliptical galaxy M87. For this and all other black holes observed so far, only a consistent quantum theory can explain what happens in their inside regions. Image credit: NASA, JPL-Caltech, Event Horizon Telescope Collaboration.

$$T_U = \frac{\hbar a}{2\pi k_B c} \approx 4.05 \times 10^{-25} \, a \left[\frac{m}{s^2} \right] \, K. \tag{2.12}$$

One immediately recognizes the similarity with (2.10), with a replaced by K. The reason for the appearance of this temperature is the fact that there is no unique vacuum (and thus no unique particle concept) for non-inertial observers in flat space time.

If black holes have a temperature, they also have an entropy, which is given by the 'Bekenstein–Hawking expression'

$$S_{BH} = \frac{k_B A c^3}{4 G \hbar} \equiv \frac{k_B A}{(2 l_P)^2}, \tag{2.13}$$

where A denotes the area of the black hole's event horizon. In the Schwarzschild case, we can express the entropy as

$$S_{BH} \approx 1.07 \times 10^{77} k_B \left(\frac{M}{M_\odot} \right)^2. \tag{2.14}$$

S_{BH} is indeed much greater than the entropy of the star that collapsed to form the black hole. The entropy of the Sun, for example, is given approximately by $S_\odot \approx 10^{57} k_B$, whereas the entropy of a solar-mass black hole is about $10^{77} k_B$, which is 20 orders of magnitude larger. All the above expressions contain the fundamental units c, G, $\hbar$ and thus point towards the need for constructing a quantum theory of gravity. Such a theory should be able to provide a microscopic interpretation of the entropy formula (2.13).

Besides black holes, quantum effects are also important in cosmology. Assuming that the Universe underwent an (almost) exponential expansion at a very early state (a phase called *inflation*), density perturbations of matter and gravity (gravitons, see below) are generated out of quantum vacuum fluctuations. All the structure in the Universe (galaxies and clusters of galaxies) is believed to arise from these perturbations. The power spectrum of these density perturbations (also called 'scalar modes') can be derived to read

$$P_S = \frac{1}{\pi} (t_P H)^2 \varepsilon^{-1} \approx 2 \times 10^{-9}, \tag{2.15}$$

where ε is a 'slow-roll parameter' that is peculiar to the chosen model of inflation, and H is the Hubble parameter (expansion rate) of the Universe during inflation. One recognizes the explicit appearance of the Planck time t_P, equation (2.6), in this formula. The power spectrum of these density fluctuations is recognized in the anisotropies of the cosmic microwave background (CMB) radiation; see figure 2.19. The number 2×10^{-9} on the right-hand side of (2.15) comes from observations.

All of what has been said so far points towards the need for a quantum theory of gravity. But how can such a theory be constructed? The first attempts date back to work done in 1939 by Léon Rosenfeld, who was then an assistant to Wolfgang Pauli in Zürich. In two papers, he pioneered two approaches: the 'covariant approach' and

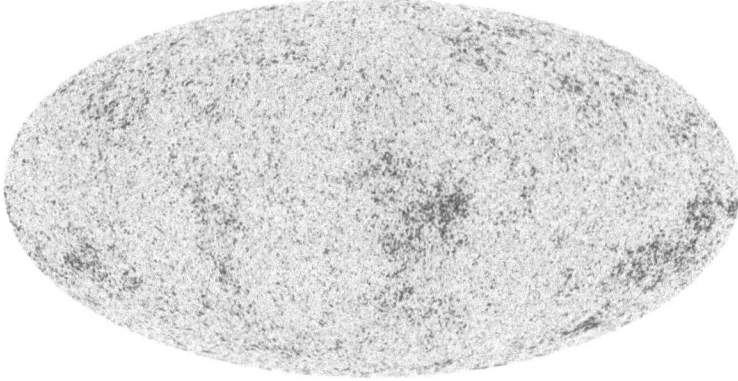

Figure 2.19. Anisotropy spectrum of the Cosmic Microwave Background (CMB). Image credit: ESA/Planck Collaboration.

the 'canonical approach'. Both approaches aim at the construction of a quantum version of general relativity. What is the status of these approaches?

The covariant approach has its name from the fact that a four-dimensional (covariant) formalism is employed throughout. In most cases, this formalism makes use of path integrals (in which, according to the superposition principle, four-dimensional spacetimes are summed over); see the appendix. Similar to the photon in quantum electrodynamics, a particle is identified as the mediator of the quantum-gravitational field, the *graviton*. It is massless, but has spin 2 (whereas the photon has spin 1). That it is indeed massless is indirectly confirmed by the detection of gravitational waves—they move with speed of light c. From this, the LIGO and Virgo collaborations report a limit of the graviton mass $m_g \lesssim 7.7 \times 10^{-23}$ eV. As remarked above, gravitons (also called 'tensor modes') are generated from the vacuum during an inflationary phase of the early Universe. Similar to the density spectrum (2.15), one can derive for them the power spectrum

$$P_T(k) = \frac{16}{\pi}(t_P H)^2. \tag{2.16}$$

A central quantity is the ratio between tensor and scalar modes,

$$r := \frac{P_T}{P_S} = 16\varepsilon. \tag{2.17}$$

So far, no observations have indicated a non-vanishing value for r. Observing such a value would constitute a direct test of quantum gravity at the linearized level.

As with all relevant quantum field theories, also the covariant quantization of general relativity exhibits divergences. But there is a major difference to the situation in the Standard Model. Whereas the perturbation theory for the SM is renormalizable, this does not apply for gravity. It is thus *not* possible to absorb divergent terms into a finite number of observable parameters; at each order of the perturbation theory, new types of divergences appear, and one would need infinitely many parameters to absorb them, rendering the theory useless. But the question arises whether higher terms in the perturbation expansion are indeed relevant. They come in powers of the parameter

$$\frac{GE^2}{\hbar c^5} \equiv \left(\frac{E}{m_\mathrm{P}}\right)^2 \sim 10^{-32}, \tag{2.18}$$

where E is the relevant observation energy, here taken to be 14 TeV, the energy of the planned LHC-upgrade. This is a very small parameter, so perturbation theory should in principle be extremely accurate. One could thus adopt the point of view that quantum general relativity is an *effective field theory* only, that is, a theory that is anyway valid only below a certain energy and must be replaced by a more fundamental, potentially renormalizable, or finite theory above that energy. An approach that makes use of standard quantum field theory up to the Planck scale is *asymptotic safety*. In this, G and Λ are not constants, but (as is typical for quantum field theory) variables that depend on energy. They may approach non-trivial fixed points in the limit of high energy and thus lead to a viable theory of quantum field theory at all scales. It is imaginable that the scale dependence of G could mimic Dark Matter; in this case, it would be hopeless to look for new particles as constitutes of Dark Matter.

To calculate quantum-gravitational path integrals is far from trivial and definitely not possible analytically. For this reason, computer methods are heavily used. One promising approach is *dynamical triangulation* which bears this name because the spacetimes to be summed over in the path integral are discretized into tetrahedra. This leads to interesting results about the possible microstructure of space time.

One candidate for a finite quantum field theory of gravity is supergravity, which combines SUSY with gravity more precisely; more precisely, a particular version called $N = 8$ supergravity. Heroic calculations over many years have shown that there are no divergences in the first orders of perturbation theory. Whether this continues to hold at higher orders and, moreover, whether this holds at all orders, is far from clear. Only new, so far unknown, principles can be responsible for this theory to be finite.

A candidate for a finite theory of quantum gravity of a very different nature is superstring theory (or M-theory). In the limit of small energy, the above covariant perturbation theory is recovered, but at higher energies, string theory is of a very different nature. Actually, its fundamental entities are not only one-dimensional entities as the name suggests, but higher-dimensional objects such as branes. Moreover, the theory makes essential use of a higher-dimensional space time (with 10 or 11 as the number of dimensions). The theory is not a direct quantization of gravity—quantum gravity appears only in certain limits as an emergent theory. In contrast to theories of quantum general relativity, string theory has the ambition to provide a unified quantum theory of all interactions (the TOE mentioned above). Such a theory should also allow to understand the origin of mass in Nature. One aspect of this is the *hierarchy problem*. We observe widely separated mass scales— neutrino masses (~ 0.01 eV), electron mass (~ 0.5 MeV), and top-quark mass (≈ 173 GeV), all of which are much smaller than the Planck mass (2.7). So far, the origin of this hierarchy is not understood. It is not clear whether there is new physics between the SM energy scale (as exemplified by the Higgs and the top-quark mass) and the Planck scale.

Out of string theory and the discussion of black holes grew insights about a possible relation between quantum-gravity theories and a class of field theories called conformal quantum field theories. The latter are defined on the boundary of the spacetime region in which the former are formulated. This is known as gauge/gravity duality, holographic principle, or AdS/CFT conjecture (see, e.g., [75]). Some claim that it will play a fundamental role in a full theory of quantum gravity.

The alternative to covariant quantization is the canonical (or Hamiltonian) approach. The procedure is here similar to the procedure in quantum mechanics where one construct quantum operators for positions, momenta, and other variables. This includes also the quantum version of the energy called Hamilton operator. In quantum mechanics, the Hamilton operator generates time evolution by the Schrödinger equation. In quantum gravity, the situation is different. Instead of the Schrödinger equation, one has *constraints*—the Hamiltonian (and other functions) are constrained to vanish. This is connected with the disappearance of spacetime at the fundamental level. Spacetime in general relativity is the analogue of a particle trajectory in mechanics; so after quantization space-time disappears in the same way as the trajectory disappears (recall the indeterminacy relations)—only space remains. This is sometimes referred to as the 'problem of time', although it is a direct consequence of the quantum formalism as applied to gravity. It is connected with the fact that already the classical theory has no fixed background, so there is no such background available to serve for the quantization of fields—different from the situation with the non-gravitational quantum fields of the SM. Background independence is one of the main obstacles on the route to quantum gravity.

If one uses the standard metric variables of Einstein's theory, one arrives at quantum geometrodynamics with the Wheeler–DeWitt equation as its central equation. Due to mathematical problems, the full equation remains poorly understood, but it can be applied to problems in cosmology and for black holes. An alternative formulation makes use of variables that show some resemblance with the gauge fields used in the SM. It is known under the name loop quantum gravity. At the kinematic level (before the constraints are imposed), it is well understood, but the exact construction of the Hamiltonian constraints and the recovery of quantum field theory in curved space time present problems. Applications of loop quantum gravity also include cosmology and black holes.

An important feature of the Wheeler–DeWitt approach is the possibility of building a bridge (at least at a formal level) from quantum gravity to quantum field theory in curved space time. In this way, spacetime (and, in particular, time) emerges as an approximate concept. This procedure is similar to the recovery of geometric optics ('light rays') from fundamental wave optics. In this, the separation of scales (the separation of Planck mass from masses of the Standard Model) is crucial[8]. This

---

[8] A review of this and other conceptual issues can be found in Kiefer [6, 7, 65] and the references therein.

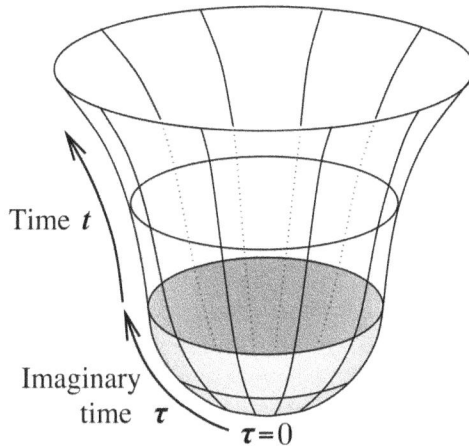

Figure 2.20. Hartle–Hawking instanton: the dominant contribution to the Euclidean path integral is assumed to be from half of a four-sphere attached to a part of de Sitter space. Reprinted with permission from [65], 2012, Oxford University Press.

emergence of time can also be described in the covariant approach. Using the method of path integrals, Hartle and Hawking constructed a certain four-dimensional geometry that elucidates the emergence of time by attaching a Euclidean ('timeless') geometry to a Lorentzian one. This 'Hartle–Hawking' instanton is shown in figure 2.20. It is frequently discussed in the application of quantum gravity to cosmology (quantum cosmology).

The Wheeler–DeWitt equation has a very peculiar structure. It is asymmetric with respect to the size of three-dimensional space and may thus allow to understand the origin of the arrow of time from fundamentally timeless quantum gravity [76]: there is an increase of entropy with increasing size of the Universe.

Besides the approaches already mentioned, there are a variety of others, and only space prevents me from discussing them in more detail. Many of these other theories make use of a discrete structure, either fundamentally imposed or derived from other principles. The reader may wish to consult Oriti [77] for more details.

2.4.2 Challenges and opportunities in the Horizon 2050

2.4.2.1 Theoretical challenges and opportunities

What can or should we expect in the coming decades? Physics is an experimental science. There can only be progress if we have testable predictions that can falsify a given approach and discriminate between different approaches. To derive such predictions is one of the main theoretical challenges.

It makes sense to distinguish between predictions at the linearized level and at the full level. The linearized level of quantum general relativity also follows from unified theories such as superstring theory, so tests at that level are very general. Looking at atomic physics, one can calculate the transition rate from an excited state to the ground state by emission of a graviton. In one example ([65], p 40) this gives a lifetime τ of the excited state as big as

$$\tau \approx 5.6 \times 10^{31} \text{ years.} \qquad (2.19)$$

It thus seems forever impossible to observe such a transition. One should, however, not forget that the predicted lifetime of a proton in the simplest unified theory of particle physics (the minimal SU(5) theory) is about 10^{32} years, which one was able to falsify in the Super-Kamiokande experiment in Japan; it turned out that the proton has a lifetime of at least about 10^{34} years. The problem with (2.19) is that this decay is drowned in electromagnetic transitions, which are very fast. But if one could identify transitions in atomic or molecular physics that emit photons at no or low rate, there may be the option to observe gravitonic emissions in, for example, thin interstellar clouds. To the best of my knowledge, however, no one so far has attempted to identify and calculate such processes.

The power spectra (2.15) and (2.16) are, in a certain sense, already effects of linearized quantum gravity. The reason for this claim is that the calculation makes use of variables that combine gravitational (metric) and matter variables in a quantum sense. This is confirmed by the appearance of the Planck time t_P in these expressions. Calculations have also been performed to derive corrections to these expressions by going beyond the linear approximation. This has been achieved in particular for the canonical theory in both the geometrodynamic and the loop version. The corrections are proportional to the inverse Planck-mass squared and turn out to be too tiny to be observable at present. Similar correction terms should appear for the power spectra of galaxy distributions; so far, however, calculations of such terms do not seem to exist. Quite generally, one would expect that the first signatures of quantum gravity come from small effects. This was the case for quantum electrodynamics, where the theoretical understanding and the successful observation of the Lamb shift in atomic spectral lines led to the general acceptance of the theory.

A second major challenge is the construction of a viable full quantum theory of gravity, preferable one that gives a unified description of all interactions. On the one hand, it is not clear whether one can construct a separate quantum theory of gravity alone, without unification. Asymptotic safety may provide an example of a stand-alone theory, but most likely, such a separate theory would be an effective theory, one that is valid only below a certain energy scale. This would be sufficient for calculating small effects, but would lack an understanding of quantum gravity at the fundamental level. On the other hand, it is far from obvious that the programme of reductionism will continue to work and that a 'theory of everything' can be found. Superstring theory, the main candidate for such a theory so far, has not proven successful in the last 50 years.

The case of superstring theory also exhibits a deep general dilemma. One might expect that a really fundamental theory would enable one to predict most of the fundamental constants of Nature from a small number of parameters. One important example is (2.9), which sets the scale at which structures in the Universe appear, and which string theory cannot predict so far. But since one knows that only a very fine-tuned set of physical parameters (masses, coupling constants, etc) allow the existence of a Universe such as ours and the formation of life, this would leave the open question of why this is so. If, on the other hand, the fundamental theory does not lead to such a prediction and if, moreover, all possible parameter values are allowed in the world (which would then constitute a kind of 'multiverse'), it would leave us only with the *anthropic principle* as a way to understand the Universe (see, e.g., [78]). It may, of course, happen that we have a mixture of the two cases, so that most constants are determined by the fundamental theory and a few (such as the cosmological constant and the Higgs mass) can only be determined anthropically. A decision about this dilemma is one of the most important theoretical challenges, if not *the* most important one.

We have remarked above that general relativity is incomplete because it predicts the occurrence of space-time singularities. The general expectation is that a quantum theory of gravity will avoid singularities. The present state of quantum gravity approaches is not mature enough to enable the proof of theorems, but preliminary investigations in various approaches indicate that singularity-free quantum solutions can indeed be constructed. It is one of the main theoretical challenges of the next decades to clarify the situation and get a clear and mathematical precise picture of the conditions under which singularity avoidance follows. This would also throw light on one important open question in the classical theory—*cosmic censorship* (see, e.g., [79]). Black holes such as the one in figure 2.18 are characterized by the presence of a horizon from behind which no information can escape to external observers. The singularity predicted by general relativity is thus hidden. The hypothesis of cosmic censorship states that all singularities arising from a realistic gravitational collapse are hidden by a horizon, thus preventing the singularity from being 'naked'. Singularity avoidance from quantum gravity would immediately lead to the non-existence of hidden *and* naked singularities and would thus prove cosmic censorship to be true in a trivial sense.

2.4.2.2 Observational challenges and opportunities

Progress in quantum gravity can eventually only come from observations and experiments. As we have seen, quantum effects of gravity are usually small and become dominant only at the Planck scale. Laboratory experiments thus may look hopeless. One can try to generate superpositions of gravitational fields in the sense mentioned in connection with figure 2.17, but it is unlikely that this could enable one to discriminate between different approaches. One may also use laboratory experiments to decide whether the superposition principle is violated for gravitational fields as advocated, for example, in Penrose [79]. The main obstacle in this is to avoid standard decoherence effects from environmental degrees of freedom [66]. Laboratory experiments are also useful to test acoustic analogies to the Hawking and Unruhe effects, from which insight relevant for quantum gravity may be drawn.

The main observational input should thus come from astrophysics and cosmology, but also from particle physics. For this to be successful, large international collaborations are typically needed. We have already mentioned the anisotropy spectrum for the CMB, which was precisely measured by international projects such as PLANCK, WMAP, BOOMERANG, and others. Whether quantum gravity effects can be seen in future projects of this kind remains open. A major step would be the identification of a non-vanishing value for the r-parameter (2.17), from which the existence of gravitons could be inferred.

Another important class of experiments are gravitational-wave experiments. They are not designed primarily for quantum-gravity effects, but they may be helpful also in this respect by detecting, for example, a stochastic background of gravitons from the early Universe. One project is the Laser Interferometer Space Antenna (LISA) scheduled for launch in 2034[9]. A planned terrestrial project is the Einstein Telescope (ET) scheduled for starting observations in 2035[10].

Aside from cosmology, black holes are perhaps the most important objects for exploring quantum gravity experimentally. Due to Hawking evaporation, black holes have a finite lifetime. Taking into account the emission of photons and gravitons only, the lifetime of a (Schwarzschild) black hole under Hawking radiation is (see, e.g., [80])

$$\tau_{\mathrm{BH}} \approx 8895 \left(\frac{M_0}{m_{\mathrm{P}}} \right)^3 t_{\mathrm{P}} \approx 1.159 \times 10^{67} \left(\frac{M_0}{M_\odot} \right)^3 \text{years.} \qquad (2.20)$$

It is obvious that this lifetime is much too long to enable observations for astrophysical black holes. This would only be possible if small black holes exist, which most likely can only result from large density fluctuations in the early Universe—for this reason they are called *primordial black holes*. So far, observations gave only upper limits on their number and on the rate for their final evaporation. Since gamma rays are emitted in the final phase, gamma-ray telescopes are crucial for their detection (e.g., the Fermi Gamma-ray Space Telescope launched in 2008)[11]. There are also speculations about the presence of a primordial black hole with the size of a grapefruit in the Solar System ('Planet X'); whether this is really the case must be checked by future observations, such as by the upcoming Vera C. Rubin Observatory in Chile[12].

Hawking's calculations that led him to the black-hole temperature (2.10) break down when the mass of the black holes approaches the Planck mass (2.7). This means that the final phase can only be understood from a full theory of quantum gravity (beyond the approximation of small correction terms). Observations may then shed light on the 'information-loss problem', that is, whether the radiation remains thermal up to the very end (and may thus lead to loss of information about the initial state) or not.

Quantum-gravity effects may also be seen in particle accelerators. This may be due, for example, to the existence of higher dimensions or due to the presence of supersymmetry. So far, no hints for this or other quantum-gravity related effects were found at the LHC[13] or other machines. The upgrade High Luminosity Large Hadron Collider (HL-LHC) is planned to start operation in 2029. Plans for various other big machines scheduled for operation before 2040 exist.

---

[9] https://www.lisamission.org
[10] http://www.et-gw.eu
[11] https://fermi.gsfc.nasa.gov
[12] https://www.lsst.org
[13] https://home.cern/science/accelerators/large-hadron-collider

2.4.2.3 A brief outlook on the year 2050

There is a quote attributed to Mark Twain—'Prediction is difficult—particularly when it involves the future'—which definitely also applies to predictions about the status of quantum gravity in 2050. Looking 30 years back (my postdoc years), most of the present quantum-gravity approaches did exist, some of them already for a while. Since then, there has been progress in both the mathematical formulation and the conceptual picture, but no final breakthrough was achieved. A hypothetical researcher time travelling from 1991 to 2021 would have no problems to follow the current literature on quantum gravity. But what about the next 30 years?

An optimistic picture would perhaps look as follows. We have a leading candidate for a quantum theory of gravity that provides an explanation of the cosmological constant (more generally, Dark Energy) and perhaps Dark Matter. It predicts testable effects for quantum-gravitational correction terms to power spectra of galaxies and the CMB and sheds light on the final phase of black-hole evaporation. Gravitons are observed as relics from the early Universe and in the form of tensor modes from the CMB. Primordial black holes are observed and their final phase can be studied in detail. Ideally, this theory should give a unified description of gravity and the other interactions.

A pessimistic version would look very differently. We still work on essentially the same approaches to quantum gravity as today and see no possibilities for testing them. The abovementioned projects for the 2030s and 2040s turn out to be very successful for astronomy and particle physics, but fail to shed light on quantum gravity. Already in 1964, Richard Feynman wrote (see [81]): 'The age in which we live is the age in which we are discovering the fundamental laws of nature, and that day will never come again. It is very exciting, it is marvellous, but this excitement will have to go.' What he means is that there are limits to performing experiments for fundamental physics coming from their sheer size and financial needs, and that these limits may appear rather soon. Still, I think, at least the next 30 years should remain exciting, and perhaps major progress, theoretically and empirically, will emerge from a totally unexpected side....

2.4.3 Appendix

In this appendix, I shall summarize some formulae which were omitted in the main text. For a clear and concise account of classical (Newtonian and Einsteinian) gravity I refer to Carlip [82].

The famous inverse-square law of Newtonian gravity reads

$$\mathbf{F} = -\frac{GM_1M_2}{r^2}\hat{\mathbf{r}}. \tag{2.21}$$

This force can be derived from a potential Φ, which obeys Poisson's equation

$$\nabla^2\Phi = 4\pi G\rho, \tag{2.22}$$

where ρ is the matter density.

In general relativity, Poisson's equation is replaced by the Einstein field equations

$$R_{\mu\nu} - \frac{1}{2}g_{\mu\nu}R + \Lambda g_{\mu\nu} = \frac{8\pi G}{c^4}T_{\mu\nu}, \tag{2.23}$$

which are of a non-linear nature (a gravitational field generates again a gravitational field, and so on). A fundamental role is played by the metric $g_{\mu\nu}$, which instead of the one function Φ in the Newtonian case contains ten functions. The physical dimension of the energy–momentum tensor $T_{\mu\nu}$ is energy density (energy per volume), which is equal to force per area (stress). Einstein once spoke of the left-hand side as marble (because of its geometric nature) and the right-hand side as timber (because of the non-geometric nature of matter fields). In fact, $T_{\mu\nu}$ contains the fields of the SM. Because these fields are quantum operators, the Einstein equations cannot hold exactly but must be modified by an appropriate quantum equation.

Covariant quantum gravity can be defined by a path integral P, which contains a sum over all permissible metrics $g_{\mu\nu}$ and over all non-gravitational fields ϕ,

$$P = \int \mathscr{D}g_{\mu\nu}\mathscr{D}\phi \exp\left(\frac{i}{\hbar}S\right), \tag{2.24}$$

where S denotes the total action of the system. In the canonical approach, one has constraints which are also fulfilled by the path integral, building in this way a bridge between the two approaches.

2.5 What is the Universe made of? Searching for dark energy/matter

2.5.1 Dark matter—general overview

Jochen Schieck[1,2]
[1]Institut für Hochenergiephysik der Österreichischen Akademie der Wissenschaften, Wien, Austria
[2]Technische Universität Wien, Austria

2.5.1.1 Introduction

Understanding the nature of dark matter is one of the big unsolved questions of modern physics. The problem was first discussed in detail in the first half of the 20th century, and almost a century later the problem is still hotly debated. However, since then, significant progress has been made, mainly by confirming and constraining dark matter with different approaches at different scales [83]. We can now better classify and narrow down the dark matter problem. We consistently observe more gravitational pull than the pull we expect from visible matter. The various measurements all point to the same dark matter abundance, which is about five times more prominent than the visible matter, which we describe by the SM of particle physics [84]. However, the underlying nature of dark matter is still not understood.

The Swiss astronomer Fritz Zwicky [85] is frequently cited as a pioneer in dark matter. In the 1930s, he studied the dynamics of the Coma galaxy cluster, and he found a clear mismatch between the observed mass and the gravitational force generated by the mass required to prevent the Coma galaxy cluster from tearing apart. The gravitational force generated by the observed mass cannot compensate for the centrifugal forces that the galaxies experience due to their speed.

The conclusion that the lack of gravity stems from unobserved matter, particularly a new particle, is currently the best approach. This assumption is very well-founded and consistent with all our observations, but it stays an assumption. A different approach motivates a solution to the dark matter problem by changing the way we understand gravity [86]. An explanation of all observations with this approach, however, is far more challenging. In any case, gravity plays an essential role in the understanding of the origin of dark matter. We have an excellent theory that describes gravity at large scales called general relativity [87]. The Standard Model (SM) of particle physics, however, does not include gravity (see, e.g., [88]). The forces represented in the SM are much stronger compared to gravity, and a particle theory that formulates all forces, including gravity, in a standard description still not exists. A better description of dark matter, or even its discovery, could provide new clues to a combined standard definition of nature that contains all four fundamental forces.

The particle character of dark matter as a solution to the dark matter problem leads to the postulation of at least one new particle. The SM of particle physics provides no particle dark matter candidate. The various astrophysical observations of dark matter allow us to narrow down its properties. First of all, the particle has to be massive to generate gravity. Secondly, the particle cannot take part in the electromagnetic interaction; it has to be dark. We know that dark matter has to travel much slower than the speed of light, often referred to as dark matter being 'cold' [89]. The measurement of its content in the Universe is consistent during the different steps of the evolution of the Universe and dark matter has to be stable or have at least a lifetime comparable to the Universe's lifetime [90]. The fact that particle dark matter has not been observed yet means that interaction with ordinary matter is weak, but hopefully stronger than the gravitational interaction. A dark matter particle that interacts via gravitation only would be impossible to observe directly in any experimental setup.

The quest for dark matter is also a prime example of a topic bringing together astronomy, astroparticle physics, cosmology, and particle physics. While particle physics explores the fundamental particles and their interactions at the smallest scale, cosmology describes the Universe's evolution—the most extensive scale we are aware of. In astronomy, celestial bodies are measured and their properties are determined. Astrophysics uses physics, especially particle physics, to explain astronomical observations. Observations of dark matter at the largest cosmological scale point towards the existence of a new microscopic particle. Particle physics studies at the smallest scale allow us to discover and characterize the new particles, which gives feedback for understanding the Universe at the largest scales and astronomical observations—an excellent example of how the various disciplines are mutually beneficial.

The discovery of a dark matter particle that gives rise to five times more matter than the currently known matter would lead to a substantial change in the understanding of nature. Humanity will realize that the nature that we experience with our senses is only a fraction of the matter of the Universe. Dark matter observation would open an entirely new research field focusing on this new particle's characterization and the related interactions.

2.5.1.2 Observation of dark matter

Since the discovery of dark matter in the first half of the last century, several different observations based on different methods confirmed the existence of dark matter. In the following section, we will briefly present the various dark matter observations and discuss their impact for a deeper understanding of dark matter. We will start with the systematic study of galaxy rotation curves, an example that provides a very illustrative picture of the dark matter problem. The mismatch between the gravitational pull and the observed matter was already discussed in the literature before the study. However, only the measurement of the rotation curves of the velocity of stars at large distances from the galactic center convinced the scientific community of the presence of dark matter. In the 70s, a new tool became available, and the American Astronomer Vera Rubin and collaborators used it to measure the velocity distribution of stars within the Andromeda galaxy, the galaxy closest to our home galaxy [91]. The new spectrograph developed by her collaborator Kent Ford Jr allowed the measurement of the characteristic 21 cm spectral line from hydrogen, and therefore the velocity of the stars away from the center of the Galaxy. Like a stone circulating on a string and whose centrifugal force is compensated by the string, the force of gravity keeps the stars in their orbit around the galactic center. Contrary to expectations, the velocity distribution does not decrease with increasing distance from the Galaxy center but somewhat flattens to constant values. The determined velocity distribution away from the galactic center is higher than the one expected from the visible mass. Additional forces have to act on the stars to keep them in orbit. Later on, the flattened-out velocity distributions were observed in other galaxies besides the Andromeda galaxy as well, and the case for missing matter in galaxies became even stronger. Embedding the Galaxy in a halo of invisible matter that extends up to large radii was proposed as a possible solution to the problem of excessive velocity distributions—a halo made of dark matter. The Andromeda galaxy is about 2.5 million light years away, and these measurements originate from processes as old as 2.5 million years. Recent measurements of the Milky Way return a similar picture and point out that our local galaxy is also embedded into a halo of dark matter. Figure 2.21 shows the velocity distribution of the planets within our Solar System and the one for a galaxy embedded in a dark matter halo. While within our Solar System, the gravitation pull is mainly determined by the Sun's mass, the dominant part of the gravitational pull in the Galaxy comes from matter which is not visible, dark matter.

Today the most accurate measurement of the dark matter content in the Universe comes from a measurement of the photons from the cosmic microwave background (see [92, 93] for a review and references therein). With the Universe's expansion after the Big Bang, the Universe continues to cool down, changing the energy of the particles. The particles are in thermal equilibrium with the Universe, which means they have the same temperature and energy as the Universe's average temperature. About 380 000 years after the Big Bang, the thermal energy of the particles is low enough to capture electrons in the electromagnetic field of protons such that

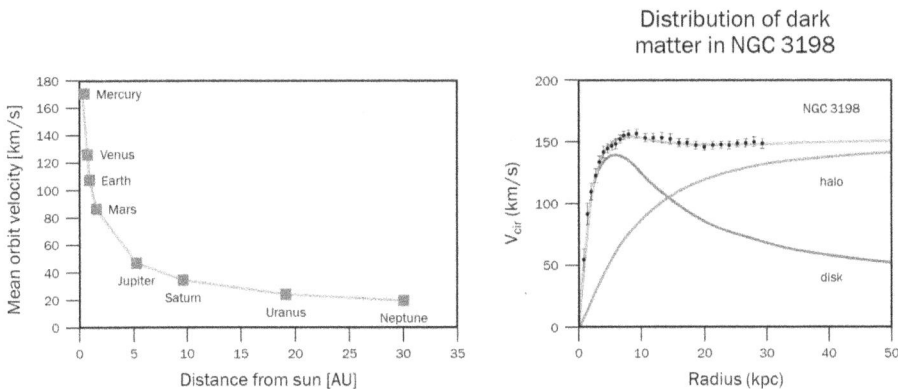

Figure 2.21. (Left) Velocity distribution of the planets orbiting around the Sun. The gravitational field is completely dominated by the Sun and the orbital velocity decreases with the distance from the Sun. (Right) The points represent the measurement of the velocity distribution of stars within the galaxy NGC 3198 [125]. For large distances the distribution flattens out and is not decreasing as expected from the visible mass indicated by the line named 'disk'. To explain the additional gravitational pull, to keep the stars on the orbit, the Galaxy has to be embedded in a halo of dark matter, indicated by the line name 'halo'. Only the combination of the gravitational pull originating from the disk and the halo can explain the observation.

hydrogen atoms can be formed, and the Universe undergoes a transition from a plasma with free charged particles to a gas of neutral hydrogen atoms. Suddenly the Universe becomes transparent to light since the charged particles are bound to neutral atoms. Photons no longer interact with scattering centers and suddenly propagate freely in space. The cosmic microwave photons represent an energy imprint at the the transition from an opaque to a transparent Universe. The measurement of these photons and a detailed subsequent analysis provides a precise determination of the Universe's matter content at the time of decoupling of the photons from the plasma. The average photon energy returns information about the average temperature when the plasma is transformed to a gas. In a gravitational field, the energy of a photon is slightly modified, as predicted by the general theory of relativity. Precision measurements of the photon energy therefore allow a mapping of the gravitational potential. Before the transition to the neutral hydrogen gas, the matter oscillates driven by the gravitational pull and radiation pressure from electromagnetic interactions. The analysis of these oscillations allows a precise measurement of the luminous matter and the non-luminous matter content of the Universe, dark matter. All matter feels the gravitational attraction, but dark matter only feels the attraction and no repulsion from the radiation pressure.

Currently, the most precise measurement of the energy and matter composition of the Universe comes from an analysis of cosmic microwave background data collected with the ESA Planck space observatory, with 26.8% [94] of the total energy density attributed to dark matter (figure 2.22).

At the transition time from the plasma to the neutral gas, the mass distribution in the Universe is almost uniform, and there are only minor density fluctuations of the gravitational field of the order of 0.001%. How did the nearly uniform Universe evolve in a Universe with matter accumulations and voids as we observe it today? The structure we observe today is the result of matter moving in the gravitational field of the Universe [95]. Sophisticated simulations can reproduce a matter structure

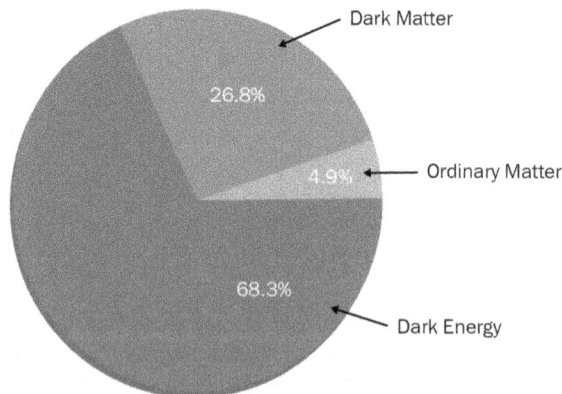

Figure 2.22. The pie chart shows the percentage of the various matter and energy contributions in the Universe as determined with data from the ESA Planck space observatory. Of the total energy and mass content, 26.8% is attributed to dark matter [94].

very similar to the one observed and, in addition, predict its evolution during the expansion of the Universe (see [96] for a review and references therein). However, the structure observed today can only be produced with dark matter included in the simulation. Visible matter only is not able to reproduce the Universe as we observe it today. In addition to the need for dark matter, we can also learn something about the dark matter properties. Dark matter must be cold, which means that it moves significantly slower than the speed of light. Otherwise, the smallest structures formed would be washed out in an early stage, and no galaxies and clusters of galaxies can form [89]. An example of the evolution of a dark matter distribution as expected from simulation based on cold dark matter is shown in figure 2.23. The similarity with observations is evident.

We know that other matter must exist from the observations discussed above, a matter that does not interact electromagnetically. However, how do we know that this matter is not made of a neutral accumulation of baryonic matter, which we know from the SM of particle physics?

A detailed study of the first nuclei generated about 3 min after the Big Bang, the so-called Big Bang nucleosynthesis, allows a precise determination of the abundance of light nuclei in the Universe [97, 98]. From this number, we can infer the Universe's total baryonic content of about 4, 9%, which is significantly lower

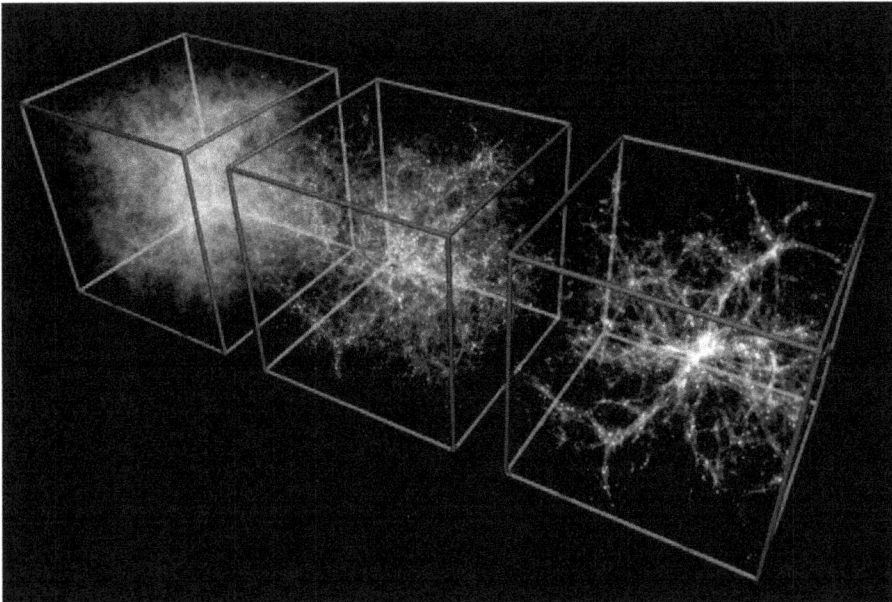

Figure 2.23. The picture shows a 3D view of the cosmic structure's evolution. The left box corresponds to the matter distribution of a part of the Universe with the age of one billion years (redshift $z = 6$), the middle one corresponds to about 3.3 billion years (redshift $z = 3$) and the right on as we would expect it today (redshift $z = 0$). The yellow areas correspond to stellar material [126] (Credit: Volker Springel, Max-Planck-Institute for Astrophysics), copyright (2008) reprinted by permission from Springer Nature.

compared to the total dark matter component [84]. Besides being invisible, dark matter is also not baryonic.

Dark matter has to exist to explain the observations presented above. All measurements agree that there is about five times more dark matter than the matter we know from the SM of particle physics. From the observations discussed above, we have some understanding of dark matter. The different measurements at different time scales during the evolution of the Universe are consistent. Dark matter consisting of one or multiple new, unobserved particles is currently the best working hypothesis for the search for dark matter.

We also have a good understanding of the dark matter in our own neighborhood, the Milky Way. Measurements show a dark matter density that corresponds roughly to the mass of a proton per cubic centimeter, which means that gravitational effects from visible matter dominate the Earth's atmosphere. The Solar System moves with a circular speed of about 230 km s^{-1} relative to the dark matter halo. Assuming a certain particle dark matter mass, we can then calculate the expected flux of dark matter particles here on Earth. The rotation of the Earth around the Sun of about 30 km s^{-1} leads to a seasonal modulation of the flux of about 5%–10%. An observation of an annual modulated signal originating from the movement of the Earth would be a clear dark matter signal [99].

2.5.1.3 Cold dark matter and open small-scale issues

On very large scales, the structure of the matter distribution can be very well described by simulations based on the Λ CDM model. The Λ CDM model is the cosmological model that provides the best description of our Universe's evolution. It offers an excellent illustration of all cosmological observations, and it is today's SM of cosmology. The Greek letter Λ refers to the cosmological constant, reflecting the so-called dark energy, the dominant part of the Universe's energy content, only discovered in 1998 (see figure 2.22). Another primary input to the Λ CDM model is cold dark matter, the central part of the understanding of several observations. For smaller scales, typically at the galactic and the sub-galactic scales, differences appear between the expectation from a collisionless Λ CDM dark matter simulation and the astronomical observations. These discrepancies are closely related to structure formation processes and to the corresponding matter being involved. A deeper understanding of astrophysical processes inside these regions is required to further improve modeling at sub-galactic scales. We already know that these simulations are incomplete, and contributions from baryonic processes, for example, are currently the subject of ongoing research. At large scales, the structure formation is entirely dominated by dark matter-caused dynamics; however, at smaller scales, baryonic processes play an increasing role and influence structure formation. Besides baryonic processes, specific properties of dark matter alter the evolution at small scales. Coherence effects of very light dark matter candidates could alter the structure at sub-galactic scales. Dark matter candidates could be warm and not cold, still in agreement with the expectations from structure formation and pointing towards a lighter dark matter particle, or dark matter might interact strongly with each other, leading to a different dynamical behavior in regions with sizable dark matter densities (see [100] for a review and references therein). Simulations using dark matter self-interactions on galactic and sub-galactic scales or with dark matter coherence effects might therefore guide the particle physics search for dark matter (see [96] for a review and references therein).

2.5.1.4 Approaches for solving the dark matter problem

Cosmology deals with our Universe's origin, evolution, and fate, and represents studies on the largest structure known to us. In addition to cosmology, which covers the largest scales known to us, the SM of particle physics plays a decisive role in unraveling the origin of dark matter. The SM of particle physics is one of the most telling physics theories ever developed and provides precision predictions with an agreement of several digits with experimental observations. It describes three of the known forces: the strong, the electromagnetic, and the weak force. Gravity, the by far weakest force, is not part of the SM of particle physics, which can explain all measurements on microscopic scales; however, it cannot be complete and is considered an effective theory of a more comprehensive theory. While explaining everything on a tiny scale, no particle in the SM could act as a dark matter candidate. In principle, the neutrino fulfills most requirements, but it nevertheless cannot act as a dark matter candidate. The SM neutrino is 'hot' and in contradiction to the condition of being a slow or 'cold' dark matter particle is necessary to explain the structure of the Universe we currently observe.

We are now in a situation where we have to have two excellent theories that describe our observations with very high accuracy. The Λ CDM model is based on general relativity, and relativistic quantum mechanics is the crucial ingredient to the SM of particle physics. We have a clear evidence for dark matter in the Λ CDM model, triggering an extended research program in particle physics. The potential discovery of a dark matter particle would have a significant impact on cosmology and therefore the Λ CDM model. Dark matter acts as a link between these two research areas dealing with the two most extreme length scales we are aware of—the largest and the smallest one.

Astrophysical observations demand the existence of dark matter; however, information on a concrete particle dark matter candidate is sparse. In particular, the mass and the size of the interaction of dark matter with SM particles are more or less unknown. It is impossible to cover the entire dark matter search region with a single experiment. In addition, any potential signal needs to be confirmed by another independent experiment and methodology before any final conclusion on particle dark matter can be drawn.

The search has to be guided by theory to set priorities for certain mass regions and interactions. New theoretical developments profit from previous work and guiding principles of physics, like symmetries and scales. These developments are complemented by new approaches, tackling the problem with alternative scenarios. A close collaboration between theory and experiment, particle physics, cosmology, and astronomy is essential to advance the field. Figure 2.24 shows the expected mass and the predicted interaction cross-section with SM particles for different dark matter models, guiding the different experimental approaches.

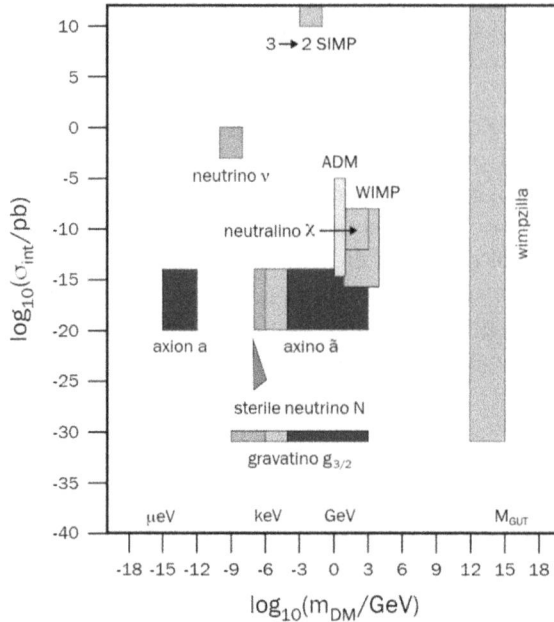

Figure 2.24. Possible dark matter models and its expectation for the corresponding mass and interaction cross-section. Adapted from [101], copyright (2015) with permission from Elsevier.

2.5.1.5 Particle dark matter candidates beyond the existing SM particle zoo

The mass range for dark matter candidates spans over several orders of magnitude. The lower mass limit is simply given by the space needed for each dark matter candidate, combined with the given dark matter relic density representing the amount of dark matter still remaining from the Big Bang. The fundamental properties of the particle, being a boson or a fermion, play a significant role. The upper limit is determined by the Universe's heaviest and most compact objects. Black holes produced during the earliest stage of the Universe ('primordial black holes') fulfill all requirements of being a dark matter candidate.

Particle dark matter candidates can be classified by different criteria. We already mentioned the classification via the temperature of the dark matter particle, like cold, warm, or hot dark matter, which reflects the particle's speed. Dark matter candidates can also be classified by the way how the dark matter relic density comes about. Thermally produced dark matter is such a possibility for the classification. All particles, including dark matter particles, are generated during the Big Bang. The particles are in thermal equilibrium with the temperature of the Universe. Dark matter particles can annihilate into SM particles, and vice versa, the temperature of the Universe determines the energy of the process. As the Universe expands, it cools down. Below a certain temperature, there is no longer enough energy available for the production of heavy dark matter particles, but dark matter can still annihilate into lighter SM particles. The Universe continues to expand and the dark matter density thins out. The annihilation process also becomes increasingly unlikely until it almost comes to a standstill and the amount of relic dark matter roughly stays constant. The expansion and cooling of the Universe leads to a freeze-out of dark matter particles from the thermal bath of the Universe. The remaining dark matter relic density is related to the properties of the interaction strength and the mass of the dark matter particle.

The most prominent thermally produced dark matter candidate is the so-called WIMP, a weakly interacting massive particle [102]. In this freeze-out scenario, the observed dark matter relic density can be obtained with a particle mass and weak interaction cross-section similar to the electroweak scale of the SM. The conjunction of these physical quantities is not regarded as a coincidence and motivates a class of models, leading to targeted searches for dark matter. Thermally produced WIMP-like dark matter particles are expected to have a mass between a few GeV up to several hundreds of TeV.

The supersymmetric extension of the Standard Model (SUSY) is based on the introduction of a new symmetry, which relates fermions and bosons—each particle of the SM gets assigned a supersymmetric partner. The supersymmetric partners of the neutral gauge bosons and the Higgs Boson of the SM can mix and form new particles, the so-called neutralinos. Under a certain symmetry, the lightest of these neutralinos has to be stable and has the expected characteristic of a WIMP particle precisely and is therefore a prime dark matter candidate. Despite an intensive search, SUSY and the neutralino have sill eluded discovery. However, the SUSY parameter space is large and this attractive theory is not ruled out yet.

Particles with an interaction strength to SM particles much smaller than the weak interaction will never reach thermal equilibrium and therefore cannot freeze-out as discussed above. These feebly interacting massive dark matter particles (FIMPs) are produced in a non-thermal process, like via decays of primordial heavier particles. This so-called freeze-in mechanism can also lead to the correct dark matter relic density with a similar mass range as WIMP particles. Only the interaction strength is much weaker, which leads to a more difficult experimental search.

Another non-thermally produced dark matter candidate is the so-called Axion, with a mass region between meV down to peV. A thermally produced dark matter candidate in that mass range would be 'hot', traveling with relativistic velocities, and in contradiction to the observed structure formation. Despite their lightness Axions are not 'hot'. Originally, they were introduced to solve another problem of the SM of particles physics, the so-called strong *CP*-Problem, and it turns out that Axions could also act as dark matter. Within the strong sector of the SM, a fundamental symmetry, the CP-symmetry, should be violated, but it is not. The non-existence of this broken symmetry can be explained by introducing a new particle, the Axion. Inspired by its properties, particles with similar properties can be introduced, so-called Axion-like particles (ALPs). While they are suited to explain the dark matter problem, they do not solve the strong *CP*-Problem [84].

There is also the idea to introduce a dark twin brother of the electromagnetic photon. This dark photon could act as a messenger between the visible and the invisible dark sector, or it could be dark matter itself. The SM neutrino has almost all the required dark matter particle properties, but as a thermally produced light particle it behaves like a hot dark matter candidate, which is in contradiction to structure formation. In the SM, each particle exists in two different orientations, so-called chirality states. The neutrino only interacts via the weak force and is therefore the only fundamental particle in which only one state of chirality can interact with other particles. The non-interacting neutrino chirality state, often referred as a 'sterile' neutrino, might be significantly heavier and a valid dark matter candidate. This scenario could also explain why normal neutrinos are so light in the first place and solve two open questions at once.

At even smaller dark matter mass scales, fuzzy dark matter models are introduced as a solution to the dark matter problem [103]. At these tiny mass scales, it is more appropriate to interpret particles as waves. The corresponding wavelengths are as large as known astronomical scales and coherence effects have to be taken into account. Fuzzy dark matter models offer a solution to the problems at sub-galactic scales discussed in the context of the Λ CDM dark matter model.

All models discussed above introduce a single new particle. However, with increasing knowledge about the particle dark matter problem, a solution with a single dark matter particle only is getting more and more challenging to defend. The answer might be rather a whole dark sector with several dark matter particles and fields (see [104] for a review and references therein). The interaction of the dark sector with the visible world might be feeble, even gravitational only. This could explain why the dark sector has so far eluded discovery and would make experimental discovery almost impossible.

2.5.1.6 Dark matter models beyond the particle hypotheses
In addition to the postulation of new dark matter particles, alternative explanations for the dark matter problem are discussed. Neutrinos are excluded as candidates for dark matter, but any neutral state built up from SM particles could be considered as well. The amount of SM matter is very well constrained by either the measurement of the total baryonic content of the Universe via the Big Bang nucleosynthesis or the analysis of the cosmic microwave background. Non-luminous astrophysical objects, MACHOs—'massive astrophysical compact halo objects', like neutron stars or dwarf stars—could also contribute to the dark matter density [105]. Primordial black holes are MACHOs and in particular the recent observation of mergers of medium-size black holes through gravitational waves intensified the discussion of primordial black holes, generated from gravitational collapses in the early Universe, as possible dark matter candidates and insensitive to the measurements from Big Bang nucleosynthesis. A MACHO or black hole only explanation for dark matter is strongly constrained, but not entirely excluded. A contribution to dark matter for a small MACHO mass range is still possible, and additional studies are ongoing (see [106, 107] for a review and references therein).

Modified Newtonian dynamics (MOND) tries to explain dark matter observations, in particular the mismatch between the observed matter and the corresponding gravitational pull, by modifying Newton's law of universal gravitation [86]. Several measurements of dark matter could be traced back to a deviation from the expected known $1/r$ gravitational behavior for large distances. An increased gravitational attraction for large distances would keep the systems from tearing apart.

While MOND modifies Newton's law of universal gravitation, extensions towards the inclusion of the underlying principle into general relativity exist as well [108]. While the solution of the dark matter problem with a modification of gravity or Newton's laws is elegant, a consistent explanation of all dark matter observations at different scales is much more challenging than introducing an additional particle dark matter candidate [109].

2.5.1.7 Search for dark matter

The observation of dark matter is undisputed; the range of proposed solutions is extensive. The existence of a new or several unobserved particles to explain the additional gravitational pull is currently a major direction of research. The difference in interaction strength with ordinary matter and in mass scales leads naturally to many different dark matter experiments focusing on different search regions. The interaction rate together with the involved energy allow us to draw conclusions about the abundance and the mass of dark matter particle candidates. To gather more information several possible approaches are used. Dark matter particles can annihilate to ordinary matter particles (so-called indirect detection; see [110] for a review and references therein), or vice versa, standard matter particles can annihilate to dark matter particles (so-called production; see [111, 115] for a review and references therein). Dark matter particles can also scatter from ordinary matter particles, including transfer of energy to the SM particle (so-called direct detection; see [112] for a review and references therein). The different interactions are sketched in figure 2.25. In principle, one can also gain new insight from the interaction between dark matter particles only, but a complete characterization or discovery of a dark matter particle with this process only is impossible.

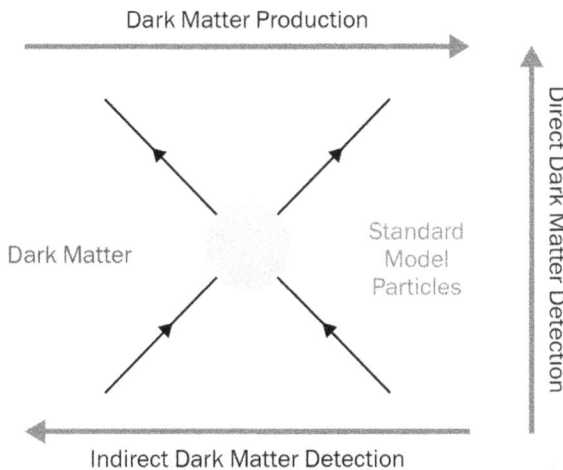

Figure 2.25. Sketch of possible interaction scenarios between dark matter particles and ordinary matter. The arrow indicates the direction of time, leading to three different interaction scenarios [84].

The annihilation of dark matter particles into SM particles is enhanced in areas with large dark matter density. This process is identical to the mechanism that would lead to a freeze-out of dark matter particles in the early Universe and to the relic dark matter content we observe today. Possible regions with enhanced dark matter density are the center of the Sun, the center of our Galaxy, or dwarf galaxies, which are expected to have a particular high dark matter content. The search is based on observing the annihilation products, leading to certain signatures, which can be well distinguished from background processes. This method includes the search for dark matter decay products, like antiparticles, which rarely exist otherwise. For a review and the references for the experiments mentioned below see [110]. Space-based experiments look for an excess of antimatter with energy distributions that indicate the annihilation of dark matter. The AMS-02 experiment at the International Space Station or the PAMELA experiment reconstruct the charge, identity, and the momentum of cosmic particles and determines the fraction of anti-particles over particles. The annihilation of a pair of dark matter particles would leave a clear signature in the anti-particle over particle spectrum. Another dark matter annihilation signature would be an increased number of dark matter annihilation products that originate from a region with increased dark matter density. These decay products must reach the observer as undisturbed as possible. Charged particles, however, are deflected by the galactic magnetic fields, and tracing back to the source is impossible. Therefore, neutral particles, such as photons or neutrinos, are an ideal way to look for dark matter decay products and their directional origin. The interpretation of indirect detection crucially depends on the understanding of the underlying astrophysical processes, like the propagation of matter and anti-matter in space or the dark matter density of potential dark matter sources. Photon detectors searching for dark matter signals are operating either as satellites in space, like the Fermi Gamma-ray Space Telescope or as Large Air Cherenkov telescopes, located at high altitudes, like the H.E.S.S. telescope in Namibia or the MAGIC telescope on the Canary islands. These telescopes detect Cherenkov light being emitted from extensive electromagnetic showers, initiated by high-energetic photons hitting the atmosphere of the Earth. Even larger detectors are necessary to detect weakly interacting neutrinos (e.g., Ice Cube at the South pole), searching for an excess of neutrinos originating from potential dark matter decays.

If dark matter annihilates into ordinary matter, the latter can also be converted into dark matter by the same underlying process—equivalent to turning around the time arrow of the process as indicated in figure 2.25. Additional kinetic energy is necessary to generate the mass of heavy dark matter particles. Since the main detection principle of particle detectors relies on electromagnetic processes, dark matter particles do not leave any signature in any particle physics experiment. However, energy and momentum conservation help to overcome this problem. Accelerator-based experiments have a well-defined initial state, and momentum or energy carried away by dark matter particles leads to energetically unbalanced final states with an unmistakable signature. Any observation of unbalanced energy or momentum proves the involvement of new particles which do not interact electro-magnetically. These studies do not allow us to make a statement about the relic dark matter density and give only limited information of the lifetime of the particle. The

LHC at the European Center for Particle Physics (CERN) is a prime accelerator to search for dark matter particles. Protons with the highest energies collide at the two multi-purpose experiments, ATLAS and CMS, to produce dark matter particles in a controlled environment. Besides the LHC experiments, dark matter is also searched at other accelerators. Experiments with electron–positron collisions, like Belle II, or experiments at beam-dump facilities are also set up to search for missing energy and momentum signatures, pointing towards the production of dark matter particles. For a review and the references see [111, 115].

The working principle of experiments for direct dark matter detection is based on the fact that the Earth moves through the dark matter halo of our Galaxy. Dark matter particles from the halo can scatter with SM particles, and the measurement of the recoil energy of this scattering process leads to a featureless, exponential energy distribution. Due to the constant dark matter density on Earth, the expected particle flux for light dark matter particles is more prominent than for heavier particles, leading a lower signal rate for heavy candidates. Taking into account the movement of the Earth around the Sun leads to well-defined annual modulation of the flux, indicating a clear dark matter signal [113]. The expected interaction strength between dark matter particles and ordinary matter is at most weak and despite the high flux of dark matter particles only few scattering events are expected [114]. Dark matter particles can scatter with the nucleus or the electrons of the target, and from the detection rate the interaction cross-section between dark matter and ordinary matter can be determined. Due to the low interaction rate the experiment has to be shielded from any potential background sources, like interactions induced by cosmic ray events. For this reason, these direct detection experiments are operated in underground laboratories, and for the construction of the experiment, material with very low natural radioactivity is used [112]. The energy transfer of the scattering event is deposited in the detector and allows to infer the particle dark matter mass. The capability to measure smallest energies is crucial in order to obtain sensitivities down to the smallest dark matter masses. Solid-state experiments, like Edelweiss, SuperCDMS, or CRESST, currently provide the best sensitivity towards smallest energy depositions and give the best limit for sub-GeV dark matter particles. For dark matter particles in the GeV mass range, experiments based on noble gases such as XENON1T, PandaX, or LUX provide the best sensitivity. A summary of the limits from ongoing direct dark matter searches is shown in figure 2.26 [84]. For a review and references see [112, 115].

Axions and ALPs interact weakly with matter and radiation. However, due to the lightness of these particles any energy deposition from scattering events is below any direct detection threshold and different experimental approaches have to be used. The main search strategy follows the two-photon interaction properties. In a strong magnetic field, Axions or ALPs can transform into a photon, and the produced photon can then be detected. The characteristic two-photon vertex is also used to generate Axions or ALPs artificially by sending a light source through a strong magnetic field. These artificially generated Axions or ALPs are sent through a light tight obstacle and another strong magnetic field is used to re-transform them back to photons, which can then be detected [116].

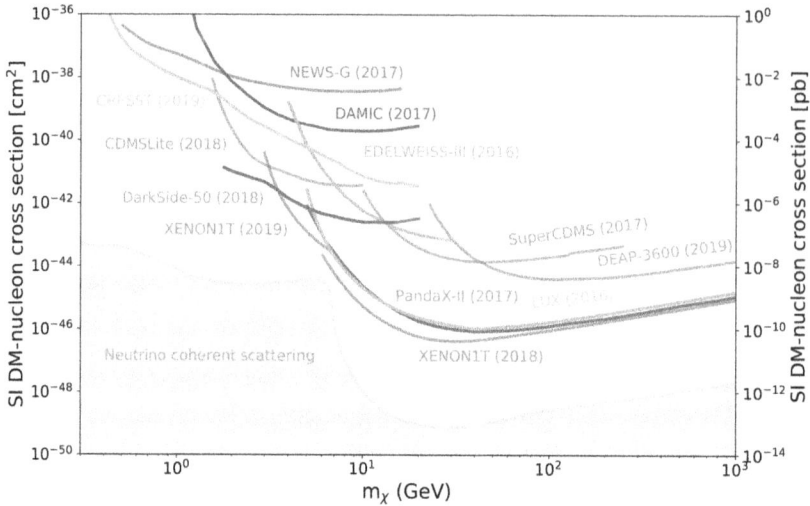

Figure 2.26. Overview of limits for direct dark matter searches performed by various experiments. The interaction strength of the dark matter particle with ordinary matter as a function of the mass of the dark matter particle is shown. The regions above the colored lines are excluded by the experiment. The blue shaded area at the bottom is the parameter range where neutrinos from Supernovae, atmospheric neutrinos, or neutrinos from the Sun leave a similar signature in the experiment. Reproduced from [84] CC BY 4.0.

The discovery of gravitational waves [117] opens a complete new possibility to study dark matter phenomena. While all measurements described above use non-gravitational interactions between dark matter and ordinary matter, gravitational waves will allow dark matter to be studied directly via gravity—the fundamental interaction all large-scale dark matter observations are based on. Besides large objects, like primordial black holes as potential dark matter candidates, gravitational waves can also make statements about possible microscopic dark matter particles. Scientific observations about the connection between dark matter and gravitational waves just started and still leave a lot of room for future theoretical and experimental studies [118].

2.5.1.8 Challenges and opportunities

Despite intense searches during the last decades, the origin of the additional gravitational pull caused by dark matter is still unknown. Intensive experimental investigations are carried out, but none has provided evidence for an undisputed particle dark matter discovery. Several theoretical approaches exist to solve the dark matter problem based on experience, scientific intuition, and extrapolation from existing knowledge. These theories are well thought out and justified to be scrutinized by different experiments. In the end, only the experimental results can give the direction towards the solution of the dark matter problem. A challenge for the scientific community is to propose suitable experiments to draw the correct conclusions from the results.

The search for a new particle as the solution to the dark matter solution provides a unique opportunity to bridge the largest and the smallest scales of fundamental physics. The proof of a dark matter particle would provide leverage to tackle significant conceptual challenges of physics. While gravity and the resulting dynamics at the largest scales are theoretically well described by general relativity, the smallest scales we can experimentally access are described by quantum mechanics and special relativity. Cosmology is based on general relativity and the SM of particle physics on relativistic quantum mechanics. Dark matter would open an opportunity to link these two fundamental theories. We know that at tremendous energies, at the so-called Planck Scale, these two theories can no longer be treated separately, and gravitational interactions have to influence quantum mechanics and vice versa. However, the required energies cannot be reached experimentally, now and not soon, and we need different input to relate these two fundamental concepts. The question for dark matter, driven by gravitational observation and the observation of a new dark matter particle, might open the possibility to give first insights into this problem.

The amount of theoretical models explaining dark matter is large and an extensive discussion of the scientific community's final confirmation and broad acceptance is expected. The fact that dark matter is reliably observed by various astronomical observations, combined with very little information on its origin, poses quite strong expectations on its scientific proof. Any potential dark matter explanation, being of particle character or any other solution, must consistently explain all existing dark matter observations at the same time. A great advantage arises from the fact that dark matter is searched by using different orthogonal experimental methods based on different assumptions. Some experimental approaches are based on the fact that we are embedded in a dark matter halo and any positive signal in direct detection experiments allows us to conclude on the relic dark matter density and provides a direct connection to astrophysical measurements. The production and observation of dark matter particles in an accelerator-based experiment is entirely independent of astrophysical conditions and provides additional input and consistency checks. A single approach can hardly solve degeneracies for specific fundamental parameters. Observations with different experimental methods are required to establish a consistent picture of dark matter and to dissolve degeneracies [119]. The establishment of dark matter measurements by different scientific approaches will provide a strict path towards the claim of discovery. The Initiative for Dark Matter in Europe and beyond (IDMeu)[14] is such a joint venture of the astroparticle physics community (represented by APPEC), the nuclear physics community (represented by NUPPEC), and the particle physics community (represented by ECFA) to tackle together the question of dark matter. This initiative includes joint scientific events, collaborative software tools as well as outreach activities.

Open science will play an increasingly important role in the future. This refers not only to open access of published results but even more to the open access to the underlying data used to produce the individual scientific results. Publicly available

---

[14] http://www.idmeu.org

data from different experimental sources will encourage collaboration between the respective disciplines. Scientists not being directly involved in generating these data will be able to provide input to their interpretation. Studies with cross-experiment data are becoming increasingly crucial for developing new knowledge.

The discovery will not only solve a long-standing problem of modern physics, it will rather establish a completely new research field, working on the detailed characterization of this new phenomena, including new insights in the underlying theories of quantum mechanics, general and special relativity.

The quest for dark matter has been ongoing for almost a hundred years with significantly increased interest during the last decades. The theoretical solutions to the dark matter problem push the experimental requirements more and more to the limits. The searches are carried out using plausible and target-oriented assumptions. It must be always kept in mind at all times what the evidence-based requirements and what reasonable assumptions are; the latter has to be questioned occasionally. Technical challenges are tempting, but the experimental search must always be driven by physics and not by technical challenges and possibilities.

A very close interaction between theory and experiment at all times is crucial for progress. Theory input is necessary to define new dark matter search strategies and for the interpretation of data from existing experiments in light of recent findings. In particular, theoretical studies are of utmost importance for bringing together experimental results from the different approaches. These studies include particle physics calculations for predicting and interpreting interactions with matter as well as theoretical astrophysical processes for the distribution and interaction of dark matter in the Universe.

2.5.1.9 Technical challenges, technology development, and knowledge transfer

Like all other fundamental sciences, technology drives the experimental progress in the field, and physics requirements from the research will drive technology. The experimental searches will advance into a new parameter space that can only be investigated with the help of new or improved experimental methods. The expected signatures for the different experimental approaches are known. There is also a good understanding of potential background sources, complicating the measurement of a dark matter signal. In addition, there will be 'unknown unknowns,' unexpected signals, which might lead to a misinterpretation of the result.

For direct dark matter detection, the current technologies will soon reach the so-called neutrino floor. An irreducible background of neutrinos from the Sun, hadronic interactions in the atmosphere, and supernovae will leave the same signature in the experiment as the one expected from dark matter particles. A discrimination between dark matter- and neutrino-induced events is no longer possible. To continue the search for dark matter with the highest sensitivity, the technology of the experiments must be adapted and further improved. Like the direction of the dark matter particle flux, additional information needs to be reconstructed to enhance the sensitivity for dark matter particles further.

Experimental searches for dark matter require cutting-edge technology, and these technological challenges can only be tackled together in extensive international collaborations. These experimental challenges for the various approaches differ significantly, from space missions for indirect dark matter searches, the construction of large-scale accelerators for the production of dark matter particles to the detection of smallest energy depositions from dark matter in our Milky Way, requiring an experiment operating in an extremely low background environment. The scale of the experiments is diverse, from tabletop experiments to massive experiments operated by large international collaborations. The challenge is to find the right balance between the approaches at different scales to ensure that all areas are covered and equipped with the necessary resources. Sometimes, these experiments are so extensive that only a few, or even only one experiment, can be implemented worldwide. The development, construction, and operation of such experiments pose new challenges for science. Financing across national borders is increasingly becoming the norm in large-scale equipment research. The scientific exchange across borders promotes scientific progress and encourages the interaction of young scientists and thus opens up the possibility that the best scientists in their field can work together on the most pressing questions. For some dark matter-related approaches, the work in large collaborations is unavoidable. The scientific work in large collaboration poses significant challenges for the scientific community, like evaluating the scientific performance of individual scientists in a large group. While scientific excellence is a crucial ingredient for recognizing personal achievements, successful work in large collaborations requires additional skills beyond the traditional assessment criteria, like coordination skills or empathy.

In addition to increasing the detector volume, improving the sensitivity to detect the smallest energy deposits plays an important role to cover more dark matter models. The conventional detection methods soon reach their limits, and new technologies and detection principles must be developed to overcome them. One possible step will be the development and application of quantum sensors that use quantitative phenomena, such as interferometry or entanglement. With this technology, the future search for dark matter will extend beyond particle and astroparticle physics towards new areas like quantum technology, further expanding the interdisciplinary nature of this topic [120].

Experiments for the direct search for dark matter must be shielded from cosmic radiation and have to be carried out in underground laboratories. With a growing number and ever-more extensive experiments, the need for laboratory space increases. The construction and operation of underground laboratories are complex, and it will be more challenging to meet all requests for laboratory space in the future. Worldwide there are only few suitable underground laboratories. Italy hosts the largest general-purpose underground laboratory in operation, the Laboratorio Nazionale del Gran Sasso (LNGS) [121]. More underground laboratories are currently under construction worldwide, and these laboratories will continue to be critical infrastructure for conducting dark matter experiments. Soon the largest and deepest underground laboratory will be operated by China.

The experiences gained during the research and development in the laboratory have to be transferred to mini-series to build these large-scale experiments. These quantities can no longer be produced by research laboratories alone, and often the production has to be carried out together with industry. Some technological challenges can only be solved together with industry and need close collaboration right from the beginning. The technical difficulties and the knowledge developed in developing new cutting-edge technologies for dark matter experiments are passed on to the industry. For example, the Astroparticle Physics European Consortium (APPEC) regularly organizes joint workshops between academia and industry[15]. In the past, developments driven by dark matter research led to successful technology transfers to society and the establishment of spin-off companies. Many developments in the field of particle detection find their way into medical technology.

Another example is the development and construction of radiation detectors based on noble gases, developed in technologies to search for dark matter. This technology development led to the spin-off now producing and selling radiation detectors[16]. The astroparticle community organizes regular joint meetings between scientists and industry on dedicated experimental topics to enhance scientific exchange further and increase knowledge transfer. The technology developed for dark matter experiments also inspires other related scientific disciplines. Scientists working on sub-GeV dark matter searches used the technology to investigate coherent neutrino scattering [122]. This technology can potentially enable tabletop experiments for neutrino physics, allowing studies of nuclear reactors without intervening directly in the reactor [123].

However, one key aspect of knowledge transfer is the education of young scientists. Young researchers, in particular students at various levels, from bachelor to PhD students, work on different dark matter-related projects as part of their training. The search for dark matter is undoubtedly one of the most exciting questions of modern physics and thus attracts many young people. The academic job market is too limited to offer all students a future in science, and most are leaving to work in the industry. The dark matter thus fulfills two critical aspects—the attraction and inspiration of young people for science and technology and an ideal, international environment to acquire all the skills necessary for a successful career in industry. Besides the science and technology aspects, the global environment of the dark matter research field, similar to other sciences, and the necessity to operate in large collaborations are beneficial for students' education.

The discovery of dark matter will open a new research field working on the detailed characterization of this recent phenomenon. Dark matter is not a minor add-on to ordinary matter; it is five times more abundant and is expected to play an essential role in the Universe's evolution. Unraveling the origin of dark matter will provide a new insight to our fundamental understanding of nature.

---

[15] https://www.appec.org/implementation/technology
[16] https://www.arktis-detectors.com

2.5.2 Dark energy

Emmanuel N Saridakis[1]
[1]National Observatory of Athens Lofos Nymfon, 11852 Athens, Greece

We provide a review on the dark energy, namely the unknown component that triggers the accelerated expansion of the Universe and comprises 70% of its energy content.

2.5.2.1 Introduction

In the history of science in general, and in the history of physics and astronomy in particular, the role of observations in changing our view has been crucial. In the Aristotelian-Ptolemaic cosmological and physical paradigm the Earth is spherical, motionless, and exists in the center of the Universe, around which revolve the spheres of the planets, Sun, and fixed stars (the Earth is composed of four elements and their combinations: Earth, Water, Fire, and Air, while the spheres are made of a perfectly transparent substance known as 'quintessence'—the 'fifth' element). It remained the absolute physical model for more than 1500 years, and actually it was the most long-lived scientific system in history.

It was only after the 11th century AD, where various Arab and Persian scholars incorporating observations performed in Maragha observatory, started putting into doubt the details of the paradigm such as the epicycles and the Earth's non-rotation. And despite the theoretical considerations of Copernicus, who was based on Aristarchos of Samos, it was only after the detailed observations made by Tycho Brahe, Johannes Kepler, and Galileo Galilei that the paradigm shift was established. In a similar way, despite the successes of Newtonian-Keplerian astronomical and physical model, a new paradigm shift was made necessary after the observations of the precession of Mercury's perihelion, which could not be explained in the previous framework.

The following decades were characterized by a significant advance in the quality and quantity of observations, leading to corrections and improvements in the new cosmological paradigm. Astronomers discovered extra galaxies beyond the Milky Way that are moreover moving away from each other (this was deduced through the Doppler effect, namely the shift of their emitted radiation wavelength towards larger values). Since every galaxy moves away from every galaxy, one can establish the framework of an expanding and cooling Universe originating from a primordial super-dense and super-hot state. This 'Big Bang' theory offered verified quantitative predictions (e.g., the abundance of primordial elements and the cosmic microwave background radiation) and was able to describe all observations.

However, theoretical investigation revealed that it might have some theoretical 'problems' (or at least issues whose explanation was not 'natural'), such as the horizon, the flatness, and the magnetic monopole ones. Thus, after 1980 the phase of inflation was established as a necessary ingredient of the cosmological paradigm at its early stages. Finally, in the last decades the cosmological paradigm underwent through another modification in order to incorporate the 'indirect observation' of the dark matter sector (see section 2.5.1). Hence, in the mid-1980s a concordance cosmological paradigm had been well established, namely an expanding Universe governed by General Relativity, in which one has all particles of the SM of particle physics (proton, neutrons, electrons, photons, and neutrinos) plus the unknown sector of Cold Dark Matter (CDM).

2.5.2.2 Accelerated expansion

The only open question was to determine the rate in which the Universe expansion is slowing down. If it was too small then the Universe would expand forever with reducing speed, while if it was sufficiently large then the Universe expansion would stop at a finite time, after which it should re-collapse resulting eventually in a Big Crunch. Finally, if the Universe expansion had a very specific value in-between, the Universe would expand with a speed that would asymptotically become zero at infinite time. The fact that the expansion should be decelerating was considered undoubtful, since gravity is an attractive force and hence the clusters of any form of matter (baryonic or dark) would tend to either re-collapse or move mutually away with lower and lower speed.

In the mid-1990s two Collaborations, namely the Supernova Cosmology Project and the High-z Supernova Search Team, started to measure in detail the 'deceleration parameter' of the Universe in order to provide a definitive answer to the above question. The way to do that was the following: if we find objects in the sky for which we know their moving-away speed, as well as their distance (and hence how long ago their image corresponds to, since their light needs some time to reach us) we can determine the expansion rate of the Universe at various times and hence deduce the 'deceleration parameter'.

Measuring speeds is quite easy in cosmology due to the Doppler effect and the redshift of the observed spectrum mentioned above. Measuring small distances is also quite easy due to the parallax phenomenon (perceived change in position of a relatively nearby object seen from two different places of the Earth's orbit) and simple geometrical calculations, but measuring large distances can be challenging. Nevertheless, one can still measure large distances if he can find objects emitting light with known flux. In particular, knowing how much radiation the object emits, observing how much radiation reaches us, and considering the inverse-square law for the reduction of light intensity with distance, we can calculate the object's distance. Fortunately, such 'standard candles' exist, and they are a particular class of supernovae explosions called Type Ia (these are believed to be caused by the core collapse of white dwarf stars when they accrete material to take them over a particular limit). Such objects have the same luminosity, although they are at different distances, and hence even if that luminosity is not known it will appear merely as an overall scaling factor.

Observing 50 such Type Ia supernovae in 1998 the Project teams resulted in a conclusion that was completely unexpected. The supernovae were less bright than expected, which implies that light had traveled more time in order to reach us, and hence the Universe expansion is not slowing down but on the contrary it is accelerating! This result was indeed verified later on by observations of completely different origin, namely the structure of the spectrum of the cosmic microwave background (CMB) radiation (there is an almost isotropic black-body microwave photon radiation coming to us from all points of the sky, with a temperature of around 2.7 K, which nevertheless has very small anisotropies that provide valuable information), the baryonic acoustic oscillations (BAO) (fluctuations in the density of the baryonic matter caused by acoustic density waves in the primordial plasma of the early Universe), etc, and this cross-verification gave to the leaders of the supernovae Collaborations Saul Perlmutter, Adam Riess, Brian P Schmidt the Nobel Price in Physics in 2011 [124].

The results delivered a major surprise to cosmologists. None of the usual cosmological models was able to explain the observed luminosity distance curve, called the apparent magnitude-redshift diagram, presented in figure 2.27. Speaking in quantitative terms, introducing the density parameters for the various components of the Universe (i.e., Ω_i for the 'ith' component), observations indicate that all known SM sectors (heavy elements, stars, free hydrogen, and helium, neutrinos,

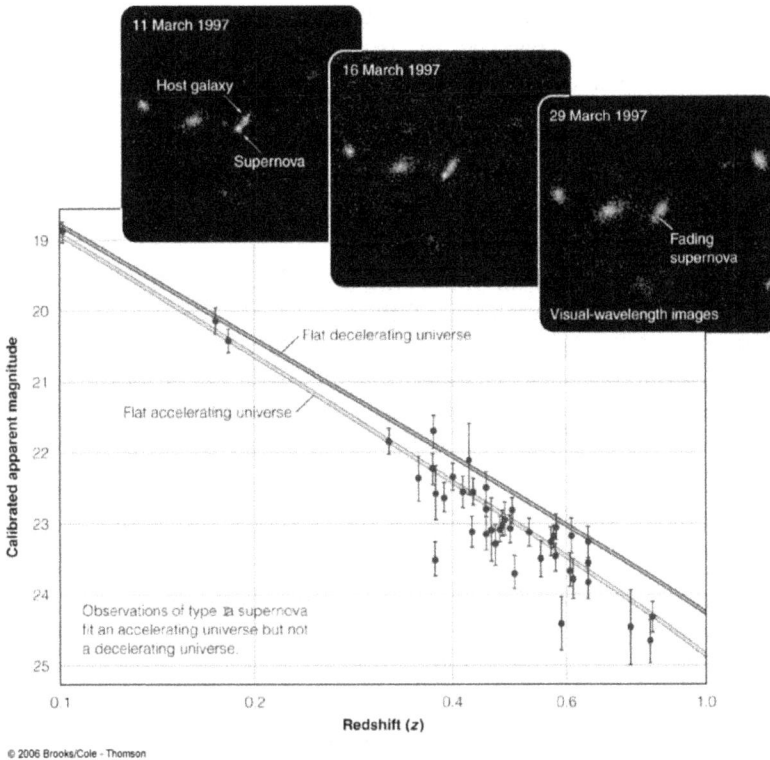

© 2006 Brooks/Cole - Thomson

Figure 2.27. The apparent magnitude-redshift diagram for 50 Type Ia supernovae elaborated by the Supernova Cosmology Project and the High-z Supernova Search Team, i.e. the observed luminosity distance curve as a function of the redshift (larger redshifts correspond to larger velocities, thus larger distances, thus earlier times). The data show that the expansion rate of the Universe is currently larger than it used to be in the recent cosmological past, and thus the expansion is accelerating.

photons, etc) constitute only 5% of the total Universe content, cold dark matter constitutes approximately 25%, and the remaining approximately 70% corresponds to this unknown cause of acceleration. The Universe is not only accelerating, but the source of acceleration is by far its dominant sector. So what can it be?

2.5.2.3 ΛCDM concordance model

When Einstein was formulating the theory of general relativity, astronomers believed that the Universe was static. General relativity could not permit this, simply because it gives rise to attractive gravitational interactions and hence a static Universe should collapse under its own gravitational attraction. In order to allow for a static Universe, in 1917 he proposed a change to the equations by introducing a 'cosmological constant', called Λ, which at cosmological scales could compensate the gravitational attraction. However, when the Universe was found to be expanding, and thus general relativity could describe it without the need of such a term, Einstein removed it famously calling it the 'greatest blunder' of his life.

Interestingly enough, the cosmological constant Λ term is exactly what is needed in order to describe the accelerating expansion of the Universe. Since it compensates the attractive gravitational interaction, it can have exactly this effect. Equivalently, viewing it from the fluid perspective, the cosmological constant is completely different than any other form of matter, since it corresponds to negative pressure, which is what would be needed to obtain accelerated expansion. Hence, after the 1998 discovery, the cosmological constant was proudly re-introduced to cosmology as the main component of the Universe, giving rise to the current concordance model, the so-called ΛCDM paradigm. Once again, observations had played a crucial role in leading to a paradigm shift.

Now, in order to describe an expanding Universe in the framework of general relativity we introduce the concept of the four-dimensional space time. This is imposed to be homogeneous and isotropic in order to comply with the 'Cosmological Principle' (which states that at large scales the Universe is the 'same' everywhere). Hence, we use the Friedmann-Robertson-Walker (FRW) metric,

$$ds^2 = -c^2dt^2 + a^2(t)\left[\frac{dr^2}{1 - kr^2} + r^2(d\theta^2 + \sin^2\theta \, d\phi^2)\right], \qquad (2.25)$$

in which t is time, r, θ, ϕ are the spatial dimensions in spherical coordinates, c is the speed of light, and $a(t)$ is the 'scale factor', the single quantity that quantifies the 'size' of the Universe (in an expanding Universe $a(t)$ is an increasing function of time). Finally, the parameter k accounts for the geometrical features of the three-dimensional space, and it is 0 for 'flat' space, $+1$ for 'closed' space, and -1 for 'open' space.

The essence of general relativity is that 'matter tells space time how to curve and the curved space time tells matter how to move'. More accurately, the distribution of the material that constitutes the Universe determines its evolution. Hence, according to the Standard Model of Cosmology, namely the ΛCDM paradigm, the Universe evolution is determined by the two Friedmann equations, which provide the differential equation for the scale factor [124]:

$$\left[\frac{\dot{a}(t)}{a(t)}\right]^2 = \frac{8\pi G}{3}\sum_i \rho_i(t) - \frac{kc^2}{a(t)^2} + \frac{\Lambda c^2}{3}, \tag{2.26}$$

$$\frac{\ddot{a}(t)}{a(t)} = -\frac{4\pi G}{3}\left[\sum_i \rho_i(t) + \frac{\sum_i p_i(t)}{3c^2}\right] + \frac{\Lambda c^2}{3}. \tag{2.27}$$

In these equations ρ_i and p_i are respectively the energy density and pressure of the various sectors that constitute the Universe material (baryonic matter, dark matter, photon radiation, neutrino radiation), which at large scales in a homogeneous and isotropic Universe depend only on time (which is consistent with the metric choice (2.25)). Moreover, G is the usual Newton's constant, which determines the strength of gravitational interaction. Finally, the equations close by considering the 'equation of state' of the various sectors, namely the expression $p \equiv p(\rho)$ that relates the pressure with the energy density.

The cosmological constant can perfectly describe quantitatively the observed acceleration of the Universe expansion. But what is its physical nature and more importantly why does it have the value it has? Although in mathematical terms Λ is just a constant that is introduced in the equations of general relativity, from the physical point of view it corresponds to the energy of space, or equivalently the vacuum energy. In any field theory one can estimate the vacuum energy following basic theoretical steps. The problem is that these basic field-theoretical estimations of the value of the zero-point energy (i.e., of the cosmological constant) lead to a number that (depending on the cutoff and other factors) is around 120 orders of magnitude larger than its observed value, which is found to be $\sim 7 \times 10^{-30}$ g cm^{-3}. This 'largest discrepancy between theory and experiment in all of science' is the renowned 'cosmological constant problem'.

As we can see, although Λ CDM paradigm can describe the Universe evolution and features, it conceptually suffers from the aforementioned problem on the nature and value of the cosmological constant. Seeing from the fluid point of view, according to Friedmann equations (2.26) and (2.27) the cosmological constant corresponds to a 'fluid' with energy density $\rho_\Lambda = \frac{c^2}{8\pi G}\Lambda$ and negative pressure $p_\Lambda = -\frac{c^2}{8\pi G}\Lambda$, and thus with equation-of-state parameter $w_\Lambda \equiv p_\Lambda/\rho_\Lambda = -1$.

2.5.2.4 Dark energy models

Having the above in mind, and trying to come closer to a microphysical description of the cosmological constant, or more fundamentally to the source that drives the Universe acceleration, one can substitute the cosmological constant by the more general concept of 'dark energy'. This umbrella term accounts for all possible explanations of the Universe late-time acceleration, either the cosmological constant itself or any other alternative. The term is adequate to describe the main features of this unknown source of the Universe acceleration, namely that it is a kind of 'energy' and that it is 'dark', in the sense that it does not interact with electromagnetism (in a similar way that dark matter is a form of matter which is 'dark').

In order for a mechanism, field, fluid etc, to be the dark energy, it should satisfy some necessary requirements. First of all it should not interact with Standard Model particles, including photons, or if it does the interaction cross-section should be extremely small; otherwise, we should have observed it. Additionally, seeing from the fluid point of view, dark energy should correspond to negative pressure, and in particular its effective equation-of-state parameter today should be close to -1 (i.e., to the cosmological constant value), since this is what observations show. In particular, in figure 2.28 we present the observational constraints on the value of the dark-energy equation-of-state parameter at the present Universe w_0, versus the matter density parameter Ω_m (thus the dark-energy density parameter is approximately $1 - \Omega_m$). Any dark energy model should satisfy these constraints. Hence, one could follow two main ways in order to obtain the above basic features: Construct models with known microphysical behavior, or construct phenomenological, effective models, with unknown microphysics but efficient to play the role of dark energy.

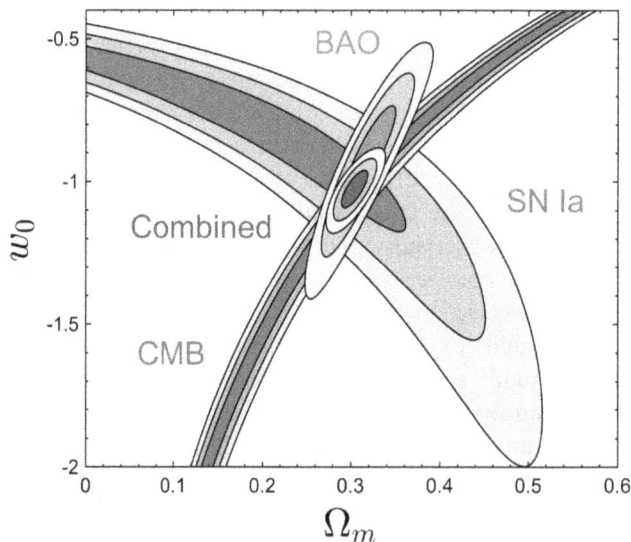

Figure 2.28. Constraints on the dark-energy equation-of-state parameter at the present Universe w_0, versus the matter density parameter Ω_m, arising from various observational datasets, namely from supernovae Type Ia (SNIa), cosmic microwave background (CMB) radiation, and baryonic acoustic oscillations (BAO) (see text) and their combination. The contours correspond to 68.3%, 95.4%, and 99.7% (1σ, 2σ and 3σ) likelihood, respectively, and one assumes a spatially-flat universe. Reproduced from [147], copyright IOP Publishing Ltd. All rights reserved.

The most well-studied model that can be a candidate for dark energy is a simple scalar field (similarly to the case of the inflaton scalar field, which is the most well-studied mechanism for the inflation realization in the early Universe). In particular, the whole Universe is filled with a scalar field ϕ whose dynamics drives the Universe acceleration. This scalar field is not a part of the SM of particle physics, and it is called 'quintessence' (Aristotle's revenge one could say!). Hence, one introduces the Lagrangian

$$\mathcal{L}_{\text{quint}} \equiv -\frac{1}{2}\partial^{\mu}\phi\partial_{\mu}\phi - V(\phi), \qquad (2.28)$$

with μ taking the four coordinate values 0, 1, 2, 3, and where $V(\phi)$ is the potential of the quintessence field. Note that no interaction term between ϕ and SM particles is introduced, as required. In the cosmological FRW metric (2.25), in which the quintessence field depends only on time, the above Lagrangian leads to the equation of motion:

$$\ddot{\phi}(t) + 3\frac{\dot{a}(t)}{a(t)}\dot{\phi}(t) + \frac{dV(\phi)}{d\phi} = 0. \qquad (2.29)$$

Additionally, the energy density and pressure of the scalar field, respectively, are

$$\rho_{\text{quint}} = \frac{\dot{\phi}^2}{2} + V(\phi) \qquad (2.30)$$

$$p_{\text{quint}} = \frac{\dot{\phi}^2}{2} - V(\phi). \qquad (2.31)$$

Hence, the Friedmann equations in a quintessence Universe will be (2.26) and (2.27), but instead of the cosmological constant terms ρ_{Λ} and p_{Λ} one should use the quantities above. Finally, the equation-of-state parameter of quintessence dark energy is simply

$$w_{\text{quint}} = \frac{\frac{\dot{\phi}^2}{2} - V(\phi)}{\frac{\dot{\phi}^2}{2} + V(\phi)}. \qquad (2.32)$$

Interestingly enough, by choosing suitably the quintessence potential (e.g., cases that lead to $V(\phi) \gg \dot{\phi}^2$), we can obtain $w_{\text{quint}} \approx -1$ and in this case the quintessence field plays the role of dark energy (i.e., it drives the Universe acceleration). We mention that although the scenario at hand is quantitatively similar to the cosmological constant, physically it is radically different since now the dark energy density is not the vacuum energy but the energy density of a scalar field, and thus there is no cosmological constant problem, and no 120-orders-of-magnitude error. Nevertheless, we should say here that if one wants to describe quantitatively the Universe acceleration then he should use potentials that in particle terms correspond to quintessence mass of the order of the cosmological constant, namely $\sim 10^{-33}$eV. Hence, we do not face the cosmological constant problem but we do face a huge hierarchy problem, namely we need to explain why in the Universe we have a massive particle that is tens orders of magnitude lighter than all the others (some people call this 're-parametrization of our ignorance').

Inspired by the above simple dark energy scenario, in the literature there appeared hundreds of dark energy models based on field Lagrangians. For instance, one can consider a large number of possible potentials, he can assume non-minimal interactions of the field to gravity, consider generalized kinetic terms (the so-called K-essence models), consider the case where the scalar field is tachyonic, phantom, or dilaton, use more than one scalar fields, use vector fields, etc. Recently, the most general scalar-field dark-energy models were investigated, the so-called Horndeski theories, or Generalized Galileon theories, in which one has many possibilities and many parameters and free functions of the field in order to successfully describe the cosmological data.

One different approach to the dark energy is the classic historical approach of introducing an unknown 'fluid'. In particular, since we know the basic requirements that dark energy sector needs to fulfill at macroscopic-cosmological level, we can introduce by hand a fluid with the specific phenomenological macroscopic properties, without caring about its microphysical description. Since a fluid in a homogeneous and isotropic Universe is determined just by its equation-of-state parameter (at least for the simple 'barotropic' fluids used in cosmology), one just introduces a desired form for the equation-of-state parameter of the dark-energy fluid, namely w_{DE}, in order to phenomenologically fit the data. In figure 2.29 we depict the

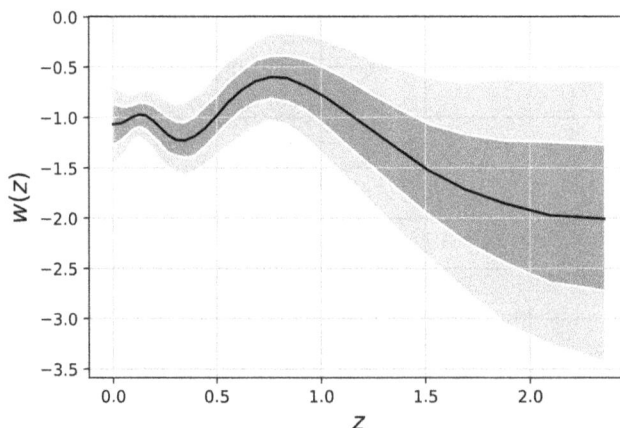

Figure 2.29. The data-driven reconstructed dark-energy equation-of-state parameter w_{DE} as a function of redshift z. Today $z = 0$, while $z = 1$ corresponds to approximately 8 Gyrs ago, and $z = 2.5$ to approximately 11 Gyrs ago. The contours correspond to 68.3% and 95.4% (1σ and 2σ) likelihood respectively, while the black curve denotes the best-fit value. Reprinted by permission from Springer Nature [148], copyright (2017).

data-driven reconstructed w_{DE} as a function of the redshift. Hence, any dark energy phenomenological model should satisfy these constraints.

Definitely, as we discussed above, the cosmological constant itself lies within this effective fluid description, with $w_{DE} = \text{const} = -1$. The first extension would be to consider dark-energy fluids with $w_{DE} = \text{const} = w_0$, the so-called wCDM dark energy models. One could then proceed to a large variety of w_{DE} parametrizations, like the Chevallier–Polarski–Linder (CPL), Linder, Linear, Logarithmic, oscillating, etc, ones, in which w_{DE} is considered to be time-varying (equivalently scale-factor-varying or redshift-varying) with a particular form. For instance, in the CPL parametrization one assumes that

$$w_{DE} = w_0 + w_a(1 - a),\tag{2.33}$$

with a the Universe scale factor and w_0, w_a two parameters. Alternatively, one could assume that the dark energy fluid has richer properties, satisfying, for instance, the equation of state class of the Chaplygin gas, whose pressure depends on energy density as

$$p_{Cg} = -A\rho_{Cg}^{-\alpha},\tag{2.34}$$

with A, α two parameters. All these effective dark energy models remain at the phenomenological level, without offering an explanation for the microphysical nature of dark energy. However, they are very efficient in describing the dark energy features, as well as the cosmological data, and thus they are extensively used.

Let us now refer to a completely different approach to dark energy and Universe acceleration that has attracted a huge amount of research effort the last two decades. The essence of general relativity, which lies in the foundation of cosmology, is that the Universe evolution (i.e., the dynamics of space time) is determined by the distribution of the Universe content. In the dark energy approaches described above the strategy was to introduce an extra component in the Universe content, with suitable properties that can change the space-time dynamics in the desired way. However, one could follow a different strategy: instead of assuming that there are extra components in the Universe, to consider that the gravitational interaction is not general relativity, but a modified/extended one, which will lead to different space-time dynamics and thus to different Universe evolution.

Hence, in the modified gravity approach to dark energy, the latter arises in an effective way due to the extended gravitational interaction. In particular, one constructs new gravitational theories, which of course accept general relativity as a particular limit (in a similar way that general relativity tends to Newtonian gravity at a particular limit), but which in general have extra degree(s) of freedom and richer dynamics that can trigger the acceleration of the Universe without the need of extra fields, fluids, particles, etc. Additionally, contrary to the phenomenological/fluid approaches to dark energy, in the modified gravity approach one does obtain a microphysical description and a full explanation, which is exactly the modified gravity.

Having in mind the structure of general relativity, by changing it in various ways one can obtain numerous classes of modified/extended gravity. For instance, one can generalize the Einstein–Hilbert action through the addition of extra geometrical terms and their possible couplings to scalars and vectors such as in tensor-vector-scalar (TeVeS) theories. Alternatively, one may consider extra vector or tensor degrees of freedom, such as in massive or bimetric gravity. Moreover, one can use richer geometries, e.g. incorporating torsion, non-metricity, and non-commutativity, or consider extra dimensions, such as in Braneworld models or in string theory, etc. In all these theories and models, in the end of the day one chooses the details, as well as the model parameters, in order to be able to obtain effective dark energy features in agreement with observations. Actually, since cosmology (amongst others) is our laboratory for gravity, confrontation with detailed observational data is a necessary and important test for every gravitational theory. Even if a gravitational theory is constructed with different goals and motivations than to describe dark energy, at the end of the day it should be confronted with Universe acceleration data just to ensure that it does not lead to wrong effective dark energy features.

2.5.2.5 Conclusions

Let us close this section with a historical apposition. In the second half of the 19th century, astronomers started to observe differences in the precession of Mercury's orbit comparing to the predicted behavior. This led Urbain Le Verrier in 1859 to propose that this was a result of another, extra, planet between Mercury and Sun, namely the 'Vulcan' (note that Verrier in 1846 was able to describe the discrepancies with Uranus's orbit by predicting through purely theoretical calculations the existence and current position of a new planet, namely Neptune, which was indeed discovered by Johann Gottfried Galle at exactly the predicted coordinates). However, the answer to the problem of Mercury's precession was not an extra planet, a planet that we had not seen, and thus in a sense a 'dark' planet, the answer was modified gravity, namely general relativity. This new gravitational theory had different foundations, physical interpretation, and mathematical structure from Newtonian gravity; nevertheless at the level of predictions it was served as its modification, exhibiting the latter as a particular limit and offering corrections that were larger at scales that had just then started to become accessible by technological advance.

Interestingly enough, from the beginning of the 21st century and until now, we are in a similar situation that physicists were in the the beginning of the 20th century. Technological advance made it possible to access scales and perform accurate observations that suggest modification of the concordance cosmological paradigm. In general, we have two main ways to explain it: either we will consider new forms of material that in the framework of general relativity will do the job, or we will modify the theory of gravity itself. In figure 2.30 we present a schematic classification of all such dark energy approaches. Will the final outcome be similar to the one of the previous century, namely modified gravity, or it will it be an unknown, for the moment, 'dark' material? The combined theoretical and observational research will provide the answer. Stay tuned!

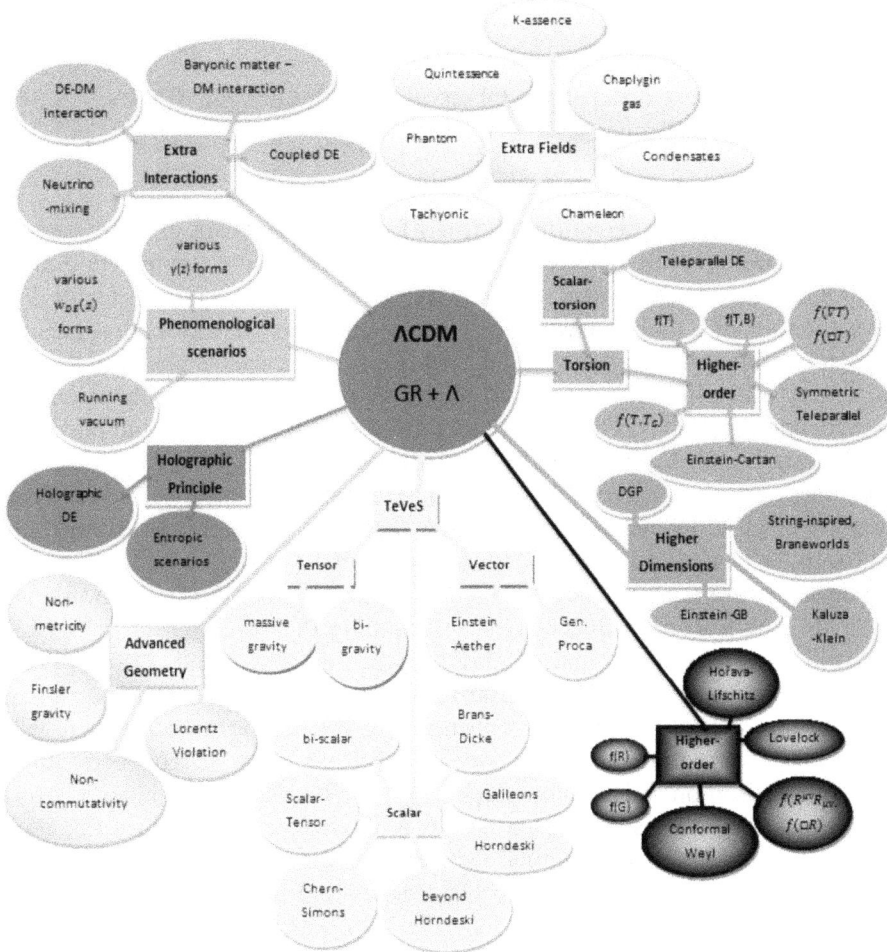

Figure 2.30. A schematic, not unique, classification of the dark energy approaches, representing the current status of cosmological research. These models arise from various extensions and modifications of the concordance paradigm of ΛCDM cosmology, which is based on general relativity (GR), cosmological constant (Λ), cold dark matter (CDM), and all particles of the Standard Model of particle physics. Details can be found in [149].

Acknowledgment

This section is partially supported by the National Natural Science Foundation of China and by the University of Science and Technology of China Fellowship for International Cooperation.

2.6 A gravitational universe: black holes and gravitation waves

Nelson Christensen[1]

[1]Artemis, Université Côte d'Azur, Observatoire Côte d'Azur, CNRS, CS 34229, 06304 Nice Cedex 4, France

A century after their prediction by Albert Einstein, numerous gravitational-wave events have been directly detected by the two Michelson interferometers that form the LIGO in the United States, and the European interferometric detector Virgo. This is the start of a global network of gravitational-wave detectors that begins the era of gravitational-wave astronomy. The majority of the events observed have been from binary black hole mergers. Two binary neutron star mergers have also been observed, one of which was in coincidence with a gamma-ray burst, and the remnant *kilonova* was found. Gravitational waves are the best way to get information on black holes, and test general relativity. The current gravitational-wave detector network will grow to include the KAGRA detector in Japan, and a third LIGO detector to be located in India. The third generation of ground-based gravitational-wave detectors are presently being designed, with the hope to be operational in the mid-2030s. The European Space Agency is leading the development of the LISA, which will operate in space, and observe a lower frequency band. LISA is planned for launch also in the mid-2030s. Pulsar timing arrays will likely soon be observing gravitational waves at even lower frequencies. In the decades to come there will be observations of gravitational waves over a broad spectrum of frequencies, thereby giving information on stellar-mass black holes, intermediate-mass black holes, and supermassive black holes. Presented here is a description of the plans for gravitational-wave detection in the coming decades, and what can be learned from these observations, especially pertaining to black holes and what they tell us about the Universe.

2.6.1 Introduction

A century after their prediction by Albert Einstein [127–129], gravitational waves, produced from a binary black hole merger, were directly detected for the first time by the two Michelson interferometers that form the LIGO [130]. LIGO, along with the European detector Virgo, are the start of a global network of gravitational-wave detectors that begin the era of gravitational-wave astronomy. Together LIGO and Virgo have announced the detection of 50 gravitational-wave events from their first three observational runs in the advanced detector era (O1, O2, and the first half of O3, O3a), namely 48 binary black hole mergers and two binary neutron star mergers [131]. These second-generation ground-based interferometric gravitational-wave detectors are referred to as Advanced LIGO [132] and Advanced Virgo [133]. With arm lengths of 3 km for Virgo and 4 km for LIGO, these detectors can measure a relative displacement of the interferometers' arm lengths of 10^{-18} m.

Presently numerous collaborations are building and operating second-generation interferometers in order to detect gravitational waves. Advanced LIGO in the

United States consists of two interferometers located in Livingston, Louisiana, and Hanford, Washington. Advanced LIGO started observations in 2015, and will be working over the coming years to achieve its design sensitivity, with the goal to reach it by 2025 [134]. The European Advanced Virgo interferometer is near Pisa, Italy [133]. Virgo started acquiring data in 2017, and will also be aiming for its target sensitivity in the coming years. GEO 600, a German–British collaboration, is a 600 m detector near Hanover, Germany [135], and is currently operational. KAGRA is the Japanese 129 km interferometer that is presently under commissioning, and will commence observations in 2022 [136]. There will be a third 4 km LIGO interferometer, LIGO-India [137], located in India, with the goal to be operational in the coming years. All of the km-length detectors will be attempting to detect gravitational waves with frequencies from 20 Hz up to a few kHz. The higher the mass of a binary system, the lower the frequency of the gravitational-wave signal. The second-generation gravitational-wave detectors will observe lighter binaries, like binary neutron star systems with individual component masses around $1.4\,M_{\odot}$ ($M_{\odot}$ represents the mass of our Sun), up to binary black hole systems with total masses up to hundreds of $M_{\odot}$.

In Europe there are plans to build a third-generation gravitational-wave detector, the Einstein Telescope [138], while in the United States there are plans for the Cosmic Explorer [139]. These detectors would have a better sensitivity over the present second-generation detectors by a factor of 10, thereby seeing 1000 more of the Universe. The low-frequency sensitivity will be pushed down to 5 Hz, and maybe even 1 Hz, which will allow for the observation of more massive binary black hole systems. Intense research and development is presently ongoing so as to allow for these detectors to observe in the mid-2030s. The Einstein Telescope will be in a triangular configuration, 10 km arm-length, with three low-frequency detectors and three high-frequency detectors. In order to minimize seismic and anthropogenic noise, the Einstein Telescope is planned to be 100–300 m below ground. The Cosmic Explorer will have an L-shape (like the present LIGO and Virgo), but gain in sensitivity with a possible arm-length of 40 km; this detector will be on the surface of the Earth. The third-generation detectors will measure binary black hole systems up to a few thousand $M_{\odot}$.

The LISA will be a gravitational-wave detector in space [140]. It will consist of three satellites, forming an equilateral triangle, with arm lengths of 2.5 million km. Laser beams propagating between each satellite create three gravitational-wave detectors. LISA will orbit the Sun, but 20° behind the Earth's orbit. LISA will observe gravitational waves in the 10^{-5} to 1 Hz band. LISA will measure binary black hole systems up to a few million $M_{\odot}$.

Pulsars, or spinning neutron stars, emit radio signal in a very stable fashion, like clock. The received timing of the pulses can be perturbed due to gravitational waves. As such, Pulsar Timing Arrays are being used to search for gravitational waves at very low frequencies, 10^{-9}–10^{-6} Hz. Binary systems containing supermasssive black holes would emit in this frequency band. The NANOGrav collaboration could already be observing a first hint of a gravitational-wave signal [141]. There is a growing effort around the world to monitor pulsars for pulsar timing gravitational-wave detection. In addition to existing collaborations in the USA, Europe, and Australia, the upcoming radio telescope network, the Square Kilometer Array, will also contribute to pulsar timing observations in the decades to come [142–144]. Pulsar timing will measure binary black hole systems up to a few billion $M_\odot$.

2.6.2 Gravitational waves

An accelerating electric charge produces electromagnetic radiation—light. It should come as no surprise that an accelerating mass produces gravitational light, namely gravitational radiation (or gravitational waves). In 1888 Heinrich Hertz had the luxury to produce and detect electromagnetic radiation in his laboratory. There will be no such luck with gravitational waves because gravity is an extremely weak force.

Albert Einstein postulated the existence of gravitational waves in 1916, and Joe Taylor and Joel Weisberg [145] indirectly confirmed their existence through observations of the orbital decay of the binary pulsar 1913+16 system. The direct detection of gravitational waves has been difficult, and has literally taken decades of tedious experimental work to accomplish. The only possibility for producing detectable gravitational waves comes from extremely massive objects accelerating up to relativistic velocities. The gravitational waves that have been detected so far have come from the coalescence of binary neutron star or binary black hole systems. For example, the first direct detection of a gravitational-wave event, GW150914 [130], was produced by the merger of a 29 $M_\odot$ black hole and a 36 $M_\odot$ black hole some 1.3×10^9 light years away. The total energy radiated in gravitational waves was equivalent to 3 $M_\odot\, c^2$, with a peak luminosity of 3.6×10^{56} ergs s^{-1}, or about 10^{23} times the luminosity of the Sun.

Other possibly detectable gravitational-wave sources are also astrophysical: core-collapse supernovae, pulsars, neutron star—black hole binary systems, newly formed black holes, or even early Universe inflation. The observation of these types of events would be extremely significant for contributing to knowledge in astrophysics and cosmology. Gravitational waves from the Big Bang would provide unique information of the Universe at its earliest moments. Observations of core-collapse supernovae will yield a gravitational snapshot of these extreme cataclysmic events. Pulsars are neutron stars that can spin on their axes at frequencies up to hundreds of Hertz, and the signals from these objects will help to decipher their characteristics. Gravitational waves from the final stages of coalescing binary neutron stars could help to accurately determine the size of these objects and the equation of state of nuclear matter; they would also help to explain the mechanism that produces short gamma-ray bursts. The observation of black hole formation from these binary systems, and the ringdown of the newly formed black hole as it approaches a perfectly spherical shape, would be the *coup de grâce* for the debate on black hole existence, and the ultimate triumph for general relativity.

Advanced LIGO [132] and Advanced Virgo [133] are second-generation interferometric gravitational-wave detectors. Initial LIGO and Virgo conducted observations from 2002 through 2010. Advanced LIGO and Advanced Virgo will ultimately have better sensitivities, by a factor of 10, over their initial designs. They search for gravitational waves from 20 Hz up to a few kHz. Their target sensitivities will allow them to observe signals from the coalescence of binary neutron star systems (1.4 $M_\odot$– 1.4 $M_\odot$) out to distances past 300 Mpc for Advanced LIGO and past 200 Mpc for Advanced Virgo. The mergers of more massive binary black holes systems will extend much farther. Already from their first three observational runs Advanced LIGO and Advanced LIGO have detected gravitational waves from 48 binary black hole coalescences, and two binary neutron star inspirals [131].

2.6.2.1 Some general relativity

Electromagnetic radiation has an electric field transverse to the direction of propagation, and a charged particle interacting with the radiation will experience a force. Similarly, gravitational waves will produce a transverse force on massive objects, a tidal force. Explained via general relativity it is more accurate to say that gravitational waves will deform the fabric of space time. Just like electromagnetic radiation there are two polarizations for gravitational waves. Let us imagine a linearly polarized gravitational wave propagating in the z-direction, $h(z, t) = h_{0+}e^{i(kz-\omega t)}$. The fabric of space is stretched due to the strain created by the gravitational wave. Consider a length L_0 of space along the x-axis. In the presence of the gravitational wave the length oscillates like

$$L(t) = L_0 + \frac{h_{0+}L_0}{2} \cos(\omega t)$$

hence there is a change in its length of

$$\Delta L_x = \frac{h_{0+}L_0}{2} \cos(\omega t).$$

A similar length L_0 of the y-axis oscillates, like

$$\Delta L_y = -\frac{h_{0+}L_0}{2} \cos(\omega t).$$

One axis stretches while the perpendicular one contracts, and then vice versa, as the wave propagates through. Consider the relative change of the lengths of the two axes (at $t = 0$),

$$\Delta L = \Delta L_x - \Delta L_y = h_{0+}L_0,$$

or

$$h_{0+} = \frac{\Delta L}{L_0}.$$

So the amplitude of a gravitational wave is the amount of strain that it produces on space time. The other gravitational wave polarization ($h_{0\times}$) produces a strain on axes 45° from (x,y). Imagine some astrophysical event produces a gravitational wave that has amplitude h_{0+} on Earth; in order to detect a small distance displacement ΔL one should have a detector that spans a large length L_0. The first gravitational wave observed by LIGO had an amplitude of $h \sim 10^{-21}$ with a frequency at peak gravitational-wave strain of 150 Hz [130]. The magnitude of a gravitational wave falls off as $1/r$, so it will be impossible to observe events that are too far away. However, when the detectors' sensitivity is improved by a factor of n, the rate of signals should grow as n^3 (the increase of the observable volume of the Universe). This is because the gravitational-wave detectors measure signals from all directions; they cannot be pointed, but reside in a fixed position on the surface of the Earth.

A Michelson interferometer can measure small phase differences between the light in the two arms. Therefore, this type of interferometer can turn the length variations of the arms produced by a gravitational wave into changes in the interference pattern of the light exiting the system. This was the basis of the idea from which modern laser interferometric gravitational-wave detectors have evolved. Imagine a gravitational wave of amplitude h is incident on an interferometer. The change in the arm length will be $\Delta L \sim h L_0$, so in order to optimize the sensitivity it is advantageous to make the interferometer arm length L_0 as large as possible. The Advanced LIGO and Advanced Virgo detectors will measure distance displacements that are of order $\Delta L \sim 10^{-18}$ m or smaller, much smaller than an atomic nucleus [146]. The recent observation of gravitational waves has been one of the most spectacular accomplishments in experimental physics, and has been greeted with much excitement across the globe.

The optical systems for a laser interferometric gravitatational-wave detector are quite complex. Figure 2.31 displays the optical set-up for the Advanced LIGO detector. Every photon that enters the detector can be thought of as a *meter stick* that is used to measure the length difference between the two arms. The statitics of repeated measurements then implies that the more photons used, the more measurements that are made, and ultimately the better the detection statistic. Hence, high-power lasers are used. Also, various schemes are employed to *recycle* the light to build up power and the signal. Because these interferometers are so large, 4 km for LIGO and 3 km for Virgo, the required lasers beams are large in size. Figures 2.32 and 2.33 show photographs of actual optical components. Even though they are large in size, they still must be prestine in their optical qualities. This is a major technological challenge for gravitational-wave detectors.

Figure 2.34, top, presents an aerial view of the LIGO site at Hanford, Washington State in the USA. The magnitude of the 4 km system is apparent. Figure 2.34, bottom, displays the Virgo detector with its 3 km, located near Pisa, Italy.

2.6.3 Black holes

Black holes are predicted by general relativity. They are so dense that nothing can escape from them, including light. Far from being a purely theoretical concept, their presence has been observed in our Galaxy and Universe via numerous means. For example, the presence of black holes can be inferred from x-ray observations [150]. Advanced LIGO [132] and Advanced Virgo [133] have now directly observed gravitational waves from stellar-mass binary-black-hole systems [151]. With the observation of GW190521, the birth of a $142 M_\odot$ intermediate mass black hole has

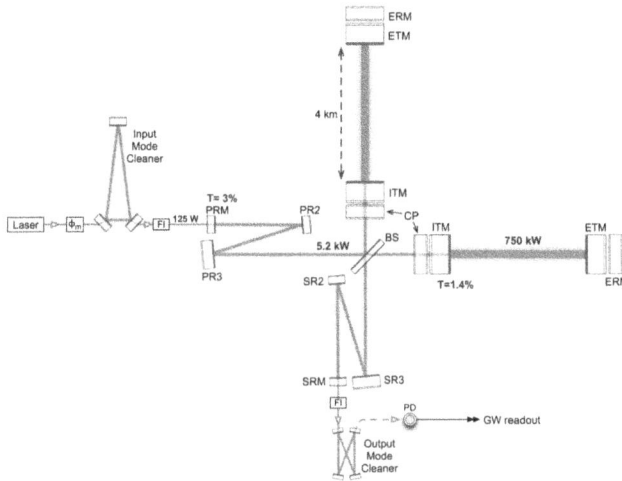

Figure 2.31. The advanced LIGO optical system. The laser (~200 W) light propagates from the stabilized laser through a phase modulator (ϕ_m) to the input mode cleaner, then through a Faraday Isolator (FI) to the power recycling mirror (PRM). The folding mirrors, PR2 and PR3, direct the light to the beamsplittler (BS) input of the interferometer. Note that approximately 125 W of light impinges upon the power recycling mirror, resulting in 5.2 kW at the input port to the beam splitter. The Fabry–Perot cavities are formed with the input test mass (ITM), which is coupled to a compensation plate (CP), and the end test mass (ETM) which is coupled to a end reaction mass (ERM). The Fabry–Perot cavities will contain 750 kW of light power. Note too that the output signal from the interferometer can itself be recycled and amplified at specific frequencies, dependent on the reflectivity of the signal recycling mirror (SRM); SR2 and SR3 are folding mirrors. The output beam also has its spatial features cleaned with the output mode cleaner before the light falls upon a photodetector (PD). Reproduced from [132], copyright IOP Publishing Ltd. All rights reserved.

Figure 2.32. A picture of the input test mass (mirror R_1) for Advanced LIGO within its vibration isolation suspension system. The fused silica component is 40 kg, 34 cm in diameter and 20 cm thick. Photograph courtesy of LIGO/Caltech/MIT.

Figure 2.33. Advanced Virgo optics. Left: The power recycling substrate (35 cm in diameter, left) beside the beam splitter of 55 cm in diameter. Right: One of the Fabry–Perot cavity mirrors, which also serves as a test mass for the gravitational wave detector. Reproduced from [133], copyright IOP Publishing Ltd. All rights reserved.

Figure 2.34. Top: Aerial view of the LIGO Hanford, Washington site. The vacuum enclosure at Hanford contains the 4 km interferometer. Photograph courtesy of LIGO/Caltech/MIT. Bottom: Aerial view of the Virgo detector, with 3 km arms, located near Pisa, Italy. Photograph courtesy of the European Gravitational Observatory.

been observed [152, 153]. Observations of the orbital dynamics of stars in the center of the Milky Way indicate that there is a supermassive black hole of mass $4 \times 10^6 \, M_\odot$ [154, 155]. The Event Horizon Telescope has recently imaged the shadow of a $6.5 \times 10^9 \, M_\odot$ supermassive black hole in the center of the galaxy M87 [156]; see figure 2.36.

While black holes are a consequence of general relativity, they were not initially recognized as we consider them today. Soon after Einstein published his description of general relativity in 1915 [127], Schwarzschild quickly followed up with the solution for the metric about a spherical body [157]. The concept of a black hole was not immediately apparent from the solution, and it took many decades for the modern interpretation to emerge [158]. Other interesting features of black holes were subsequently deciphered. A proposition known as *cosmic censorship* states that there should be no naked singularities; specifically, we should not be able to view from afar the point where the curvature of space time for the black hole goes to infinity. As such, the magnitude of the dimensionless spin vector, $\overrightarrow{\chi} = \overrightarrow{S}c/(GM^2)$, where M is the mass of the black hole and $\overrightarrow{S}$ is its spin angular momentum, must not exceed 1 [159, 160]. The *no-hair theorem* says that black holes are described by only three parameters; their mass, spin angular momentum, and charge [161, 162]. Finally, black holes obey laws of thermodynamics, and radiate as a black-body with temperature $T = \frac{\hbar c^3}{8\pi k_B G M}$, where $\hbar$ is the reduced Planck's constant, c is the speed of light, k_B is Boltzmann's constant, G is Newton's constant, and M is the black hole mass [163, 164].

2.6.4 Gravitational-wave detections and what they tell us about black holes

On September 14, 2015, at 09:50:45 UTC a gravitational wave was detected directly for the first time. The gravitational wave was first observed at the LIGO Livingston Observatory (Louisiana), and then 7 ms later at the LIGO Hanford Observations (Washington). An online transient search algorithm identified the signal in 3 min. An offline examination of the data using a template-based search for compact binary coalescence signals identified the gravitational wave with a signal-to-noise ration of 24 [130]. Parameter estimation routines were used to determine that the gravitational-wave signal was emitted from the merger of two black holes with masses of 36 $M_\odot$ and 29 $M_\odot$. The newly created black hole had a mass of 62 $M_\odot$, meaning that the total energy of gravitational wave emitted was equivalent to 3 $M_\odot c^2$. The system was 1.3 billions light years away from us when it merged.

The measured gravitational-wave signal, GW150914, from the two LIGO detectors is displayed in figure 2.35 [130]. The peak amplitude of GW150914 is $h \sim 10^{-21}$ which corresponds to a displacement of the interferometers' arms of $\Delta L \sim 2 \times 10^{-18}$ m. The exquisite sensitivity of these interferometers can be seen from these numbers. In addition to GW150914, during Advanced LIGO's first observing run two other gravitational-wave events were observed (also stellar-mass binary black hole mergers) [165, 166].

In their first three observing runs, O1, O2, and O3 (first half O3a, with results from the second half O3b still to be announced), Advanced LIGO, and Advanced Virgo reported the observations of 50 gravitational-wave events; 48 from binary black hole mergers and two from binary neutron star mergers [131]. The first reported three-detector (the two LIGO detectors plus Virgo) observation of gravitational waves was GW170814, a binary black hole coalescence [167]. The binary

Hanford, Washington (H1)

Livingston, Louisiana (L1)

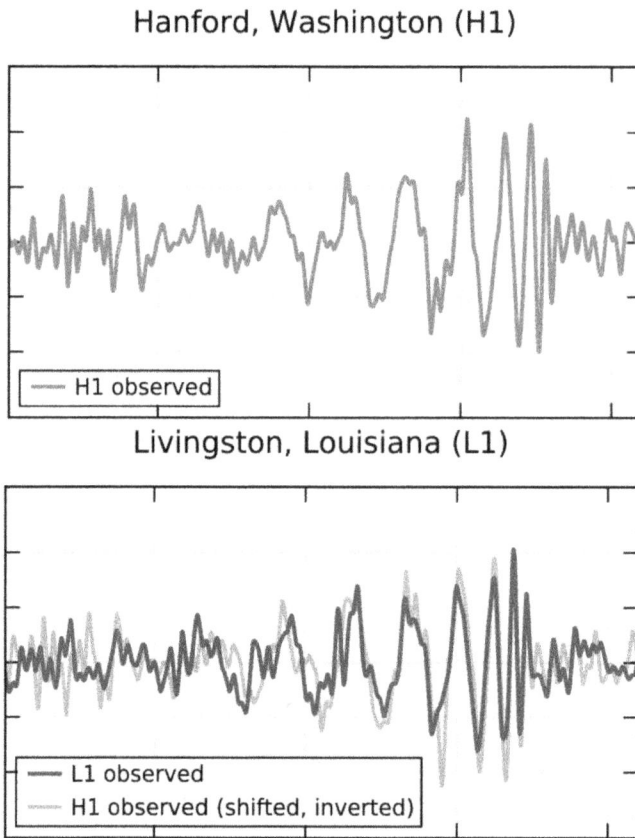

Figure 2.35. The measured gravitational-wave signal GW150914 as observed at the two LIGO interferometric detectors. The data has been bandpass filtered (35–350 Hz), and the gravitational-wave signal is clearly observable by eye. Top: The signal as observed from the LIGO Hanford detector. Bottom: The signal as observed from the LIGO Livingston detector (blue). In addition, the Hanford signal (red) is superimposed after it has been displaced by 7 ms and inverted (due to the relative orientation of the two detectors). The similarity of the two measured signals is clearly visible. Figures [130] courtesy of the LIGO Open Science Center (losc.ligo.org).

neutron star event, GW170817 [168], was also observed by the three LIGO and Virgo detectors, and as such, the position of the event could be estimated to a spot size of 28 deg^2 on the sky and a distance of 40^{+8}_{-14} Mpc. There was also a gamma-ray signal observed 1.7 s after the merger time of the binary neutron star system [169, 170]. From the gravitational-wave and gamma-ray position estimates it was possible to find the location of the source of these signals [171]. GW170817 signaled the birth of gravitational-wave multimessenger astronomy.

2.6.4.1 Binary black holes

With so many observed binary black hole mergers it is possible to study the statistical distributions of the physical parameters of the black holes. This includes the masses, the spins, and their distance. These parameters are estimated via Bayesian methods [172, 173]. The frequency with which the observations are made, and the distances to the sources allow for an estimation of the merger rates. For example, the LIGO-Virgo observations to date predict a binary black hole merger rate of 23.9 Gpc^{-3} yr^{-1}, where a Gpc is 3.3 billion light years. This means that the binary black hole mergers that LIGO-Virgo are observing (stellar mass) are happening about once every 4min in the observable Universe. For comparison, the rate of binary neutron star mergers is estimated to be 320 Gpc^{-3} yr^{-1}. Further observations will help to reveal how the merger rate of binary black holes varies with redshift (or their occurrence as a function of the age of the Universe). The present LIGO-Virgo observations indicate the the binary black hole merger rate does increase as one goes back in time, but not as fast as the rate at which stars form looking back in the history of the Universe. Knowing better the binary black hole merger rate over the history of the Universe will provide important information as to how these systems were formed, and even information on the evolution of massive systems, such as galaxies [174].

2.6.4.1.1 *Formation in the field*

The observation of the spins of the initial component black holes, and the resulting statistical distribution, can provide information on the possible formation mechanism. The field formation consists of two initial massive stars in a binary system. One star's life ends, and it forms a black hole. Material is then pulled off of the second star, into the black hole. A common envelope of material then forms about the pair, and a close orbit is produced. Eventually the core of the second star collapses to a black hole [175, 176]. The binary black hole's orbit decays via subsequent gravitational-wave emission, and ultimately there is the merger into a new and larger black hole.

For the binary black holes that have formed by field formation the spins of the initial black holes are more likely to be aligned with the orbital angular momentum. This scenario is consistent with many, but not all, of the LIGO-Virgo observed binary black hole mergers.

However, the reality of stellar physics can make this formation scenario difficult for more massive black holes. The formation by stellar processes of black holes in the $\sim 64 - 135 M_{\odot}$ range is prohibited by a process known as the (pulsational) pair-instability supernova [177, 178]. For these masses, the star's core becomes unstable, and the star disintegrates, leaving no remnant behind. The recent detection by LIGO-Virgo of GW190521 had the initial black hole masses at $85 M_{\odot}$ and $66 M_{\odot}$ [152, 153]. These masses are in tension with stellar formation, especially for the $85 M_{\odot}$ black hole. Multiple formation processes are probably necessary to describe all of the events that LIGO-Virgo have observed.

It is also important to consider the lower mass limit for black holes, especially for those formed via stellar processes. LIGO and Virgo have observed a unique system GW190814 Abbott:2020khf, with initial component masses of $23.2 M_{\odot}$ and $2.59 M_{\odot}$. This small secondary mass is difficult to explain, and an important question is whether this is a neutron star or a black hole. Neutron stars with masses above $2 M_{\odot}$

are also difficult to construe. Possibly this small mass seen in the GW190814 signal could be a black hole formed by the merger of two neutron stars, and would be at the lower limit for black hole masses formed by stellar processes.

2.6.4.1.2 Dynamic formation

Dense regions, like the center of galaxies or in globular clusters, could be environments where black holes have multiple interactions with one another, mergers happen, and more massive black holes are formed. This is the other major formation scenario to explain the binary black holes that LIGO-Virgo are observing. With such dynamic formations the spins of the initial black holes can be expected to be randomly distributed, independent from one another, and independent of the orbital angular momentum [179, 180]. LIGO and Virgo see evidence for some binary black hole systems where the initial component masses have spins parallel to the orbital plane (or perpendicular to the orbital angular momentum).

When the spins are perpendicular to the orbital angular momentum, there is spin–orbit coupling (similar to atomic systems), and this will cause a precession of the orbital plane. This was the case with the very massive GW190521 system, where there were indications of such a precession [152, 153]. The statistical studies of all the LIGO-Virgo binary black hole observations to date show the presence of orbital precession. This, plus the character of the mass distribution at high masses, supports the assumption that some of the observed binary black hole systems have been formed by dynamical processes [174].

2.6.4.1.3 Primordial black holes

The observations of numerous binary black hole mergers has generated much discussion as to whether black holes might be an explanation to the missing mass, or dark matter, problem. About 5% of the mass–energy of the Universe is baryonic matter (protons and neutrons are baryons), namely the material that makes up the elements (H, He, C, O, N, etc) that we are familiar with. However, there is much evidence that there is missing mass, and missing energy. It is estimated that 27% of the mass energy in the Universe is dark (unobserved) and non-baryonic mass.

The numerous binary black hole mergers observed by LIGO and Virgo has renewed interest in the possibility that black holes could be formed in the early Universe, namely primordial black holes [181, 182], and these could contribute to the missing mass [183, 184]. It has been theorized that cosmic strings could also be responsible for primoridal black hole formation [185]. The possibility that black holes could be dark matter has been constrained by microlensing observations, and the structure of the cosmic microwave background. However, a window in allowable masses exists, from $30M_\odot$ to $100M_\odot$, consistent with the LIGO-Virgo observations. Some studies claim that the black holes observed by LIGO-Virgo could account for all of the dark matter [186]. For the LIGO-Virgo observations, primordial black holes in binary systems would have their spins randomly distributed, independent of the orbital angular momentum of the binary system [187].

If black holes were formed in the early Universe it is expected that they would be created with a distribution of masses. LIGO and Virgo are conducting searches especially targeting compact binary systems where the component masses are under $1 M_\odot$ [188]. If such systems were to be found, it would be difficult to explain the presence of a compact object less than $1 M_\odot$. This would be too small to be a neutron star, so presumably it would be a black hole. In such a case the best assumption would be that the black hole was formed in the early Universe, namely a primordial black hole. Continuing searches for such sub-solar mass sources by LIGO-Virgo, and future third-generation ground-based detectors, will be critical in determining the existence of primordial black holes.

The work of Jacob Bekenstein showed that black holes should have entropy [189], which was followed by Stephen Hawking who showed that such an object with entropy would have a temperature, and would therefore radiate like a black body [164]. According to the derivation by Hawking, a black hole formed in the early Universe would have evaporated by now (universe age of about 14 billion years [189]) if the initial mass was 1.7×10^{11} kg or less. Such a decay would release many photons and neutrinos, especially at the final moments; such an explosion has not been observed. The cosmic microwave background has a temperature of 2.7°, and if one considers a thermal equilibrium between Hawking radiation and the acquisition of energy from the cosmic microwave background, then the mass limit for primordial black holes would be approximately 5×10^{22} kg, very similar to the mass of the moon.

2.6.5 The path to supermassive black holes

The centers of most galaxies seem to have a supermassive black hole, of millions of solar mass, or even much larger. Our Milky Way Galaxy contains a black hole of $4 \times 10^6 \, M_\odot$ [154, 155]. The Event Horizon Telescope has recently imaged the shadow of a $6.5 \times 10^9 \, M_\odot$ supermassive black hole in the center of the galaxy M87; see figure 2.36 [156]. An important question is how these supermassive black holes were formed, and what role they play in the formation of galaxies. One assumption is that there were very massive stars in the early Universe that quickly became black holes with masses of tens to hundreds of solar masses. These black holes then consumed the material in its environment (gas, stars, other black holes) and grew [190]. Another scenario has dense regions of gas and dark matter collapsing directly to a black hole of hundreds of thousands of solar masses, without even forming stars. These would then be the seeds for the supermassive black holes in the center of galaxies [191].

However, it is difficult to explain very large black holes that exist in a short time after the Big Bang. For example, the quasar ULAS J1120+0641 was estimated to have a central black hole mass of $8 \times 10^8 M_\odot$ at just 690 million years after the Big Bang [192]. Another quasar, J0313-1806, is estimated to have a black hole with mass $1.6 \times 10^9 M_\odot$ at just 670 million years after the Big Bang [193]. Such large masses early in the Universe are difficult to explain. Gravitational-wave observations of supermassive binary black hole systems with pulsar timing and LISA will hopefully provide more information about these systems.

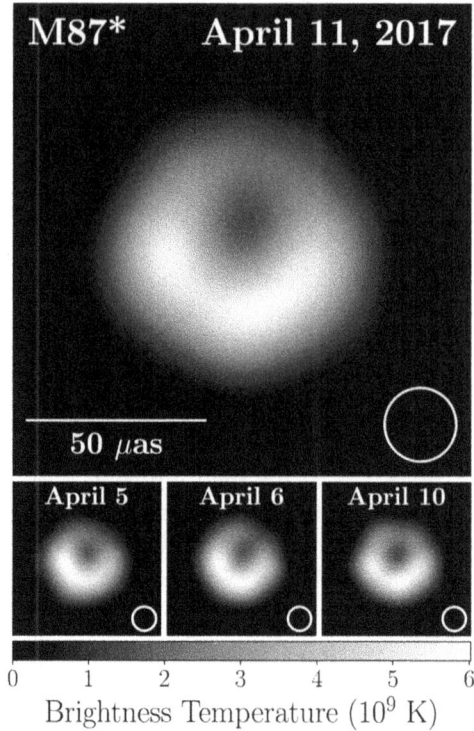

Figure 2.36. The Event Horizon Telescope image of the black hole in the center of the galaxy M87. Radio telescope observations were combined to display the shadow of the supermassive black hole at the center of the Galaxy. Reroduced from [132], copyright IOP Publishing Ltd. All rights reserved.

2.6.5.1 Intermediate mass black holes

There is much observational evidence for stellar-mass black holes and supermassive black holes. However, there are relatively few observations of intermediate-mass black holes, $100 M_{\odot}$ to $100\,000 M_{\odot}$, the bridge from stellar mass to supermassive black holes. It will be important for our understanding of the formation of supermassive black holes to accurately measure the presence of intermediate mass black holes in the Universe. Gravitational-wave observations are likely to be the best way to do this.

The direct observation by LIGO-Virgo of the birth of a $142 M_{\odot}$ black hole displayed how intermediate mass black holes can be observed and measured [152, 153]. The other observations of intermediate mass black holes are sparse, and indirect. For example, evidence comes from the dynamics of the central regions of galaxies and star clusters [194]. LIGO and Virgo gravitational-wave observations should provide more observations up to a few hundred solar masses, third-generation detectors like the Einstein Telescope and Cosmic Explorer could observe systems up to a few thousands of solar mass, while the space-based LISA could extend the observations through hundreds of thousands to a few millions of solar mass. Gravitational-wave observations truly give the best means to measure the distribution of intermediate mass black holes, and build the bridge between stellar mass and supermassive black holes.

2.6.6 Testing general relativity

The observations of gravitational waves offer an important means to test general relativity, especially from regimes where gravity is extremely strong. Important tests have already been made with the LIGO–Virgo observations for binary black holes and binary neutron stars. That general relativity is correct is the basic assumption in these tests. The data can be analyzed to see if the total signal is consistent with general relativity. Comparisons can also be made to alternative models of gravity. At some level it is expected that quantum mechanics must play a part in the correct description of gravity, and gravitational-wave observations are being used to look for these quantum effects.

2.6.6.1 Signal residual test
One test that has been done for binary black hole merger observations is the signal residual test. Bayesian parameter estimation routines [172, 173] use general relativity to model the signal. The estimate of the signal is then subtracted from the data, and an examination is done to see if the residual is consistent with LIGO-Virgo detector noise [195]. None of the LIGO-Virgo observations have failed this test. See figure 2.37 for an example of this test for binary black hole merger GW170104 [196].

2.6.6.2 Gravitational-wave polarization
Gravitational waves have two polarizations. This is similar to light, which also has two polarizations. The two LIGO detectors are essentially aligned with respect to each other, although separated by 3000 km. As such, the two detectors tend to respond to the same polarization. The first three-detector observations were made when the European detector, Virgo, joined the network. Due to its large distance displacement from the American detectors, and its orientation, it is able to measure polarization content from gravitational waves that complements that observed by the LIGO detectors. This was first demonstrated with the three-detector observations of binary black hole-produced signals GW170814 [167] and GW170818 [151]. General relativity predicts a type of polarization, tensor, that is different from electromagnetic polarization, vector. Alternative theories of gravity predict vector and scalar polarizations, in addition to the tensor polarization. The polarization content of the binary black hole-produced gravitational-wave signals observed by the LIGO-Virgo three-detector network are consistent with general relativity. This type of polarization test will become more stringent once the worldwide gravitational-wave network expands to include KAGRA [197] in Japan and LIGO-India [137].

2.6.6.3 Speed of gravity
It is predicted by general relativity that gravitational waves travel at the speed of light. In addition, it is thought that short gamma-ray bursts are created by the

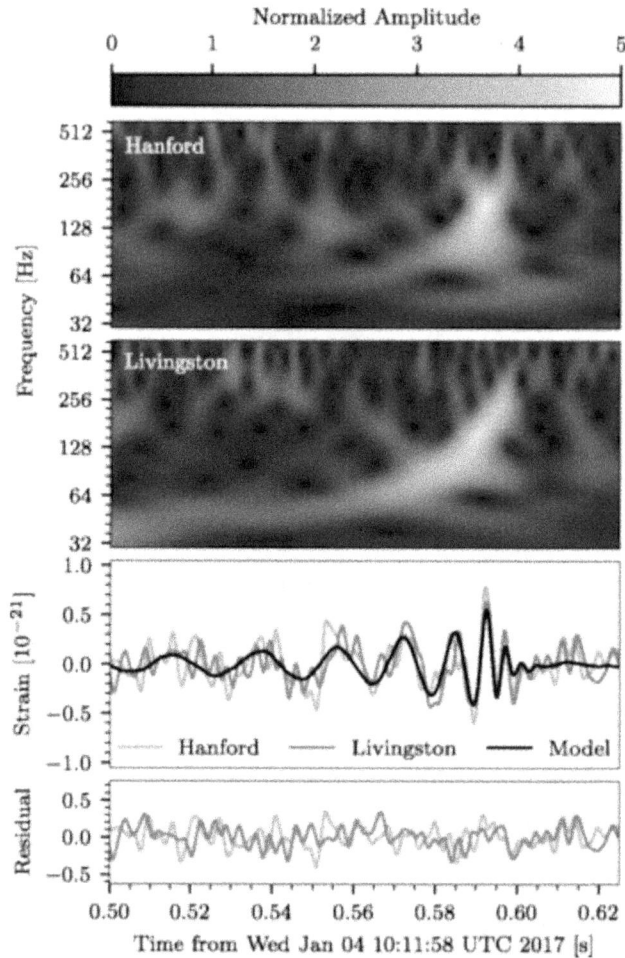

Figure 2.37. Gravitational-wave signal GW170104 observed by the two LIGO detectors. The time-frequency representations of the signal as observed in the LIGO Hanford (top panel) and the LIGO Livingston (second panel) detectors. The third panel shows the data time series from Hanford (orange) and Livingston (blue), along with the waveform best describing the signal (black). The bottom panel is the result when the waveform model is subtracted from the two data streams, leaving Gaussian noise and no indication of a remaining signal. See [196] for more details. Reproduced from [196] CC BY 4.0.

coalescence of binary neutron stars. These predictions were verified when gravitational waves from a binary neutron star merger, GW170817 [198], were seen 1.74 s before the gamma-ray burst, GRB 170817A [199]. This was detected by two satellite gamma-ray detectors, the Anti-Coincidence Shield for the Spectrometer for the International Gamma-Ray Astrophysics Laboratory (INTEGRAL) [200], and the Fermi Gamma-ray Burst Monitor (Fermi/GBM) [201]. The gravitational-wave signal from LIGO and the gamma-ray signals from INTEGRAL and Fermi/GBM are shown in figure 2.38. The difference between the gravitational-wave merger time and the gamma-ray arrival time of 1.7 s can be seen.

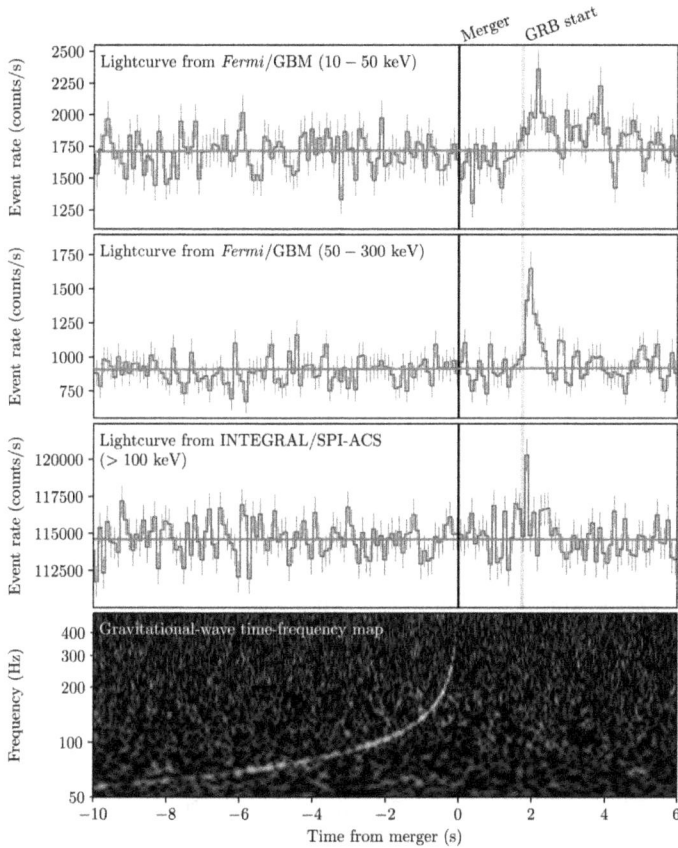

Figure 2.38. The LIGO observed gravitational-wave signal GW170817 and the Fermi/GBM [201] and INTEGRAL [200] observations of gamma-ray burst GRB 170817A. See [199] for more details. Reproduced from [199] CC BY 4.0.

From the gravitational-wave signal alone it is possible to estimate the distance (technically the luminosity distance) to the location of the binary merger. For GW170817, the estimate is 40^{+8}_{-14} Mpc [198], or about 130 million light years. The subsequent searches for an optical counterpart to this event succeeded, with the kilonova found in the galaxy NGC 4993. This observation also showed that the formation of the heaviest elements truly does come from neutron star mergers.

The difference between the arrival times of the gamma rays and the gravitational waves, along with the measurement of the distance, gives the information needed to calculate the difference in the speeds of propagation.

For making the comparison between the speed of gravity v_g, and the speed of light c, LIGO-Virgo used the most conservative distance estimate, namely the lower limit of $D = 26$ Mpc. One does not know, however, the time of the emission for the gamma rays relative to the time of the merger of the neutron stars. While it could very likely correspond to the 1.74 s observation, one can assume that the gravitational waves and gamma rays were emitted simultaneously. As such, the speed of gravity would be greater than the speed of light, and $\Delta v = c - v_g$, $\Delta v/c \approx -3 \times 10^{-15}$. On the other hand, a conservative assumption would be that the gamma emission occurred 10 s after the merger. For this scenario $\Delta v/c \approx 7 \times 10^{-16}$. The gravitational-wave and gamma-ray data set a limit for the difference between v_g and c of [199]

$$-3 \times 10^{-15} \leqslant \frac{\Delta v}{c} \leqslant +7 \times 10^{-16}. \tag{2.35}$$

2.6.6.4 Binary black hole inspiral—merger—ringdown

The gravitational-wave signal from a binary black hole merger has different regimes that correspond to the different physical effects from the source. The assumption is that general relativity provides the description of all these processes. When the two black holes are orbiting one another, energy is lost by gravitational-wave emission, and the orbit decays. There is an increase in the orbital- and gravitational-wave frequencies, and the amplitude of the gravitational wave grows. This is the so-called inspiral, or *chirp* signal. Eventually the system arrives at the innermost stable circular orbit; general relativity does not allow for stable circular orbits for objects too close to each other. At this point the masses fall toward one another, the merger part of the signal. Finally the two black holes meet. By the no-hair theorem, the final black hole can only described by its mass, spin, and electric charge. When the two black holes meet it will resemble something like a peanut shell, but in the end the final remnant black hole must approach a stable configuration. It does this by oscillating, ringing like a bell, losing energy by gravitational-wave emission.

LIGO and Virgo have tested general relativity by comparing the physical parameters estimated from the inspiral part of the signal, with those from the merger-ringdown part. This test has been done with a number of binary black hole merger signals. To date the parameter estimation for the different parts of the signal give consistent results, and hence no deviations from general relativity were observed [202].

2.6.6.5 Echoes and quantum gravity

There has been much research and discussion pertaining to applying quantum mechanics to gravity. For example, it has been theorized that the area of a black hole event horizon is quantized [203]. It is further speculated that this could create a scenario where gravitational waves produced during a binary black hole merger could reflect off of the event horizon of the newly formed black hole [203]. This would produce gravitational-wave *echoes*, and multiple signals would be subsequently observed in gravitational-wave detectors [204]. It has been claimed that the gravitational-wave signal GW150914 [205] had observable echoes [206]. This claim has been challenged by others, however [207]. As LIGO and Virgo (and eventually KAGRA) improve on their sensitivities, it will be important to look for such echo signals in the binary black hole-produced gravitational-wave signals.

2.6.7 Hubble constant

As already noted, one can estimate the luminosity distance to a coalescing compact binary gravitational-wave source. If the source can be found and the redshift measured (giving the velocity) it is then possible to estimate the Hubble constant, the measure of the present expansion rate of the Universe [208, 209]. The definition of the Hubble constant, H_0 comes from

$$v = H_0 d, \tag{2.36}$$

where d is the distance to the source and v is the velocity for the source moving away from us because of the Universe expanding. This gravitational-wave approach would complement the existing methods of measuring the Hubble constant using the cosmic microwave background data [210, 213] or supernovae observations [211]. There is presently a tension between the cosmic microwave background and supernova estimates of the Hubble constant, and so using gravitational waves could help to clarify the description of the expansion of the Universe.

For the binary neutron star-produced gravitational-wave signal GW170817 it was possible to identify the source in the Galaxy NGC 4993, and then make a redshift measurement and obtain the velocity.

The measurement of the Hubble constant was $H_0 = 70^{+12}_{-8}$ km s^{-1} Mpc^{-1} [212]. This observation is consistent with the results from the other means used to estimate the Hubble constant. Cosmic microwave background observations give $H_0 = 67.74 \pm 0.46$ km s^{-1} Mpc^{-1} [213], and supernovae measurements have $H_0 = 73.24 \pm 1.74$ km s^{-1} Mpc^{-1} [211]. Note that the uncertainty for the gravitational-wave Hubble constant measurement is quite large compared to the other two methods. However, with order of 50–100 such binary neutron star gravitational-wave observations the error should be reduced to the level of the other methods, and could help to explain the tension between the cosmic microwave background and supernovae estimations.

2.6.8 LIGO, Virgo, and KAGRA observation schedules

The LIGO and Virgo detectors sensitivity improvements are described in a document that is regularly updated [214]. The arrival of the KAGRA gravitational-wave detector to the worldwide network is expected in 2022 [136], and is the next important step for worldwide gravitational-wave observations. KAGRA, located in Japan, will observe with LIGO and Virgo at for the fourth observing run O4 in 2022. Note too that a third LIGO detector, located in India, will join the worldwide gravitational-wave network in 2027 [197]. The LIGO-Virgo-KAGRA observing scenario calls for a succession of data taking periods separated with detector upgrades and commissioning periods to reach the detectors' target sensitivities. Advanced LIGO and Advanced Virgo recently completed their third observational run O3 at the end of March, 2020. lIGO and Virgo release their data to the public after a proprietary period; the data can be downloaded from the Gravitational Wave Open Science Center [215].

LIGO, Virgo, GEO, and KAGRA have created a new type of telescope to peer into the heavens. With every new means of looking at the sky there has come unexpected discoveries. This has started with the unexpected observation of gravitational waves produced by binary black hole systems with tens of solar masses. Physicists do know that there will be other signals that they can predict. Binary systems containing neutron stars, for example. It is suspected that short gamma-ray bursts come from the coalescence of binary neutron stars, or neutron star—black hole binary systems. The binary neutron star merger producing GW170817 proved this assumption [199]. A core collapse supernova will produce a burst of gravitational waves that will hopefully rise above the noise. Pulsars, or neutron stars spinning about their axes at rates sometimes exceeding hundreds of revolutions per second, will produce continuous sinusoidal signals that can be seen by integrating for sufficient lengths of time. Gravitational waves produced by the Big Bang will produce a background stochastic noise that can possibly be extracted by correlating the outputs from two or more detectors. These are exciting physics results that will come through tremendous experimental effort. The exciting initial observations of gravitational waves have been made, but it is just the beginning of a new astronomy. The second-generation gravitational-wave detectors started detection in 2015 and will continue through the 2020s. The third-generation detectors should arrive in the 2030s, alongside the space-based LISA detector. By 2050 we should have a tremendously more complete view of neutron stars, black holes, and our Universe through gravitational-wave observations from these detectors. Plans will certainly be developing for future improvements.

2.6.9 Challenges and opportunities

To date ground-based gravitational-wave detection has already achieved remarkable success. The second-generation detectors, Advanced LIGO [132] and Advanced Virgo [133], have already announced 50 gravitational-wave observations [131]. More observations will soon be announced as the results from the second half of the third observing run, O3b, are presented. The arrival of the Japanese detector KAGRA [136] for the fourth observing run O4 in 2022 will be a tremendous improvement for the worldwide detector network. With the LIGO-India detector [137] commencing observations around 2027, the 2020s looks very bright for continued gravitational-wave observations. The gravitational-wave observations have given us new insight into the physics associated with neutron stars, black holes, short gamma-ray bursts, kilonovae, tests of general relativity, and cosmology.

2.6.9.1 Future gravitational-wave detectors
These second-generation gravitational-wave detectors observe gravitational waves from about 20 Hz up to the kHz regime. The low-frequency cutoff determines the possible observable mass range for binary black hole systems. The second-generation detector are able to observe binary black hole systems with masses up to a few hundred solar masses.

The third-generation detectors will have sensitivities that are about 10 times better than the second generation. This will allow them to observe 1000 times more of the Universe. Importantly, a goal for the third-generation detectors is to have a lower sensitivity of about 5 Hz (although if the technology advances, this might be pushed to 1 Hz). This will allow for observing binary black hole systems with masses of a few thousands of solar masses. This is a mass range for intermediate mass black holes that is relatively unknown. The ability to measure the population of such binary black hole systems will be extremely important for our understanding of the black hole mass spectrum in the Universe.

In Europe the Einstein Telescope [138] is the proposed third-generation gravitational-wave detector. Whereas the second-generation detectors are Michelson interferometers in a 90° L-shape configuration, the Einstein Telescope will have the configuration of an equilateral triangle with 10 km long arms. There will actually be six interferometers, each with arms at 60°. There will be three interferometers optimized for low-frequency observations, from 1 to 250 Hz. The mirrors for the low-frequency interferometers will be cooled to 10–20 K in order to reduce thermal noise. The Einstein Telescope will then have three high-frequency interferometers, operating from 10 Hz to 10 kHz. For the high-frequency interferometers the mirrors will be at room temperature. To reduce seismic and anthropogenic noise, the Einstein Telescope will be 100–300 m underground. In order to achieve its sensitivity goals much technology needs to be further developed. Researchers in Europe, and around the world, are actively conducting research in a number of areas in order to meet the Einstein Telescope requirements. These include high-power and stabilized lasers (>500 W), mirror coatings with low thermal noise but able to tolerate the 3 MW of light stored in the arms of the high-frequency detectors, cryogenic mirrors for the low-frequency detectors, sophisticated seismic isolation systems for the interferometer mirrors, vacuum systems to accommodate the six interferometers with arm lengths of 10 km, and an extensive underground tunnel network to accommodate the detectors. Bringing the Einstein Telescope to its proposed sensitivity will require tremendous scientific and engineering developments, but the prospects are realistic. The scientific payoffs from the observations will be tremendous, which is the motivation for developing the third-generation detectors. The goal is to have the Einstein Telescope operating by the mid-2030s. Figure 2.39 shows a simplified picture of the Einstein Telescope.

The American led third-generation gravitational-wave detector project is known as Cosmic Explorer [139]. One of the ways that Cosmic Explorer hopes to achieve an improvement in sensitivity by a factor of 10 is to make an L-shaped Michelson interferometer with 40 km arms. This is to compared with the 4 km arms for the LIGO detectors. There will be technological advancements over the current LIGO detectors, but the bulk of the sensitivity gain comes from the arm-length increase. Cosmic Explorer will be on the surface of the Earth, but a difficulty concerns the curvature of the Earth over this distance. In addition, a vacuum system with two 40 km arms will be a challenge. A simple artistic representation of Cosmic Explored is given in figure 2.40.

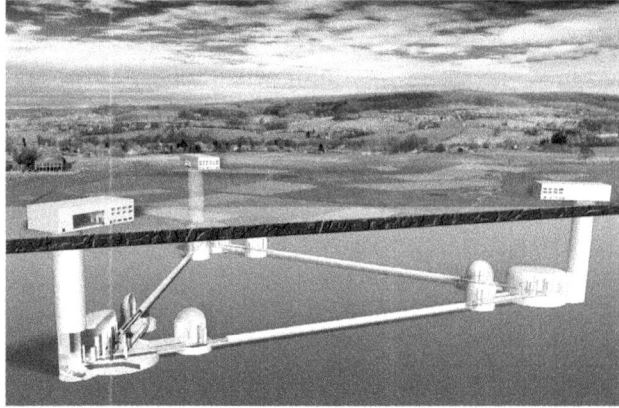

Figure 2.39. The Einstein Telescope will be located in Europe, have six actual interferometric gravitational-wave detectors with arm lengths of 10 km, and located 100–300 m below ground. Figure from the Einstein Telescope collaboration.

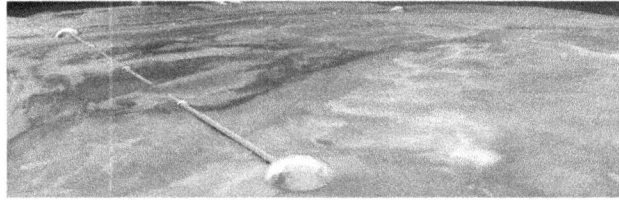

Figure 2.40. The Cosmic Explorer will be located in the United States. It will be an L-shaped interferometric gravitational-wave detector with arm lengths of 40 km, and located the surface of the Earth. Figure from the Cosmic Explorer collaboration.

The third-generation detectors will be expensive, with costs easily exceeding a billion Euros. Tremendous scientific and technical planning will be necessary. The importance of the scientific goals also will need to have universal recognition from the broader physics community. While initial observing is planned for the mid-2030s, the infrastructures created are expected to last for many decades. It is expected that upgrades will be made to specific elements in the detectors, but the vacuum systems and tunnels need to survive for many decades. As LIGO and Virgo have already seen, maintaining such large vacuum systems is not trivial, and problems do arise.

The LISA mission is led by the European Space Agency, but also supported by NASA [140]. The launch is currently planned for 2034. The nominal duration for the mission is 4 years of data acquisition, but this could be extended to 10 years. LISA will be searching for gravitational waves in the 10^{-4}–0.1 Hz band, but it may be possible to extract useful information out to a band of 10^{-5}–1 Hz. The three LISA satellites will form an equilateral triangle with arm lengths of 2.5 million km; see figure 2.41. Each satellite will have two test masses in free fall. The optical system will measure the distance between the masses in the different satellites to an accuracy

Figure 2.41. A simple figure displaying the three LISA satellites, separated from one another by 2.5 million km. The LISA constellation will orbit the Sun, 20° behind the orbit of the Earth. Figure from NASA.

of pico-meters (10^{-12} m) over the 2.5 million km total distance. There will be of order 1 Watt of laser light exiting the satellite, but only a few hundred pico-Watts of light will be detected after this long distance. The beam will have diverged significantly. As such, it will be impossible to make a true Michelson interferometer, like LIGO or Virgo. LISA will use other optical techniques to measure the small distance displacements. But like LIGO and Virgo, the gravitational wave will make a change to the detector arm lengths, which will induce an optical phase shift in the laser light, which will be detected. Unlike LIGO, Virgo, and the other ground-based detectors, LISA must work on the first attempt. There will be no means for human intervention after the launch. The physics necessary for gravitational-wave detection by LISA are already established; it is a question as to whether such a detector can be properly engineered to work as expected.

Much of the technology for making accurate distance measurements between test masses in space has already been demonstrated by the LISA Pathfinder Mission [216, 217]. This mission was launched on December 3, 2015, conducted experiments for 576 days, and was deactivated on June 30, 2017. The experiment consisted of two test masses in the satellite. There were two cubes (46 mm per side), 1.928 kg, made of a gold–platinum alloy. One mass was in a true free fall, with the satellite fixing its position about the mass. The second mass was electrostatically controlled to remain at a fixed distance separation of 376 mm. The differential acceleration between two freely falling test masses in space was measured to sufficient accuracy to demonstrate that the LISA mission could succeed in its quest to measure gravitational waves in space. This measurement was made in the observational band of LISA. Figure 2.42 displays the LISA Pathfinder and its optical system.

In the coming years gravitational-wave searches will continue to be conducted by pulsar timing. Radio telescope networks around the world will be observing pulsars. The Square Kilometer Array is currently being constructed in South Africa and Australia, with the final network assembly coming online in 2027 or possibly later.

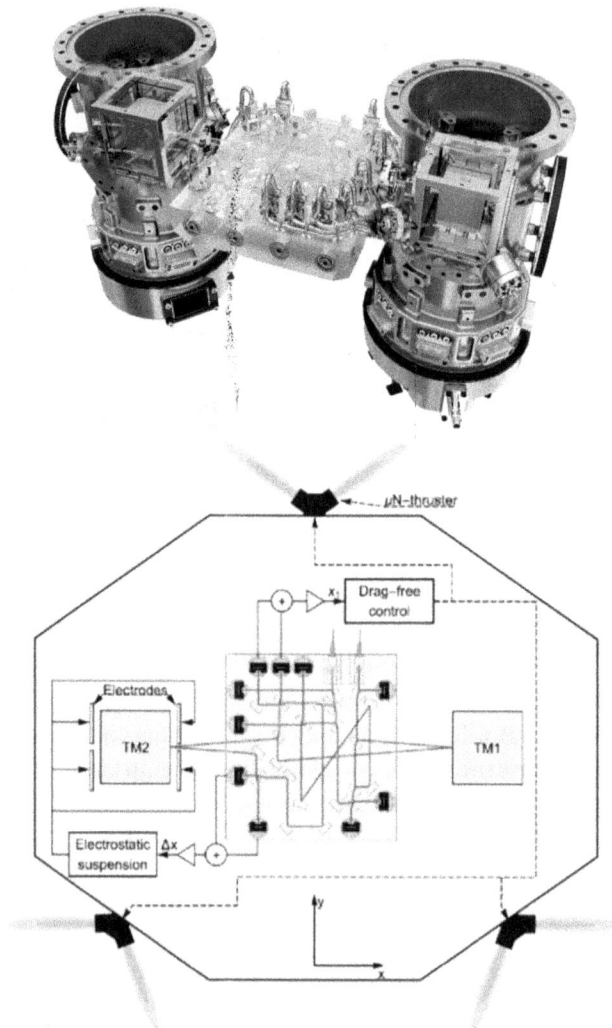

Figure 2.42. Top: Representation of the LISA Pathfinder experiment. Note the two masses and the laser interferometric system to measure the distance between them. Bottom: Schematic of the spacecraft, test masses, and optical system for measuring the distance between the masses. Top figure from the European Space Agency, bottom figure from [216] CC BY 4.0.

The Square Kilometer Array will augment other radio observations and search for gravitational waves in the 10^{-9}–10^{-6} Hz frequency band [142–144].

The 2.7° cosmic microwave background is the remnant of the thermal radiation present when free protons and electrons formed neutral atoms 370 000 years after the Big Bang. Gravitational waves produced in the Big Bang, during inflation, should have left and imprint in the distribution of the polarization of the cosmic microwave background. Numerous observations are being conducted, and will continue to be conducted, to observe this signature of gravitational waves and hopefully gain knowledge of the parameters describing inflation [218].

The coming decades will be seeing extensive observations of gravitational waves over a broad frequency range. This should propel our knowledge of astrophysics and cosmology.

2.6.9.2 Future science results from gravitational-wave observations
The 2030s will see the arrival of the third-generation ground-based gravitational-wave detectors, the Einstein Telescope, and Cosmic Explorer. Similar to LIGO, Virgo, and KAGRA, these detectors will continue to search for gravitational-wave signals from binary neutron star mergers, binary black hole mergers, supernovae, pulsars, a stochastic gravitational wave background, and exotic objects such as cosmic strings. With an improvement in sensitivity by a factor of 10, the ability of third-generation detectors to observe these signals is greatly improved.

Binary neutron stars are relatively light objects with a total mass of around $3M_\odot$. Still, the third-generation detectors will be able to detect the gravitational waves from a binary neutron star merger out to a redshift of $z \approx 2$ (when the Universe had an age of 3.3 billion years), and maybe even $z \approx 3$ (when the Universe had an age of 2.2 billion years). For comparison, the age of the Universe is approximately 13.8 billion years [179, 213]. It is estimated that the third-generation detectors could detect 70 000 binary neutron star merger events per year [219]. The large number of observations, including many with large signal-to-noise ratio, should tell us much about the nuclear physics associated with neutron stars, including their equation of state. This will happen when tidal effects change the shape of the neutron stars before merger, thereby changing the gravitational-wave signal.

Some of the observed binary neutron star mergers will probably also be seen in electromagnetic observations, be they short gamma-ray bursts or kilonova identification. Multimessenger astronomy will greatly benefit from the large number of these observations. This depends, of course, on there being sufficient electromagnetic observatories, especially gamma-ray detection on satellites. Coincident observations of high-energy neutrinos will also be very informative. These observations will tell us more about the physics of short gamma-ray bursts, the formation of gamma-ray jets, what part neutrinos play in these mergers, and the formation of heavy elements in the Universe.

Joint gravitational-wave and electromagnetic observations will also help to reduce the uncertainty in the estimation of the Hubble constant. The gravitational-wave based Hubble constant measurement will be as statistically accurate as other methods. This will certainly be important in explaining the tension between the Hubble constant measurement from observations of the cosmic microwave background [213] or supernovae [211]. The accurate determination of the Hubble constant is critical for our ability to describe the Universe.

Being more massive, binary black hole events will be easier for the third-generation detectors to observe, and so they will be seen even further back in the Universe. In fact, almost every binary black hole merger in the observable Universe with masses up to a few thousand solar masses will be detected by the third-generation detectors. This will amount to around a million observations per year. This will provide the means to do detailed statistical studies of the observations. This will include measurements of the merger rate as a function of total mass, component spins, and redshift (distance). This should provide the means to understand the formation channels for black holes, from stellar based, to dynamical black hole mergers in dense environments. Any evidence of primordial black hole formation would be revolutionary for cosmology. The important information about intermediate mass black holes, from 100 solar mass up to thousands of solar mass, will be extremely important for understanding black holes in the Universe, and making the bridge between stellar-mass black holes and supermassive black holes.

The observation of gravitational waves produced during the Big Bang would provide cosmological information that is impossible to obtain by any other means [220]. This cosmologically produced stochastic gravitational-wave background could provide data about inflation, phase transitions in the early Universe, or the presence of exotic objects like cosmic strings. However, the merger of binary neutron stars and binary black holes over the history of the Universe will create another stochastic gravitational-wave background that will obscure the observation of the cosmological background. Techniques have been proposed to separate these different backgrounds using data from third-generation detectors [221, 222].

However, the third-generation detectors will be so sensitive that they will directly detect virtually all binary black mergers in the Universe, and most binary black hole mergers. As such, these events can be removed from the data, and a search for the cosmologically produced stochastic gravitational-wave background can be done without this astrophysically produced foreground. This could lead to an important discovery, or setting stringent limits on the cosmological gravitational-wave background [223, 224].

While the third-generation gravitational-wave detectors will begin observations in the mid 2030s, they will continue observing well past 2050. For example, the proposal for the Einstein Telescope is to have an infrastructure that will last for at least 50 years. We can expect that the technology for gravitational-wave detection will progress, and important scientific observations will happen of the decades.

The LISA Mission will be launched in the mid 2030s, with a plan for 4 years of observations, with a possible extension to 10 years. Within the observation band of 10^{-4}–0.1 Hz LISA will be expected to observe thousands of signals. The science case for LISA is extremely strong.

Starting closest to us, LISA will tell us much about compact binary stars in the Milky Way. This will include observations of binaries comprised of white dwarfs, neutron stars, and stellar-mass black holes. The observation of how the tidal disruption of the objects affects the orbital evolution will be important. Some galactic binary systems will be observable with gravitational waves and electromagnetic observations. The comparison of these observations will be important for understanding tidal interactions and other features than can perturb the orbit.

LISA will be sensitive to binary black merger signals for systems with a total mass up to a few million solar mass. The merger of such systems will be observable out to the earliest moments of the Universe. Specifically, these mergers will be observed out to redshifts of $z \approx 10$ (when the Universe had an age of 500 million years) to $z \approx 20$ (when the Universe had an age of 200 million years). This will be extremely important as it can help us understand the starting process for the the formation of the supermassive black holes that we now observe in the centers of galaxies. Mapping out the black hole masses as a function of redshift,as well as the merger rates as function of redshift, will also be critical for our understanding of structure and its formation in the Universe. The observation of an electromagnetic signal in coincidence with an observed binary black hole merger would also provide information about the environment occupied by these black holes.

LISA should be able to observe gravitational-wave signals from extreme mass ratio inspirals. This would be when a stellar-mass black hole ($\sim 50 M_\odot$) orbits around and then plunges into a $10^5 M_\odot$ to $10^6 M_\odot$ black hole. Such a signal would be given information about the environment about large black holes, and the rate that they consume smaller black holes. The redshift dependence of such mergers would also help to understand how black holes grow in size throughout the history of the Universe.

The stellar-mass black holes mergers that have been observed by LIGO and Virgo could also be seen by LISA. In some cases, it might be possible for LISA to observe the inspiral signal from a binary black hole months to years before the merger would be observed by ground-based gravitational-wave detectors [225]. Such an observation across numerous frequency decades would provide for interesting tests of general relativity. While the merger signal is outside of the LISA observing band, the observation of stellar-mass black hole binaries would contribute to our knowledge of such systems, and possibly also contribute to deciphering how such systems are formed.

LISA's observations of binary black hole mergers will allow for testing general relativity and alternative theories of gravity in multiple ways. The observation of a post-merger signal from binary black hole merger, namely the ringdown signal from the newly formed black hole, will help to confirm whether general relativity truly is the theory that describes black holes. It would also help to confirm that these objects are, in fact, black holes. Gravitational waves from extreme mass ratio inspirals also have a more complicated structure than when the two masses are approximately equal. Comparing such signals to models will be another test of general relativity. These signals can also be used to confirm that gravitational-waves travel at the speed of light. The presence of dark matter (such as axions) in the environment of an extreme mass ratio inspiral could also affect the gravitational-wave signal.

Just as LIGO and Virgo have already shown that the use of gravitational waves to measure the Hubble constant is possible, LISA will also contribute to this important measurement of the expansion of the Universe. This will be especially true when the source can be found, for example, with an electromagnetic counterpart. Any electromagnetic signal from a binary black hole merger would provide critical information on the environment about the black hole. Such observations at very high redshift would also give information on the dark energy content of the Universe, as the Hubble constant does change of the history of the Universe (so it is not really a constant, as per its name).

LISA will also attempt to measure the stochastic gravitational-wave background [220]. It will certainly observe a stochastic background made by binary black hole mergers throughout the Universe. There will also be a foreground signal from the binary systems galaxy. In order to try to measure a cosmologically produced stochastic background methods must be developed to simultaneously characterize the galactic foreground [226], and the binary black hole produced background [227]. In the LISA observational band a cosmologically produced stochastic background could be produced by phase transitions in the early Universe or cosmic strings. The gravitational waves produced from inflation are likely too small for LISA to observe.

LISA will be observing thousands of signals and have the opportunity to make significant discoveries. See [140] for a more complete description of the scientific goals for LISA.

Pulsar timing observations in the decades to come will also be contributing important science. At the very low frequencies (nanoHertz) there is the opportunity to observe gravitational waves from supermassive black hole binaries, with masses of billions of solar mass [228]. The addition of the Square Kilometer Array for the radio observation of pulsars will be an important contribution for the future [142–144].

Gravitational-wave astronomy started in earnest in 2015 with the first direct detection of gravitational waves by LIGO. In the five years since LIGO and Virgo have announced 50 events, with many more arriving in the years to come. The future looks bright for gravitational-wave science in the decades to come. With detectors with even better sensitivities, and operating in many observation bands, the scientific output should accelerate. However, pushing the detection technology is not easy, and will require substantial work and resources.

Acknowledgments

The production of this article was supported by funds from the Observatoire Côte d'Azur, and U.S. National Science Foundation grant PHY-1806990.

2.7 Stars, the Sun, and planetary systems as physics laboratories

Patrick Eggenberger[1]
[1]University of Geneva, Geneva, Switzerland

2.7.1 General overview

As a very simple and general definition, stars can be seen as *spheres of hot gas*. While such a definition may seem at first sight to be too simplistic to be meaningful, it already contains some key aspects of the physical nature of stars. First, the spherical shape of stars directly reveals the dominant role played by gravity for these objects. This is of course directly related to the important mass of these objects that enables gravity to dominate over electric forces, which are responsible for the shape of the objects that we encounter in our daily life. This evolution of the shape of the objects according to their sizes as a result of the competition between gravity and electric forces is visible when observing the different celestial bodies of our Solar System. One indeed notes that the small objects of the Solar System like asteroids exhibit various shapes, which differ from the spherical symmetry that characterizes the more massive objects like the Moon, the planets, or the Sun.

2.7.2 The mechanical equilibrium of a star

While the first part of the simple definition of stars as spheres of hot gas underlines the dominant role played by gravity in these objects, the second part underlines the need to have another force in order to compensate for the contraction that would be induced by gravity alone. A key element of stellar evolution is indeed related to the fact that during most of their lifetime, stars are in hydrostatic equilibrium. In order to reach this equilibrium state, the gravity force acting on a fluid element inside the star must be compensated by a force of the same amplitude but acting outwards. This is where the second part of the definition related to hot gas comes into play. Indeed, this force is directly related to the properties of the matter in stellar interiors: this is the pressure of the gas, and even more precisely the variation of this pressure with the radial distance inside a star that is able to sustain a star against gravity. This constitutes the first basic equation of stellar structure, the equation of hydrostatic equilibrium:

$$\frac{\mathrm{d}P}{\mathrm{d}r} = -\rho \frac{GM_r}{r^2}, \qquad (2.37)$$

where the radius r is defined as the distance to the stellar center, P is the pressure, ρ the density, G the constant of gravitation, and M_r the total mass inside a sphere of radius r. This equation is directly obtained by expressing the difference of pressure dP in a thin shell inside a spherically symmetric star between a radius r and $r + dr$. Recalling that the local gravity $g = \frac{GM_r}{r^2}$, one sees that this equation simply expresses that at any point inside a star, the pressure gradient must sustain the stellar matter against gravity in order to reach an equilibrium state.

While the first basic equation of stellar structure equation (2.37) constitutes a very simplified version of the conservation of angular momentum, the second basic equation of stellar structure is simply obtained by expressing the conservation of the mass:

$$\frac{\mathrm{d} M_r}{\mathrm{d} r} = 4\pi r^2 \rho. \qquad (2.38)$$

It is interesting to notice that in theses two equations (2.37) and (2.38), only the pressure P and the density ρ appear explicitly and not the temperature T. To solve the equations of stellar structure, one needs to know the microphysics and in particular the equation of state of the stellar matter that expresses the relation between the three physical parameters P, T, and ρ:

$$\frac{\mathrm{d} \rho}{\rho} = \alpha \frac{\mathrm{d} P}{P} - \delta \frac{\mathrm{d} T}{T}, \qquad (2.39)$$

with $\alpha = (\frac{\partial \ln \rho}{\partial \ln P})_{T, \mu}$, $\delta = -(\frac{\partial \ln \rho}{\partial \ln T})_{P, \mu}$ and μ the mean molecular weight (i.e., the average number of atomic mass units per particle). The physics describing the state of the stellar matter is included in the different values assigned to α and δ. For instance, in the case of a perfect gas, α and δ are equal to one, while in the case of a degenerate non-relativistic gas, $\alpha = 3/5$ and $\delta = 0$. In the latter case, the density depends then only on the pressure and not on the temperature. As mentioned above, equations (2.37) and (2.38) form then a complete set of equations describing the internal structure of the star without any need to consider the thermal part of the problem. However, the pressure generally also depends on the temperature in stellar interiors as illustrated by the perfect gas case that is more representative of the state of the stellar matter during the main duration of the evolution of a star (the degenerate case describing the evolved phases of evolution of a star). In this case, there is an important coupling between the mechanical and the thermal state in stellar interiors.

2.7.3 The energy of a star

We are now interested in discussing in more detail the second part of our simple definition of stars as spheres of *hot gas*. As discussed above, a gradient of pressure is necessary in stellar interiors to ensure hydrostatic equilibrium. In the general case of an equation of state where the pressure depends both on the density and temperature (as for instance in the perfect gas case), this results in a gradient of temperature with a higher temperature in the central part of the star than at its surface. For instance, in the solar case, a central temperature of about 15 million of Kelvin (K) is reached, while the surface is characterized by a much lower temperature of about 5800 K. These differences of temperature lead to a transport of energy from the central layers to the surface: a star shines as a direct consequence of hydrostatic equilibrium and its lifetime then directly depends on the reservoir of energy available to maintain its luminosity.

This leads us to consider the different sources of energy available for a star. A first source of energy for a star consists in simply extracting this energy from its gravitational potential through a global contraction. Based on the simple equations derived in the preceding section, we can discuss the impact of a uniform contraction of a star. As a result of such a contraction, there is a change of pressure dP and of radius dr. From equation (2.37) expressing hydrostatic equilibrium, these two variations are then related by:

$$\frac{dP}{P} = -4\frac{dr}{r}.\tag{2.40}$$

Similarly, using equation (2.38) that expresses the conservation of the mass, the variation of density $d\rho$ is then related to the one of the radius by:

$$\frac{d\rho}{\rho} = -3\frac{dr}{r}.\tag{2.41}$$

Combining these two equations and using equation (2.39) for the general expression of the equation of state leads to the following relation between the variations of temperature and density during a slow uniform contraction of a star:

$$\mathrm{d}\ln T = \frac{4\alpha - 3}{3\delta}\mathrm{d}\ln\rho.\tag{2.42}$$

In the perfect gas case representative of stellar matter in non-degenerate conditions, the coefficients α and δ are equal to one so that a slope of 1/3 is obtained from equation (2.42) between the variations in temperature and density. This situation corresponds to the case of a star that extracts its energy from its gravitational potential only. In order to estimate the lifetime of a star based on this sole source of energy, one can simply divide its total gravitational energy by its luminosity L to obtain:

$$t_{\mathrm{KH}} \cong \frac{GM^2}{RL},\tag{2.43}$$

where M, R, and L are the total mass, radius, and luminosity of the star, respectively. The timescale t_{KH} is referred to as the Kelvin–Helmholtz timescale and corresponds to the lifetime of a star that would only be able to produce its luminosity at the sole expense of its gravitational energy. By introducing numerical values in equation (2.43) a typical duration of about 30 million of years is obtained for the Sun. In this context of the study of the lifetime of a star, the interest of a multi-disciplinary approach can be nicely illustrated. Indeed, in the case of the Sun, the lifetime predicted by the Kelvin–Helmholtz timescale can be directly compared to other constraints available for the age of the Solar System. In particular, a comparison with geophysical constraints coming from the dating of the oldest rocks analyzed on Earth reveals that this gravitational timescale is much too short to be compatible with the geophysical understanding of the evolution of our planet. This disagreement reveals that another important source of energy is at work in stellar interiors. This additional source of energy is not needed to explain the luminosities of the stars (since as seen above they can be easily reproduced from the sole gravitational energy) but to explain how these luminosities can be maintained over long durations and in particular over billion of years instead of a few million of years as predicted by the Kelvin–Helmholtz timescale.

In order to discuss the possible additional sources of energy in stars, we first notice that the slow uniform gravitational contraction discussed above leads to a simultaneous increase in density and temperature in central stellar layers under non-degenerate conditions (see equation (2.42) with $\alpha = \delta = 1$ for the perfect gas case). When the stellar matter is in degenerate conditions, the situation is completely different and a contraction can then lead to a cooling of the matter instead of an increase in the temperature. This is of course directly related to the equation of state of a completely degenerate non-relativistic gas, and in particular to the fact that in this case the density depends only on the pressure (and not on the temperature) with $P \propto \rho^{5/3}$ (equation (2.42) with $\alpha = 3/5$ and $\delta = 0$). For a perfect gas, $P \propto \rho T$, so that the transition between the non-degenerate and degenerate cases occurs when the two pressures are equal (i.e., when $\rho T \propto \rho^{5/3}$). This leads to $T \propto \rho^{2/3}$ and hence to a slope of 2/3 for the transition between the non-degenerate and degenerate conditions that can be compared to the slope of 1/3 obtained above for a slow contraction in non-degenerate conditions. This means that a star will begin its evolution in non-degenerate conditions with its central values of temperature and density evolving along this 1/3 line, but that at a given point of its evolution, the frontier defined by the slope of 2/3 will be reached and the central stellar layers will then evolve in degenerate condition (see figure 2.43). Close to this transition from non-degenerate to degenerate conditions, the coefficients α and δ will then evolve from the value of 1 to tend to the values of 3/5 and 0 that characterize the completely degenerate case. The value of α becomes lower than 4/3 before the value of δ reaches 0; consequently, the coefficient of proportionality between the variations in temperature and density given by equation (2.42) becomes negative. A contraction is then found to produce a cooling of the stellar matter in the central regions so that an increase in the central temperature can no longer be obtained in degenerate conditions.

This different behaviour in non-degenerate and degenerate conditions is of fundamental interest in the discussion of the additional sources of energy needed in stellar interiors and for the evolution of a star in general. Indeed, thermonuclear reactions can produce this additional energy in stars. For these reactions to occur, specific physical conditions are needed and in particular high values of temperatures and densities must be reached. As discussed above, these conditions of high densities and temperatures can be obtained in central stellar layers from a slow contraction in non-degenerate conditions. A star then begins its evolution by extracting its energy solely from its gravitational potential on a characteristic timescale given by the Kelvin–Helmholtz timescale (equation (2.43)). This leads to a continuous increase of both the central temperature and density until the values needed for thermonuclear reactions to occur are met.

The first reaction to take place in stellar cores is the hydrogen burning during which the production of energy is obtained from nuclear binding energy by

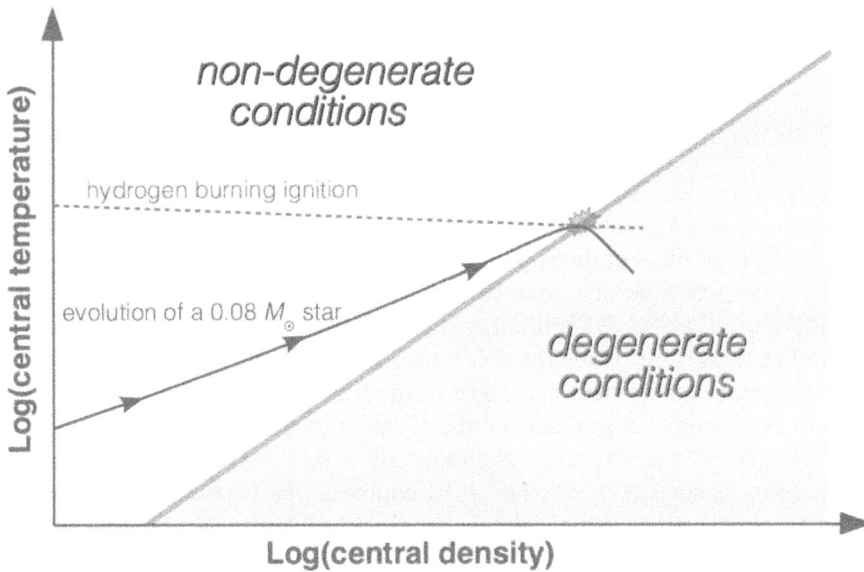

Figure 2.43. Illustration of the change in the central temperature and pressure during the evolution of a star (continuous blue line with a slope of about 1/3 in such a diagram). The transition from the perfect gas to the degenerate region (red shaded area) is indicated by the red line with a slope of about 2/3. The dashed line shows the values of temperature and density needed for the ignition of core hydrogen burning.

transforming four hydrogen nuclei into one helium nucleus. Hydrogen burning can occur through two main reactions: the proton–proton chain and the CNO cycle. For the proton–proton chain, two protons are first needed to form a deuterium nucleus, which then interacts with another proton to form ^{3}He. The last step in the formation of ^{4}He can be done in different ways, with either the interaction of two ^{3}He nuclei to obtain ^{4}He (referred to as the proton–proton1 chain), or the interaction with existing ^{4}He nuclei (referred to as the proton–proton2 and 3 chains). Another possibility for hydrogen burning is referred to as the CNO cycle. In this case, isotopes of carbon, nitrogen, and oxygen are needed in order to transform hydrogen into helium. Hydrogen burning through the proton–proton chain is found to dominate over the CNO cycle at relatively low temperatures (lower than about 1.5×10^7 K), while hydrogen burning through the CNO cycle dominates at higher temperatures.

An estimate of the characteristic timescale related to thermonuclear reactions as the source of energy of a star can then be obtained from this transformation of hydrogen into helium. A relative mass defect of about 0.7% is obtained for the transformation of four protons into one ^{4}He nucleus. The corresponding timescale is then simply obtained by dividing the total energy produced by nuclear reactions by the luminosity of the star. If nuclear reactions would be able to occur in the whole interior of a star of total mass M, this energy would then simply be equal to 0.007 Mc^2, where c corresponds to the speed of light. The values of temperature and density are, however, high enough for hydrogen burning to occur only in the central regions of a star. This leads to the introduction of a factor q (lower than 1) in the preceding expression to account for this fact. The characteristic nuclear timescale is then simply expressed as

$$t_{nuclear} \cong 0.007 \frac{qMc^2}{L}. \qquad (2.44)$$

Introducing numerical values corresponding to the solar case in this equation together with $q = 0.1$ (thus assuming that nuclear reactions are only at work in the 10% of the total mass of the star corresponding to the central layers), one obtains a characteristic timescale of about 10 billion years. This is much longer than the typical lifetime of about 30 million years estimated from the sole gravitational energy and enables to reconcile the theoretical modeling of the evolution of the Sun with the different constraints on the age of the Solar System and of our planet. In addition to providing an important source of stellar energy, thermonuclear reactions play a key role in the chemical evolution of a star by changing the chemical composition in its central layers. This is of course at the basis of stellar nucleosynthesis and in particular of the creation of chemical elements until the iron-peak elements.

Similarly to the cases of slow contractions in non-degenerate and degenerate conditions discussed above, the impact of thermonuclear reactions is also completely different depending on the state of the stellar matter. In non-degenerate conditions, the nuclear energy production in a star is self-regulated. Indeed, when more energy is produced by nuclear reactions in the central regions, an expansion of these layers will follow as a result of the increase in temperature. This expansion will then produce a decrease of the temperature and a corresponding decrease in the energy production rate by nuclear reactions. In degenerate conditions, an increase in temperature will also occur in the stellar core as a result of an excess of energy production, but this will not lead to an increase in pressure due to the particular equation of state of a degenerate gas with the pressure that does not depend on the temperature. As a result, no regulating effect is at work in degenerate conditions so that nuclear reactions are unstable and will continue to produce an increasing amount of energy that will finally result in a flash or an explosion. Conversely, nuclear reactions are stable in non-degenerate conditions due to the dependency of the pressure on temperature.

2.7.4 Evolution as a function of the initial mass

The simple physical aspects of stellar structure discussed above enable to draw a global view of the different evolutions of stars as a function of their initial masses. This can be understood from figure 2.43. We begin by recalling the two key points discussed in the preceding section. First, a star will approximately evolve along a line with a slope of 1/3 in such a plot in non-degenerate conditions (see equation (2.42)) and this leads to the simultaneous increase in both temperature and pressure in the stellar core. When sufficiently high values of temperature and pressure are reached, nuclear burning can be ignited. Second, the separation between non-degenerate and degenerate conditions is given by a line with a slope of 2/3 in figure 2.43 so that a star globally evolves towards degenerate conditions. Once the star enters this domain of degenerate gas, a contraction does not result anymore in an increase in the temperature, but leads instead to a cooling of the central layers, so that the conditions for nuclear burning ignition cannot be reached anymore.

• Brown dwarfs and planets

As can be seen in figure 2.43, there exists a minimal initial mass for an object to reach central physical conditions that enable the ignition of hydrogen burning. This is directly related to the moment when the star will enter the degenerate phase. If an object is not massive enough, the slow gravitational contraction occurring on a Kelvin–Helmholtz timescale will increase its central temperature under non-degenerate conditions, but without being able to reach a sufficiently high value to start hydrogen burning before entering the degenerate phase. As discussed above, once central layers are in degenerate conditions, there is no possibility for a contraction to increase further the central temperature so that physical conditions favorable to the fusion of hydrogen cannot be reached. The mass limit for this to happen is typically of about $0.08 M_\odot$ (see figure 2.43); this corresponds to the minimal mass for a star, and objects with initial masses below $0.08 M_\odot$ are brown dwarfs ans planets.

Interestingly, there is also a minimal mass that can be reached for the formation of these objects following the contraction and the subsequent fragmentation of an interstellar cloud. This minimal mass results from the need to radiate away the energy related to the contraction of the cloud in order for fragmentation to proceed. This leads to a minimal mass of about $0.01 M_\odot$ for an object to be formed from the initial contraction of a cloud. This fragmentation limit is lower than the limit of about $0.08 M_\odot$ discussed above for hydrogen burning to be ignited in non-degenerate conditions. The mass intervalle between about 0.01 and $0.08 M_\odot$ is typically the domain of brown dwarfs. This fragmentation limit also shows that objects with masses lower than about $0.01 M_\odot$ cannot be formed by fragmentation. This means that another formation mechanism is needed for planets. The latter are formed by collision and accumulation of rocks, ice, and gas in protostellar disks surrounding young forming stars.

• Low-mass stars

When the initial mass of the object is larger than about $0.08 M_\odot$, hydrogen burning can take place in the central layers and one leaves the domain of substellar objects to enter the stellar domain. As discussed in the preceding section, the typical timescale corresponding to this new source of nuclear energy is much larger than the one associated with the energy released by the sole gravitational contraction. Once hydrogen burning is ignited, the star is then in a long-standing evolutionary phase referred to as the main-sequence phase. This phase will last as long as hydrogen is present in the central stellar layers and will simply end once all the central hydrogen has been transformed into helium by nuclear reactions.

After the main-sequence phase, the central layers of the star contract to extract energy from the gravitational potential. This situation is then similar to the one preceding the ignition of hydrogen burning, with the contraction resulting in an increase in the central stellar temperature as long as central layers are in non-degenerate conditions. This leads to another mass limit that is defined by the possibility for a star to reach a sufficiently high central temperature for helium burning ignition before the helium core becomes too degenerate. This limit is found around $0.5M_\odot$ and stars with an initial mass lower than this value will then be able to ignite hydrogen burning, but not helium burning.

- Solar-type stars: the Sun as a reference star

Stars with initial masses above about $0.5M_\odot$ will then experience both hydrogen burning and helium burning. Our Sun is the perfect representative of stars evolving on the main sequence in this mass range. Owing to its proximity, the global and internal physical properties of the Sun can be determined with exquisite precision, which makes the Sun a fundamental reference for the modelling of the complex physical processes acting in stellar interiors. For instance, a precise determination of the mass and radius can be achieved for the Sun. Interestingly, this can be directly used to obtain a simple estimate of the values of the pressure and temperature in the solar core. To illustrate this, we first approximate the left-hand side of the equation expressing hydrostatic equilibrium (equation (2.37)) by:

$$\left|\frac{dP}{dr}\right| \cong \frac{P_\odot^{\text{Core}} - P_\odot^{\text{Surface}}}{R_\odot} \cong \frac{P_\odot^{\text{Core}}}{R_\odot}. \tag{2.45}$$

Similarly, the right-hand side of equation (2.37) can be simply estimated using rough average values for the mass, radius, and density in the solar interior based on the total mass and radius of the Sun: $M_r \cong M_\odot/2$, $r \cong R_\odot/2$ and a mean density $\rho \cong 3M_\odot/(4\pi R_\odot^3)$:

$$\rho\frac{GM_r}{r^2}r^2 \cong \frac{3}{2\pi}\frac{GM_\odot^2}{R_\odot^5}. \tag{2.46}$$

Using the equation of hydrostatic equilibrium (equation (2.37)) with the approximations given in equations (2.45) and (2.46) leads to the following estimate for the pressure in the central layers of the Sun:

$$P_\odot^{\text{Core}} \cong \frac{3}{2\pi}\frac{GM_\odot^2}{R_\odot^4}. \tag{2.47}$$

Introducing numerical values for the solar mass and radius in this expression leads to a central pressure in the Sun of about 5.4×10^{15} g s^{-2} cm^{-1}. An estimate of the central temperature of the Sun can be directly obtained from equation (2.47) by adopting an equation of state. The hypothesis of a perfect gas is well suited for a main-sequence star like the Sun and in this case the pressure P is related to density ρ and temperature T by:

$$P = \frac{k}{\mu m_u}\rho T, \tag{2.48}$$

where m_u is the atomic mass unit and k corresponds to the Boltzmann constant. Combining equations (2.47) and (2.48) (using again that $\rho \cong 3M_\odot/(4\pi R_\odot^3)$) leads to the following estimate for the temperature of the solar core:

$$T_\odot^{Core} \cong 2\frac{\mu m_u}{k}\frac{GM_\odot}{R_\odot}. \tag{2.49}$$

The chemical composition of the Sun being largely dominated by hydrogen, we can simply approximate the value of the mean molecular weight μ by the one of pure ionized hydrogen ($\mu = 1/2$) to obtain a temperature of about 2.3×10^7 K for the central layers of the Sun. While this value for the central temperature of the Sun is obtained from very crude approximations, it is nevertheless in relatively good agreement with the more realistic determination of 1.5×10^7 K based on the computation of solar models.

The fact that the properties of the Sun can be precisely determined (see next section for the determination of the solar internal properties) is also of prime interest to discuss the internal structure of stars and in particular the relation between this internal structure and the transfer of energy. Indeed, above the limit of about 0.5 $M_\odot$, the whole stellar interior is not fully convective and both radiative and convective zones are present depending on the local physical conditions. In the specific case of the Sun, the internal region is radiative below about 70% of the total solar radius, while the remaining external layers constitutes the convective envelope. This leads us to briefly discuss here the internal transport of energy in a star. We first discuss the radiative transfer of energy, which is the transport of energy by photons. Energy is produced by nuclear reactions in the central layers and we thus consider a photon emitted in the stellar core. This photon will then travel on a typical distance between two interactions that is called the mean-free path. In the solar case, the length of this mean-free path is of the order of 10 mm in the deep interior and increases to about 1 cm in the envelope. This typical length for the travel of a photon between two interactions is thus much smaller than the solar radius: a photon emitted in the core will then undergo a very large number of interactions in its way to the surface that will only be reached after a time of the order of the Kelvin–Helmholtz timescale. One can also compare this mean free path to the typical variation of temperature in stellar interiors. Such a variation can be very simply estimated from the solar properties, and in particular from the temperature in the solar core obtained above (equation (2.49)):

$$\left|\frac{dT}{dr}\right| \cong \frac{T_\odot^{Core} - T_\odot^{Surface}}{R_\odot} \cong 10^{-4} \text{ K cm}^{-1}. \tag{2.50}$$

Adopting a value of the mean-free path of 1 cm, one then obtains that the variation of temperature is typically only of about 10^{-4} K during the travel of a photon between two interactions. This very small variation compared to typical temperatures of million of K in the solar interior is at the basis of the fundamental assumption of local thermodynamic equilibrium in stellar interiors, which implies that the radiation field at a given radius r inside a star can be simply approximated by the radiation emitted by a black body using the local temperature $T(r)$ at this radius r. Transport of energy by radiation is present in the whole stellar interior. However, depending on the different physical conditions characterizing stellar interiors, the efficiency of energy transport by radiation can be found to be insufficient to transfer a given heat excess. This happens when the opacity of the stellar matter is too high to enable an energy transfer by radiation alone. In this case, convection is active in order to transport efficiently energy through turbulent motions. As mentioned above, low-mass stars with initial masses below about $0.5 M_\odot$ are fully convective, while more massive stars like the Sun are characterized by a radiative core and a convective envelope. The depth of this convective envelope decreases when the initial mass of the star increases, due to the corresponding increase in temperature and related decrease in the opacity of the external layers of the star. For a solar chemical composition, this convective envelope disappears for stars with initial masses typically larger than about $1.5 M_\odot$, which exhibit radiative envelopes. Concerning central layers, a convective core appears for main-sequence stars with initial masses typically larger than about $1.2 M_\odot$ (for a solar chemical composition) as a result of the transition from hydrogen burning through the proton–proton chains to hydrogen burning through the CNO cycle discussed in the preceding section.

The end of the main-sequence phase corresponds to the exhaustion of hydrogen in the central layers of these stars. At this stage, the physical conditions in the core are not favorable for the ignition of helium-burning: in particular the central temperature is still lower than the typical value of about 10^8 K required for helium burning. Consequently, there is a contraction of the central layers, where no nuclear reactions are taking place due to the lack of hydrogen fuel, together with hydrogen burning in more external layers where hydrogen is still present. This post-main sequence phase is then characterized by the contraction of the central layers on a Kelvin–Helmholtz timescale and the simultaneous expansion of the envelope that leads to a rapid decrease in the surface temperature of the star that becomes a red giant star. The subsequent evolution of the star as a red giant follows the contraction of the central layers and the corresponding increase in luminosity together with the expansion of the envelope. Due to contraction, the central values of density and temperature continue to increase: the density then becomes high enough for the gas to become degenerate, while the central temperature is high enough for helium burning to be ignited in the core. As discussed above, nuclear reactions are unstable in degenerate conditions: this ignition of helium burning leads then to an important increase in temperature on a very short timescale called the helium flash. However, this flash does not result in the disruption of the star, but is able to remove the degeneracy in the central stellar layers, and core helium burning can then proceed in non-degenerate conditions. The star is then in another long-standing evolutionary phase, the core helium burning phase, during which energy is obtained from the triple alpha reaction, which is the formation of ^{12}C from three ^{4}He nuclei. This phase ends once helium fuel is exhausted in the central stellar layers. The star evolves then on the asymptotic giant branch and will end up as a white dwarf for which the degeneracy pressure of electrons is able to counteract gravity.

• Intermediate-mass stars

As discussed above, stars with masses typically larger than about $1.2M_\odot$ (for a solar chemical composition) exhibit a convective core while evolving on the main sequence. For these stars, hydrogen burning takes place in well-mixed cores, because of the efficient transport of chemicals by the turbulent convective motions. This results in helium core of significant mass at the end of the main sequence, with an increase of the mass of the core with the initial mass of the star. For stars with an initial mass typically larger than about $2.3M_\odot$ (for a solar chemical composition), the contraction of this helium core after the main sequence always occurs in non-degenerate conditions until the ignition of core helium burning. Contrary to lower-mass stars, these stars will not experience the helium flash and this difference marks the transition to intermediate-mass stars. As for lower-mass stars, the ignition of core helium burning in intermediate-mass stars stops the contraction of the central layers and the star remains in this long-lasting phase until exhaustion of helium in the core. After the core-helium burning phase, stars with masses lower than about $8M_\odot$ evolve on the asymptotic giant branch and end up as white dwarfs. For more massive intermediate-mass stars with initial masses typically in the range $8-10M_\odot$, the evolution after the core-helium burning phase is more complicated. Depending on key physical parameters as the efficiency of mass-loss at the surface of the star or the modelling of convection, the end point of evolution could be a white dwarf or the collapse of the star due to electron capture.

• Massive stars

Stars more massive than about $10M_\odot$ (always for a solar chemical composition) are massive enough to be able to ignite all the different nuclear burning phases in their central layers during their evolution. This marks the transition from intermediate-mass to massive stars. Similarly to lower-mass stars, these massive stars will begin with core-hydrogen burning and core-helium-burning, but then the contraction of their central layers will continue in order to be able to successively ignite carbon burning, neon photodisintegration, oxygen, and finally silicon burning. After these different phases of nuclear burning, an iron core is present, which corresponds to the end of this series, because no exothermic fusion is then possible. At this stage, the internal structure of the star is characterized by different layers exhibiting heavier elements when the distance to the stellar centre decreases. This characteristic 'onion-skin' chemical structure directly results from the successive nuclear reactions that were at work in the stellar core as the central temperature increased during the evolution of the star. This shows that heavy elements until the iron-peak elements are natural products of nuclear reactions occurring in the core of massive stars. Of course, chemical elements heavier than iron are not produced in the central layers of stars, but can be processed during the supernova explosion that characterized the end of the evolution of a massive star. These supernova explosions will then result in the chemical enrichment of the interstellar medium, and successive generations of stars and planets can then be formed from an interstellar medium that contains a higher and higher fraction of heavy elements. The nature of the remnant of a massive star depends on the final mass of the iron core. If this mass is lower than a given threshold (typically of about $2 - 3M_\odot$), a neutron star is formed. Similarly to white dwarfs where the degeneracy pressure of electrons is able to counterbalance gravity, the pressure of degenerate neutrons is able to counteract gravity in the case of a neutron star. However, if the mass of the iron core is larger than this threshold, the degeneracy pressure of neutrons is not anymore able to stop the collapse, which results in the formation of a black hole.

2.7.5 Challenges and opportunities

In the preceding section, the basic physical ingredients needed to model stars were briefly described together with their evolution according to their initial mass. These results rely on models obtained by solving the equations of stellar structure. The computations of these models require of course many different physical inputs, going from microphysics with the equation of state and opacity of stellar matter or nuclear reaction rates, to macro-physics with the modelling of convection and magneto-hydrodynamic transport mechanisms in stably stratified radiative regions.

Observational constraints are then of fundamental importance in order to progress in the theoretical modelling of these various physical processes taking place in stellar interiors. Global stellar properties observed either in photometry such as the luminosity of the star or in spectroscopy with the surface temperature and surface gravity of the star are the first observational constraints used for a comparison with theoretical predictions. In addition to these global properties, surface chemical abundances can be obtained by spectroscopic measurements. The aim of stellar models is then to be able to reproduce these different surface properties either for one specific well-studied target or for a large population of stars.

The comparisons of theoretical models with observed surface properties are of course particularly useful to test different physical inputs of the models, but suffer from a key limitation: it's very difficult to constrain the theoretical modelling of a whole internal structure of a star or a planet, and hence the various physical processes at work in the deep layers of these objects, by only having access to its surface properties. In this respect, spectroscopic determinations of surface chemical composition can nevertheless be particularly useful for the modelling of transport processes in radiative zones. This is for instance the case of light elements like lithium and beryllium in solar-type stars and of nitrogen surface abundance in more massive stars that can reveal the transport of material processed by the CNO cycle from the stellar core to the surface. We can also enlight here the specific cases of intermediate-mass stars exhibiting peculiar surface abundances and of the massive Wolf-Rayet stars. The former stars reveal the role played by radiative forces on the relative surface abundances of chemical elements, while the latter stars nicely illustrate the key impact of mass loss on the evolution of massive stars. Indeed, strong mass loss, either due to stellar winds in single stars or to mass transfer in binary stars, are able to remove the surface layers of Wolf-Rayet stars, so that the chemical composition of internal layers, which has been modified by the different nuclear reactions, can then be observed at their surface.

While observing surface properties for these specific targets can bring some indirect constraints on stellar interiors, the main challenge consists of course in obtaining direct observational constraints on the internal physical properties of stars. To address this key question, we recall that the internal structure of the Earth can be studied thanks to the analysis of seismic waves. These waves can either be artificially generated by hitting the ground at fixed frequencies in order to probe the structure of layers close to the surface, or be produced by an earthquake and then travel through deeper layers of the Earth to give some information about the internal structure of our planet. By analogy to the seismic probing of our planet, one can then wonder whether a 'starquake' could happen and the analysis of the generated waves be used to probe the internal structure of a star.

2.7.6 The Sun: the solar modelling problem

In the case of the Sun, this question has been beautifully answered by the observation and analysis of the so-called solar five-minute oscillations. To understand how a 'sunquake' can happen, we recall that convection is active in the envelope of the Sun in order to efficiently transfer energy in these external layers (see previous section). Due to these turbulent convective motions, waves can be generated and will then travel in the solar interior. These travelling waves can then give raise to standing waves, or global oscillation modes, whose properties depend on the internal structure and composition of the Sun. The characterization and analysis of these solar oscillations can then be used to probe the interior of the Sun similarly to the analysis of seismic waves in the case of the Earth: this is the goal of helioseismology, which is the study of solar oscillations.

The analysis of solar oscillations has led to key results regarding the internal structure of the Sun [229, 230]. Thanks to precise determinations of the frequencies of pressure modes (global oscillation modes related to acoustic waves), the solar sound speed and density profiles have been determined. These helioseismic data have also enabled a precise determination of the location of the transition from the radiative interior of the Sun to the convective envelope [231], as well as the determination of the abundance of helium in the solar envelope [232–234]. In addition to the well-determined global parameters of the Sun (mass, radius, luminosity, and age), these helioseismic measurements constitute of course fundamental constraints for theoretical solar models. Interestingly, standard solar models (i.e., models that do not take into account hydrodynamic or magnetic transport processes) based on the solar photospheric abundances determined from spectroscopy in the 90s [235] were found to be in fairly good agreement with these different observational constraints [236]. In particular, the location of the base of the solar convective zone as well as the helium content of the solar envelope predicted by these standard models were in perfect agreement with the helioseismic determinations. Moreover, the sound speed and density profiles deduced from helioseismic data were satisfactorily reproduced by these solar models. The success of these standard solar models based on the solar surface abundances available in the 90s then suggested that our global understanding of the Sun was quite good.

This situation completely changed about 15 years ago with the redetermination of the solar surface abundances and the significant decrease by about 25% of the heavy element abundances of the Sun [237, 238]. This important change in the spectroscopic determination results from the use of three-dimensional models of the atmosphere (instead of the previous one-dimensional models), from a new consideration of the selection of the solar spectral lines and from the inclusion of effects related to non local thermodynamic equilibrium. Standard solar models computed by accounting for these revised photospheric abundances were then found to be unable to correctly reproduce the helioseismic constraints. In particular, the transition from the internal radiative zone to the convective solar envelope is predicted to be to close to the surface compared to helioseismic data, while the predicted helium surface abundance is too low compared to the helioseismic determination. Moreover, the sound speed and density profiles deduced from helioseismology are not correctly reproduced anymore [239–241]. This important tension between models based on the revised solar abundances and helioseismic constraints is referred to as the solar modelling problem. One of the main challenges concerning the Sun at the horizon 2050 will then be related to the study of the origins of this tension between solar abundances and helioseismic measurements with the final goal of solving the solar modelling problem.

Another key result of helioseismology is the determination of the internal rotation profile of the Sun. These measurements reveal an approximately uniform rotation in the radiative interior of the Sun with a rapid transition to differential rotation as a function of latitude in the solar convective envelope. These two key features of the internal rotation of the Sun have always been challenging to reproduce. Indeed, solar models that account for hydrodynamic transport processes related to rotation-induced instabilities predict a strong radial differential rotation in the solar radiative interior, which is in contradiction with the almost uniform rotation inferred from helioseismic measurements [242]. This shows that additional dynamical physical processes are missing in the solar interior, in particular regarding the efficient transport of angular momentum in the radiative zone. The effects of magnetic fields are of prime interest in this context. For instance, the strong coupling ensured by large-scale fossil fields can be invoked to reproduce the flat rotation profile in the radiative interior of the Sun [243–248]. However, such a scenario has some difficulties to simultaneously account for the second key observational constraint, which is the sharp transition from the approximately uniform rotation in the radiative interior to differential rotation in the convective zone [249, 250]. The effects of magnetic instabilities could then play an important role. In particular, the Tayler instability [251] could be at the basis of a small-scale dynamo leading to an efficient transport of angular momentum in stellar radiative zones [252]. Interestingly, such a process can simultaneously account for the approximately uniform rotation in the radiative interior of the Sun and the transition to differential rotation in its convective envelope [252, 253]. However, the existence of this kind of process has to be confirmed by simulations performed under realistic stellar conditions [249]. Obtaining numerical simulations of these complex hydrodynamic and magnetic processes under realistic stellar conditions is another main challenge for the years to come.

The different results obtained so far by helioseismology have been derived from the study of pressure oscillation modes. A big observational challenge for the Sun at the horizon 2050 is related to the detection of the solar gravity oscillation modes. This is a long and difficult quest owing to the extreme difficulty of detecting these very low-amplitude oscillation modes. Characterizing solar gravity modes would enable the determination of the rotation rates in the core of the Sun, which is not feasible from the rotational splittings of pressure modes. This is of course particularly interesting to better understand the specific case of the internal rotation of the Sun, but also for stars in general by directly constraining angular momentum transport in radiative interiors where strong chemical gradients able to inhibit the transport efficiency by turbulent processes are present [254]. In addition to the rotational properties of the central layers of the Sun, gravity modes would of course also offer an interesting possibility to better constrain the chemical mixture and opacities in the solar core that would perfectly complement the constraints coming from solar neutrinos [255].

2.7.7 From the Sun to distant stars

The beautiful successes obtained from the study of oscillation modes in the case of the Sun motivated the detection of solar-like oscillations for other stars. Using the same seismic techniques for more distant stars is of course of prime interest to probe the internal structure of stars with properties that can be quite different from the Sun: for instance, stars with different masses and chemical compositions, stars at various evolutionary stages, or even physical processes not at work in the Sun, such as mixing by convective cores. Having a direct observational access to the internal properties of different stars is needed to progress in our understanding of stellar evolution.

A first important issue to achieve this goal is related to the difficulty of detecting these oscillations for a distant star. Indeed, contrary to the solar case, another star is seen as a point source and its disc is not resolved. This has a direct consequence on the type of oscillation modes that can be detected for a distant star. Figure 2.44 illustrates the deformation of the surface corresponding to different oscillation modes. These modes can be identified with the degree l and azimuthal order m of the spherical harmonics, where l and m indicate the total number of nodal lines at the surface and the number of these lines that intersect the equator, respectively. While all these oscillation modes can be detected for the Sun, only the lowest-degree modes can be observed for a distant stars (typically modes with $l \leqslant 3$) due to the geometrical average of the deformation on the stellar disc. Fortunately, these low-degree oscillation modes are of particular interest, because they penetrate deeply in stellar interiors and can then bring key constraints on both global stellar properties (e.g., masses and radii) and properties of the stellar cores (e.g., central hydrogen abundance during the main sequence, size of convective cores). A second observational difficulty is related to the very low amplitudes of solar-like oscillations

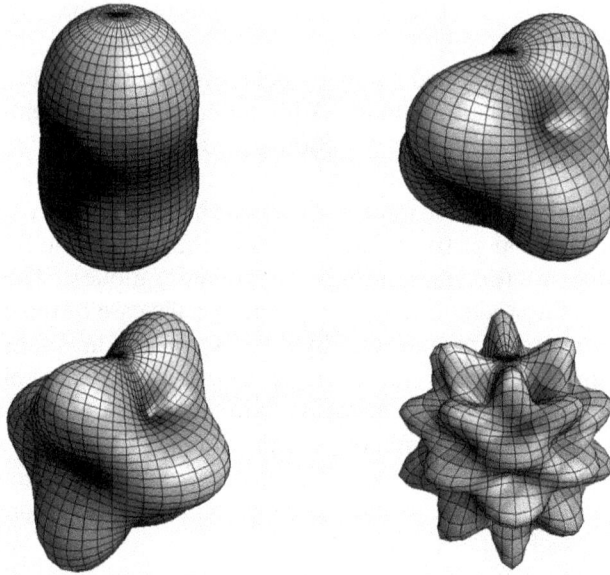

Figure 2.44. Deformation of the stellar surface due to global oscillations of the star. The amplitudes of these oscillations are largely exaggerated for illustrative purpose. Oscillations modes with 2 nodal lines at the surface ($l = 2$) and none of this nodal line that intersects the equator ($m = 0$) is shown in the top-left panel. A mode with 3 nodal lines at the surface ($l = 3$) and 2 nodal lines that intersect the equator ($m = 2$) is shown in the top-right panel. Oscillation modes with $l = 4$, $m = 2$ and $l = 10$, $m = 4$ are shown in the bottom left and right panels, respectively. (Courtesy of Sébastien Salmon, University of Geneva.)

(typically about 20 cm s^{-1} in radial velocity for the Sun): these oscillations can then only be detected in radial velocity from the ground by using high-accuracy spectrographs or from space in photometry in order to avoid the variations of luminosities induced by atmospheric turbulence.

The observational effort aiming at detecting solar-like oscillations for distant stars has led to the characterization of these oscillations for a large number of stars with various properties, in particular thanks to the space-based missions CoRoT, *Kepler* and TESS. The study of these stellar oscillations, or asteroseismology, can then bring different type of constraints for stellar evolution, from the determination of the global properties for a large population of stars to the detailed study of the internal structure of specific targets for which the most precise data are available. For instance, the masses and ages can be precisely determined for a large number of stars thanks to asteroseismology, which opens new perspectives for our understanding of stellar populations of the Galaxy and hence of the chemical evolution of the Milky Way [256, 257]. An important challenge for the years to come will be to extend these results in the metal poor regime and to dense fields of stars such as globular clusters [258].

Another key challenge for stellar physics is to progress in the modelling of the complex dynamical processes at work in stellar interiors. A first long-standing problem is of course related to the modelling of convection. The combined development of sophisticated numerical simulations of convection [259] and of asteroseismic constraints on the size of convective cores and shape of the border of these cores [260] is very promising to achieve significant progresses in this direction. A second key issue in stellar physics is related to the modelling of transport of angular momentum and chemicals in radiative stellar interiors. The observation of solar-like oscillations for subgiant and red giant stars has led to the measurements of rotational splittings of mixed oscillation modes. This has enabled the determination of the core rotation rates for these stars [247, 261–263]. The comparison with theoretical models has revealed that purely hydrodynamic processes do not provide an efficient enough transport of angular momentum to reproduce the observed core rotation rates [264–266] and that the efficiency of the missing transport processes can be precisely quantified thanks to these seismic data [267, 268]. This constitutes a unique opportunity to reveal the physical nature of these undetermined transport mechanisms in stellar radiative zone, which will be one of the main challenge of stellar physics for the years to come.

2.7.8 The star–planet connection

With the first detection of a planet around a solar-type star [269] and the following observations of numerous planets around various stars other than the Sun, the study of exoplanets rapidly became a central topic of modern astrophysics. These exoplanets show a great varieties of masses and sizes going from small rocky cores to massive gaseous planets on different kind of orbits (http://exoplanet.eu). Exoplanetary science is thus a very vast topic and we will only briefly touch here on the fundamental link between the stellar physical processes discussed above and the physics of the planets. With thousands of exoplanets with precisely characterized properties observed so far, one important challenge for the coming years is indeed related to the understanding of the evolution of these planetary systems studied as a whole.

A first obvious link between stars and planets is of course related to the need to have precise knowledge of the host-stars properties to derive the planetary parameters. For instance, a determination of the mass and the radius of the host star is needed to obtain the planetary mass from radial-velocity measurements and the planetary radius from transit observations, respectively. In order to study the evolution of planetary systems, one also needs a precise determination of the age of these systems. This is a particularly challenging goal, which can be successfully achieved thanks to the capability of asteroseismology to probe the deep stellar layers in order to reveal the central hydrogen abundance (i.e., the quantity of nuclear fuel still available for a main-sequence planet-host star). The observation of solar-like oscillations is then a powerful means to precisely determine the properties of the planet-host star (in particular its mass, radius, and age) needed to fully characterize the planetary system. This complementarity between asteroseismic and exoplanetary observations is well illustrated with space-based missions aiming at simultaneously detecting transiting exoplanets and performing asteroseismic measurements to characterize planetary systems such as TESS [270] or the future PLATO mission [271].

Being able to precisely determine the current state of a planetary system is the first mandatory step to study its past and future evolution. The different interactions between the host star and the planets have then to be taken into account to study the evolution of the system. A first important interaction between a star and a planet is the tidal interaction. To follow the orbital evolution of a planet due to tides implies to account for the transfer of angular momentum between the planetary orbit and the spin angular momentum of the star. The efficiency of tidal dissipation is also sensitive to the convective and radiative structure of stellar interiors as well as to the internal rotation of the star. Sophisticated rotating stellar models able to correctly predict the internal transport of angular momentum and the lost of angular momentum due to magnetized winds are then needed to coherently follow the rotational evolution of the host star together with the orbital evolution of the planet. This is another illustration of the link between exoplanetary science and stellar physics, in particular regarding the challenging question of the modelling of transport processes in stellar interiors discussed in the previous section. Developing such sophisticated models of rotating stars coupled with the orbital evolution of the planets due to tides constitutes one of the challenging questions to be addressed in the coming years to understand the evolution of planetary systems; preliminary attempts have thus been made in this direction [272–274].

Interestingly, these rotating stellar models including magnetic effects for the host star are also needed to study another fundamental star–planet interaction, namely the impact of the radiation emitted by the star on the planet properties. This is of particular relevance for the evolution of the high-energy flux received by a planet that will directly influence the photo-evaporation and the physical properties of its atmosphere [275, 276]. For solar-type and low-mass stars, the high-energy flux emitted by a star is indeed closely related to the rotational properties of its convective envelope, which determines the intensities of the magnetic fields that can be produced by a dynamo. The high-energy flux received by a planet can then change by orders of magnitudes during the evolution of a planetary system as a result of the rotational evolution of the host star [277, 278]. This evolution of the high-energy flux has to be coherently taken into account to study the evolution of planetary systems due to the photo-evaporation of the planets [240–281]. This nicely illustrates the complementarity between the physical description of stars and planets and the need for a multi-disciplinary approach to tackle the challenging questions related to the evolution of planetary systems at the horizon 2050.

2.8 Physics of the Earth's interior

Emmanuel Dormy[1]
[1] CNRS & Ecole Normale Supérieure, Paris, France

2.8.1 General overview

The Earth is, until now, the only planet we know on which lifeforms have developed and grown for several billions of years. Perhaps more surprisingly, it is also considered to be a living planet. By this we mean that it spontaneously evolves with time. Of course, we all know about the geothermal gradient: the temperature within the Earth increases with increasing depth. There are two reasons for that: first the Earth, very hot when it formed, is slowly cooling down with time, and second the presence of low radioactivity of its deep-seated rocks provides a source of internal heating.

When the Earth formed, it differentiated: the very heavy elements went to the centre, a place we now call the core of the Earth, mainly formed of iron (see figure 2.45). The comparatively lighter rocks floated and formed a deep layer

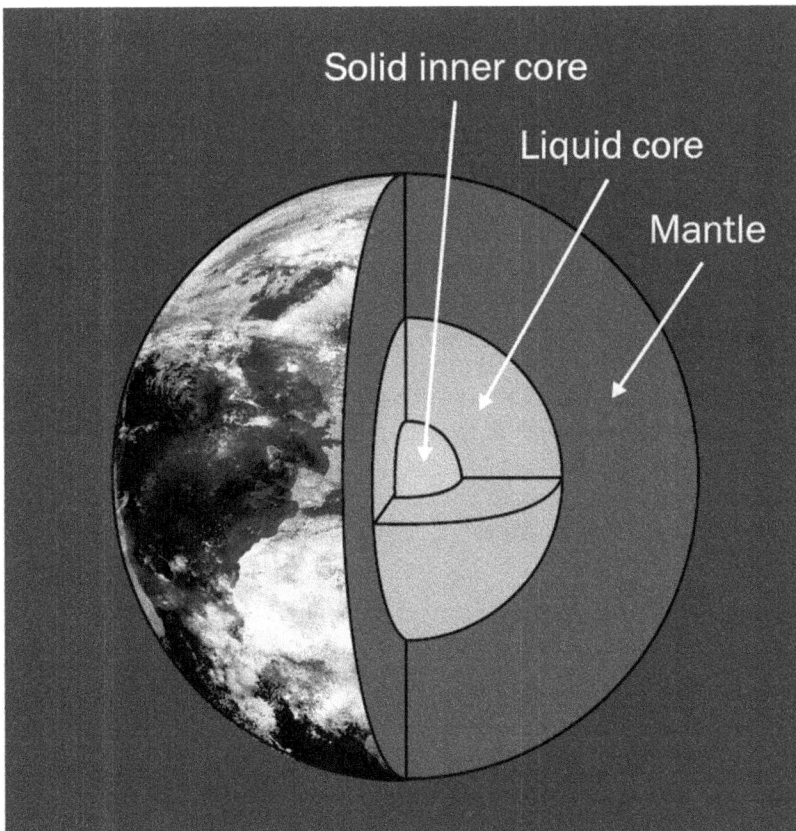

Figure 2.45. Cross-section of the Earth highlighting the simplified inner structure of our planet.

surrounding the core: the so-called mantle. The radioactivity in the mantle heats the rocks so as to keep them in a ductile state that allows the development of large-scale convective motion (through diffusion creep and dislocation creep). This slow convection is responsible at the surface for plate tectonics, the formation of mountains, earthquakes, and volcanism! About 3000 km below the surface, the core of the Earth is so hot that the iron is in a liquid phase, except for its very central part, where solid iron is met, as the pressure increase does more than compensate for the increase in temperature. Motions in the Earth liquid core are responsible for the Earth magnetic field, without which the compass would be of little use to find north.

During its long history, the geography of the Earth, its oceans, the composition of its atmosphere, its climate, and its biosphere (all forms of life) have undergone significant changes. The evolution of the Earth during its 4.5 billion years of existence is spectacular. It is only in the last century that we have begun to perceive the physical phenomena that have constantly altered it. Geophysics offers a venue for many branches of physics. When studying the interior of the Earth, the parameters are often extreme (in terms of pressure or temperature, for example) and the range of scales involved is (as we will see) phenomenal. The main specificity of this discipline being that theory, experiments, and simulations are completed by field work and *in situ* observations, which constitute the ultimate constraint over all other approaches. Of course, direct observations are limited to the surface of the Earth and a combination of physical insight, experiments, and indirect observations is needed to understand the processes at work within our planet.

In this short presentation, I will focus on three, and only three, open challenges in the physics of the Earth's interior. Such a choice is of course necessarily partial, yet I do hope that these questions can shed light on some aspects of research in geophysics and will motivate some of the readers to take an interest in the broad field of geophysics. Advanced texts such as the book by Turcotte and Schubert [282] can be referred to in order to develop a more complete picture of the current state of knowledge in this field.

The three subjects I wish to highlight concern the origin of the Earth magnetic field, of volcanic eruptions, and of earthquakes. If a motivation was needed to justify this choice, it would be only fair to say that the Earth magnetic field has played a major role through the centuries in the exploration of our planet and the discoveries of new worlds overseas; and since its appearance on Earth, mankind has been subjected to a changing environment, which includes deadly catastrophes such as volcanic eruptions and earthquakes.

2.8.2 The origin of the Earth's magnetic field

We have all played at least once in our life with a compass. The systematic orientation of the magnetised needle in the ambient field is mesmerising. The physical understanding of this observation is of course well established: the needle orientates itself according to the ambient magnetic field, which has both a strength and a direction. What is not so well understood, however, is where this field is coming from.

As we explained above, whereas the mantle is formed of silicate rocks, comparable to those found on the surface, the Earth's core is a spherical huge metallic ocean, mainly made of molten iron. The core probably formed 'rapidly' (in a geological sense, i.e., in 'only' some 100 million years) by differentiation. The temperatures in the core are over 3000 °C at the base of the mantle and about 5000 °C near the inner core. These temperatures keep the metal molten in spite of the considerable pressure, which varies from 1.3 to 3.5 million atmospheres, the latter being reached some 5200 km below the surface. At this depth, despite temperatures over 5000 °C, the pressure forces the crystallisation of a solid inner core.

At such high temperatures, permanent magnetism is not possible because of thermal agitation. This phenomenon was studied by Pierre Curie during his PhD thesis in 1895. The critical temperature at which permanent magnetism is lost is now known as the 'Curie temperature'. It is of the order of 800 °C for iron. Permanent magnetism in the core of the Earth can thus be ruled out. Only cold material, close to the surface can be magnetised (this is known as the crustal field, which is rather weak and small-scale).

Yet the Earth exhibits a large-scale magnetic field, known as the main field, which can only be of internal origin. The rocky mantle being—to a good approximation— insulating, the magnetic field, as measured at the surface, can be downward continued to the core-mantle boundary using a harmonic potential. Since the core is free of permanent magnetism, the magnetic field we reconstruct at its surface is the signature of electrical currents crossing the core. The key open question lies in the generation mechanism for these electrical currents.

The electrical currents, and thus the internal magnetic field, are most likely due to a magnetohydrodynamic instability in the liquid core known as 'self-exciting dynamo action'. This instability is one of the rare examples of a mechanism acting at a scale which is neither very small (say nanophysics), nor very large (say cosmology), and which still lacks a proper physical understanding. How does a conducting fluid spontaneously transfer part of its kinetic energy into electrical energy?

The origin of the kinetic energy in the core is rather well understood. Since its formation, some 4.5 billion years ago, our planet has continuously been cooling down, by evacuating the heat it contains to the surface of the planet where temperatures are lower. In this process, heat is partly carried from the inner depths by thermal convection, that is to say by the transport of buoyant hot material toward the surface and of cold material to lower depths. In the mantle, this convective motion is rather slow. In the core, which is characterised by a kinematic viscosity comparable to that of water at ambient temperature and pressure, it is on the contrary very fast (again in a geological sense!). The flow is estimated to be of the order of a meter per hour. This may not seem terribly large at first, but it is almost a million times faster than the velocities of tectonic plates at the surface!

Besides thermal convection, a second source of energy is available to drive motions in the liquid core as the Earth is cooling down. High-pressure physics and geochemistry allowed to estimate that the liquid core of our planet is not made of pure iron. Instead, it contains approximately 80% in mass of iron, 5% of nickel (with a density comparable to iron), and of 15% 'light elements' including possibly silicon, sulphur, and oxygen. The density of the liquid core is lower by approximately 10% than that of pure iron under the same physical conditions. The solid inner core solidifies and grows as the planet is slowly cooling down. In this process, which is not at the eutectic, only pure iron solidifies and the lighter elements are released at the base of the liquid core. They will rise and mix through buoyancy in a way similar to thermal convection, known as 'solutal convection'.

We know that an electric current appears by induction within a conductor moving in the presence of a magnetic field. We also know that an electric current generates a magnetic field. Sir Joseph Larmor, then Lucasian Professor of mathematics in Cambridge, proposed a little bit over a century ago a mechanism by which these two phenomena could be combined to yield an instability. This is to say that, if the conditions are met, the motion of the liquid metal in the presence of electrical currents in the core could act to regenerate these very currents against ohmic losses.

However, these 'conditions' are yet to be understood (we refer the reader to the book by Moffatt and Dormy [283] for a more advanced discussion). There exist, of course, theoretical bounds such as lower bounds on the strength of the flow (as measured by the magnetic Reynolds number) in order to be able to sustain the magnetic field. There are also theoretical results ruling out field amplification for too simple geometries (e.g., Zeldovich's theorem, ruling out dynamo action from two-dimensional flows, and Cowling's theorem for axisymmetric fields). But the detailed conditions for dynamo action are still challenging.

This is made even more complicated by the rotation of the Earth (which is remarkably fast: one spin a day!). The effect of the Coriolis force on the flow is determinant, as highlighted by the fact the axis of the large-scale field is nearly aligned with the axis of rotation of the planet (a decisive property for the use of a compass!). The day is a very short timescale compared, for example, to the time needed for the weak viscosity of the fluid to slow down the motion. It follows that very sharp boundary layers will form at the boundaries of the liquid core. These can be estimated to be about a meter thick, to be compared with the 3400 km radius of the core. Despite the complexity of the non-linear equations describing the induction mechanisms in the core, some theoretical advances were made possible thanks to the disparity of scales involved in the process.

Back in 1955, Eugene Parker introduced the idea that small-scale motions, influenced by rotation, will be of a helical nature (a property of rapidly rotating flows) and that 'cyclonic events' could be incorporated by an averaging procedure in equations for the mean magnetic field. But it was only some ten years later that the mathematical formalism was properly derived independently by Braginskii (1964) [284] and by Steenbeck, Krause, and Rädler (1966) [285].

The essential idea is that an average can then be taken to estimate the large-scale field resulting from small helicoidal vortices. It is found to be non-vanishing. In this two-scale analytical approach, some randomness is assumed on the small-scale flow, and the large-scale field (known as 'mean field') is constructed on the basis of this assumption. Braginskii considered the axisymmetric field as 'large-scale' and non-axisymmetric fluctuations as small scales.

The magnetic Reynolds number relevant to this scale separation can be shown to involve the square root of the product of the large scale and the small scale (i.e., the geometric mean of the two length-scales) [283]. Because of this, mean-field effects are possible down to a very small scale. In the liquid core of the Earth, the flow down to a scale of a few 100 m can contribute to the large-scale magnetic field generation. Such mean-field models usually assume either no-flow at the large-scale, or more simply a spectral gap between the small-scale dynamo vortices and the large-scale flow and field. This formalism has been used for decades, and was until recently the only way to tackle this challenging problem. The full non-linear set of equations not being prone to a complete analytical treatment.

Since 1995, direct numerical simulation has taken the lead. The full set of equations can be successfully discretised on a computer (usually computers working in parallel). The outcome is spectacular magnetic field lines rendering in three-dimensional visualisation (see figure 2.46), Earth-like magnetic fields, and occasionally reversals of the field polarity. Yet, it should be stressed that all of these models have to rely on non-physical filtering of the small scales (either through unrealistic parameters, or through additional terms in the equations). This does not come as a surprise when one realises the formidable disparity of scales involved (the size of the core being, for example, about a million times that of the boundary layers we

Figure 2.46. The celebrated Glatzmaier–Roberts model for the geodynamo. One of the first numerical models introduced in 1995. Computers and numerical models have improved since, but not to the point of closing the formidable gap between the smallest scales of the flow relevant to magnetic field generation and the size of the core. Reprinted from [286], Copyright (1995), with permission from Elsevier.

mentioned above). A proper understanding of the various scales at play and their interaction is sadly still missing despite the use of some of the largest computers currently available.

It is not at all easy to perform experiments of self-exciting dynamo action. Only three experiments have been successful so far. The first two in Karlsruhe and Riga involving severely constrained flows. The third one in Cadarache, allowing for an unconstrained turbulent flow, has highlighted Earth-like polarity reversals when the symmetry between the forcing blades was broken [287]. However, it needed ferromagnetic blades to operate.

It is then interesting to consider other planets to try and gain an insight from their magnetic fields. Some planetary magnetic fields are similar, such as Jupiter and the Earth (despite the fact Jupiter is a gaseous planet, with a very different inner structure). In other cases they are very different. Uranus and Neptune, for example, both exhibit an asymmetric multipolar field, whereas Saturn presents a nearly axisymmetric field. There is thus no simple picture to be drawn from such comparisons. The latter geometry raising interesting questions, since Cowling's theorem prevents dynamo action to sustain a purely axisymmetric field.

2.8.3 The physics of volcanic eruptions

Let us now turn our attention to convection in the mantle and some of its interactions with us: volcanoes. Contrary to our first problem, which was purely academic and not directly related to society concerns (see [288] for a discussion), this problem addresses a life-threatening hazard. We cannot discuss the physics of volcanoes without asking the question: to what extent can volcanic eruptions be predicted?

Volcanoes are one of the signatures at the Earth's surface of the deep-seated thermal convection in the Earth's mantle. We have already mentioned three 'layers' in the Earth's interior: the inner core, the liquid outer core and the mantle. We should also introduce the thin shell at the surface: the Earth's crust, with thickness ranging from some 5–10 km under the oceans and 30–50 km under continents. Together with a thin portion of the upper mantle, it forms the Lithosphere (i.e., the 'rigid' outer layer of the Earth). Because of the underlying thermal convection in the mantle, it is broken into tectonic plates (the physical laws governing the sizes of these plates also constitute a challenge in geophysics).

A volcano is formed of one or more magma reservoirs that are connected to the surface by small pipes that contain a liquid, the magma, formed of molten silicate. The magma is produced by a deep source, located in the Earth's mantle (remember the core is formed of liquid metal, not magma!), or near the surface in the crust. Some volcanoes, known as 'hot spots', originate from very deep in the mantle (probably plumes of melted material emerging from the base of the mantle), while others are located at plates boundary and are of shallower origin.

The volume of a volcanic eruption is much greater than can be sustained by the flow through the deep plumbing system that brings the magma to the surface. A volcanic eruption is therefore only possible when a sufficient quantity of magma has accumulated in a reservoir, the so-called magma chamber. These reservoirs are formed because of the migration of magma to the surface, due to buoyancy. This migration stops in the vicinity of the surface, when the magma reaches rocks of a lower density. The magma then stops rising and starts forming the reservoir. The magma chamber does more than just storing the magma. During the period of stay in the chamber, the magma changes properties, in terms of crystallinity, viscosity, dissolved water content as the magma cools and partly solidifies. Crystallization is fractional, and the residual liquid chemically evolves and is enriched in volatile elements, such as carbon dioxide, sulphur dioxide, or hydrogen sulfide. These changes in the magma are essential, because they modify its physical properties and thus determine the type of the next eruption. The content in gazes of course depends on the volcano. In general, volcanoes associated with convergent plate boundaries tend to emit more water vapour than those at hot spots or at divergent plate boundaries (because of the seawater trapped in the subduction).

An eruption occurs when the magma chamber fails. This occurs when the overpressure in the chamber exceeds the strength of the rocks surrounding it. This is, of course, very hard to predict, if only because it is not possible to know the exact level of stress within a volcano.

When an eruption occurs and the magma moves up to the surface, the flow of magma is characterised by a complex rheology. Besides, the pressure drop causes both exsolution (i.e., the separation of the various constituents previously dissolved in a homogeneous phase) of volatiles and expansion of the gas phase present.

It turns out that the nature of the magma is essential in explaining the various observed eruptive behaviours. A given volcano, and even a given eruption, can evolve between different eruptive regimes. Some being far more dangerous than others and the changes being very difficult to forecast. The quantity of gas in the mixture, of course, depends on the initial content of the magma in the reservoir, but also on that quantity of gas lost during the ascent through the permeable walls and fractured rocks. The total content of volatile elements thus depends on the rising velocity. If the magma rises very quickly, gas losses can safely be neglected. If, on the contrary, the rising velocity is low, these losses can become very significant. Since the rate of ascent itself depends on the gas fraction, the non-linear coupling induces a strong sensitivity to small changes in the initial conditions in the reservoir.

The regimes can be classified in increasing proportion of gas. If the gas content is low, the magma contains a few gas bubbles in suspension. This is the regime corresponding to effusive lava flows. At a higher gas fraction, an intermittent rate is obtained, where the gas is collected in pockets whose diameter are close to that of the volcano conduit. With even more gas, the gas forms a continuous jet and the walls of the conduit are then covered with a thin layer of magma. The eruption then consists of high-velocity turbulent jets carrying fine droplets of magma. This is the so-called explosive regime. In this explosive regime, the volcanic jet can follow remarkably different dynamical evolutions depending on small variations in its mass discharge rate and in the amount of free gas in the eruptive mixture.

In the so-called 'Plinian' eruptions, the jet is buoyant enough for the eruptive columns to go as high as the stratosphere, where it spreads horizontally. However, if the turbulent jet is not light enough, it can collapse. This is due to the nature of the jet, but also to the amount of air being entrapped as the rising panache expands [289, 290]. If the jet collapses, it falls back on the slope of the volcano at formidable speeds. This is known as a pyroclastic flow. The dynamics of the flow is similar in many aspects (including its speed of some 100 km h^{-1}) to an avalanche, but its temperature is about 1000 °C. Pyroclastic flows are the most terrible of all volcanic hazards. When the conditions are close to those of collapse, the eruption can be transitional: a buoyant column can rise to high altitudes until it suddenly collapses (see figure 2.47).

Pyroclastic flows are notoriously life threatening. Famous examples of catastrophic events in the past include the terrible eruption of Mount Vesuvius in 79 AD which destroyed, among others, the city of Pompeii and caused a huge number of deaths (including the famous Pliny the Elder), or that of Mount Unzen, in Japan on June 3, 1991 which caused the death of two well-trained volcanologists, Katia and Maurice Krafft.

It may be surprising at first to note how the volatile content and the micro-bubbles forming in the magma will have such decisive large-scale consequences. Minute variations in the pressure of the reservoir can induce variations in velocity, pressure, and gas content, and shift from an explosive regime to an effusive regime. Initially, small bubbles are not buoyant enough to rise quickly through the viscous magma. However, as the magma ascends, the bubbles will grow, both because of expansion as the pressure drops, and because the solubility of volatiles in the magma decreases causing more gas to exsolve. This results in a complex rheology which prevents classical solutions of equations of fluid mechanics to describe the flow of magma. The presence of bubbles will drive a faster upward-propagation. What happens next largely depends on the gas concentration and on the viscosity of the magma. The bubbles may rise and merge to form bigger bubbles, or they may be stuck in the magma and slowly connect to form some sort of network. The

Figure 2.47. A Plinian column (on the left) characterised by a buoyant jet extending to high altitudes, and a collapsing fountain producing a devastating pyroclastic flow (on the right) which is not buoyant enough to rise high in the atmosphere and collapses on the flanks of the volcano.

overpressure in the conduit caused by these gases is directly related to the explosivity of an eruption [291].

Volcanoes do not only occur on Earth. The surface of Venus is covered by many volcanic structures. They are probably not active, though a recent study indicates that some of them may have been active during the past few million years [292]. The most active extra-terrestrial volcanoes were reported on moons, namely on Io (Jupiter's moon), Triton (Neptune's moon), and Enceladus (Saturn's moon). Io was the first moon for which evidence of active volcanism on a body of the Solar System other than Earth was obtained, some 40 years ago. It is now regarded as the most volcanically active body in the Solar System.

2.8.4 Earthquakes

Let us now turn to yet another consequence of mantle convection at the surface of our planet: earthquakes. Earthquakes are, without a doubt, among the most devastating natural phenomena on Earth. They are also intimately related to our knowledge of the interior of our planet. Seismology has allowed the use of earthquakes as an indirect means of observing the Earth's inner depth (we would not know of the internal structure of the Earth introduced above, if it was not for seismology).

Earthquakes are a direct consequence of plate tectonics. Tectonics is driven by thermal convection in the mantle and by the fact the colder and thus heavier plates want to sink. Gravity pulls the slab downward in the form of cold downwelling. Seismic tomography reveals that some slabs appear to stop sinking and are horizontally deflected at a depth of some 660 km (i.e., the transition between the upper-mantle and the lower-mantle, corresponding to a mineral phase transition, that of Olivine). Other slabs are sinking further down in the mantle.

Earthquakes occur at plate boundaries, where the plates converge, diverge, or slide along each other. Brittle deformation causes the formation of faults in the lithosphere. Earthquakes are caused by a sudden slip on an active fault: tectonic plates are slowly moving, but they tend to 'stick' at their edges due to friction with neighbouring plates. Stress builds up and at some stage becomes strong enough to overcome friction. An earthquake then occurs, releasing a large amount of energy in the form of waves.

While predicting earthquakes remains exceedingly challenging, our physical understanding of these phenomena has significantly progressed since the under-standing, over a century ago, that they result from the brutal sliding of rocks on faults, which are regions of weakness of the Earth's crust. Fault motion can be viewed as frictional sliding on a fault plane. From a mechanical point of view, friction will evolve as a result of slip (i.e., the relative displacement of the two sides of the fault plane), the velocity, and the history of contact.

The very nature of the trigger makes it extremely difficult to predict the likely ground motions during future earthquakes. It is currently impossible to predict their time, location, and magnitude. Observations of precursor phenomena are numerous but involve a significant risk of false alarms. This should be contrasted with the good efficiency of early warning systems, which rely on the finite propagation speed of seismic waves to provide real-time assessment of the hazard, and notification of people in the endangered area. Fatalities in many parts of the world have also been reduced through the development of building codes designed to withstand strong earthquakes.

To understand the formation of earthquakes, we must think of the tectonic stress on a fault, slowly building up, until it reaches the critical stress necessary for failure. The frictional stress is essential as it controls the seismic motion. An earthquake can occur only if friction decreases rapidly with slip; in other words, if the 'dynamic friction' is much less than the 'static friction', a process often referred to as 'slip weakening'. In that case, the two walls of an earthquake fault are either sticking to each other or sliding over each other. They can be thought as rough surfaces with irregular contact points. The resulting sliding motion then takes a form of relaxation oscillations, common in non-linear physics, and known in this case as 'stick-slip'.

This is probably best understood using a physical analog: the slider-block model. In this model, originally introduced by Burridge and Knopoff [293] and often used to describe earthquake dynamics, a block is attached to a spring, which is pulled over a surface with a constant force. The friction between the block and the surface prevents the block from sliding. The area of the frictional surface in contact with the block is then analogous to the fault. As the spring is pulled, tension slowly builds up in the spring. At some stage the tension will be such that the critical stress is reached and the block will slip. On most surfaces, the block will simply gently slide at constant velocity. However, if we meet a velocity-weakening of the friction force, that is to say a decrease of friction with velocity, a 'stick-slip' behaviour will occur. The block will slide for a distance and then stop again as the tension in the spring has dropped too much (in the same way, after an earthquake occurs, it yields a sudden stress drop). The tension on the spring will then build again until the next event. A new 'earthquake' cycle subsequently begins.

This experiment is easily performed using a standard soap in a paper box as 'sliding block' and a yoga mat or any rubber mat as supporting surface. The spring can conveniently be replaced by one of these surgical masks that have sadly filled our lives in the recent years. The experiment is illustrated in figure 2.48(a).

More realistically, small irregularities at the surface will involve local variations of the critical stress, loading rate, and stress drop. So that the stick-slip cycle will not be periodic. Burridge and Knopoff went beyond this and introduced chains of such sliding blocks, all coupled to nearest neighbours by elastic springs (see figure 2.48(b)) and then a two-dimensional network of such blocks interconnected with springs. They performed both laboratory experiments and numerical simulations. Remarkably, their model accounts for the redistribution of stresses associated with a sliding event. A sequence of event occurs, much like 'aftershocks' in real earthquakes. This simplified physical system also produces a power law probability density function for event size, in agreement with observations of real earthquakes known as the Gutenberg-Richter law [293, 294]. Such confrontation of physical models to observations is an essential part in the understanding of geophysical processes [295].

Burridge and Knopoff's model is now over 50 years old. It cannot be considered as new! Yet, it highlights very clearly one of the key challenges in understanding

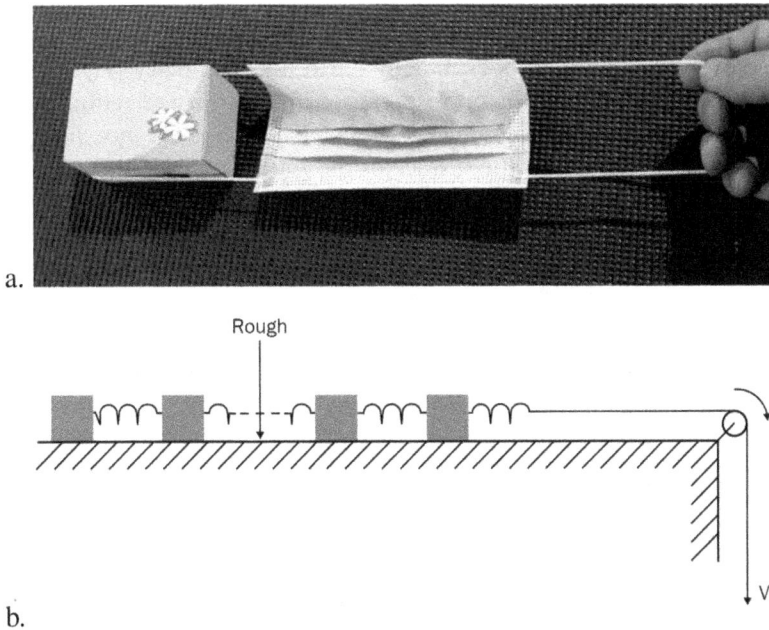

Figure 2.48. A sliding block model for earthquake, highlighting the importance of small-scale contacts in large-scale events. Realised with a single block and low-cost equipment (a) and in the Burridge and Knopoff (1967) one-dimensional chain model (b).

earthquakes: the very complex mechanisms at their source and the importance of small-scale heterogeneities in the rupture process. In real earthquakes, ruptures in a fault with heterogeneous stress and rupture resistance distributions are very complex. The rupture front itself is not given. It can change directions because of small-scale heterogeneities. It will follow regions of strong stress concentration and avoid those with lower stress or higher rupture resistance. This dynamics will essentially depend on small-scale heterogeneities.

This complex behaviour of the fault network, governed by small-scale heterogeneities, hinders an accurate *a priori* estimate of the energy being released. Trying to forecast when a propagating rupture will or will not stop appears nearly impossible. The dynamics are controlled by terribly small-scale irregularities in the fault. In order to be able to estimate potential ground motions that may occur due to rupture on an active fault, one would need a remarkably precise description of the state of the rocks, both in terms of stress and resistance. The resulting complex rupture and earthquake clustering appear essential to achieving a better understanding of earthquakes [296].

Earthquakes can be viewed as terrible 'large-scale events' that appear to depend crucially on the dynamics of very small-scales heterogeneities within the rocks.

'Quakes' are not a specificity of the Earth. They were also recorded on our Moon some 50 years ago by Apollo's seismometers. These are known as 'moonquakes'. On Mars, the first 'marsquake' was recorded by the Insight mission in 2019, thus revealing the inner structure of the planet.

2.8.5 Challenges and opportunities

Only three of the many open questions in Earth and planetary physics have been highlighted here. This, of course, corresponds to an arbitrary selection. Many other issues offer a challenge to researchers. These include but are not limited to: the formation of tectonic plates at the surface of the Earth from mantle convection and the parameters controlling their characteristic sizes; why Earth exhibits plate tectonics but not other planets; the necessary ingredients for magnetic polarity reversals; the formation of banded structures in giant gaseous planets and what sets the depth of the associated flow; how the strong zonal winds in these bands might interact with flow in the electrically conducting interior and affect the magnetic field of those planets. These are only examples of the wide variety of open questions which offer a venue for physical investigation.

I thought nonetheless that these three questions had some merits to be put forward. They highlight both the wide variety of open questions in Earth sciences, ranging from fundamental physical processes (such as the dynamo instability) to phenomena of direct societal concern (such as volcanic hazard and earthquakes) and the impressive variety of scales involved in many geophysical processes.

Those three problems have more in common than may appear at first sight. They all belong the the broad field of non-linear physics. In all three problems, the basic physics is, to a large extent, reasonably well understood. It has been theoretically described, sometimes more than a century ago. The large-scale dynamics, however, is highly non-linear and hardly predictable.

The large-scale non-linear physics can often be efficiently tackled through high-performance numerical simulations. These have made impressive progresses over the last decades with the development of high-performance computing. Increasingly complex physics can be accounted for, with ever more realism and higher resolutions. Yet the problems associated with extreme ranges of scale remain challenging.

In many cases, current numerical simulations will have to rely on some parameterisation of the small scales. Small-scale dynamics, out of reach of current simulations, have to be either ignored or parameterised using ad hoc simplifying assumptions. Such a separation of scale is a very common approach in geophysics, including external geophysics. For example, the microphysics of droplet formation is essential to large-scale atmospheric phenomena such as tropical cyclones and has to be parameterised.

In view of the rapidly increasing computational power, it may be tempting to simply sit and wait for faster computers and increased numerical resolution to close this gap. Yet the disparity of scale is such, in many geophysical problems, that the wait could still be long. Some new physical insight is more likely to yield progress in the years to come. In many cases, numerical work does not outrun theory, but rather complements it or sometimes provides new ideas to be tested theoretically.

It should be stressed that the importance of small-scale physics goes here far beyond the theory of deterministic chaos. It is well established that a tiny perturbation of a non-linear system can yield a drastic change in behaviour. In the above examples, and in many others, however, the small scales are an intrinsic part of the physical mechanism. They cannot be viewed as some small-scale perturbations of a large-scale dynamics. Instead, they are an essential part in the dynamics itself.

The reason for which these open problems involve such a wide variety of scales is probably that, had it not been the case, they would have been solved long ago! Such an extreme range of scale is of course not restricted to geophysics, but it is surprisingly common in Earth sciences.

When facing such a wide range of scale, a physicist will usually rely on separation of scale to handle it. This is the usual theoretical formalism to apprehend such a wide range of phenomena, and it is a very powerful tool. Yet, it usually assumes a 'spectral gap' between the small scales and the large scales or in other words an irrelevance of the intermediate scales. Sadly this approximation is often not justified.

Rather than a simple coupling of the small- and large-scale physics, some new developments in the years to come are probably needed to offer a better description of physics of the 'intermediate scale'. Beyond the standard separation of scale, it becomes increasingly evident that some physical problems are associated with a huge variety of scales, all relevant to understanding the large-scale behaviour. The physics of the 'intermediate-scale' is of course challenging as it necessarily involve a combination of the non-linearities typical of large-scales and of the complex small-scale physics.

Acknowledgments

I would like to thank a few colleagues and friends whom have shared with me, over the years, their understanding of some of the problems mentioned above. In alphabetical order: Claude Jaupart, Keith Moffatt, Dmitri Pissarenko, Andrew Soward, and Albert Tarantola.

References

[1] Director: Mark Levinson 2013 Particle Fever (Documentary) (http://particlefever.com/)
[2] Kajita T and McDonald A B 2015 The Nobel Prize in physics was awarded for the discovery of neutrino oscillations, which shows that neutrinos have mass [online]
[3] Bednyakov A V, Kniehl B A, Pikelner A F and Veretin O L 2015 Stability of the electroweak vacuum: gauge independence and advanced precision *Phys. Rev. Lett.* **115** 201802
[4] Grange J *et al* 2015 Muon (G-2) technical design report 1 arXiv:1501.06858
[5] Rizzo T 2012 The BSM Zoo *Technical Report* (https://www-conf.slac.stanford.edu/ssi/2012/lectures.asp).
[6] Alimena J *et al* 2020 Searching for long-lived particles beyond the standard model at the large hadron collider *J. Phys. G: Nucl. Part. Phys.* **47** 090501
[7] *LHCb Trigger and Online Upgrade Technical Design Report* 5 2014
[8] Data Parking and Data Scouting at the CMS Experiment 9 2012
[9] Yoo J H 2019 The milliQan experiment: search for milli-charged particles at the LHC *PoS* ICHEP2018:520
[10] Lubatti H *et al* 2020 Explore the lifetime frontier with MATHUSLA *JINST* **15** C06026

[11] Ariga A *et al* 2019 FASER: ForwArd Search ExpeRiment at the LHC. 1 arXiv:1901.04468

[12] The LHCf CollaborationAdriani O *et al* 2008 The LHCf detector at the CERN Large Hadron Collider *J. Instrum.* **3** S08006–6

[13] The TOTEM CollaborationAnelli G *et al* 2008 The TOTEM experiment at the CERN large hadron collider *J. Instrum.* **3** S08007–7

[14] Pinfold J *et al* 2009 Technical design report of the MoEDAL experiment. 6

[15] Akerib D S *et al* 2015 LUX-ZEPLIN (LZ) conceptual design report. 9 arXiv:1509.02910

[16] Aprile E *et al* 2005 The XENON dark matter search experiment *New Astron. Rev.* **49** 289–95

[17] European Strategy Group 2020 Update of the European strategy for particle physics *Technical Report, Geneva* 2020 (https://cds.cern.ch/record/2720129)

[18] *The NuPECC Long Range Plan 2017* (http://nupecc.org/pub/lrp17/lrp2017.pdf)

[19] *The NuPECC Long Range Plan Brochure* (https://nupecc.org/pub/lrp17/nupecc_lrp_brochure_2017.pdf)

[20] *The 2015 Long Range Plan for Nuclear Science* (https://science.osti.gov/~/media/np/nsac/pdf/2015LRP/2015_LRPNS_091815.pdf)

[21] Durante M *et al* 2019 All the fun of the FAIR: fundamental physics at the facility for antiproton and ion research *Phys. Scr.* **94** 033001

[22] *JINR Long-Term Development Strategy up to 2030 and Beyond* (http://jinr.ru/posts/jinr-long-term-development-strategy-up-to-2030-and-beyond/)

[23] Gales S *et al* 2018 The extreme light infrastructure—nuclear physics (ELI-NP) facility: new horizons in physics with 10 PW ultra-intense lasers and 20 MeV brilliant gamma beams *Rep. Prog. Phys.* **81** 094301

[24] Savalle A *et al* 2018 SPIRAL2: une sonde de nouvelle génération pour explorer la matière nucléaire *Reflets Phys.* **59** 11

[25] Jokinen A 2014 The Jyväskylä Accelerator Laboratory *Nucl. Phys. News* **24** 4

[26] De Angelis G *et al* 2016 The SPES radioactive ion beam project of LNL: status and perspectives *EPJ Web Conf.* **107** 01001

[27] Calabretta L *et al* 2017 Overview of the future upgrade of the INFN-LNS superconducting cyclotron *Mod. Phys. Lett.* A **32** 1740009 (Special issue on cyclotrons and their applications)

[28] Colonna N *et al* 2018 Neutron physics with accelerators *Prog. Part. Nucl. Phys.* **101** 177

[29] Ashford M *et al* 2020 Exploratory study for the production of Sc beams at the ISOL facility of MYRRHA preliminary thermal investigations *Nucl. Instrum. Methods Phys. Res.* B **463** 244

[30] Warren S *et al* 2020 Offline 2, ISOLDE's target, laser and beams development facility *Nucl. Instrum. Methods Phys. Res.* B **463** 115

[31] Kuchi V *et al* 2020 High efficiency ISOL system to produce neutron deficient short-lived alkali RIBs on GANIL/SPIRAL 1 facility *Nucl. Instrum. Methods Phys. Res.* B **463** 163

[32] Minaya E *et al* 2020 New program for measuring masses of silver isotopes near the $N = 82$ shell closure with MLLTRAP at ALTO *Nucl. Instrum. Methods Phys. Res.* B **463** 315

[33] Kaminki G *et al* 2020 Status of the new fragment separator ACCULINA-2 and first experiments *Nucl. Instrum. Methods Phys. Res.* B **463** 504

[34] Berger N *et al* 2021 The MESA experimental program: a laboratory for precision physics with electron scattering at low energy *Nucl. Phys. News* **31** 5

[35] Facility for Rare Isotope Beams 2021 Rare isotopes aplenty at FRIB *CERN Cour.* https://frib.msu.edu/

[36] Shimizu Y *et al* 2020 Database of radioactive isotope beams produced at the BigRIPS separator *Nucl. Instrum. Methods Phys. Res.* B **463** 158

[37] Marchetto M *et al* 2020 Status of the CANREB high-resolution separator at TRIUMF *Nucl. Instrum. Methods Phys. Res.* B **463** 227

[38] Kim Y J 2020 Current status of experimental facilities at RAON *Nucl. Instrum. Methods Phys. Res.* B **463** 408

[39] Burkert V D 2018 Jefferson Lab at 12 GeV: the science program *Annu. Rev. Nucl. Part. Sci.* **68** 405–28

[40] Ohnishi H *et al* 2020 Hadron Physics at J-Park *Prog. Part. Nucl. Phys.* **113** 103773

[41] Drischl C *et al* 2021 Towards grounding nuclear physics in QCD *Prog. Part. Nucl. Phys.* **121** 103888

[42] Arbuzova A *et al* 2020 On the physics potential to study the gluon content of proton and deuteron at NICA SPD *Prog. Part. Nucl. Phys.* **113** 103858

[43] Peset C 2021 The proton radius (puzzle?) and its relatives *Prog. Part. Nucl. Phys.* **121** 103901

[44] Ireland D G *et al* 2020 Photoproduction reactions and non-strange baryon spectroscopy *Prog. Part. Nucl. Phys.* **111** 103752

[45] Constantinou M *et al* 2021 Parton distributions and lattice-QCD calculations: toward 3D structure *Prog. Part. Nucl. Phys.* **121** 103908

[46] Bartmann W *et al* 2018 The ELENA facility *Philos. Trans. A: Math. Phys. Eng. Sci.* **376** 20170266

[47] Tino G M *et al* 2020 Precision gravity tests and the Einstein equivalence principle *Prog. Part. Nucl. Phys.* **112** 103762

[48] Dong X and Greco V 2019 Heavy quark production and properties of Quark–Gluon plasma *Prog. Part. Nucl. Phys.* **104** 97 and the ALICE Publications https://alice-publications.web.cern.ch/publications

[49] ALICE Collaboration 2020 Unveiling the strong interaction among hadrons at the LHC *Nature* **588** 232–8

[50] Chapon E *et al* 2022 Prospects for quarkonium studies at the high-luminosity LHC *Prog. Part. Nucl. Phys.* **122** 103906

[51] Oganessian Y T *et al* 2017 Superheavy nuclei: from predictions to discovery *Phys. Scr.* **92** 023003

[52] Wilson J N *et al* 2021 Angular momentum generation in nuclear fission *Nature* **590** 566–70

[53] Giovinazzo J *et al* 2021 4D-imaging of drip-line radioactivity by detecting proton emission from Ni pictured with ACTAR TPC *Nat. Commun.* **12** 4805

[54] Mougeot M *et al* 2021 Mass measurements of 99–101In challenge *ab initio* nuclear theory of the nuclide 100Sn *Nat. Phys.* **17** 1099

[55] Leimbach D *et al* 2020 The electron affinity of astatine *Nat. Commun.* **11** 3824

[56] Burgio G F *et al* 2021 Neutron stars and the nuclear equation of state **120** 103879

[57] Arnould M and Goriely S 2020 Astronuclear physics: a tale of the atomic nuclei in the skies *Prog. Part. Nucl. Phys.* **112** 103766

[58] Mossa V *et al* 2020 The baryon density of the Universe from an improved rate of deuterium burning *Nature* **587** 210–3

[59] The Borexino Collaboration 2020 Experimental evidence of neutrinos produced in the CNO fusion cycle in the Sun *Nature* **587** 577–82

[60] Korten W *et al* 2020 Physics opportunities with the advanced gamma tracking array: AGATA *Eur. Phys. J.* A **56** 137

[61] Bracco A *et al* 2021 Gamma spectroscopy with AGATA in its first phases: new insights in nuclear excitations along the nuclear chart *Prog. Part. Nucl. Phys.* **121** 103887 and AGATA Publication list: https://agata.org/

[62] Fallon P *et al* 2016 GRETINA and its early science *Rev. Nucl. Part. Sci.* **66** 321 and GRETINA Publication list; http://gretina.lbl.gov/publications

[63] Milner R G 2021 The electron ion collider: the 21st-century electron microscope for the study of the fundamental structure of matter *MIT Physics Annual 2020 and Nuclear Physics News*

[64] EicC working group 2021 EicC white paper *Front. Phys.* **16** 64701

[65] Kiefer C 2012 *Quantum Gravity* (Oxford: Oxford University Press) 3rd edn

[66] Schlosshauer M 2007 *Decoherence and the Quantum-to-Classical Transition* (Berlin: Springer)

[67] Joos E, Zeh H D, Kiefer C, Giulini D, Kupsch J and Stamatescu I-O 2003 *Decoherence and the Appearance of a Classical World in Quantum Theory* 2nd edn (Berlin: Springer)

[68] DeWitt C 1957 *Proc. of the Conf. on the Role of Gravitation in Physics (University of North Carolina, Chapel Hill, January 18–23, 1957)* WADC Technical Report 57–216 (unpublished). These proceedings have recently been edited inD Rickles and C M DeWitt Edition Open Sources, http://www.edition-open-sources.org/sources/5/

[69] Bose S, Mazumdar A, Morley G W, Ulricht H and Toroš M 2017 Spin entanglement witness for quantum gravity *Phys. Rev. Lett.* **119** 240401 6

[70] Marletto C and Vedral V 2017 Witness gravity's quantum side in the lab *Nature* **547** 156–8 see also the Correspondence in *Nature* 549

[71] Albers M, Kiefer C and Reginatto M 2008 Measurement analysis and quantum gravity *Phys. Rev.* D **78** 064051 17

[72] Hehl F W and Lämmerzahl C 2019 Physical dimensions/units and universal constants: their invariance in special and general relativity *Ann. Phys. (Berlin)* **531** 1800407 10

[73] Carlip S 2001 Quantum gravity: a progress report *Rep. Prog. Phys.* **64** 885–942

[74] Woodard R P 2009 How far are we from the quantum theory of gravity? *Rep. Prog. Phys.* **72** 126002 42

[75] Kiefer C 2013 Conceptual problems in quantum gravity and quantum cosmology *ISRN Math. Phys.* **2013** 509316

[76] Zeh H D 2007 *The Physical Basis of the Direction of Time* 5th edn (Berlin: Springer)

[77] Oriti D (ed) 2009 *Approaches to Quantum Gravity* (Cambridge: Cambridge University Press)

[78] Carr B (ed) 2007 *Universe or Multiverse?* (Cambridge: Cambridge University Press)

[79] Penrose R 2007 *The Road to Reality: A Complete Guide to the Laws of the Universe* (New York: Vintage)

[80] Page D N 2013 Time dependence of Hawking radiation entropy *J. Cosmol. Astropart. Phys.* **9** 28 28

[81] Feynman R 1990 *The Character of Physical Law* (Cambridge, MA: MIT Press)

[82] Carlip S 2019 *General Relativity: A Concise Introduction* (Oxford: Oxford University Press)

[83] Bertone G and Hooper D 2018 History of dark matter *Rev. Mod. Phys.* **90** 045002

[84] Zyla P A *et al* 2020 Review of particle physics *Prog. Theor. Exp. Phys.* **2020** 083C01

[85] Zwicky F 1933 Die Rotverschiebung von extragalaktischen Nebeln *Helv. Phys. Acta* **6** 110 127

[86] Milgrom M 1983 A modification of the Newtonian dynamics as a possible alternative to the hidden mass hypothesis *Astrophys. J.* **270** 365 370

[87] Einstein A 1916 The foundation of the general theory of relativity *Ann. Phys.* **49** 769–822

[88] Gaillard M K, Grannis P D and Sciulli F J 1999 The standard model of particle physics *Rev. Mod. Phys.* **71** S96–S111

[89] Blumenthal G R, Faber S M, Primack J R and Rees M J 1984 Formation of galaxies and large scale structure with cold dark matter *Nature* **311** 517–25

[90] Audren B, Lesgourgues J, Mangano G, Serpico P D and Tram T 2014 Strongest model-independent bound on the lifetime of dark matter *J. Cosmol. Astropart. Phys.* **12** 028

[91] Rubin V C and Ford W K 1970 Rotation of the Andromeda nebula from a spectroscopic survey of emission regions *Astrophys. J.* **159** 379–403

[92] White M J, Scott D and Silk J 1994 Anisotropies in the cosmic microwave background *Ann. Rev. Astron. Astrophys.* **32** 319–70

[93] Wayne Hu and Dodelson S 2002 Cosmic microwave background anisotropies *Ann. Rev. Astron. Astrophys.* **40** 171–216

[94] Ade P A R *et al* 2016 Planck 2015 results. XIII. Cosmological parameters *Astron. Astrophys.* **594** A13

[95] Lifshitz E 1946 Republication of: on the gravitational stability of the expanding universe *J. Phys. (USSR)* **10** 116

[96] Vogelsberger M, Marinacci F, Torrey P and Puchwein E 2020 Cosmological simulations of galaxy formation *Nature Rev. Phys.* **2** 42–66

[97] Alpher R A, Follin J W and Herman R C 1953 Physical conditions in the initial stages of the expanding universe *Phys. Rev.* **92** 1347–61

[98] Wagoner R V, Fowler W A and Hoyle F 1967 On the synthesis of elements at very high temperatures *Astrophys. J.* **148** 3–49

[99] Cerdeno D G and Green. A M 2010 Direct detection of WIMPs arXiv:1002.1912 [astro-ph. CO] 2

[100] Bullock J S and Boylan-Kolchin M 2017 Small-scale challenges to the Λ CDM paradigm *Ann. Rev. Astron. Astrophys.* **55** 343–87

[101] Baer H, Choi K-Y, Kim J E and Roszkowski L 2015 Dark matter production in the early Universe: beyond the thermal WIMP paradigm *Phys. Rept.* **555** 1 60

[102] Steigman G and Turner M S 1985 Cosmological constraints on the properties of weakly interacting massive particles *Nucl. Phys.* B **253** 375–86

[103] Hu W, Rennan B and Gruzinov A 2000 Cold and fuzzy dark matter *Phys. Rev. Lett.* **85** 1158–61

[104] Hooper D, Leane R K, Tsai Y-D, Wegsman S and Witte. S J 2020 A systematic study of hidden sector dark matter:application to the gamma-ray and antiproton excesses *J. High Energy Phys.* **07** 163

[105] Griest K 1991 Galactic microlensing as a method of detecting massive compact halo objects *Astrophys. J.* **366** 412–21

[106] Villanueva-Domingo P, Mena O and Palomares-Ruiz S 2021 A brief review on primordial black holes as dark matter *Front. Astron. Space Sci.* **8** 87

[107] Carr B, Kohri K, Sendouda Y and Yokoyama J 2021 Constraints on primordial black holes *Rept. Prog. Phys.* **84** 116902

[108] Bekenstein J D 2005 Modified gravity vs dark matter: relativistic theory for MOND *PoS* JHW2004:012

[109] McGaugh S S 2015 A tale of two paradigms: the mutual incommensurability of Λ CDM and MOND *Can. J. Phys.* **93** 250–9

[110] Gaskins J M 2016 A review of indirect searches for particle dark matter *Contemp. Phys.* **57** 496–525

[111] Kahlhoefer F 2017 Review of LHC dark matter searches *Int. J. Mod. Phys.* A **32** 1730006

[112] Teresa Marrodán UndagoitiaRauch L 2016 Dark matter direct-detection experiments *J. Phys.* G Nucl. Phys. **43** 013001

[113] Silk J *et al* 2010 *Particle Dark Matter: Observations, Models and Searches* (Cambridge: Cambridge University Press)

[114] Lewin J D and Smith P F 1996 Review of mathematics, numerical factors, and corrections for dark matter experiments based on elastic nuclear recoil *Astropart. Phys.* **6** 87 112

[115] Agrawal P *et al* 2021 Feebly-interacting particles: FIPs 2020 workshop report *Eur. Phys. J.* C **81** 1015

[116] Marsh D J E 2016 Axion cosmology *Phys. Rept.* **643** 1 79

[117] Abbott B P *et al* 2016 Observation of gravitational waves from a binary black hole merger *Phys. Rev. Lett.* **116** 061102

[118] Bertone G *et al* 202 Gravitational wave probes of dark matter: challenges and opportunities *SciPost Phys. Core* **3** 007

[119] Arbey A and Mahmoudi F 2021 Dark matter and the early Universe: a review *Prog. Part. Nucl. Phys.* **119** 103865

[120] Carney D *et al* 2021 Mechanical quantum sensing in the search for dark matter *Quantum Sci. Technol.* **6** 024002

[121] Donzelli M and Carpineti M 2017 The underground laboratories global research infra-structures *Case Studies Report, GSO—Group of Senior Officials on Global Research Infrastructures, August, 2017. Presented to the G7 Science Ministers' Meeting (Turin, 27–28 September)*

[122] Angloher G *et al* 2019 Exploring with NUCLEUS at the Chooz nuclear power plant *Eur. Phys. J.* C **79** 1018

[123] Bowen M and Huber P 2020 Reactor neutrino applications and coherent elastic neutrino nucleus scattering *Phys. Rev.* D **102** 053008

[124] Dodelson S and Schmidt F 2020 *Modern Cosmology* (London: Academic)

[125] van Albada T S, Bahcall J N, Begeman K and Sancisi R 1985 The Distribution of dark matter in the spiral galaxy NGC-3198 *Astrophys. J.* **295** 305 313

[126] Springel V, White S D M, Frenk C S, Navarro J F, Jenkins A, Vogelsberger M, Wang J, Ludlow A and Helmi A 2008 A blueprint for detecting supersymmetric dark matter in the Galactic halo arXiv:0809.0894 [astro-ph] 9

[127] Einstein A 1916 Die Grundlage der allgemeinen Relativitätstheorie *Ann. Phys. (Berlin)* **49** 769

[128] Einstein A 1916 Sitzungsber *Preuss. Akad. Wiss. Berlin* **1** 688

[129] Einstein A 1918 Sitzungsberichte *Preuss. Akad. Wiss. Berlin* **1** 154

[130] Abbott B P *et al* (LIGO Scientific Collaboration and Virgo Collaboration) 2016 *Phys. Rev. Lett.* **116** 061102

[131] Abbott R *et al* 2020 (LIGO Scientific, Virgo) (*Preprint*)

[132] Aasi J *et al* 2015 Characterization of the LIGO detectors during their sixth science run *Class. Quantum Grav.* **32** 074001

[133] Acernese F *et al* 2015 Advanced Virgo: a second-generation interferometric gravitational wave detector *Class. Quantum Grav.* **32** 024001

[134] Abbott B P *et al* 2020 Prospects for observing and localizing gravitational-wave transients with Advanced LIGO, Advanced Virgo and KAGRA *Living Rev. Rel.* **23** 3 (KAGRA, LIGO Scientific, Virgo)

[135] Affeldt C *et al* 2014 Advanced techniques in GEO 600 *Class. Quantum Grav.* **31** 224002

[136] Somiya K 2012 Detector configuration of KAGRA—the Japanese cryogenic gravitational-wave detector *Class. Quantum Grav.* **29** 124007

[137] Unnikrishnan C S 2013 IndIGO and LIGO-India: scope and plans for gravitational wave research and precision metrology in India *Int. J. Mod. Phys.* D **22** 1341010

[138] Punturo M *et al* 2010 The Einstein Telescope: a third-generation gravitational wave observatory *Class. Quantum Grav.* **27** 194002

[139] Reitze D *et al* 2019 Cosmic explorer: the U.S. contribution to gravitational-wave astronomy beyond LIGO *Bull. Am. Astron. Soc.* **51** 141 (*Preprint*)

[140] Amaro-Seoane P *et al* 2017 Laser interferometer space antenna arXiv:1702.00786 [astro-ph. IM] *ArXiv e-prints* (*Preprint*)

[141] Arzoumanian Z *et al* 2020 The NANOGrav 12.5 yr data set: search for an isotropic stochastic gravitational-wave background *Astrophys. J. Lett.* **905** L34 (*Preprint*)

[142] Smits R, Kramer M, Stappers B, Lorimer D, Cordes J and Faulkner A 2009 Pulsar searches and timing with the square kilometre array *Astron. Astrophys.* **493** 1161–70

[143] Lazio T 2013 The Square Kilometre Array pulsar timing array *Class. Quantum Grav.* **30** 4011

[144] Stappers B W, Keane E F, Kramer M, Possenti A and Stairs I H 2018 The prospects of pulsar timing with new-generation radio telescopes and the Square Kilometre Array *Phil. Trans. R. Soc.* A **376** 20170293

[145] Taylor J and Weisberg J 1989 Further experimental tests of relativistic gravity using the binary pulsar PSR 1913+16 *Astrophys. J.* **345** 434

[146] Abbott B P *et al* 2016 GW150914: the advanced LIGO detectors in the era of first discoveries *Phys. Rev. Lett.* **116** 131103 (LIGO Scientific, Virgo) (*Preprint*)

[147] Huterer D and Shafer D L 2018 Dark energy two decades after: observables, probes, consistency tests *Rept. Prog. Phys.* **81** 016901 [arXiv:1709.01091 [astro-ph.CO]]

[148] Zhao G B *et al* 2017 Dynamical dark energy in light of the latest observations *Nat. Astron* **1** 627–32 [arXiv:1701.08165 [astro-ph.CO]]

[149] Saridakis E N *et al* [CANTATA] Modified gravity and cosmology: an update by the CANTATA network [arXiv:2105.12582 [gr-qc]]

[150] Fender R and Belloni T 2012 Stellar-mass black holes and ultraluminous X-ray sources *Science* **337** 540 (*Preprint*)

[151] Abbott B P *et al* 2019 GWTC-1: a gravitational-wave transient catalog of compact binary mergers observed by LIGO and Virgo during the first and second observing runs *Phys. Rev.* **X9** 031040 (LIGO Scientific, Virgo)(*Preprint*)

[152] Abbott R *et al* 2020 GW190521: a binary black hole merger with a total mass of 150 $M_\odot$ *Phys. Rev. Lett.* **125** 101102 (LIGO Scientific, Virgo)(*Preprint*)

[153] Abbott R *et al* 2020 Properties and astrophysical implications of the 150 M$_\odot$ binary black hole merger GW190521 *Astrophys. J. Lett.* **900** L13 (LIGO Scientific, Virgo)(*Preprint*)

[154] Ghez A M *et al* 2008 Measuring distance and properties of the Milky Way's central supermassive black hole with stellar orbits *Astrophys. J.* **689** 1044–62 (*Preprint*)

[155] Genzel R, Eisenhauer F and Gillessen S 2010 The Galactic Center massive black hole and nuclear star cluster *Rev. Mod. Phys.* **82** 3121–95 (*Preprint*)

[156] Akiyama K *et al* 2019 First M87 event horizon telescope results. I. The shadow of the supermassive black hole *Astrophys. J.* **875** L1 (Event Horizon Telescope)(*Preprint*)

[157] Schwarzschild K 1916 On the gravitational field of a mass point according to Einstein's theory *Sitzungsber. Preuss. Akad. Wiss. Berlin (Berlin)* **189** 96

[158] Penrose R 1965 Gravitational collapse and space-time singularities *Phys. Rev. Lett.* **14** 57–9

[159] Penrose R 1969 Gravitational collapse: the role of general relativity *Riv. Nuovo Cim.* **1** 252

[160] Wald R M 1999 *Gravitational Collapse and Cosmic Censorship* (Dordrecht: Springer) pp 69–86

[161] Israel W 1967 Event horizons in static vacuum space-times *Phys. Rev.* **164** 1776–9

[162] Carter B 1971 Axisymmetric black hole has only two degrees of freedom *Phys. Rev. Lett.* **26** 331–3

[163] Bekenstein J D 1972 Black holes and the second law *Lett. Nuovo Cim.* **4** 737–40

[164] Hawking S W 1974 Black hole explosions? *Nature* **248** 30–1

[165] Abbott B PLIGO Scientific Collaboration and Virgo Collaboration *et al* 2016 GW151226: observation of gravitational waves from a 22-solar-mass binary black hole coalescence *Phys. Rev. Lett.* **116** 241103

[166] Abbott B PLIGO Scientific Collaboration and Virgo Collaboration *et al* 2016 Binary black hole mergers in the first advanced LIGO observing run *Phys. Rev. X* **6** 041015

[167] Abbott B PLIGO Scientific Collaboration and Virgo Collaboration *et al* 2017 GW170814: a three-detector observation of gravitational waves from a binary black hole coalescence *Phys. Rev. Lett.* **119** 141101

[168] Abbott B PLIGO Scientific Collaboration and Virgo Collaboration *et al* 2017 GW170817: observation of gravitational waves from a binary neutron star inspiral *Phys. Rev. Lett.* **119** 161101

[169] Abbott B PVirgo, Fermi-GBM, INTEGRAL, LIGO Scientific *et al* 2017 Gravitational waves and gamma-rays from a binary neutron star merger: GW170817 and GRB 170817A *Astrophys. J.* **848** L13

[170] Goldstein A *et al* 2017 An ordinary short gamma-ray burst with extraordinary implications: *Fermi*-GBM detection of GRB 170817A *Astrophys. J.* **848** L14 (*Preprint*)

[171] Abbott B P *et al* 2017 Multi-messenger observations of a binary neutron star merger *Astrophys. J.* **848** L12

[172] Christensen N and Meyer R 1998 Markov chain Monte Carlo methods for Bayesian gravitational radiation data analysis *Phys. Rev.* D **58** 082001

[173] Veitch J *et al* 2015 Parameter estimation for compact binaries with ground-based gravitational-wave observations using the LALInference software library *Phys. Rev.* D **91** 042003 (*Preprint*)

[174] Abbott RLIGO Scientific, Virgo *et al* 2020 Population properties of compact objects from the second LIGO-Virgo gravitational-wave transient catalog arXiv:2010.14533 [astro-ph. HE] (*Preprint*)

[175] Belczynski K *et al* 2020 Evolutionary roads leading to low effective spins, high black hole masses, and O1/O2 rates for LIGO/Virgo binary black holes *Astron. Astrophys.* **636** A104 (*Preprint*)

[176] Ivanova N *et al* 2013 Common envelope evolution: where we stand and how we can move forward *Astron. Astrophys. Rev.* **21** 59 (*Preprint*)

[177] Spera M and Mapelli M 2017 Very massive stars, pair-instability supernovae and intermediate-mass black holes with the SEVN code *Mon. Not. Roy. Astron. Soc.* **470** 4739–49 (*Preprint*)

[178] Farmer R, Renzo M, de Mink S E, Marchant P and Justham S 2019 Mind the gap: the location of the lower edge of the pair-instability supernova black hole mass gap *Astrophys. J.* **887** 53

[179] Rodriguez C L, Zevin M, Pankow C, Kalogera V and Rasio F A 2016 Illuminating black hole binary formation channels with spins in advanced LIGO *Astrophys. J. Lett.* **832** L2 (*Preprint*)

[180] Vitale S, Lynch R, Sturani R and Graff P 2017 Use of gravitational waves to probe the formation channels of compact binaries *Class. Quant. Grav.* **34** 03LT01 (*Preprint*)

[181] Carr B J and Hawking S W 1974 Black holes in the early Universe *Mon. Not. R. Astron. Soc.* **168** 399–415 (*Preprint*)

[182] Carr B, Kühnel F and Sandstad M 2016 Primordial black holes as dark matter *Phys. Rev. D* **94** 083504

[183] Bird S, Cholis I, Muñoz J B, Ali-Hamoud Y, Kamionkowski M, Kovetz E D, Raccanelli A and Riess A G 2016 Did LIGO detect dark matter? *Phys. Rev. Lett.* **116** 201301 (*Preprint*)

[184] Sasaki M, Suyama T, Tanaka T and Yokoyama S 2016 Primordial black hole scenario for the gravitational-wave event GW150914 *Phys. Rev. Lett.* **117** 061101
Erratum 2018 Primordial black hole scenario for the gravitational-wave event GW150914 *Phys. Rev. Lett.* **121** 059901 (*Preprint*)

[185] Jenkins A C and Sakellariadou M 2020 Primordial black holes from cusp collapse on cosmic strings arXiv:2006.16249 [astro-ph.CO] (*Preprint*)

[186] Boehm C, Kobakhidze A, O'Hare C A J, Picker Z S C and Sakellariadou M 2020 Eliminating the LIGO bounds on primordial black hole dark matter arXiv:2008.10743 [astro-ph.CO] (*Preprint*)

[187] Fernandez N and Profumo S 2019 Unraveling the origin of black holes from effective spin measurements with LIGO-Virgo *J. Cosmol. Astropart. Phys.* **08** 022 (*Preprint*)

[188] Abbott B P LIGO Scientific, Virgo *et al* 2019 Search for subsolar mass ultracompact binaries in advanced LIGO's second observing run *Phys. Rev. Lett.* **123** 161102 (*Preprint*)

[189] Knox L, Christensen N and Skordis C 2001 The age of the Universe and the cosmological constant determined from cosmic microwave background anisotropy measurements *Astrophys. J. Lett.* **563** L95–8 (*Preprint*)

[190] Kulier A, Ostriker J P, Natarajan P, Lackner C N and Cen R 2015 Understanding black hole mass assembly via accretion and mergers at late times in cosmological simulations *Astrophys. J.* **799** 178 (*Preprint*)

[191] Begelman M C, Volonteri M and Rees M J 2006 Formation of supermassive black holes by direct collapse in pre-galactic haloes *Mon. Not. Roy. Astron. Soc.* **370** 289–98 (*Preprint*)

[192] Banados E *et al* 2018 An 800-million-solar-mass black hole in a significantly neutral Universe at a redshift of 7.5 *Nature* **553** 473–6 (*Preprint*)

[193] Wang F *et al* 2021 A luminous quasar at redshift 7.642 *Astrophys. J. Lett.* **907** L1 (*Preprint*)

[194] Greene J E, Strader J and Ho L C 2020 Intermediate-mass black holes *Annu. Rev. Astron. Astrophys.* **58** 257–312 (*Preprint*)

[195] Abbott B PLIGO Scientific, Virgo *et al* 2020 A guide to LIGO–Virgo detector noise and extraction of transient gravitational-wave signals *Class. Quant. Grav.* **37** 055002

[196] Abbott B PLIGO Scientific Collaboration, Virgo Collaboration *et al* 2017 GW170104: observation of a 50-solar-mass binary black hole coalescence at redshift 0.2 *Phys. Rev. Lett.* **118** 221101

[197] Unnikrishnan C S 2013 IndIGO and LIGO-India: scope and plans for gravitational wave research and precision metrology in India *Int. J. Mod. Phys.* D **22** 1341010 (*Preprint*)

[198] Abbott B PLIGO Scientific Collaboration, Virgo Collaboration *et al* 2017 GW170817: observation of gravitational waves from a binary neutron star inspiral *Phys. Rev. Lett.* **119** 161101 (*Preprint*)

[199] Abbott B PLIGO Scientific Collaboration, Virgo Collaboration, Fermi-GBM, INTEGRAL *et al* 2017 Gravitational waves and gamma-rays from a binary neutron star merger: GW170817 and GRB 170817A *Astrophys. J.* **848** L13 (*Preprint*)

[200] Savchenko V *et al* 2017 *INTEGRAL* detection of the first prompt gamma-ray signal coincident with the gravitational-wave event GW170817 *Astrophys. J.* **848** L15 (*Preprint*)

[201] Goldstein A *et al* 2017 An ordinary short gamma-ray burst with extraordinary implications: *Fermi*-GBM detection of GRB 170817A *Astrophys. J. Lett.* **848** L14 (*Preprint*)

[202] Abbott B PLIGO Scientific Collaboration, Virgo Collaboration *et al* 2019 Tests of general relativity with the binary black hole signals from the LIGO-Virgo catalog GWTC-1 *Phys. Rev.* **D100** 104036 (*Preprint*)

[203] Bekenstein J D and Mukhanov V F 1995 Spectroscopy of the quantum black hole *Phys. Lett.* B **360** 7–12 (*Preprint*)

[204] Cardoso V, Foit V F and Kleban M 2019 Gravitational wave echoes from black hole area quantization *J. Cosmol. Astropart. Phys.* **2019** 6

[205] Abbott B PLIGO Scientific, Virgo *et al* 2016 Observation of gravitational waves from a binary black hole merger *Phys. Rev. Lett.* **116** 061102 (*Preprint*)

[206] Abedi J, Dykaar H and Afshordi N 2017 Echoes from the abyss: tentative evidence for Planck-scale structure at black hole horizons *Phys. Rev.* D **96** 082004 (*Preprint*)

[207] Westerweck J, Nielsen A B, Fischer-Birnholtz O, Cabero M, Capano C, Dent T, Krishnan B, Meadors G and Nitz A H 2018 Low significance of evidence for black hole echoes in gravitational wave data *Phys. Rev.* D **97** 124037

[208] Schutz B F 1986 Determining the Hubble constant from gravitational wave observations *Nature* **323** 310–1

[209] Nissanke S, Holz D E, Hughes S A, Dalal N and Sievers J L 2010 Exploring short gamma-ray bursts as gravitational-wave standard sirens *Astrophys. J.* **725** 496–514 (*Preprint*)

[210] Christensen N, Meyer R, Knox L and Luey B 2001 Bayesian methods for cosmological parameter estimation from cosmic microwave background measurements *Class. Quantum Grav.* **18** 2677–88

[211] Riess A G *et al* 2016 A 2.4% determination of the local value of the Hubble constant *Astrophys. J.* **826** 56 (*Preprint*)

[212] Abbott B PLIGO Scientific, Virgo, 1M2H, Dark Energy Camera GW-E, DES, DLT40, Las Cumbres Observatory, VINROUGE, MASTER *et al* 2017 Hungary rewards highly cited scientists with bonus grants *Nature* **551** 425–6 (*Preprint*)

[213] Ade P A R *et al* 2016 *Planck* 2015 results *Astron. Astrophys.* **594** A13 (*Preprint*)

[214] Abbott B PKAGRA, LIGO Scientific Collaboration and Virgo Collaboration *et al* 2019 Prospects for observing and localizing gravitational-wave transients with Advanced LIGO, Advanced Virgo and **KAGRA** *Living Rev. Relativ.* **23** 3 (*Preprint*)

[215] https://gw-openscience.org

[216] Armano M *et al* 2016 Sub-femto-*g* free fall for space-based gravitational wave observatories: LISA pathfinder results *Phys. Rev. Lett.* **116** 231101

[217] Armano M *et al* 2018 Beyond the required LISA free-fall performance: new LISA pathfinder results down to 20 μHz *Phys. Rev. Lett.* **120** 061101

[218] Kamionkowski M and Kovetz E D 2016 The quest for B modes from inflationary gravitational waves *Annu. Rev. Astron. Astrophys.* **54** 227–69 (*Preprint*)

[219] Maggiore M *et al* 2020 Science case for the Einstein telescope *J. Cosmol. Astropart. Phys.* **03** 050 (*Preprint*)

[220] Christensen N 2019 Stochastic gravitational wave backgrounds *Rep. Prog. Phys.* **82** 016903

[221] Biscoveanu S, Talbot C, Thrane E and Smith R 2020 Measuring the primordial gravitational-wave background in the presence of astrophysical foregrounds *Phys. Rev. Lett.* **125** 241101

[222] Martinovic K, Meyers P M, Sakellariadou M and Christensen N 2021 Simultaneous estimation of astrophysical and cosmological stochastic gravitational-wave backgrounds with terrestrial detectors *Phys. Rev.* D **103** 043023

[223] Regimbau T, Evans M, Christensen N, Katsavounidis E, Sathyaprakash B and Vitale S 2017 Digging deeper: observing primordial gravitational waves below the binary-black-hole-produced stochastic background *Phys. Rev. Lett.* **118** 151105

[224] Sachdev S, Regimbau T and Sathyaprakash B 2020 Subtracting compact binary foreground sources to reveal primordial gravitational-wave backgrounds *Phys. Rev.* D **102** 024051 (*Preprint*)

[225] Sesana A 2016 Prospects for multiband gravitational-wave astronomy after GW150914 *Phys. Rev. Lett.* **116** 231102 (*Preprint*)

[226] Adams M R and Cornish N J 2014 Detecting a stochastic gravitational wave background in the presence of a galactic foreground and instrument noise *Phys. Rev.* D **89** 022001 (*Preprint*)

[227] Boileau G, Christensen N, Meyer R and Cornish N J 2020 Spectral separation of the stochastic gravitational-wave background for LISA: observing both cosmological and astrophysical backgrounds arXiv:2011.05055 (*Preprint*) arXiv e-prints

[228] Mingarelli C M F 2019 Improving binary millisecond pulsar distances with Gaia arXiv:1812.06262 [astro-ph.IM] (*Preprint*)

[229] Buldgen G, Salmon S and Noels A 2019 Progress in global helioseismology: a new light on the solar modelling problem and its implications for solar-like stars *Front. Astron. Space Sci.* **6** 42

[230] Christensen-Dalsgaard J 2021 Solar structure and evolution *Living Rev. Sol. Phys.* **18** 2

[231] Basu S and Antia H M 1997 Seismic measurement of the depth of the solar convection zone *Mon. Not. R. Astron. Soc.* **287** 189–98

[232] Basu S and Antia H M 1995 Helium abundance in the solar envelope *Mon. Not. R. Astron. Soc.* **276** 1402–8

[233] Vorontsov S V, Baturin V A, Ayukov S V and Gryaznov V K 2013 Helioseismic calibration of the equation of state and chemical composition in the solar convective envelope *Mon. Not. R. Astron. Soc.* **430** 1636–52

[234] Vorontsov S V, Baturin V A and Pamiatnykh A A 1991 Seismological measurement of solar helium abundance *Nature* **349** 49–51

[235] Grevesse N and Noels A 1992 Cosmic abundances of the elements *Origin and Evolution of the Elements: Proceedings of a Symposium in Honour of H. Reeves held in Paris, June 22–25, 1993* ed N Prantzos, E Vangioni-Flam, M Casse, N Prantzos, E Vangioni-Flam and M Casse (Cambridge: Cambridge University Press) p 14

[236] Christensen-Dalsgaard J *et al* 1996 The current state of solar modeling *Science* **272** 1286

[237] Asplund M, Grevesse N and Sauval A J 2005 The solar chemical composition *Cosmic Abundances as Records of Stellar Evolution and Nucleosynthesis, of Astronomical Society of the Pacific Conf. Series* ed III Barnes, G Thomas and F N Bash (San Francisco, CA: Astronomical Society of the Pacific) vol 336 p 25

[238] Asplund M, Grevesse N, Jacques Sauval A and Scott. P 2009 The chemical composition of the Sun *Annu. Rev. Astron. Astrophys.* **47** 481–522

[239] Basu S and Antia H M 2008 Helioseismology and solar abundances *Phys. Rep.* **457** 217–83

[240] Pezzotti C, Eggenberger P, Buldgen G, Meynet G, Bourrier V and Mordasini C 2021 Revisiting Kepler-444. II. Rotational, orbital, and high-energy fluxes evolution of the system *Astron. Astrophys.* **650** A108

[241] Serenelli A M, Basu S, Ferguson J W and Asplund M 2009 New solar composition: the problem with solar models revisited *Astrophys. Lett.* **705** L123–7

[242] Pinsonneault M H, Kawaler S D, Sofia S and Demarque P 1989 Evolutionary models of the rotating sun *Astrophys. J.* **338** 424–52

[243] Charbonneau P and MacGregor K B 1993 Angular momentum transport in magnetized stellar radiative zones. II. The solar spin-down *Astrophys. J.* **417** 762

[244] Gough D O and McIntyre M E 1998 Inevitability of a magnetic field in the Sun's radiative interior *Nature* **394** 755–7

[245] Mestel L 1953 Rotation and stellar evolution *Mon. Not. R. Astron. Soc.* **113** 716

[246] Mestel L and Weiss N O 1987 Magnetic fields and non-uniform rotation in stellar radiative zones *Mon. Not. R. Astron. Soc.* **226** 123–35

[247] Mosser B *et al* 2012 Spin down of the core rotation in red giants *Astron. Astrophys.* **548** A10

[248] Spada F, Lanzafame A C and Lanza A F 2010 A semi-analytic approach to angular momentum transport in stellar radiative interiors *Mon. Not. R. Astron. Soc.* **404** 660

[249] Braithwaite J and Spruit H C 2017 Magnetic fields in non-convective regions of stars *R. Soc. Open Sci.* **4** 160271

[250] Brun A S and Zahn J-P 2006 Magnetic confinement of the solar tachocline *Astron. Astrophys.* **457** 665–74

[251] Tayler R J 1973 The adiabatic stability of stars containing magnetic fields-I. Toroidal fields *Mon. Not. R. Astron. Soc.* **161** 365

[252] Spruit H C 2002 Dynamo action by differential rotation in a stably stratified stellar interior *Astron. Astrophys.* **381** 923–32

[253] Eggenberger P, Maeder A and Meynet G 2005 Stellar evolution with rotation and magnetic fields. IV. The solar rotation profile *Astron. Astrophys.* **440** L9–L12

[254] Eggenberger P, Buldgen G and Salmon S J A J 2019 Rotation rate of the solar core as a key constraint to magnetic angular momentum transport in stellar interiors *Astron. Astrophys.* **626** L1

[255] Salmon S, Buldgen G, Noels A, Eggenberger P, Scuflaire R and Meynet G 2021 Standard solar models: perspective from updated solar neutrino fluxes and gravity-mode period spacing *Astron. Astrophys.* **651** A106

[256] Miglio A *et al* 2013 Galactic archaeology: mapping and dating stellar populations with asteroseismology of red-giant stars *Mon. Not. R. Astron. Soc.* **429** 423–8

[257] Miglio A, Montalbán J, Baudin F, Eggenberger P, Noels A, Hekker S, De Ridder J, Weiss W and Baglin A 2009 Probing populations of red giants in the galactic disk with CoRoT *Astron. Astrophys.* **503** L21–4

[258] Miglio A *et al* 2021 HAYDN *Exp. Astron.* **51** 963–1001

[259] Arnett W D, Meakin C, Viallet M, Campbell S W, Lattanzio J C and Mocák M 2015 Beyond mixing-length theory: a step toward 321D *Astrophys. J.* **809** 30

[260] Aerts C 2021 Probing the interior physics of stars through asteroseismology *Rev. Mod. Phys.* **93** 015001

[261] Beck P G *et al* 2012 Fast core rotation in red-giant stars as revealed by gravity-dominated mixed modes *Nature* **481** 55–7

[262] Deheuvels S *et al* 2012 Seismic evidence for a rapidly rotating core in a lower-giant-branch star observed with kepler *Astrophys. J.* **756** 19

[263] Gehan C, Mosser B, Michel E, Samadi R and Kallinger T 2018 Core rotation braking on the red giant branch for various mass ranges *Astron. Astrophys.* **616** A24

[264] Ceillier T, Eggenberger P, Garca R A and Mathis S 2013 Understanding angular momentum transport in red giants: the case of KIC 7341231 *Astron. Astrophys.* **555** A54

[265] Eggenberger P, Montalbán J and Miglio A 2012 Angular momentum transport in stellar interiors constrained by rotational splittings of mixed modes in red giants *Astron. Astrophys.* **544** L4

[266] Marques J P *et al* 2013 Seismic diagnostics for transport of angular momentum in stars. I. Rotational splittings from the pre-main sequence to the red-giant branch *Astron. Astrophys.* **549** A74

[267] Eggenberger P *et al* 2019 Asteroseismology of evolved stars to constrain the internal transport of angular momentum. I. Efficiency of transport during the subgiant phase *Astron. Astrophys.* **621** A66

[268] Eggenberger P *et al* 2017 Constraining the efficiency of angular momentum transport with asteroseismology of red giants: the effect of stellar mass *Astron. Astrophys.* **599** A18

[269] Mayor M and Queloz D 1995 A Jupiter-mass companion to a solar-type star *Nature* **378** 355–9

[270] Ricker G R *et al* 2015 Transiting Exoplanet Survey Satellite (TESS) *J. Astron. Teles., Instrum., Syst.* **1** 014003

[271] Rüdiger G and Kitchatinov L L 1996 The internal solar rotation in its spin-down history *Astrophys. J.* **466** 1078

[272] Privitera G, Meynet G, Eggenberger P, Vidotto A A, Villaver E and Bianda M 2016 Star–planet interactions. I. Stellar rotation and planetary orbits *Astron. Astrophys.* **591** A45

[273] Privitera G, Meynet G, Eggenberger P, Vidotto A A, Villaver E and Bianda M 2016 Star–planet interactions. II. Is planet engulfment the origin of fast rotating red giants? *Astron. Astrophys.* **593** A128

[274] Rao S, Meynet G, Eggenberger P, Haemmerlé L, Privitera G, Georgy C, Ekström S and Mordasini C 2018 Star–planet interactions. V. Dynamical and equilibrium tides in convective zones *Astron. Astrophys.* **618** A18

[275] Lopez E D and Fortney J J 2013 The role of core mass in controlling evaporation: the Kepler radius distribution and the Kepler-36 density dichotomy *Astrophys. J.* **776** 2

[276] Owen J E and Lai D 2018 Photoevaporation and high-eccentricity migration created the sub-Jovian desert *Mon. Not. R. Astron. Soc.* **479** 5012–21

[277] Johnstone C P, Bartel M and Güdel M 2021 The active lives of stars: a complete description of the rotation and XUV evolution of F, G, K, and M dwarfs *Astrophys. J.* **649** A96

[278] Tu L, Johnstone C P, Güdel M and Lammer H 2015 The extreme ultraviolet and X-ray Sun in time: high-energy evolutionary tracks of a solar-like star *Astrophys. J.* **577** L3

[279] Johnstone C P, Güdel M, Stökl A, Lammer H, Tu L, Kislyakova K G, Lüftinger T, Odert P, Erkaev N V and Dorfi E A 2015 The evolution of stellar rotation and the hydrogen atmospheres of habitable-zone terrestrial planets *Astrophys. Lett.* **815** L12

[280] Kubyshkina D *et al* 2019 The Kepler-11 system: evolution of the stellar high-energy emission and initial planetary atmospheric mass fractions *Astron. Astrophys.* **632** A65

[281] Rao S, Pezzotti C, Meynet G, Eggenberger P, Buldgen G, Mordasini C, Bourrier V, Ekström S and Georgy C 2021 Star–planet interactions VI. Tides, stellar activity, and planetary evaporation *Astron. Astrophys.* **651** A50

[282] Turcotte D and Schubert J 2014 *Geodynamics* (London: Cambridge University Press)

[283] Moffatt K and Dormy E 2019 *Self-Exciting Fluid Dynamos* (London: Cambridge University Press)

[284] Braginskii S I 1964 Kinematic models of the Earth's hydromagnetic dynamo *Geomagn. Aeron.* **4** 572–83

[285] Steenbeck M, Krause F and Radler K-H 1966 Berechnung der mittleren Lorentz-Feldstärke für ein elektrisch leitendes Medium in turbulenter, durch Coriolis-Kräfte beeinflusster Bewegung *Z. Naturforsch.* **21a** 369–76 (English translation: Roberts and Stix (1971) pp 29–47)

[286] Glatzmaier G A and Roberts P H 1995 A three-dimensional convective dynamo solution with rotating and finitely conducting inner core and mantle *Phys. Earth Planet. Inter.* **91** 63–75

[287] Pétrélis F, Fauve S, Dormy E and Valet J-P 2009 Simple mechanism for reversals of Earth's magnetic field *Phys. Rev. Lett.* **102** 144503

[288] Dormy E 2006 The origin of the Earth's magnetic field: fundamental or environmental research? *Europhys. News* **37** 2

[289] Kaminski E and Jaupart C 1998 The size distribution of pyroclasts and the fragmentation sequence in explosive volcanic eruptions *J. Geophys. Res.* **103** 29759–79

[290] Michaud-Dubuy A, Carazzo G, Kaminski E and Girault F 2002 A revisit of the role of gas entrapment on the stability conditions of explosive volcanic columns *J. Volcanol. Geotherm. Res.* **357** 349–61

[291] Geshi N, Browning J and Kusumoto S 2020 Magmatic overpressures, volatile exsolution and potential explosivity of fissure eruptions inferred via dike aspect ratios *Sci. Rep.* **10** 9406

[292] D'Incecco P, Filiberto J, López I, Gorinov D and Idunn Mons K G 2021 Evidence for ongoing volcano-tectonic activity and atmospheric implications on Venus *Planet. Sci. J.* **2** 215

[293] Burridge R and Knopoff L 1967 Model and theoretical seismicity *Bull. Seismol. Soc. Am.* **57** 341

[294] Clancy I and Corcoran D 2006 Burridge-Knopoff model: exploration of dynamic phases *Phys. Rev. E* **73** 046115

[295] Pétrélis F, Chanard K, Schubnel A and Hatano T 2023 Earthquake magnitude distribution and aftershocks: A statistical geometry explanation *Phys. Rev. E* **107** 034132

[296] Morell K D, Styron R, Stirlin M, Griffin J, Archuleta R and Onur T 2020 Seismic hazard analyses from geologic and geomorphic data: current and future challenges *Tectonics* **39** e2018TC005365

IOP Publishing

EPS Grand Challenges
Physics for Society in the Horizon 2050

Mairi Sakellariadou, Claudia-Elisabeth Wulz, Kees van Der Beek, Felix Ritort, Bart van Tiggelen, Ralph Assmann, Giulio Cerullo, Luisa Cifarelli, Carlos Hidalgo, Felicia Barbato, Christian Beck, Christophe Rossel and Luc van Dyck

Chapter 3

Matter and waves

Marco Baldovin, Antoine Browaeys, José María De Teresa, Claudia Draxl, Frédéric Druon, Giacomo Gradenigo, Jean-Jacques Greffet, Franck Lépine, Jan Lüning, Lucia Reining, Pascal Salières, Pierre Seneor, Luis Silva, Thomas Tschentscher, Kees van Der Beek, Antje Vollmer and Angelo Vulpiani

3.1 Introduction

Kees van Der Beek[1]
[1]Institut Polytechnique de Paris, Paris, France

The scales of length, time, and energy that are intermediate between the infinitely small and the infinitely large define the world we live in and that we experience. The relevant fundamental forces that act on these scales are, for nearly all phenomena, the gravitational, and the electromagnetic forces. At the energy scales that characterise our world, the relevant physical approaches take atoms, ions, and electrons as fundamental building blocks of matter, and photons as those of light. This chapter considers how these constituents, in the unimaginable vastness of their numbers and the fantastic variety of their possible arrangements, lead to the astounding complexity of our world and the dazzling phenomena it harbours. The world we experience is also the world on which we may intervene. When done in a controlled manner, this intervention—or experiment—belongs to the realm of science and technology. But even when we do not strive to control, our actions are still controlled by the physical workings of the world's constituent building blocks and their interactions.

When one comes to think of it, an atom itself is a hugely complex system. Mathematics abdicates when faced with the problem of providing an exact description of more than three interacting objects; all atoms are in this situation: quarks make up hadrons, hadrons make up the atomic nuclei, and nuclei and, for most elements, numerous electrons make up the atom. The internal organisation of

doi:10.1088/978-0-7503-6342-6ch3 3-1

the atom is the result of 'correlations'. The simultaneous presence of several and often many interacting particles lead to the organisation of matter. This is even truer when one assembles atoms into molecules, atoms, and molecules into liquids and solids, and liquid, and solid components into complex systems. What we witness is the result of this internal, physical organisation, the forces that objects around us exert on each other and the phenomena we witness as the net result of this organisation. The myriad particles involved and the manner in which they can interact and made to interact vow for this complexity.

Section 3.2, 'Quantum many-body systems and emerging phenomena', treats this organisation of matter and of the excitations the constituents matter can undergo when stimulated by light or other, impinging, particles, how we can understand it, and how we may describe it. It describes how highly complex behaviour can emerge from the interaction of many constituent particles, and how the quantum mechanical nature of our world determines the nature of the objects that constitute it, even though we may not at all be aware of such. It also describes when a quantum description is needed, that is, when the conditions of observation are such that coherent interactions and propagation of light and particles are guaranteed. When the quantum mechanical coherence of light and matter is disturbed in the process at hand, we may resort to what has become known as the 'classical description'.

Section 3.3, 'The search for new materials', details how our—even limited—understanding of the organisation of matter can help us fashion materials and tools needed to tackle the world's challenges and problems. While in principle there is nothing new—did not the smithies and rock-hewers of prehistoric and early historic times draw on their experience to fashion stone and metal tools?—the formidable scientific progress made in the last two centuries allows for mankind to imagine and make materials according to need, in a sustainable and controlled manner, taking into account availability of material and energetic resources, and designing those material assemblies and those processes that are most appropriate for a given application from the many millions of materials that nature would allow. Nowhere does the force of science have an impact as great as when matter and light are manipulated on the smallest of scales—the nanometer (i.e., 10^{-9} m) or below. Section 3.4, 'Manipulating photons and atoms: photonics and nanophysics', explains how today's and tomorrow's technologies allow us to dive into the smallest length scales defining materials and the systems they can build and, indeed, back into the quantum realm. It explains how interactions between particles manifest themselves differently depending on the length scale we are working on, how the quantum mechanical nature of these interactions can be brought out, and how both can be put to work to define entirely new tools and systems.

What goes for matter also goes for light, and what goes for space also goes for time. Section 3.5 on 'Extreme light' provides an account of the astounding advances made in fashioning extremely bright light beams, extremely short light pulses, or a combination of both. Extreme brightness allows us to examine the structure of matter, its organisation, and its response to excitation in the very finest details—for it is those details that most often give away the fundamentals at stake in determining the nature of the studied object in the first place. Ultra-short and bright pulses allow one to make 'movies of matter'. Much as stroboscopic illumination of moving beings can reveal the gestures in motion, pico- and femtosecond illumination of matter unveil the dance of electrons in matter: how matter is excited, how matter reacts, how matter transforms. Extremely spectacular results with high-stake implications in many fields (chemistry, biology, medicine, etc) are to be expected here. Moreover, being able to intervene at the rhythm of matter itself allows us to reorganise it, yielding yet another tool for the creation of new (quantum) states of matter designed as tools to face the challenges of humankind.

Finally, section 3.6 'Systems with numerous degrees of freedom' reflects on the multiplicity of interactions in systems that are not necessarily characterized atomic length and timescales and quantum coherence. This opens the way for the most complex of organisations of matter as we know it: life itself, to be treated in chapter 4.

3.2 Quantum many-body systems and emergent phenomena

Lucia Reining[1]
[1]Laboratoire des Solides Irradiés, CNRS, École Polytechnique, CEA/DRF/ IRAMIS, Institut Polytechnique de Paris, F-91128 Palaiseau, France European Theoretical Spectroscopy Facility (ETSF)

The development of theoretical and experimental approaches to understanding many-body effects in both soft and hard condensed matter, but also in dilute matter, atomic, and molecular systems, as well as in atomic nuclei, remains one of the key challenges in physics today. A proper description of the effects of interaction and inter-particle correlation is key to the understanding of materials properties, their quantum states, as well as the outcome of experiments that probe them, modern spectroscopies, scattering, or collisions. Beyond phenomenology, the adequate description of quantum many-body phenomena will allow the understanding of emergent states of matter and new physics associated with these. Recent advances have concerned novel types of charge and spin order, new magnetic and super-conductive systems (see section 3.3), as well as multiferroic systems, the switching between different orders by the tuning of a parameter (e.g., pressure, density, electronic excitation by radiation) but also the absence of order in a number of key systems. The advent of new states of matter will open the way to novel applications in the field of electronics, thermodynamics, sensors, biology (see section 3.3.2.3.4), health and medicine, and it will provide us with new understanding that may go well beyond condensed matter. In this section, we will introduce the main concepts and challenges and explain their consequences on materials science.

3.2.1 General overview

The quantum many-body problem has been one of the most challenging topics of physics for decades. In spite of much progress, all evidence indicates that it will still remain a challenge for a while, both intellectually and because of its potential impact on applications. It turns an old subject, condensed matter theory, into an exciting adventure of unpredictable outcome and potentially highest impact. If we want to understand what is measured in laboratory spectroscopy experiments, if we want to draw maximum benefit from large instruments such as synchrotrons, if we want to use known materials for new purposes or design new materials to solve long-standing problems, we must adapt our thinking to the many possible stories written by a system that is composed of numerous interacting particles in a quantum state.

In order to appreciate the depth and breadth of the problem and its implications, let us first analyse its ingredients. What is *the quantum many-body problem?*, or, to be more precise, what is *the quantum interacting many-body problem?* and *why does it lead to new emergent phenomena?* We will explain each term, and immediately point out why each of the ingredients is so important, both for the whole and for particular applications. Further reading can be found in the references; we cite texts that give an overview, that conveniently illustrate a particular aspect, or that are particularly accessible, whereas we do not claim to cite all, and not even a significant fraction of, the first or the most forefront works in the field.

3.2.2 Quantum

Our mind is classical, which means that we describe the world in terms of our observations: an object is in a given place, or it moves from one place to another with a given speed that may change with time. This makes it difficult to understand or even simply to describe the quantum world, where an object is in a given state, or where it changes state with time. For simplicity, let us think that our problem is static. We may, for example, have two positive charges. Let us say, two protons (depicted by the blue balls in figure 3.1) that create a potential (depicted by the blue lines) which is attractive for negative charges. In the classical world, if we place a negative charge (red ball in the upper row) directly into one minimum of the potential, it will stay there. The two possibilities (charge on the right and charge on the left) are shown on the left and right side, respectively. If we place the charge not in the minimum, but somewhere in that region, it will fall into a minimum, and then it will oscillate around that minimum by converting potential into kinetic energy, if there is not too much friction. In the quantum world (treating for the moment the potential as given), we place the negative charge into a quantum state. If we put the charge into an eigenstate, which is a special state of the system, it will remain there forever. We could also put it into another state that is a mixture, a superposition, of eigenstates: in that case, the wavefunction will oscillate. So, the eigenstates play the role of the minimum positions, and other states somehow play the role of places away from the minima—somehow, because the states do not at all, in general, correspond to a position. This is depicted in the bottom row of figure 3.1, where the red shaded regions represent the eigenstates. To be precise, the red line is the

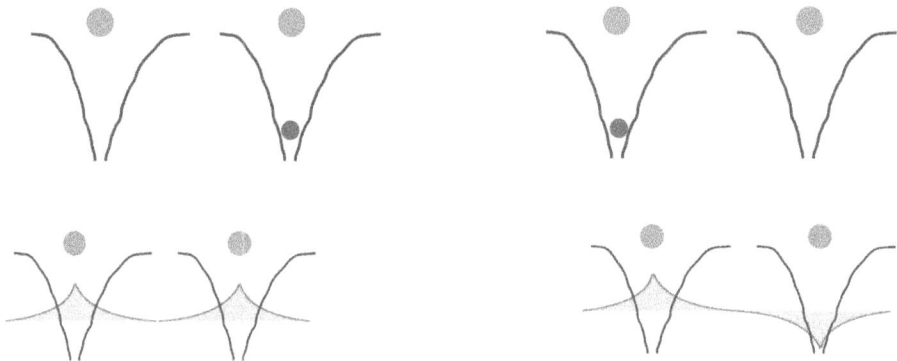

Figure 3.1. Schematic view of a system with two protons and one electron, a model for the H_2^+ molecule. Light blue balls, the two positive charges (e.g., protons). Blue lines: attractive potential created by the protons. Red ball in the upper row: the electron, understood as a classical particle. It can be captured by the left or right potential minimum with the same probability. Red line with shaded area in the lower row: the quantum state represents a probability amplitude. The probability to find the electron on the right or left side is the square of the wavefunction, and it is the same in the two quantum states, since the shaded area indicates the regions of positive and negative sign.

wavefunction of a given state, and there are two of them (on the left and right, respectively). The wavefunction is a probability amplitude, and its square gives the probability to find the particle in a given place. Both states are extended over the whole region: we cannot see such thing as a well-defined position. There is, instead, a well-defined shape of the wavefunction.

Of course, one may object that looking at the picture it is clear that we can build a state, superposition of the two eigenstates, which privileges a position: this state is the simple sum (appropriately normalized) of the wavefunction in the left panel and the one in the right panel. The sum cancels the portion of the wavefunction in the right minimum, so now the square of this new wavefunction is localized on the left: our particle is confined on the left side. However, because the superposition is not an eigenstate, there is oscillatory behavior in which the electron explores the possible realisations of the superposition state, and some time later we will find the charge on the right side. Moreover, the localization is not perfect, meaning that even at the start we do not know *exactly* where the charge is, and at intermediate times, the wavefunction is again delocalized over both potential wells.

This simple example illustrates the difference between the information of classical and of quantum mechanics: classical mechanics tells in which position you are, quantum mechanics tells in which state you are. It is unambiguous about the state-information, but when you ask the question about position, there are only probabilities.

We now have a choice: we can accept this fact and follow the mathematical rules given by quantum mechanics to calculate the properties of the system using wavefunctions. Or we can try to project the quantum world onto our classical world in a way that allows us to interpret and maybe even predict what is going on. Usually one will do both, but here we will avoid mathematics and try to give a feeling of how one may look at the situation. Of course, many books have been written about the interpretation of quantum mechanics, and the debates are fascinating; here our purpose is just to highlight certain aspects. In particular, if we want to use a classical way to look at the problem—for example, we would like to talk about positions, we will have to admit that not knowing where a particle is forces us to ask the question *what if?* For example, suppose we add a second negative charge to our model system. If it gets close to the first one, there will be a strong Coulomb repulsion energy. But since we do not have a well-defined and ever-lasting position, we have to consider different scenarios, where the two charges come close, or where they do not. The outcome of an experiment will be the sum of all possible stories of what could happen, so in this case, we should measure at least two peaks in a spectroscopy experiment: one at an energy where the particles meet, one where they do not. This is a first challenge: to take into account in our reasoning, and in the theory that we build, all or at least the most important, of the possible stories that a system can write.

Why is this important? The importance of this fact is not limited to esoteric questions of quantum mechanics. It determines our daily life. As an example, take two helium atoms. They both have a nucleus in the center and a spherical wavefunction of two electrons around it. These are two charge neutral and perfectly symmetric objects, and in principle they should not interact (i.e., they should not 'see' each other via the Coulomb interaction). However, this consideration interprets the square of the spherical wavefunction as a classical charge distribution that would stay the same forever. This classical interpretation (which is used, for example, in one of the first approximations to the interacting electron problem, the one by Hartree [1]) tries to give charges a well-defined position, and it does not take into account the probability aspect. To respect quantum mechanics, we have to write the possible stories of these objects, meaning we have to consider all possible positions of the two electrons of one atom (with a probability that is spherically symmetric) and the reaction of the electrons of the other atom to this particular distribution. Now, as soon as the distribution that we consider is not symmetric, the other atom starts to see a dipole or higher multipoles, so it will itself polarize. This fact, schematized in the left panel of figure 3.2, leads to an attractive potential, so the atoms bind together. The effect is called the van der Waals dispersion interaction. Note that the dipole–dipole interaction or the coupling of oscillations also exist in classical physics, but of course only when the classical picture predicts that electro-static dipoles or oscillating charges are present. The van der Waals dispersion interaction is a quantum effect, because it is due to the coupling of fluctuations that exist only on the quantum level.

One may not care whether helium atoms can bind or not (although it is interesting to know that solid helium can exist), but one should care about this effect in general. The weak van der Waals dispersion interaction is often important in biology, for example, where it is in particular an ingredient of protein folding. The left panel of figure 3.2 shows results of a computer simulation of an alanine helix structure [2]. In the computer, one can switch off the van der Waals dispersion interaction artificially. The simulation shows that after some time, the resulting conformation is completely different.

This is but one example. One may imagine how rich the world becomes by writing all the possible stories. 'Quantum' or 'Quantum Materials' has become a hot topic

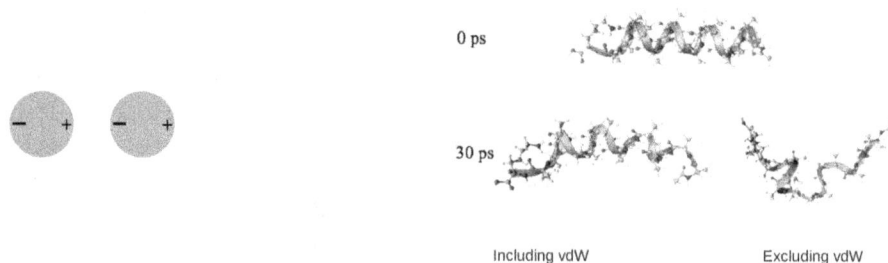

Figure 3.2. Quantum mechanical aspects of binding. Left: schematic picture of fluctuating dipoles in neutral atoms or molecules, leading to weak bonding. This is a pure quantum effect. Right: alanine helix structures in alanine polypeptides. The α-helix structure is favoured by the weak van der Waals dispersion interaction. In experiment, it is stable up to 725 K. The figure shows snapshots from computer simulations using density functional molecular dynamics, at the start (top) and after 30 ps (bottom). At 700 K inclusion of the van der Waals contribution preserves the α-helix (bottom left), whereas the structure opens when this contribution is neglected (bottom right). Results from reference [2] with permission from APS, figure in right panel from [3] with permission from Cambridge University Press.

over the last years [4]. The terms are maybe not always perfectly well defined, but they are essentially saying that one is interested in materials that have very particular properties stemming from the fact that quantum mechanics does not fix classical observables, but quantum mechanical states.

3.2.3 Interacting

In the discussion above, we have already anticipated that particles interact. This is true in the classical and in the quantum world. We can see it all around us. Take a drop falling on a water surface: as depicted on the left side of figure 3.3, the water reacts and concentric waves form. This means that the water 'sees' the drop; in other words, the water molecules that form the drop and that form the water interact. Both objects, the drop and the water, are affected by the interaction, and the result of the drop falling on the surface is different from having simply more water at the end.

Another good example is the coupled motion of two pendula. A single pendulum has its well-known regular left-right oscillatory motion. When we couple two identical pendula by a weak spring, as schematized on the right side of figure 3.3, each of the two pendula oscillates with an amplitude that changes over time, and there are moments where one of the two even stands still. The amplitudes of the right and left pendula are out of phase, so when the left one stands still the right one moves strongly, and vice versa.

Why is this important? The fact that two or more objects in interaction behave differently than just the sum of the two is of utmost importance. Take chemistry: it is not rare to hear people asking the question *carbon and oxygen molecules are not toxic, so how can carbon monoxide be toxic?* To scientists, this may seem a naive question, but it is not naive when asked by someone who is not used to thinking about effects of interactions. We have to imagine pictures like the one in figure 3.4, where carbon atoms (blue), oxygen molecules (yellow), and carbon monoxyde (violet/yellow gradients) are schematized. We have to explain that two oxygen atoms influence each other, so if we break the molecule into two atoms there is a first change, and then, oxygen influences the carbon atom and carbon influences oxygen, so together they form a new and completely different object. The fact that a composed object is potentially completely different from the sum of its constituents is sometimes phrased as *emergent behavior*, because new properties seem to emerge without obvious explanation. It implies, for example, that one has to be extremely careful when evaluating the impact of chemicals, such as drugs or pesticides, because

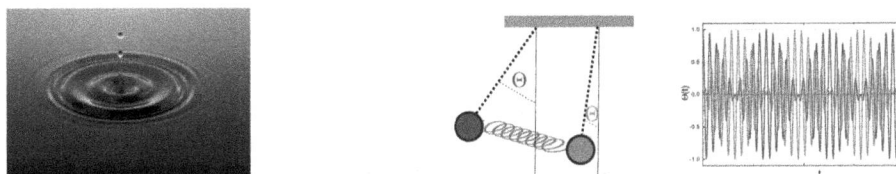

Figure 3.3. Left: Water drop falling on water surface. Because of the interaction between water molecules, the small drop has a visible effect over a large range. Right: Two coupled pendula. The motion of the coupled object is more complicated than the motion of each pendulum alone.

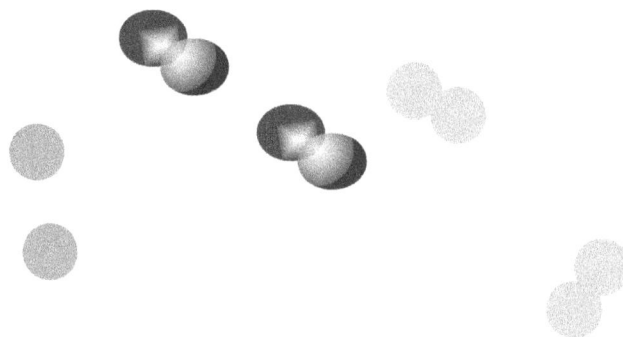

Figure 3.4. Neither carbon nor oxygen molecules are toxic, but CO is: interactions can change everything.

investigating them in a situation where they are isolated from other drugs or chemicals and from the environment where they will be used may lead to dramatically wrong conclusions, positive or negative.

It also means that experiments are in general difficult to interpret. For example, in a photoemission experiment ultraviolet (UV) light or x-rays are sent on a material and their energy absorbed by electrons, which are then ejected from the material and their energy measured in a detector [5]. In a naive picture one would think that each electron in the materials has its own energy, so we would measure a spectrum with sharp peaks. Instead, such spectra are in general very complex, with broad peaks and many extra features, called satellites and incoherent background, that cannot be explained by considering the material as a sum of independent nuclei and electrons [3]. Figure 3.5 shows the photoemission spectrum of a very simple metal, aluminum [6]. The horizontal axis represents the momentum of the electrons in the crystal, and the vertical axis, energy. The relatively sharp lines close to the top are the energy bands that one would naively expect for electrons sitting in the allowed energy levels of aluminum. These can be well described by relatively simple band structure calculations, where interaction effects only lead to quantitative, not qualitative, changes. All the rest of the spectrum is exclusively due to interactions, either between electrons and nuclei that vibrate around their ideal crystal position because of temperature, or, most importantly, among the electrons. In more complex systems, the situation is correspondingly even more intricate, and a close collaboration between experiment and theory is often necessary to extract sound information from the measurements. Photoemission is one of the key techniques used to study interaction effects in materials, because the spectra display these effects in such a dramatic way.

3.2.4 Quantum and interacting

If we adopt for a moment the point of view that quantum mechanics leads us to write all possible stories, and if we take into account at the same time that particles are interacting, an additional difficulty arises: to describe the state of a system, we cannot treat particles separately. The *what if?*—aspect that was worked out in the first example of H_2^+ implies that while the state of the system with all its particles is expressed by the wavefunction, this wavefunction cannot be simply the product of

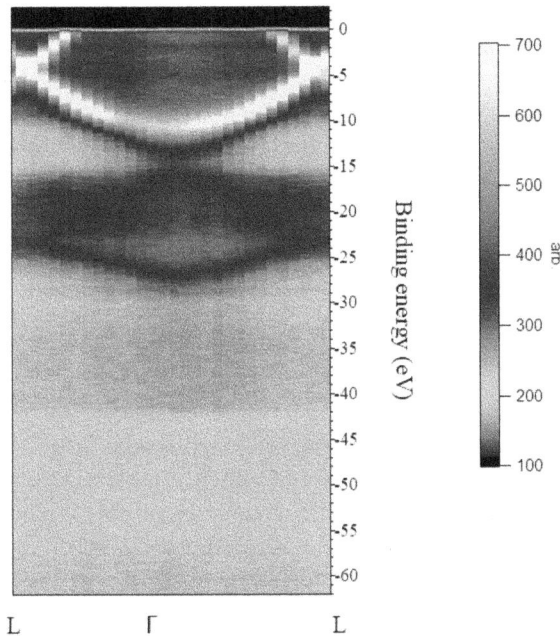

Figure 3.5. Angle-resolved photoemission spectrum of bulk aluminum. The relatively sharp lines close to the top are the energy bands that one would naively expect for electrons sitting in the allowed energy levels of aluminum. All the rest is exclusively due to interactions. Results from [6], figure courtesy of F Sirotti.

single-particle wavefunctions. If it were, then the probability to realize a certain event would be the product of independent probabilities from the state of each single particle, but asking *what if?* means we must have conditional probabilities. In other words, in classical mechanics the state of a static system is well described by telling separately where each particle is, but in a quantum many-body system, we cannot consider particles one by one, or mathematically speaking, the wavefunction is not a product of the wavefunctions of individual particles.

Why is this important? Due to this fact, strictly speaking, even when they are very far away particles are not independent. This *entanglement* leads to some of the most discussed aspects of quantum mechanics and to some of the most recent promising applications, in particular, in quantum computing [7]. It is also the major difficulty when one tries to develop efficient numerical approaches to predict materials properties, because treating all particles at the same time—writing all possible stories of all particles, with all their *what if?*—is simply impossible [8]. The box summarizes ways to deal with the problem that are used in practice; more information and references can be found in [3, 9].

3.2.4.1 What do we do in practice?
Experimentally,

- *Scattering experiments* such as x-ray diffraction or neutron scattering detect patterns of charge and spin.

- *Spectroscopy* probes the electronic structure and response. Prototypical examples are direct and inverse photoemission and scanning tunneling spectroscopies that give insight about the density of electronic states as a function of energy, optical measurements such as reflection or ellipsometry experiments probing electronic excitations in the visible and UV range, or inelastic x-ray scattering and electron energy loss measurements, which can access excitations that are visible with higher momentum. Another way to make otherwise undetectable excitations visible are resonant spectroscopies such as resonant photoemission or resonant inelastic x-ray scattering. Spectroscopic experiments can be resolved in time, up to attosecond resolution.

Theoretically,

- *The solution of models*, such as the Hubbard model or the Anderson impurity model, yields precious insight about general tendencies and can also predict parameter ranges where interesting phenomena may occur.
- *Wavefunction-based approaches* approximate the many-body wavefunction while still yielding a good description of observables. Prototypical examples are stochastic (Quantum Monte Carlo) approaches and expansions in finite basis sets such as Configuration Interaction.
- *Density functional theory* (DFT) relies on the fact that the density of a system in its lowest energy (ground) state contains in principle all the information necessary to predict ground state observables. First principles DFT calculations are today widely used. They are very successful, in spite of the fact that in most cases it is not known *in which way* observables depend on the density; in other words, the functionals are unknown and must be approximated. DFT has an extension to the time-dependent case (TDDFT). In linear and higher-order response, this also yields compact expressions to describe spectroscopy.
- *Approaches based on Green's functions*, similarly to DFT, avoid wavefunctions by expressing observables in terms of more compact quantities. One-body and higher-order Green's functions are more complicated objects than the density, but much more compact than the many-body wavefunctions. Also in this case, most functionals are unknown. Major approximations are perturbation expansions in the Coulomb interaction (Many-Body Perturbation Theory) and expansions around a local problem, such as Dynamical Mean Field Theory. Green's functions describe the propagation of particles. They are therefore helpful for finding approximations that are intuitive, because they naturally write the possible stories of the system.

3.2.5 Many-body

In the examples above we were looking at systems composed of few objects, such as two hydrogen atoms or two pendula, and others composed of many molecules or electrons and nuclei, such as liquid water or bulk aluminum. The drop-on-the-water observation is stunning, if you think about it. Liquid water is a disordered ensemble of a huge number of molecules, but when the drop hits the surface, these molecules somehow coordinate and give rise to a beautifully symmetric pattern. Such a collective behaviour of a many-body system is not determined if you just have all the detailed information of every one of its particles; it is determined if, to this, one adds the interaction and information such as the average distance *between* particles. Collective effects are very important for the understanding of materials, and they also happen at the quantum level. Probably the most well-studied collective effect of the many-electron system is the plasmon. Plasmons are collective oscillations of an electron gas and have been first studied theoretically for a homogeneous distribution of electrons [10, 11], but they also occur in real, inhomogeneous materials. The plasmon frequency is the resonance frequency of an electron gas, and it depends on its density and shape. Real materials show regions of different densities and can therefore have plasmon spectra with more than one resonance. Moreover, as in classical mechanics the double, triple, etc, frequency can be excited [12]. Plasmons are measured as sharp spectral peaks in experiments such as inelastic x-ray scattering, where an x-ray scatters from the sample but loses energy and momentum to excitations of the material [13], or electron energy loss spectroscopy, where an electron beam (e.g., in an electron microscope) is transmitted or reflected, again losing energy and momentum by exciting the material [14].

Why is this important? Because collective motion gives rise to sharp spectral features corresponding to long-lived excitations, they are interesting for applications. Plasmons, for example, have created an own field of research called plasmonics [15]. It is mostly concerned with the nanoscale, where it considers resonances of metallic particles (see, e.g., the example of photography below) or at metal-dielectric interfaces. It aims to establish the generation, manipulation, and detection of optical signals. Potential applications are found in optical communication, microscopy, and more general imaging, for example, of biological materials.

The response of a system at resonance is strong. This is used in string instruments, where the body and the air inside the sound box have characteristic resonance frequencies that determine the sound of the instrument. The response can be so strong that it becomes destructive. This may be a problem, for example, for the stability of buildings or vehicles, but it can also be used on purpose, for example, to destroy tumors. More generally, the violent response of materials at certain frequencies allows one to change their properties in a targeted way, for example, using visible light.

A nice example is the first colour photography by E Becquerel. His technique did not meet great success. The colours could be captured (see his first attempt to capture the colours of the rainbow in figure 3.6), but the process could not be stopped, so the photography has to be kept in the dark to avoid further evolution. Nevertheless, the approach he proposed is interesting, and the origin of the colours in this material has long been debated, not least because a better understanding can result in better conservation of the fragile pictures. A collaboration of researchers and the National Museum for Natural History in Paris has found that the explanation is given by plasmon resonances [16]: Becquerel's material is composed

Figure 3.6. *Solar spectrum*, Edmond Becquerel 1848, Musée Nicéphore Niépce, Chalons-sur-Saône (France).

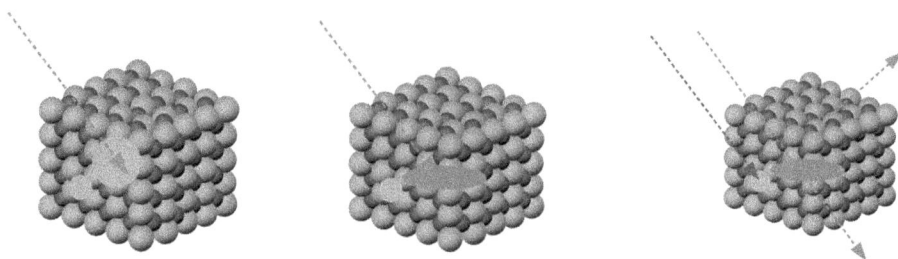

Figure 3.7. Schematic view of a photochromic material. When light of a certain wavelength is absorbed, the absorbing object may undergo changes. Subsequently, light of that same wavelength will be transmitted and reflected.

of silver chloride crystals that contain small, nano-sized, silver particles of various sizes. This is schematically represented by the crystal (green/blue) and two different nanoparticles (light blue circles) in the leftmost part of figure 3.7. According to the size of the nanoparticles and their shape, their plasmon resonance frequencies differ. Therefore, they absorb light of different colours. In the beginning of the process, all resonance frequencies are present and the visible light is completely absorbed: the material appears to be black, whereas pure silver chloride would be white. Now suppose you shine red light (represented by the red arrow) on it: this will be absorbed by those nanoparticles that have a resonance frequency corresponding to the wavelength of red light. The violent reaction of the electron system perturbs these nanoparticles to the point that they explode or, at least, change shape (middle panel). Visible light (represented by the red and blue arrows in the right panel) is now absorbed at all frequencies except red, since the corresponding particles no longer exist. Red light is transmitted and reflected, and therefore the object is seen as red in the places where the nanoparticles have absorbed red light. Such changes form the principle of photochromic materials, which change colour when exposed to light. In many applications, such as automatically darkening sun-glasses, one needs a reversible process. For example, some organic molecules are photochromic because they respond to UV light by changing shape and subsequently absorb light also in the visible range. When there is no longer any UV radiation, they return to their original shape.

In large (bulk) materials, typical plasmon frequencies of the weakly bound outer valence electrons lie well above the range of visible light. However, another kind of many-body resonance exists at lower frequencies, often in the visible range: these are excitons. As in the case of plasmons, one cannot understand excitons within an independent-particle picture, in which electrons absorb photons and transit between single-particle states independently. Instead, the whole system of the order of 10^{23} electrons is promoted from its initial state, which, as pointed out earlier, corresponds to a wavefunction describing all particles at the same time, to a new many-body state. Whereas it is impossible to give a simple picture of the initial and final states, one can more easily describe the final state *with respect to* the initial one. This is schematized in figure 3.8: an electron (red) is displaced and leaves a hole (white) behind. The electron carries a negative charge, and the hole, or 'missing' electron, acts correspondingly like a positive charge, so it creates an attractive potential for the electron. If the interaction is strong enough, the electron is captured in this potential and we have a bound electron–hole pair called exciton. The whole system participates to this situation, in particular, all electrons contribute to screen the Coulomb interaction between electron and hole, so the interaction responsible for the exciton binding depends on the material. The pair of a positive and a negative charge has some similarity with a hydrogen atom. Therefore, bound excitons create very characteristic features in optical absorption experiments, sometimes extremely similar to the Rydberg series of optical absorption of the hydrogen atom. The right panel of figure 3.8 shows an example.

Excitons are important in many applications, such as solar cells, or photo-catalysis, as illustrated in figure 3.9. One can easily imagine why: on one side, when an electron and a hole travel together they form an overall neutral object, which reduces the interaction with the lattice compared to a charge travelling alone. On the other side, when one wants to separate the positive and negative charges, as is the case, for example, in photovoltaics, strong exciton binding can be an obstacle. Other

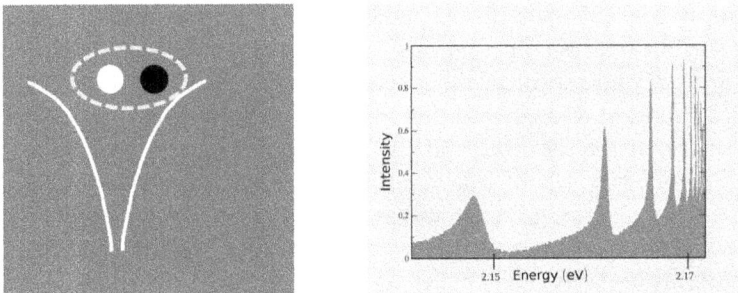

Figure 3.8. An exciton is a bound electron–hole pair. Left, schematic representation of an exciton. Right, schematic view on the absorption spectrum of copper oxide Cu_2O. The typical hydrogen-like series of bound exciton peaks is observed. Figure representing results published in [17].

Figure 3.9. Excitons are important in photovoltaics (left illustration) or photocatalysis (right illustration), where electron–hole pairs are created by light absorption and where the transport, separation of charges, and migration of electrons and holes to the desired places is crucial. Theory can help to get insight into these complex processes.

examples of applications where excitons play a role are UV excitonic lasers, tuneable UV photodetectors, or LEDs.

3.2.6 Tiny differences, large effects: symmetry breaking

One of the reasons why the waves in figure 3.3 do not surprise us is because they are concentric, meaning, they are perfectly symmetric. We expect that in the absence of a clear reason no particular direction is privileged. In quantum mechanics, this translates into the fact that a Hamiltonian (the quantum mechanical operator describing the system) with certain symmetries will also lead to properties with that symmetry. For example, in figure 3.1 the potential is left-right symmetric, and the two eigenstates are, respectively, symmetric and anti-symmetric, so the square of the wavefunction is symmetric in both cases. We would expect that this is true independently of the distance between the two atoms, and this is indeed so on paper. In reality, the two hydrogen atoms are not alone, but surrounded by other objects—or even the vacuum—in a continuously changing universe, which leads to fluctuations of the potential. These may be very small, and their effect negligible. However, while we separate the protons more and more, the energy difference between the two eigenstates of the molecule becomes smaller and smaller, and eventually smaller than the fluctuations. At this stage, a fluctuation that breaks the symmetry will win and localize the electron on the side with lower energy. Even when the fluctuation ceases and the molecule experiences again a symmetric potential, such that the localized state of the electron is no longer an eigenstate of the system, it will stay where it is for a long time, because the oscillation of the composite quantum state that follows has a period that is inversely proportional to the energy difference between the eigenstates. For an observer who is not aware of the tiny fluctuations, this will appear as if the system had decided to break the symmetry spontaneously. This *spontaneous symmetry breaking* is therefore a consequence of degeneracy, which means energy differences between eigenstates becoming smaller than typical fluctuations. After all, this is not too mysterious:

even in the classical world, such things occur. For example, if you place a ball exactly on top of a fixed sphere, it will stay there, because there is no reason to fall down in any particular direction. However, the slightest disturbance is enough to make the ball fall.

Why is this important? In the above classical example, it is important to note that it is impossible to predict in which direction the ball will move, unless you have perfect control over the fluctuations of the surroundings of the sphere, for example, the vibrations of the table where you have put it. What you can predict, on the other hand, is that it will eventually fall down. At this stage, one may understand that this can trigger fundamental discussions about the question whether one can ever predict what will happen to a system even if one knows all its constituents and fundamental interactions (i.e., the reductionalist hypothesis), or whether other ways to approach the problem, or even additional knowledge, are needed [18]. To understand the practical consequences for materials, we first have to see how situations of degeneracy and spontaneous symmetry breaking can come into play in large systems.

3.2.7 More can be very different

More is different—with this famous expression P W Anderson [19] meant to stress the ongoing excitement of condensed matter theory, and the fact that one can discover radically new effects even when all constituents have been known for a long time. This is actually something that even children know—in the movie Finding Nemo, for example, a school class of small fish learns how to self-organize in order to simulate one large fish that could frighten an enemy [20].

One would immediately object that small fish can think and decide to do this, while electrons or protons cannot. Still, they form patterns. Take, for example, a crystal: if you start from empty space and throw in a bunch of atoms, at low temperature they may form a crystal. In a large homogeneous environment this crystal would have the same probability to sit in whatsoever place—this is a very degenerate problem. However, the crystal will 'decide' to form somewhere. Of course, space is usually not empty and homogeneous, but even if it were, the crystal position would be pinned by the small fluctuations as introduced earlier. Electrons can also form crystals, although this is less common. A homogeneous electron gas at very low density crystallizes; this is called a Wigner crystal [21]. The left panel of figure 3.10 shows the electron density that one obtains by an analytic calculation on paper for many electrons in a completely flat potential, taking into account the translational invariance: it shows no feature whatsoever. The result in the right panel was obtained for the same system numerically, using the Quantum Monte Carlo method, a stochastic approach to optimizing the many-body electronic ground state wavefunction by minimizing its energy [3]. One can clearly see a crystal structure. The difference between the results of the analytical and the numerical calculation is huge. This does not mean that one of them is wrong; they are both correct. However, the analytic calculation is done in unrealistic conditions, because the tiny fluctuations of the environment are missing. Therefore, the system cannot break the

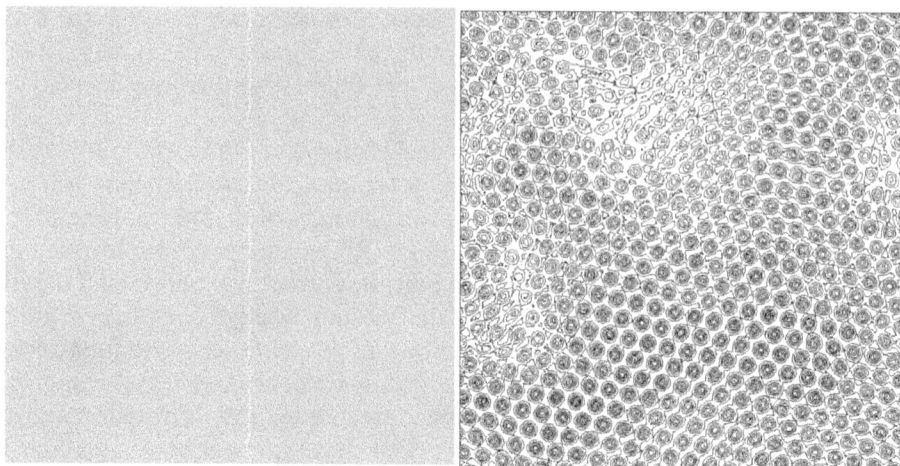

Figure 3.10. A Wigner crystal breaks the symmetry of a homogeneous electron distribution. Left: Analytic solution for the electron density in a homogeneous potential. Right: Solution obtained by path-integral Quantum Monte Carlo calculations. Figure by D Ceperley, published in [3], copyright (2016) with permisssion from Cambridge University Press.

symmetry. The calculation, instead, involuntarily contains fluctuations, through initial conditions of the simulation, its finite length, and the random number generator that is used. In a sense, one would not call this realistic either: why should this simulation noise have anything to do with a natural environment? Indeed, if one repeats the very same calculation, one will most likely obtain a completely different picture. Or, to be precise, one will find the crystal in a different position and with a different orientation. However, it will be the same crystal, with the same shape of the unit cell and the same periodicity. This means that the particular realization is meaningless, but the pattern is meaningful.

There is still a piece missing in the puzzle: the example of the H_2^+ molecule cannot explain the formation of a regular pattern as the one we observe here, and one would think that tiny irregular local fluctuations should lead to an irregular picture. Somehow the information of how to organize is spread throughout the whole system: the density fluctuations are correlated. Indeed, one can define a correlation function that gives the probability to find a density change in one place together with another density change in a second place [3]. For the Wigner crystal, one finds that this correlation function diverges for a wavevector leading to the density changes that have the periodic pattern seen in figure 3.10. There is a direct link to the H_2^+, though: the probability amplitudes calculated in such a correlation function are inversely proportional to energy differences, so divergence of the correlation function is directly related to degeneracy. Degeneracy, in turn, is likely to arise in infinitely extended systems where the energy spectrum becomes continuous in the so-called thermodynamic limit. The correlation functions express the capability of the system to respond to an external perturbation, and a divergence implies that even the most tiny fluctuation will lead to a sizeable response—right as the Wigner crystal example shows.

Why is this important? There are many examples of the formation of patterns in many-body systems: in ferromagnetic materials, small magnetic moments are aligned in a given direction, and there is no apparent reason why the moments point in this, and not in opposite direction: ferromagnetism breaks a mirror symmetry. In antiferromagnetic materials, small magnetic moments are periodically aligned in alternating direction, and this can break the translational invariance of the ideal crystal structure. Charge density waves can stabilize patterns in materials, similar to the Wigner crystal in the homogeneous electron gas. Nematic order occurs in liquids, but also in some superconductors: in this case, translational invariance is not broken, but the electron cloud is deformed locally, for example, around an atom, such that some of the original rotational symmetries around each atom are broken and the material may become anisotropic. These are symmetry breakings that one can visualize quite easily. But there are other, less evident symmetry breakings that lead to exotic behaviour of great interest. A superconductor, for example, breaks electromagnetic gauge symmetry, and this gives the material extraordinary properties, the most well known being zero resistance at non-zero temperature. Similarly, superfluids can creep up walls like magic.

One can easily understand that these drastic changes of the material are very important for applications. *More is different* can open new avenues to design materials with non-toxic, earth-abundant elements, to replace compounds with harmful properties or at the origin of conflicts. Finally, symmetry and symmetry breaking is an important concept all over physics. Therefore, the study of symmetry-broken phases of condensed matter may also shine light on fields where experimental data are more difficult to obtain, and condensed matter analogues for high-energy physics is a fascinating topic [22].

What to look at? From what we have seen up to now, it becomes clear that the goal cannot be to describe every possible story of every single particle in materials with a number of particles of the order of 10^{23}, neither theoretically nor experimentally: it would not be possible, and even it it were possible, it would not be meaningful. What is important are not the details of a single realization, but averages and patterns, in the widest sense. The former are the expectation values that are measured experimentally with good statistics, and calculated using increasingly complex and accurate approaches from quantum chemistry, stochastic methods, or functional approaches such as density functional theory or functionals of Green's functions.

Most often materials are interesting even in the absence of challenging symmetry breaking: the interaction between electrons and of electrons with the lattice still determines their properties. Some of them can be explained in terms of single-particle features that are simply renormalized by the presence of all particles. Others are due to collective effects such as plasmons or excitons, or such as magnons, which are collective spin excitations, or phonons, stemming from lattice vibrations. These are excitations that appear as quite sharp features in the spectra, which means that they live a long time before they decay into the continuum of incoherent excitations of the many-body system [3]. This means that their characteristics to some extent resemble those of independent particles, although they are different from the bare electrons and nuclei that constitute the system. They are therefore called *quasiparticles* [23]. This is an important concept: while the behaviour of a material cannot be understood in terms of the motion of each of its constituent particles, it can be understood in terms of these quasiparticles. We have seen examples above: the sharp bands of aluminum correspond to quasiparticles, where the removal of an electron from the system can be described in terms of the hole left behind. This hole is surrounded by a cloud of electrons that are attracted by its effective positive charge, an effect called screening, which changes the energies but still leaves the material with a clearly visible band structure. The importance of these quasiparticles reflects itself in the fact that they have launched new directions of research and applications that are named after them, such as plasmonics [15] that is based on plasmons and that was already mentioned above, excitonics [24] where ultra-small optical switches, solar energy production, or low-consumption solid-state light sources are based on excitons, or magnonics [25], where, for example, magnetic materials are to be used in miniaturized programmable devices. Even for complex materials, much can be understood in these terms, especially by combining experiments with electronic structure calculations.

In the more difficult case, when degeneracies determine the materials properties, still much can be done: fluctuations are naturally present in experiment, and symmetry breaking can be induced in calculations by adding small perturbations [26]. However, the tendency of a material to organize in a pattern can be more directly understood by looking at correlation functions: in an antiferromagnet, for example, the spin–spin correlation function would tell us that every spin prefers to have an opposite spin as next neighbour, whereas in a ferromagnet, a neighbour of same spin is preferred. The Wigner crystal, in turn, can be deduced from a density–density correlation function. It is important to note that the 'will' to self-organize is already encoded in such a correlation function that would be calculated or measured in the symmetry-unbroken state, and it is therefore independent of any particular realization of the symmetry breaking. The tendency to organize appears as a divergence of the correlation function at a parameter that corresponds to the pattern, for example, a wavevector related to the period of a translationally broken symmetry. As pointed out earlier, this divergence, in turn, stems from division by an energy difference that vanishes, hence from degeneracy. Correlation functions actually help us out of the dilemma: we may admit that it is impossible to predict a particular Wigner crystal or to reproduce the very same Wigner crystal by two subsequent experiments, but it is possible and pertinent to measure or calculate the density–density correlation function and therefore, what kind of pattern will be formed—even with pen and paper, in principle.

3.2.8 Challenges and opportunities

No doubt, this is a fertile field, where new discoveries are made every day. Above, we have outlined the complexity of the quantum many-body problem. Most importantly, we have explained to which extent potential surprises and yet unpredicted properties are inherent in its very nature. In particular, emergent phenomena due to spontaneous symmetry breaking are not exotic singularities, but common in condensed matter. Their importance for fundamental insight and technological applications has been widely acknowledged—maybe the fact that a large portion of what used to be called *condensed matter physics* is now called the *study of quantum materials* [27] is just reflecting that fact. It should not make us forget that even 'simple' metals or semiconductors may have interesting properties and lead to crucial technological breakthrough.

In the following, we will not try to predict what we cannot even dream of today. We shall simply remind the reader that we can only be ready for new discoveries, and be humble enough to admit that some ideas or observations put forward recently may seem of modest interest right now, but could have huge impact tomorrow. We have described the basic ingredients above in simple terms, using prototypical illustrations, but the consequences in real materials are so multifaceted that it is impossible to give an exhaustive summary. Valuable roadmaps exist and are regularly updated, where one can monitor progress and have a glimpse at the possible future (see, e.g., [28]).

What we can do is, however, to ask questions. Many new materials will be synthesized, and in particular the combination of different materials is a vast playground. So, for example, we can ask *How can we use the two-dimensional electron gas that forms at the interface between two insulating oxides* [29]*?* or *How much is still to be discovered in low-dimensional systems, in particular quasi one-dimensional ones, knowing that the effective Coulomb interaction is stronger in lower dimensions?* And *What about combinations of such low-dimensional systems?* For example, layers of two-dimensional crystals are often kept together by the weak van der Waals dispersion interaction explained above. Therefore, they can be stacked in many ways. The stacking can be twisted, and we know today that such twisted systems at 'magic' angles have very unconventional properties [30–33]. On the other hand, still new classes of materials are discovered and/or characterized, leading to questions such as *What will be future research in, and applications of, spin liquids* [34], where, for example, magnetic excitations with unconventional statistics may be found. Spin liquids belong to the wider class of topological materials, a field of research that by itself is not very recent [35], but the importance of which is increasingly acknowledged [36]. Topological phases, such as topological insulators [37], though clearly distinct from other phases of matter, cannot be described by conventional spontaneous symmetry breaking. Still, they are distinct phases and can be described by an order parameter which plays the role of conventional symmetry-breaking order parameters (e.g., magnetization is the order parameter when the spins order). In topological phases the order parameters are certain integrals over the momentum frequency space that do not depend on details of the integrand, which makes it more difficult to imagine than quantities such as the magnetization. So, *what other order parameters and related phases may we find in the future?* Also, superconductivity remains in the focus of interest. Superconductivity in cuprates has been studied since the 1980s, but still, new features are discovered, such as a pattern of the charge density (called charge order, a reminder of the Wigner crystal) that has been measured using different techniques such as inelastic x-ray scattering [38] and that does not come with spin order, which is quite unusual. *What is the relation between different* *forms of symmetry breaking in superconductors?* is a question that is not yet satisfactorily answered. At the same time, superconductors are used to push quantum technologies, and there is a dream to develop quantum computing based on superconducting qbits [39]. *Can materials science lead to the breakthrough of quantum computing?*

While this admittedly very partial list of questions develops, and while it is clear that every single research group or even every single researcher would have at least one more important question to suggest, some broad directions of search can be distilled from the discussion above.

The search for new quasiparticles: We have pointed out the importance of quasiparticles, such as plasmons or excitons, for the understanding of materials and for whole classes of applications. There are in principle infinitely many possibilities for such collective effects, and quasiparticles with very exotic properties may exist. Hunting for such quasiparticles is a lively field of research. For example, zero-energy modes which exist at the ends of one-dimensional conductors that are in a topological superconducting state [40], called Majorana fermions, have attracted much effort. Also skyrmions, chiral magnetic objects that are expected to be stable, localized in a few nanometers, and easily manipulable by weak electric or magnetic stimuli, are intensively studied for their promises ranging from low-power applications [41] to neuromorphic computing [42].

The exploration of new couplings: 'More is different' always applies, when one brings together particles such as electrons and nuclei, when one combines different atoms, different crystalline layers, or other pieces of matter, and also when one couples matter to photons. A strongly coupled photon–electron system can give rise to new quasiparticles that are detected in experiment. A convenient theoretical formulation is the Floquet approach [43]. The strong electron–photon coupling can, for example, lead to replica of the energy bands called Floquet bands, which are found in experiment [44]. One can understand them in a similar way as the replica of the energy bands in aluminum that can be observed in figure 3.5, the only difference being that there, the hole coupled to plasmons, whereas here, the coupling is to photons. Both can be described by a model of fermion–boson coupling, but of course, with different parameters. The most puzzling observation is probably that even a molecule in vacuum is coupled to vacuum fluctuations, which can be observed when putting the molecule into a cavity. In that case, the coupling to something that is seemingly *nothing* (but that actually consists of photons) can significantly alter the properties of the molecule [45]. This motivates the search for new couplings, involving potentially more than just two ingredients, with infinite possibilities of combining pieces of matter and electrons, holes, plasmons, excitons, magnons, lattice vibrations (phonons), photons, etc.

A new dimension: time: Time adds a new dimension to possible studies, and the typical timescales of lattice vibrations (picoseconds) and electrons (femto- and attoseconds) are now accessible experimentally. Experimental developments such as free-electron lasers are tools that have yet been used only to a fraction of their potential, and that will without doubt challenge our understanding. Time-dependent perturbations can be used both to probe and to create: they can drive a system out of equilibrium, and it can be very instructive to see how the system evolves. They may also allow one to bring a system into a new state; one prominent example is light-induced superconductivity [46]. Theoretical efforts, for example, with new developments in the framework of TDDFT or non-equilibrium Green's function theory and real-time propagation of Green's functions, are already paying off, especially with joint experimental–theoretical studies, and open new persepctives such as coherent manipulation of the excitonic properties on ultrafast timescales [47].

Additional parameters: Many more parameters influence materials: for example, these can be pressure, an environment such as a substrate or a solvent, external electric or magnetic fields, doping, or temperature. Often it is not easy to take this additional information into account, but in order to exploit the full range of possibilities for understanding and designing materials, we have to face the challenge. Changing the parameters can lead to new observations, such as phase transitions (e.g., a transition from one crystal structure to another). Temperature is a good example: often calculations suppose that the lattice is frozen and that the electrons are in their ground state, whereas in reality there is thermal motion and the electrons occupy also higher states with a thermal distribution. This may require to take disorder into account, entropy, exchange of energy between electrons and lattice vibrations, thermal fluctuation of magnetic moments, etc, on a level that is more than just taking averages. For example, some materials above a certain critical temperature appear to be metallic when one neglects magnetic moments by considering that they are zero on average, whereas these materials are paramagnetic insulators in nature, and they can only be described correctly by taking the fluctuating magnetic moments into account [3].

Order parameters: As we have seen, while symmetry is a fundamental constraint underlying all domains of physics, symmetry breaking occurs (though for good reasons) and leads to the most amazing phenomena. Much more remains to be discovered and understood. Investigations are already ongoing on topics such as understanding the interplay of distortions and superconductivity [48], or the formation of time crystals [49], where a periodicity in time of a driven system is superseeded by a different time periodicity. We will have to be creative and imagine new forms of symmetry breaking, with new order parameters (including those not related to global symmetry breaking such as the topological ones, see in the questions above) that can bring us to unexplored territory. It will also be interesting to find new ways to suppress certain symmetry breakings, since different forms of symmetry breaking, such as magnetism and superconductivity, can be in competition. From the theoretical and numerical point of view, the example of the Wigner crystal shouldn't give the impression that having a noisy method or computer is enough to find everything that might possibly occur, since the computational choices that are made still restrict the range of what one could possibly find. We will have to improve theory for the calculation of correlation functions, and set up correlation functions for new kinds of fluctuations. We will also have to include potential symmetry breaking in the functional approaches, which is especially delicate when going to infinite systems, where the order of limits may seem to be a mathematical subtlety, but strongly impacts the physics that one may find.

Machine learning and data science: Experiments, theory, and computation see constant progress, which should be acknowledged. However, recently a new tool has emerged that may well change our way of working in a fundamental way and therefore merits special mention. Computers can not only perform electronic structure calculations, but they can handle large amount of data in an efficient way. This leads to several possibilities that find exponentially growing use today and may well completely transform our field by the horizon 2050. First, databases allow us to profit from existing knowledge and in such a way avoid huge waste [50, 51]. Today, materials databases exist with millions of entries, where one can find information about a given material, or materials with desired properties. Second, computers can perform nonlinear interpolations and extrapolations in huge parameter spaces. This is the essence of machine learning, which allows us to predict properties of materials knowing the properties of a set of other materials [52]. Third, they can recognize patterns and correlations, which, although the computer does not 'know' the reason for correlations, can be of tremendous help for modeling, developing theory (functionals), and classifying materials to make predictions [53]. Many difficulties remain, and in particular in materials science, the absence of clean, unified, and abundant data is a challenge. Nevertheless, machine learning can be a plus also for experimental data, in particular imaging [51], even today, and the future is promising. Needless to say, be it with or without machine learning, computational materials design is an increasingly important branch of condensed matter physics, and collaboration with mathematicians and computer scientists will be needed in order to use the expanding computing capacities in the best possible and responsible way.

Much of our discussion reflects the very essence of science: trying to discover what is not yet known, and not even thought of. To do so, we have to continuously find new ways to think and imagine, and we have to be ready for surprises. As a final remark, we may be proud of the fact that the investigation of materials on a quantum level can respond to many of the grand challenges of humankind, such as the development of renewable energies, or of new drugs. However, the complexity of the field may also serve to remind us that scientists have the mission to learn and transmit to society ways to structure, understand, and solve problems that are too difficult for any of us. And that, yes, 'more is different', and we will solve those problems only collectively, or not at all.

Stimulating discussions with many members of the Palaiseau Theoretical Spectroscopy Group are gratefully acknowledged.

3.3 The search for new materials

Claudia Draxl[1] and José María de Teresa[2]
[1]Professor of Physics at the Humboldt-Universität zu Berlin, Berlin, Germany
[2]Research Professor at CSIC and Leader of the Nanofabrication and Advanced Microscopies Group at the Institute of Nanoscience and Materials of Aragón (INMA, CSIC-University of Zaragoza), Zaragoza, Spain

3.3.1 General overview

Ever since *Homo sapiens* started to create tools and develop technologies, new materials that could improve the quality of life have been discovered and deliberately fashioned. New elements, metals, alloys, molecular assemblies, and compounds have contributed to create additional functionalities or implement better physical properties. Today as well, there is an intense research activity aiming to create new substances, expanding the *palette* of materials available for higher efficiency and quality of life.

Apart from the material composition, the dimensions and microstructure of materials are crucial for the physical properties they exhibit, especially when the nanometer scale is approached. As a consequence, it is important to control the conditions by which crystals, polycrystals, or amorphous states of a bulk material are synthesized, as well as to investigate their properties not only in bulk form but also when grown as thin films, nanoparticles, nanowires, aerosols, etc.

In certain cases, the discovery of a new material or a new form of a material entails its prompt technological application, which explains the healthy state of research on new materials. Sometimes, a targeted application can guide what material to search for, whereas in other serendipitous cases, a new material brings about unexpected applications and technological developments.

In the following sections, the three main aspects of state-of-the art research on new materials will be addressed: synthesis, physical properties, and applications. Moreover, we shall discuss aspects of data-centric materials science. The focus will first be on the present understanding of the topic; in the second part of this chapter the challenges and opportunities on the Horizon 2050 will be tackled. As materials research is an extremely wide area, we will focus on some—subjectively chosen—examples rather than covering the whole field.

3.3.2 Present understanding and applications

3.3.2.1 Synthesis

Once a researcher, guided by, perhaps, theoretical predictions, by previous experience, by experiments, or by intuition has decided to prepare a new material, it is very important to choose an appropriate synthesis method. This task is not obvious. To synthesize new materials, one can rely on either physical or chemical methods, or a combination thereof—the variety of existing growth methods is immense. An optimized synthesis method will depend on multiple variables such as the equipment available or the budget, the required bulk, thin-film, or nanostructured form of the material, its single-crystalline, polycrystalline, or amorphous nature, the desired throughput, the accuracy of the stoichiometry that is needed, and so forth. Moreover, if the new material is intended to be used commercially, one should also consider other requirements such as the cleanliness and sustainability of the synthesis method, the potential for large-scale synthesis, the availability of the starting materials, etc. Given the breadth of the topic and space limitations, we shall restrict the discussion to the most common physical methods for the synthesis of new materials and provide a few examples.

For bulk materials, most of the synthesis routes are based on chemical methods, but in some particular examples physical routes can be more convenient. For example, the semiconductor industry relies on single-crystal Si wafers that are grown by means of the Czochralski method. In this method, the necessary elements are molten and the surface of the liquid is put in contact with a small crystalline seed that is pulled and serves as a guide for obtaining a large single crystal [54]. More generally, the use of molten metal fluxes allows the synthesis of novel materials in the form of single crystals, which are ideal to investigate the intrinsic physical properties of a new material [55]. A recent example is the synthesis of a new class of Fe-based superconductors [56]. The arc-melting growth method is of interest for the synthesis of ingots of metallic materials. In this method, stoichiometric amounts of the required elements are molten by the electrical arc discharge from terminals submitted to a high voltage. In this way, multielement alloys can be grown; a recent prominent example is that of high-entropy alloy systems [57]. The growing interest for the fabrication of materials with arbitrary shapes and resolution down to the micrometer range and for manufacturing objects that cannot be machined or assembled otherwise has boosted the use of additive printing techniques [58].

In many applications, the new material will need to be grown in the form of a thin film of sub-micrometer thickness. Such films are usually not self-supporting, therefore synthesis is directly performed on a support substrate. A wide variety of physical techniques are available to achieve the required stoichiometry, crystallinity, and roughness; these include sputtering, thermal, or electron-beam evaporation, molecular beam epitaxy (MBE), pulsed laser deposition (PLD), and physico-chemical techniques such as chemical vapour deposition (CVD) and atomic layer deposition (ALD). Each of these techniques has proven to be very successful in the growth of particular thin films.

We will provide a few examples to give the reader the flavour of the topic. Sputtering techniques use targets composed of the elements or compounds required for synthesis. A plasma maintained in the deposition cell provides for ion bombardment of the targets, which results in the sputtering of their constituent elements and their subsequent deposition on the substrate. In some cases elements present in the plasma such as oxygen or nitrogen are incorporated in the process. This is the general technique of choice in the magnetic storage- (memory) and magnetic sensor-industry [59]. Thermal or electron-beam evaporation is convenient to produce thin films from small molecules [60] or from elements and compounds; these can even be mixed after annealing processes [61]. On the other hand, MBE uses the evaporation of elementary components under ultra-high vacuum conditions to achieve an accurate stoichiometry, which has been extremely useful to grow high-quality semiconductors [62] and devices based on multilayers of high-quality semiconductors [63]. The PLD technique, which uses a laser to vaporize material from a target onto a substrate under high-vacuum conditions, is frequently used to grow high-quality multicomponent oxide films [64]. CVD, where appropriate precursor gases are mixed on a chamber and made to react on a heated substrate, is a widely used growth technique for films [65]; it has recently allowed the synthesis of graphene films on large areas, thus opening the route for many applications [66]. Finally, ALD, where precursor gases are brought into the chamber sequentially and react to grow a film layer by layer with great control, has found applications in the semiconductor industry to grow high-dielectric-constant gate oxides [67].

In those cases where nano-structured materials (nanowires, nanoparticles, flakes, etc) need to be synthesized, it is possible to either grow the material in the form of a thin film and subsequently carry out a nanolithography process to pattern it into small structures [68] or to directly grow it in its nano-structured form (see figure 3.11 with ZnO nanostructures as an example). In general, for direct growth of the nanostructures, chemical synthesis routes are used. However, other methods benefitting from physical–chemical phenomena have been used. For example, the Vapour–Liquid–Solid (VLS) technique, in which a metal nanoparticle such as Au acts as a catalyser and forms a liquid droplet that contains the incoming gas to produce a growing nanowire [69], is very convenient to synthesize high-quality Si nanowires or other types of semiconducting, oxide, and nitride nanowires [70]. Another popular nanowire growth method is electrodeposition on appropriate membranes with small tubular structures [71], which allows the exploration of

Figure 3.11. Different morphologies of ZnO. From left to right and from top to bottom: Quantum dot. Nanotubes. Nanowires. Nanobelts. Nanoring. Nanocombs. Tetrapod. Nanoflowers. Hollow spheres. Sponge-like film. Nanosphere. Nanoplates. Copyright (2017) John Wiley & Sons [75].

new and multicomponent stochiometries [72]. Moreover, focused ion- or electron beam-induced deposition techniques (FIBID or FEBID), where a precursor gas is delivered onto the substrate surface and decomposed by an impinging focused electron or ion beam, produce nano-deposits with metallic, magnetic, optical, or superconducting functionality [73]. Moreover, FEBID and FIBID techniques allow for the growth of three-dimensional nanostructures in a single step, which is convenient to explore novel physical effects [74].

Interestingly, it is possible to exfoliate certain materials and produce flakes down to thicknesses of an atomic monolayer or a few atomic layers. In this way, graphene flakes were grown and studied for the first time, opening the important research field of 2D materials [76]. Moreover, deterministic transfer techniques allow the synthesis of multilayer-like systems with various flakes one on top of another, including control of the rotation angle of the flakes [77].

It is also worth discussing some emerging synthesis techniques for new materials not mentioned so far. For instance, self-organization and self-assembly are useful strategies to synthesize unconventional nano-patterns, as shown by the example of polymethyl methacrylate–polystyrene block copolymers [78]. Also, the combination of various synthesis and/or patterning techniques is common to fabricate novel structures and devices [79]. In this direction, smart combinations of materials and structures can give rise to metamaterials with new physical properties, such as the ability to cloak objects from electromagnetic fields [80]. The synthesis of topological materials (section 3.2) is also an intense field of research guided by theoretical predictions [81]. In such materials, symmetries in the electronic structure produce robustness against smooth topological variations and the materials show unusual physical properties. Thus, bulk, thin films, and nanowires of topological materials have been recently synthesized and their properties thoroughly studied [82].

Beyond the technical aspects of materials synthesis, we call the reader's attention to the important aspect of sustainability of the synthesis process, especially if it is intended to be commercially used on a large scale. First, the synthesis process should be clean, avoiding the release of contaminant products. Second, scarce materials should be avoided and recycling processes developed. For example, the use of rare-earth materials in permanent magnets raises important questions regarding strategic dependence on a single provider; alternative materials and strategies are under investigation [83]. Similarly, the increasing dependence on Li for its use in batteries and the difficulty in recycling them has raised concerns on the sustainability of this technology [84].

3.3.2.2 Physical properties of new (or improved) materials and their applications
It is not only our inherent human nature that makes us strive for steady advancements in all aspects of our life. Fast progress is also owed to the pressing needs to meet the enormous challenges arising from the world's tremendously increasing energy consumption and environmental problems, to a large extent caused by the rapid growth of the population and of industry in the absence of sustainability concepts or collective responsibility of the current 'throwaway society'. Materials are an essential part of our prosperity and lifestyle. In fact, there is hardly any sector

where they do not play a crucial role, be it energy, environment, information technology (IT), mobility, or health. Sustainable energy involves photovoltaics, thermoelectricity, batteries, catalysis, solid-state lighting, and superconductivity. IT is concerned with electronic devices, data storage, sensors, switches, touch screens, and alike. Mobility and infrastructure are based on *robust* materials, in terms of being strong, hard, non-corrosive, heat-resistant, or inflammable. This also applies to various tools. Finally, the health sector is dealing with biocompatible, non-toxic components for implants or drug delivery, but is equally concerned with a huge variety of medical instruments that, in turn, not only require proper materials for their functions but also, again, electronics and IT. Below, we illustrate the crucial nature of materials development with a few examples, discussing where we are now and where improvements are needed.

The key components in electronics are semiconductors. These are materials that typically exhibit a band gap (range of forbidden electron energies) in the infrared (IR) and visible part of the energy spectrum. Their behaviour is tuned by the selective addition of impurity atoms. These introduce 'free' charge carriers (i.e., not bound to a particular atom) that make the material electrically conducting, in the form of either negatively charged electrons (n-doping) or positively charged holes (p-doping, see section 3.2). The material's electrical conductivity can also be controlled by external stimuli, such as electric or magnetic fields, by exposure to light or heat, or by mechanical deformation. The prime materials in electronic devices are silicon, germanium, and gallium arsenide; silicon is by far the most common. It is used, for instance, in MOSFETs (metal–oxide–semiconductor field-effect transistors) of which no less than several trillion have been made to date. Despite this obvious success, and their widespread application in light-absorbing or -emitting devices, neither Ge nor Si are optimal materials for the semiconductor industry, because of the 'indirect' nature of their band gaps. The excitation of electrons to higher energy levels in these materials requires the electron to interact not only with light (photons) to gain energy, but also with lattice vibrations (phonons, see section 3.2) in order to gain or lose momentum. Such indirect processes happen at much lower rates and are thus less efficient than direct processes. Despite this disadvantage, Si is hard to beat on the market as it is abundant on the Earth's surface and thus rather cheap; there are well-established processing techniques that ensure its high quality and purity. The total production volume of silicon worldwide in 2020 amounted to about eight million metric tons (https://www.statista.com/topics/1959/silicon/).

Inorganic semiconductors rule today's electronics and optoelectronics industry. While GaAs is the main player in red-light emitting devices and lasers, GaN plays the same role for the blue and green colours—and, as a consequence, for the generation of white light. In 2014, Isamu Akasaki, Hiroshi Amano, and Shuji Nakamura were awarded the Nobel Prize for Physics for *'for the invention of efficient blue light-emitting diodes which has enabled bright and energy-saving white light sources'. A long process* requiring many years of in-depth investigations towards a full understanding and a full development of the material preceded its use in such applications. Today, LEDs are omnipresent in TV screens and cover walls and buildings as giant advertisements. Most important, LEDs are a true game changer towards more efficient lighting devices, compared to traditional lightbulbs. The latter waste about 90% of the energy as heat could be considered 'heating devices' rather than light sources. Nevertheless, ample room for improvement remains, as LEDs too should become more efficient.

On the other hand, organic semiconductors have been studied for several decades as potential replacements for inorganic components. Besides their extraordinary electro-optical properties, they combine several advantages such as mechanical flexibility, a huge variety of building blocks, and low production costs. Therefore, all kinds of opto-electronic devices have been demonstrated. On the downside are their inferior thermal stability and low charge-carrier mobilities. Therefore, this class of materials is only used on a large scale in very distinct applications such as organic light-emitting diodes (OLEDs). How to compensate for major disadvantages by forming organic/inorganic hybrid interfaces will be discussed in section 3.3.3.

Over 40% of current energy consumption is wasted in the form of heat, be it in industrial ovens, chemical plants, exhaust pipes, engines, or computers. The thermo-electric effect allows the conversion of heat into electricity using thermoelectric generators. Thus, such devices have great potential to reduce the world's energy consumption. However, today's thermoelectric materials are not efficient enough to make this concept economically viable. Designing high-performance thermoelectric materials that could boost their widespread deployment is a difficult task. Since the thermoelectric efficiency depends on the ratio of charge and heat transport, the goal is to maximize the former and minimize the latter at the same time. As the enhancement of one goes typically hand in hand with that of the other, a real breakthrough is a great challenge. One way out of this dilemma is the development of complex crystals. Such compounds typically have a cage-like crystal structure formed by atoms with covalent bonds to ensure a high performance in their electronic properties, while guest atoms inside their cavities ensure low thermal transport. Typical examples for such inclusion compounds are skutterudites and clathrates. Such structures offer a playground for tuning their properties to the optimal performance. Still, their figure of merit ZT—a measure for the usefulness of the material in thermoelectric applications—hardly exceeds values of about 1, whereas a value larger than 2 would be required for use on a large scale. Another approach to reaching high ZTs is the reduction of dimensionality or nanostructuring via superlattices, quantum dots, nanowires, and nanocomposites [85–88]. When the distance between nanostructured interfaces is shorter than the mean free path of phonons but still larger than that of the electron, the engineered interfaces enhance phonon scattering, thereby impeding heat transport, without affecting electron-mediated electrical transport. For the quantum-well superlattices Bi_2Te_3/Sb_2Te_3, a high ZT of 2.4 has been achieved [89]. More recently, simpler structures consisting of weakly bound 2D layers have gained a lot of attention. For example, $SnSe_2$ shows a remarkable low thermal conductivity of 0.4 W m^{-1} K^{-1} and a ZT of 2.6 [90]. These properties are realized over a wide temperature range.

It is not only the direct consumption of fuel and electricity that needs to be reduced to facilitate the transition to a carbon-neutral economy. Consider how much energy is lost because all our means of transportation are heavy, being mainly made of steel. Clearly, heavy is considered synonymous to robust and strong, which is a requirement for cars, trains, and airplanes to be safe. On the other hand, strength is not the only prerequisite. It is their ductility that prevents materials from breaking. How to make materials extremely strong and simultaneously ductile when, typically, one property can only be improved at the expense of the other? This question has triggered generations of researchers and engineers. The situation is getting even more puzzling when one also needs to reduce weight (e.g., by reducing thickness). Airplanes, cars, and trains would consume so much less energy if they were lighter. One solution of this problem may be found in high-entropy alloys, a currently booming field of research [91].

Another solution lies in materials that involve only light chemical elements such as aluminium and titanium. Aluminium, being super light, lacks strength and is also extremely expensive to produce—in terms of energy consumption. Titanium—more than an order of magnitude less abundant than Al—is, on the positive side, heat-resistant, tough (but weaker than heat-treated steel), and non-magnetic, and also shows low thermal expansion. On the downside, it is difficult and expensive to manufacture. The intermetallic compounds TiAl, lightweight and resistant to oxidation, find use in aircrafts, jet engines, and automobiles. Its γ-phase has excellent mechanical properties and is oxidation- and corrosion resistant, to temperatures well above 600 °C. It is therefore discussed as a possible replacement for traditional Ni-based superalloys. In this context, we briefly address another problem. Turbine blades operate at temperatures higher than their melting point. Therefore, they require coatings. Such 100 μm to 2 mm thick layers of thermally insulating materials reduce the temperature of the underlying metal, thereby also providing protection against oxidation and corrosion by high-temperature gases. The prototypical material here is Y-stabilized ZrO_2. Since this material undergoes a phase transformation at elevated temperatures, active research is currently dedicated to adequate replacements.

3.3.2.3 Selected emerging applications and the need for new materials
As shown in the previous section, new materials have applications in numerous fields. In the present section, we will focus our attention only on emerging applications related to the digital world as well as energy. In the first case, we have chosen emergent technologies that largely depend on new or improved materials and allow for improved or energy-efficient computing and sensing. In the second case, we have chosen materials that could help methods that lower our carbon footprint such as solar energy, energy harvesting, and thermoelectricity. Without being exhaustive, this approach should permit the reader to appreciate how new materials can impact our future lives.

3.3.2.3.1 Quantum computing, neuromorphic computing
Quantum computing has the potential to tackle difficult or long computational problems that are tedious, intractable, or impossible using classical computation [92]. Recent demonstrations even suggest that quantum supremacy has been achieved in a particular case [93]. Various technologies have been put forward to

perform quantum computation, but the one based on superconducting qubits is so far the most promising [94]. Other proposals for quantum computing make use of topological materials, photonic structures, molecular magnets, etc. This is an open and exciting field of research fed by the use of new or improved materials.

Inspired by the low energy consumption of the human brain (20 W), an emergent field of research is neuromorphic computing. Here, the use of materials and devices that mimic the behaviour of neurons and synapses is pursued. Neuronal computation is performed through the generation of spikes, depending on the integrated input coming from other neurons. The synapses, or connections between neurons, contribute to the computation by changing their connection strength as a result of neuronal activity, which is known as synaptic plasticity. Synaptic plasticity is the mechanism that is believed to underlie learning and memory of the biological brain [95]. Materials with phase change, memresistors, ferroelectric materials, spintronic oscillators, etc, are currently being studied for this application [96] and proof-of-concept results indicate the potential of this technology for language recognition [97].

3.3.2.3.2 Low-power, flexible, high-frequency electronics

One of the main paradigms of the 21st century society is the existence of electronic devices and their connectivity. The underlying hardware supporting this technology is silicon-based microelectronics, which is sustained by two main pillars: the materials involved in microelectronics (semiconducting, metallic, insulating, magnetic, etc) and the lithography techniques allowing one to reach the required small dimensions (below 10 nm). These techniques comprise optical and electron-beam lithography, focused ion beams, and nanoimprinting, among others. Future developments in electronics are expected to come, within Moore's approach, from improvements in the materials used or in the lithography processes; alternatively, they can come from new applications of microelectronic chips (More-than-Moore), or from different approaches beyond standard paradigms of the semiconductor industry (Other-than-Moore) [98]. Within the new context of More-than-Moore and Other-than-Moore approaches, intense research is being carried out to target novel electronic devices harnessing low power, flexible, and high-frequency properties [99]. Whereas low power is required to comply with today's and tomorrow's energy requirements, flexibility is needed in wearable and implantable devices; high-frequency response allows for fast data collection, treatment, and communication. Given their intrinsic flexibility and low power consumption due to their one-atom-thick nature, as well as the absence of the working-frequency limitations of Si, Graphene, and other two-dimensional (2D) materials such as transition-metal dichalcogenides (TMDs) are expected to excel in these applications [100]. An example is shown in figure 3.12, where a device based on graphene transistors has been implanted in a rat brain for mapping its neuronal activity.

3.3.2.3.3 Ubiquitous sensing and in situ and operando characterization

Today's digital society and modern industry require ubiquitous sensing. Many sensors are based on physical (i.e., optical, magnetic, electrical, mechanical, etc) effects. They can be carried by objects (vehicles, electrical appliances, etc) or by

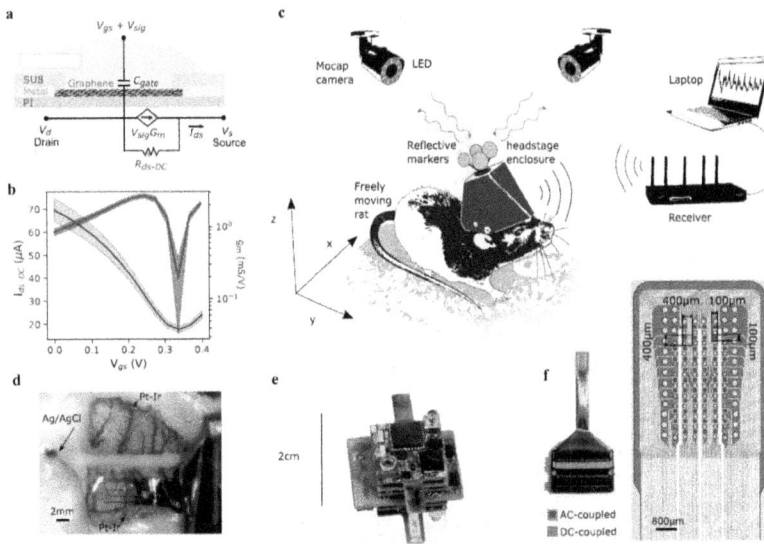

Figure 3.12. (a) Schematic of the graphene-based transistor and its equivalent circuit. (b) Transistor characteristics. (c) Illustration of the rat with the untethered recording system implanted. (d) Graphene transistor array device implanted in the rat cortex. (e) Photograph of the wireless headstage. (f) Photograph of the graphene transistor array mounted on a customized connector and zoomed image of the probe active area (right). This figure has been obtained from reference [101] and is subject to a free license under a Creative Commons Attribution 4.0 International License.

humans (cell phones, smart watches, etc), placed in an industrial environment (for the monitoring of manufacturing, fabrication by robotic machines, etc), in the workplace, or at home (for purposes of security, safety, comfort, etc) [102]. A particular requirement of sensors is their capability to give an accurate output in an energy-efficient way [103]. If we take the example of CO_2 monitoring, extremely important in closed spaces during the COVID-19 pandemic, technology has evolved from bulky, heavy sensors to small portable sensors that are fabricated in clean rooms in a manner similar to microchips. Using the same physical principles as the old bulky CO_2 sensors (i.e., the absorption of infrared light by CO_2 molecules), new materials for semiconductor diodes providing emission of light at the required wavelength, for miniaturized waveguides coated with efficient light reflection, and for solid-state photodiodes have paved the way for these modern CO_2 sensors (see figure 3.13).

3.3.2.3.4 Materials for improved energy production

Clean and renewable energy sources are increasingly used in the energy mix of most countries. New materials can help us find more efficient and cleaner processes for energy production and reduce CO_2 emissions (https://www.elsevier.com/connect/net-zero-report). This hot topic is also very broad, so, we will just mention a few examples.

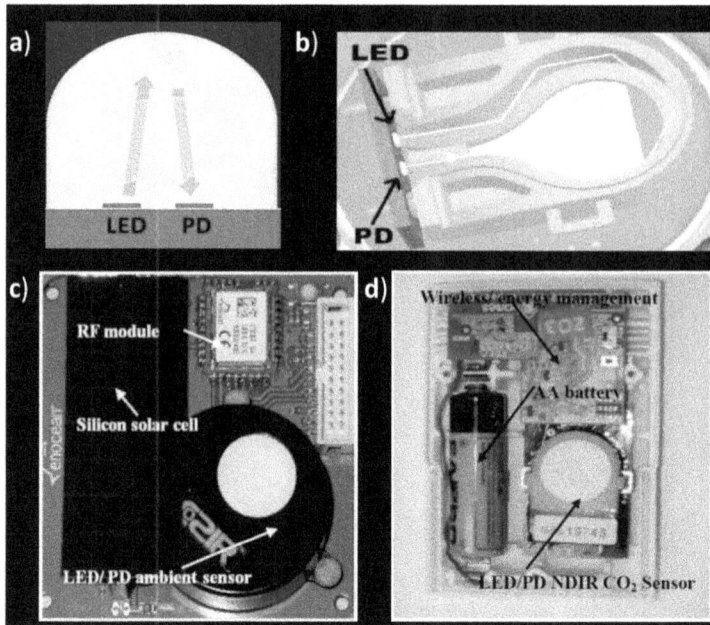

Figure 3.13. (a) Simple design of a CO_2 sensor, with a single reflection. The light emitted by the light emitting diode (LED) is absorbed at particular infrared wavelength by CO_2, which leads to a decrease in the light intensity detected by the photodetectors (PD). (b) Optimized design of the pathway travelled by light, which includes multiple reflections between LED and PD. (c) Wireless self-powering LED/PD-based CO_2 sensor. (d) Battery-based LED/PD-based CO_2 sensor. Images have been taken from the article reference [104] through Creative Commons CC BY license.

Solar cells for photovoltaic energy production have a quite long history. We draw the reader's attention to the well-known chart provided by NREL (https://www.nrel.gov/pv/cell-efficiency.html) that shows the efficiency of solar cells over the years. Also in this area, silicon has long dominated the market. In the last two decades, a great many new or improved materials are investigated for their application in solar cells, photovoltaic cells, solar absorbers, heat storage materials, photocatalysis, light trapping, solar concentrators, and so forth. A huge boost was achieved in recent years with the advent of hybrid halide perovskites that, so far, keep breaking all records in terms of efficiency [105]. Unless replacement elements are found, the widespread application of these materials is likely to be halted because they contain the toxic element lead. Another example is that of electro-chromic materials. These change the absorption of light under small electric fields and are used as chromogenic materials [106].

As described in the previous section, an interesting way to produce energy is to use the waste heat of other energy sources to feed thermoelectric devices, which relies on materials with good electrical conduction but bad heat transport, with great expectations from nanostructured materials and new compounds [107]. Energy harvesting from friction, vibration, RF sources, etc, is also intensively studied to power small (electronic) devices [108]. Future energy sources such as fusion energy call for the investigation of materials that could stand the stress and hard radiation inside the fusion reactor [109]. This topic will be later discussed in more detail.

3.3.2.4 Data-centric materials research

While *Make and Measure*—synthesizing materials, characterizing them with various experimental probes, and computing and analyzing their properties—represents the current state-of-the art of materials research, there is an urgent need of speeding up the process by complementary approaches. '*Twice as fast at a fraction of the cost compared to traditional methods*' was also the aim of President Obama's Materials Genome Initiative (MGI) (https://www.mgi.gov/about), launched in 2011. In fact, data-centric science—combining emerging techniques of machine learning and other methods of artificial intelligence (AI) with the Big Data that the community is producing on an everyday basis—is currently establishing the fourth pillar of materials research. This is evidenced by the increasing number of publications and the advent of new journals emphasizing the importance of data, but also by the many workshops and sessions at international conferences dedicated to this topic. To mention an example, the 2021 spring conference of the American Physical Society (APS) counted 51 sessions with either *Data Science*, *Machine Learning*, *Deep Learning*, *or Artificial Intelligence* in the session title. One might get the impression here that data-centric materials science is already fully established. How huge the challenges truly are will be discussed below.

A popular approach among data-driven efforts is high-throughput screening (HTS). It has led, for instance, to the establishment of large-scale US-based computational data collections (http://aflow.org, https://materialsproject.org, http://oqmd.org) already before the MGI; and all over the world, there is a rapidly growing number of other small and large databases (see also [110]), examples being the Computational 2D Materials Database (C2DB) [111] (https://cmr.fysik.dtu.dk/c2db/c2db); the HybriD^3 materials database (https://materials.hybrid3.duke.edu)—a collection of experimental and theoretical halide-perovskite data; the experimental metal–organic framework database CoRE MOF [112], just to mention a few examples. Systematically replacing atomic species in known materials or combining them in new ways such to create novel structures offers a huge playground for exploring materials and their properties before they are even synthesized. There are also more and more *experimental* initiatives of this kind, examples being the high-throughput platforms for catalysis (https://www.hte-company.com/en/company/high-throughput-experimentation) or photovoltaics (https://data.nrel.gov/submissions/75, https://www.hi-ern.de/hi-ern/HighThroughputPV/node.html). We note that HTS has by no means always been a computational approach. The most famous example may be the ammonia catalyst demonstrated by Fritz Haber in 1909; this was followed by systematic testing of about 20 000 materials by Alwin Mittasch in Carl Bosch's group at BASF.

However, all these efforts are insufficient when it comes to exploring the infinite amount of possible materials that can be created, considering the building blocks provided by the periodic table of elements (PTE) and the many ways of combining them, in terms of composition and configuration. To reach the ambitious goal—

phrased in reference [113]—of creating 'materials maps' from which one can read in what region of 'materials space' one is likely to find promising candidates for a given application, HTS and Big Data must be accomplished by novel AI concepts and tools. This issue will be addressed below.

First, we discuss the current limitations of data-centric materials research (i.e., the facts that are hampering fast success). A first problem concerns scientific bias introduced by our publication culture. Published success stories show what works and may lead us to comprehend why. However, this procedure alone does not provide us with the full picture (i.e., it does not allow us to understand why alternative approaches or materials *do not work*). This problem is particularly critical in materials synthesis. Even publicly accessible databases such as Landolt–Börnstein (SpringerMaterials, https://materials.springer.com), which provide comprehensive information about materials properties, lack information on synthesis. A recent effort is placed on scanning literature for synthesis recipes and parameters [114]. However, this information cannot be complete as unsuccessful synthesis attempts are typically not published.

The second showstopper concerns the amount and description of shared research data. The materials science community is steadily producing enormous volumes of valuable data, be it with the diverse experimental techniques and instruments or the large variety of computational methods. Eventually, papers are published that report on the gained knowledge in a very compact form. Most of the information inherent in the data is, typically, 'forgotten' and might even be discarded rather than published and made available for future use. However, even if data are kept, they are often insufficiently described, rendering them of little value. This problem is less severe in theoretical research. As long as the input and output files and the computer programme (and its version) are known, calculations can be verified. Nevertheless, for comparison with other results, the meaning of the data must be uniquely described in terms of metadata. The in-depth annotation of measured data is an even bigger issue as the full information on the sample (including preparation), the apparatus, and the measurement type and conditions is required.

Sharing and publishing research data also implies 'sustainability'. For instance, doubling of identical calculations or measurements can be avoided, leaving time and resources for work beyond. This topic brings us back to HTS. Typically, those materials that fulfil the required criteria are investigated and well characterized, while all others are not further pursued. Obviously, while they may not be useful for the purpose of the particular research pursued, they may be extremely valuable in other contexts.

To harvest all this currently disregarded information, research data—but also workflows and tools—must follow the FAIR principles [115]. What does FAIR—Findable, Accessible, Interoperable, and Reusable—mean in the context of our field? Certainly, publicly available data stores fulfil the *F* and to a large extent also the *A* as uploaded data can be inspected and, possibly, also downloaded. This is the prerequisite also for the *R*—re-using or re-purposing data. The latter refers to what was mentioned above, namely that research results can be useful for a purpose that is different from the original intention. The *I* (interoperability) is the most critical and largely unresolved issue. This applies in particular when data from different sources are brought together. For the experimental characterization, there exist a few benchmark databases, an example being the *EELS Atlas* (https://eels. info/atlas). However, many more well-characterized data obtained from state-of-the-art instruments would be needed. On the theory side, reproducibility [116] and benchmarking for solids [117–120] have become an issue only rather recently.

An early data infrastructure, following FAIR principles even before the term FAIR was introduced is the NOMAD Laboratory (https://nomad-lab.eu). Being an open platform for sharing data within the entire community, it is different from other data collections in computational materials science. Information about its services, comprising the NOMAD Repository, the Encyclopædia, and the AI Toolkit, can be found in [121, 122]. Its extension to data from sample synthesis and experimental characterization is described elsewhere [123]. Likewise, the Open Access to Research (OAR) data infrastructure at NIST [124] is built to allow NIST scientists and others to share research data using standards and best practices adopted in the scientific community. Given the pressing needs, many such initiatives are currently established or will follow.

3.3.3 Challenges and opportunities on the Horizon 2050

As discussed in the previous section, in the next years and decades, a great many bottlenecks must be overcome as far as materials exploration and development is concerned. Clearly, innovative solutions and ideas and therefore '*thinking out of the box*' is required. Which are the most promising material classes for what kind of applications? A first example is that of nanostructured and composite materials; these will certainly help us tailor and tune desired properties. The introduction of interfaces allows us to exploit the advantages of pristine components while getting rid of their disadvantages. This may apply to the transition from pure organic electronics to hybrid electronics (e.g., in light-harvesting or light-emitting device, where, one may want to make use of the superior light–matter interaction of organic components with the advantageous higher charge-carrier mobilities and weaker electron–hole binding of inorganic semiconductors). This concept of combining the best of two worlds applies to other interfaces and nanostructures as well. In addition, data-driven approaches (see above) is expected to bring 'outliers' to light (i.e., materials that have not been considered for a given application so far). These may well be 'simpler' materials and therefore less involved as far as synthesis is concerned.

The development of technologies based on new or improved materials will surely have a great impact in our lives on the Horizon 2050, no matter the technology at hand: information technologies, transport, energy production, health, and more. In what follows, we shall focus our attention on materials for a digital and energy-efficient world in 2050. Unfortunately, many topics will remain unaddressed, but we think that most of the profound changes in the coming decades will come from the increasingly digitalized society and the evolution of our means of producing energy.

3.3.3.1 Materials for a digital world

The control of information (storage, processing, and transmission) shapes every aspect of our daily life, thus permeating cultural and social changes. The digitalization of information, which started around 50 years ago, is accelerating and it is reasonable to suppose that by 2050 the relevant information in our society will be digital and in most cases only digital. A multi- and cross-disciplinary approach is needed to cover the present challenges in this field, ranging from more technological aspects to social ones. The current digital transformation is enabled by developments in physics and engineering and concerns several fields including electronics, optics, material science, and quantum technologies. Today's challenges include sustainable and energy-efficient electronics, integrated photonics with new functionalities, quantum computing, machine learning, and operation within the Internet of Things [125].

Materials for ubiquitous sensing. The future is that of an omni-connected society, where sensors will play a key role and represent an unavoidable interface amongst humans, objects, and the environment (see also section 3.3.2.3.3). Sensors are needed for monitoring of the environment (gas concentrations, radiation, etc), for energy-efficient control of autonomous cars and drones, for the precise localisation of persons and objects, for safety and security applications, and so on. Sensors are more efficient if they are integrated with the electronics handling the obtained data into a single device (system), which requires micro/nano-fabrication techniques. Sensing matter at the micro- and nano-scale allows, on the one hand, to probe new physical states endowed with novel properties and allowing the development of new devices and, on the other hand, the detection of biological objects and processes with high spatial and temporal resolution. In addition, new tools such as artificial intelligence and machine learning can help design a new generation of smart nano-sensors that process the data and communicate them more efficiently in order to provide a faster and energy-saving response. In sensors, one of the most important constituents (if not the most important) is the material that shows the functional response: optical, electrical, magnetic, mechanical, etc. Also, the material into, or onto which the sensor is integrated (the substrate) is of utmost importance; their physical properties determine their applicability: flexibility, optical transparency, heat or electrical conductivity, etc. These are some of the challenges and opportunities in this field on the horizon 2050:

- *Functional properties of materials.* With respect to electrical conductors, good electrical contacts and materials at the nanoscale are required in the pursuit of ever-increasing miniaturization of electronic devices. This opens opportunities to achieve improved metallic contacts by resorting to new materials such as graphene and other 2D materials that do not suffer from electromigration issues when their dimensions are decreased to the nanoscale. As for magnetic materials, in the case of permanent magnets it is important to decrease our dependence on rare-earth elements, all the while not losing performance. In nano-magnetism and spintronics applications, the challenge is to find suitable magnetic materials and architectures that decrease power consumption. The purpose of spin-orbitronics is to obtain materials or combinations of materials that produce topological magnetic structures (skyrmions, etc) or useful three-dimensional magnetic textures. In the field of nano-optics, superb opportunities exist for the fabrication of smaller optical components; even if the light wavelength is intrinsically large (in the order of half a micrometer), it is possible to exploit plasmonic effects to reduce the working wavelength to smaller values (see also section 3.2.5).

- *Integration of materials in suitable electronics platforms.* Flexible electronics is a very important development field for electronic devices on clothes, on our skin or inside our body (such as neuro-chips). In this application, the added value of other substrate properties such as optical transparency or electrical conductivity is very relevant. The realization of hybrid fabrication processes combining functional materials with semiconductor-based platforms, which are generally very strict with respect to compatible materials, offers important opportunities.

- *Working range of materials.* The tendency in telecommunications is to work at high frequencies, which opens the need for (conductive, magnetic, optical, etc) materials that are (energy-)efficient at GHz and THz frequencies. Here, graphene and other 2D materials as well as superconducting materials are very promising. For superconducting materials, the Holy Grail is to find superconductivity at room temperature and ambient pressure, perhaps in materials where it is today unsuspected. Extending the temperature range in which sensors work is very important in applications where substantial heating is produced; one cannot forget that the increasing electronic miniaturization is generally accompanied by a large Joule heating in a small volume space.

- *Biocompatibility of materials.* Whereas for *in vitro* and point-of-care biosensors any material can be used, biosensors that are to be implanted inside the body or will be in contact with the skin must be biocompatible. This poses strict requirements on the materials used, especially in the case of implanted biosensors that track disease markers. Given the overall requirements of biosensors in terms of sensitivity and specificity, this implies significant challenges. For example, the biosensor must be able to operate within the therapeutic range of the target substance whilst in the presence of complex solutions (e.g., interstitial fluid or blood). It must be biostable and biocompatible, since negative immune reactions may cause the device to become non-functional. It must be self-sufficient in terms of power supply and control from external devices; as far as data transmission is concerned: the signal output transmitted to an external communication device should arrive in a meaningful form for ease of use for the patient/clinician [126].

- *Energy efficiency of materials.* Ubiquitous sensing has a pitfall: the need for energy to power the sensors, to process the data, and to transmit them. Power can be provided to the sensor by batteries, through energy harvesting, or via wireless means. Thus, improved materials for batteries and energy harvesters will remain necessary for decades to come. In addition to the intense search for materials that could serve as efficient thermoelectric convertors that was already mentioned, similar searches are conducted for materials for power supply and for detection.

- *Scarcity of materials.* There is plenty of room for the substitution of scarce/strategic materials with more abundant ones, provided that the performance of the sensors is not jeopardized. Just to have an overall picture, a mobile phone contains up to 64 elements, many of which could be unavailable in a few years (https://www.bbc.com/future/article/2014031414-the-worlds-scarcest-material). Besides the long-standing problem of the scarcity of rare earths for permanent magnets and that of Li for batteries, some heavily used metals are also scarce, such as Ta, Co, Pd, Cu, In, Pt, Ag, Rh, Au, Al, etc. In this topic, another interesting avenue is the development of recycling processes that could reduce this problem.

Materials for quantum technologies (see also section 3.2). Quantum technologies comprise quantum computing, quantum communication, and quantum sensing and hold potential for paradigm shifts across several disciplines. Huge research investments have been announced in this field by the main players in technology, which include the most important technological companies as well as entire nations and clusters of nations. The treasure hunt is on for a disruptive technology that can beat semiconductor-based computation in specific applications. Although promises in this field have already existed for a few decades, the current state-of-the-art in this discipline as well as the incoming investments are expected to pay off by 2050.

- *Quantum computing.* The challenge is to develop robust platforms with a sufficient number of qubits allowing fault-tolerant computation. Although superconducting platforms based on Josephson Junctions are currently leading the race, as recently demonstrated by Google [39], an intense research using other platforms also exists. In particular, semiconductor-based platforms are very promising, such as single-spin qubits hosted in quantum dots. Another emerging platform is that of topological qubits based on Majorana quasiparticles occurring on topological superconductors [127]. Basic research is also underway regarding the use of molecular and magnetically based qubits [125]. In all these platforms, one can foresee a tremendous research effort in the next decades to implement new or improved materials and their integration in optimized electronic platforms.

- *Quantum communications.* Today, we already use a quantum effect whenever we use a navigation system: an atomic clock. In the future, our communications will be private thanks to quantum encryption, the technical name of which is quantum key distribution (QKD). This technology is protected by the very laws of physics: the mere fact of observing a quantum object perturbs it in an irreparable way (section 3.2). The span of current QKD systems is limited by the transparency of optical fibres and typically reaches one hundred km, which opens opportunities for improved optical fibres and optical platforms that could extend the working range of this technology. By 2050, it is expected that various ambitions of quantum connectivity will be achieved [128].

- *Quantum sensing.* In our tireless effort to probe matter with the highest sensitivity and the highest spatial and temporal resolution, quantum sensors are expected to contribute enormously. For instance, the most advanced detectors of magnetic flux today are quantum ones, such as SQUIDs (super-conducting quantum interference devices) or SNVM (scanning nitrogen-vacancy microscopy) [129]. Sensors able to detect such magnetic fields with nanometre spatial resolution enable powerful applications, ranging from the detection of magnetic resonance signals from individual electron or nuclear spins in complex biological molecules to readout of classical or quantum bits of information encoded in an electron or nuclear spin memory [130].

Materials for neurotechnologies. Interfacing with the brain (brain–machine interfaces, BMIs) is one of the most exciting opportunities of today's technologies, but there are many challenges to overcome before the technology is ready for broad use. For certain applications, the external detection of the weak electric fields produced by the brain activity will be sufficient, such as in videogames, communication with our mobile phone, treatment of certain brain disorders, or neurostimulation (https://www.bitbrain.com/#). However, implanted BMIs are more powerful and even have the potential to produced enhanced human beings (https://neuralink.com/). Whereas advancements in BMI for electrophysiology, neurochemical sensing, neuromodulation, and optogenetics are revolutionizing scientific understanding of the brain and enabling treatments, there are many gaps in the technology. The grand challenge in neural interface engineering is to seamlessly integrate the interface between neurobiology and engineered technology to record from and modulate neurons over chronic timescales. However, the biological inflammatory response to implants, neural degeneration, and long-term material stability diminishes the quality of the interface over time [131]. Recent advances in functional materials are aimed at engineering solutions for chronic neural interfaces, yet, the development and deployment of neural interfaces designed from novel materials have introduced new challenges that have been largely unaddressed. Many aspects of this topic are related to the use of suitable materials, such as the optimization of the individual electrodes and probes, including their softness and flexibility, and other critical multidimensional interactions between different physical properties of the device that contribute to overall performance and biocompatibility. Before regulatory approval for use in human patients is achievable, the behaviour of these new materials and of the overall device must be addressed. By 2050, we predict that significant progress will have taken place and some of these BMIs will be available for broad use. Hopefully, this disruptive technology will develop following appropriate regulations (neurorights) [132] that avoid the existence of neuronal paradises (https://nanofab-deteresa.com/science-popularization/).

3.3.3.2 Materials for energy production (see also section 3.3.2.3.4)
Crucially important for the production of chemicals and fuels is the sustainable production of hydrogen. For this, photo-catalytic water splitting could become a key technology. So far, limitations in suitable catalyst materials render this process unfeasible. To trigger the chemical reaction requires a stable semiconductor to

harvest the solar energy as well as an efficient catalyst to split the water molecules and evolve hydrogen and oxygen from the electron–hole pairs that are created by sunlight. As such, identification of a better catalyst for water splitting has enormous potential in carbon-dioxide management and hydrogen-based energy technology.

On the longer-term perspective, our cars and trucks may run on hydrogen. For the time being, society and industry rely more on electric drive. Thus, batteries are a critical player here (see section 3.3.3.1).

Also, we should not forget the energy losses caused by transmission which amount to about 6% in average in Europe. Superconductors would be the ideal solution as they transport current without losses; these materials can operate so far, however, only at low temperatures and/or high pressures. Yet, we are far from room-temperature superconductivity at ambient conditions, but the huge progress in identifying materials with high superconducting transition temperatures and high critical currents over the last years has given rise to big hopes. The highest temperatures so far have been achieved in pressurized superconducting hydrides (see, e.g., [133] and references therein).

Making current technologies more energy efficient by improving or replacing the underlying materials will be essential for the period leading up to 2050 and beyond. We should, however, also look ahead towards a possible additional source of energy —nuclear fusion. The idea is to replicate the fusion processes of the Sun to create energy on the Earth. To this extent, the thermonuclear fusion reactor ITER is built as a collaborative megaproject of (in alphabetic order) China, the European Union, India, Japan, Russia, South Korea, and the USA; other partners being the UK, Switzerland, Australia, Canada, Kazakhstan, and Thailand. ITER will have the capability to produce 500 MW of power with a necessary input power of 50 MW for more than 300 s [134]. This reactor, is, however, not built for energy supply but for technology demonstration and can be considered as the world's largest plasma physics experiment. Besides being an incredible engineering effort, it also poses severe challenges to materials science. Interestingly, construction of the ITER complex started in 2013, while the choice of most crucial materials was not made. Issues here are, for instance, temperature and radioactive contamination. Components must withstand extremely high temperatures up to 1200 °C, or even possibly 3000 °C. An obvious candidate here is tungsten, exhibiting a melting point of 3422 °C. A severe drawback of this material is that it turns brittle above room temperature. Therefore, for instance, doping elements must be found to remedy the situation. As a very promising candidate for the reactor walls, carbon-based materials like CC fiber composites or SiC were discussed. Unfortunately, it turned out that they would quickly absorb all tritium from the plasma, making the reactor walls highly contaminated. Overall, the case is still not settled today.

3.3.3.3 New knowledge from research data
What do we need to enhance our research by the fourth paradigm? These are basically two aspects: FAIR data and the corresponding data infrastructure, including tools for processing, storing, and accessing data; and novel AI methods

that allow us to turn data into knowledge. All this is in the making, but we are still far from a real breakthrough.

Let us focus on FAIR data first. It is clear that the big picture can only be realized when bringing together data from different sources, essentially, from all over the world. This comes with challenges with respect to all *4V of Big Data*. The incredible *Volume* of materials data requires a distributed data infrastructure that allows for accessing its content with standardized protocols. A first example for a specification of a common REST API has been developed by the OPTIMADE consortium (https://www.optimade.org/) [135]. Such tools are indispensable as the community, and every sub-community, is using and building very different tools in their daily research. This means that, in the first place, we need to overcome the issue of *Variety* without forcing people to drastically change their habits but rather supporting them with appropriate tools. Variety not only concerns different instruments and measurement modes for one specific type of experiment or one specific calculation method that can be applied by different software packages. Overall, we are hit by variety as the community is dealing with an enormous number of different theoretical and experimental characterization techniques that should be dealt with on equal footing. This variety comes with *Veracity*, the uncertainty in the data. Veracity also has different faces. On the one hand, calculations can be based on one or another method and approximation; likewise, different instruments may have different resolutions, and measurements can be carried out at different temperatures. On the other hand, even measured with the same instrument under the same conditions, different samples may give rise to different results. Finally, theory and experiment need to be reconciled. In contrast to other fields, like high-energy physics, *Velocity* may be least critical in materials research. The emerging new (time-resolved) measurement techniques producing enormous amounts of data in short time, may pose, however, challenges in the future.

From the above, it is clear that veracity and variety largely hamper interoperability if the data are not fully annotated. If data can be misinterpreted because their meaning and quality are not known or not considered, even the most innovative AI method can be misled. Therefore, the description of the data by rich metadata is *key* for interoperability and for the success of data-centric approaches. Metadata should capture all parameters that may influence the results. The establishment of metadata schema for each synthesis route, experimental probe, and theoretical approach, possibly connected by a 'materials ontology', is the most critical step towards FAIR handling of all materials science data. An example for how such data infrastructure can be built by the community is described in reference [123].

Let us now turn to the role of AI. Machine learning methods are applied to many problems of materials research (see, e.g., [136, 137]) to predict stable structures and materials properties. Most of them are very successful as long as it concerns interpolation of data or classification problems. To go beyond interpolation, dedicated data analysis and AI tools are yet to be developed. Most crucial thereby is the representation of a material in terms of *descriptors*—the most relevant parameters behind a certain property or function [138–141]. Obviously, having such methods in hand along with the right data is key for their success. Right in this context means certainly Big, as the complex interplay of interactions taking place in materials can only be learned and trends can only be identified from a large amount of data. Right, however, also means that the data need to be well characterized (as noted above) and need to carry the required information. In that sense, just increasing the amount of data does not help. How to obtain and choose those very precious data that enable exceptional findings is another big challenge.

To summarize, as often stated, research data are a goldmine of the 21st century. Turning it into gold, however, means refining the feedstock and enhancing it by novel tools. On the Horizon 2050, big steps toward this goal can and will be realized.

3.3.4 Final considerations

Role of theory. Last but not least, we should consider the role of theory, which is invaluable for the characterization of materials as well as for data-centric approaches. For instance, a big bottleneck in superconductivity research is that a full theoretical understanding of unconventional superconductivity is lacking [142]. Theoretical concepts overall, and in particular, *ab initio* methods, have advanced during the last decades as a stable pillar of materials research. Density-Functional Theory, Green-function-based methods, (*ab initio*) Molecular Dynamics, and Monte Carlo Simulations are state-of-the art techniques that are able to describe and predict materials properties with high accuracy. For details, we refer to the section 3.2 by L Reining. We note though that significant advances in methodology are needed to arrive at a fully quantitative description such to enhance their predictive power. While first principles theory is well advanced in capturing electron–electron, electron–lattice, electron–hole, etc interactions very well, their interplay is often less well understood. This is partly owing to the very complex formalisms and partly to the computational costs that render such calculations infeasible for more complex and even so for rather simple materials. Here, also new computing technology may be one way out.

Societal aspects. There are practical and political aspects to be considered. It is not always the best material that provides the best solution to a problem. We always need to keep a bigger picture in mind; for instance, the whole life cycle of a product and the side effects it may cause. Many questions may arise here: Is it better to use cheap and easy to produce, but imperfect materials than the best possible, most efficient ones that are expensive? Are good products recyclable? If yes, are, for instance, contaminations of an alloy with all kinds of elements and deviations from the desired composition still acceptable such to be able to produce the product in mind? Can toxic elements be tolerated if they are kept in the circle such to avoid contamination by waste? Is industry reluctant to switch to new materials as a redesign of production lines costs enormous amounts of money that has to be compensated by higher gain and/or possible competitive advantage? On the political side, a big issue is the dependence on third countries: Just to give one example, China is by far the world's largest producer of silicon. Around 5.4 million metric tons of silicon were produced in China in 2020, which accounted for about two-thirds of the global silicon production that year. In 2017, the European Union identified 27 critical raw materials among 78 candidates to be critical [143]. This situation requires rethinking in view of our global goals and even more careful assessment of what material to choose for what purpose.

Acknowledgments

CD acknowledges helpful discussions with J Spitaler, L Romaner, and R Pippan and funding from the DFG through the NFDI consortium FAIRmat, project 460197019. JMDT acknowledges funding from Gobierno de Aragón (grant E13_23R)

3.4 Manipulating photons and atoms: photonics and nanophysics

Jean-Jacques Greffet[1], Antoine Browaeys[1], Frédéric Druon1 and Pierre Seneor[2]

[1]Université Paris-Saclay, Institut d'Optique Graduate School, CNRS, Laboratoire Charles Fabry, Palaiseau, France

[2]Université Paris-Saclay, Thales, CNRS, Unité Mixte de Physique, Palaiseau, France

Our knowledge on the physics of atoms and light has been revolutionized by the advent of quantum mechanics during the first 30 years of the 20th century. It was Planck who first postulated the fundamental principle that energy exchange between light and matter can only be provided by electromagnetic energy quanta, a postulate needed to explain the form of the high-energy spectrum emitted by incandescent blackbodies. The very existence of the photon with quantized electromagnetic energy was the basis of the explanation of the photoelectric effect given by Einstein, an achievement for which he was awarded the Nobel Prize. The origin of the narrow spectral lines of atoms was finally understood in the framework of quantum theory. After these first achievements, quantum mechanics became the basis for understanding molecules and solid-state properties with seminal contributions in the middle of the 20th century. During all these years, quantum mechanics was used to describe physical systems with a macroscopic number of particles, on the order of the Avogadro number.

A revolution started in the mid 1980s: *understanding and manipulating light and matter at the nanoscale and even at the level of single atoms and single photons.* The road there was long and full of obstacles, but was adorned with many achievements in the last 30 years, celebrated by over 20 Nobel Prizes in physics and chemistry. We will briefly sketch this long journey, highlighting a few examples for the sake of brevity. Let us start with a description of the state of the art in the middle of the 20th century.

3.4.1 The landscape in the 1960s

To realize why manipulating individual atoms and photons is a revolution—after all, in the mid-20th century quantum mechanics was already believed to be the theory of single quantum objects—let us quote a few sentences written in 1952 by Erwin Schrödinger, one of the founding fathers of quantum mechanics [144]:

> … it is fair to state that we are not experimenting with single particles, any more than we can raise Ichthyosauria in the zoo.

> … we never experiment with just one electron or atom or (small) molecule. In thought-experiments we sometimes assume that we do; this invariably entails ridiculous consequences …

A few years later, in 1960, Richard Feynman gave a famous speech at a meeting of the American Physical society entitled '*There is plenty of room at the bottom*' [145]. He argued that the physics at the scale of only hundreds or thousands of atoms had not yet been explored. It was a genuine scientific *terra incognita* with real challenges for new physics and many practical applications.

> Atoms on a small scale behave like nothing on a large scale, for they satisfy the laws of quantum mechanics. So, as we go down and fiddle around with the atoms down there, we are working with different laws, and we can expect to do different things.

> The principles of physics, as far as I can see, do not speak against the possibility of manoeuvring things atom by atom. It is not an attempt to violate any laws; it is something that can be done; but, in practice, it has not been done because we are too big.

In 2022, physicists perform experiments with single atoms emitting single photons despite the fact that physicists are still too big. They invented advanced tools to manipulate single atoms, to detect single electrons and single photons. They build materials by depositing atomic layers. This may sound like science fiction happening only in advanced laboratories, but the fabrication of materials by depositing individual monoatomic layers is state of the art in today's industry. What was deemed impossible and ridiculous in 1952, 'futuristic' in 1960, has become today the basis of widely spread technology. Every person using a GPS or checking the time on its smart phone is relying on atomic clocks, every person storing information on a hard disk or using a LED relies on devices fabricated with nanometer-scale precision.

3.4.2 Nanophysics: there is plenty of room at the bottom

3.4.2.1 What are nanosciences and nanotechnologies?

The pioneering vision put forward by Richard Feynman in 1960 became what is known today as nanotechnology. LEDs and hard disks are examples of what has been called nanotechnology. Why do we speak of nanotechnologies and never of micro-technologies or mega-technologies? What is so special about the nanometer scale?

As pointed out by Feynman, an important part of the answer is that quantum mechanical rules come into play. This is definitely the basis of all quantum technologies. Yet, so many nanoscale devices exist where quantum is not the key word. Why not? To address that question a simple example is enlightening. When measuring the electrical resistance R of a wire with a length L, the measurement can be fitted with a simple Ohm's law depending on L. This formula is valid if L is 1 km, 1 m or 1 μm. However, as soon as the length becomes smaller than 100 nm[1], the wire

---

[1] One nanometer—1 nm or 10^{-9} m—is equal to one billionth of a meter.

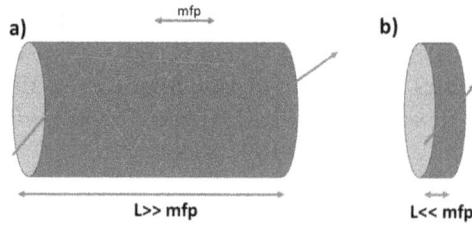

Figure 3.14. Illustration of diffusive (a) versus ballistic electron (b).

enters the nanoscale regime and the formula does not work anymore. Why is this? It turns out that when electrons move in a metal, they typically have collisions with defects or phonons and the mean displacement between two collisions (the so-called mean free path) is on the order of 100 nm. Hence, when a wire becomes smaller than 100 nm, the electrons can travel ballistically, that is, without collisions (see figure 3.14). In this regime, their fate is described by quantum mechanics: electrons cannot be viewed any longer as particles bouncing back and forth; they have to be described as a wave trapped in a wire which plays the role of an optical fibre. This intermediate regime between the microscopic and the macroscopic world is denoted by the mesoscopic regime.

Let us take another simple example: the colour of a gold layer. Depositing a layer of gold on a substrate will make it bright and yellow independent of the layer thickness. But if this thickness becomes smaller than 100 nm, gold is no longer opaque and becomes semitransparent and green in transmission. Here, we compare the thickness of the layer gold with the attenuation length (called the skin depth) of light in a metal. Nanoscale gold nanorods may be either blue or green. This behaviour is not due to quantum effects.

From these two examples, we conclude that the behaviour of a physical system changes when its size is reduced as compared to some typical length of a physical phenomenon. This often happens at the nanoscale but not always. This often happens because of quantum effects but not always. Nanoscience is the study of all new laws at the nanoscale. Nanotechnology uses these new laws to design new devices.

During the last 30 years, physicists and chemists have learnt how to control light–matter interactions at the nanoscale and even at the atomic scale and have learnt how to design new devices using them. In what follows, a few examples will illustrate these ideas.

3.4.2.2 Seeing the atoms: novel microscopies

Looking at ever smaller objects has always been a challenge in physics. In the 19th century, optical microscopy was well understood and the resolution limit derived by Ernst Karl Abbe provided a fundamental limit in terms of the wavelength used. Hence, two tiny particles separated by 200 nm can be distinguished if a wavelength of 400 nm or less is used.

The introduction of electron microscopes was a revolution. These microscopes are based on the fact that in quantum mechanics, electrons are described by a wave. It was possible to develop lenses for electron beams and assemble them to make a microscope. The tremendous advantage is that the electron wavelength can be made much smaller than 200 nm by accelerating the electron. Ruska received the 1986 Nobel Prize for this invention. Another Nobel Prize was given in 2017 in chemistry to Dubochet, Frank, and Henderson who managed to adapt the electron microscope to image molecules of biological interest noninvasively (i.e., without destroying them). Their technique, now known as cryo-electron microscopy, provides a sharp three-dimensional image with atomic resolution of biomolecules such as proteins and surfaces of viruses.

The 1986 Nobel Prize was shared by Binnig and Rohrer who introduced a totally different design of a microscopy that allowed observing and manipulating single atoms. The basic idea is very simple: a tip attached to a cantilever is brought in close distance to an electrically conducting surface. Before the tip actually touches the sample surface, a tiny current is established between the tip and the surface through the insulating vacuum by means of the so-called tunnelling effect, a purely quantum phenomenon. An image of the surface is then produced by scanning the tip above the surface, while keeping the current intensity constant and recording the vertical position of the tip. This technique, known as scanning tunnelling microscopy (STM), reveals the structure of the crystal with atomic resolution. It was the first time that individual atoms could be directly observed. A few years later, it was shown that atoms could be moved one by one, and then observed as shown in figure 3.15 [146].

The STM technique had initially been envisioned as a means to study the electronic properties of surfaces rather than to use it as an imaging technique. As a

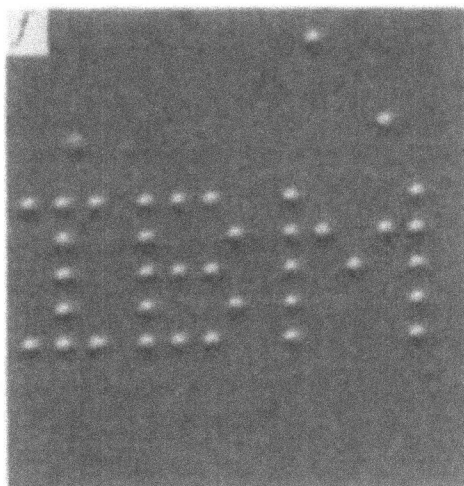

Figure 3.15. Image showing Xe atoms on a Ni surface. The atomic structure of Ni is not resolved. The size of a letter is 5 nm from top to bottom. Reproduced from [144] with permission from Oxford University Press.

matter of fact, the atomic resolution was quite unexpected. This observation triggered the development of a large number of novel microscopy designs, known as near-field microscopes based on scanning tips brought close to surfaces to detect different signals such as forces (atomic force microscopy with many variants such as magnetic force, electrostatic forces, Kelvin probes, etc), heat flux (scanning thermal microscopy), etc. These instruments can be used to measure different physical quantities with nanometer resolution. While the dream of developing nanotechnology had been put forward by Feynman in the 1960s, its real implementation by the scientific community was boosted by the advent of near-field microscopies that provided some of the tools needed to measure and manipulate at the nanoscale. It is the development of these tools that, among others, enabled to a large extent the development of nanosciences and nanotechnology.

3.4.2.3 Detecting single molecules and super-resolution imaging

The imaging techniques described above are based on electrons and forces. They do not use light. Hence, the Abbe resolution limit is not beaten but circumvented. Today, several types of optical microscopy can produce images with resolution much better than the Abbe limit. The first breakthrough was the demonstration of an image performed with a sub-wavelength aperture in a metallic screen. This aperture was used to illuminate very locally a sample so that light transmitted or reflected could only originate from this small area. With this scheme, the resolution does not depend on the wavelength but only on the size of the aperture. This simple idea was first demonstrated experimentally in 1972 by Ash and Nicolls, using microwaves [147]. Yet, the implementation of the idea in the wavelength regime of visible light was deemed to be impossible as it implied bringing a nanoscale aperture at a nanoscale distance from the sample. The first experimental evidence was reported in the 1980s by D Pohl [148] in the visible using a scanning tip to hold a tiny aperture producing a light spot on the order of 100 nm. The Scanning Near-Field Optical Microscope was born.

It was later shown that imaging beyond Abbe's resolution limit could even be achieved without using a near-field technique. The idea is based on the possibility of detecting light emitted by a single molecule. While the image of a point is blurred and has a spatial extension on the order of half a wavelength, it is possible to determine its centre position with much better accuracy. By operating in conditions where the molecules can be detected separately, it becomes possible to locate their positions. If they are attached to an object, its structure can be reconstructed. Betzig [149], Hell [149], and Moerner [150] were awarded the 2014 Nobel Prize in Chemistry for the development of super-resolved fluorescence microscopy. These techniques can be designed in standard microscopes objectives and do not require scanning a tip. They have become very popular for biological applications (figure 3.16).

As we mentioned, a key asset for these techniques is the ability to detect optically a single molecule using its fluorescence [152]. Very sensitive detectors have been developed so that even a single photon can be detected. Hence, detecting a molecule standing alone in vacuum is not so difficult. The formidable challenge is to distinguish the faint signal emitted by a single molecule in a solid or a liquid

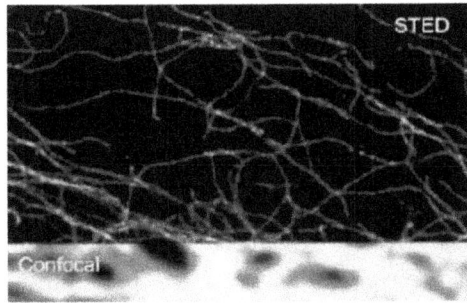

Figure 3.16. Image of a cellular cytoskeleton with the STED super-resolution technique (top) and a diffraction limited confocal microscope (bottom). Adapted from [151], SPIE (1963), with permission from Nobel Foundation.

because its contribution has to be distinguished from the background due to billions of surrounding molecules. This was achieved at low temperature with molecules embedded in a matrix taking advantage of the fact that the interactions with the local environment shifts the fluorescence line so that each molecule emit at a different frequency with a very narrow spectral width. By combining a narrow spectral filter with a microscopic technique isolating light coming from a tiny volume, it is possible to reduce the background and isolate the light emitted by single molecules.

3.4.2.4 Manipulating single atoms with light. Cooling and optical tweezers

Is it possible to grab a single atom and move it around? It can be done with light using so-called optical tweezers. Arthur Ashkin (Nobel Prize 2018) predicted and demonstrated that a dielectric particle can be captured at the focus of a laser beam that acts as an optical trap [153]. It is then possible to move the beam to displace the particle trapped at its focus. This technique was later extended to single atoms in a dilute vapour of atoms. Figure 3.17 shows an Eiffel tower made of single Rubidium atoms spatially arranged using focused lasers. This method can be used to explore the interactions between atoms and forms the basis of a platform needed for the development of quantum computers.

Another important example of using light to apply a force on an atom is the technique of laser cooling. The idea is to reduce the velocity of an atom by making it absorb photons propagating in the direction opposite to its movement. Due to the recoil associated with each frontal collision, the atom experiences on average a drag force so that its kinetic energy is rapidly reduced. This technique is called laser cooling. With this type of technique, atoms can be almost stopped and stored in a trap. Further cooling techniques can be applied and temperatures as low as microKelvins can be attained. The Nobel Prize 1997 was awarded for the prediction and development of these techniques. A further development enabled by laser cooling was the observation of a so-called Bose–Einstein condensate, a state of matter predicted already in 1924 but never observed before, in which thousands of atoms all occupy the exact same quantum state. This offers the possibility to fashion

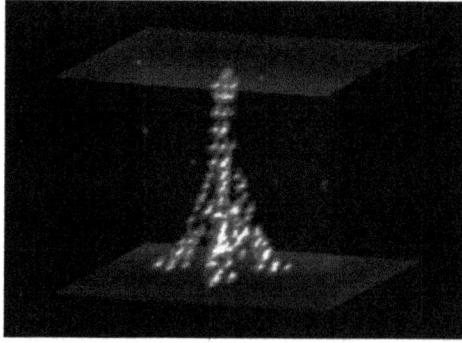

Figure 3.17. Image of an Eiffel tower made of N rubidium atoms trapped by optical tweezers. The height of this atomic Eiffel tower is 90 μm. Courtesy of A Browaeys.

an 'atom laser' out of atoms, such as was done with photons, with a minimum uncertainty in energy and momentum.

While these techniques were initially developed by curiosity-driven research, they have paved the way for novel and future technologies. They are at the heart of new quantum technologies with numerous applications such as atomic clocks, enabling accurate positioning systems, or ultrasensitive gravitational sensors.

3.4.2.5 The advent of spintronics: electronics using the electron spin

The discovery of giant magneto-resistance has been a revolution in electronics. For the first time in the history of electricity, the electrical resistance was tailored through the spin of the electrons. Electrons are elementary particles with a mass, a charge, and a spin. The spin is an intrinsic quantum property of the electron. While quantum in nature, it is also the ultimate constituent of the magnetism observed every day in magnets at the macroscopic scale. In a normal metal or a semiconductor, the electrical resistance, originating from collisions, depends on the charge and on the mass but not on the spin of the electron. However, if the conducting medium is magnetic, such as is the case of iron, for example, a propagating electron experiences collisions that now depend on its spin orientation. This leads to the electrical current in the magnetic medium being spin polarized and, hence, the creation of an electronic spin source. One may then take advantage of this effect to control the electron flow and hence the electrical resistance. This effect, named giant magneto-resistance (GMR), was first observed by A Fert and P Grünberg in 1988. They were awarded the Nobel Prize in 2007. As is often the case in nanoscience, this breakthrough in fundamental physics was made possible because of the development of a new technology. Here, it was molecular beam epitaxy, or MBE (see section 3.3.2.1) that allowed one to produce magnetic multilayers superposing high-quality layers of different materials, only a few atoms thick. In turn, this breakthrough in fundamental physics became the starting point for novel applications. The discovery of GMR has found a major application in data storage and has contributed to the digital age and big data revolution, which is ever expanding with data centre-based cloud applications. Indeed, today digital information is still stored in hard disks drives using the orientation of the magnetisation of nanometer-scale magnetic materials. Remarkably, hard disks relying on GMR-based technology for the read-out the orientation of magnetic domains were released in 1997, less than ten years after the discovery of the effect.

On a broader perspective, GMR has been the starting point of a new type of electronics based on the spin in addition to the electronic charge. This field is called spintronics [154, 155].

3.4.3 Photonics: let there be light

Until the middle of the 20th century, light sources had slowly evolved from candles to incandescent light bulbs and fluorescent lamps. A first revolution came in the 1960s, with the advent of the laser. For the first time, it was possible to produce a coherent light source, i.e., a source such that all the energy is concentrated in a given direction and emitted at a very well-defined frequency, due to the light propagating perfectly in phase. Since then, light has become a powerful tool to probe material properties and lasers have played a key role in the subsequent development of physics and chemistry. They also became the basis of many technologies, ranging from cutting and soldering materials to the design of telecommunication through optical fibres.

In the quest of mastering light emission, three breakthroughs were achieved in the last 30 years. The first was the ability to *emit light at a single-photon level* in a controlled way. This is much more than doing optics with extremely dimmed light. It is about manipulating quantum properties with light, and has paved the way for an entirely new field that started in the 1970s: quantum optics. This field is now mature in the sense that the textbook experiments reported in the 1980s can now be reproduced in laboratory classes. A major output of quantum optics has been to understand the implications of the concept of quantum entanglement, and to initiate a second 'quantum revolution'. This strange feature is the cornerstone of all applications of quantum physics to quantum computing, and will be explained in more detail below. Today's state-of-the art in this quantum revolution is undoubtedly at the level that will allow the development of the quantum technologies of the future. A second break-through is the development of *semiconductor light sources*. The development of solid-state devices requires exquisite control of materials (section 3.3.2.1). This was made possible by a major effort in material sciences and nanofabrication. A ubiquitous consequence of this great success is the development of LEDs with unprecedented energy conversion efficiency. The third breakthrough, started in the 1980s, is the ability to generate *extremely short light pulses* (section 3.5.3). This enables one to observe ultrafast phenomena that were impossible to study before. One can now follow, in real time, atomic dynamics and chemical reactions. Progress has been astonishing: from picosecond pulses in the late 1970s, one has gone to femtosecond pulses in the 1980s, and finally attoseconds in the late 1990s. Another consequence of short light pulses is the possibility to generate extremely large laser powers during short times. This paves the way for many applications ranging from laser surgery to new particle accelerators (see section 3.5).

3.4.3.1 Single photons and single atoms

Controlling light at the level of a few photons forms the basis of quantum optics. If the concept of quantization of the energy of light was first introduced by Planck and Einstein in the beginning of the 20th century to explain the spectrum of blackbody radiation, it is only in the 1960s that a general framework of quantum optics was finally introduced by Glauber (Nobel Prize 2005), largely motivated by a new generation of experiments made possible by the introduction of the laser. The challenge of the last 30 years in quantum optics has been the control of light with very few photons and the exploration of their unique properties. We will briefly quote three examples here: the experimental observation of wave-particle duality, the generation of entangled photons, and the realization of atomic systems interacting with a single photon.

In the late 1980s, it became possible to generate single photons and observe wave-particle duality. Reconciling the wave model of light with the particle point of view had been a major issue in the early days of quantum mechanics. A spectacular experiment, performed in 1985, was to observe interferences with single photons, which highlights one of the central concepts of quantum physics, that is, the superposition principle. This concept states that a quantum object can be in different states at the same time. Its validity has been tested extensively, and although quite counterintuitive, it is one of quantum physics' cornerstones. Single photons are ideal to illustrate both particle-wave duality and the superposition principle.

Producing single photons, which are truly quantum objects, was an experimental challenge before 1980 but has now become nearly routine. The basic scheme consists of exciting an atom with a short laser pulse. The atom then releases its energy with a time delay given by the atom lifetime, on the order of a nanosecond. A lens collects the emitted photons, which can then be sent into an interferometer and then detected at the output by single-photon counter (that also had to be developed). Over the years, photon sources have improved considerably. Starting from faint atomic beams where one tried to isolate the light coming from a single atom, or the use of exactly one atom trapped in a laser beam, one has gone to the use of bright quantum dots (see figure 3.18) that finally turned out to be the most practical solution. It is now possible to produce solid-state sources emitting single photons into optical fibres on demand.

As mentioned above, the concept of interference is crucial in quantum physics, but there is more to it than wave-particle duality. Take, for example, a beam splitter that reflects 50% of the incident power and transmits 50%. Send two photons, one in each input port, as shown in figure 3.19. Amazingly enough, the two photons always exit the beam splitter in the same direction. This observation, called the Hong–Ou–Mandel effect, cannot be understood either by considering the photons as pure waves or by pure particles! This effect only occurs when the two photons are in a quantum state in which they are truly indistinguishable. The Hong–Ou–Mandel effect is now often used as a diagnostic for single-photon sources to check that they really produced exactly identical photons.

The superposition principle leading to interferences is the first important concept that one encounters when learning quantum mechanics. Applied to, for example,

Figure 3.18. Single quantum dots in a semiconducting cavity heterostructure.

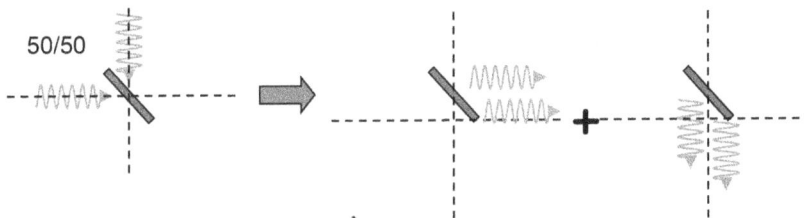

Figure 3.19. Principle of the Hong–Ou–Mandel experiment.

two photons that can each be in two polarization states (horizontal: H or vertical: V), it leads to the existence of quantum correlations. To illustrate this concept, we consider a superposition state where both photons 1 and 2 have a vertical polarization (1V, 2V) or both photons have a horizontal polarization (1H, 2H). Such a state is called an entangled state. Measuring the polarization of the first photon will result in random observations of horizontal and vertical polarization. However, once this measurement is performed, the state is no longer a superposition but either (1V, 2V) if the result was V or (1H, 2H) if the result was H. Hence, measuring the polarization of the second photon yields the same result as for the first one. In such an entangled state, the polarization of one photon is not defined, but the state exhibits strong correlations between the two photons. What is amazing is that those correlations are stronger than any classical correlations in a sense that was made clear by J Bell in 1964. From the philosophical debate between Bohr and Einstein, which revealed the intrinsic and surprising absence of reality in quantum mechanics, we have reached the stage where entanglement has become a fundamental resource for quantum technologies, following experimental tests starting in the early 1980s, for which J Clauser, A Aspect and Z Zeilinger were awarded the 2022 Nobel prize in physics.

One of the most fundamental physical processes is the interaction between light (i.e., a train of photons) and atoms. As soon as experimental techniques were devised that made it possible to isolate individual atoms, it became tempting to study these processes at the level of just one atom interacting with just one photon. These studies started in the late 1980s and led to the Nobel Prize of Serge Haroche in 2012. The basic idea was to send an attenuated beam of excited atoms through a microwave cavity, which isolates a single mode of the field. With the cavity initially empty, the excited atom drops the photon in the cavity, and is able to reabsorb it after the photon has bounced onto the cavity mirrors, leading to a periodic exchange of energy (called Rabi oscillation) between the atom and the cavity involving exactly one photon and one atom! Pushing the technique, it was even possible to entangle two atoms via the single photon of the cavity.

3.4.3.2 Semiconductor light sources

The advent of lasers and semiconductor light sources has deeply modified our playground. Telecommunication infrastructures rely heavily on the interplay between photons and semiconductors. On the one hand, processors manipulate electrons, and information is exchanged via electrical currents that circulate within the processor. On the other hand, information exchange over large distances is achieved using photons propagating in optical fibres. Sending light over hundreds of km was made possible by the development of extremely low absorption materials (2009 Nobel Prize C K Kao). It is thus critical for information devices to be able to convert, with high quality, an optical signal into an electrical current and vice versa. Another fundamental issue is the energy required for lighting. The incandescent light bulbs that have been used for decades typically deliver 3 W of optical power at the cost of 100 W of electrical power (see also section 3.3.2.1).

Both optical communication and lighting have been impacted by the development of light-emitting diodes. An LED is a remarkable device that provides energy to an electron and converts this electrical energy into light with efficiency larger than 30% for commercially available lamps. This was made possible by significant progress in materials science, which has enabled the design of materials with high-purity and high-light emission efficiency. For instance, GaAs is a very good emitter whereas silicon is a very poor emitter of light. In order to convert electric energy to light inside this material, an electric current is injected. Light can be emitted by an electron within a very thin region only. In order to favour the emission of light, so-called nanoscale heterostructures have been introduced (Nobel Prize 2009 awarded to Z I Alferov and H Kroemer). They can be viewed as traps capturing the electrons in those regions where light emission can take place. These heterostructures are made of alternate layers of semiconductors with thicknesses on the order of 10 nm. This type of nanotechnology is the key to efficient lighting. For decades, LEDs emitted red light but could not produce the much more desired white light because no material was known to emit in the blue part of the spectrum. This finally became possible with the advent of GaN, a semiconductor with a relatively large gap of 3.4 eV, more than three times the gap of Silicon. (Nobel Prize 2014 awarded to I Akasaki, H Amano, and S Nakamura.)

Artificial atoms called quantum dots were discovered in the 1980s (2023 chemistry Nobel prize, M Bawendi, L Brus, A Yekimov). Physically, they are realised as semiconductor nanocrystals with a typical radius on the order of a few nanometers. They can be grown at the interface between two semiconductors or be synthesized chemically as semiconductor nanocrystals in a solution. These nanocrystals emit light and behave as artificial atoms. Remarkably, by tuning their size, it is possible to tune the emission frequency with high accuracy. They are currently used in display technology to produce high-quality colour rendering.

3.4.3.3 Extreme light: attosecond and petawatt

Imaging on a very short timescale is fundamental to understand the processes involved in matter at molecular and atomic scales. Just like high-speed photography has facilitated the capture of the movement of a horse at the turn of the 19th century, one can now capture the movement of an electron by illuminating it with ultrashort pulses. In the 1980s, the development of ultrafast lasers producing flashes of light on the femtosecond (fs) scale revolutionized our understanding of molecular vibrations. This work has provided the basis for femtochemistry research and led to Ahmed Zewail receiving the Nobel Prize for Chemistry in 1999 (see section 3.5).

Since then, ultrafast lasers have continued to improve and electronic motion in atoms, molecules, or solids can now be observed *in situ*. A significant step was taken to reach attosecond (as) resolution (an attosecond duration amounts to one second divided by a billion and again by a billion), an achievement recognized by the 2023 Nobel prize in physics awarded to Anne L'Huillier, Pierre Agostini and Ferenc Krausz. This was made possible by an additional nonlinear process producing x-rays and ultraviolet (UV) light via the generation of high harmonics (HHG) using intense ultrafast lasers. This nonlinear process is based on a second advantage of ultrafast lasers: the ability to concentrate a large number of photons over ultrashort times. This amount of energy (from micro-Joules to hundreds of Joules) compressed to an ultra-short duration (about 20 fs) gives rise to extreme peak powers, typically in the range of giga watts (GW) and even peta watts (PW). If the laser is focused as well, light intensities are typically in the range of 10^{13}–10^{23} W cm^{-2}. This increase in the intensity of ultrafast lasers has been made possible by a technological breakthrough. The laser intensity is increased by propagation through an amplifying medium. Yet, beyond some threshold, the amplifying medium is damaged so that further amplification is not possible. To circumvent this limit, Gérard Mourou and Dona Strickland introduced a technique consisting of stretching the pulse duration before sending it through the amplifier and then compressing it in time as it propagates in a vacuum so that there is no damage threshold. They were awarded the Nobel Prize in 2018 for this invention, known as chirped pulse amplification (CPA).

The development of intense and femtosecond-laser pulses has enabled new sources producing attosecond pulses by sending them on gas atoms. Ultrafast pulses interact with gas atoms in the following way: First, the electric field pulls an electron out of the atom via the tunnelling effect; during the next half-period of the laser's electric field, this electron is accelerated back to the ion. Finally, the recombination of the fast electron with its ion results in the emission of an attosecond pulse in the

XUV. At the current state of the art, the shortest light pulses have a duration of 43 attoseconds, which is the shortest physical event ever produced in the laboratory.

Increasing the laser intensity even more allows for other interesting applications. For example, the ultrafast ionization enabled by femtosecond-laser/matter interactions can create an athermal ablation process that leads to precise micromachining. This athermal micromachining can be used for industrial and medical applications such as LASIK (laser-assisted *in situ* keratomileusis). Pushing the laser energy even further may result in what is called ultra-high intensity laser-matter interaction, in which the electromagnetic field is so intense that the acceleration of the electrons leads to relativistic velocities. In these relativistic regimes, the electron can be accelerated in the laser-wakefield (LWFA) up to the GeV range in relatively short distances since the electric field is about tens of GeV m^{-1}. Moreover, when an ultraintense laser interacts with a solid target, this ultrafast bunch of electrons creates a high-space-charge-field zone in which protons and ions can be accelerated. Finally, due to the betatron effect, the relativistic and nonlinear motion of the electrons creates x-rays and gamma rays. All these effects in the relativistic (10^{18} W cm^{-2}) to ultra-relativist (10^{23} W cm^{-2}) regimes for laser-matter interaction can then lead to remarkable ultrafast 'secondary' sources of light and particles.

3.4.4 Challenges and opportunities

In summary, during the last 30 years, the physics community has explored the light–matter interaction at the nanoscale, in the regime where few photons and few atoms are involved. It is now possible to observe and manipulate single atoms and single photons. It has also explored regimes of interaction at extremely short times and at very large powers. Theoretical tools are now available to model these systems. But this is by no means the end of the road. A challenge for the years to come is to increase the size of the systems while keeping them under control. Increasing the size of the systems from a few particles to thousands or billions of particles introduces complexity. Complexity has often been regarded as a source of disorder and noise. But if complexity can be controlled, it may become an opportunity. Complexity of classical systems allows us to revisit optical devices and components. Complexity of quantum systems may offer an opportunity to take advantage of the exponentially growing number of degrees of freedom of a quantum system.

3.4.4.1 *Classical complexity: complex media and metasurfaces*

A good example of a complex system is light propagation in a cloud. The latter consists of a myriad of water droplets randomly located. As the sunlight illuminates a cloud, the light is scattered by these droplets and the transmitted light becomes isotropic. When using a laser beam to illuminate a slab containing randomly located particles of TiO_2 forming a white powder, the transmitted intensity shows fluctuations known as speckle (figure 3.20(a)). They are due to interference between the light emitted by the particles which behave as secondary sources with random phase. Although a microscopic theory is available, solving the equations becomes computationally very challenging when the number of particles increases. Even more

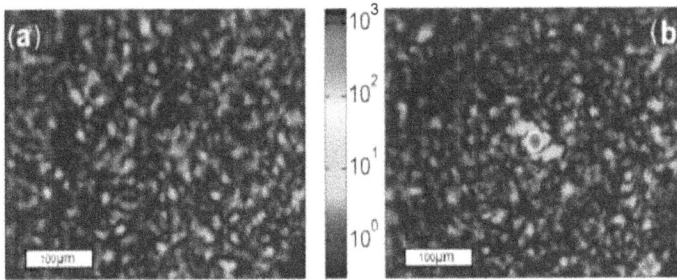

Figure 3.20. (a) Light transmission of a laser beam through a strongly scattering system. As expected, a properly shaped beam can be focused through the sample as seen in (b). Adapted from [156] with permission from OSA, CC BY 3.0.

challenging is the idea of taking advantage of such a complex medium to control light. Figure 3.20(a) suggests that a large number of particles just generates a random pattern. Yet, complexity can be viewed as an opportunity to benefit from many degrees of freedom to control light–matter interaction. Taming complexity to take advantage of this large number of degrees of freedom is a challenge.

A beautiful and counterintuitive example is the possibility of focusing light through a random medium made of TiO_2 particles forming a white powder as seen in figure 3.20(b). The key to achieving such unexpected behaviour is to try to shape the phase of each pixel of the incident beam such that it compensates the random action of the slab. Surprisingly, it is possible to obtain a nicely focused beam out of a random system as opposed to our daily experience: You can't see through the fog.

If this is possible, why not use a very complex system to design optical devices instead of always using perfect mirrors, lenses, and beam splitters? Not only the direction but also the spectrum and the polarization of the light have to be controlled. The motivation to use complexity is that a complex system contains a very large number of degrees of freedom (e.g., the positions of the particles) that all control the propagation of the light. This leads to a paradigm shift. Currently, optical systems use optical components such as lenses, mirrors, polarisers, and prisms. Each component has a single function so that an optical system is obtained by adding several independent components. The art of optical engineering is to efficiently design these multicomponent systems. The new paradigm is completely different. It amounts to taking a slab of material and considering removing material at arbitrary places and replacing with other materials using numerical simulations to find out if some combination of the modifications could end up in a useful system. The success of this approach obviously relies heavily on numerical solutions.

Let us give two more examples of this type of approach. Figure 3.21 shows a recently developed method to separate three beams with different wavelengths propagating in a silicon wire (called a waveguide) on a substrate. We want to guide them into three different waveguides. Classically, one would use a combination of lenses and prisms to achieve this goal. Here, a rectangular intermediate region with micrometer size is defined (see figure 3.21(b)). To achieve the required objective, silicon is replaced by silicon dioxide in some areas, shown in white. The figure shows

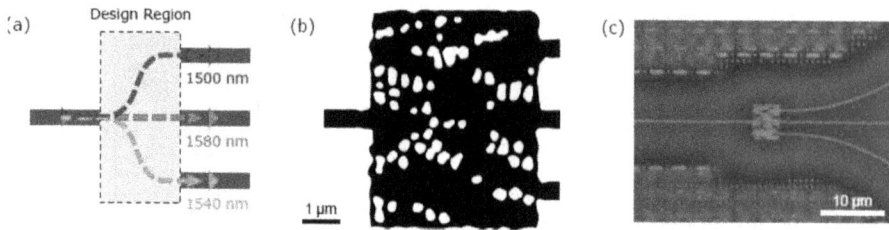

Figure 3.21. Example of the numerical design of a microscale optical system separating different spectral waves and funneling them into different waveguides. Adapted from [149], SPIE (1963) with permission from ACS.

the result of a purely numerical optimization solving Maxwell equations, which rule the fate of light. Although at first glance, the distribution of the white areas is reminiscent of droplets scattering light in all directions, the system is remarkably efficient at sorting out the different frequencies and focusing them at the right points. The fabricated sample is shown in figure 3.21(c). The length scale shows that the system is extremely compact. It fits inside a square of 40 μm, roughly the size of a hair's diameter.

A second example of intelligent complexity to control light with extremely compact systems is a metasurface. It consists of an array of resonant scatterers distributed over a surface. These metasurfaces have a thickness on the order of a micrometer. They can be used to mimic optical components and, more importantly, combinations of optical components (i.e., optical systems). A major advantage is to be able to design the ultrathin optical systems that are the premises of a new generation of optical systems. In this section, we have considered systems that have a number of parameters that is very large but can still be described with a deterministic approach. We have referred to them as complex using the layman meaning of the word. In physics, a complex system is beyond any deterministic calculation. An example of complex behaviour is the phase transition (see section 3.2) between liquid water and ice. At the melting point, an ensemble of water molecules can move from a very ordered crystal form to a disordered liquid form, spontaneously. This type of transition phenomena is observed in many fields of physics. It is often beyond our current simulation possibilities. However, this situation could change with quantum computers.

3.4.4.2 From quantum complexity to quantum technologies

One of the ultimate goals of physics is to understand matter and waves and more generally to understand the natural phenomena that one observes in Nature. Today, our understanding of the laws of physics is based on quantum mechanics, which allows the superposition of states. It is a huge challenge to predict the behaviour of N interacting quantum particles even if we assume that each particle only has two possible quantum states denoted 0 or 1. In practice, the 0 or 1 could be the polarization states of a photon (V or H), or two atomic states of an atom or the spin states of an electron. A possible state for the system of N particles is thus a list of N values (0, 0, 0, 1, 0, 1, 1, ...) specifying the state of each of them. Owing to the superposition principle, the quantum state of the N particles is a superposition of all the 2^N possible states with a complex amplitude for each of them. Hence, a single quantum state of a physical system is defined by 2^N complex numbers. This exponential scaling means that beyond typically 50 particles we cannot even store all these numbers, on any computer we have today or that we can hope to develop soon, thus thwarting the numerical solution of Schrödinger's equation beyond this number. Simply adding one extra particle doubles the time of a calculation! But a single milligram of matter already contains 10^{20} atoms—therefore, any *ab initio* calculation is thus completely out of reach! Over the 100 years that have lapsed since the beginning of quantum physics, physicists have of course devised sophisticated approximations methods to understand and capture complexity, such as the spontaneous replica breaking method in statistical physics by Giorgio Parisi and rewarded by the Nobel Prize of 2022. They obtained great successes in doing that, but many situations remain very hard to understand. This is the case, for example, of high-T_c superconductivity, many-body Anderson localization, the magnetic properties of oxides, or the conduction properties of materials. As the understanding of all these cases could lead to important technological developments, our impossibility to model them *ab initio* is problematic, etc. Usually, the failure of the approximation methods in such situations originates from the very strong interactions between the particles.

The idea to move forward was once again proposed by Richard Feynman in the early 1980s: We need a quantum machine to solve a quantum problem! If the machine is ruled by quantum physics, the superposition principle tells us that N atoms with two states can encode 2^N numbers, and we would need 2^N classical bits of information! Feynman's suggestion was the birth of the concept of quantum computer that is today actively investigated experimentally and theoretically. It relies on the superposition principle, quantum interference, and entanglement. Indeed, as the quantum state of N two-state particles encode 2^N numbers at the same time owing to the superposition principle, a quantum calculation on this state would also operate on the 2^N numbers simultaneously. However, one would then also obtain a superposition of all the results! This is where quantum interferences play a role: the quantum algorithm is designed such that all the paths interfere destructively, except the one leading to the desired result. In doing so the quantum computer generates a huge amount of entanglement.

Building such a quantum computer that operates on thousands of quantum bits (the superposition of two states) is a formidable task, as one has to fight decoherence (i.e., the tendency of any superposition state to collapse into one of the states of the superposition, and then behave classically). This decoherence originates from the fact that any system is not perfectly isolated but does interact (even very weakly) with its environment. Unfortunately, decoherence gets worse when the number of particles increases. Strategies to mitigate the role of decoherence are being devised at the same time as physicists are developing better-controlled quantum systems. These systems can be photons, atoms, and ions that are 'naturally quantum', but physicists have also designed what are called artificial atoms. A prominent example is a superconducting loop, which consists of a huge number of atoms, but which behaves as a two-state 'real' atom: the current through the loop can flow clockwise (state 1), anti-clockwise (state 2), or both at the same time in the same way a photon can be in a superposition of two polarizations. Today's systems already contain a few dozen quantum bits and roadmaps predict several thousands soon.

Even if, in terms of technological complexity, the quest for a quantum computer resembles the challenge of landing a man on the Moon and could take decades to reach, the quantum machines that exist in the laboratory today are already at the stage where they can be useful. For example, we have mentioned that a quantum computer generates a large amount of entanglement: this entanglement can be used as a resource to improve the accuracy of atomic clocks by many orders of magnitude, leading to extremely precise sensors or geo-localisation devices. Moreover, the elementary quantum processors already available are solving open scientific questions related to the conductivity of materials or the magnetic properties of matter. Amazingly enough, they can also be used to optimize industrial processes: finding the lowest energy state of an ensemble of interacting particles turns out to be equivalent to minimizing a cost function. Applications in logistic, finance, and in the optimisation of industrial processes are already actively being developed!

3.4.4.3 Extreme light

The emblematic challenge for ultrahigh-power lasers is to reach the so-called Schwinger intensity of 10^{29} W cm^{-2} at which electron–positron pairs can be created from the vacuum subject to the ultraintense electromagnetic fields. To achieve this value, several orders of magnitude in intensity need to be gained. One technological breakthrough would consist of generating high-energy attosecond pulses by sending a multi-PW laser onto a solid target, thereby creating a surface plasma mirror oscillating at a speed close to that of light. This relativistically oscillating mirror induces a Doppler effect on the reflected laser beam, which temporally compresses parts of the laser field, producing attosecond x-rays, thus reducing the duration and potential focus of the beam and finally allowing access to tremendous intensities (see section 3.5).

Currently, 100 PW lasers are under development. These lasers (compact compared to a synchrotron) pave the way, with new laser-acceleration concepts, to the possibility of reaching 100 GeV, and even TeV for acceleration lengths of a few meters instead of several tens of km today.

In order to study light–matter interaction with increased temporal and spatial resolution, attosecond pulses operating at shorter wavelengths in the XUV range are needed. Moreover, extending the XUV range will open up new applications such as in biology where the 200–400 eV photon energy is targeted since this energy band corresponds to a water-transparent window. Another important issue concerning the generation of XUVs attosecond pulses concerns the low photon flux. This is crucial for extending the range of application of these relatively compact XUV coherent sources such as for photoemission spectroscopy with pulses of 10–50 eV or photo-lithography with pulses of 100 eV.

3.4.4.4 Spintronics: toward all-spin systems

Sparked by the initial discovery of GMR, spintronics now pushes to provide new energy-conserving solutions to the existing silicon technology known as CMOS (complementary metal-oxide semiconductor). A good example of the potential of spintronics is the ultrafast and low-power non-volatile magnetic random access memories (MRAMs) based on the tunnelling of spins—a quantum property where electrons can pass through walls when these are thin enough (i.e., of the order of few atoms) between magnetic layers. We then speak of magnetic tunnel junctions. It is also possible to envision a sustainable post-CMOS technology based on spin as the information carrier. Thanks to magnetism's intrinsic non-volatility[2], it could deliver quantum information (a single spin is the perfect quantum variable) using ultrafast and low-power electronics while giving access to stochastic and neuromorphic computing, ultimately offering compatibility with optical transmission by exploiting the properties of circularly polarized light.

Achieving all-spin logics integration is one of the key remaining challenges. It requires the ability to transport and process spin information. Indeed, while spin information can be stored on the long term through magnetism (a magnet stays on a fridge forever) it quickly vanishes when carried away for processing: this is the spintronics paradox. Along this way, several paths are now being actively followed; these include the recently discovered 2D materials for efficient electronic spin transport, or, again, the development of radiofrequency communication between oscillator spin devices via spin wave technologies (magnonics).

---

[2] Note here the key importance of stabilizing ultimately small magnetic structures. This is one of the very active research fields of magnetism with magnetic textures such as Skyrmions that resembles a hedgehog of spins that is said to be 'topologically protected'.

3.5 Extreme light

Franck Lépine[1], Jan Lüning[2], Pascal Salières[3], Luis O Silva[4], Thomas Tschentscher[5]
and Antje Vollmer[2]
[1]CNRS-ILM-Lyon, Lyon, France
[2]Helmholtz Zentrum, Berlin, Germany
[3]Université Paris-Saclay, CEA, LIDYL, Gif-sur-Yvette, France
[4]Instituto Superior Técnico, Universidade de Lisboa, Lisbon, Portugal
[5]European XFEL, Hamburg, Germany

Ever since the discovery of x-rays by W G Röntgen in December 1895, the use of energetic radiation to interrogate matter has been a key tool for science, industry, medicine, and society at large. The invention of the laser by Theodore H Maiman on May 16, 1960 has increased the impact and applications of the light–matter interaction manifold still. Over the decades, light sources have been developed to access ever-shorter wavelengths, higher energies, higher powers, shorter pulse durations, and higher repetition rates. Such light sources are electron-accelerator based synchrotron radiation and free-electron laser sources, as well as laser-based attosecond and ultrahigh-intensity sources, considered 'extreme' with respect to what had been possible only two or three decades before today (2022).

It is safe to say that the development of these extreme sources of light, providing radiation across the electromagnetic spectrum, has been a critical ingredient in solving scientific and societal problems over the last century, and that this will continue for many decades and even centuries to come. *'We are in a climate and environmental emergency'* is a quote from Frans Timmermans, the Executive Vice-President of the European Commission in 2019 [157]. With climate change, sustainable energy supply, environment, biodiversity, clean water, food, health, digital transformation, and pandemics, humankind is confronted with unprecedented challenges. Light sources and radiation facilities, operating as national/international research units, each providing light with unique properties, are highly versatile tools that continue to prove their importance for an ever-expanding range of scientific fields. Starting from physics, chemistry, materials science, biology, and life sciences, they now encompass fields such as environmental science and the study of artefacts of cultural interest. The manifold examples of breakthrough experiments [158] advancing our knowledge on new energy sources, on the environment, and in the sector of health and medicine clearly demonstrate how extreme light sources instigate strategies and provide solutions to societal challenges.

Below, we present the realisation and prospects for extreme light sources, as well as their scientific and societal impact (note that single-photon sources are discussed in section 1.2.5). In order to respond to the current and forthcoming societal challenges as well as to lead the harvest of new knowledge in many fields, it is imperative that Europe as a leading industrialised area stays on the forefront of the development of new methods and techniques and maintains its own array of cutting-edge analytical research tools.

3.5.1 Synchrotron radiation

Synchrotron radiation facilities have proven to provide eminent and highly relevant analytical tools to advance our understanding of materials and, more generally, to find solutions for pressing societal challenges [158]. This is also reflected by the spread and use of the facilities themselves. Starting from three synchrotron radiation facilities worldwide in the 1960s, today, in Europe alone, fourteen synchrotron sources are operated as user facilities in ten European countries, together serving more than 25 000 users per year. In a synchrotron radiation source, a high-energy (GeV, i.e., a billion electron-volts) and finely focused electron beam propagating in an alternating magnetic field serves as a source of extremely bright radiation. The spectrum of the emitted radiation depends on the electron beam and magnetic field parameters and may range from the THz and infrared via the range of visible light, visible-to-ultraviolet (VUV) radiation, to soft and hard x-rays from which specific energies can be selected. In addition, full control of the polarization of the light is provided. At the same time, the radiation is pulsed, allowing time-resolved studies down to the picosecond (10^{-12} s) range.

The remarkable success of synchrotron radiation facilities fuels continuous interest in the further development of their capabilities. On the one hand, the increasing relevance of materials and devices that are spatially inhomogeneous on the nanometer (nm) length scale, whether due to the material's underlying physical properties or to tailored structuring, continuously drives the improvement of the spatial resolution that can be achieved. On the other hand, there is an ever-increasing user demand for new analytical capabilities, with a notable shift of interest from understanding a material's static ground state properties towards its behaviour in operation under application relevant conditions. Time-resolved experiments that probe local properties study how fundamental properties as well as properties relevant for applications evolve, thereby yielding the insight necessary for application specific tailoring of materials. This has lead to advances *in situ* or *operando* experimental techniques, experiments on liquids, on biological objects, and yet more, which, in turn, stimulate a broadening of the user communities themselves.

To address these needs, synchrotron radiation facilities continuously improve the performance of their light-generating accelerators. The much-discussed transition from the third to the fourth generation of synchrotron sources is characterised by a radical increase in brightness, which is the main performance parameter describing the beam quality of an accelerator-based light source; this has become possible with the recent advent of a novel technique, the so-called multibend achromat (MBA) storage ring design (figure 3.22). The increase in brightness goes along with an increase in the spatial coherence of the emitted light, which dramatically improves focusing capabilities and the application of novel imaging techniques. The new Max IV source in Sweden has pioneered the implementation of the MBA design. Essentially all facilities worldwide currently envision an upgrade of their accelerator to this technology [159], with the ESRF [160] having been the first to do so (in 2020), thus becoming the world's brightest synchrotron light source. As an illustration, the upgrade to a fourth-generation source increased the brightness of ESRF's x-ray

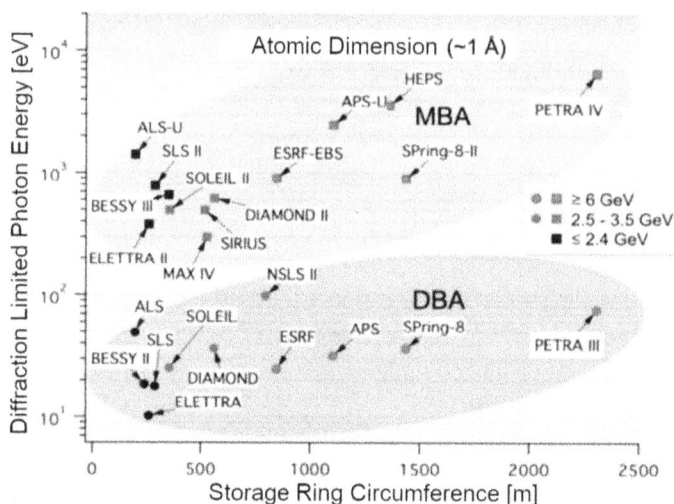

Figure 3.22. The diffraction limited photon energy is shown versus the circumference of the accelerator for synchrotron radiation facilities based on the double-bend-achromat (DBA, third generation) and the multi-bend-achromat (MBA, fourth generation) accelerator lattice design. The comparison highlights the jump-like boost in brightness and coherence gain by transition from the third to the fourth generation sources.

beam by a factor 100 over the previous upgrade completed in 2015; the source, which feeds 40 beamlines, is now 10 000 times stronger than when it was first inaugurated in 1994!

3.5.2 X-ray free-electron lasers—studying the dynamics of atomic and electronic structures

While synchrotron radiation sources had matured by around 2000, new developments allowed the conception of a new type of electron accelerator-driven x-ray light source. Compared to (synchrotron) storage rings, free-electron lasers use an electron beam with narrower collimation, a much smaller bandwidth, and a much higher peak current to create laser-like (i.e., intrinsically coherent) x-ray radiation. The free-electron laser (FEL) process uses a resonance in the x-ray generation process—the self-amplified spontaneous emission or SASE process—to produce slices in the electron bunches coherently emitting x-rays. The FEL process shows an exponential gain along many thousands of magnetic periods (so-called undulators), until saturation is reached (see figure 3.23). Due to the high peak current and the ultra-short electron bunches, FEL pulses can be very energetic (of the order of several mJ) and very short (of the order of tens of femtoseconds, or 10^{-14} s). Their wavelength is governed by the same equation as undulator radiation and can be tuned by the electron energy and the magnetic field of the undulators. The concept of FELs was proposed in the 1970s [161], but only the small emittance of modern accelerators and the single-pass self-amplified spontaneous emission or SASE principle [162] has allowed physicists to reach the conventional x-ray regime with wavelengths shorter than 0.1 nm.

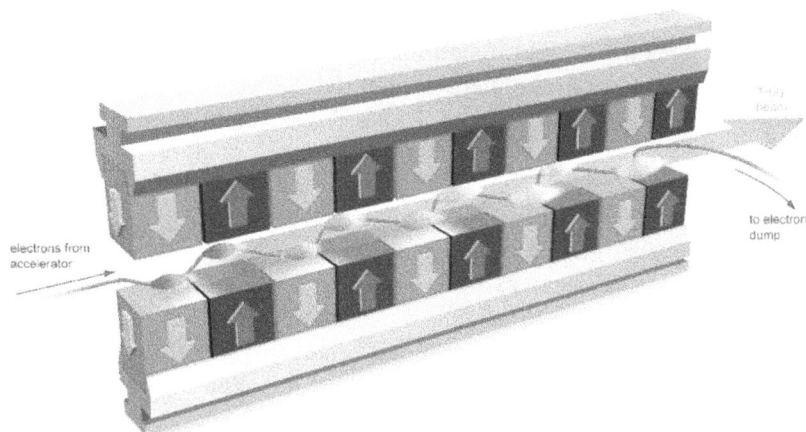

Figure 3.23. Scheme of SASE FEL process using alternating magnetic fields to initially create synchrotron radiation, which then acts back on the electron beam producing density modulations within the electron bunches. In the final section of very long undulators (typically 1000 periods; shown are five periods) the electron bunch is fully structured and electrons emit radiation coherently.

Since the slices do not emit coherently between them, the temporal coherence of x-ray FELs is not complete. The development of fully coherent sources is an area of active research and has already provided fully coherent FEL radiation for wavelengths exceeding 1 nm.

The properties of FEL radiation offer unique possibilities for following the time-dependent evolution of geometric and electronic structures of atoms, ions, molecules, clusters, nano- and bio-particles, as well as those of bulk materials. Sometimes called 'taking the molecular movie', the application of ultra-short and intense XFEL pulses allow the detection of atomic-scale movements as well as extremely rapid electronic structure changes (e.g., during a chemical reaction [163], a phase transition of a solid [164], or a magnetic recording [165]). Such rapid transient states regularly occur in reality, but cannot be observed using conventional x-ray sources. XFEL pulses can also be used to record the structure of very short-lived states, such as in matter exposed to extreme electromagnetic fields relevant to materials research and to the geo- and planetary sciences [166]. While most FEL applications apply conventional x-ray techniques, the intense and coherent FEL radiation also allows one to develop and employ new, nonlinear x-ray spectroscopy methods [167].

The first user facility based on the SASE FEL principle FLASH at DESY (Hamburg, Germany) opened to a broad scientific community in 2005. Now, about ten facilities for the wavelength regime from the VUV to very hard x-ray radiation operate worldwide and serve a growing community of scientists. Their operation is similar to that of synchrotron radiation facilities, and the scientific applications and user communities employ the complementary features of x-ray radiation provided by these two types of extreme light sources. The international network of FEL facilities collaborations with the goal to expand scientific applications and to further develop FEL sources and techniques. The most important of these developments are the increased application of high repetition rate FELs with pulse rates from 0.1 to above 1 MHz [168], the development of methods to provide attosecond (10^{-18} s) pulse duration and attosecond pulse trains [169–171], the generation of fully coherent FEL pulses [172], as well as the development of nonlinear x-ray spectroscopy [173]. Some of the first FEL facilities are in the process of upgrading in order to improve the quality of FEL radiation. In addition, a second generation of FEL facilities is either under construction or has been proposed. Already, present-day facilities are characterised by an extreme data rate, related to the fact that each x-ray pulse basically corresponds to a full experiment for which the corresponding scattering and emission signals are recorded. With increasing repetition rates and more complex and larger x-ray detection schemes this issue will become more and more important.

3.5.3 Attosecond light sources

The quest for ultrashort light sources has been triggered by the dream to follow and control physical/chemical/biological processes at the level of the fundamental constituents of matter, from molecules and atoms, down to electrons. Tabletop lasers have been instrumental in this endeavour. Indeed, the rapid development in laser technology from 1960 has made it possible to produce shorter and shorter light pulses, down to a few femtoseconds in the 1980s, reaching the single-cycle limit in the visible/infrared (IR) range. Such ultrashort laser pulses allow the observation of atomic motion within molecules, which has spawned the field of femtochemistry [174] and was rewarded with a Nobel Prize to A Zewail in 1999. In order to follow the electronic motion within atoms, molecules, and solids, attosecond pulses are required[3]. The production of attosecond flashes of light is confronted with a fundamental difficulty. Namely, the Fourier transform link between time and frequency implies a very broad spectrum extending from the visible to the extreme ultraviolet (XUV), with different spectral frequencies all in phase (i.e., one requires coherent broadband XUV radiation). A solution was identified in the 1990s: the interaction of intense femtosecond IR laser pulses with atomic gases results in a huge broadening of the laser spectrum due to strongly nonlinear optical processes in the form of the generation of a whole set of high-order harmonics of the laser frequency [175, 176]. As these harmonics inherit the coherence of the fundamental laser, the Fourier conditions are met: there is a parallel increase in central frequency/ bandwidth and a concomitant decrease in pulse duration. The clarification of the underlying physics [177–180] confirmed the possibility to generate attosecond pulses, and soon after, at the dawn of the 21st century, attosecond metrology techniques required for such short pulses were developed [181, 182].

At the time of writing, huge progress has been accomplished in both the generation and the characterization of isolated attosecond pulses [183, 184]. While

---

[3] To give an idea of the involved timescale, in the Bohr model of the hydrogen atom, the characteristic time associated with the journey of the electron on the first orbit around the nucleus is 150 attoseconds

all the initial studies were performed using Titanium:Sapphire laser drivers (λ = 0.8 μm), the recent use of few-cycle driving pulses in the mid-IR (λ = 1–2 μm) has yielded XUV pulse durations as short as 50 attoseconds [185, 186]; the central frequency was also increased to reach the so-called 'water window' [187]. Using a λ = 3.9 μm driving wavelength, the high harmonic generation process was extended to orders greater than 5000, reaching a 1600 eV photon energy and providing a record 700 eV spectral bandwidth [188]. While the latter would be compatible with the generation of pulses as short as 2 attoseconds, close to the zeptosecond (10^{-21} s) range, these are still to be demonstrated, and will require significant progress both in spectral phase management, and in metrology techniques. Coherently mixing IR and mid-IR pulses allows sub-cycle control of the driving light wave, providing μJ energies in a single attosecond pulse centred at 30 eV photon energy [189]. The corresponding gigawatt (GW) peak power results in focused intensities in the 10^{13}–10^{14} W cm^{-2} range, large enough to induce multiphoton transitions. In terms of repetition rate and average power, bright prospects are brought by the new Ytterbium laser technology. As compared to the typical 1–10 kHz repetition rate of standard Titanium:Sapphire lasers, Yb lasers hold the promise of attosecond sources with repetition rates in excess of 100 kHz and average powers reaching 0.1 μW eV^{-1} at 90 eV [190]. Finally, advanced control over the spatial, temporal, and polarisation/angular momentum properties of the attosecond beams has been achieved.

These exquisitely controlled tabletop attosecond sources have spread in laboratories all around the world, with Europe keeping the lead after having pioneered the field (as illustrated by the 2023 Physics Nobel Prize awarded to A L'Huillier, P Agostini and F Krausz). The European attosecond community has been structured thanks to a number of (Marie Curie) European Networks that have trained young researchers, starting as early as 1993. Furthermore, the cluster of European laser facilities Laserlab-Europe encompasses several attosecond facilities offering beam time to international users since 2006. Finally, important investments worldwide are currently devoted to the construction and operation of large-scale attosecond facilities, such as the European Extreme Light Infrastructure—Attosecond Light Pulse Source (ELI-ALPS) in Szeged-Hungary, the NSF National Extreme Ultrafast Science Facility (NeXUS) at Ohio State-USA, and a number of new projects are blooming (e.g., in China).

In parallel, two new sources of attosecond pulses have emerged. First are the x-ray FELs [170, 171] that provide extremely powerful—100 GW—tunable pulses extending to the x-ray spectral region and discussed above. Second, high harmonic beams generated from the laser–plasma interaction at very high intensity ($>10^{18}$ W cm^{-2} instead of typically 10^{14} W cm^{-2} for gas harmonics, see discussion below) present promising properties, in particular for high-energy pulses. Both attosecond pulse trains [191] and isolated pulses [192] have been demonstrated for these plasma sources.

3.5.4 Attosecond applications—imaging at the attosecond scale

The stability of matter is the result of the complex interplay between interacting microscopic constituents: electrons and nuclei. The complexity involved has its origin in the electronic correlations (see section 3.2) that determine how electrons interact quantum mechanically with each other. The difficulty to account for electronic correlations theoretically limits the accurate prediction of physical and chemical properties. While the description of static properties of matter has made enormous progress, our knowledge on how electronic correlations influences out-of-equilibrium properties is still in its infancy. Attosecond science allows one to measure and manipulate electrons on the quantum level, using the coherent properties of light. Over the past 20 years, a large effort was dedicated to the investigation of small quantum systems (atoms and small molecules) in the gas phase [193]. This has constituted a first playground to test state-of-the-art quantum theories and has allowed the investigation of the stability of excited electronic states [194].

Accessing processes on the attosecond timescale has revealed the complexity of seemingly trivial issues. One example is the well-known photoelectric effect, with the question 'how much time does it take for an electron to escape an atom, a molecule or a surface, after the absorption of a photon?' This simple question hides profound interrogations on the quantum mechanical properties of matter, on causality, and on the concept of measurement in quantum mechanics, time not being an observable. It involves discussion of the 'Wigner delay' semi-classical concept and its connection to the scattering phase acquired by the electron. Experimentally, the temporal dynamics of the scattering process has become accessible in atoms, molecules, and solids [195]. Recent investigations include the study of resonance effects [196–198], of single vs. multiphoton transitions [199], of electron scattering in nanoparticles [200] as well as in liquids [201]. The high coherence of the attosecond sources gives direct access to the phase and amplitude of the ejected electron wave packets, allowing the dynamical reconstruction of the formation of quantized structures such as Fano resonances [202], or resonant two-photon transitions [203] (figure 3.24).

Attosecond molecular physics has emerged as a major field of interest, with, notably, the attosecond control of molecular reactions using the coherent manipulation of their electronic degrees of freedom [204]. The first demonstration was performed on H_2 through the quantum manipulation of electronic localization following molecular dissociation [205]. Light controlled with attosecond precision has now allowed the modification of the absorption and dissociation of larger molecules such as N_2, CO_2, and C_2H_4 [206], as well as to trigger coherent charge

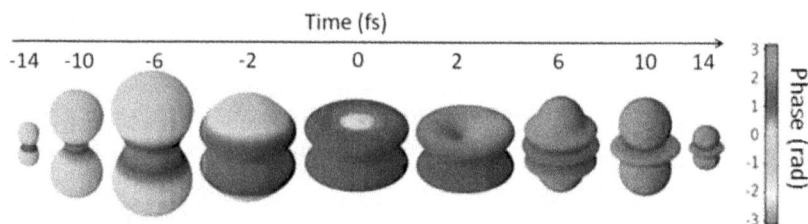

Figure 3.24. Space-time movie of the 2-photon XUV+IR ionization of a helium atom through the intermediate 1s3p and 1s4p bound states [204]: time dependence of the complex-valued atomic ionization rate (amplitude in polar coordinates, color-coded phase).

oscillations across the molecular backbone of amino-acids such as phenylalanine [207]. The modification of molecular reactivity by directly acting on the electronic localization questions the role played by electronic correlations [208] and the complex interaction between electrons and nuclei. In these experiments, the electronic charges are manipulated faster than the timescale on which any nuclear motion can occur, so that the nuclear structure has to re-adapt to the new charge configuration. This new paradigm was called 'charge driven reactivity'. In the past years, tremendous work has been done to reveal the role of coherent hole dynamics [207], non-adiabatic relaxation [209], and structural rearrangements [210].

While attosecond pump–probe schemes initially used electron spectroscopy, other promising spectroscopic techniques are currently being developed. For instance, attosecond transient-absorption spectroscopy has been performed on dense jets and solids, with demonstrated control of the absorption processes using the interaction with strong field-controlled IR pulses. It has thus been possible, for example, to drive insulator-to-conductor transitions in crystals [211] and to study exciton dynamics [212].

3.5.5 Laser–plasma interaction and the ultra-relativistic limit

3.5.5.1 The present

The invention of chirped pulsed amplification, or CPA (figure 3.25), recognized with the 2018 Nobel Prize awarded to Gérard Mourou and Donna Strickland [213], has triggered an exponential increase in laser intensity and a concomitant decrease in the achievable brevity of pulse duration, with a celerity that recalls Moore's law in electronic transistors.

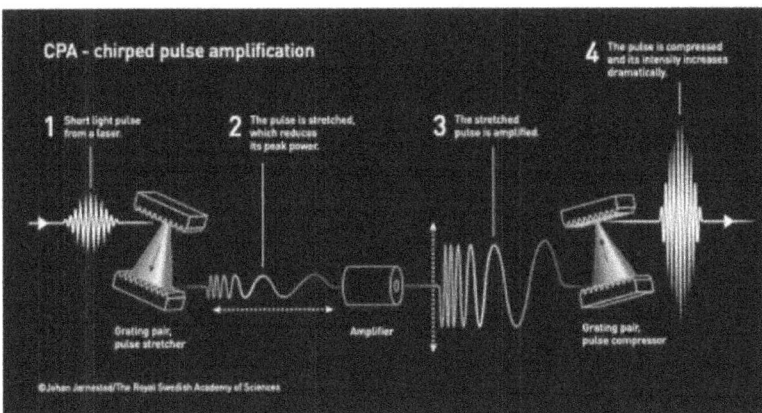

Figure 3.25. Chirped pulsed amplification (CPA)—a short laser pulse (left) is stretched spatially by a pair of gratings with a well-defined spectral encoding (low frequencies first, high frequencies of the pulse in the back of the stretched pulse). The longer pulse can then be safely amplified without damaging the amplifier and/or significant nonlinear effects. The amplified stretched pulse can then be compressed with another pair of gratings, and the frequencies properly combined to recover the original short pulse length.

As the laser intensity increases, novel regimes of physics can be explored in existing laser laboratories worldwide [214]. Notably, matter irradiated by intense laser light becomes fully ionized and forms a plasma. At intensities in excess of 10^{23} W cm^{-2} [215], the electrons in the plasma quiver in the laser field with velocities very close to the speed of light; the plasma response to the laser light is highly nonlinear and determined by special relativity.

At the same time, we are witnessing a golden age for plasma kinetic simulations. Taking advantage of the largest supercomputers in the world, these are able to capture the physics in these novel relativistic and ultra-relativistic regimes, and complement experimental advances with unprecedented detailed synthetic diagnostics. This virtuous combination has propelled exciting advances in laser–plasma interactions.

The most prominent are associated with the development of the entirely new concept of compact advanced laser–plasma accelerators (figure 3.26) [216]. Here, an intense laser pulse propagating through a transparent plasma excites a trail of plasma oscillations with their associated periodic, intense electric field. These oscillations propagate close to the speed of light, much in the way a boat drives a wake in a lake. The intense electric field oscillations accelerate the charged particles of the plasma. Thus, recent experiments have delivered electron beams of energy close to 10 GeV, propelling charges of up to a fraction of a nanoCoulomb (nC, or 10^{-9} C) in pulses of duration of a mere few femtoseconds. Thus, laser–plasma

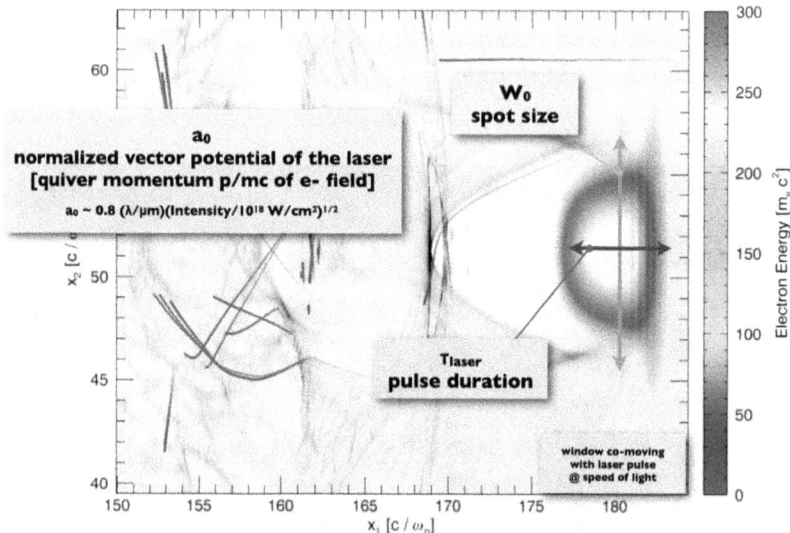

Figure 3.26. Schematic of laser wakefield acceleration. A strong laser pulse (depicted at the right of the simulation window in red) propagates in the plasma—the radiation pressure is strong enough to expel the electrons from the wakefield region, forming a bubble/electron cavity. The ions in the plasma column thus create an electric field that pulls electrons in, closing the bubble, but that also provides electron acceleration. Inside the bubble, electrons (coloured according to their energy) can accelerate and gain energy. Image courtesy of Samuel F Martins GoLP/IPFN/Instituto Superior Tecnico.

acceleration has demonstrated ultrahigh accelerating gradients in excess of 1 GeV cm^{-1}. The properties of electron beams generated in this manner are continuously optimized. In particular, high-quality beams with energy spread $\Delta E/E$ as low as 0.1% have been demonstrated. Critical advances on laser technology (such as stability, reproducibility, and repetition rate) and on targets (plasma channels of controlled length and density) have been (and are) the main stepping stones for further advances.

Electron beams generated by the laser–plasma interaction are now routinely used as compact drivers for betatron sources of (incoherent) x-rays [217]. These sources provide a unique source for high-resolution, high-contrast coherent diffraction imaging with a wide range of applications going from biology and biomedicine to materials science (figure 3.27). A demonstration of free-electron lasing driven by a laser–plasma accelerator in the x-ray range [211] has just been achieved, holding the promise to further increase the impact of laser–plasma accelerators as compact drivers for ultrafast (duration of a few fs) x-ray sources unleashing a wide range of applications across many scientific fields.

Advances on ion acceleration have been steady with the exploration of several laser-driven proton/ion acceleration mechanisms and their optimization. For ion acceleration, the laser hits a solid (or near solid) density target where its energy is predominantly absorbed by the electrons. The dynamics of the hot electrons then generates electric fields, via different mechanisms, that can pick up and accelerate the protons. Many conceptual studies have been performed, targeting proton therapy with these unique beams, and addressing the distinct features of ultrashort, but high-intensity, doses. As laser technology progresses and novel targets are engineered it is expected that one of the key milestones for proton therapy, proton acceleration to energies in the 200–250 MeV range, can be achieved within the next few years.

Fundamental studies of laser–plasma interactions have an important impact in laser fusion. The laser energy to be deposited in the fuel must first propagate through very long transparent and semi-transparent plasmas, in which different types of plasma instabilities can develop. Progress in computer modelling has underscored

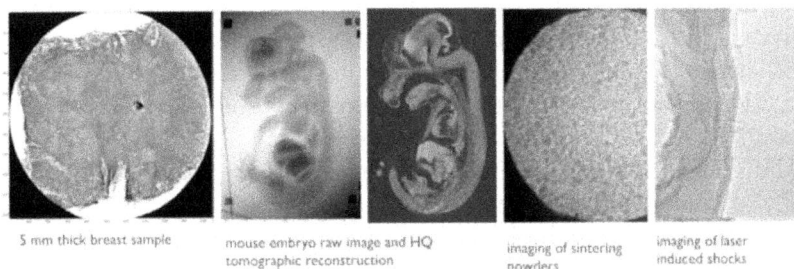

5 mm thick breast sample   mouse embryo raw image and HQ tomographic reconstruction   imaging of sintering powders   imaging of laser induced shocks

Figure 3.27. Imaging with betratron-generated x-rays, illustrating several applications of betatron sources [218] probing different targets/materials. Image courtesy of Nelson Lopes (Imperial College, London and GoLP/IPFN/Instituto Superior Tecnico). Reprinted from [219], copyright (2020) by permission from Springer Nature.

the importance of controlling deleterious laser–plasma instabilities by careful engineering of the spectral content of the lasers. This will become even more relevant as laser fusion moves to the burning plasma and ignition domain, triggered by the recent results at the Lawrence Livermore National Laboratory announced in late 2022, and recently published. These conditions provide an important test bed for burning plasma conditions, of relevance for any other fusion device, but also can give additional momentum to the more energy-efficient direct-drive laser fusion, based on shock ignition, and other advanced concepts such as fast ignition.

The first steps on the manipulation of light beyond the damage threshold of conventional materials have been taken. These most notably involve plasma mirrors that are now routinely used to improve the intensity contrast of lasers. Extensive theoretical and numerical work has explored the combination of intense lasers and plasmas. We cite laser–surface interactions and relativistic plasmonics and novel routes, based on strong coupling with the plasma, to amplify or to focus laser light in order to achieve ultra-high intensities in excess of 10^{26} W cm^{-2}.

These advances are touching other fields of physics. The confluence of high-intensity lasers, relativistic dynamics, and optics allows the study of laser–plasma interactions with exotic beams. These feature, for example, a phase velocity opposite the group velocity, or a superluminal moving focus. The detailed control of the phase of the laser light, as well as its orbital angular momentum, is opening novel directions for fundamental studies and for laser–plasma accelerators and fine-tuned secondary sources.

Exciting developments at the interface between high-intensity lasers and plasma astrophysics are starting to shed light on fundamental processes in astrophysics. This emerging field of *laboratory astrophysics* has seen important developments in recent years. One example concerns shock waves. These are pervasive in the Universe and are thought to be the site for the acceleration of cosmic rays, the most energetic particles in the Universe. These shock waves are collisionless, in the sense that the mean free path of the particles is much longer than the shock wave thickness—the shock structure is determined by collective plasma processes. High-intensity lasers provide a unique tool to drive these shocks in the laboratory, generating collisionless conditions and sub- to near-relativistic plasma flows. Recent experiments, supported by extensive theoretical and numerical work, have demonstrated the formation of laser-driven collisionless shock waves, spanning from electrostatic to fully electro-magnetic shocks (figure 3.28). The exploration of the acceleration mechanisms in these experiments has provided first hints on possible proton acceleration and Fermi acceleration— i.e., the key mechanism for cosmic ray acceleration—in laboratory shocks [219]. The first studies on magnetized plasmas driven by lasers demonstrate a possibility of studying jets, magnetogenesis, and magnetized turbulence, providing yet stronger connections with plasma astrophysics.

For many years, electron–positron plasmas have been addressed theoretically and numerically, motivated by their relevance in high-energy astrophysics (most notably associated with pulsars), and also due to their (apparent) simplicity. Recent experiments resorting to high-intensity lasers have not only generated copious amounts of electrons and positrons but have also been able to produce a quasi-neutral

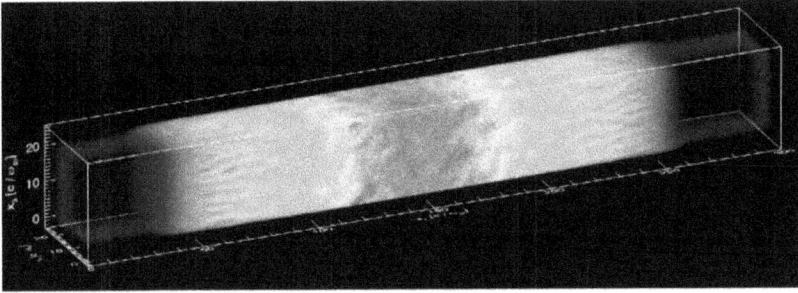

Figure 3.28. Numerical simulation of a laser-driven collisionless shock driven by two ablated plasmas, originating from the left-hand side and the right-hand side of the simulation window. In the centre, the compressed density, permeated by magnetic filaments, originates the collisionless shocks propagating from the central region (red corresponds to the regions of higher density—downstream, while blue represents the lower region—upstream region of the shock). Image courtesy of Frederico Fiuza (SLAC/Stanford University).

electron–positron plasma plume. The interaction of energetic electrons (accelerated via the laser wakefield) with a solid target of high mass number ('high Z') atoms generates gamma rays, via bremsstrahlung. In turn, these can generate electron–positron pairs through scattering on the high Z nuclei. The road is now open to the experimental study of plasma properties of relativistic fireballs, thought to be a key ingredient to the understanding of gamma ray bursters, including the onset of plasma instabilities, the self-consistent generation of magnetic fields, radiation from fireballs, and the formation of collisionless relativistic shocks.

These studies highlight the convergence of high-intensity lasers, quantum electrodynamics, and plasma physics. These opportunities have triggered several experimental directions, as described previously, but have also motivated new theoretical directions. As the laser intensity increases, a plethora of quantum electrodynamics processes starts to play an important role. The quantum electrodynamics of ultrahigh-intensity laser fields poses highly challenging theoretical questions [220] —ranging from radiation reaction, multiphoton processes in the ultra-relativistic regime, to vacuum birefringence—but the very possibility to generate very large numbers of secondary particles (e.g., electron–positron pairs, gamma rays) via cascade or avalanche processes can trigger further QED processes and might provide an alternative route to the generation of even higher density electron–positron(-photon) plasmas, with densities comparable to those of solids. Collective plasma dynamics in such conditions is unexplored and has strong connections with extreme astrophysical phenomena. Laser experiments are on the brink of reaching the quantum regime. Experiments on the collision of high-intensity lasers and electron beams have started to explore the transition between radiation reaction in the classical regime and in the quantum regime. It is clear, however, that this is a first step in the pursuit of even more extreme conditions, where fully developed electron–positron cascades can develop, providing a test bed for fundamental processes of relevance in astrophysics (figure 3.29).

Planned or commissioned laser facilities will move into these exciting new directions, pursuing both fundamental science and applications across multiple

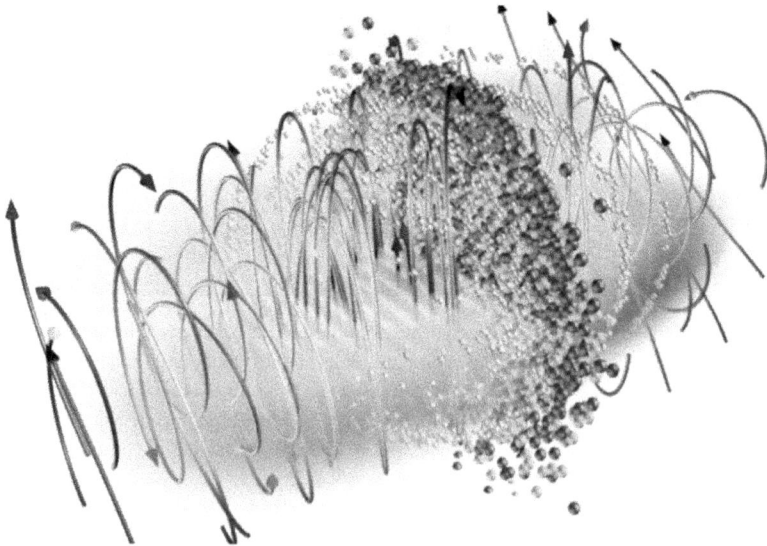

Figure 3.29. Electron–positron cascades driven by two counter-propagating ultraintense laser pulses (field lines of the beating structure are depicted as well as the amplitude of the beating pattern in the plane). In the interaction region the strongly accelerated electrons (green) radiate photons (in yellow) that in turn interact with the ultrastrong laser field and generate copious amounts of pairs (electrons—green and positron—red) in a cascade process. Image courtesy of Thomas Grismayer GoLP/IPFN/Instituto Superior Tecnico.

disciplines. These facilities include the EU-wide Extreme Light Infrastructure, Apollon (France), EPAC (UK), or, more focused on laser fusion, the Laser MJ (France) or the EU-wide effort on laser–plasma accelerators EUPRAXIA or HIPER+ on laser fusion.

3.5.6 Challenges and opportunities

3.5.6.1 Synchrotrons and free-electron lasers

About 30 years before today (2022) x-ray FELs and their scientific use had not yet been proposed and synchrotron radiation sources had just started into a new era. Looking 30 years ahead, it is certain that electron accelerator-based synchrotron radiation sources and x-ray FELs will have developed into x-ray light sources with full control of all kinds of radiation parameters in a mature way.

The jump in source performance provided by the MBA synchrotron source technology will provide the basis for entirely new experimental capabilities, enabling new discoveries and insights in a broad range of scientific fields, serving existing as well as new research communities. Higher brightness allows for unprecedented improvement in spatial resolution for microscopy, imaging, tomography, and diffraction, enabling novel types of experiments, and drastically speeding up experiments. Better performance and higher throughput will be beneficial for both the quality and the quantity of experimental results and with this for the amount of created knowledge.

Therefore, after ESRF, all other European facilities, such as the Spanish synchrotron ALBA ('dawn', inaugurated in 2010, with eight beamlines), BESSYII at the Helmholtz Zentrum in Berlin (with 46 beamholes), the DIAMOND [221] light source in Didcott (UK), ELETTRA [222] (Italy), PETRA III [223] (Jordan), the Swiss Light Source [224], and SOLEIL [225] (France) (in alphabetical order) are each planning their upgrades so as to deliver radiation with a brightness improved by at least two orders of magnitude. Similar projects are pursued by the leading facilities outside of Europe.

The improvement of the coherence of synchrotron light concomitant with MBA will allow the application of 3D microscopy with nanometer spatial resolution as a routine analytical tool, using dedicated sample environments for *in situ* and *operando* studies, in a broad range of research fields, with time resolution as a fourth dimension.

X-ray FELs will provide extreme light with wavelengths down to 0.02 nm, pulse durations ranging from attoseconds to picoseconds, arbitrary pulse shapes, and full coherence that users will be able to select. These features will enable new applications in a vast range of scientific domains. In addition, alongside attosecond and femto-second lasers, laser-driven Compact-FELs will complement accelerator-driven facilities and will contribute to the worldwide network of FEL infrastructures, large laser installations, and synchrotron radiation sources. Compact-FELs may feature reduced power and peak performances, but their reduced size, operation effort, and cost will allow to use them for specific applications or to install them in regions without access to x-ray sources. There is a possibility that Compact-FELs will become the preferred technology for the UV to soft x-ray (>1 nm) spectral range.

The current development of applying new data science technologies both to the operation and experiments of both synchrotron and FEL facilities will have a huge impact, similar to the evolution of the instruments themselves. In addition to improving photon beam parameters, rigorous digitalization of the facilities and the experiment offers unique opportunities to improve efficiency. Applying artificial intelligence will allow the operation of more stable, more performant, and more energy-efficient light sources. Software developments will not only mitigate today's increasing challenges to deal with huge data rates and volumes, but they will likely change our way of doing science: data algorithms will be looking for patterns in huge datasets, and will find new correlations, which will become experimental observations.

The implementation of remote access to the infrastructures is a very important step towards making the facilities even more open, accessible, and inclusive, to a much wider user community, and thus increase their relevance. Recording, storing, and instantaneous data analysis will provide real-time guidance for experiments, ideally coupled with autonomous decision taking. The growing scientific community and the expansion of research fields also presents synchrotron radiation and FEL facilities with new challenges. Tailored sample environments, multimodal measurements, and offline instruments complementing synchrotron radiation-based techniques allow for higher throughput and more efficient experiments, thus increasing the attractiveness and outreach of synchrotron research even further.

3.5.6.2 Attosecond light sources

Predicting the properties of attosecond sources on the Horizon 2050 is challenging. Thirty years ago, the high-order harmonic generation process had just been observed, x-ray FELs did not exist yet, and the production of attosecond pulses was not even envisioned. The rate at which advances are currently being made means both the ultrafast laser and x-ray FEL fields are in full evolution, meaning that different complementary paradigms are evolving rapidly. All in all, we may expect that in 2050 attosecond light sources are available as routine tools for scientific research; tabletop laser sources and large-scale accelerator-driven x-ray FELs will, together, provide a wide spectrum of attosecond light source properties, including a wide wavelength spectrum from the XUV to very hard x-rays, pulse energies in the millijoule range, high repetition rates, full coherence, and polarization control, as well as access to the entire attosecond range and even to pulse durations entering the zeptosecond regime. A particularly interesting recent prospect is opened by tabletop laser sources relying on the generation of attosecond pulses by solid materials such as crystals instead of gas jets or cells [184]. This may lead to further integration and compactness, key to cost-efficient highly reliable systems that could then spread in laboratories and industries and be used by researchers/engineers from other fields. Attosecond techniques could become an everyday tool for applications in chemistry, materials science, or spectroscopy labs in general. In parallel, the accelerator-driven FELs will offer the most extreme characteristics for pioneering scientific studies, in particular requiring high flux. The best possible vision is doubtlessly that which is fuelled by outstanding scientific challenges, as we now outline.

3.5.6.3 Taking movies of molecules and real materials as they function

As to the most important scientific applications of ultrafast x-ray sources in decades to come, we first note that we are currently witnessing a phase of exploratory experiments, during which new methods and techniques are developed and tested on prototypical or highly standard sample systems. The results of these experiments largely serve to qualify experimental results. Their success will lead to a maturing of methods and techniques and to their application to less standard, more complex, but also more relevant scientific or technological problems. Therefore, in 30 years, one is likely to see many instruments specialized in specific methods and applying these to specific classes of material samples. At the same time, the development of methods will continue in order to enable new applications of extreme, ultrafast x-ray light sources.

A highly promising scientific application of ultrafast x-ray sources will be to take molecular movies in the very broadest sense. Employing the unique properties of x-ray FELs to adjust spatial and temporal resolution, such movies will allow one to follow atomic motion and electronic excitations from the shortest (attosecond) time regime to the rearrangement of macroscopic numbers of atoms in the millisecond regime. Probed samples will range from small molecules, via complex bio-systems, to real materials, and investigations will be performed under *operando* conditions. Understanding and describing a probed volume at all length scales, from the atomic to the macroscopic regime, and being able to dial in to the relevant voxels will employ techniques developed at synchrotron sources today, and will imply a huge data management and software challenge. Materials' research will benefit the most from these movies, as they will allow to develop a profound understanding of the dynamic processes responsible for specific materials' characteristics. Such will contribute to the development of new energy conversion materials, catalytic systems, and entire systems such as batteries. The investigation of the dynamics and behaviour of hard materials will also benefit from this 'four-dimensional atomic resolution microscopy' (time being the fourth dimension). A very important area of application will be the investigation of stochastic processes such as crack propagation, defect formation, or crystallisation. These require special techniques to observe the spontaneous and usually non-ergodic events. Nonetheless, the majority of experiments will apply deterministic and tailored drive pulses with THz to x-ray wavelengths, available at each FEL instrument or multicolour FEL operation to initiate and probe specific dynamic processes of samples mimicking nature.

Another field where x-ray FELs will have a big impact is the investigation of the nature and dynamics of liquids allowing one to understand their behaviour and physical properties, and thereby allowing one to design new liquids with designed physical properties. The emergence of fully coherent FEL radiation and the development of nonlinear x-ray spectroscopy in the next decades will open the route towards new scientific applications which in 30 years may just transit from the prototypical phase to applications with direct relevance to societal challenges. In particular, in the area of quantum optics completely new insights are to be expected. With the transition to mature FEL applications for specific scientific problems the presence of industrial applications of x-ray FELs will have significantly increased compared to today. Following trials to use FEL for 13.6 nm lithography during the last decade, x-ray FELs might turn out to be the light source providing powerful, stable, and coherent radiation enabling lithography in the regime of 1 nm or even below.

3.5.6.4 *Future applications of attophysics*

Like femtosecond technology before, we expect that attosecond technologies and their applications will swiftly evolve in the coming years [167]. Femtosecond science has demonstrated the new possibilities offered by ultrashort broadband coherent light pulses and has progressively evolved from the investigation of basic concepts such as the dynamics of nuclear wave packets to increasingly more complex spectroscopic approaches, including multidimensional spectroscopy and coherent imaging, and has progressively found industrial applications such as surgery and in material structuring. At the time of writing, attosecond science has unveiled new questions on the nature and dynamics of all phases of matter, from atoms and molecules in the gas phase, to liquids and solids. Augmented spectroscopies and the control of material properties and chemical reactions have already emerged as addressing very fundamental aspects of quantum mechanics. They have pushed experimental performances and precision, broadened the community of interest, and touched possible applications.

Overall, manipulating light with ultrahigh accuracy has clearly pushed our understanding and capacity to control nature and create new generations of technologies. The prime requirements have been to adapt light technologies and spectroscopic methods to the frontier of new temporal resolutions. In the next decades, small quantum systems could be interrogated with unprecedented accuracy in order to perform 'complete experiments' where, electrons, nuclei, and photons could be all measured, together with their 3D properties (position, velocity, spin, etc). We foresee experiments in which the complete, phase, and amplitude-resolved reconstruction of electronic and nuclear wave packets and the elucidation of their properties over the wide spectral range from UV to x-ray, including coincidence between these particles, will be possible. Such would present a major test for quantum theories, and would provide crucial information for a universal description of electron correlations. It would provide a quantum mechanical view of non-equilibrium processes occurring at microscopic scales.

The use of attosecond pulses to study and control molecules will be further pushed towards larger and more elaborate complexes, including isolated fragile biomolecules, metal-complexes, molecules in an environment, and exotic ions. This opens the perspective to investigate and control elementary processes in complex biomolecules such as proteins, in which radiation damage could be investigated at its ultimate timescale. By doing so, attosecond science will address questions beyond proof of principle, and connect to research fields such as radiolysis, photocatalysis, and biochemistry. It will be possible to control phase transitions in molecular materials and the material bulk by acting coherently on electrons and holes. The recent application of ultrashort XUV pulses to materials has emerged as a new frontier, and one may expect that the usual technologies that are available at synchrotrons or at femtosecond labs will also be used and coupled to attosecond light sources in order to address correlated materials, topological insulators, complex nanostructures, molecules in liquid samples and so forth. Proof-of-principle experiments are already being carried out at time of writing using electron energy- and momentum spectroscopy (ARPES, 3D PEEM). While already used with femtosecond and UV pulses, they might soon become also common tools in attosecond science, providing detailed dynamical information on molecules and nanostructures at interfaces with e.g., metallic surfaces. Applications will cover spintronics as well as plasmonics, in which 3D fields will be resolved with attosecond precision. Combining light control with attosecond precision and microscopy techniques will allow ultrafast microscopy in the XUV or x-ray domain with ultrahigh time and space resolution. Proposals are also emerging in which atto-second light control will allow laser machining with higher (nano-scale) spatial resolution.

Attosecond science offers the capacity to manipulate quantum properties of matter. Because attosecond pulses can act on the timescale where the atomic nuclei are essentially frozen, one might expect that it will teach us how to manipulate matter using its quantum properties on ultrashort timescales, even in complex systems. This might open new areas where quantum states are not only measured but also manipulated, and electron quantum currents in vacuum or in a mesoscopic or nanoscopic circuits controlled, thus providing new tools for quantum computation, quantum microscopy, and quantum detectors in general. These developments will address increasingly specialized and diverse research topics, for which attosecond technology will become a tool among others. Here again, the complementarity between tabletop sources and large-scale installations will play an important role in disseminating attosecond techniques and contributing to these research programs. Finally, with the progress made in attosecond sources and technologies, zeptosecond resolution will be reachable, which opens the possibility to track particles on sub-Angstrom dimensions, thus, perhaps, connecting attosecond physics to nuclear physics.

3.5.6.5 Ultraintense beams and laser–plasma accelerators

In the next 30 years, laser technology at the intensity frontier will continue to evolve, moving towards higher focused intensities (beyond 10^{24} W cm^{-2}), higher pulse energies (in excess of several MJ), shorter pulse durations (of a few fs or lower), and a wider spectrum of laser wavelengths (moving towards the mid-infrared). The combination of these laser parameters allows for a broad range of novel physics and applications across many fields of science. Furthermore, it is now very likely, as demonstrated by ongoing projects or planned facilities, that an even stronger impact can be achieved from the combination/co-location of ultraintense laser beams and high-intensity charged particle beams, either generated from conventional accelerators or from laser–plasma accelerators, where beams in excess of 10 GeV in bunch durations of, typically, dozens of femtoseconds, will be within reach of the next generation of lasers. This combination opens the way for even more exciting physics, associated with intense sources of low-energy photons, high-energy (secondary) photons, and multi-GeV electron beams. State-of-the-art numerical tools, critical to the pursuit of extreme light-plasma studies, will strongly benefit from progress in machine learning, through the convergence of data from experiments and simulations, as well as quantum computing, as novel algorithms are deployed to future quantum computers.

3.5.6.6 Secondary sources and applications

Many experimental advances will rely on fully developed strategies for laser control and delivery of secondary sources on demand, with fine-tuned control of secondary sources with unique properties. The availability in the laboratory of such sources, with the potential for co-location of synchronous generation, of high-brightness photons, high-energy electron, proton, and neutron beams, until now only available at large-scale facilities, will revolutionize a broad range of fields, from fundamental science to applications.

3.5.6.7 Plasma optics

Taming plasma properties for the manipulation of light is a critical and challenging aspect of the road to ultrahigh intensities. Conventional materials cannot sustain the intensities of state-of-the-art lasers. Therefore, optics at ultrahigh intensities will have to rely on plasmas to steer, focus, and shape light. Relativistic plasmonics will thus become a central question to address the manipulation of extreme light. As we envisage moving to even higher laser intensities, even the laser amplification process itself will have to rely on the nonlinear properties of the plasma to transfer light from long lasers pulses to ultrashort and ultrahigh laser intensities pulses, via Raman or Brillouin amplification. Further manipulation of the topological properties of light with angular momentum resorting to the unique topological plasma response provides an additional exciting direction, with deep and fundamental consequences for laser–plasma interactions and their applications.

3.5.6.8 Quantum-electrodynamics; applications to particle- and astrophysics
Lasers (and particle beams) at the intensity frontier will first provide a unique stage to explore fundamental quantum electro-dynamical phenomena in the novel regime of multiphoton processes, driven by (a very large number of) low- energy photons. These will also provide a unique path to the exploration of the transition from the classical to quantum physics in the relativistic regime. The envisaged intensities can eventually approach the regime in which it has been conjectured that quantum electrodynamics might 'break down'; a strongly non-perturbative field theory would then be needed, which would move into QED regimes that remain hitherto theoretically and experimentally unexplored.

The density of the electron–positron plasma generated at such laser intensities via QED processes (from laser-beam or laser-target collisions) will exceed the plasma density of solids and will therefore itself affect the interaction of light with the target and the plasma and laser dynamics. A new fundamental direction for laser–plasma interactions will be opened up, in which QED processes determine collective plasma dynamics, and even brighter gamma ray and electron–positron sources can be designed for a multitude of applications. It is even possible to conceive the generation and exploration of secondary sources of pions and muons generated via laser–plasma interactions. This is now only possible at a few very large accelerator facilities worldwide.

Similar electromagnetic intensities and extreme electron–positron plasmas are also present in extreme astrophysical scenarios such as gamma ray bursts, pulsars, magnetars, and black holes. Therefore, laser–plasma interactions in the ultrarelativistic limit opens a novel path for the convergence of ultrahigh laser intensity physics with laboratory astrophysics in the high-energy regime, and with ultrastrong (electro) magnetic fields, where numerical tools and experiments commonly associated with laboratory experiments can shed light on some of the fundamental processes underpinning the scientific mysteries of the most exotic objects in the Universe.

As the laser energy increases and laser facilities are complemented with ultrastrong magnetic fields or relativistic particle beams, the next 30 years will also witness further developments on laboratory astrophysics, driven by high-intensity/ high-energy lasers, of collisionless shock waves and their physics and impact on cosmic ray acceleration, magnetized relativistic flows and the formation of jets, fundamental magnetogenesis processes, and the long-term evolution of turbulence in magnetized plasma, a fundamental question underpinning many astrophysical questions.

Plasma acceleration is expected to reach two key milestones. For electrons, the realm of high-energy physics will be within reach, with high-energy, high-current beams, high-brightness beams, and moderate repetition rate (kHz), determined by the laser technology. For protons, acceleration beyond 200 MeV will further place laser–plasma interactions as a route for compact proton accelerators for proton therapy and other applications.

As the world addresses climate change, the quest for clean energy sources is at the forefront of the scientific and technology challenges of the 21st century. One of the possible directions, still requiring major scientific and technological advances and the development of new large-scale facilities, is laser fusion, or its combination with other clean technologies. This is a very long-term, very high-risk endeavor but humanity just cannot leave a direction filled with outstanding scientific and technological challenges and a possible path to achieve a commercial clean energy source unexplored.

3.6 Systems with numerous degrees of freedom

Marco Baldovin[1,2], Giacomo Gradenigo[3] and Angelo Vulpiani[4]
[1]Université Paris-Saclay, Orsay, France
[2]CNR - Instute for Complex Systems, Rome, Italy
[3]Gran Sasso Science Institute, L'Aquila, Italy
[4]Università 'Sapienza', Rome, Italy

3.6.1 Introduction

Hamiltonian systems have great relevance in many fields, from celestial mechanics to fluids and plasma physics. Not even to mention the importance of the Hamiltonian formalism for quantum mechanics and quantum field theory [226]. The aim of this contribution is to discuss the relation between the dynamics of Hamiltonian systems with many degrees of freedom and the foundations of statistical mechanics [227–229]. In particular, we want to address the problem of thermalization in integrable and nearly integrable systems, questioning the role played by chaos in this processes. The problem of thermalization of high-dimensional integrable Hamiltonian systems is, in our opinion, a truly non-academic one and is today of groundbreaking importance for one main reason: it has a direct connection with the problem of quantum systems thermalization. The analogy between the problem of thermalization in high-dimensional integrable Hamiltonian systems and thermalization of quantum many-body systems would make it possible to exploit the rich conceptual and methodological framework already developed for classical systems for the investigation of the quantum ones. The point of view we want to promote here is that the presence or absence of thermal equilibrium is not an *absolute* property of the system but depends on the set of variables chosen to represent it. We will present some examples supporting the idea that the notion of thermal state strongly depends on the choice of the variables used to describe the system. One could ask at this point what makes the study of thermalization still such an interesting problem. The answer is that the most recent technologies allow now for manipulation of quantum mesoscopic systems in an uprecedented way, so that many foundational questions gain renewed importance and are today crucial for experimental purposes. Last but not least, the interplay between the absence of thermalization in quantum many-body systems and the processing (storage/transmission) of information in quantum devices is of paramount importance for all applications to quantum computing.

The discussion develops through sections as follows: in section 3.6.2 the standard viewpoint, according to which chaos plays a key role to guarantee thermalization, is summarized; in section 3.6.3 we present three counter-examples to the above view and provide a more refined characterization of the central point in the 'thermalization problem', namely the choice of the variables used to describe the system; in section 3.6.4 we discuss the analogies between the problem of thermalization in classical and quantum mechanics and we compare the scenarios established, respectively, by the Khinchin's (classical) ergodic theorem [230] and the quantum ergodic theorem by Von Neumann [231–233]. An account of open questions and future perspectives follows at the end.

3.6.2 The role of Kolmogorov–Arnold–Moser (KAM) and chaos for the timescales

It is usually believed that ergodicity is a necessary condition to observe thermal behaviour in a given Hamiltonian system [229]. The ergodic hypothesis, first introduced by Boltzmann, states that a Hamiltonian system with energy E, during its motion, will visit the whole hypersurface of constant energy E in the phase space; in each region of the hypersurface the system will spend an amount of time which is proportional to the volume of the region itself (i.e., its Lebesgue measure). In modern terms, the assumption of ergodicity amounts to asking for the phase space to not be partitioned in disjoint components which are invariant under time evolution [227, 234]. Indeed, if dynamics can explore the whole hypersurface, the latter cannot be divided into measurable regions that are invariant under time evolution without resulting in a contradiction. The existence of conserved quantities different from energy implies in turn the existence of other invariant hypersurfaces, which prevent the system from being ergodic.

The claim that thermalization is related to ergodicity summarizes the viewpoint of those who think that *invariant manifolds* in phase space, which work as *obstructions* to ergodicity, are the main obstacle to thermal behaviour. The problem of '*thermalization*' can be thus reformulated as the determination of the existence of integrals of motion other than energy. This was solved by Poincaré in his celebrated work on the three-body problem [227, 228]. Consider the following Hamiltonian

$$H = H_0(I) + \varepsilon H_I(I, \theta), \tag{3.1}$$

where H depends on a set of 'angles' $\{\theta_n\}$, with $n = 1, \dots, N$, and on their conjugate momenta, the 'actions' $\{I_n\}$. Actions and angles are the canonical coordinates which are typically chosen to describe the motion of an integrable system, which is the limiting case $\varepsilon = 0$ of the above Hamiltonian. In that case, the actions are constant in time (since the angles do not appear in the Hamiltonian) and the angles increase with constant velocities. There are therefore N conservation laws and the motion takes place on N-dimensional invariant tori.

If ε is small but non-vanishing, we are in an 'almost' integrable situation. Poincaré showed that, for $\varepsilon \neq 0$, excluding specific cases, there are no conservation laws apart from the trivial ones (i.e., energy and total momentum). In 1923 Fermi generalized Poincaré's theorem, showing that, if $N > 2$, it is not possible to have a foliation of phase space in invariant surfaces of dimension $2N - 2$ embedded in the constant-energy hypersurface of dimension $2N - 1$. From this result Fermi argued that generic Hamiltonian systems are ergodic as soon as $\varepsilon \neq 0$. At that time it was believed that global invariant manifolds were the main obstruction to thermalization; as a consequence, Fermi's mathematical proof appeared to imply that good thermal behaviour should have been expected for non-integrable systems. It was then a numerical experiment realized by Fermi himself in collaboration with J Pasta, S Ulam, and M Tsingou (who did not appear among the authors) which showed that this was not the case [235]. Fermi and coworkers studied a chain of weakly nonlinear oscillators, finding that the system, despite being non-integrable, was not showing relaxation to a thermal state within reasonable times when initialized in a very atypical condition. This observation raised the problem that while the absence of globally conserved quantities guarantees that no partitioning of phase space takes place, on the other hand, it does not tell anything on the *timescales* to reach equilibrium [236]. A first understanding of the slow FPUT timescales from the perspective of phase-space geometry came from the celebrated Kolmogorov–Arnold–Moser (KAM) theorem, sketched by A N Kolmogorov already in 1954 and completed later [227, 228]. The theorem says that for any value of $\varepsilon \neq 0$, even very small, some tori of the unperturbed system, the so-called resonant ones, are completely destroyed, and this prevents the existence of analytical integrals of motion. Despite this, if ε is small, most tori, slightly deformed, survive; thus the perturbed system (for 'non-pathological' initial conditions) has a behaviour quite similar to an integrable one.

3.6.2.1 KAM scenario, chaos, and slow timescales

After the discovery of the KAM theory it became clear that also in weakly nonlinear systems the phase space is characterized by the presence of invariant manifolds which, even if not partitioning it into disjoint components, might crucially slow down the dynamics. The foundational problem of statistical mechanics thus turned from *'do we have a partitioning of phase space?'* to *'how long does it take to thermalize in the large-N limit?'*.

A crucial role to answer this question was then played by the notion of chaos. Very roughly, chaos means exponential divergence of initially nearby trajectories in phase space. The rate of divergence τ_{div} of nearby orbits is the inverse of a quantity known as the first (maximal) Lyapunov exponent [237] λ_1: $\tau_{\text{div}} = 1/\lambda_1$. Of course, the behaviour of nearby trajectories may, and actually does, depend on the phase space region they start from. In fact, for a system with N degrees of freedom one finds N independent Lyapunov exponents: in the multiplicity of Lyapunov exponents is encoded the property that the rate of divergence of nearby orbits has fluctuations in phase space. This is indeed the picture which comes from both the KAM theorem and the evidence of many numerical experiments: at variance with dissipative systems (characterized by a sharp threshold from regular to chaotic behavior) Hamiltonian systems are characterized at any energy scale by the coexistence of regular and chaotic trajectories. In particular, one has in mind that fast thermalization takes place in chaotic regions while the slow timescales are due to the slow intra-region diffusion. The slow process is thus controlled by the location and structure of invariant manifolds which survive in the presence of nonlinearity. The whole issue of determining on which timescale the system exhibits thermal properties boils down to estimating the timescale of diffusion across chaotic regions in phase space: this is the so-called Arnold's diffusion.

3.6.2.2 Arnold's diffusion

Arnold's diffusion is the name attached to the slow diffusion taking place in a phase space characterized by the coexistence of chaotic regions and regular ones [227, 228]. Typically, chaotic regions in high-dimensional phase spaces do not have a smooth geometry and are intertwined in a very intricate manner with invariant manifolds.

Diffusion across the whole phase space is driven by chaotic regions, but their complicated geometry, typically fractal, slows down diffusion at the same time [238]. From this perspective, let us try to have a look at the scenario for diffusion in phase space in the case of weakly non-integrable systems in the limit of large N [239–241]. Is it compatible with the appearance of a thermal state?

Let us consider the generic Hamiltonian for a weakly nonintegral system. This is a system with N degrees of freedom where, after introducing the Liouville–Arnold theorem's action-angle variables $I_n(t)$, $\theta_n(t)$, the symplectic dynamics is defined as follows:

$$\theta_n(t+1) = \theta_n(t) + I_n(t), \mod 2\pi, \tag{3.2}$$

$$I_n(t+1) = I_n(t) + \varepsilon \, \partial F(\theta(t+1)) \partial \theta_n(t+1), \mod 2\pi \tag{3.3}$$

where $n = 1, ..., N$, $F(\theta) = F(\{\theta_i ; 1 \leqslant i \leqslant n\})$ is the term which breaks integrability and ε is a dimensionless coefficient which we assume to be small. Let us note that a symplectic map with N degrees of freedom can be seen as the Poincaré section of a Hamiltonian system with $N + 1$ degrees of freedom. The key point for the purpose of our discussion is to understand what happens to invariant manifolds in the large-N limit. Several numerical works (see, e.g., [239]) showed that the volume of regular regions decreases very fast at large N. In particular, the normalized measure $P(N, \varepsilon)$ of the phase space, where the numerically computed first (maximal) Lyapunov exponent λ_1 is very small, goes to zero exponentially with N:

$$P(N, \varepsilon) \sim e^{-a(\varepsilon)N} . \tag{3.4}$$

At a first glance such a result sounds quite positive for the possibility to build the statistical mechanics on dynamical bases, since, no matter how small ε is, the probability to end up in a non-chaotic region ($\lambda_1 \leqslant 0$) becomes exponentially small with the size of the system.

At this stage the problem of dynamical justification of statistical mechanics may seem to be solved. On the one hand, it is true that a set of manifolds of finite measure which are invariant along the dynamics can always be found, even for non-integrable systems (KAM theorem), and this invariance may possibly spoil the relaxation to equilibrium; on the other hand, however, very reasonable estimates also assure that in the large-N limit, the probability to end up in a non-chaotic region is negligible.

But this is not the end of the story. Indeed, the presence of chaos is not sufficient to avoid the long timescale correlations in the motion of the system. From a mere analysis in terms of Lyapunov exponents it is indeed only possible to conclude that *almost* all trajectories have the *'good'* statistical properties we expect in statistical mechanics. The main reason is that the time $\tau_{\mathrm{Lyap}} = 1/\lambda_1$, which is related to the trajectory instability, is not the unique relevant characteristic time of the dynamics. There might be more complicated collective mechanisms which prevent a fast thermalization of the system which are not traced by the Lyapunov spectrum. Think for instance to the mechanism of ergodicity breaking in systems such as spin glasses [242].

The trajectories of an N-dimensional system lie on a $(2N - 1)$ dimensional hypersurface in the phase space (i.e., the phase-space region with constant energy), while the KAM tori have dimension N. If $N = 2$, a 1D torus can thus separate the 2D hypersurface it lies on into an 'inner' and an 'outer' portion. As soon as $N \geqslant 3$, instead, a constant-energy hypersurface cannot be separated into disjoint components by a torus and, as we have said, ergodicity is guaranteed asymptotically from the point of view of dynamics. A trajectory initially closed to a KAM torus may visit any region of the energy hypersurface, *but*, since the compenetration of chaotic regions and invariant manifolds typically follows the pattern of a fractal geometry, the diffusion across the isoenergetic hypersurface is usually very slow; as we said, such a phenomenon is called Arnol'd diffusion [227, 240]. The important and difficult problem is therefore to understand the 'speed' of the Arnold diffusion, i.e., the time behaviour of

$$\langle |\mathcal{I}(t) - \mathcal{I}(0)|^2 \rangle = \sum_{n=1}^{N} \langle |\mathcal{I}_n(t) - \mathcal{I}_n(0)|^2 \rangle \tag{3.5}$$

where the $\langle \rangle$ denotes the averages over the initial conditions. There are some theoretical bounds for $\langle |\mathcal{I}(t) - \mathcal{I}(0)|^2 \rangle$, as well as several accurate numerical simulations, which suggest a rather slow 'anomalous diffusion':

$$\langle |\mathcal{I}(t) - \mathcal{I}(0)|^2 \rangle \sim D \, t^\nu, \; \nu \leqslant 1 \tag{3.6}$$

where D and ν depend on the system's parameters. We can say that in a generic Hamiltonian system Arnold diffusion is present and, for small ε, it is very weak. It can thus happen that different trajectories, even with a rather large Lyapunov exponent, maintain memory of their initial conditions for considerably long times. See, for instance, the results of [239] discussed below in section 3.6.3. The existence of Arnold's diffusion is a hint that chaos is sometimes not sufficient to guarantee thermalization.

3.6.2.3 Lyapunov exponents in the large-N limit
Although we are going to depart from this point of view, let us put on the table all evidences that chaos is apparently a good property to justify a statistical mechanics approach. For instance, in a chaotic system it is possible to define an entropy directly from the Lyapunov spectrum [237]. In any Hamiltonian (symplectic) system with N degrees of freedom we have $2N$ Lyapunov exponents $\lambda_1 \geqslant \lambda_2 \geqslant \cdots \lambda_{2N-1} \geqslant \lambda_{2N}$ which obey a mirror rule, i.e.,

$$\lambda_{2N} = -\lambda_1, \; \lambda_{2N-1} = -\lambda_2, \quad \ldots \tag{3.7}$$

and the Kolmogorov–Sinai entropy is

$$H = \sum_{n=1}^{N} \lambda_n . \tag{3.8}$$

Regarding statistical mechanics, the natural question is about the asymptotic behaviour of $\{\lambda_n\}$ for $N \gg 1$. There is clear numerical evidence [243, 244] and analytical results for a few special cases [245] that at large N one has

$$\lambda_n = \lambda^* f(n/N) \qquad (3.9)$$

where $\lambda^* = \lim_{N\to\infty}\lambda_1$ as $f(0) = 1$. Such a result implies that the Kolmogorov–Sinai entropy equation (3.8) is proportional to the number of degrees of freedom:

$$H \simeq h^* N$$

where $h^* = \lambda^* \int_0^1 f(x)dx$ does not depend on N. The extensivity of H in the thermodynamic limit resonates with the properties of entropy in statistical mechanics, but it should not be overestimated, as we will discuss in the next section.

3.6.3 Challenging the role of chaos

The previous section was entirely dedicated to present evidences in favour of chaos being a sufficient condition for an equilibrium behaviour in the large-N limit and the possibility to apply a statistical mechanics description. In the present section we present counter-examples to this point of view. These examples suggest that in the large-N limit a statistical mechanics description holds for almost all Hamiltonian systems irrespective of the presence of chaos. First, we will quote two results showing that, even in the presence of finite Lyapunov exponents, there are clear signatures of (weak) ergodicity breaking and of the impossibility to establish a statistical mechanics description on finite timescales [239, 246]. Conversely, we can also present the examples of an integrable system, the Toda chain [247], which shows thermalization on short timescales notwithstanding all Lyapunov exponents are zero. Clearly, in the case of an integrable system, thermalization takes place with respect to some observables and not with respect to others. But this is the point of the whole discussion we want to promote: thermal behaviour is not a matter of dynamics being regular or chaotic, it is just a matter of choosing the description of the system in terms of the appropriate variables. And, as we will see, this is also the case for quantum mechanics (which has important similarities with classical integrable systems); almost all choices of canonical variables are good while only very few are not good. In the case of an integrable system, in order to detect thermalization the canonical variables which diagonalize the Hamiltonian must be, of course, avoided.

3.6.3.1 Weak ergodicity breaking in coupled symplectic maps
We start our list of 'counter-examples' to the idea that chaotic properties guarantee fast thermalization by recalling the results of [239]. In [239] the authors studied the ergodic properties of a large number of coupled symplectic maps (3.1) and (3.2) where one considers periodic boundary conditions, i.e., $\theta_{N+1} = \theta_1$, $I_{N+1} = I_1$, and nearest-neighbour coupling:

$$F(\theta) = \sum_{n=1}^{N} \cos(\theta_{n+1} - \theta_n) \, .$$

represented by following system of discrete time update equations:

$$\theta_n(\tau + 1) = \theta_n(\tau) + I_n(\tau) \quad (\mathrm{mod}\ 2\pi)$$

$$I_n(\tau + 1) = I_n(\tau) + \varepsilon \, \frac{\partial F}{\partial \theta_n}[\mathbf{q}(\tau + 1)] \tag{3.10}$$

Skipping all the details of the analysis let us move straight to the main results of [239]. The authors perform a numerical mesure of the Lyapunov spectrum and provide an estimate for the size of 'regular regions', which is a way to refer to the measure of KAM tori, by looking at the fraction of Lyapunov exponents compatible with zero within error bars.

As anticipated above, reference [239] shows that the volume of phase space filled with invariant manifolds decreases exponentially with system size, as shown in figure 3.30. This is a remarkable evidence that in the large-N limit the probability that thermalization is obstructed by KAM tori is exponentially (in system size) small. The system has good chaotic properties. This notwithstanding, at the same time there are clear signatures of a phenomenon analogous to the so-called *weak-ergodicity breaking* of disordered systems. The terminology weak-ergodicity breaking is jargon indicating the situations where, despite relaxation being always achieved asymptotically, at any *finite* time memory of the initial conditions is preserved [248]. Let us define as t_w the time elapsed since the beginning of the dynamics, $t_w + \tau$ is a later time, and $C(t_w, t_w + \tau)$ is the time auto-correlation

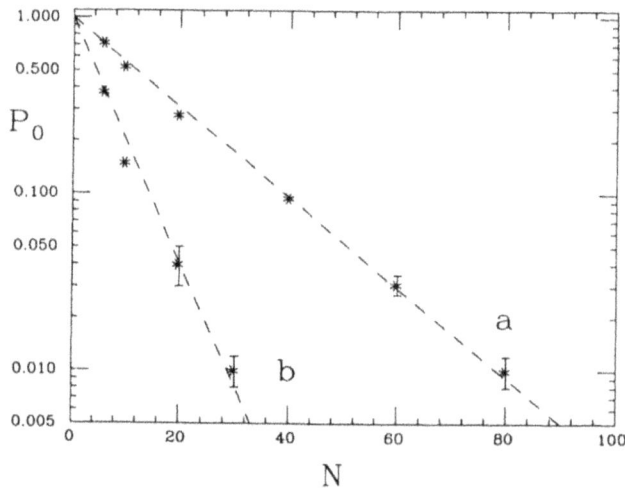

Figure 3.30. Fraction of maximal Lyapunov exponents smaller than 0.002 as a function of N for (a) $\varepsilon = 0.025$ and (b) $\varepsilon = 0.05$, $T = 10^5$, averaged over $\mathcal{N} = 500$ initial conditions. Reprinted figure with permission from [239], Copyright (1991) by the American Physical Society.

(averaged over an ensemble of initial conditions) of some relevant observable of the system. Weak ergodicity breaking is then expressed by the condition

$$\lim_{\tau \to \infty} C(t_w, t_w + \tau) = 0 \quad \forall \ t_w$$

$$C(t_w, t_w + \tau) \neq C(\tilde{t}_w, \tilde{t}_w + \tau) \quad \forall \ \tau < \infty \ \& \ t_w \neq \tilde{t}_w \qquad (3.11)$$

The dependence on waiting times t_w encoded in equation (3.11) represents the fact that we do not find only one characteristic timescale emerging at the macroscopic level (which is the standard behaviour at thermodynamic equilibrium): on the contrary, a broad distribution of timescales can be appreciated even macroscopically. In jargon, we say that the probability distribution of characteristic timescales is *not self-averaging* (i.e., it cannot be approximated with a Dirac delta in the thermodynamic limit). A clear signature of a situation with the features of weak ergodicity breaking is revealed by the study of how the distribution of the maximum Lyapunov spectrum depends on the system size [239]. While the probability distribution of the exponents peaks at a value of order $\mathcal{O}(1)$ with respect to N, still the scaling of the variance is anomalous (i.e., it decays much slower than $1/\sqrt{N}$): this is the distinguishing feature of a probability distribution which is not self-averaging. One of the main results of [239] is, in fact, that the variance of the Lyapunov exponent distribution, in particular its asymptotic estimate $\sigma_\infty(\varepsilon, N)$, scales as

$$\sigma_\infty(\varepsilon, N) \sim \frac{1}{N^{b(\varepsilon)}} \qquad (3.12)$$

where $b(\varepsilon) \sim \sqrt{\varepsilon}$, so that for small ε, $\sigma_\infty(\varepsilon, N)$ is much larger than $1/\sqrt{N}$.

$$\sigma_\infty(\varepsilon, N) \sim \frac{1}{N^{a(\varepsilon)}} \gg \frac{1}{N^{1/2}}. \qquad (3.13)$$

since

$$a(\varepsilon) \sim \sqrt{\varepsilon}. \qquad (3.14)$$

3.6.3.2 *High-temperature features in coupled rotators*
Another example of a system which does not show thermal behaviour despite having positive Lyapunov exponents is represented by the coupled rotators at high energy studied in [246]. The Hamiltonian of this system reads as:

$$\mathcal{H}(\mathbf{j}, \mathbf{p}) = \sum_{i=1}^{N} \frac{\pi_i^2}{2} + \varepsilon \sum_{i=1}^{N} (1 - \cos(\varphi_{i+1} - \varphi_i)), \qquad (3.15)$$

where φ_i are angular variables and π_i are their conjugate momenta. The specific heat can be easily computed from the partition function $\mathcal{Z}_N$ as

$$C_V = \frac{\beta^2}{N} \frac{\partial^2}{\partial \beta^2} \log \mathcal{Z}_N(\beta), \qquad (3.16)$$

with $\beta = (k_B T)^{-1}$, k_B being Boltzmann's constant and T the temperature. In particular, the expression of the specific heat reads in terms of a modified Bessel function as

$$C_V = \frac{1}{2} + \beta^2 (1 - \frac{1}{\beta} \frac{I_1(\varepsilon\beta)}{I_0(\varepsilon\beta)} - \left[\frac{I_1(\varepsilon\beta)}{I_0(\varepsilon\beta)}\right]^2). \tag{3.17}$$

It is possible to compare the analytical prediction of equation (3.17) with numerical simulations by estimating in the latter the specific heat from the fluctuations of energy in a given subsystem of M rotators, with $1 \ll M \ll N$:

$$C_V = \frac{1}{MT^2}[\langle \mathcal{H}_M^2 \rangle - \langle \mathcal{H}_M \rangle^2] \tag{3.18}$$

where $\mathcal{H}_M$ is the Hamiltonian of the chosen subsystem and the 'ensemble' average in equation (3.18) includes averaging over initial conditions and averaging along the symplectic dynamics for each initial condition. The result of the comparison is presented in figure 3.31. From the figure it is clear that, while at small and intermediate energies the canonical ensemble prediction for the specific heat equation (3.17) matches the quantity (3.18) observed in numerical simulations, it fails at high energies. What is remarkable is that in the high-temperature regime where statistical mechanics fails, the value of the largest Lyapunov exponent is even

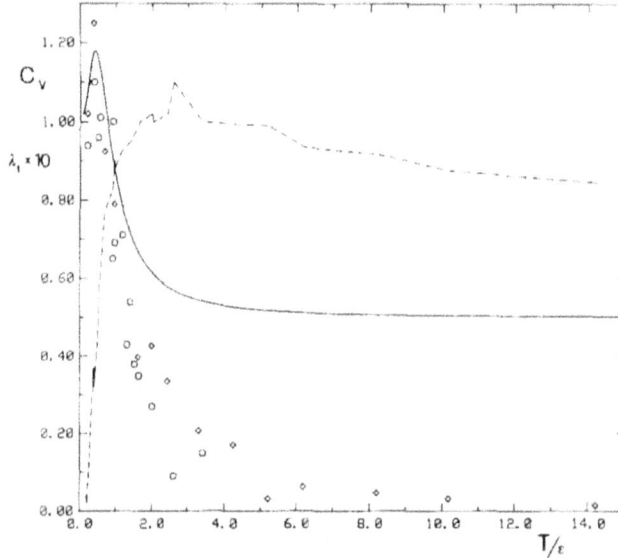

Figure 3.31. Specific heat versus temperature in the rotator model. The continuous line represents the analytic prediction from equation (3.17). Empty symbols represent the results of numerical simulations, respectively, for a subsystem with $M = 10$ (circles) and $M = 20$ (diamonds) rotators, in a chain of $N = 100$ and $N = 400$ particles, respectively. The dashed line is the maximal Lyapunov exponent as a function of the temperature (specific energy) measured in units of the coupling constant ε [see equation (3.15)]. Reprinted by permission from Springer Nature [246], copyright (1987).

larger than in the intermediate-small energy regime where the equilibrium prediction works well. This is one of the strongest hints from the past literature that the chaoticity of orbits has nothing to do with the foundations of statistical mechanics. In practice, it happens that for rotators a sort of 'effectively integrable' regime arises at high energy. In this regime each rotator is spinning very fast, something which guarantees the chaoticity of orbits, but the individual degrees of freedom do not interact with each other, which causes the breakdown of thermal properties of the system. This very simple mechanism is a concrete example (and often examples are more convincing than arguments) of how absence of thermalization and chaos can be simultaneously present. In the next example, the role of chaos for thermalization will be challenged even further: we are going to present the case of a system which, despite being *all Lyapunov exponents equal to zero*, presents fast relaxation to equilibrium when the appropriate variables are considered.

3.6.3.3 Toda lattice: thermalization of an integrable system
We present here the example of an integrable system which shows very good thermalization properties. In some sense we find that the statement 'the system has thermalized' or 'the system has not thermalized' depends, for Hamiltonian systems, on the choice of canonical coordinates in the same way as the statement 'a body is moving' or 'a body is at rest' depend on the choice of a reference frame. In this respect, it is true that, if a system is integrable in the sense of the Liouville–Arnold theorem and we choose to represent it in terms of the corresponding action-angle variables, we will never observe thermalization. But there are infinitely many other choices of canonical coordinates which allow one to detect a good degree of thermalization. Let us be more specific about this and recall the salient results presented in [249] on the Toda chain.

It is well known that the Toda lattice,

$$\mathcal{H}(q, p) = \sum_{i=1}^{N} \frac{p_i^2}{2} + \sum_{i=0}^{N} V(q_{i+1} - q_i), \qquad (3.19)$$

where $V(x)$ is the Toda potential,

$$V(x) = \exp(-x) + x - 1, \qquad (3.20)$$

admits a complete set of independent integrals of motion, as shown by Henon in his paper [247]. This result can also be understood in light of Flashka's proof of the existence of a Lax pair related to the Toda dynamics [250]. The explicit form of such integrals of motion is rather involved, and their physical meaning is not transparent; one may wonder whether the system is able to reach thermalization under a different canonical description, such as the one which is provided by the Fourier modes,

$$Q_k = \sum_{i=1}^{N} \frac{\sqrt{2}\, q_i}{\sqrt{N+1}} \sin\left(\frac{\pi i k}{N+1}\right), \quad P_k = \sum_{i=1}^{N} \frac{\sqrt{2}\, p_i}{\sqrt{N+1}} \sin\left(\frac{\pi i k}{N+1}\right). \qquad (3.21)$$

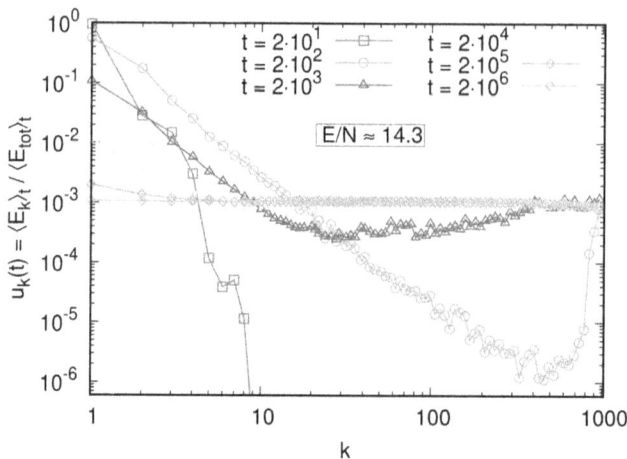

Figure 3.32. Normalized harmonic energy as a function of the Fourier mode number k, for different values of the averaging time, in a Toda dynamics (3.19) with FPUT initial conditions. Here $N = 1023$, total energy $E = 14.3$.

A typical test which can be performed, inspired by the FPUT numerical experiment, is realized by considering an initial condition in which the first Fourier mode $k = 1$ is excited, and $Q_k = P_k = 0$ for all $k > 1$; the system is evolved according to its Hamiltonian dynamics, and the average energy corresponding to each normal mode,

$$\langle E_k \rangle_t = \int_0^t \left[\frac{P_k^2}{2} + \omega_k^2 \frac{Q_k^2}{2} \right] dt, \quad \omega_k = 2 \sin\left(\frac{\pi k}{2(N+1)} \right), \qquad (3.22)$$

is studied as a function of time. It is worth noticing that in the limit of small energy E, the Toda Hamiltonian is well approximated by a harmonic chain in which the P_k's are conserved quantities; it is therefore not surprising that, for $E/N \ll 1$, it takes very long times to observe equipartition of harmonic energy between the different modes (figure 3.32). This ergodicity-breaking phenomenology has been studied, for instance, in reference [236] to underline relevant similarities with the FPUT dynamics.

The scenario completely changes when specific energies of order $\approx \mathcal{O}(1)$ or larger are considered. In figure 3.31 the distribution of the harmonic energy among the different Fourier modes is shown for several values of the averaging time t; no ergodicity breaking can be observed, and in finite times the harmonic energy reaches equipartition.

3.6.4 From classical to quantum: open research directions

After the examples of the previous section we feel urged at this point to draw some conclusions. We started our discussion from wondering how much having a phase space uniformly filled by chaotic regions is relevant for relaxation to equilibrium and we came to the conclusion, drawn from the last set of examples, that in practice chaos does not seem to be the crucial point for relaxation to equilibrium. What is then the correct way to frame the 'foundation of statistical mechanics' problem? It seems that the correct paradigm to understand the results of section 3.6.3 is the one established by the ergodic theorem by Khinchin [230]: in order to guarantee a thermal behaviour, i.e., that dynamical averages correspond to ensemble averages, it is sufficient to consider appropriate observables for large enough systems. That is, the regime $N \gg 1$ is mandatory. As anticipated in section 3.6.3 by our numerical results, one finds that the hypothesis under which the Khinchin ergodic theorem is valid are for instance fulfilled even by integrable systems. In a nutshell we can summarize his approach by saying that it is possible to show the practical validity of the ergodic hypothesis when the following three conditions are fulfilled:

- the number of the degrees of freedom is very large
- we limit the interest to 'suitable' observables
- we allow for a failure of the equivalence of the time average and the ensemble average for initial conditions in a 'small region'

The above points will be clarified in the following.

3.6.4.1 The ergodic problem in Khinchin's perspective

Let us see in practice what the theorem says. Consider a Hamiltonian system $\mathcal{H}(q, p)$ with canonical variables $(q, p) = (q_1, p_1, \ldots, q_N, p_N)$. Let $\Phi_t(q, p) = (q(t), p(t))$ be the flow under the symplectic dynamics generated by $\mathcal{H}(q, p)$. The time average for given initial datum $x = (q, p)$ is defined as

$$\overline{f(x)} = \frac{1}{T} \int_0^T dt\, f(\Phi_t(x)) \tag{3.23}$$

while the ensemble average is defined with respect to an invariant measure $\mu(q, p)$ in phase space:

$$\langle f \rangle = \int d\mu(q, p) f(q, p). \tag{3.24}$$

Let us define as *sum function* any function reading as

$$f(x) = \sum_{i=1}^{N} f_i(q_i, p_i) \tag{3.25}$$

For such functions Khinchin [230] was able to show that for a random choice of the initial datum x the probability that the time average in equation (3.23) and the ensemble average in equation (3.24) are different is small in N, that is,

$$p\left(\left| \frac{\overline{f(x)} - \langle f \rangle}{\langle f \rangle} \right| \geqslant \frac{C_1}{N^{1/4}} \right) < \frac{C_2}{N^{1/4}}, \tag{3.26}$$

where C_1 and C_2 are constants with respect to N. This is the well-known Khinchin's theorem. We can note that many, but not all, relevant macroscopic observables are sum functions. While the original result of Khinchin was for non-interacting system, i.e., systems with Hamiltonian:

$$\mathcal{H}(q, p) = \sum_{i=1}^{N} h_1(q_i, p_i) \tag{3.27}$$

Mazur and van der Linden [251] were able to generalize it to the more physically interesting case of (weakly) interacting particles:

$$H = \sum_{i=1}^{N} h_1(q_i, p_i) + \sum_{i<j} V(|q_i - q_j|) \tag{3.28}$$

From the results by Khinchin, Mazur, and van der Linden we have the following scenario: although the ergodic hypothesis mathematically does not hold, it is 'physically' valid if we are tolerant, namely if we accept that in systems with $N \gg 1$ ergodicity can fail in regions sampled with probability of order $O(N^{-1/4})$ (i.e., vanishing in the limit $N \to \infty$). Let us stress that the dynamics has a marginal role, while the very relevant ingredient is the large number of particles.

Whereas it is possible, on the one hand, to emphasize some limitations of the Khinchin ergodic theorem, as for instance the fact that it does not tell anything about the timescale to be waited for in order to have equation (3.26) reasonably true, let us stress here its goals. For instance, let us highlight the fact that the Khinchin ergodic theorem guarantees the thermalization even of *integrable* systems! In fact, if we consider the Toda model discussed in the previous section, the conditions under which the Khinchin ergodic theorem was first demonstrated perfectly apply to it. In fact, integrability, proved first by Hénon in 1974 [247], guarantees the existence of N action-angle variables such that the Hamiltonian reads as:

$$\mathcal{H}(\mathcal{I}, \mathbf{f}) = \sum_{i=1}^{N} h_1(\mathcal{I}_i, \phi_i), \tag{3.29}$$

which is precisely the *non-interacting-type* of Hamiltonian considered in first instance by Khinchin. The results presented in [249] on the fast thermalization of the Toda model have been in fact proposed to re-establish with the support of numerical evidence the assertion, hidden between the lines of the Khinchin theorem and perhaps overlooked so far, that even integrable systems do thermalize in the large-N limit. The phase-space of Toda chain is in fact completely foliated in invariant tori. But, according to the Khinchin theorem and our numerical results, this foliation of phase space in regular regions is not an obstacle as long as the *'thermalization'* of sum functions is considered (for almost all initial data in the limit $N \to \infty$).

3.6.4.2 Quantum counterpart: Von Neumann's ergodic theorem

The observation that *'even integrable systems thermalize well in the large-N limit'* becomes particularly relevant as soon as we regard, in the limit of large N, the dynamics of a classical integrable system as the *'classical analog'* of quantum system dynamics. In quantum systems one finds in fact the same 'ergodic problem' of classical mechanics: is it possible to replace time averages with ensemble averages?

Let us say a few words on the quantum formalism in order to point out the similarities between the ergodic theorem by Khinchin and the one by Von Neumann for quantum mechanics. As it is well known, due to the self-adjointness of the Hamiltonian operator, any wavevector $|\psi\rangle$ can be expanded on the Hamiltonian eigenvectors basis:

$$|\psi\rangle = \sum_{\alpha \in \text{Sp}(\hat{H})} c_\alpha |\alpha\rangle, \tag{3.30}$$

where $\text{Sp}(\hat{H})$ denotes the discrete spectrum of the Hamiltonian operator $\hat{H}$. The projection on a limited set of eigenvalues defines the quantum microcanonical ensemble. Since it is reasonable to assume that also in a quantum system the total energy of an N-particle system is known with finite precision (i.e., usually we know that it takes values within a finite shell $I_E \in [E - \delta E, E + \delta E]$ with $E \sim N$ and $\delta E \sim \sqrt{N}$), one defines the microcanonical density matrix as the projector on the eigenstates pertaining to that shell:

$$\hat{\rho}_E = \frac{1}{\mathcal{N}_E} \sum_{\alpha | \varepsilon_\alpha \in I_E} |\alpha\rangle \langle \alpha|, \tag{3.31}$$

where $\mathcal{N}_E$ is the number of eigenvalues in the shell. The microcanonical expectation of the observable thus reads:

$$\langle \hat{O} \rangle_E = \text{Tr}[\hat{\rho}_E \hat{O}] = \frac{1}{\mathcal{N}_E} \sum_{\alpha | \varepsilon_\alpha \in I_E} \langle \alpha | \hat{O} | \alpha \rangle . \tag{3.32}$$

Clearly in the limit $N \to \infty$ where the eigenvalues are densely distributed on the real line one has that even in a finite shell $\mathcal{N}_E \sim N$. By preparing a system in the initial state $|\psi_0\rangle$ the expectation value of a given observable $\hat{O}$ at time t reads as:

$$\langle \hat{O}(t) \rangle_{\psi_0} = \langle \psi_0 | e^{i\hat{H}t/\hbar} \hat{O} e^{-i\hat{H}t/\hbar} |\psi_0\rangle = \langle \psi_0(t) | \hat{O} | \psi_0(t) \rangle \tag{3.33}$$

Quantum ergodicity amounts then to the following equivalence between dynamical and ensemble averages:

$$\langle \hat{O} \rangle_E = \lim_{T \to \infty} \frac{1}{T} \int_0^T dt \, \langle \hat{O}(t) \rangle_{\psi_0}. \tag{3.34}$$

The reader can easily convince himself of the fact that, if for *almost all* times the expectation of $\hat{O}(t)$ on the initial state $|\psi_0\rangle$ is *typical*, namely one has $\langle \hat{O} \rangle_E \approx \langle \hat{O}(t) \rangle_{\psi_0}$, then the quantum ergodicity property as stated in equation (3.34) is realized. The fact that for almost all times $\langle \hat{O} \rangle_E \approx \langle \hat{O}(t) \rangle_{\psi_0}$ is ususally called 'normal typicality': it is discussed thoroughly in [232]. Similarly to the scenario later proposed by Kinchin for classical systems, already in 1929 Von Neumann proposed a quantum ergodic theorem which proves the *typicality* of $\langle \hat{O}(t) \rangle$ without making any reference to *quantum chaos* properties. The definition of quantum chaos, which will be formalized much later [252], is to have a system characterized by a Hamiltonian operator, $\hat{H}|\alpha\rangle = \varepsilon_\alpha|\alpha\rangle$, such that its eigenvectors $|\alpha\rangle$ behave as random structureless vectors in any basis. This property, notwithstanding the different formalism of quantum and classical mechanics, has a deep analogy with the definition of classical chaos [228, 253]: a quantum system is said to be chaotic when a small perturbation of the Hamiltonian, $\hat{H} + \hat{\lambda}$, produces totally uncorrelated vectors, $\langle \varepsilon_\alpha | \varepsilon_\alpha + \lambda_\alpha \rangle \sim 1\sqrt{N}$, in the same manner that a small shift in initial conditions produces totally uncorrelated trajectories in classical chaotic systems.

Quite remarkably, Von Neumann's ergodic theorem does not make any explicit assumption of the above kind on the Hamiltonian's eigenvalues structure. Here follows a short account of the theorem; mathematical details can be found in the recent translation from German of the original paper [231] and in [232, 233]. Let $\mathscr{D}$ be the dimensionality of the energy shell $I_E = [E - \delta E, E + \delta E]$, namely $\mathscr{D} = \dim(\mathcal{H}_E)$, where $\mathcal{H}_E$ is the Hilbert space spanned by the eigenvectors $|\alpha\rangle$ such that $\varepsilon_\alpha \in I_E$, and define a decomposition of $\mathcal{H}_E$ in orthogonal subspaces $\mathcal{H}_\nu$ each of dimension d_ν:

$$\mathcal{H}_E = \bigoplus_\nu \mathcal{H}_\nu \quad \dim(\mathcal{H}_\nu) = d_\nu \quad \sum_\nu d_\nu = \mathscr{D} \tag{3.35}$$

Then define $\hat{P}_\nu$ as the projector on subspace $\mathcal{H}_\nu$. It is mandatory to consider the large system size limit where $1 \ll d_\nu \ll \mathscr{D}$. It is then demonstrated that, under quite generic assumptions on the original Hamiltonian $\hat{H}$ and on the orthogonal decomposition $\mathcal{H}_E = \bigoplus_\nu \mathcal{H}_\nu$, for *every wavefunction* $|\psi_0\rangle \in \mathcal{H}_E$ and for *almost all* times one has normal typicality, i.e.,

$$\langle \psi_0 | \hat{P}_\nu(t) | \psi_0 \rangle = \mathrm{Tr}[\hat{\rho}_E \hat{P}_\nu], \tag{3.36}$$

where $\hat{\rho}_E$ is the microcanonical density matrix in equation (3.31). A crucial role is played by the assumption that the dimensions d_ν of the orthogonal subspaces of $\mathcal{H}_E$ are macroscopically large (i.e., $d_\nu \gg 1$.) In this sense the projectors P_ν correspond to macroscopic observables. From this point of view, Von Neumann's quantum ergodic theorem is constrained by the same key hypothesis of Khinchin's theorem: the limit of a very large number of degrees of freedom. At the same time, Von Neumann's theorem does not make any claim on chaotic properties of the eigenspectrum, in the very same way as the Khinchin's theorem does not make any claim on chaotic properties of trajectories.

For completeness we also need to mention what is regarded today as the 'modern' version of Von Neumann's theorem, namely the celebrated eigenstate thermalization hypothesis (ETH) [254]. By expanding the expression of $\langle \hat{O}(t) \rangle_{\psi_0}$ as

$$\langle \hat{O}(t) \rangle_{\psi_0} = \sum_{\alpha \in \mathrm{Sp}\hat{H}} |c_\alpha|^2 \, O_{\alpha\alpha} + \sum_{\alpha \neq \beta} e^{i(E_\alpha - E_\beta)t/\hbar} \, c_\alpha^* c_\beta \, O_{\alpha\beta}. \tag{3.37}$$

it is not difficult to figure out that ergodicity, as expressed in equation (3.34), is guaranteed if suitable hypotheses are made for the matrix elements $O_{\alpha\beta}$. For instance, it can be assumed that:

- The off-diagonal matrix elements are exponentially small in system size, $O_{\alpha\neq\beta} \sim e^{-S(E_0)/k_B}$ with S being the microcanonical entropy and $E_0 = \langle\psi_0|\hat{H}|\psi_0\rangle$.
- The diagonal elements are a function of the initial-condition energy, $O_{\alpha\alpha} = O(E_0)$.

The hypothesis $h.\,a$ guarantees that *for large enough systems* relaxation to a stationary state is achieved within reasonable time, still leaving open the possibility that stationarity is different from thermal equilibrium. In fact, it is thanks to the exponentially small size of non-diagonal matrix elements that in the large-N limit one does not need to wait the astronomically large times needed for dephasing in order to have

$$\frac{1}{T}\int_0^T dt \sum_{\alpha\neq\beta} e^{i(E_\alpha-E_\beta)t/\hbar} \, c_\alpha^* c_\beta \, O_{\alpha\beta} \approx 0 \qquad (3.38)$$

Hypothesis $h.\,b$ then guarantees that relaxation is towards a state well characterized macroscopically, i.e., a state which depends solely on the energy of the Hilbert space vector $|\psi_0\rangle$ and not on the extensive number of coefficients c_α:

$$\sum_{\alpha\in\mathrm{Sp}\hat{H}} |c_\alpha|^2 \, O_{\alpha\alpha} = O(E_0) \cdot \sum_{\alpha\in\mathrm{Sp}\hat{H}} |c_\alpha|^2 = O(E_0) \qquad (3.39)$$

Though inspired from the behaviour of (quantum) chaotic systems, ETH is clearly a *different* property since it makes no claim on the structure of energy eigenvectors. For this reason and also because a necessary condition for ETH to be effective is a large-N limit, ETH is rather close in spirit to Von Neumann's theorem. Then, we must say that a complete understanding of the reciprocal implications of quantum chaos and the ETH is still an open issue which deserves further investigations. In this respect let us recall the purpose of this chapter, namely to furnish reasons to believe that an investigation aimed at challenging the role of chaos in the thermalization of both quantum and classical systems is an interesting and timely subject. For instance, a deeper understanding of the mechanisms which trigger, or prevent, thermalization in quantum systems is crucial to estimating the possibilities of having working scalable quantum technologies, like quantum computers and quantum sensors.

An interesting point of view which emerged from the results on the Toda model discussed in section 3.6.3.3 and which can be traced back to both Khinchin and Von Neumann's theorem is the following: the presence (or absence) of thermalization is a property pertaining to a given choice of observables and cannot be stated in general (i.e., solely on the basis of the behaviour of trajectories or the structure of energy eigenvectors). We have in mind the choice of the projectors P_ν in Von Neumann's theorem and the choice of sum functions in Khinchin's theorem. This perspective of considering 'thermal equilibrium as a matter of observables choice' is in our opinion a point of view that deserves careful investigation.

3.6.4.3 Summary and perspectives

Let us try to summarize the main aspects here discussed. At first, we stressed that the ergodic approach, even with some caveats, appears the natural way to use probability in a deterministic context. Assuming ergodicity, it is possible to obtain an empirical notion of probability which is an objective property of the trajectory. An important aspect often not considered is that both in experiments and numerical computation one deals with a unique system with many degrees of freedom, and not with an ensemble of systems. According to the point of view of Boltzmann (and the developments by Khinchin, Mazur, and van der Linden) it is rather natural to conclude that, at the conceptual level, the only physically consistent way to accumulate a statistics is in terms of time averages following the time evolution of the system. At the same time ergodicity is a very demanding property and, since in its definition it requires to consider the infinite time limit, physically it is not very accessible.

We then presented strong evidence from a numerical study of high-dimensional Hamiltonian systems that chaos is neither a necessary nor a sufficient ingredient to guarantee the validity of equilibrium statistical mechanics for classical systems. On the other hand, even when chaos is very weak (or absent), we have shown examples of a good agreement between time and ensemble averages [249]. The perspective emerging from our study is that the choice of variables is crucial to saying whether a system has thermalized or not. We have found that this point of view emphasizes the commonalities between classical and quantum mechanics for what concerns the ergodic problem. This is particularly clear by comparing the assumptions and conclusions of the ergodic theorems of Khinchin and Von Neumann, where, without *any* assumption on the chaotic nature of dynamics, it is shown that for *general enough observables* a system has good ergodic properties even in the case where interactions are absent, provided that the system is large enough.

This brought us to underline as a possibly relevant research line the one dedicated to find 'classical analogs' of thermalization problems in quantum systems and to study such problems with the conceptual tools and the numerical techniques developed for classical Hamiltonian systems.

The ability to control and exactly predict the behavior of quantum systems is in fact of extreme relevance, in particular for the great bet presently made by the worldwide scientific community on quantum computers. In particular, understanding the mechanisms preventing thermalization might certainly help to improve the function of quantum processors and to devise better quantum algorithms.

In summary, we tried to present convincing motivations in favour of a renewed interest towards the foundations of quantum and statistical mechanics. This is something which in our opinion should be pursued while approaching the one century anniversary of the two seminal papers by Heisenberg [255] and Schrödinger [256] on quantum mechanics. How much over the past 100 years has the quantum mechanics revolution influenced not only the understanding of the microscopic world, but even the thermodynamic properties of macroscopic systems? This is a key question for future research.

References

[1] Hartree D R 1928 The wave mechanics of an atom with non-Coulombic central field: parts I, II, III *Proc. Cambridge Phil. Soc.* **24** 89,111,426

[2] Tkatchenko A, Rossi M, Blum V, Ireta J and Scheffler M 2011 Unraveling the stability of polypeptide helices: critical role of van der waals interactions *Phys. Rev. Lett.* **106** 118102

[3] Martin R M, Reining L and Ceperley D M 2016 *Interacting Electrons* (Cambridge: Cambridge University Press)

[4] Keimer B and Moore J E 2017 The physics of quantum materials *Nat. Phys.* **13** 1045–55

[5] Hüfner S 2003 *Photoemission Spectroscopy: Principles and Applications* (Berlin: Springer)

[6] Sky Zhou J *et al* 2020 Unraveling intrinsic correlation effects with angle-resolved photoemission spectroscopy *Proc. Natl Acad. Sci.* **117** 28596–602

[7] Fitzgerald R 2000 What really gives a quantum computer its power? *Phys. Today* **53** 20–2

[8] Kohn W 1999 Nobel lecture: electronic structure of matter—wave functions and density functionals *Rev. Mod. Phys.* **71** 1253–66

[9] Martin R M 2020 *Electronic Structure: Basic Theory and Practical Methods* 2nd edn (Cambridge: Cambridge University Press)

[10] Bohm D and Pines D 1953 A collective description of electron interactions. 3. Coulomb interactions in a degenerate electron gas *Phys. Rev.* **92** 609–25

[11] Pines D and Bohm D 1952 A collective description of electron interactions.2. Collective vs individual particle aspects of the interactions *Phys. Rev.* **85** 338–53

[12] Huotari S, Sternemann C, Schülke W, Sturm K, Lustfeld H, Sternemann H, Volmer M, Gusarov A, Müller H and Monaco G 2008 Electron-density dependence of double-plasmon excitations in simple metals *Phys. Rev.* B **77** 195125

[13] Schülke W 2007 Electron dynamics by inelastic x-ray scattering *Oxford Series on Synchrotron Radiation* (Oxford: Oxford University Press)

[14] Egerton R F 2009 Electron energy-loss spectroscopy in the TEM *Rep. Prog. Phys.* **72** 016502

[15] Yu H, Peng Y, Yang Y and Li Z-Y 2019 Plasmon-enhanced light–matter interactions and applications *npj Comput. Mater.* **5** 45

[16] de Seauve V, Languille M-A, Kociak M, Belin S, Ablett J, Andraud C, StÃ©phan O, Rueff J-P, Fonda E and Lavédrine B 2020 Spectroscopies and electron microscopies unravel the origin of the first colour photographs *Angew. Chem. Int. Ed.* **59** 9113–9

[17] Kazimierczuk T, Fröhlich D, Scheel S, Stolz H and Bayer M 2014 Giant Rydberg excitons in the copper oxide Cu_2O *Nature* **514** 343–7

[18] Laughlin R B 2005 *A Different Universe: Reinventing Physics from the Bottom Down* (New York: Basic Books)

[19] Anderson P W 1972 More is different *Science* **177** 393–6

[20] Unkrich L and Stanton A 2003 Finding Nemo (Pixar Animation Studios)

[21] Wigner E 1934 On the interaction of electrons in metals *Phys. Rev.* **46** 1002–11

[22] Novoselov K, Geim A and Morozov S *et al* 2005 Two-dimensional gas of massless dirac fermions in graphene *Nature* **438** 197–200

[23] Pines D 1997 *The Many Body Problem (Advanced Book Classics, originally published in 1961)* (Reading, MA: Addison-Wesley)

[24] David P 2009 Excitonics heats up *Nat. Photonics* **3** 604–4

[25] Kruglyak V V, Demokritov S O and Grundler D 2010 Magnonics *J. Phys.* D **43** 264001

[26] Zhao X-G, Wang Z, Malyi O I and Zunger A 2021 Effect of static local distortions vs. dynamic motions on the stability and band gaps of cubic oxide and halide perovskites *Mater. Today* **49** 107–22

[27] Editorial 2016 The rise of quantum materials *Nat. Phys.* **12** 105

[28] Giustino F *et al* 2020 The 2021 quantum materials roadmap *J. Phys.: Mater.* **3** 042006

[29] Ohtomo A and Hwang H Y 2004 A high-mobility electron gas at the $LaAlO_3$/$SrTiO_3$ heterointerface *Nature* **427** 423–6

[30] Bistritzer R and MacDonald A H 2011 Moiré bands in twisted double-layer graphene *Proc. Natl. Acad. Sci.* **108** 12233–7

[31] Cao Y *et al* 2018 Correlated insulator behaviour at half-filling in magic-angle graphene superlattices *Nature* **556** 80–4

[32] Cao Y, Fatemi V, Fang S, Watanabe K, Taniguchi T, Kaxiras E and Jarillo-Herrero P 2018 Unconventional superconductivity in magic-angle graphene superlattices *Nature* **556** 43–50

[33] Lopes dos Santos J M B, Peres N M R and Castro Neto A H 2007 Graphene bilayer with a twist: electronic structure *Phys. Rev. Lett.* **99** 256802

[34] Savary L and Balents L 2016 Quantum spin liquids: a review *Rep. Prog. Phys.* **80** 016502

[35] Volkov B A and Pankratov O A 1985 Two-dimensional massless electrons in an inverted contact *JETP Lett.* **42** 178

[36] Wang J and Zhang S-C 2017 Topological states of condensed matter *Nat. Mater.* **16** 1062–7

[37] Qi X-L and Zhang S-C 2011 Topological insulators and superconductors *Rev. Mod. Phys.* **83** 1057–110

[38] Ghiringhelli G *et al* 2012 Long-range incommensurate charge fluctuations in (Y,Nd) $Ba_2Cu_3O_{6+x}$ *Science* **337** 821–5

[39] Arute F *et al* 2019 Quantum supremacy using a programmable superconducting processor *Nature* **574** 505–10

[40] Lejinse M and Flensberg K 2012 Introduction to topological superconductivity and Majorana fermions *Semicond. Sci. Technol.* **27** 124003

[41] Zázvorka J *et al* 2019 Thermal skyrmion diffusion used in a reshuffler device *Nat. Nanotechnol.* **14** 658–61

[42] Song K M *et al* 2020 Skyrmion-based artificial synapses for neuromorphic computing *Nat. Electron.* **3** 148–55

[43] Bukov M, D'Alessio L and Polkovnikov A 2015 Universal high-frequency behavior of periodically driven systems: from dynamical stabilization to floquet engineering *Adv. Phys.* **64** 139–226

[44] Mahmood F, Chan C-K, Alpichshev Z, Gardner D, Lee Y, Lee P A and Gedik N 2016 Selective scattering between Floquet–Bloch and Volkov states in the topological insulator Bi_2Se_3 arXiv:1512.05714 [cond-mat.mes-hall]

[45] Flick J, Ruggenthaler M, Appel H and Rubio A 2017 Atoms and molecules in cavities, from weak to strong coupling in quantum-electrodynamics (QED) chemistry *Proc. Natl Acad. Sci.* **114** 3026–34

[46] Fausti D, Tobey R I, Dean N, Kaiser S, Dienst A, Hoffmann M C, Pyon S, Takayama T, Takagi H and Cavalleri A 2011 Light-induced superconductivity in a stripe-ordered cuprate *Science* **331** 189–91

[47] Mor S, Gosetti V, Molina-Sánchez A, Sangalli D, Achilli S, Agekyan V F, Franceschini P, Giannetti C, Sangaletti L and Pagliara S 2021 Photoinduced modulation of the excitonic

resonance via coupling with coherent phonons in a layered semiconductor *Phys. Rev. Res.* **3** 043175

[48] Sachdev S 2003 Colloquium: order and quantum phase transitions in the cuprate superconductors *Rev. Mod. Phys.* **75** 913–32

[49] Sacha K and Zakrzewski J 2017 Time crystals: a review *Rep. Prog. Phys.* **81** 016401

[50] Vergniory M G, Elcoro L, Felser C, Regnault N, Andrei Bernevig B and Wang Z 2019 A complete catalogue of high-quality topological materials *Nature* **566** 480–5

[51] Zhang T, Jiang Y, Song Z, Huang H, He Y, Fang Z, Weng H and Fang C 2019 Catalogue of topological electronic materials *Nature* **566** 475–9

[52] Schmidt J, Marques M R G, Botti S and Marques M A L 2019 Recent advances and applications of machine learning in solid-state materials science *NPJ Comput. Mater.* **5** 83

[53] Schleder G R, Padilha A C M, Acosta C M, Costa M and Fazzio A 2019 From DFT to machine learning: recent approaches to materials science—a review *J. Phys.: Mater.* **2** 032001

[54] Canfield P C and Fisk Z 1992 Growth of single crystals from metallic fluxes *Phil. Mag.* B **65** 117

[55] Kanatzidis M G *et al* 2005 The metal flux: a preparative tool for the exploration of intermetallic compounds *Angew. Chem. Int. Ed.* **44** 6996

[56] Stewart G R 2011 Superconductivity in iron compounds *Rev. Mod. Phys.* **83** 1589

[57] Tong C-J *et al* 2005 Microstructure characterization of Al$_x$CoCrCuFeNi high-entropy alloy system with multiprincipal elements *Metall. Mater. Trans.* A **36** 881

[58] Ligon S C *et al* 2017 Polymers for 3D printing and customized additive manufacturing *Chem. Rev.* **117** 10212

[59] Parkin S S P *et al* 2004 Giant tunnelling magnetoresistance at room temperature with MgO (100) tunnel barriers *Nat. Mater.* **3** 862

[60] Hoppe H *et al* 2004 Organic solar cells: an overview *J. Mater. Res.* **19** 1924

[61] Priyadarshini P *et al* 2021 Structural and optoelectronic properties change in Bi/In$_2$Se$_3$ heterostructure films by thermal annealing and laser irradiation *J. Appl. Phys.* **129** 223101

[62] Strite S and Morkoç H 1992 GaN, AlN, and InN: a review *J. Vac. Sci. Technol.* B **10** 1237

[63] Faist J *et al* 1994 Quantum cascade laser *Science* **264** 553

[64] Zheng H *et al* 2004 Multiferroic BaTiO$_3$–CoFe$_2$O$_4$ nanostructures *Science* **303** 661

[65] Özgür Ü *et al* 2005 A comprehensive review of ZnO materials and devices *J. Appl. Phys.* **98** 041301

[66] Li X *et al* 2009 Large-area synthesis of high-quality and uniform graphene films on copper foils *Science* **324** 1312

[67] George S M 2010 Atomic layer deposition: an overview *Chem. Rev.* **110** 111

[68] De Teresa J M (ed) 2020 *Nanofabrication: Nanolithography Techniques and Their Applications* (Bristol: IOP Publishing)

[69] Wagner R S and Ellis W C 1964 Vapor–liquid–solid mechanism of single crystal growth *Appl. Phys. Lett.* **4** 89

[70] Law M *et al* 2004 Semiconductor nanowires and nanotubes *Ann. Rev. Mater. Res.* **34** 83

[71] Nietsch K *et al* 2000 Uniform nickel deposition into ordered alumina pores by pulsed electrodeposition *Adv. Mater.* **12** 582

[72] Fert A and Piraux L 1999 Magnetic nanowires *J. Magn. Magn. Mater.* **200** 338

[73] Utke I *et al* 2008 Gas-assisted focused electron beam and ion beam processing and fabrication *J. Vac. Sci. Technol.* B **26** 1197

[74] Córdoba R *et al* 2019 Three-dimensional superconducting nanohelices grown by He^+-focused-ion-beam direct writing *Nano Lett.* **19** 8597

[75] Laurenti M *et al* 2017 Surface engineering of nanostructured ZnO surfaces *Adv. Mater. Interfaces* **4** 1600758

[76] Novoselov K S *et al* 2004 Electric field effect in atomically thin carbon films *Science* **306** 666

[77] Cao Y *et al* 2018 Unconventional superconductivity in magic-angle graphene superlattices *Nature* **556** 43

[78] Lopes W *et al* 2001 Hierarchical self-assembly of metal nanostructures on diblock copolymer scaffolds *Nature* **414** 735

[79] Lu W and Lieber C 2007 Nanoelectronics from the bottom up *Nature* **6** 841

[80] Pendry J B *et al* 2006 Controlling electromagnetic fields *Science* **312** 1780

[81] Bradlyn B *et al* 2017 Topological quantum chemistry *Nature* **547** 298

[82] Xu S-Y *et al* 2015 Discovery of a Weyl fermion semimetal and topological Fermi arcs *Science* **349** 613

[83] Cui J *et al* 2018 Current progress and future challenges in rare-earth-free permanent magnets *Acta Mater.* **158** 118

[84] Larcher D and Tarascon J M 2015 Towards greener and more sustainable batteries for electrical energy storage *Nat. Chem.* **7** 19

[85] Chen Z-G *et al* 2012 Nanostructured thermoelectric materials: current research and future challenge *Prog. Nat. Sci. Mater. Int* **22** 535

[86] Dresselhaus M S *et al* 2007 New directions for low-dimensional thermoelectric materials *Adv. Mater.* **19** 1043

[87] Bux S K *et al* 2010 Nanostructured materials for thermoelectric applications *Chem. Commun.* **46** 8311

[88] Szczech J R *et al* 2011 Enhancement of the thermoelectric properties in nanoscale and nanostructured materials *J. Mater. Chem.* **21** 4037

[89] Venkatasubramanian R *et al* 2001 Thin-film thermoelectric devices with high room-temperature figures of merit *Nature* **413** 597

[90] Zhao L-D *et al* 2014 Ultralow thermal conductivity and high thermoelectric figure of merit in SnSe crystals *Nature* **508** 373

[91] George E P *et al* 2019 High- entropy alloys *Nat. Rev. Mater.* **4** 515

[92] Bennett C H and DiVincenzo D P 2000 Quantum information and computation *Nature* **404** 247

[93] Arute F 2019 Quantum supremacy using a programmable superconducting processor *Nature* **574** 505

[94] Wendin G 2017 Quantum information processing with superconducting circuits: a review *Rep. Prog. Phys.* **80** 106001

[95] Kuzum D *et al* 2013 Synaptic electronics: materials, devices and applications *Nanotechnology* **24** 382001

[96] Saïghi S *et al* 2015 Plasticity in memristive devices for spiking neural networks *Front. Neurosci.* **9** 51

[97] Torrejón J *et al* 2017 Neuromorphic computing with nanoscale spintronic oscillators *Nature* **547** 428

[98] Clark R *et al* 2018 Perspective: new process technologies required for future devices and scaling *APL Mater.* **6** 058203

[99] Nomura K *et al* 2004 Room-temperature fabrication of transparent flexible thin-film transistors using amorphous oxide semiconductors *Nature* **432** 488

[100] Kim K S *et al* 2009 Large-scale pattern growth of graphene films for stretchable transparent electrodes *Nature* **457** 7230

[101] Garcia-Cortadella R *et al* 2021 Graphene active sensor arrays for long-term and wireless mapping of wide frequency band epicortical brain activity *Nat. Commun.* **12** 211

[102] Baselt D R *et al* 1998 A biosensor based on magnetoresistance technology *Biosens. Bioelectron.* **13** 731

[103] Klauk H *et al* 2007 Ultralow-power organic complementary circuits *Nature* **445** 745

[104] Gibson D and MacGregor C 2013 A novel solid state non-dispersive infrared CO_2 gas sensor compatible with wireless and portable deployment *Sensors* **13** 7079

[105] Yin W-J *et al* 2015 Halide perovskite materials for solar cells: a theoretical review *J. Mater. Chem.* A **3** 8926

[106] Lampert C M 2004 Chromogenic smart materials *Mater. Today* **7** 28

[107] Sootsman J R *et al* 2009 New and old concepts in thermoelectric materials *Angew. Chem., Int. Ed.* **48** 8616

[108] Harb A 2011 Energy harvesting: state-of-the-art *Renew. Energy* **36** 2641

[109] Ehrlich K 2001 Materials research towards a fusion reactor *Fusion Eng. Des.* **56–7** 71

[110] Himanen L *et al* 2019 Data-driven materials science: status, challenges, and perspectives *Adv. Sci.* **6** 1900808

[111] Haastrup S *et al* 2018 The computational 2D materials database: high-throughput modeling and discovery of atomically thin crystals *2D Mater.* **5** 042002

[112] Chung Y G *et al* 2019 Advances, updates, and analytics for the computation-ready, experimental metal–organic framework database: CoRE MOF 2019 *J. Chem. Eng. Data* **64** 5985

[113] Draxl C and Scheffler M 2019 *Handbook of Materials Modeling* ed W Andreoni and S Yip (Cham: Springer)

[114] Kim E *et al* 2017 Materials synthesis insights from scientific literature via text extraction and machine learning *Chem. Mater.* **29** 9436

[115] Wilkinson M D *et al* 2016 The FAIR Guiding Principles for scientific data management and stewardship *Sci. Data* **3** 160018

[116] Lejaeghere K *et al* 2016 Reproducibility in density functional theory calculations of solids *Science* **351** aad3000

[117] Gulans A *et al* 2018 Microhartree precision in density functional theory calculations *Phys. Rev.* B **97** 161105(R)

[118] Jensenet S R *et al* 2020 Polarized Gaussian basis sets from one-electron ions *J. Chem. Phys. Lett.* **152** 134108

[119] Nabok D *et al* 2016 Accurate all-electron G_0W_0 quasiparticle energies employing the full-potential augmented plane-wave method *Phys. Rev.* B **94** 035418

[120] Rangel T *et al* 2020 Reproducibility in G_0W_0 calculations for solids *Comput. Phys. Commun.* **255** 107242

[121] Draxl C and Scheffler M 2018 NOMAD: the FAIR concept for big data-driven materials science *MRS Bull.* **43** 676

[122] Draxl C and Scheffler M 2019 The NOMAD laboratory: from data sharing to artificial intelligence *J. Phys.: Mater.* **2** 036001

[123] Scheffler M *et al* 2022 FAIR data enabling new horizons for materials research *Nature* **604** 635

[124] Greene G *et al* 2019 Building open access to research (OAR) data infrastructure at NIST *CODATA Data Sci. J.* **18** 30

[125] Zambrini R and Rius G 2021 *Digital and Complex Information* (CSIC)

[126] Gray M *et al* 2018 Implantable biosensors and their contribution to the future of precision medicine *Vet. J.* **239** 21

[127] Qi X L 2011 Topological insulators and superconductors *Rev. Mod. Phys.* **83** 1057

[128] Kimble H J 2008 The quantum internet *Nature* **453** 1023

[129] Marchiori E *et al* 2022 Nanoscale magnetic field imaging for 2D materials *Nat. Rev.* **4** 49

[130] Maze J R *et al* 2008 Nanoscale magnetic sensing with an individual electronic spin in diamond *Nature* **455** 644

[131] Wellman S M *et al* 2018 A materials roadmap to functional neural interface design *Adv. Funct. Mater.* **28** 1701269

[132] Goering S *et al* 2021 Recommendations for responsible development and application of neurotechnologies *Neuroethics* **14** 365

[133] Errea I 2022 Superconducting hydrides on a quantum landscape *J. Phys.: Condens. Matter* **34** 231501

[134] Hawryluk R and Zohm H 2019 The challenge and promise of studying burning plasmas *Phys. Today* **72** 34

[135] Andersen C W *et al* 2021 OPTIMADE, an API for exchanging materials data *Sci. Data* **8** 217

[136] Tanaka I, Rajan K and Wolverton C 2018 Data-centric science for materials innovation *MRS Bull.* **43** 659

[137] Botti R S and Marques M 2021 Roadmap on machine learning in electronic structure *Electron. Struct.* **4** 023004 2021

[138] Ghiringhelli L M, Vybiral J, Levchenko S V, Draxl C and Scheffler M 2015 Big data of materials science: critical role of the descriptor *Phys. Rev. Lett.* **114** 105503

[139] Isayev O *et al* 2017 Universal fragment descriptors for predicting properties of inorganic crystals *Nat. Commun.* **8** 15679

[140] Jäckle M, Helmbrecht K, Smits M, Stottmeister D and Groß A 2018 Self-diffusion barriers: possible descriptors for dendrite growth in batteries? *Energy Environ. Sci.* **11** 3400

[141] Ward L, Agrawal A, Choudhary A and Wolverton C 2016 A general-purpose machine learning framework for predicting properties of inorganic materials *npj Comput. Mater.* **2** 16028

[142] Zhou X *et al* 2021 High- temperature superconductivity *Nat. Rev. Phys.* **3** 462

[143] European Commission 2020 Study on the EU's list of critical raw materials *Final Report*

[144] Schrödinger E 1952 Are there quantum jumps? (Part II) *Br. J. Philos. Sci.* **III** 233

[145] Feynman R P 1960 There is plenty of room at the bottom *Eng. Sci.* **XXIII** 22

[146] Eigler D M and Schweizer E K 1990 Positioning single atoms with a scanning tunnelling microscope *Nature* **344** 524

[147] Ash E A and Nicholls G 1972 Super-resolution aperture scanning microscope *Nature* **237** 510

[148] Pohl D W, Denk W and Lanz M 1984 Optical stethoscopy: image recording with resolution l/20 *Appl. Phys. Lett.* **44** 651

[149] Betzig E *Single Molecules, Cells, and Super-Resolution Optics* (Nobel Lectures) https://nobelprize.org/prizes/chemistry/2014/betzig/lecture/

[150] Moerner W E *Single-molecule Spectroscopy, Imaging and Photocontrol: Foundations for Super-Resolution Microscopy* (Nobel Lecture) www.nobelprize.org/prizes/chemistry/2014/moerner/lecture/

[151] Hell S 2014 *Nanoscopy with Focused Light* (Nobel Lectures) www.nobelprize.org/prizes/chemistry/2014/Hell/lecture/

[152] Orrit M and Bernard J 1990 Single pentacene molecules detected by fluorescence excitation in a p-terphenyl crystal *Phys. Rev. Lett.* **65** 2716

[153] Ashkin A 1970 Acceleration and trapping of particles by radiation pressure *Phys. Rev. Lett.* **24** 156

[154] Fert A *The Origin, Development and Future of Spintronics* (Nobel Lectures) www.nobelprize.org/prizes/physics/2007/fert/lecture/

[155] Grünberg P A *From Spinwaves to Giant Magnetoresistance (GMR) and Beyond* (Nobel Lecture) www.nobelprize.org/prizes/physics/2007/grunberg/lecture/

[156] Vellekoop I M and Mosk A P 2007 Focussing coherent light through opaque scattering media *Opt. Lett.* **32** 2309

[157] European Commission 2019 The European Green Deal sets out how to make Europe the first climate-neutral continent by 2050, boosting the economy, improving people's health and quality of life, caring for nature, and leaving no one behind (press release) https://ec.europa.eu/commission/presscorner/detail/en/IP_19_6691

[158] See books such asFan C and Zhao Z 2018 *Synchrotron Radiation Applications, or Synchrotron Radiation in Materials Science: Light Sources, Techniques, and Applications* (Wiley)

[159] Eriksson M 2016 The multi-bend achromat storage rings *AIP Conf. Proc.* **1741** 020001

[160] European Synchrotron Radiation Facility (https://esrf.eu/about/upgrade)

[161] Madey J M J 1971 Stimulated emission of bremsstrahlung in a periodic magnetic field *J. Appl. Phys.* **42** 1906

[162] Pellegrini C, Marinelli A and Reiche S 2016 The physics of x-ray free-electron lasers *Rev. Mod. Phys.* **88** 015006

[163] Kim J G *et al* 2020 Mapping the emergence of molecular vibrations mediating bond formation *Nature* **582** 520

[164] Buzzi M, Först M and Cavalleri A 2019 Measuring non-equilibrium dynamics in complex solids with ultrashort X-ray pulses *Phil. Trans. R. Soc.* A **377** 20170478

[165] Büttner F *et al* 2021 Observation of fluctuation-mediated picosecond nucleation of a topological phase *Nat. Mater.* **20** 30

[166] Cerantola V *et al* 2021 New frontiers in extreme conditions science at synchrotrons and free electron lasers *J. Phys.: Condens. Matter* **33** 274003

[167] Young L *et al* 2018 Roadmap of ultrafast x-ray atomic and molecular physics *J. Phys. B: At. Mol. Opt. Phys.* **51** 032003

[168] Tschentscher Th 2023 Investigating ultrafast structural dynamics using high repetition rate x-ray FEL radiation at European XFEL *Eur. Phys. J. Plus* **138** 274

[169] Huang S *et al* 2017 Generating single-spike hard x-ray pulses with nonlinear bunch compression in free-electron lasers *Phys. Rev. Lett.* **119** 154801

[170] Duris J *et al* 2020 Tunable isolated attosecond x-ray pulses with gigawatt peak power from a free-electron laser *Nat. Photonics* **14** 30–6

[171] Praveen Kumar M *et al* 2020 Attosecond pulse shaping using a seeded free-electron laser *Nature* **578** 386–91

[172] Huang Z and Ruth R R 2006 Fully coherent x-ray pulses from a regenerative-amplifier free-electron laser *Phys. Rev. Lett.* **96** 144801
Kwang-je Kim Y, Shvyd'ko and Reiche S 2008 A proposal for an x-ray free-electron laser oscillator with an energy-recovery linac *Phys. Rev. Lett.* **100** 244802

[173] Rouxel J R *et al* 2021 Hard X-ray transient grating spectroscopy on bismuth germanate *Nat. Photonics* **15** 499

[174] Zewail A H 2000 Femtochemistry: atomic-scale dynamics of the chemical bond *J. Phys. Chem.* A **104** 5660–94

[175] Ferray M, L'Huillier A, Li X F, Lompre L A, Mainfray G and Manus C 1988 Multiple harmonic conversion of 1064 nm radiation in rare gases *J. Phys.* B **21** L31–5

[176] McPherson A, Gibson G, Jara H, Johann U, Luk T S, McIntyre I A, Boyer K and Rhodes C K 1987 Studies of multiphoton production of vacuum-ultraviolet radiation in the rare gases *J. Opt. Soc. Am.* B **4** 595–601

[177] Schafer K J, Yang B, DiMauro L F and Kulander K C 1993 Above threshold ionization beyond the high harmonic cutoff *Phys. Rev. Lett.* **70** 1599–602

[178] Corkum P B 1993 Plasma perspective on strong field multiphoton ionization *Phys. Rev. Lett.* **71** 1994–7

[179] Lewenstein M, Balcou P, Ivanov M Y, L'Huillier A and Corkum P B 1994 Theory of high-harmonic generation by low-frequency laser fields *Phys. Rev.* A **49** 2117–32

[180] Salières P *et al* 2001 Feynman's path-integral approach for intense-laser-atom interactions *Science* **292** 902–5

[181] Krausz F and Ivanov M 2009 Attosecond physics *Rev. Mod. Phys.* **81** 163–234

[182] Paul P M *et al* 2001 Observation of a train of attosecond pulses from high harmonic generation *Science* **292** 1689–92

[183] Chang Z 2016 *Fundamentals of Attosecond Optics* (Boca Raton, FL: CRC Press)

[184] Li J, Lu J, Chew A, Han S, Li J, Wu Y, Wang H, Ghimire S and Chang Z 2020 Attosecond science based on high harmonic generation from gases and solids *Nat. Commun.* **11** 2748

[185] Li J *et al* 2017 53-Attosecond X-ray pulses reach the carbon k-edge *Nat. Commun.* **8** 186

[186] Gaumnitz T, Jain A, Pertot Y, Huppert M, Jordan I, Ardana-Lamas F and Wörner H J 2017 Streaking of 43-attosecond soft-X-ray pulses generated by a passively cep-stable mid-infrared driver *Opt. Express* **25** 27506–18

[187] Johnson A S *et al* 2018 High-flux soft x-ray harmonic generation from ionization-shaped few-cycle laser pulses *Sci. Adv.* **4** eaar3761

[188] Popmintchev T *et al* 2012 Bright coherent ultrahigh harmonics in the kev x-ray regime from mid-infrared femtosecond lasers *Science* **336** 1287–91

[189] Takahashi E J, Lan P, Mücke O D, Nabekawa Y and Midorikawa K 2013 Attosecond nonlinear optics using gigawatt-scale isolated attosecond pulses *Nat. Commun.* **4** 2691

[190] Klas R, Eschen W, Kirsche A, Rothhardt J and Limpert J 2020 Generation of coherent broadband high photon flux continua in the XUV with a subtwo- cycle fiber laser *Opt. Express* **28** 6188–96

[191] Nomura Y *et al* 2009 Attosecond phase locking of harmonics emitted from laser-produced plasmas *Nat. Phys.* **5** 124–8

[192] Wheeler J A, Borot A, Monchocé S, Vincenti H, Ricci A, Malvache A, Lopez-Martens R and Quéré F 2012 Attosecond lighthouses from plasma mirrors *Nat. Photonics* **6** 829–33

[193] Lépine F, Sansone G and Vrakking M J J 2013 Molecular applications of attosecond laser pulses *Chem. Phys. Lett.* **578** 1–14

[194] Drescher M, Hentschel M, Kienberger R, Uiberacker M, Yakovlev V and Scrinzi A 2002 Time-resolved atomic inner-shell spectroscopy *Nature* **419** 803–7

[195] Pazourek R, Nagele S and Burgdörfer J 2015 Attosecond chronoscopy of photoemission *Rev. Mod. Phys.* **87** 765

[196] Kotur M *et al* 2016 Spectral phase measurement of a Fano resonance using tunable attosecond pulses *Nat. Commun.* **7** 10566

[197] Nandi S *et al* 2020 Attosecond timing of electron emission from a molecular shape resonance *Sci. Adv.* **6** eaba7762

[198] Cattaneo L, Pedrelli L, Bello R Y, Palacios A, Keathley P D, Martín F and Keller U 2022 Isolating attosecond electron dynamics in molecules where nuclei move fast *Phys. Rev. Lett.* **128** 063001

[199] You D *et al* 2020 New method for measuring angle-resolved phases in photoemission *Phys. Rev.* X **10** 031070

[200] Seiffert L *et al* 2017 Attosecond chronoscopy of electron scattering in dielectric nano-particles *Nat. Phys. (N.Y.)* **13** 766–70

[201] Jordan I, Huppert M, Rattenbacher D, Peper M, Jelovina D, Perry C, von Conta A, Schild A and Wörner H J 2020 Attosecond spectroscopy of liquid water *Science* **369** 974–9

[202] Gruson V *et al* 2016 Attosecond dynamics through a Fano resonance: monitoring the birth of a photoelectron *Science* **354** 734–8

[203] Autuori A *et al* 2022 Anisotropic dynamics of two-photon ionization: an attosecond movie of photoemission *Sci. Adv.* **8** eabl7594

[204] Lépine F, Ivanov M Y and Vrakking M J J 2014 Attosecond molecular dynamics: fact or fiction? *Nat. Photonics* **8** 195–204

[205] Sansone G *et al* 2010 Electron localization following attosecond molecular photoionization *Nature* **465** 763–6

[206] Neidel C *et al* 2013 Probing time-dependent molecular dipoles on the attosecond time scale *Phys. Rev. Lett.* **111** 033001

[207] Calegari F *et al* 2014 Ultrafast electron dynamics in phenylalanine initiated by attosecond pulses *Science* **346** 336–9

[208] Barillot T *et al* 2021 Correlation-driven transient hole dynamics resolved in space and time in the isopropanol molecule *Phys. Rev.* X **11** 031048

[209] Hervé M *et al* 2021 Ultrafast dynamics of correlation bands following XUV molecular photoionization *Nat. Phys. (N.Y.)* **17** 327–31

[210] Attar A R, Bhattacherjee A, Pemmaraju C D, Schnorr K, Closser K D, Prendergast D and Leone S R 2017 Femtosecond x-ray spectroscopy of an electrocyclic ring-opening reaction *Science* **356** 54–9

[211] Schultze M *et al* 2013 Controlling dielectrics with the electric field of light *Nature* **493** 75–8

[212] Moulet A, Bertrand J B, Klostermann T, Guggenmos A, Karpowicz N and Goulielmakis E 2017 Soft x-ray excitonics *Science* **357** 1134–8

[213] The Nobel Prize organisation 2018 https://nobelprize.org/prizes/physics/2018/summary/

[214] Danson C *et al* 2019 Petawatt and exawatt class lasers worldwide *High Power Laser Sci. Eng.* **7** E54

[215] Yoon J W, Y G K, Choi I W, Sung J H, Lee H W, Lee S K and Nam C H 2021 Realization of laser intensity over 10^{23}W/cm^2 *Optica* **8** 630–5

[216] Joshi C 2021 New ways to smash particles *Sci. Am.* **55**

[217] Kneip S *et al* 2010 Bright spatially coherent synchrotron X-rays from a table-top source *Nat. Phys.* **6** 980–3

[218] Fiuza F, Swadling G F and Grassi A *et al* 2020 Electron acceleration in laboratory-produced turbulent collisionless shocks *Nat. Phys.* **16** 916–20

[219] Wang W, Feng K and Ke L *et al* 2021 Free-electron lasing at 27 nanometres based on a laser wakefield accelerator *Nature* **595** 516–20

[220] For a popular account seeSilva L O 2017 Boiling the vacuum: in silico plasmas under extreme conditions in the laboratory and in astrophysics *Europhys. News* **48** 34–7
for comprehensive reviews, seeMarklund M and Shukla P K 2006 Nonlinear collective effects in photon-photon and photon-plasma interactions *Rev. Mod. Phys.* **78** 591
Di Piazza A, Müller C, Hatsagortsyan K Z and Keitel C H 2012 Extremely high-intensity laser interactions with fundamental quantum systems *Rev. Mod. Phys.* **84** 1177
Gonoskov A, Blackburn T G, Marklund M and Bulanov S S 2022 Charged particle motion and radiation in strong electromagnetic fields arXiv:2107.02161

[221] Diamond Light Source 2022 https://diamond.ac.uk/Home/About/Vision/Diamond-II.html

[222] Elettra Sincrotrone Trieste 2023 https://elettra.eu/images/Documents/ELETTRA%20Machine/Elettra2/Elettra_2.0_CDR_abridged.pdf

[223] Deutsches Elektronen-Synchrotron DESY 2019 PETRA IV, Upgrade of PETRA III to the Ultimate 3D X-ray Microscope, Conceptual Design Report https://bib-pubdb1.desy.de/record/426140/files/DESY-PETRAIV-Conceptual-Design-Report.pdf

[224] Paul Scherrer Institute 2022 SLS 2.0 https://psi.ch/en/sls2-0

[225] *Source Optimisée de Lumière d'Energie Intermédiaire du LURE* (https://synchrotron-soleil.fr/en/news/conceptual-design-report-soleil-upgrade)

[226] Lee T D 1981 *Particle Physics and Introduction to Field Theory* (New York: Harwood Academic Publishers)

[227] Castiglione P, Falcioni M, Lesne A and Vulpiani A 2008 *Chaos and Coarse Graining in Statistical Mechanics* (Cambridge: Cambridge University Press)

[228] Cencini M, Cecconi F and Vulpiani A 2009 *Chaos: From Simple Models to Complex Systems* (Singapore: World Scientific)

[229] Oono Y 2013 *The Nonlinear World* (Berlin: Springer)

[230] Khinchin A I 1949 *Mathematical Foundations of Statistical Mechanics* (New York: Dover)

[231] Von Neumann J 2010 Proof of the Ergodic theorem and H-theorem in quantum mechanics *Eur. Phys. J.* H **35** 201

[232] Goldstein S, Lebowitz J L, Mastrodonato C, Tumulka R and Zanghì N 2010 Normal typicality and von Neumann's quantum ergodic theorem *Proc. R. Soc.* A **466** 3203–24

[233] Goldstein S, Lebowitz J L, Tumulka R and Zanghì N 2010 Long-time behavior of macroscopic quantum systems *Eur. Phys. J.* H **35** 173

[234] Emch G and Liu C 2002 *The Logic of Thermo-Statistical Physics* (Berlin: Springer)

[235] Gallavotti G (ed) 2008 *The Fermi-Pasta-Ulam Problem: A Status Report* (Berlin: Springer)

[236] Benettin G, Chrisodoulidi H and Ponno A 2013 The Fermi–Pasta–Ulam problem and its underlying integrable dynamics *J. Stat. Phys.* **152** 195–212

[237] Pikovsky A and Politi A 2016 *Lyapunov Exponents, A Tool to Explore Complex Dynamics* (Cambridge: Cambridge University Press)

[238] Kaneko K and Bagley R J 1985 Arnold diffusion, ergodicity and intermittency in a coupled standard mapping *Phys. Lett.* A **110** 435

[239] Falcioni M, Marini Bettolo Marconi U and Vulpiani A 1991 Ergodic properties of high-dimensional symplectic maps *Phys. Rev.* A **44** 2263

[240] Yamagishi J F and Kaneko K 2020 Chaos with a high-dimensional torus *Phys. Rev. Res.* **2** 023044

[241] Hurd L, Grebogi C and Ott E 1994 On the tendency toward ergodicity with increasing number of degrees of freedom in Hamiltonian systems *Hamiltonian Mechanics* ed J Siemenis (New York: Plenum) p 123

[242] Mézard M, Parisi G and Virasoro M-A 1987 *Spin Glass Theory and Beyond* (Singapore: World Scientific)

[243] Kaneko K and Konishi T 1987 Transition, ergodicity and Lyapunov spectra of Hamiltonian dynamical systems *J. Phys. Soc. Japan* **56** 2993

[244] Livi R, Politi A and Ruffo S 1986 Distribution of characteristic exponents in the thermodynamic limit *J. Phys. A: Math. Gen.* **19** 2033

[245] Bunimovich L A and Sinai G 1993 Statistical mechanics of coupled map lattices *Theory and Applications of Coupled Map Lattices* ed K Kaneko (New York: Wiley) p 169

[246] Livi R, Pettini M, Ruffo S and Vulpiani A 1987 Chaotic behavior in nonlinear Hamiltonian systems and equilibrium statistical mechanics *J. Stat. Phys.* **48** 539

[247] Hénon M 1974 Integrals of the Toda lattice *Phys. Rev.* B **9** 1921

[248] Cugliandolo L F and Kurchan J 1995 Weak ergodicity breaking in mean-field spin-glass models *Philos. Mag.* B **71** 501–14

[249] Baldovin M, Vulpiani A and Gradenigo G 2020 Statistical mechanics of an integrable system arXiv:2009.06556

[250] Flaschka H 1974 The Toda lattice. II. Existence of integrals *Phys. Rev.* B **9** 1924–5

[251] Mazur P and van der Linden J 1963 Asymptotic form of the structure function for real systems *J. Math. Phys.* **4** 271

[252] Bohigas O, Giannoni M and Schmit C 1984 Characterization of chaotic quantum spectra and universality of level fluctuations law *Phys. Rev. Lett.* **52** 1

[253] Vulpiani A 1994 Determinismo e Caos *Carocci Editore* (Roma)

[254] Rigol M and Srednicki M 2012 Alternatives to Eigenstate thermalization *Phys. Rev. Lett.* **108** 110601

[255] Heisenberg W 1925 Über quantentheoretische Umdeutung kinematischer undmechanischer Beziehungen *Z. Phys.* **33** 879

[256] Schrödinger E 1926 An ondulatory theory of the mechanics of atoms and molecules *Phys. Rev.* **28** 1049

IOP Publishing

EPS Grand Challenges

Physics for Society in the Horizon 2050

Mairi Sakellariadou, Claudia-Elisabeth Wulz, Kees van Der Beek, Felix Ritort, Bart van Tiggelen, Ralph Assmann, Giulio Cerullo, Luisa Cifarelli, Carlos Hidalgo, Felicia Barbato, Christian Beck, Christophe Rossel and Luc van Dyck

Chapter 4

Physics for understanding life

Patricia Bassereau, Angelo Cangelosi, Jitka Čejková, Carlos Gershenson, Raymond Goldstein, Zita Martins, Sarah Matthews, Felix Ritort, Sara Seager, Bart Van Tiggelen, David Vernon and Frances Westall

4.1 Introduction

Felix Ritort[1] and Bart Van Tiggelen[2]
[1]University of Barcelona, Barcelona, Spain
[2]Laboratoire de Physique et de Modélisation des Milieux Condensés, University Grenoble Alpes/CNRS, Grenoble, France

Infinitely many challenges exist for physics and science in general to understand life and human behaviour. Many of them were identified already a long time ago: What is the nature of the human mind? Does life exist elsewhere in the Universe? How did life emerge on Earth? Did molecules essential for life come out of space? The huge progress in physics, chemistry, biology, computer science, and astrophysics over the last decennia has made it clear that answers to these complex questions have become within reach. They will emerge from the next generation of scientists, forming collaborations using advanced technologies where frontiers between disciplines will eventually disappear.

We currently have a lack of knowledge of the early environmental conditions on Earth when life emerged, especially because no well-preserved crust is known to exist older than 3.5 billion years. Hopefully, a remnant of the ancient crust may be discovered under the Antarctic ice cover. Removing the uncertainties around the forcing of climate by the young Sun is a second major challenge. The Sun–Earth connection must have played an essential role in the prebiotic chemistry, before life was created. Finally, we are the first generation able to truly search for extraterrestrial life. Today we can send landers to targets in the Solar System and remotely probe the atmospheres of exoplanets for biosignatures. Many new fly and probe missions are under study or even under design. Understanding how life is created on Earth and unveiling life elsewhere is a tremendous challenge for mankind that will change drastically our vision of the Universe.

In the past decades, physics has faced the challenge of unravelling hidden principles that explain the marvelous complexity of living matter. While biology and biochemistry have made huge steps over the past two centuries, a quantitative understanding of life itself remains a formidable challenge. Life is permanently out-of-equilibrium, and despite the major progress in physical and chemical sciences, we still do not understand how to produce life from raw materials. Is it reasonable to expect the advent of new physical laws or principles emerging from biological studies? Is biology going to unveil a deep nexus with physics one day, as chemistry already did a century ago? Finally, does natural evolution have the flavor of physical law, or is it just conditional for life to exist? Almost sure, all these questions will be at the core of physics research in this century.

Artificial intelligence and artificial life aim at answering to these questions posing new and provocative challenges for the so-called fourth industrial revolution. Can we build machines that solve tasks like humans do? Can we empower machines and robots with brain-inspired algorithms? Will we ever understand how intelligence and consciousness work and how we deal with uncertainty? Besides designing machines and algorithms that behave as living beings, we would like to implement living matter in the lab. Physics offers uncountable possibilities to build artificial wet life. Compared to living cells, synthetic cells exhibit limited properties and function-alities, yet there may be ways to organize living matter beyond what we see in nature. A remarkable feat would be to develop artificial autonomous systems with open-ended evolution, the feature that makes animate matter so different from inanimate matter.

This chapter presents a collection of articles on these four topics by exposing the main challenges that lie ahead in these exciting fields.

4.2 Searching for life in the Universe: what is our place in the Universe?

Sara Seager[1] and Zita Martins[2]

[1]Massachusetts Institute of Technology, 77 Massachusetts Ave., Cambridge, MA 02139, USA

[2] Centro de Química Estrutural, Institute of Molecular Sciences and Department of Chemical Engineering, Instituto Superior Técnico, Universidade de Lisboa, 1049-001 Lisboa, Portugal

For thousands of years, inspired by the star-filled dark night sky, people have wondered what lies beyond Earth. Today, the search for signs of life is a key factor in modern-day planetary exploration, both for *in situ* exploration of our own Solar System's planets and moons, and by remote sensing via telescopes of exoplanets orbiting nearby stars. Planetary *in situ* measurements enable us to search directly for organics and even life itself. The holy grail of detection of life in our Solar System would be to detect a 'fingerprint' of present or/past extraterrestrial life, either in the atmosphere or regolith of planets, or in the oceans of the icy moons of Jupiter and Saturn. Beyond our Solar System, we aim to detect a gas in an exoplanetary atmosphere that might be attributed to life. A suitable 'biosignature gas' must be able to accumulate in an atmosphere against atmospheric radicals and other sinks, have strong atmospheric spectral features, and have limited abiological false positives. This review summarizes the organic compounds that are representative of potential biosignatures in planets and moons, and how to distinguish these from prebiotic chemistry, it recalls the growing list of potential biosignature gases, as well as the next generation of telescopes under construction or in development that have the capability to detect biosignature gases in exoplanet atmospheres.

4.2.1 General overview

4.2.1.1 Introduction and motivation through the lens of time

As a species we have wondered about life beyond Earth for millennia, since at least the time of the Greek philosophers (e.g., [1]). Hundreds of years ago some accepted the concept of life elsewhere. A few of the Founding Fathers of the United States of America are quoted with public statements in the late 1700s saying that stars were suns with planetary systems[1] [1]. After Bessel had invented new tools for astrometry and quantified distances to stars, US school children in the mid-1800s Midwest were reportedly taught not only about those stellar distances but also that there are humans out there in planetary systems surrounding those stars [2].

Over a century ago, astronomers laid the foundations of the astronomical search for life by remote sensing. Arcichovsky [3] proposed looking for vegetation

---

[1] '*The probability, therefore, is that each of these fixed stars is also a Sun, round which another system of worlds or planets, though too remote for us to discover, performs its revolutions, as our system of worlds does round our central Sun.*' Thomas Paine, in [1].

signatures in the Moon's Earthshine as a reference case for vegetation or chlorophyll searches on other planets. In 1930, the famous astronomer Jeans described the concept of oxygen as an atmospheric biosignature gas[2].

The search for signs of life on Mars also began over one hundred years ago with the sensational reports of canals on Mars [5, 6]. About half a century later, the search for signs of life by way of remote sensing spectroscopic analysis yielded a report of the detection of the 3.4 micron absorption of vegetation on Mars' surface [7, 8], which was later found to be deuterated water (HDO) in the Earth's atmosphere [9]. In the 1960s and 1970s, the first spacecrafts, the Mariners, took images of Mars on orbit, revealing several extinct craters and volcanoes, but no canals or vegetation [10]. Hence, ideas about canals and vegetation on Mars were completely dropped.

The *in situ* search for life beyond Earth began in earnest with the Mars Viking lander missions [11]. No organic compounds were detected on the Martian regolith above a threshold level [11]. This was due to two reasons: (i) technological problems, as the gas chromatography–mass spectrometer (GC–MS) onboard the spacecraft was not able to analyze and detect organic molecules in the regolith [12], and (ii) the location of the search—on the surface of Mars—where several reactions (that were unknown at the time) destroy organic compounds. These destructive reactions may happen because of UV radiation [13–22], cosmic rays [23, 24] and oxidation species [25], such as perchlorates. Indeed, perchlorates were detected directly on the Mars regolith by the Wet Chemistry Laboratory (WCL) of the Phoenix spacecraft, and indirectly by the Thermal Evolved Gas Analyser, with the analysis indicating the thermal decomposition of perchlorate [26, 27].

While today sample return and *in situ* search for ancient and present-day life has become within reach for Solar System bodies (including Mars, Europa, Enceladus, and Titan), the search for signs of life on exoplanets will remain in the domain of remote sensing. We know of thousands of planets beyond our Solar System, orbiting stars other than the Sun, called exoplanets (figure 4.1; [28]). Nearly every star is expected to have planets of some kind. While Solar System copies are hard to find due to observation selection bias, planetary systems like our own appear to be rare. Figure 4.1 shows that the observation of an Earth analog (an Earth-size planet in an Earth-like orbit around a sun-like star) remains technologically out of reach. Hence, today's search for habitable worlds and signs of life is focused on a planet category easier to discover and to observe than an Earth analog, such as Earth-sized or larger planets orbiting small red dwarf stars. A planet orbiting a red dwarf has a larger signal with currently favoured detection techniques than a planet of the same size orbiting a sun-sized star or even larger. Nonetheless, committees dating back to the 1960s discussed space mission designs to search for Earth analogs. NASA's

---

[2] '*It seems at first somewhat surprising that oxygen figures so largely in the Earth's atmosphere, in view of its readiness to enter into chemical combination with other substances. We know, however, that vegetation is continually discharging oxygen into the atmosphere, and it has often been suggested that the oxygen of the Earth's atmosphere may be mainly or entirely of vegetable origin. If so, the presence or absence of oxygen in the atmosphere of other planets should shew whether vegetation similar to that we have on earth exists on those planets or not.*' [4].

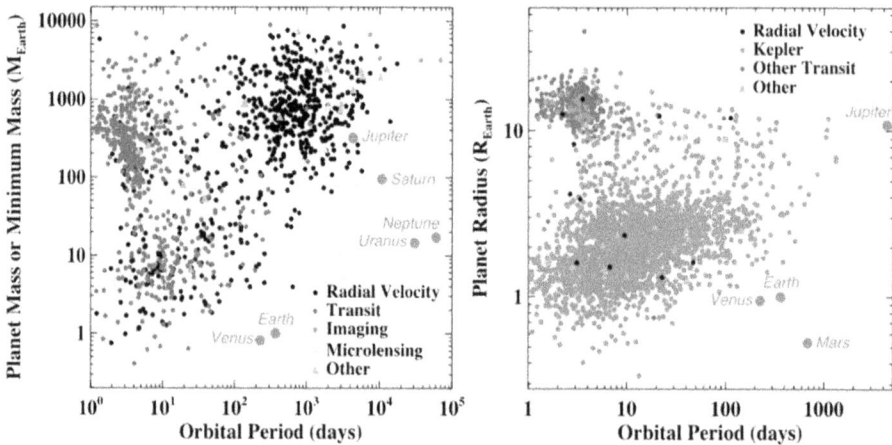

Figure 4.1. The distributions of known exoplanets as a function of their mass and period (left) and radius and period (right). The observed semi-major axis is converted to period via Kepler's third law. The color coding denotes the method by which planets were detected. Some of the features in these diagrams are real, but many are due to the current observational selection effects. Notably, the ~4300 confirmed exoplanets have an astonishing diversity in mass, size, and orbital period. Earth analogs are isolated on this log–log diagram, supporting the point that they are technologically just out of reach. Figure reproduced from [29].

Terrestrial Planet Finder studies[3] in the early 2000s led to the space-based, high-contrast, direct imaging mission concepts of today that will be able to find and identify a true Earth analog.

The study of exoplanetary atmospheres relies on remote sensing, blending decades of solar-system-planet foundational work with new techniques developed to cope with the exoplanet atmospheres' low spectral resolution and the lack of spatial or vertical resolution. With our first generation capability of observing atmospheres of rocky exoplanets of suitable temperatures for life, eventually, the goal is to search for 'biosignature gases', i.e., gases produced by life that can accumulate in exoplanetary atmospheres to remotely detectable levels. The main assumption we make is that life elsewhere uses a chemistry as life on Earth does, to extract, store, and release energy for metabolism, and in this process generates waste products as biosignature gases. Astronomical spectroscopy begun with the detection of dark lines in the solar spectrum by Baker in 1801 and continues today with stellar, planetary, and exoplanetary spectroscopy of ever-increasing sophistication. The search for signs of extraterrestrial life will continue for the next century and beyond.

4.2.1.2 Origin of life, habitability, and signs of life

Since antiquity, the question of the origin of life has been under discussion. In the beginning of the 20th century, Oparin and Haldane [30–32] suggested that the atmosphere of the early Earth was strongly reduced, mainly composed of methane,

---

[3] http://science.jpl.nasa.gov/projects/TPF/

ammonia, water vapor, and molecular hydrogen [33]. According to Oparin and Haldane, spark discharges in this atmosphere would generate organic compounds, which would later accumulate in the ocean of the primitive Earth. Amino acids were then experimentally synthesized in the laboratory by Stanley Miller using equipment that mimicked the atmospheric and ocean conditions of the primitive Earth suggested by Oparin and Haldane [34, 35]. However, towards the end of the century, new data (e.g., geology information, composition of volcanic gases, models of accretion of the Earth) showed that the atmosphere of the primitive Earth was not reduced, but was composed of nitrogen, carbon dioxide, and water vapor [36–40], so that this could not be the mechanism to produce significant amounts of organic molecules [41–45]. Presently, the exact time and conditions associated with the origin of life on Earth are the subject of intensive debates in the scientific community (e.g., [46–57], section 4.6). Despite not knowing the exact mechanism of formation, the first cell—the basic unit of life—must have needed water and organic molecules, either formed abiotically on Earth or exogenously delivered [58]. The origin of life was undoubtedly not unique to our planet, and it makes sense to search for habitability conditions in other worlds.

The search for habitable worlds in our Solar System (figure 4.2) focuses on planets and moons that have three main requirements: The presence of (1) a source of energy, (2) elements essential for life that support a geochemical mechanism, and (3) liquid water [59]. Liquid water is a solvent that is able to solubilize ions (cations and anions), which is fundamental to generate ionic gradients between the inside and outside the cells. In addition, water is also expected to be the most abundant liquid in the cosmos [60]. In our Solar System, Mars is considered to have the right habitability conditions as it was much wetter in its past, during the Noachian and early Hesperian Eons [61, 62]. Furthermore, icy moons of our Solar System are modelled to have rocky cores [63, 64], whose interaction with the subsurface ocean is a fundamental condition to generate geochemical reactions. Europa and Enceladus

Figure 4.2. Potentially habitable planets or moons in our Solar System. Not to scale. Planet images credit, Venus: JAXA, others: NASA.

are amongst the icy moons in which water/rocky core interactions are present [65, 66], thus increasing the potential for finding signatures of life at these locations.

We may, on the whole, exclude giant exoplanets from the list of life-supporting planet types as there is no solid surface below their atmospheres but rather a compressed phase of hydrogen and helium at temperatures far too high to support covalent bonds. Water vapor in a small rocky exoplanet atmosphere still is a key gas in our search for habitable worlds.

Organic matter is widespread throughout the Universe, so to successfully detect extraterrestrial life and its biosignatures, it is necessary to first distinguish between abiotic matter and signatures of past or present life. Abiotic matter that may have played a role in prebiotic chemistry just before life originated in our Solar System may be studied by laboratory analysis of organic residues produced by simulated photo- and thermo-processing of icy mixtures. They are considered as analogues for the organic material synthesized in interstellar or circumstellar icy grains [67–72], laboratory analysis of meteorites, micrometeorites, interplanetary dust particles (IDPs) and Ultra-Carbonaceous Antarctic Micrometeorites (UCAMMs) [73–85], on-site analysis of small planetary bodies (comets and asteroids), and laboratory analysis of sample return missions from these small planetary bodies (see [86, 87] and references therein).

A biosignature is a substance and/or a chemical pattern whose origin requires a biological agent [88, 89]. A cell is made up of i) proteins, which in turn are made up of amino acids, ii) genetic material, which is made up of nucleotides which in turn are made up of nucleobases, ribose, and phosphates, and iii) a membrane consisting of amphiphilic molecules spontaneously self-assembled into vesicles in water. Proteins, genetic material, i.e., RNA, and DNA would be ideal biosignatures, but they easily degrade under oxidizing and radiation conditions [90]. Furthermore, and with very few exceptions, life on Earth uses L-amino acids and D-sugars, while on meteorites mixtures of D- and L-amino acids have been detected (for a review [82, 83]). So chirality may be used as a tool to distinguish between abiotic organic molecules and products of past or present life. In addition, stable carbon, nitrogen, hydrogen, and sulfur isotope data of individual organic molecules may also be used to make this distinction. Abiotic molecules formed in extraterrestrial environments have been found to be enriched towards the heavy isotopes such as deuterium, ^{13}carbon, and ^{15}N [82, 83]). A likely biosignature deriving from molecular attributes of amino acids and sugars would require to exhibit chiral asymmetry, as well as a simple molecular distribution with structural isomer preference, and a light isotopic composition [91]. A summary of targeted biosignatures and their level of priority for future space missions in our Solar System is provided by Parnell *et al* [92].

To search for signs of life in an exoplanetary atmosphere, we make the assumption that like on Earth, life elsewhere will use chemistry to extract energy from the environment, to store energy, and will use the energy in a way that might generate a by-product gas. In other words, we assume that life elsewhere produces by-product 'biosignature gases'. The biosignature gases must accumulate in an exoplanet atmosphere to levels that astronomers can detect using next-generation ground- and space-based telescopes. The general consensus is to look out for gases that are present in too high amounts for them to be produced by abiotic chemistry, and that are also incompatible with their atmospheric and planetary environment. A touted example is that the simultaneous presence of CH_4 and O_2 could be a clear indicator of life, because CH_4 (a reduced gas) would be rapidly destroyed by reactions with the types of molecules present in a highly oxidized environment. However, both were never seen to be simultaneously present at high enough levels on Earth for this idea to be validated [93, 94]. Researchers are working hard to come up with many scenarios where a particular gas could be attributed to life given the context of a certain planetary environment. Part of this flourishing research area is complicated by false positives. This happens, for instance, when a biosignature gas is in reality produced by abiotic processes. For a comprehensive review see [95]. A key challenge is to overcome the problem that many gases of real interest may not be detectable and that those that can be detected may not be robustly associated with life without excluding false-positive scenarios.

There is a growing list of gases that are being considered as candidate biosignatures. These gases are input in computer programs to simulate their survival against destructive photochemistry and to assess their detectability by remote observations with future telescopes. The gases include dimethyl sulfide (DMS or $(CH_3)_2S$) [96–98]; nitrous oxide (N_2O) [99]; methyl chloride (CH_3Cl) [100, 101]; phosphine (PH_3) [102]; isoprene (C_5H_8) [103]; and ammonia NH_3 [104, 105] and other amines [104]. The unique approach by Seager *et al* (2016) aims to exhaustively sort through all classes of potential biosignature gases in order to find new biosignatures. More details can be found in reviews [95, 106, 107].

4.2.1.3 Solar System planets and moons: current status, and future plans
The planet Venus seems to be an unlikely abode for life due to the massive CO_2 greenhouse atmosphere creating surface temperatures of 700 K, too hot for complex molecules to survive, and thus too hot for life of any kind. However, for over half a century [108], people have speculated that life may exist in an aerial biosphere, populating the Venus atmosphere's clouds at altitudes of 48–60 km above the surface where temperatures are suitable for life. The clouds are permanent and cover Venus completely. Yet, the Venus cloud environment is undeniably harsh (for a review see [109]), both very dry and very acidic. The clouds are composed of liquid sulfuric acid which are more than ten orders of magnitude more acidic than the most acidic environment on Earth [110, 111]. A number of intriguing atmospheric anomalies on Venus have persisted for decades and may or may not be connected to life. Some are robust, including the unidentified but very strong ultraviolet absorber (e.g., [112]); the *in situ* detection of O_2 by two different probes, Pioneer 13 [113], and Venera 14 [114]; and the anomalously low abundance of H_2O and SO_2 [115]. Other anomalies are still tentative, such as the controversial detection of PH_3 from ground-based radio telescopes [116–118] that is also found in re-analyzed data from the Pioneer 13 probe [119], and the suggested detection of NH_3 (Pioneer 13: Mogul *et al* 2021, and Venera 8: [120]). It is interesting to note that the last entry probes were done nearly four decades ago: the Pioneer Venus probes in 1979 (e.g., [121]) and the Russian VEGA balloons in 1985 (e.g., [122]). Currently, the Japanese

Aerospace Exploration Agency (JAXA) Venus Akatsuki is orbiting Venus, and the European Space Agency (ESA) Venus Express orbited for nearly a decade until 2015. Both have returned valuable data on Venus' atmospheric dynamics and composition as well as on the cloud composition and structure. A number of missions are either being proposed, are under study, or are planned by a range of countries including the USA, Russia, Europe, Japan, and India.

The surface of Mars contains oxidized species such as perchlorates and chlorates (e.g., chlorinated hydrocarbons), which were detected by the Sample Analysis at Mars (SAM) of the Mars Science Laboratory's (MSL) in the Gale Crater and Sheepbed mudstone at Yellowknife Bay [123–127]. These molecules destroy any potential biosignature, and therefore, it is not likely to detect life on the surface of Mars, even if it was there. In February 2021, the Mars2020 mission—which will investigate the past habitability conditions, the potential for past life on Mars, and the preservation of biosignatures—landed the rover Perseverance and its helicopter Ingenuity on Mars. The mission will store samples that will later be returned to Earth on a future Mars sample return mission. ESA and the Russian space agency Roscosmos have launched the ExoMars Trace Gas Orbiter (TGO) into the orbit of Mars. ESA will launch the ExoMars rover—named Rosalind Franklin—to search for biosignatures at various depths of Mars up to 2m, but it is unlikely that this will happen before 2028 [128]. Since April 2018 the TGO has been analyzing the atmosphere of Mars to establish the presence of methane and other trace gases [129]. The previous numerous reports of CH_4 come from various instruments, including three different large ground-based telescopes (e.g., [130]), the Prime Focus Spectrograph (PFS) instrument on the Mars Orbital Express (e.g., [131, 132]), the thermal emission spectrometer (TES) onboard the Mars Global Surveyor (e.g., [133, 134]), and from the Tunable Laser Spectrometer (TLS) component of the Sample Analysis at Mars (SAM) instrument on the Mars Curiosity Rover [135, 136]. However, observations from the TGO did not find any methane in any hemispheres beyond 0.05 parts per billion by volume [137]. The presence of CH_4 can only be reconciled with the non-detection from the TGO if concentrated amounts produced locally at the near-surface are destroyed by an unknown process that can rapidly remove or sequester methane from the lower atmosphere before it spreads globally [137]. If methane is indeed present, it is not known (see figure 4.3) if the source is geological in origin, due to early Mars serpentinization reactions leading to the formation of hydrogen and methane, or possibly even biological, for instance produced by methane-producing microorganisms.

The favorable habitability conditions of the Jovian and Saturnian icy moons increase the probability for these moons to accomodate life. The Saturnian moon Enceladus contains a silicate core and an ocean beneath its surface [138, 139]. Indirect analysis of this ocean by the Cassini space mission showed the existence of a mixture of small (below 50 atomic mass units) and complex macromolecular organic compounds (beyond 200 atomic mass units) [140, 141]. The European Space Agency is planning to launch in the next decade the Jupiter Icy Moon Explorer (JUICE) to explore the Jovian system—Ganymede, Callisto, and Europa—and its organic content related to biosignatures in Europa [142].

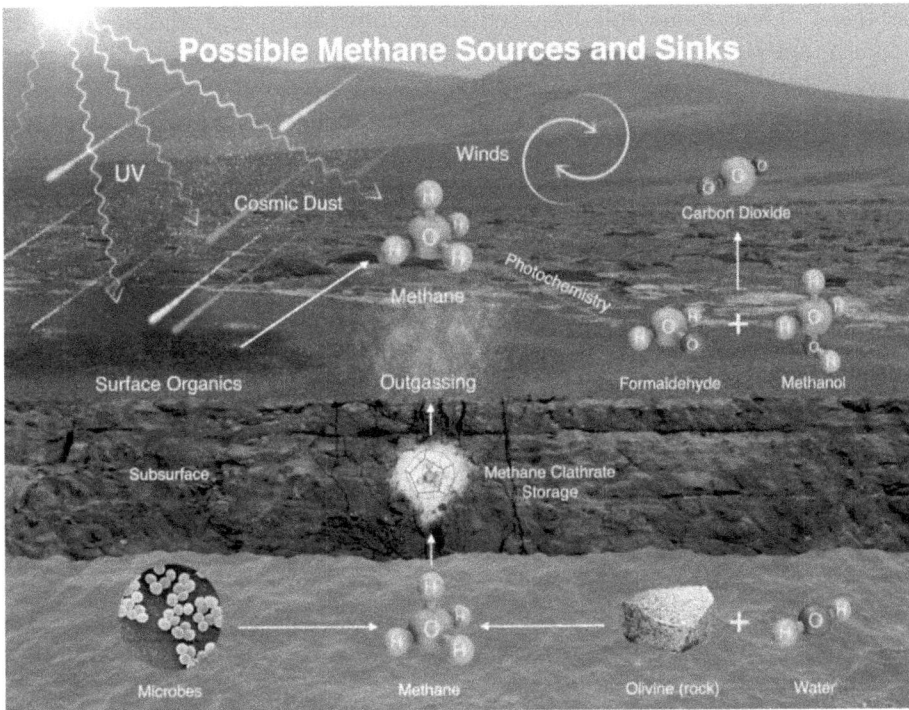

Figure 4.3. Possible sources and sinks of Mars' atmospheric methane (biological and/or geological). Credit NASA/JPL-Caltech/SAM-GSFC/University of Michigan. https://www.jpl.nasa.gov/images/possible-methane-sources-and-sinks.

4.2.1.4 Building blocks of life on small Solar System bodies

The building blocks of life have been analyzed and detected in several small Solar System bodies, such as comets and asteroids. The comet Halley contains an organic mantle composed of highly unsaturated compounds [143]. In the beginning of the 21st century the Stardust mission collected particles of comet 81P/Wild 2, which were returned to Earth. Analysis showed deuterium and ^{15}N excesses in some of the comet particles, with both aromatic and non-aromatic compounds [144, 145], including refractory material with highly aromatic organic matter [146]. Glycine (with a carbon isotopic composition of 29‰ ± 6‰), methylamine, and ethylamine were detected in the particles of comet 81P/Wild 2 [145, 147]. The dust particles of comet 67P/Churyumov–Gerasimenko were analyzed *in situ* by the Rosetta space mission and showed solid organic matter similar to the insoluble organic matter (IOM) of carbonaceous chondrites [148]. The nucleus of the comet 67P/Churyumov–Gerasimenko contained non-volatile organic macromolecular materials [149], carbon-rich species (e.g., alcohols, carbonyls, amines, nitriles, amides, isocyanates, the polymer polyoxymethylene [150, 151], phosphorus, and glycine) [152].

Samples were collected from asteroids and returned to Earth to be analyzed. The 25143 Itokawa near-Earth asteroid was explored by the Hayabusa spacecraft from the Japanese Aerospace Exploration Agency (JAXA) [153], and some of its particles were found to be carbonaceous, also containing nitrogen, oxygen, and trace amounts of fluorine and sulfur [154]. More recently, the Hayabusa2 mission touched down on the near-Earth C-type asteroid 162173 Ryugu, and successfully returned samples to Earth in December 2020. A variety of molecules were detected and identified [155–157]. In 2023 scientists received samples from another asteroid, the B-type asteroid 101955 Bennu, which was visited by the OSIRIS-Rex space mission [158]. Both asteroids and comets have delivered significant amounts of extraterrestrial organic molecules to Earth, which may have contributed to the feedstock of organic matter necessary for the first living organisms to be created on our planet.

4.2.1.5 Exoplanets: current status, and future plans

The current status of the search for signs of exoplanetary life is an exciting one. We are on the brink of the launch of a brand-new space telescope, one that has been many decades in the making. The James Webb Space Telescope (JWST; [159]) was launched in December 2021. JWST is an internationally supported mission, primarily NASA, with contributions from the ESA and the Canadian Space Agency (CSA). JWST is an infrared telescope that has more than seven times the collecting area of the Hubble Space Telescope and is orbiting more than 1 million miles from Earth at the L2 Earth–Sun gravitational balance point, an ideal vantage point for observational astronomy because it is away from the contaminating light and heat of Earth.

Because JWST is a general-purpose observatory, it will be used by hundreds of astronomers to study transiting exoplanets and their atmospheres. Transiting planets are a set of exoplanets with fortuitously aligned orbits such that the planets pass in front of their stars as seen from our viewpoint (figure 4.4). During transit, the starlight shines through the planet's atmosphere. Some wavelengths of light make it through

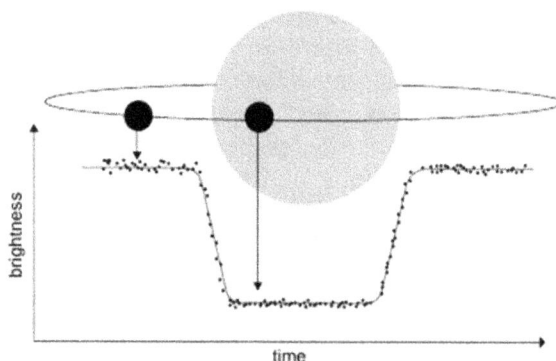

Figure 4.4. Transit planet schematic. Transiting planets are those that orbit in front of the star as seen from our viewpoint and cause a tiny drop in the star's brightness of a very characteristic shape and duration (graph). Note that stars other than the Sun are not spatially resolved; the cartoon star and planet are for illustration only. We know of thousands of transiting exoplanets. Astronomers are able to study the atmospheres of transiting planets because during transit the exoplanets atmosphere absorbs some wavelength-specific starlight, depending on the gases present in the exoplanet atmosphere (not shown). Figure credit: https://www.open.edu/openlearn/ocw/mod/oucontent/view.php?id=67466&section=3.1.

while others are absorbed by gases—enabling us to identify gases in an exoplanet atmosphere [160]. Today, we use the Hubble Space Telescope to study mostly hot, giant exoplanets. JWST holds promise to push observations down to smaller rocky exoplanet atmospheres—the kind of planets that might host life. JWST is limited to transmission spectra of planets transiting red dwarf stars—a very common type of star much cooler and smaller than the Sun. We should keep in mind that JWST was not designed for exoplanet science, and that we do not yet know its systematic noise floor. Nonetheless, JWST is already programmed to observe the so far most favorable system of potentially habitable planets that we can observe, those orbiting the ultracool red dwarf star TRAPPIST-1 [161] provided that the planets have maintained their atmospheres against the high-energy radiation coming from the host star [162].

Beyond JWST are planned the very large next-generation ground-based telescopes now under construction. These include the Extremely Large Telescope (ELT, with 39 m aperture diameter) [163], the Thirty Meter Telescope (TMT, 30 m aperture diameter) [164], and the Giant Magellan Telescope (GMT, 20 m aperture diameter) [165, 166]. These ELTs will be able to study atmospheres of exoplanets orbiting small red dwarf stars— though not limited to transiting exoplanets. The telescopes will operate at near-infrared wavelengths and must have special instruments to block out the starlight so we can see the planet directly, a method called direct imaging. Direct imaging requires special control of the telescope optics, called adaptive optics, to counteract Earth's atmospheric turbulence that causes distortions to the images. The challenge for directly imaging exoplanets in the habitable zones of M dwarf stars is to overcome the huge brightness difference between the planet and star: the light reflected by the planet is 10^7–10^8 times fainter than the star. A very suitable planet for the ELT observations is the Earth-mass planet orbiting the star nearest to the Sun, Proxima Centauri b [167].

The lofty goal to discover an Earth analogue exoplanet is extremely challenging. By Earth analogues we mean Earth-sized planets in Earth-like orbits around Sun-like stars. The challenge is that an Earth analogue in reflected light is ~10 billion times fainter than the host star (see figure 4.5). We must be able to suppress the starlight by such a tremendous factor. Only in space, above the blurring effects of Earth's atmosphere can telescopes meet this challenge for a significant number of stars. Moreover, for space telescopes, the search for biosignature gases will not be contaminated by the same gases already present in the Earth's atmosphere.

One promising and fancy concept to suppress the starlight is Starshade, a giant specially shaped screen tens of meters in diameter that would fly in formation thousands of kilometeres in front of a space-based telescope (figure 4.6). Starshade would block out the starlight so that only planet light enters the telescope. Starshade was first conceived in the 1960s [168] and has been revisited every decade since then until the mid-2010s, when a serious plan to make Starshade a reality was formulated by a NASA Study team [169]. Today there are several Starshade mission concepts of different sizes, including the Starshade Rendezvous Mission and the Habitable Exoplanet Observatory ([28]; figure 4.7) and the projects are awaiting funding. A related concept is the internal coronagraph, a starlight suppression device inside the telescope, such as described for the Large Ultraviolet Optical

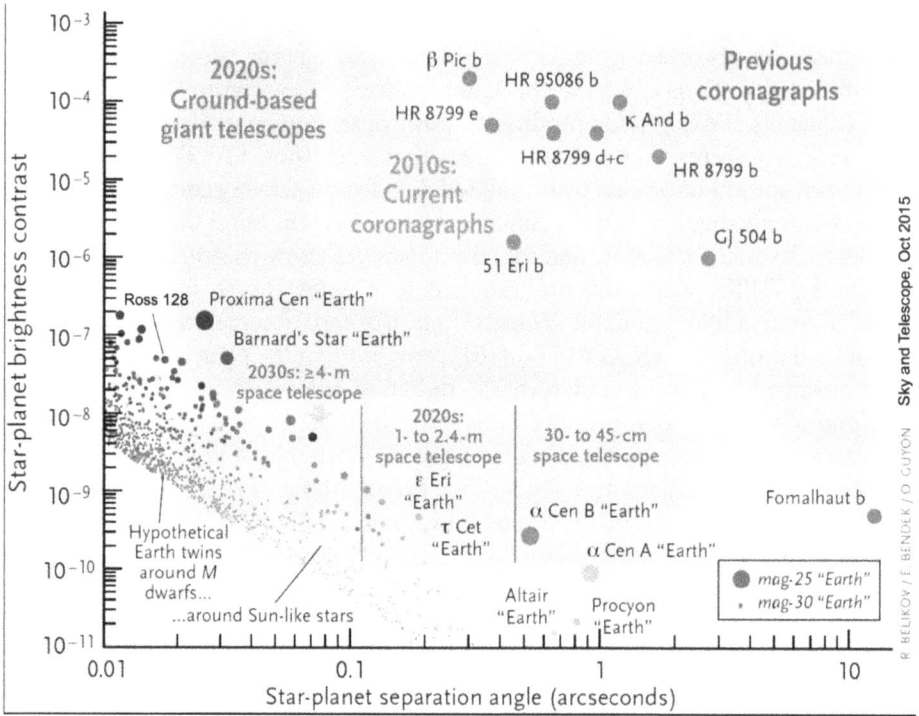

Figure 4.5. Overview of the field of exoplanet 'direct imaging' in visible and infrared light. The *x*-axis is the planet-star projected separation on the sky in angular units and the *y*-axis is the ratio of the planet's reflected light to the starlight. The point is that any telescope has to be able to suppress starlight to very significant levels for a planetary atmosphere to be detected by direct imaging. Capabilities of past, current, and future direct imaging instruments are shown. Red dots in the upper right are known as directly imaged giant exoplanets. Figure credit: Belikov, Bendek and Guyon, Sky and telescope 2015.

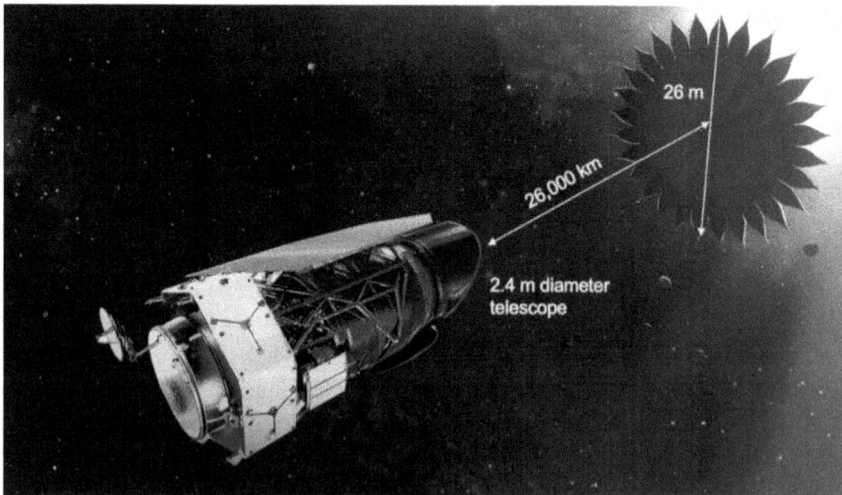

Figure 4.6. The Starshade Rendezvous Probe Mission consists of the Roman Telescope and a formation-flying starshade. The mission will enable direct imaging of the habitable zones of nearby Sun-like stars with sufficient sensitivity to detect and identify Earth-like exoplanets. Image not to scale. Credit: NASA.

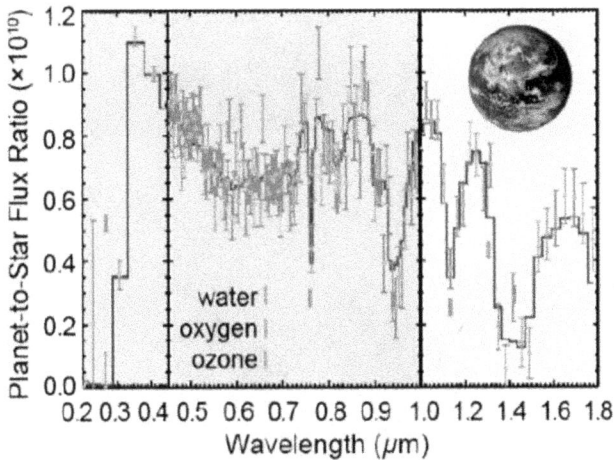

Figure 4.7. Simulated Earth-like planet spectra with the HabEx Observatory. The majority of HabEx-discovered planets will be well-characterized, with relatively high signal-to-noise ratio (SNR $\geqslant$ 10), and for a spectral resolution $R\,\lambda\,/\,\Delta\lambda$ of about 7 in the UV channel from 0.20 to 0.45 μm, $R \sim 140$ spectra in the visible channel from 0.45 to 1.0 μm, and $R \sim 40$ spectra in the IR channel from 1.0 to 1.8 μm. Figure from the HabEx Final Report [28].

Infrared Surveyor [170]. There is one hybrid Starshade coronagraph mission concept, HabEx. The bright, nearby Sun-like star tau Ceti is reported to host a multiple planet system [171, 172], including two habitable planets suitable for space-based high-contrast direct imaging.

4.2.2 Challenges and opportunities for 2050

4.2.2.1 Solar System planets and moons: 2050 and beyond
Currently there is an ongoing discussion on how best to obtain pristine samples from Solar System bodies where life may have started, either by future on-site missions or by sample-return missions. The decision will be based on time, cost, risk of contamination, and payload [87]. While on-site analysis will constrain the risk of contamination, the equipment onboard will be much more limited than what can be done in laboratories on Earth. On the other hand, sample-return missions will allow the use of a diversity of state-of-the-art equipment, but the risk of contamination of the samples is high, and curation facilities need to be in place. Even if contamination controls are implemented, minimal levels of terrestrial contamination will always occur, so that scientists must be able to distinguish between extraterrestrial organic matter and terrestrial organic matter [173]. There are several targets for 2050 and beyond, that will be briefly described below.

There is a tremendous opportunity to visit Venus with instruments with today's technology to search for signs of life or even life itself directly in the Venus atmosphere (venuscloudlife.com). Even though about three dozen space missions have already gone to Venus, there have never been any direct probes of the atmosphere since the NASA Pioneer and the Russian Venera Probes have flown

Figure 4.8. Dragonfly mission concept of entry, descent, landing, surface operations, and flight at Titan. Figure credit: NASA. https://www.nasa.gov/image-feature/dragonfly-dual-quadcopter-lander.

by nearly four decades ago. Even more exciting is the possibility of a Venus atmosphere sample-return mission of both gas atmosphere and cloud particles, to be analyzed on Earth with the most sophisticated instruments.

The Red Planet Mars is one of the main targets for the search for life in the Solar System, and several space agencies have plans to organize the first sample-return mission. This will be a unique opportunity to analyze samples from Mars—apart from Martian meteorites—with state-of-the-art equipment in the laboratory, in the hope that biosignatures will be unequivocally detected.

While it is not expected to also have sample-return missions to icy moons of our Solar System in the near future, analysis of their subsurface oceans has been performed indirectly by studying plumes that contained ice grains and vapor. In the next decades it would be ideal to develop technology in order to penetrate the surface ice and to directly analyze the subsurface ocean.

A next project is the launch of the future Dragonfly mission to Saturn's moon Titan (figure 4.8)—expected arrival 2036—that will assess its habitability conditions and prebiotic chemistry. Titan (figure 4.2) may become the main target of future life detection missions in 2050 and beyond. Potential microbial lifeforms living on Titan may use methane and other hydrocarbons present in the methane cycle [174, 175] rather than water, challenging our current ideas about the requirements for life in our Solar System.

4.2.2.2 Exoplanets: 2050 and beyond

Exoplanet astronomers are confident that the incredible pace of discovery during the last 10 years will continue into the next decade and beyond. New observatories with capabilities that have improved orders of magnitude with respect to their predecessors will open up new regions in parameter space, and computational power and tools such as AI also make huge progress.

In the search for signs of life on exoplanets, however, there are two inevitable limitations. First, despite the multitude of stars in the night sky and the hundreds of billions of stars in our own Galaxy, only stars in our solar neighbourhood (out to about 30–100 light years) are suitable targets for the search for planets with possible atmospheric biosignature gases. This is because exoplanets are adjacent to their much larger and much brighter host stars. This makes the signal of rocky exoplanet atmospheres so small that we require bright stars for observations. Because nearby stars are brighter than distant stars, our search is necessarily restricted to very nearby stars. In other words, the current generation of exoplanet astronomers is pinning its hope on the search for signs of life on what will be a very small number of habitable planets. If we do not find exoplanets with signs of life on the handful of exoplanets accessible to us in the next 10–20 years, the challenge will be how to scale up our efforts in 2050 and beyond, to access more, fainter, and more distant stars. Larger aperture space telescopes are limited to what fits within a rocket fairing, even for telescopes whose primary mirrors can be folded up and deployed on orbit. Larger apertures have been considered (e.g., the Large UV/Optical/Infrared Surveyor LUVOIR), and to design aperture sizes needed to reach even more distant stars, the next generation of astronomers will have to facilitate space-based manufacturing and space-based assembly.

The second inevitable limitation is a harsh reality: we may never know if an exoplanet atmospheric biosignature gas is indeed produced by life, or instead is produced by some unknown abiotic planetary chemistry. Indeed, this is a limitation difficult to overcome with remote sensing alone, leaving us with what we might call 'inhabited candidates for exoplanetary life'. As scientists, we are used to coping with uncertainty and unforeseen surprises, but when it comes to the search for life, would a *possible* detection of a sign of life be enough proof for scientists, and eventually for the taxpayer? If not, we need to consider brand new types of futuristic telescopes so that we might follow up in detail on any planets that show signs of life.

Ultimately, we will need to have a detailed look at any exoplanet that shows some evidence of hosting life. There are two paradigm-changing concepts under development. Both concepts, however, require a new operating timescale, at least 25 years from launch, before we get useful scientific results.

Breakthrough Starshot (https://breakthroughinitiatives.org/concept/3) is an effort funded by Breakthrough Initiatives to send thousands of tiny spacecrafts dubbed 'starchips' to Alpha Centauri B, a star in the nearest star system about four light years from Earth. Each starchip will deploy a solar sail of about 4 meter in diameter which will enable an acceleration to 20% of the speed of light, using the radiation pressure from a bank of coherent ground-based lasers with a combined power output of Gigawatts. The starchips that survive the 20-year journey will rapidly fly by Alpha Centauri B and any planets in the system. They will take images and relay them back to Earth, with a data transmission time at the speed of light equal to 4.4 years. The Starshot project has defined 19 technological challenges that are currently being worked out to make the mission possible during the next generation.

The Solar Gravitational Lens Telescope is an observatory that would operate very far from Earth at 550 AU (where 1 AU is the Earth–Sun separation), a distance to which no spacecraft has ever traveled yet. The concept is to use the Sun as a powerful gravitational lens with a focal distance of 550 AU [176–178]. The solar mass bends space such that any distant well-aligned background object will have its light rays bent towards its focal point. The background object then becomes magnified and spatially resolved. The better aligned the distant object is with the gravitational lens the higher the magnification and the better the achieved spatial resolution. Using the Sun as a gravitational lens this telescope could image exoplanetary surfaces at a resolution of 10 km. To reach 550 AU in a reasonable time, typically 25 years),the spacecraft would have to travel one order of magnitude faster than the Voyager spacecraft, the single one that has left our Solar System so far and that has taken about 45 years to reach a distance of 130 AU. Upon reaching the Sun's gravitational lens focal point, the spacecraft must follow the motion of the Sun-exoplanet focal line with a tailored trajectory in order to monitor the exoplanet. The planet image would be huge—on the order of kilometers—which is much larger than any conceivable telescope focal plane. For these and other reasons sophisticated computer algorithms would be needed to reconstruct the planet image. Because each planet has its own focal point with the Sun's gravitational lens, the telescope would be single-purpose, able to study only one exoplanet along the line of sight. Multiple telescopes could be sent into different directions, each for a pre-chosen target [178]. Despite the list of challenges, the revolutionary possibility for the direct imaging and spectroscopy of an Earth-like exoplanet up to 30 parsec away at a spatial resolution as small as 10 km makes this concept highly worth the investment.

4.2.2.3 Private sector
The number and kinds of opportunities to explore planets of the Solar System may increase with the growing involvement of private commercial space-flight companies. We have already mentioned the incredible utility of small space satellites in Earth orbit and a growing number of countries have become involved that were not traditionally space-faring nations. We get a glimpse into what could be a possible new future for Solar System exploration, that of small, focused missions visiting Solar System planets. Rocket Lab, a private company in the USA and New Zealand, has plans to launch their two-stage Electron rocket with the Photon satellite bus to Venus to drop a probe, most likely not before the end of 2024. The probe will be a light-weight (about 20 kg) with a small science instrument of around 1kg without parachute, spending just a few minutes in the Venus clouds but about an hour overall descending to the surface. Virgin Orbit's new LauncherOne, a rocket that is launched in the Earth's atmosphere from a modified Boeing 747 plane, can travel to Mars or Venus with an additional rocket stage to launch more mass than the current version. Public–private partnerships may become the norm in 2050 and beyond. Unlike national space agencies, private companies usually accept to take more risk at lower cost for novel space missions.

4.2.3 Conclusions and recommendations
The Grand Challenge of the search for life beyond Earth is ripe for huge progress. With strategic and concentrated international investment, we can accelerate the

search, and increase our chance to find life beyond Earth, if it exists. Increased and highly focused efforts and funding in the following areas will serve this goal: field work in extreme environments on Earth that serve as planetary analogues; infrastructure and fundamental research for the study of habitability, life detection, and biosignatures; development of prototype equipment for future life detection missions; planetary protection development for both *in situ* measurements and for sample-return mission, shared data, experiment hardware, and protocols published in open access databases. It is also important to attract the next generation of scientists, with active public outreach and training of the next generation scientists and engineers.

The huge progress in technology and design that has taken place since the dawn of the space age only half a century ago means that we can send landers to the many targets in the Solar System that exhibit the necessary ingredients for life, and in some cases even return a sample to Earth. The discovery and characterization of exoplanets has come a long way in the new millennium since humans have pondered the mysteries of the multitude of stars. We are lucky to be the first generation that will not just imagine but that can truly carry out the search for extraterrestrial life. It is closer than it has ever been, and, once discovered, it will put an end once and for all of the often-believed unique position of Earth and change drastically our vision of the Universe.

Acknowledgments

Centro de Química Estrutural acknowledges the financial support of Fundação para a Ciência e Tecnologia (FCT) through projects UIDB/00100/2020 and UIDP/00100/2020. Institute of Molecular Sciences acknowledges the financial support of FCT through project LA/P/0056/2020. This work was partially supported by NASA Grant 80NSSC19K0471 and Heising-Simons Foundation Grant 2018-1104.

4.3 Artificial intelligence: powering the fourth industrial revolution

Angelo Cangelosi[1] and David Vernon[2]
[1]University of Manchester, Manchester, UK
[2]Carnegie Mellon University Africa, Kigali, Rwanda

Artificial intelligence (AI) is the branch of computer science and engineering that allows us to harness the power of computing and technology to mimic and extend human intelligence. Together with ubiquitous communications and near-universal access to information, AI is driving the Fourth Industrial Revolution, ushering in an era of unprecedented and rapid change in how humans live, work, and relate to one another through the fusion of physical, digital, and biological technologies. In this article, we trace the origin and evolution of the different strands of AI and consider the implications of its pervasive presence in society, addressing some of its many applications—in medicine, robotics, the World Wide Web and social media, and sport—and their impact on society across the globe, in developed and developing countries, and the ethical issues it raises for humankind.

4.3.1 What is AI, where did it come from, and where is it taking us?

In 1960, J C R Licklider predicted a symbiotic partnership between humans and computers that would perform intellectual operations much more effectively than humans alone can perform them [179]. Today that symbiotic partnership is being realized through AI, a technology that both amplifies and extends human cognitive abilities. AI forms the foundation of the fourth industrial revolution, a revolution that is characterized by a fusion of physical, digital, and biological technologies, powered by AI and enabled by ubiquitous communication and near-universal access to information. It is irreversibly altering how humans live, work, and relate to one another [180]. At the same time, it is important to consider how to harness AI within an ethical framework that achieves economic benefits and social development for all.

The world of the AI-powered fourth industrial revolution may well be the destination, but how did AI get started? For many people, the discipline of AI has its origins in a conference held at Dartmouth College, New Hampshire, in July and August 1956. It was attended by luminaries such as John McCarthy, who coined the term artificial intelligence, Marvin Minsky, Allen Newell, Herbert Simon, and Claude Shannon, all of whom had a very significant influence on the development of AI over the next half-century. The essential position of AI at this time was that intelligence—both biological and artificial—is achieved by computations performed on internal symbolic knowledge representations, an approach referred to as 'computationalism', grounded in cognitivist psychology, and often referred to as GOFAI: good old-fashioned artificial intelligence.

However, AI has other roots in cybernetics, which is concerned with self-organization, regulation, and control [181]. In 1950, Grey Walter developed two robotic turtles, Elmer and Elsie, that could roam around a room, find a charging station, and recharge themselves. These systems were built on behaviorist psychology by using associative and reinforcement learning in relatively simple neural networks, rather than focusing on internal models and symbolic computation. Neural networks process information by propagating it through an interconnected layered network of relatively simple processing units: artificial neurons, very simplified versions of the neurons in biological brains. This approach referred to as 'connectionism', progressed in parallel with the computationalist approach over the next 60 years and more. We'll say more about computationalist symbolic AI in section 4.3.2.1 and connectionist AI in section 4.3.2.2.

From the outset, symbolic AI was concerned with producing intelligent artifacts that exhibited the versatility, flexibility, and robustness of humans in rational problem-solving. For this reason, it became known as strong AI. Despite the early optimism, strong AI proved to be very difficult to achieve. Consequently, AI techniques began to be applied in more limited domains with stronger constraints and a narrower focus. This approach became known as weak AI. Despite continual progress in both symbolic AI and connectionist AI in the 1970s and 1980s, performance on more challenging problems was disappointing and the popularity of AI waned during a period known as the AI winter.

The AI winter came to an end in the 2000s when, building on research in the late 1990s, artificial neural networks with deeper network topologies (i.e., networks with many more layers than had been used in the mid-1980s to mid-1990s) and new learning techniques were introduced, leveraging the recent availability of much greater computing power in the form of graphic processor units (GPUs) and much larger datasets to train the networks. This period also saw the development of some landmark probabilistic approaches to AI, perhaps the most celebrated of which was the Watson system from IBM (named after its founder Thomas J Watson) which won the TV show *Jeopardy!* in 2011, beating two human champions in answering rich natural language questions over a very broad domain of topics. The success of Watson was the result of probabilistic knowledge engineering that integrated many knowledge sources and exploited many techniques for search, hypothesis formulation, and hypothesis evaluation [182].

4.3.2 The nature of AI

4.3.2.1 Symbolic AI and GOFAI

One of the key historical, methodological, and epistemological approaches to AI is that of 'Symbolic AI' (i.e., GOFAI). This has its origins in the 1950s (i.e., part of the 1956 Dartmouth Workshop for the start of the AI movement) and constituted the primary, classical approach in the first 30 years of AI research, before the first AI Winter and the advent of Connectionist AI and machine learning [183].

The term 'symbolic' refers to the fact that AI algorithms and programs are based on a set of symbols and symbol manipulation processes. Two of the founding fathers of symbolic AI, Allen Newell and Herbert Simon, proposed the concept of a **Physical Symbol System**, 'a machine that produces through time an evolving collection of symbol structures' [184, p 116]. These symbols are purely formal and meaningless entities, though in practice they are normally interpreted by the programmer with a particular semantic content such as words, numbers, pictures, actions, etc. The symbolic expressions are created using logic formalisms, such as propositional logic with Boolean connectives (e.g., 'Red AND Round') or predicate calculus (e.g., 'Apple(Red, Round)'). They can also be arranged in IF–THEN production rules (e.g., 'IF apple, THEN eat'). In specific symbolic systems such as semantic networks, each node has a symbol ('Red', 'Apple', 'Fruit') with links having a label for the semantic relationship between nodes (e.g., 'IS A' or 'HAS') and hierarchical relationship between nodes. A collection of symbolic structures for a specific domain constitutes the knowledge base used by the system to reason about the problem. In general, symbols systems solve problems by using the processes of heuristic search [184], where the search for the optimal link between the problem definition and its solution must be guided by heuristics (i.e., rules of thumb that help guide the program toward the solution in an optimal way). The AI heuristic search and planning algorithms are widely used today for scheduling and logistics, data mining, games, searching the web, and planning in robotics.

An important aspect of the GOFAI approach is the idea that symbolic systems can model human intelligence. Newell and Simon [184] proposed the **Physical Symbol Systems Hypothesis**, which states that 'A Physical Symbols Systems has the necessary and sufficient means for general intelligent action' [184, p 116]. This is why GOFAI systems have been applied to modeling mathematical reasoning, natural language processing, planning, game playing, etc. A classic example of a GOFAI system is an expert system (i.e., a program that represents the knowledge of the human expert in a specific domain, using a set of IF–THEN production rules, and which can be used to offer advice to non-experts or to provide solutions to experts). Beyond the historical examples of the first expert systems, such as Mycin to support medical doctors in the diagnosis and treatment of infectious diseases, today expert systems have been developed in a wide range of domains (commercial, education, medical, and military applications), with some capable of highly complex planning on the order of tens of thousands of search steps [185].

The major strengths of GOFAI are its abilities to model hierarchical and sequential tasks, such as language processing, problem-solving, and games, and to represent knowledge bases using propositional contents and inference processes.

Some limitations of GOFAI systems are that these AI programs are brittle and they cannot learn new knowledge. This, as well as the initial strong claims about the power of symbolic systems to deal with general intelligence and any problem domain, led to the first AI Winter in the 1980s, and to the subsequent developments of connectionist and machine learning approaches, which we cover in the next section. However, significant achievements of GOFAI include the widespread use of commercial expert systems and their essential role in the games industry (to control the intelligent behavior of the virtual agents) including the historical victory of the IBM Deep Blue system in 1997 in beating the chess world champion Gasparov, and IBM Watson's victory in 2011 over two human champions in the *Jeopardy!* TV game [185].

4.3.2.2 Connectionist AI: from perceptrons to deep neural networks

Connectionist AI differs from symbolic AI in that information is processed by propagating it through a large interconnected network of relatively simple processing elements, typically implemented as artificial neural networks. They use statistical properties rather than logical rules to analyze information. Although the term *connectionist model* is usually attributed to Feldman and Ballard [186], the roots of connectionism reach back well before the computational era, with connectionist principles evident in William James' 19th-century model of associative memory [187].

Neural networks also have strong foundations in physics, as many of the mathematics concepts on neuron modeling and computation come from physics principles. For example, the Ising model (also known as the Ising–Lenz model), a mathematical model of ferromagnetism in statistical mechanics, inspired a model of associative memory [188] that was popularized by John Hopfield's recurrent neural network: the Hopfield net [189]. Boltzmann machines are variants of Hopfield nets that use stochastic rather than deterministic weight update procedures to avoid problems with the network becoming trapped in non-optimal local minima during training [190]. In the future, the principles of quantum mechanics may provide the basis for efficient neural networks [191], in particular, and for quantum AI [192], in general.

The seminal paper by McCulloch and Pitts [193], 'A logical calculus immanent in nervous activity', is regarded as the foundation of artificial neural networks and connectionism. Connectionism advanced significantly in the late 1950s with the introduction of the perceptron [194] and with the introduction of the delta rule for supervised training [195]. However, perceptron networks suffered from a severe problem: no learning algorithm existed to allow the adjustment of the weights of the connections between input units and hidden units in networks with more than two layers (i.e., multi-layered perceptrons, MLPs). In 1969, Minsky and Papert [196] showed that these perceptrons can only be trained to solve linearly separable problems and can't be trained to solve more general problems. This had a very negative influence on neural network research for over a decade, marking the beginning of a decade-long winter for connectionist AI.

Perceptron-based neural networks underwent a strong resurgence in the mid-1980s with the introduction of the backpropagation algorithm [197], which had previously been derived independently by Paul Werbos [198], among others [199]. Backpropagation finally made it feasible to train MLPs, overcoming the restriction highlighted by Minsky and Papert [196], thereby enabling MLPs to learn solutions to complex problems that are not linearly separable. This was a major breakthrough in neural network and connectionist research.

By the early 2000s, the traditional neural network approach had fallen out of favor because effective training was limited to relatively small networks, both in terms of the number of layers and the number of units per layer, due to the lack of computational resources for training and the infeasible amount of time required to train large networks. However, in the late 1990s, significant breakthroughs in deep networks, such as long short-term memory (LSTM) by Hochreiter and Schmidhuber [200] and convolutional neural networks (CNNs) by LeCun *et al* [201], heralded a new era in connectionism, although it took another ten years before they were widely adopted because of the lack of sufficiently large datasets and sufficient computational power for training. A CNN network is similar in principle to the multi-layer perceptrons of the 1980s and early 1990s but they have more layers, each of which performs a different function. In a CNN, convolution refers to the application of a filter to the data being processed by the neural network. The key feature of a CNN is that these filters are learned by the network during the training phase. This marked a significant departure from previous approaches where the filters, and the features they extracted, were the result of hand-crafted design. Consequently, CNNs can map directly from the input space (e.g., the image to be classified or the image in which you want to search for a given object, directly to the image label or the object location). For this reason, they are referred to as end-to-end systems. The first CNN was created by Yann LeCun, focusing on handwritten character recognition [201]. In 2011, AlexNet [202], a CNN with seven hidden layers, won the ImageNet Large Scale Visual Recognition Challenge.

Since then, deep neural networks have been applied successfully in many challenging applications [203, 204]. The networks have become deeper, with 22 or many more layers, and performance has improved through the use of more effective activation functions (e.g., the rectified linear unit ReLU), the use of specialized layers (e.g., pooling), more advanced learning techniques (e.g., batch normalization and dropout), techniques to overcome the problem of vanishing gradients (where the error terms become too small to effect an improvement in network performance as they are propagated back in a deep network), and a better understanding of how to adjust the system hyper-parameters during training to improve performance.

While CNNs proved their mettle with a very impressive performance in image recognition, object detection and localization, face detection, face recognition, and object tracking, new forms of recurrent neural networks proved very successful on problems that involve processing and analyzing sequences of states (e.g., in natural language), by exploiting new recurrent elements such as long short-term memory (LSTM) and gated recurrent units (GRU).

Progress has continued, with modern architectures successfully combining the power of deep CNNs and LSTMs to address problems that involve both images and language (e.g., automatic image annotation and captioning, image retrieval, and synthesis based on linguistic descriptions) [205].

Progress using deep neural networks for language understanding and generation has recently been advanced even further with the series of generative pre-trained transformer (GPT) architectures, culminating, for now, in GPT-3 [206]. This system is capable of generating natural language text that is often indistinguishable from that generated by humans.

4.3.2.3 Statistical and machine learning

A parallel development in AI in the last 20 years, with partial overlap with the AI connectionist approach, has been that of machine learning. This is the field primarily based on a variety of **statistics-based inference methods** that use large datasets to estimate (i.e., learn) the parameters of a model that has classification and predictive capabilities. This approach developed in conjunction with AI research in computer vision and speech (or more generally, pattern recognition), in robotics (e.g., reinforcement learning), and in neural networks (MLP and deep neural networks). Some people today use the terms AI and machine learning interchangeably, especially because of the big, common emphasis on deep learning. But as we will see below, machine learning keeps a distinctive emphasis on data-driven statistical inference methods.

Amongst the various inferential strategies in statistics (e.g., analogical inference, domain-specific inference, and structural inference), the bulk of machine learning uses the structural inference approach. This uses domain-general algorithms which exploit the **internal structure of the data**, rather than identifying the semantic, domain-specific, content of the data. Structural inference is the basis of most machine learning frameworks, such as the well-known methods of regression, neural networks, and Bayesian networks [207]. Given this data-centric (sometimes known as 'data-hungry') approach, the recent, easy availability of potentially unlimited data from social media and the web, and wider access to cloud-based parallel computing systems such as GPUs (which are necessary to apply computationally intensive statistical computations on large datasets) can in great part explain the recent, impressive contribution of machine learning to AI, and information technology in general.

Machine learning comprises a set of methods typically grouped into supervised and unsupervised techniques, as well as reinforcement methods. **Supervised learning** algorithms need a labeled dataset (i.e., where each data point such as an image of a dog, for example) is associated with a supervision signal or ground-truth (e.g., the category label 'dog'). The learning algorithm has to find the parameters of the model (e.g., weights of a neural network) using the error between the model's guess and the supervision label. Examples of supervised learning algorithms include MLP, CNN, and LSTM neural networks, decision trees, support vector machines, and regression. **Reinforcement learning** can be considered part of the supervised approach, but where the supervision to learn a policy (e.g., actions that should be taken when certain sensory conditions prevail) is guided by a reward function. **Unsupervised learning** algorithms do not require a labeled dataset, as they discover the regularity in the data and their organisation in separate categories. Examples of unsupervised learning include clustering algorithms such as k-means and autoencoder neural networks,

An important set of machine learning approaches is that of Bayesian learning algorithms. The general Bayesian framework is based on the intuition that the beliefs after observing some data are determined by the probability (prior probability distribution) of each possible explanation given that data. When processing a dataset, the machine learning algorithm uses the Bayesian rule to calculate the correct probability distribution over the hypotheses given that data. And given large datasets, the computations required for Bayesian learning become too difficult to be done analytically, thus the recent boost of Bayesian algorithms with easy access to parallel computational resources [207].

Machine learning in large part is responsible for the recent successful developments of AI. The most successful and widely used applications in speech recognition, computer vision, natural language processing, and robotics applications are based on deep learning and other Bayesian approaches. This includes the design of DeepMind's AlphaGo and AlphaZero systems, based on the combination of deep neural networks, reinforcement learning, and AI search algorithms, which, as we will see in section 4.3.4, were able to outperform human champions in the Go game, even without human knowledge or supervision [208].

4.3.3 Example applications

4.3.3.1 AI applications in medicine
The application of AI to medicine and healthcare has its origins in the early GOFAI developments of expert systems, such as the MYCIN for infectious diseases and DENDRAL on the discovery of chemical compounds. More recently, the advent of machine learning and its focus on learning from data has led to a resurgence of the development of medicine AI systems.

Deep learning methods, such as CNNs because of their impressive performance with 2D image recognition, have been widely used for **image-based cancer detection and diagnosis** [209]. For example, in skin cancer diagnosis, the performance of CNNs to classify biopsy-proven clinical images (e.g., malignant melanomas versus benign nevi) was on par with that of 21 board-certified dermatologists. A recent landmark achievement in medicine and biochemistry is the **AlphaFold AI model for protein folding**. This was developed by Google DeepMind and was the winner in 2020 of the biennial critical assessment of protein structure prediction (CASP) competition. AlphaFold achieved a performance similar to the results from experimental methods.

Medical AI applications bear significant **technological and ethical challenges**. One key issue is the reliance on the quality and variety of the training data, as healthcare datasets typically are sparse, noisy, heterogeneous, and time-dependent. Moreover, new methods and tools are needed to enable interactive machine learning to interface with healthcare information workflows, keeping the human in the loop [210, 211]. There are also important ethical considerations. For example, the need for explainable systems so that clinicians (both novice and expert doctors) can access causal explanations of the AI's decision-making process [212]. We return to this issue of trust and explainability in AI in section 4.3.1.

4.3.3.2 AI applications in robotics
AI is used in robotics for many purposes, including autonomous navigation, task planning, task execution, object detection, object grasping and manipulation, inspection and surveillance, social human–robot interaction, including natural language processing, facial recognition, sentiment analysis, gesture understanding, and intention recognition. It is also used in an extensive range of robots. At the time of writing, the IEEE robots website [213] features 229 robots of many different types: wheeled, legged, tracked, airborne, underwater, and humanoid, targeting

consumers, entertainment, education, research, medicine and healthcare, disaster response, service and industrial, aerospace, military and security applications, telepresence, self-driving cars, and agriculture. Perhaps the epitome of AI in robotics is the goal of creating a collaborative robot (i.e., one that can share a common goal and share the human's intentions to achieve that goal, acting jointly with the human, paying attention to what the human is doing, and, crucially, anticipating any help the human might need to complete whatever tasks she or he is working on).

One example of AI in robotics is robot-enhanced therapy [214] where **robots assist a psychotherapist working with children** with autism spectrum disorder. Under the guidance of clinical practitioners, this project developed interactive capabilities for social robots that allowed them to engage a child in clinically derived exercises. The robot can operate autonomously for limited periods under the supervision of a psychotherapist. AI plays a major part in the success of this application, specifically in its cognitive ability to interpret body movement and appearance-based cues of emotion. This allows the robot to assess the child's actions by learning to map them to therapist-specific classes of behavior. In turn, the robot also learns to map these child behaviors to appropriate robot responses, again as specified by the therapists.

4.3.3.3 AI applications in the web and social media

AI is having a tremendous impact on a variety of applications and functionalities for the web (e.g., search algorithms, music and video recommendations, automatic translation) and social media (e.g., news selection and recommendation, sentiment analysis, face recognition). Although this progress is resulting in clear benefits to people and society, it also carries important ethical considerations and risks.

AI has significantly changed the **search algorithms** for the web. For example, Google's initial PageRank algorithm (based on standard mathematical methods) has now developed a collection of search tools, such as the Hummingbird framework. This framework complements PageRank's results with RankBrain, based on machine learning algorithms for entity recognition, and the recently introduced BERT (Bidirectional Encoder Representations from Transformer), which uses a neural network for natural language processing.

Another example of the widespread use of AI and machine learning algorithms on the web is for **recommender systems, which make** recommendations for related purchases on e-commerce sites, suggestions of related news and friends in social media, and personalised recommendations in media-streaming sites and apps. In fact, 80% of movies watched on Netflix are based on AI recommendations [215]. As in other domains, deep machine learning systems have become the default algorithm.

AI applications for **face recognition** have also become widespread on the web and in social media. These algorithms can be used for image matching and people recognition (e.g., in social media photo tagging) as well as for authentication (e.g., to implement secure access in some smartphone systems). However, face recognition algorithms based on learning from datasets have important ethical implications (e.g., regarding possible biases in the data used for the training). For example, in 2018, a seminal paper by computer scientists Buolamwini and Gebru [216] demonstrated that leading facial recognition systems produced substantial disparities in the accuracy of gender classification (e.g., with error rates of up to 34.7% in the classification of darker-skinned females, whilst the maximum error rate for lighter-skinned males was 0.8%). This highlights the urgent need to address and remedy implicit bias in such systems and make sure they are based on fair, transparent, and accountable facial analysis algorithms.

4.3.3.4 AI applications in sports

While the use of statistical analysis is well-established in sports, AI is taking it to a new level. Possible applications range across the entire spectrum of activities. Barlow and Sriskandarajah [217] identified eleven applications across 17 sports that are being or will be impacted by AI. These applications include identifying talent and determining optical game strategies.

AI technologies such as computer vision have been used routinely to **assist with umpiring during games**, especially using automated ball tracking and line calling applications. For example, the Hawk-Eye system [218] visually tracks the trajectory of the ball using six high-speed cameras, the images from which are used to triangulate the ball's position over time, displaying a virtual reality trajectory of its statistically most likely path. AI can also be used for **automated generation of video highlights**, while integrated vision and natural language technology can be used for automated generation of copy for publication in print and online.

The All England Lawn Tennis Club hosts the annual tennis championship at Wimbledon and uses IBM's Watson technology to provide a variety of services, including **real-time match reports and uncovering player insights**. It also powers a voice-activated cognitive assistant 'Fred', named after the late champion Fred Perry, to help spectators find their way around the venue [219].

To identify successful game strategies, an **AI system can play against itself**, as the DeepMind AlphaGo system did, before going on to beat Lee Sedol, the winner of 18 world titles, in 2016, and achieve 60 straight wins in time-control games against top international players in 2017 [220]. Subsequently, in AlphaGo Zero, even better performance was achieved based purely on reinforcement learning without any prior supervised training. Apart from its formidable performance, what is significant about AlphaGo is that it uncovered several innovative strategies that greatly surprised expert players, demonstrating the potential for AI to augment human abilities and exceed human performance.

4.3.4 Future challenges

4.3.4.1 Collaborating with machines and robots

AI has contributed significantly to the design of intelligent control models and cognitive architectures for sensorimotor behavior (e.g., perception, navigation manipulation) and cognitive capabilities (e.g., planning, language) in robots, as we have seen in section 4.3.1. But major challenges still remain, specifically with regards to the robot being able to handle the complexity of real-world scenarios (i.e., cluttered, dynamic, unpredictable environments where objects to be grasped or obstacles to be avoided are difficult to see, can be occluded, or change their position over time). Beyond the complexity of designing skills in individual robots, a significant challenge for robots, and intelligent machines in general, is that of handling **interaction with people for collaborative tasks**, also known as human–robot interaction (HRI) or social robotics [221]. This type of interaction includes a variety of scenarios, such as joint action in a flexible manufacturing setup between a worker and a cobot (i.e., a collaborative robot), assistive robot companions for older and disabled people or in hospitals and care homes, and robot tutors for education or entertainment.

The research challenges on the use of AI for the design of social and cognitive skills for interaction include, for example, the **capability of 'intention reading' and the implementation of an artificial 'Theory of Mind'** [222]. Intention reading is the capability of the robot to detect the human user's intended goal of the joint interaction. The 'Theory of Mind' describes a more general view of intention reading, as this concerns the robot's capability to understand and predict the beliefs, desires, and goals of the person; as shown in figure 4.9. AI methods, such as Bayesian networks and deep learning, can be used to build artificial Theory-of-Mind skills in robots [222].

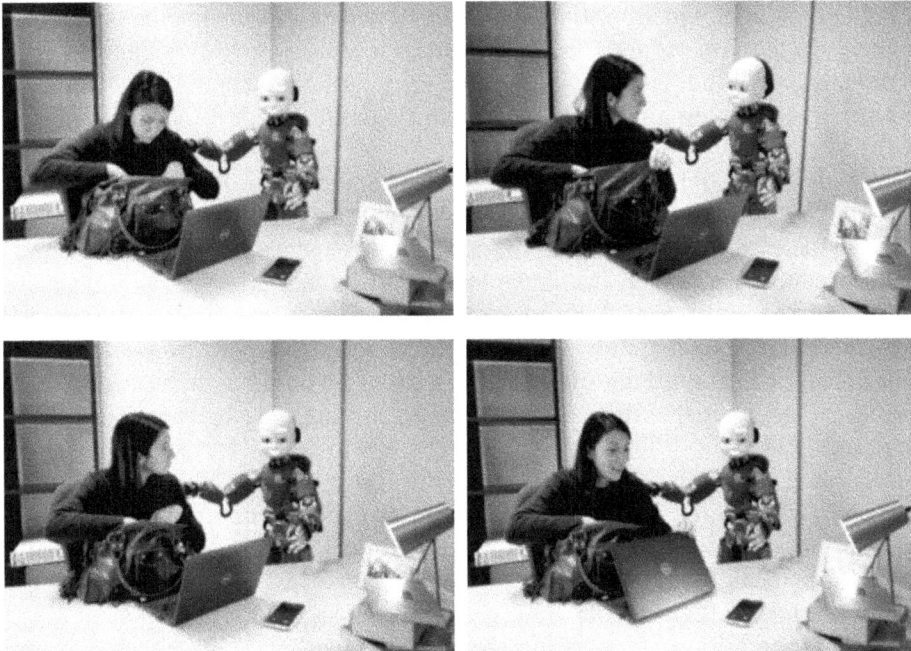

Figure 4.9. This sequence of pictures depicts a situation in which the iCub humanoid robot (www.icub.org) is interacting with a human, reading her intention to get her phone from her bag, and alerting her to the fact that it is on the desk, hidden from her by the laptop. This sequence has been staged to illustrate the future capabilities of a cognitive robot and has not yet been implemented. Reprinted from [223], copyright (2021) with permission by Springer Nature.

Another important future research direction in AI for collaborative robotics concerns the quality of the interaction (i.e., the design of **long-term and trustworthy interaction and well-being in human–robot collaboration**). Long-term interaction requires the robot to be able to engage in continuous, meaningful, and contextualised interaction over a series of interactions lasting for days, weeks, or longer. This will require the ability to recognize the person and their personality and preferences, to remember recent interactions, and to engage in empathic behavior with the person's needs [224]. Trustworthy interaction, a growing field of research, requires people's acceptance and trust of the robot's behavior and decision-making process. This is also linked to ethical issues regarding explainable AI (see section 4.3.4.3) and to the achievement of people's and robot's reciprocal theory of mind [225, 226].

4.3.4.2 Self-learning and self-programming machines

The quest for the automatic generation of programs and codes, also known as program synthesis or self-programming machines, has been one of the main challenges of AI since the outset. With the advent of machine learning approaches, AI has started to put emphasis on self-learning machines, which can learn with no or minimal supervision from humans.

This **self-programming machine** challenge has only recently received a significant boost through the combination of deep learning and NLP methods. For example, DeepCode is a code generator that uses a neural network to predict the properties of the program that can produce the outputs given specific inputs [227]. SketchAdapt is a software system that learns, without direct supervision, when to rely dynamically on pattern recognition and when to perform symbolic search for explicit reasoning [228].

Recently, the GPT-3 system [206] has been proposed for natural language generation, with the potential application to automatic program synthesis. GPT-3 is a large-scale deep learning model for natural language processing with an order of magnitude more parameters than any previous NLP model. The model can generate new text without the need for further training or task-specific fine-tuning of its parameters. GPT-3 can produce samples of news articles that human evaluators have difficulty distinguishing from articles written by humans. In addition, GPT-3 has been used for generating programs, such as the code to create the Google homepage [229].

Regarding the challenge of creating a **self-learning machine**, the first attempts to design AI systems and robots that autonomously learn without supervision from humans have recently been realized in developmental robotics (also known as autonomous mental development). This area of robotics takes inspiration from child development to design robots that go through stages of development for the incremental acquisition of sensorimotor and cognitive skills [230]. Another example of self-learning AI is the AlphaZero system mentioned earlier in which artificial agents play the game Go against each other, bootstrapping their final learning capabilities. This led to the acquisition of skills that far outperformed the skills of the best human players ([231]; see also section 4.3.3.4).

4.3.4.3 Social and ethical aspects of AI

AI can bring significant benefits to all. However, the examples we have given so far focused on applications in the developed world and, indeed most of the national strategies on AI have been created by governments in developed countries [232]. Nevertheless, the fourth industrial revolution in general, and AI in particular, is just as relevant for developing countries. For example, AI is having an increasingly positive impact in Africa, in sectors such as energy, healthcare, agriculture, public services, and financial services [233, 234]. It has the potential to drive economic growth, development, and democratization, to reduce poverty, improve education, support healthcare delivery, increase food production, improve the capacity of existing road infrastructure by increasing traffic flow, improve public services, and improve the quality of life of people with disabilities [235]. The challenge is to ensure that developing countries have access to the technology and the datasets necessary for innovation in a socially and culturally acceptable manner (i.e., the democratization of AI, and open access to AI technology by developers everywhere). It is crucially important that the fourth industrial revolution, powered by AI, occurs in a fair, ethical manner [236].

AI can also be used for negative purposes, either intentionally or unintentionally (e.g., by fomenting religious, ethnic, social, and political divisions through fake misinformation created by deep networks) [237]. Of particular concern is the issue of implicit and explicit bias in the data that are used to train the AI models, thereby resulting in discrimination against people based on gender or race. Examples of bias against dark-skinned people include facial analysis [216], pedestrian detection [238], and predicting recidivism [239]. The success of AI in the future will depend on the elimination of such bias.

4.3.4.4 Intelligence, brains, and consciousness

Why does intelligence matter? Indeed, what is intelligence? Can a robot be conscious? Let's start by answering the question: what is intelligence? There are many possible answers but the one that has the most appeal derives from the answer to a different question: why do we have brains? The neuroscientist Daniel Wolpert provides an unexpected but compelling answer. He argues that we have brains to allow us to control movement [240]. This mirrors what Francisco Varela and Umberto Maturana say about cognition: 'Cognition is effective action' [241]. From this perspective, we see intelligence as the way to be effective in our control of our movements and in the way we act in the world. The key to understanding why this is so important—and so difficult—is to see that the number of possible ways we can move and act, and the number of possible outcomes of these movements and actions, is infeasibly large if we are to consider all the possibilities and choose the best one, or even a good one. This is what Allen Newell and Herbert Simon pointed out in their Turing Award (the equivalent of the Nobel Prize for computer science) lecture: 'The task of intelligence, then, is to avert the ever-present threat of the exponential explosion of search' [184] (i.e., the search for good ways to act). Newell and Simon were referring to the search for the solution to a problem, but it amounts to the same thing. This is a satisfyingly straightforward and very practical way of understanding intelligence and the brains that give rise to intelligence.

However, brains are even better than that. They also predict the need to act and the outcome of those actions, and they do so all the time, at every instant, as we act and as we anticipate the future, milliseconds ahead, seconds ahead, hours, days, years. Indeed, it has been argued that brains are, in effect, probabilistic prediction machines, meaning that they can deal effectively with uncertainty [242].

But what then of consciousness? Can robots and intelligent machines also be conscious? Can we take AI even further and build machines with artificial consciousness? Many people think that this is a distinct possibility. Indeed, according to Paul Verschure, 'understanding the nature of consciousness is one of the grand outstanding scientific challenges' and he proposes a scientifically grounded approach to addressing the challenge of answering the question of what consciousness is and how physical systems can develop it [243].

4.3.5 Summary and conclusion

AI impacts all aspects of human activity: it automates tasks, assists with decision-making, augments and extends our cognitive capabilities, and it can even operate autonomously, if we allow it, without recourse to human oversight.

AI began as an attempt to understand and replicate human intelligence, initially taking two routes to that goal, one via connectionism and one via symbolic computationalism, reflecting their inspiration from behaviorist and constructivist psychology, respectively. These two approaches waxed and waned in their own respective ways over the decades, to be joined in the 1980s by machine learning and in the 1990s by statistical machine learning, probabilistic inference networks, and other established disciplines in computer science. Breakthroughs in deep neural network learning and deep neural network topologies, aided by very large datasets and equally large increases in processing power, yielded great success in many application domains. The symbolic knowledge representation and reasoning approach also developed rapidly, especially in cognitive architectures, as knowledge bases and ontologies increased greatly in size and sophistication and as the hybrid paradigm, combining symbolic approaches and sub-symbolic connectionist approaches, was developed (e.g., in cognitive architectures such as Soar [244], ACT-R [245], and CLARION [246], among others).

While the success of statistical machine learning in narrowly targeted applications yielded great success, it did so at the expense of losing focus on AI's original goal of understanding and replicating human-level intelligence. There has been a resurgence of interest in what is now known as Artificial General Intelligence in cognitive science and cognitive systems. Still, the ultimate goal of replicating the versatility of human cognition remains elusive and it is unclear when it will be achieved. What is certain is that the AI quest will continue and AI in its many guises will continue to permeate our lives and change them, hopefully, for the better.

In seeking to steer the path to the future, it is likely that other strands of thinking will be woven into the fabric of AI, especially concerning the trustworthiness of AI in autonomous systems (i.e., its role in serving the bigger agenda of creating self-maintaining systems that can operate robustly and prospectively in the face of uncertainty and that can continually develop through self-programming as they interact with and learn from the world and the people in it). While there is much important work yet to be done to promote the development of democratized, trustworthy, ethical AI in the developed and developing worlds, an equal challenge will be how to control the role of AI in autonomous systems, possibly conscious ones, where the relationship with humans is no longer symbiotic. We are far from that point at present. but it is likely we will reach it, and everything will change quickly when we do.

In Ernest Hemingway's novel *The Sun Also Rises* there is a dialog between two characters which goes as follows. 'How did you go bankrupt?' Bill asked. 'Two ways,' Mike said. 'Gradually and then suddenly'. And so it will be with autonomous AI. Our collective responsibility is to work together in a directed manner during the present gradual phase so that, when the full impact of AI is suddenly felt, it will be for the greater good of all humankind.

Acknowledgments

In the preparation of this section, Angelo Cangelosi's work was partially supported by the Air Force Office of Scientific Research, USAF under Award No. FA9550-19-1-7002, by the UKRI TAS Node on Trust (EP/V026682/1), and the H2020 projects PERSEO, TRAINCREASE, and eLADDA.

4.4 Artificial life: sustainable self-replicating systems

Carlos Gershenson[1] and Jitka Čejková[2]

[1]Universidad Nacional Autónoma de México, Mexico City, Mexico
[2]University of Chemistry and Technology, Prague, Czechia

4.4.1 General overview

Nature has found one method of organizing living matter, but maybe other options exist—not yet discovered—on how to create life. To study life 'as it could be' is the objective of an interdisciplinary field called Artificial Life (commonly abbreviated as ALife) [247–249]. The word 'artificial' refers to the fact that humans are involved in the creation process. The artificial lifeforms might be completely unlike natural forms of life, with different chemical compositions, and even computer programs exhibiting life-like behaviours.

ALife was established at the first 'Interdisciplinary Workshop on the Synthesis and Simulation of Living Systems' in Los Alamos in 1987 by Christopher G Langton [250]. ALife is a radically interdisciplinary field that contains biologists, computer scientists, physicists, physicians, chemists, engineers, roboticists, philosophers, artists, and representatives from many other disciplines. There are several approaches to defining ALife research. One can discriminate between soft, hard, and wet ALife (figure 4.10). 'Soft' ALife aims to create simulations or other purely digital constructions exhibiting life-like behaviour. 'Hard' ALife is related to robotics and implements life-like systems in hardware made mainly from silicon, steel, and plastic. 'Wet' ALife uses all kinds of chemicals to synthesize life-like systems in the laboratory.

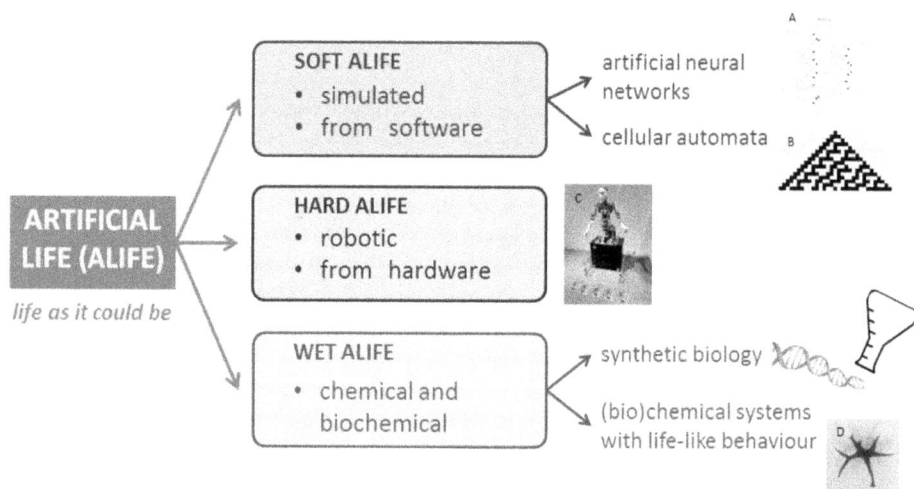

Figure 4.10. Artificial life research. (A) Artificial neural networks. (B) Cellular automaton. (C) Robot Alter 2. (D) Shape-changing decanol droplet.

Bedau *et al* [251] proposed 14 open problems in ALife in the year 2000, but none of them have been solved yet. Aguilar *et al* [249] summarized the ALife research challenges and divided them into 13 themes: origins of life, autonomy, self-organization, adaptation (evolution, development, and learning), ecology, artificial societies, behaviour, computational biology, artificial chemistries, information, living technology, art, and philosophy.

4.4.2 Soft ALife

Mathematical and computational models of living systems are naturally more abstract than robotic or chemical systems. Also, they are easier to build, so actually most ALife models are 'soft'. Still, we can discriminate between 'abstract' and 'grounded' soft ALife models. This distinction is more gradual than categorical. More abstract models focus less on physics or biology, and more on information and organization, while more physical models consider to a greater degree the actual components of living systems.

Some of the most abstract models include cellular automata [252–258], random Boolean networks [259–262] (figure 4.11), and boids [263], where there are basically space, time, and simple dynamics that may lead to complex behavior. Other abstract models have focussed on studying self-replication [264] or evolution [265–269]. Abstractly exploring the theoretical space of possibilities for living systems (necessary and sufficient conditions) has also been made at the 'chemical' level, with artificial chemistries [270, 271] and swarm chemistry [272] (figure 4.11). Another more 'grounded' strand of soft ALife involves the simulation of environments with realistic physics, either to evolve 'creatures' [273] or controllers for physical robots [274–276].

4.4.3 Hard ALife

Robots are physically situated [277, 278] and embedded [279] in their environment, so they have been useful to explore aspects of life related to behavior [280], traditionally studied by ethology [281]. There are also several soft ALife models of adaptive behavior [282, 283], dealing with physics (time, motion, inertia, gravity, hardware imperfections, etc) that already are an important challenge for ALife. As mentioned earlier, robotic controllers have usually been evolved in software that has been uploaded into hardware. This is also known as 'evolutionary robotics' [284–286]. Hard ALife models have also been used extensively to study the emergence of collective behaviors [287–291] (figure 4.12).

4.4.4 Wet ALife

Wet ALife is related to the effort of creating artificial cells in the laboratory from chemical and biochemical precursors. Living cells are the basic structural, functional, and biological units of all known living organisms. They are found in nature and produced and maintained by homeostasis, self-reproduction, and evolution [292]. In contrast, artificial cells (synthetic cells) are prepared by humans and only mimic some of the properties, functionalities, or processes of natural cells.

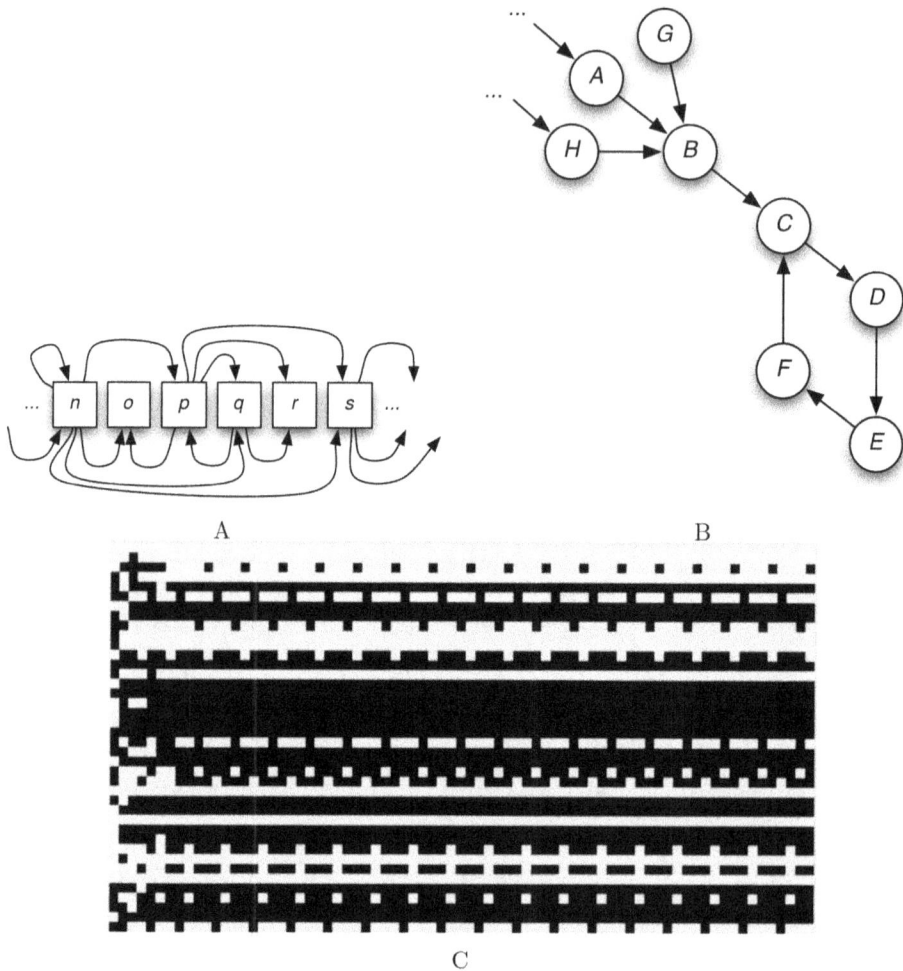

Figure 4.11. An example of a random Boolean network (RBN) [262]. (A) A structural network is formed by N Boolean nodes (can take values of zero or one) that are connected to K inputs randomly. The future state $(t + 1)$ of each node is determined by the current state (t) of its inputs following lookup tables that are also generated randomly (and then remain fixed). (B) The structural network defines a state transition network with 2^N nodes. Each state has precisely one successor, but it can have several or no predecessors. Thus, it is a dissipative system. Eventually, a visited state is repeated, and thus the network has reached an attractor. (C) Example dynamics of an RBN with $N = 40$ and $K = 2$, time flowing to the right. A random initial state converges into an attractor of period 4. A single RBN can have several attractors of different periods.

Although many laboratories are working on this task, the successful preparation of artificial cells having all features of natural living cells is still challenging. An artificial cell that can self-produce and maintain itself (a so-called autopoietic system) has not yet been demonstrated. All published papers with 'artificial cell' in the title describe usually simple particles that have at least one property in common with living cells. Nevertheless, a big challenge exists to synthesize an artificial cell having at least several properties shared by living cells. For example,

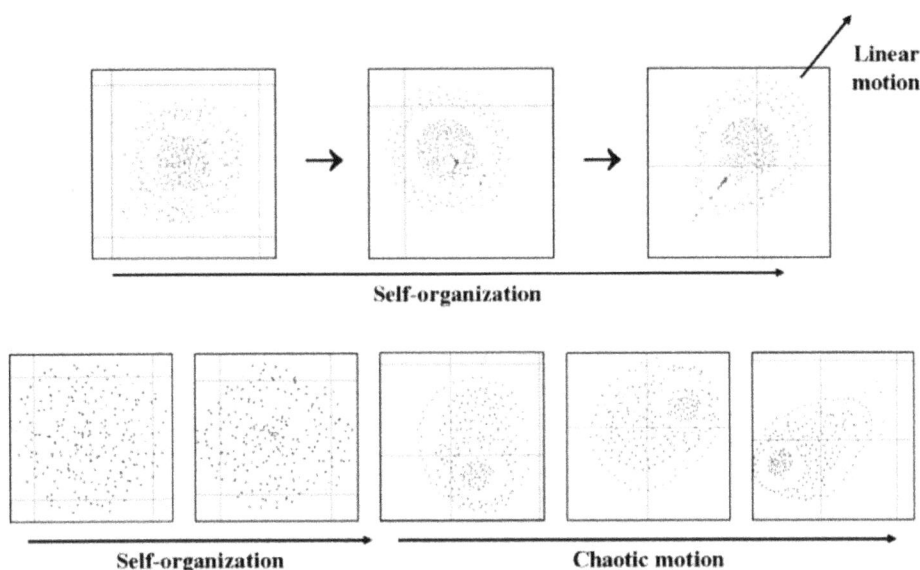

Figure 4.12. Examples of complex structures made of several different chemical species, designed using Sayama's interactive simulator [272]. The swarms self-organize from initially random states to a shape that looks like a horseshoe crab (top), or a biological cell-like structure that shows active chaotic movement after self-organization (bottom).

(i) the presence of a stable semi-permeable *membrane* that mediates the exchange of molecules, energy, and information between internal content and external environment while preserving specific identity; (ii) the possibility to sustain itself by using energy from its environment to manufacture at least some components from resources in the environment using *metabolism;* and (iii) the capacity of growth and self-replication including genetic *information.*

A simpler and less problematic approach is to synthesize protocells that are not necessarily alive, but that exhibit only a few life-like properties [293–295]. Protocells can be defined as simplified systems that mimic one or many of the morphological and functional characteristics of biological cells. Their structure and organization are usually very simple and can be orthogonal to any known living system. Protocells are used both to model artificial life and to model the origins of life. In the latter case, protocells are often considered hypothetical precursors of the first natural living cells.

In principle, two main motivations exist to create an artificial cell. One group of researchers aims to answer questions about the origin of life: they synthesize primitive cells which consist of a protocell membrane that defines a spatially localized compartment, and of genetics polymers that allow for the replication and inheritance of functional information. The aim is to create self-replicating micelles or vesicles and to observe the spontaneous Darwinian evolution of protocells in the laboratory [296, 297]. Other researchers want to prepare particles with life-like properties that mimic the behaviour of living cells, though without the ability to self-replicate or evolve. Such objects can move in their environment [298, 299], selectively exchange molecules with their surroundings in response to a local change in temperature or concentration, chemically process those molecules and either accumulate or release the product, change their shape [300, 301], and behave collectively [302, 303]. Such synthetic objects can be used for instance as smart drug delivery vehicles that release medicine *in situ*. These artificial cells could also be called chemical or liquid robots [304] (figure 4.13).

Recently, so-called Xenobots were introduced [305, 306]. Although they do not belong to traditional wet artificial life research, they should be mentioned here as a new approach on how to synthesise artificial organisms in the laboratory by using ALife tools. Xenobots were designed by an evolutionary computer algorithm and then assembled from embryonic frog cells *Xenopus laevis* (hence the name Xenobots). Whether Xenobots are 'living robots', 'living machines', or 'man-made animals' it remains a debatable question. Nevertheless, these creatures, smaller than

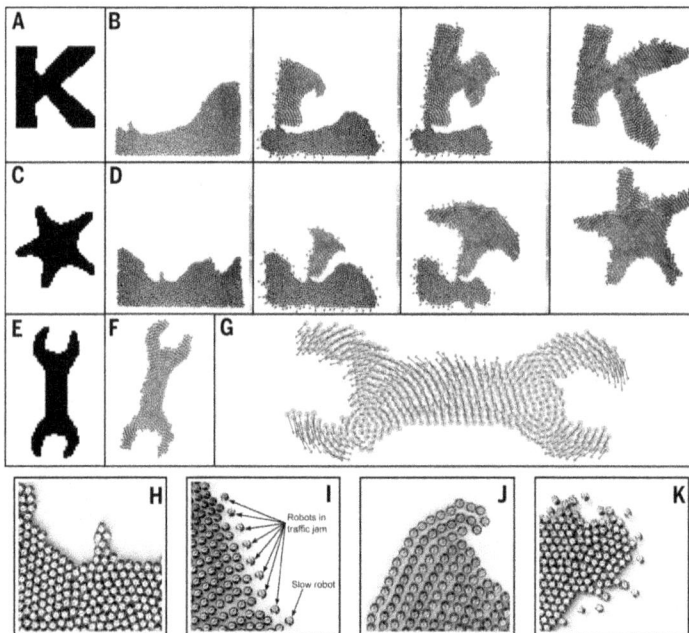

Figure 4.13. Self-assembly experiments using up to 1024 physical robots. Reprinted from [290] with permission from AAAS. (A, C, and E) Desired shapes are provided to robots as part of their program. (B and D) Self-assembly from initial starting positions of robots (left) to final self-assembled shape (right). Robots are capable of forming any simply connected shape, subject to a few constraints to allow edge-following (19). (F) Completed assembly showing global warping of the shape due to individual robot errors. (G) Accuracy of shape formation is measured by comparing the true positions of each robot (red) and each robot's internal localized position (gray). (H–K) Close-up images of starting seed robots (H), traffic backup due to a slowly moving robot (I), banded patterns of robots with equal gradient values after joining the shape (robots in each highlighted row have the same gradient value) (J), and a complex boundary formed in the initial group (dashed red line) due to erosion caused by imprecise edge-following (K).

a millimeter in size, can find potential applications in areas such as environmental remediation. At this moment, Xenobots have the ability neither to self-replicate nor to evolve, but the lifespan and the hypothetical ability to reproduce could be assessed and regulated in the future in accordance with ethical principles (figure 4.14).

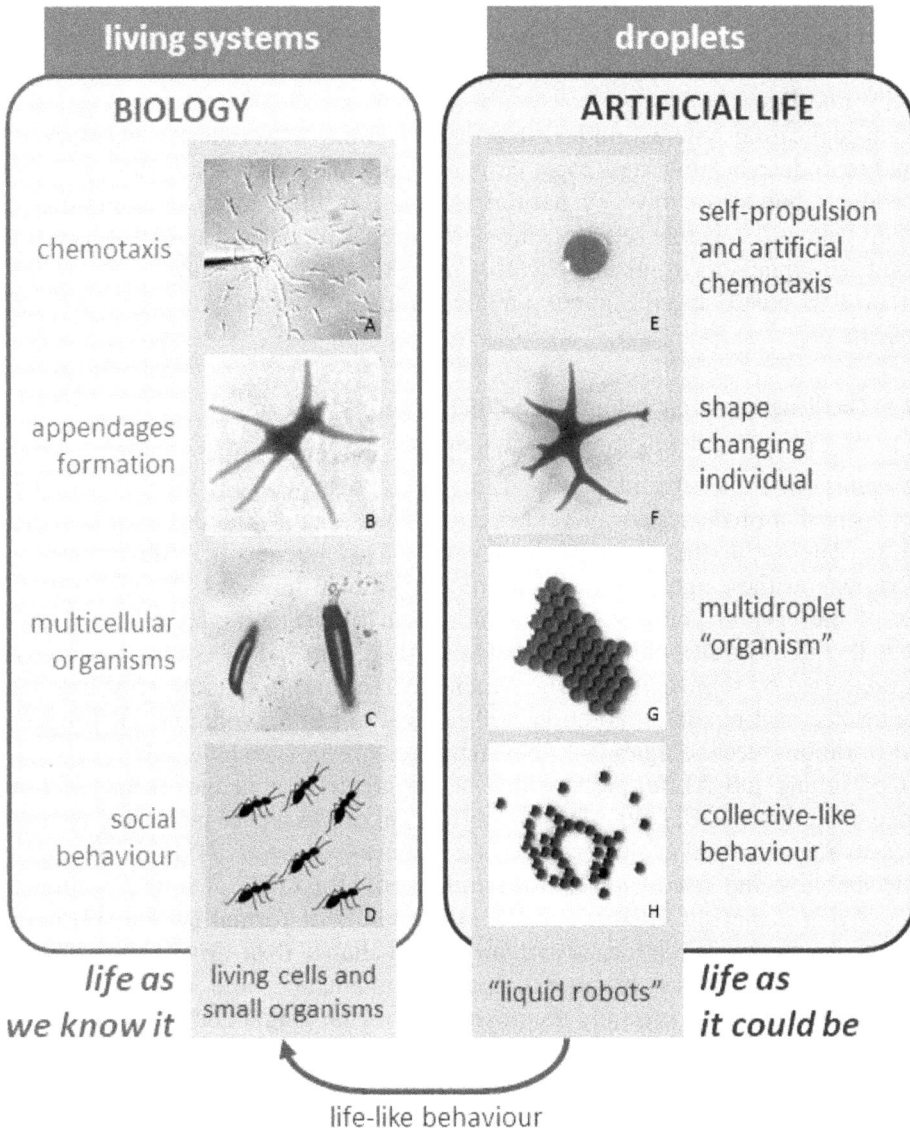

Figure 4.14. Schematic comparison of wet artificial life research on droplets to biology and studies of living systems. (A) Chemotaxis of *Dictyostelium* cells. (B) Prosthecate freshwater bacteria. (C) Multicellular slug stage of the *Dictyostelium* developmental cycle. (D) Schematic representation of ants group. (E–H) Decanol droplets were placed into an aqueous solution of sodium decanoate.

4.4.5 Self-replicating systems

Let us imagine a factory where robots are constructing other robots. We know such scenes from the movie *I, Robot* where the company U.S. Robotics produces humanoid robots, or from an island factory in the theatre play *R.U.R.—Rossum's Universal Robots*. However, these examples are fictional, and such factories have never existed in the real world and still belong to science fiction. Still, the idea of machines that beget machines was already in the air four centuries ago. A comprehensive study of the history and state of the art of self-replicating machines from a scientific and technological point of view can be found in a recent book by Taylor and Dorin [307].

Self-replicating systems can be categorized as follows: 'Standard-replicators' that could reproduce by building a copy of themselves; machines that are able not only to reproduce, but also evolve by natural selection as their living counterparts, are categorized as 'evolvable self-replicators' (evo-replicators); and so-called 'manufacturing self-replicators' (maker-replicators) have the ability not only to self-replicate, but also to create specific goods and materials as by-products when they self-replicate.

4.4.6 Challenges and opportunities: a 2050 vision

ALife will have to tackle very specific challenges in science and engineering: (a) Get the agents out of the lab and into the wild. (b) Make them able to interact with living organisms in a predictable way. (c) Make them sustainable and degradable in order not to deteriorate the already transformed environments. (d) Allow long-term operations of these agents, or even multi-generational timespans by self-reproduction of the agents. The development of sustainable technologies is thus urgently needed, and ALife could in part address that. The ALife systems should be characterized by robustness, autonomy, energy efficiency, materials recycling, local intelligence, self-repair, adaptation, self-replication, and evolution, all properties that traditional technologies lack, but living systems possess [308, 309].

Concerning **soft ALife**, perhaps one of the greatest challenges is that of open-ended evolution (OEE) [310–313]: can a program produce ever-increasing complexity (as it seems natural evolution does)? Hernández-Orozco *et al* [314] proposed that undecidability and irreducibility are requirements for OEE. It might be that the difficulty of achieving OEE is related to the limits of formal systems [315–317]. Simplifying the situation: formal systems cannot change their own axioms. This is a necessary condition for traditional logic, mathematics, and computation, but perhaps OEE requires precisely the possibility of changing axioms.

As for **hard ALife**, robots are becoming more and more sophisticated [318]. They are also gaining autonomy. However, as with the rest of artificial intelligence, all robots are specialists. They are good at performing the tasks they were designed for, but they cannot generalize and perform other activities. For example, a vacuum robot cannot paint. Not only because it lacks the appropriate hardware, but also because the software is task-specific. The so-called 'artificial general intelligence' has so far produced not more than mere speculations. Could it be that the limits of formal systems just mentioned for OEE also affect artificial intelligence in general and robots in particular? If so, can we find an alternative, to build robots that are not based on formal systems?

Perhaps the most promising is the least developed: wet ALife. If we build a system that most people agree on calling 'alive', most likely it will be from wet ALife. There are several challenges already mentioned, but there seems to be no inherent limit to building or finding alternative lifeforms, either artificial or extraterrestrial. It is a blind guess to try to say whether we will have detected or created life different from the one that evolved on Earth by 2050. Still, soft ALife and hard ALife seem to have inherent limits (derived from the limits of formal systems), so we might as well expect the most from wet ALife. In any case, it can certainly contribute to a 'general biology'.

Physics can also contribute in this direction. Already, research in self-organization [319, 320] and active matter [321] has contributed to understanding the properties of living systems. Very likely, in the next few decades, advances in physics will enhance our perspectives on what we consider to be living, how it evolved, and where it might lead.

There is still no agreed-upon definition of life. Biological systems are perhaps too complex for a sharp definition. As we have seen, ALife can help to understand the general properties of living systems. This can benefit biology and engineering, gaining insights into life and being able to build artificial systems exhibiting properties of the living.

4.5 Toward a quantitative understanding of life

Raymond Goldstein[1] and Patricia Bassereau[2]
[1]University of Cambridge, Cambridge, UK
[2]Institut Curie, Paris, France

4.5.1 General overview

More than half a century ago, physics and biology came together to understand the DNA double-helix structure, one of the most important discoveries of the 20th century. At present, the complexity of biological structure and function and the observed emergent phenomena in biology delineates a well-defined research domain where physics meets biology. Emergent phenomena include processes where larger entities exhibit properties that their simpler constituent entities (and the naive superposition of their properties) do not exhibit and, as a consequence, something new emerges from collective behaviour that could not be predicted from its constituent parts. Complexity applies to key open questions in the physics of the many-body interacting systems, but it finds its most natural setting in biology, from the synchronized dynamical behaviour of the human brain or the beating of heart cells to the concerted action across multiple scale of biomolecules and cells in tissues and organs. The cascade of energy and information across the many biological levels defines a new field of research, the thermodynamics of information, which starts at the Maxwell demon paradox and the interpretation of the second law and may lead us far beyond.

4.5.2 Physics meets biology over the last 60 years

Physics and biology have been intertwined since the dawn of modern science (figure 4.15), from Antony van Leeuwenhoek's use in 1700 of advanced optics to reveal the hidden world of microscopic life [322], to Bonaventura Corti's (1774) discovery of the persistent fluid motion inside large eukaryotic cells [323], Robert Brown's 1828 study of random motion at the microscale [324], and Theodor Engelmann's determination in 1882 of the wavelength dependence of photosynthetic activity [325]. Although at the time it might have been difficult to define the precise disciplines of each of these scientists—perhaps they were all 'natural philosophers'—in hindsight we can see clearly the way in which their discoveries impacted both biology and physics. Despite this long history of discoveries at the boundary between the two fields, and the innumerable fundamental contributions to both disciplines over the long arc of time, the field of *biological physics* as a discipline within the research enterprise of physics has only risen to great prominence since the postwar era, particularly since the mid-1980s. We are now at the point that most academic physics departments have an identifiable group in biological physics alongside those in high energy, condensed matter, atomic and astrophysics. In this introductory section we will review some of key developments over the past 60 years in order to identify what we see as key intellectual threads that run through that history, to set the stage for the forward-looking sections that follow.

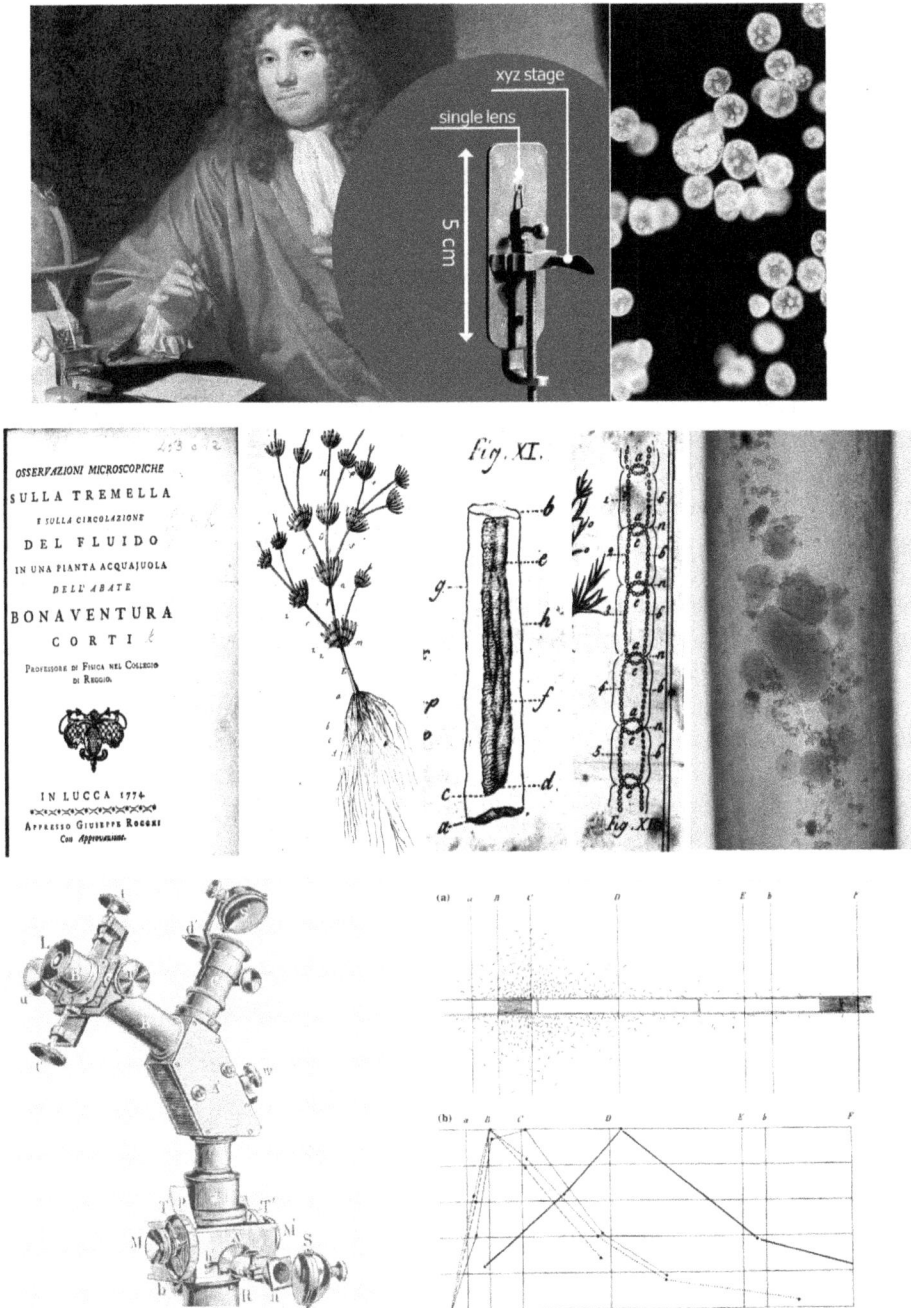

Figure 4.15. Historical connections between physics and biology. Top row: Antony vanLeeuwenhoek, his microscope, and the alga Volvox that he discovered. Middle: BonaventuraCorti's celebrated treatise on cytoplasmic streaming, with drawings of aquatic plants he studied,and a modern image of the plant Chara corallina. Bottom: The micrscope with which Theodor Engelmann's determined the action spectrum of photosynthesis by visualizing the accumulation of aerotactic bacteria along an alga illuminated by the solar spectrum. This image has been obtained by the authors from the Wikipedia website, where it is stated to have been released into the public domain. It is included within this article on that basis.

We begin by noting that each of the discoveries highlighted above was made possible by state-of-the-art *microscopy* that was able to reveal phenomena that had escaped previous notice. That technological advances often translate into scientific discoveries is a familiar pathway in science, but it takes more than just a new piece of kit to lead to true progress. Indeed, there is often a separate but equally important collective aspect of the scientific community at work, in defining the questions, bridging disciplines, and training students.

Our choice of 60 years for this overview was made to capture several of the most important immediately postwar advances at the interface of biology and physical sciences which, in an interesting historical twist, clustered around the same time (figure 4.16). These were the 1952 elucidation of the dynamics of action potentials in neurons by Hodgkin and Huxley [332] and the theoretical work by Turing in that same year showing that chemical reaction-diffusion systems can exhibit spatio-temporal patterns, followed in 1953 by the discovery of the structure of DNA by Franklin, Watson, and Crick [333, 334] using x-ray scattering methods. As in the discussion above, the work of Hodgkin and Huxley built very much on developments in experimental methods; in this case it was the invention of the 'voltage clamp' by Cole [335], a device that utilizes feedback to maintain the voltage across a membrane at a set value, that allowed the properties of ion channels in membranes to be studied as a function of (controlled) voltage. The idea of a voltage-dependent channel conductance was the key to understanding not only neuronal dynamics but 'excitable media' in general [336]. Experimental methods from physics were of course also key to the DNA structural work, which was an offshoot of the original development of x-ray scattering to determine crystal structures based on Bragg's law [337].

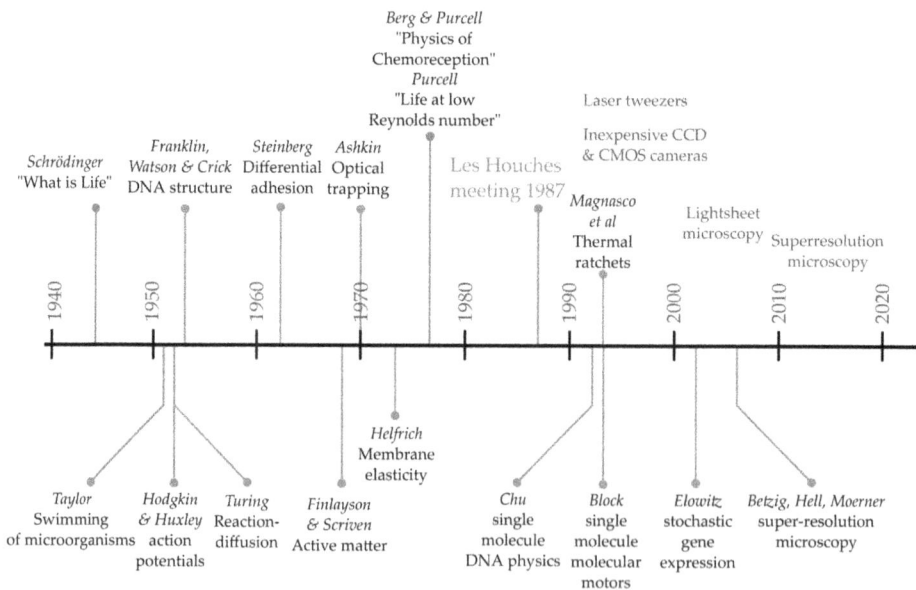

Figure 4.16. Timeline of significant developments in biological physics from the mid 20th century onwards.

Turing's work had little to do with technological advancements and his research would not generally be considered originally to be part of academic physics; it might be categorized instead as one of the earliest works in *mathematical biology* [338]. Using very simple mathematical models of two interacting chemical species, he showed that a system of chemical reactions that would, in the absence of diffusion, be linearly stable, could be rendered unstable by diffusion, leading to spatially periodic patterns. In this context, this mechanism was a revolutionary idea, although spatio-temporal symmetry breaking had already been understood in fluid mechanical contexts at least since the time of Rayleigh's work on thermal convection [339]. Yet, the relevance of the Turing mechanism to actual biological systems has been debated ever since [340], in part because in its standard form the mechanism requires the diffusivities of the two chemical species to differ by what seems to be an unphysical amount. It was physicists who provided the first experimental realizations of the Turing instability by reducing the diffusivity of one of the chemical species through binding to a gel substrate [341–343]. While it has also been unclear how to square highly regulated biological patterning with the concept of spontaneous symmetry breaking, it is important to emphasize that Turing's work introduced most of the concepts now used in the study of how biological form develops, the most important of which is that a system of reacting and diffusing chemicals can form spatio-temporal patterns [344]. It was within applied mathematics that reaction-diffusion dynamics were first applied to pattern formation in biological populations such as the spiral waves and chemotaxis exhibited by the slime mold *Dictyostelium discoideum* [345], but these ideas eventually resonated with physicists studying such pattern formation [346]. There is now ample evidence that these basic ideas are correct.

Like Turing's paper, Taylor's (1951) work [347] that offered the first explanation for the swimming of microscopic organisms in a viscous fluid sat not within physics proper, but rather in fluid dynamics (in the British school of applied mathematics). This and subsequent works by other applied mathematicians such as Lighthill [348] clarified the physics behind self-propulsion in the absence of inertia by exploiting the theoretical simplifications inherent in the dynamics of asymptotically slender filaments such as eukaryotic and prokaryotic flagella. These developments were later taken up by physicists and caught the attention of the community thanks to Purcell in his celebrated 1977 essay on 'Life at low Reynolds number' [349], and in his work with Berg [350] that highlighted physical considerations in the phenomenon of chemoreception. This is one of the many important examples in which scientists who came from more established areas within physics moved into biological physics, helping to legitimize it within the broader community.

These intellectual strands that were developing in the 1970s were very much apart from the molecular-biology-centered world of mainstream biology at the time, and indeed they were distinct from much of the field known as 'biophysics' that was focused on proteins, ion channels, and electrophysiology, but they would prove to be harbingers of the development of biological physics as a distinct discipline. Another excellent example of the biological research beyond the molecular level is that of Steinberg on the cellular organization in tissues. In his 1962 paper on this subject

[351], he formulated a differential adhesion hypothesis for the self-organization of cell types within tissues. Although the dominance of the molecular view of biology at the time meant this work gained little traction in the community, its enthusiastic uptake by physicists in the 1990s [352] led to an explosion of work on the subject of tissue biomechanics that continues to this day.

The gradual transformation of the field of solid-state physics into what we now term condensed matter physics began in the 1960s and early 1970s with the intense interest both in phase transitions and critical phenomena and also in the physics of liquid crystals. In moving away from the static description of precisely ordered solids, the field naturally began to focus on phenomena on length scales larger than molecular, and to utilize continuum descriptions such as those familiar from the Landau–Ginzburg theory for superconductivity, with particular emphasis on the formulation of scaling laws [353] in polymer physics, building on the foundational work of the theoretical chemist Flory [354]. An important strand of this research concerned the description of interfaces, as involved in phenomena such as wetting, and also in pattern formation during solidification [355]. The emphasis on continuum approaches was particularly significant in the field of liquid crystals, where the intellectual leadership of the French school led by De Gennes proved so important in creating a worldwide community of experimentalists and theorists. It was out of this field that came the seminal work of Helfrich in 1973 [356] on the elasticity of fluid membranes. The particularly simple form of this energy functional, dependent only on the shape of the membrane, allowed for an explosion of analytical and numerical work on a range of problems in membrane physics. With these advances, it was but a small step to move from the description of idealized membranes as found in lipid vesicles to real biological membranes with all their complexity.

On the experimental side, the arrival of affordable and relatively easy-to-build optical trapping setups in the early 1990s enabled quantitative experiments at the single-molecule level that were simply not possible previously. The first single-molecule experiments on the stretching of DNA in 1992 [357], combined with later theory [358], enabled the precise quantitative understanding of semiflexible polymers, opening up the understanding of aspects of chromosome structure and function in the cell. Likewise, pioneering work in 1993 on the stepping of motor proteins along biofilaments [359] revealed the stochastic nature of the motion of molecular motors and almost overnight placed their dynamics squarely within the field of non-equilibrium statistical physics. At roughly the same time, the paradigm of stochastic rectified motion on periodic potential energy landscapes (aka 'Brownian ratchets') received intense focus as the model for molecular motors [360–363] and also for the unidirectional motion exhibited by plants [364]. Together these helped to launch a subfield of stochastic thermodynamics, leading to principles for complex systems such as synthetic machines [365], including collective effects [366], with applications to such systems as hair cells in the ear [367]. From the mid-1990s onward, the advent of relatively inexpensive CCD and CMOS cameras and associated image processing techniques, including particle tracking methods [368], meant that the barrier to experimental work in this area was dramatically lowered.

Later developments in the new millennium saw a whole new generation of affordable high-speed cameras, further enabling the worldwide study of fast phenomena such as flagellar synchronization [369] and cell motility [370]. Finally, the development of soft lithography for microfluidics [371] further broadened the experimental base for studies at the microscale.

From all of the above, we see that fields that in the postwar period were considered apart from the core of physics, including fluid mechanics, transport theory, and much of continuum physics), were embraced by the biological physics community as its focus turned toward phenomena on scales from nanometers to millimeters, where so many cellular phenomena occur. It was perhaps only natural that physicists in this new era would reject [372] the view [373] that their role in biological research was simply to provide instrumentation or better intermolecular potentials in the service of the questions biologists had framed. Rather, the field has prospered precisely because physicists and biologists have joined together to pose the questions that guide the field. Finally, as the ranks of academia working on biological physics grew throughout the 1990s to the point that a new generation of PhD students was trained and themselves moved into academia, the whole field gained critical mass to the point of being among the fastest growing divisions within national physics societies.

4.5.3 Challenges and opportunities for the future

In this section we speculate on the future directions of a few of the many areas within biological physics, aiming for an overview rather than an encyclopedic account.

4.5.3.1 Soft matter and self-organization: from active matter to cell and tissues
In a remarkably insightful article in 1969 [374], Finlayson and Scriven considered the possibility of various types of hydrodynamic instabilities arising from what they termed 'active stresses'. These are contributions to the stress tensor arising from gradients of scalar fields (e.g., the concentration of some solute) that represent the conversion of chemical energy to kinetic energy, typically manifest by a pattern-forming instability. This is the essence of what we now term 'active matter'; systems in which there is injection of energy at small scales that display coherent structures on scales large compared to the microscopic constituents [375–378]. Early examples of this notion are found in new physical models for membranes that include the non-equilibrium activity of proteins that transfer ions across membranes via external sources of energy (light, ATP hydrolysis, electric fields) [379, 380]. Other examples include collections of molecular motors and microtubules [381], where the consumption of ATP by the motors powers translocation that leads to filament motion and self-organization, and in collections of self-propelled particles, in theory [382–384] and experiment [385], where coherent structures arise from hydrodynamic interactions between the organisms (figure 4.17).

These general concepts and tools were used in parallel to study the organization of living systems and of engineered ones. On the biological side, 'active membranes'

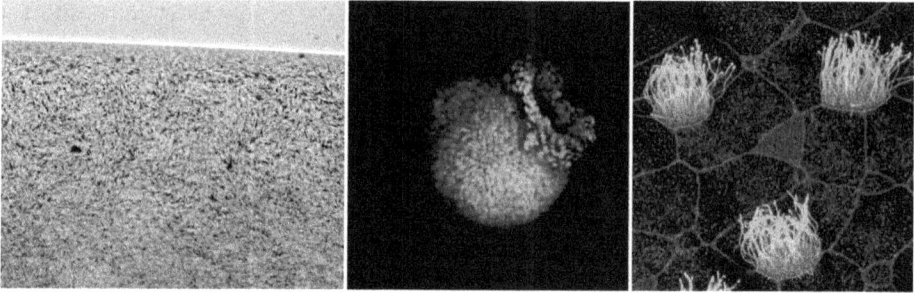

Figure 4.17. Active matter on multiple scales, from individual cells to confluent tissues. From left to right: bacterial 'turbulence' in a suspension of B. subtilis (reproduced from [326], CC BY 4.0), embryonic inversionin Volvox (reproduced from [327] CC BY 4.0), and a tissue with multiciliated cells (Mitchell Lab, Northwestern University).

have been generalized to membranes with a cortex and in contact with polymerizing actin filaments [386, 387]; they contributed to elucidate the non-equilibrium origin of the flickering of the red blood cells [388], although a consensus is still not reached. They have been applied to active gels [389, 390] made of polar dynamical cytoskeleton filaments and active crosslinkers like molecular motors that generate flows and stresses, features that do not exist in passive gels. Active gels have been instrumental for understanding actin flows in different cellular features: cell motility, blebbing, division, etc [390]. Liquid crystals principles have been extended to active nematics, smectics, or cholesterics. The power of this continuous description is that the model of active nematics can describe as well the self-organization of patterns and the flows for actively moving entities such as bacteria in films, anisotropic motile synthetic particles, animals flocks, or that of cells in living tissues during development, embryos [391]. Of course, non-equilibrium activity also affects phase separation [392]. A rapidly growing body of work has shown that liquid–liquid phase separation of multivalent assemblies occurs in the cytoplasm of cells, forming membraneless compartments [393], or in the nucleus [394]. However, Oswald ripening and thus droplet size is probably limited by active enzymatic reactions [395].

While many of the ideas and experiments on active matter systems implicitly use as a reference dilute suspensions of the motile entities, in recent years there has been a gradual shift of focus toward "confluent' tissues, typically quasi-two-dimensional sheets whose constituent cells are in space-filling contact with their neighbors. This leads to a wholly different class of phenomena and theoretical models that touch on some of the most significant issues in developmental biology. During development, tissues like epithelia are very dynamic since cells continuously die or divide and in addition, active forces are exerted at the cell–cell junctions. Thus, in spite of their morphological similarity with foams, epithelia differ strongly from them [396, 397], and non-equilibrium principles are necessary to describe their homeostasis [398, 399]. In addition, many tissues bend, fold, and even invert their topology during embryogenesis [327]. Through a combination of advances in experimental methods such as light-sheet imaging [400] and use of 'organoids' [401] to study the early stages of multicellularity, and theoretical models addressing geometric rearrangements of tissues [402], we anticipate that the near future will see significant progress in understanding many key issues in the biomechanics of development and differentiation.

So far, very little work has been done to integrate at the cell scale all active exchanges that occur between compartments, with the plasma membrane and with other cells. At a single compartment level such as the Golgi apparatus, they directly affect its shape [403]. Apart from during division, cells have to maintain their shape, area, and volume in spite of these multiple fluxes, and how they manage is a recurring issue in biology. Modeling these exchanges and understanding how homeostasis at the cellular scale is achieved remains one of the future challenges for physicists and cell biologists. The following section addresses how homeostasis is maintained in tissues.

4.5.3.2 Deciphering the physical principles of mechano-chemical networks that control homeostasis, shape, and size of living entities

In his book *On Growth and Form* [404], D'Arcy Wentworth Thompson raised the question of how physical forces contribute to determine the size and shape of living organisms and thus initiated the field of 'mechanobiology'. In addition to the non-equilibrium principles that govern cellular assemblies, mechanics also plays a key role. Indeed, cells exert, sense, and respond to external forces. For instance, the spreading velocity of cellular migration depends on the stiffness of the underlying substrate [405]. Cells exert forces on their environment generally using dynamical actin networks and the contractile actomyosin machinery. To a large extent, cell mechanosensitivity depends on proteins embedded in the plasma membranes that are linked on the extracellular side to specific ligands of the external matrix or to similar membrane proteins of a neighbouring cell in tissues. On the intracellular side, they have cryptic binding sites that unfold and allow connection to the actin cytoskeleton in a load-dependent manner. Actin structures are themselves mechanosensitive [406]. Moreover, mechanical forces trigger biochemical response and signalling pathways (i.e., cascades of biochemical reactions with positive and negative feedback loops). A revealing illustration is provided by stem cells (i.e., non-differentiated, pluripotent cells) that differentiate into very different cell types—neurons, muscle, or bones—depending on the stiffness of their micro-environment [407]. Mechanical cues are thus transduced into biochemical signals, and integrated with genetic and chemical signals to modulate diverse physiological processes. In addition, there is constant cross-talk between biochemistry and mechanics during mechanotransduction, which is itself a part of the early development since developmental genes can be switched on by internal stresses accumulated during the growth of the embryo [408]; it is also involved in cancer development [409]. Physics and bioengineering have strongly contributed to this field, in particular by developing many tools for measuring forces at all scales, from the single mechanosensitive molecules to stresses in tissues [410]. On the biochemical side, complex signalling networks have been identified in cell-extracellular matrix adhesion mediated by integrins [411, 412], or adherens-junction in cell–cell contacts [413]. Mechanosensitive channels, in particular 'piezo' channels, the main type of molecular force sensor in eukaryotes [414], are present in cell membranes; they let ions flow when they mechanically activated, which also triggers a cascade of biochemical signals, but have been less studied.

Cross-talk exists between these different signalling pathways that are all integrated at the cell level. Systems biology approaches are certainly essential in understanding these complex regulatory networks and how cells manage their mechanical interactions with their environment. But models based only on gene ON/OFF circuits are insufficient to understand how tissue integrity is preserved against mechanical stresses, extensile, or compressive, and during rearrangements of cells, and how tension homeostasis is set. The existence of cellular rearrangements in tissues implies the remodeling, destruction, or creation of the connecting structures between cells in a coordinated fashion (figure 4.18). In some rare cases, the feedback loops between mechanics and signaling are known [415]. Since ever-more force sensors are available to quantify stresses across scales, these measurements have to be integrated with the networks in cells to develop models that explain the emergence of larger-scale behavior from the interactions of their molecular components inside cells. Thus, it appears that it will still be some time before we have a final answer to D'Arcy Thompson's questions: what limits the growth and division process, and what determines the size and shape of organs or animals?

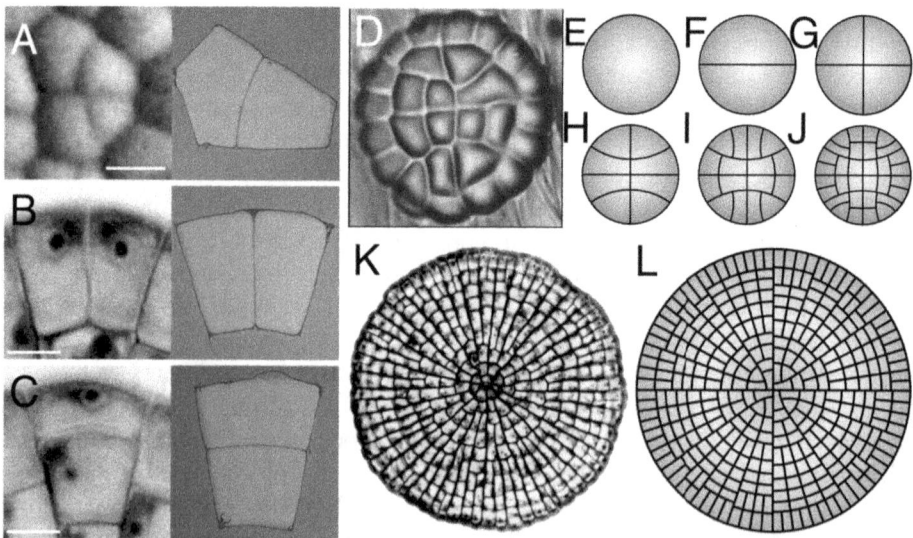

Figure 4.18. Cell division in the plant cells and similarity to shapes of soap bubble. Reproduced from [328] CC BY 4.0. (A–C) Patterns of cell division in various plants (left) compared to shapes of soap bubbles (right). (D) A large-scale geometry compared to mathematical results (E–J) in which a rule derived from soap bubble physics is iterated for uniform growth in the region. (K, L) as in (D–J), but for marginal growth.

4.5.3.3 Some perspectives on brain functions, for soft matter and computational neuroscience

Neuroscience has a singular place in biology. It is at the cross-road between disciplines, since the brain is not only an organ but controls locomotion, sensing, memory, decision making, and, at least in humans, feelings, consciousness, etc. The brain is a complex, temporally, and spatially multiscale structure. From the perspective of physics, it can be of course investigated at the cellular level. But since neurons communicate and form circuits and interconnect functional areas in the brain, it is better described as a hierarchical network (termed the 'human connectome' [416]). In addition, neurons are embedded in glial cells that protect them, but also contribute to some signaling functions. Two of the most challenging goals in science continue to be on the one hand the generation of the complete map of the neural connections in a brain and, on the other, to understand from a molecular point of view how signals are produced and transmitted and how the network builds up, in order to decipher how this incredibly complex structure can produce complex cognitive functions. The breadth of challenges for physicists of the future is too wide to list in totality; here we mention two of them, on axonal signaling and for computational neuroscience.

Neuronal cells have unique properties—mechanical, geometrical, and internal organization—and are actively studied *per se*. Strikingly, not much has been done since the work of Hodgkin and Huxley [332] to revisit their model of axonal transmission based on electrical signals. In the standard action potential model, signal propagation is achieved by the voltage-dependent opening and closing of ion channels, largely ignoring the specific physical properties of cell membranes. Only T Heimburg has challenged it, by suggesting that a lipid phase transition occurs in the membrane in the course of the action potential, increasing membrane conductance [417]. However, no study has included yet the non-equilibrium effect due to the activity of the channels we discussed above. Thus, a comprehensive model of the propagation of the electric axonal signal is still missing.

Various methods are routinely used to image whole-brain activity and detect dysfunction and disease, including x-rays (CT scan), radioactivity (PET scan), and NMR (MRI). However, to image the neuronal networks with better resolution, neurosciences have greatly benefited from the most advanced developments in microscopy and imaging: methods for imaging in diffusing media allow deep imaging in the brain, light-sheet microscopy for volumetric imaging, and optogenetics to control neuronal circuits [418]. Model animals with small brain volume or that are moderately transparent (*Drosophila*, zebrafish) have also facilitated these studies. It is now possible to follow neural algorithms in living and freely moving animals as they vary their behavior [419]. Using whole-brain functional imaging, the brain of a zebrafish can be imaged during the different stages of the decision-making process about its swimming direction [420]. Betzig and collaborators achieved the *tour de force* of imaging the whole brain of a fruit fly with molecular contrast and nanoscale resolution using combined light-sheet and expansion microscopy [421]. It is thus now possible to locate individual neurons, trace connections between them, and visualize organelles inside neurons, over large volumes of brain tissue in 3D, albeit on fixed brains. These are only a few examples showing how the field is developing with the blooming of new optical techniques, pushing the frontiers of the observations. One obvious consequence is that with these advances will come enormous quantities of data, as in many other areas of cell biology in which volumetric imaging is used. One strategy to manage and analyse such 'big data' is obviously to use artificial intelligence and deep learning methods [422] to extract meaningful information. In addition, there is now a timely opportunity for computational sciences to develop approaches based on network science to provide integrated models of interactions in neurobiological systems. In fact, a new field termed 'network neuroscience' is growing, bridging network theory and experiments [423]. One might hope that creative developments in computational sciences in theory and in functional imaging will eventually allow us to unlock the neuronal code.

4.5.3.4 *Emergence of life and physics of biological evolution: from the second law of thermodynamics to the selection of structures*

In his 1944 book *What Is Life? The Physical Aspect of the Living Cell* [424], Erwin Schrödinger stressed the apparent paradox behind life: how can living organisms maintain an organized state and grow complex structures without violating the second law of thermodynamics that predicts an evolution towards maximized entropy? He resolved it by pointing out that Earth is not an isolated system, but receives energy from the Sun, and that living systems absorb energy. Moreover, he also used thermodynamic arguments to explain why an internal organizing factor that carries information (that eventually turned out to be DNA) is necessary for living systems to develop in an organized manner and replicate faithfully.

Likewise, the emergence of life on Earth cannot be explained by the second law, but rather (in part) by far-from-equilibrium thermodynamics and the concept of dissipative structures highlighted by Prigogine and others. While this issue is covered elsewhere in this volume, we note that there is a strong school of research, by no means universally accepted, supporting the idea that the evolution of life began from 'soup' of RNA molecules before DNA appeared. Yet, the mechanisms by which the nucleotide bases and sugars could be formed beforehand by prebiotic reactions are still not elucidated. There are promising attempts to produce artificial cells [425] and a significant body of work on life-inspired and out-of-equilibrium systems at the nanoscale [426]. Compartmentalization and the appearance of membranes are key steps during evolution. Recent work on active membraneless droplets suggested that they could have formed the protocells from which cell membranes could have appeared [427]. Bottom-up reconstitution of a synthetic cell with well-characterized functional molecular entities in vesicles can also help to understand the origin of life [428]. Conversely, with a top-down approach, the Craig Venter Institute has shown it is possible to recreate artificially genomes and minimal cells [429]. A synthetic minimal organism has been reproduced *in silico* by reconstruction of a complete set of chemical reactions [430]. Hydrodynamic models have also been used to understand how DNA may have replicated in early times: laminar thermal convection, present in submarine hydrothermal vents, can very efficiently accelerate the DNA replicating polymerase chain reaction (PCR) [431], which is enhanced when the molecules are trapped in porous rocks [432]. We expect these *in vitro* approaches will become ever-more important in the future. On the modeling side, considering the complexity of this interdisciplinary problem, there is a clear need for further development of related aspects of non-equilibrium physics.

4.5.3.5 Evolution of biological complexity

High on the list of fundamental problems in biology, just behind the origin of life and the nature of consciousness, is the origin of multicellularity. While the simplest organisms to appear on earth were no doubt unicellular, eventually life evolved to become larger, in the sense of having more cells, and also more complex, dividing up life's processes into ever-more specialized cell types [433]. It has been recognized since the time of Weismann [434] in the late 19th century that a great challenge is to understand the driving forces behind the transition to multicellularity, and, as pointed out by Huxley some years later [435], to identify the biological entities on which evolution acts [435]. While there can be obvious advantages to larger and more complex organisms, such as greater motility, avoidance of predators, and larger uptake rates of nutrients, there are also metabolic costs associated with the regulatory networks that control the organism and the cellular scaffolding that holds it together [436]. Recent work has begun to address these issues using green algae [437] and choanoflagellates (figure 4.19), the closest uni- and multicellular relatives of animals [438].

The ability to track *single* living objects (bacteria, yeast, or cells) using microfluidic devices and to analyze their lineage during multiple rounds of cell division will open the way for new discoveries when these experiments will be coupled to external perturbations. Among the most promising approaches to understanding the origins of biological complexity involve the use of artificial methods to put evolutionary pressure on extant organisms. A prime example of this is recent work on yeast, in which repeated rounds of centrifugation of growing cultures, selection of

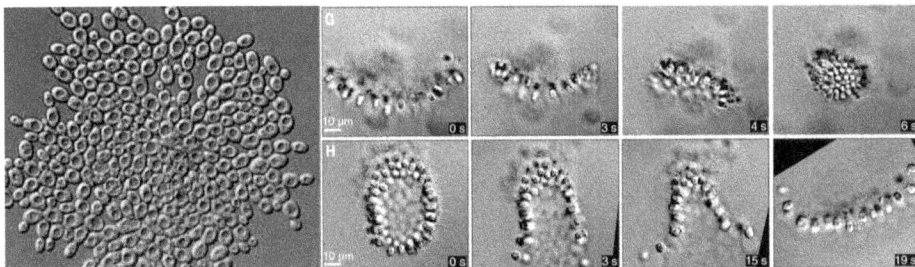

Figure 4.19. Novel multicellular organisms. Left: 'snowflake yeast'. Reproduced from [329] CC BY 4.0 formed via repeated rounds of selection for faster settling speed under centrifugation. Right: snapshots during the curvature inversion of a sheet of choanoflagellate cells comprising the organism C. flexa, as triggered by light. Reproduced from [330] CC BY 4.0.

the fastest sedimenting fraction (containing the largest organisms), and subculturing of that fraction produces 'snowflake yeast' (figure 4.19), a genuine multicellular variant [329]. This 'experimental evolution of multicellularity' enables a whole range of questions in the origins of multicellularity to be addressed, and we anticipate significant developments in this area in the coming years. It is clear that ideas from statistical physics [439, 440] will be important for the analysis of the data arising from these studies. In sum, we can see the emergence of a field centered around the physics of biological evolution, using concepts from condensed matter such as frustrated states and glasses to describe transitions during evolution [441]. With the appearance of ever-more data from advances in experimental methods due to technical developments, we can imagine that more theoretical models will be designed to understand how living systems cope with the second law to adapt to their environment.

4.5.4 The future

While predictions are always difficult, especially about the future, we offer a few final words on the greatest challenges in the field of biological physics may address in coming years. Clearly, the most fundamental, unsolved problem is the origin of life. As we have touched on here and as discussed in greater detail elsewhere in this volume, a range of highly interdisciplinary efforts appears poised to make significant progress on this problem. The issues concern the bootstrapping problem of how a truly self-replicating system can arise biochemically, and also the geophysical conditions that are amenable to such a development. Likewise, the nature of consciousness remains mysterious, but we can anticipate that the continued development of probes of neuronal structure and organization will point the way toward a deep understanding of this emergent phenomenon. In the realm of developmental biology, it is clear that the rapid explosion of experimental methods to probe cellular fate determination and global regulation combined with physical concepts regarding spatio-temporal patterning will continue apace, and we can look forward to a deeper understanding of the *regulation* of development. Even such issues as the regulation of limb size are, at this point in time, not resolved; their study in model organisms will continue to be an important research endeavor. Ultimately, we may hope that these physical methods will contribute to understanding the origins and control of the unregulated cell division that is at the heart of cancer. At the subcellular scale we still lack a comprehensive understanding of such complex machines such as the ribosome, which work in confined environments or with a limited energy supply. At the more macroscopic scale, the use of *in vitro* evolution methods will surely continue, providing a platform for the true quantitative understanding of evolution in the natural world. Coupled closely to this will be an increasing focus on what might be termed 'physical ecology', the study of communities of organisms coexisting with their natural habitat (figure 4.20). As of this writing, the world is wrestling with a global pandemic and it seems natural to expect much future research on the interplay between viruses and their hosts.

Figure 4.20. Two aspects of physical ecology. Left: marine algal blooms (green) in the Baltic Sea as seen from a European Space Agency satellite (reprinted from [331], courtesy of ESA). Right: the blue glow of bioluminescence triggered in breaking waves at a beach (photo courtesy of Gergo Rugli).

Acknowledgements

The work of P Bassereau's research group is supported by the FRM (Fondation pour la Recherche Médicale), the ANR (Agence Nationale pour la Recherche), the AFM (Association Française pour les Myopathies), the CNRS, and Institut Curie. R E Goldstein acknowledges support from the Engineering and Physical Sciences Research Council, through Established Career Fellowship EP/M017982/1, Investigator Award 207510/Z/17/Z from the Wellcome Trust, Grant 7523 from the Marine Microbiology Initiative of the Gordon and Betty Moore Foundation, and the Schlumberger Fund.

4.6 The emergence of life: the Sun–Earth connection

Sarah Matthews[1] and Frances Westall[2]
[1]Mullard Space Science Laboratory, University College London, London, UK
[2]Centre de Biophysique Moléculaire, Université d'Orléans, CNRS, Orléans, France

4.6.1 Introduction

Life on Earth is sustained by radiative emission from the Sun, as well as by heat emanating from the outer liquid core and the mantle. Indeed, it appears that this energy may have been essential for the very emergence of life on Earth. The Sun's emission varies on all timescales at which it has been observed, as a consequence of the generation, emergence, and evolution of its intrinsic magnetic field. There is increasing evidence that solar radiative changes have had an influence on Earth's climate through different mechanisms. In addition to variable radiative emission, the Sun also emits energetic particles, the flux of which changes in time depending on solar magnetic activity. Variations of the solar particle emission cause changes in the ionization of the Earth's atmosphere, its global electric circuit, and the ion-induced nucleation and condensation nuclei in the Earth's atmosphere. Understanding climate and the emergence of life on Earth thus requires knowledge of the natural variations of the radiative and particle fluxes received from the Sun. The main scientific questions are:

- What is the influence of solar radiation on the origin of life on Earth and possibly elsewhere in the Solar System and the Universe?
- How does the solar emission vary at different timescales, from seconds to centuries and millennia?
- What is the spectral dependence of the solar radiative variability?
- How does the magnetic field affect the radiative and particle emission of the Sun?
- How does solar variability affect the climate and life on Earth?

4.6.2 Challenges and opportunities

A major challenge is to get a better understanding of the environmental conditions reigning on the Earth when life emerged, including the physical characteristics of the Sun and their influence of Earth's early environment. Future missions to search for life on icy satellites, such as the NASA Europa Clipper flyby mission (2024) and the ESA JUICE flyby mission (2023) to Jupiter's moon Europa, and proposed missions to Saturn's moon Enceladus, will hopefully provide us with some indications as to whether life emerged on these satellites, vital information to understand the origin of life in general, and the exact role of solar radiation in prebiotic processes. It is now generally accepted that the recent climate change on Earth has been influenced by human activity. It is also clear that continued emissions of greenhouse gasses in the atmosphere will cause further warming and changes in all components of our climate system. Therefore, improving our knowledge of the solar radiative and particle emission and of the mechanisms behind their changes is fundamental to advance our understanding of the climate system on Earth as well as to improve the accuracy of models predicting future climate scenarios.

4.6.2.1 The influence of radiation on environmental conditions and habitability of the early Earth

Environmental conditions and therefore habitable conditions on the early Earth were dependent on a complex interplay between the early evolution of the Sun and the surface volatile envelope of the early Earth. Solar evolution must have played an important role, controlling atmospheric chemistry and temperatures and therefore the existence and state of water at the surface. It also provided the energy for certain critical prebiotic molecular reactions and may even have destroyed some molecular species. Pre-ozone levels of UV flux to the Earth's surface certainly affected processes and early metabolisms. The evolution of the Sun throughout the early period of geological history when life was getting a foothold therefore contributed to the early evolution of life. Thus, the emergence and early evolution of life must have depended a lot on the Sun and its physical environment.

4.6.2.2 Habitability

It is useful here to begin with a brief discussion of the concept of habitability and the Habitable Zone (HZ) around a star (see figure 4.21). Originally defined as the zone in which liquid water exists at the surface of a planet [442], it is now understood that other parameters, such as cloud cover, need to be taken into account and that the habitable zone can be extended. Indeed, with respect to icy satellites and exoplanets, the concept changes. Here, bodies of liquid water below icy crusts are maintained by internal planetary processes and tidal heating due to the gravitational resonance of the icy satellite with the main planet. These bodies are far outside the traditional HZ. The timing of, and the processes leading to the emergence of life on Earth were

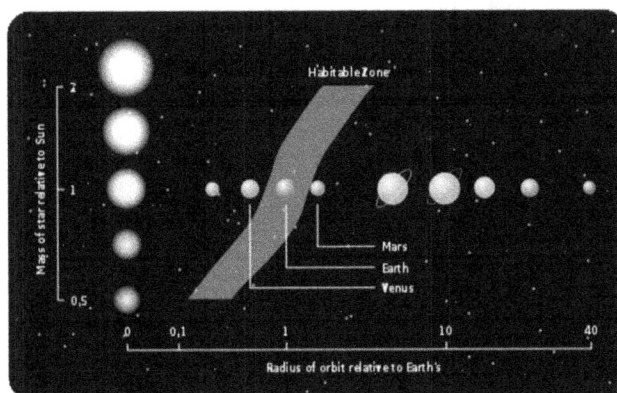

Figure 4.21. Classical view of the Habitable Zone predicated on the existence of liquid water at the surface of a planet. The planets of the Solar System, with the exception of Earth, are either too hot or too cold. This image has been obtained by the authors from the Wikipedia website, where it is stated to have been released into the public domain. It is included within this article on that basis.

predicated on the initial establishment of habitable conditions, namely the presence of liquid water at relatively low temperatures less than 120 °C, the presence of prebiotic molecules, and the availability of energy sources. However, prior to the installation of these initial conditions, the Earth had already undergone a rather violent history. Understanding when exactly the Earth became habitable is challenging because we are lacking a well-preserved rock record covering the first billion year's history of the planet: The earliest crust was thoroughly recycled through tectonic processes and impacts during the Hadean (4.56–4.0 Ga, 1 billion years ago = 1 giga annum = 1 Ga) and the Eo-Palaeoarchaean (4.0–3.2 Ga). It is thus necessary to base our understanding on theoretical considerations coming from modeling, comparative planetology, and the use of any information on the oldest, well-preserved rocks on Earth. They date back to 3.5 Ga, i.e., one billion years after the consolidation of the planet.

4.6.2.3 Early atmosphere(s) and oceans

Life arose within the ephemeral envelope of liquid and gaseous volatiles that enclosed a solid crust and mantle of silicate rocks, which in turn enclosed a core of solid and liquid iron, nickel, sulphur, and other elements, all fractionated out during the formation of the planet. The Earth consolidated from planetesimals at about 4.56 Ga. Initially heated by short-lived radio nucleotides, such as ^{28}Al, the early magma ball slowly cooled down, degassing a first, mainly H_2 atmosphere [443]. At this stage, too hot and without liquid water, the Earth was uninhabitable. One or more Moon-forming impacts reduced the Earth again to a magma ocean, either totally or at least partially, the last of which occurred between 4.9 and 4.43 Ga [444] or about 4.45 Ga according to [445]. The magma ocean rapidly crystallized out within at most a million years (see, e.g. [446]), outgassing a short-lived steam atmosphere of H_2O and CO_2 originating from an oxygenated mantle, the CO_2 outgassing earlier because of its lower solubility in magma, and H_2O later because of its higher solubility [447]. Oxygenation of the early mantle, as deduced from chondritic D/H ratios, was critical to the composition of the outgassed atmosphere. A reduced mantle (i.e., low in oxygen) would have produced an atmosphere rich in atomic hydrogen that would have been more rapidly lost to space due to interactions with the solar wind and the radiation conditions [448], whereas a CO_2-rich atmosphere would result in IR cooling of the upper atmosphere, thus decreasing atmospheric expansion and the risk of erosion by solar processes [447]. It is estimated that the early atmosphere was dense, comprising about 100 bar CO_2 and a couple of hundred bars of H_2O [449], the latter raining out as ocean.

The contribution of volatiles to the Earth's envelope is the subject of much debate. Different hypotheses oscillate between the accretion of the planet from volatile-rich planetesimals and their degassing with minimal importation from later accreted materials [450], to major importation through carbonaceous chondrites, micrometeorites, and interstellar dust particles (IDPs) [451]. The surface of the crust would have needed to be sufficiently cool for the volatiles to condense out into a liquid ocean. Models of the cooling of the crust correlated to the gradual waning of the flux of large impactors that boiled off portions of the early oceans suggest that habitable conditions, i.e., temperatures below 120 °C, could have been present as early as 4.4 Ga [452].

The composition and density of the early atmosphere were crucial to both environmental conditions and the existence of an early ocean. Both were partly influenced by the evolution of the young Sun. Extreme ultraviolet (EUV and XUV) radiation (1–103 nm) was the main source of energy and ionisation in the upper atmosphere, thus influencing atmospheric escape [453]. In particular, the rotation of the young Sun must have affected the flux of XUV coming to the Earth. Lammer *et al* [454] estimate that this flux could have been about 15 times greater than at present for a slowly rotating young Sun, and up to 150 times greater for a fastly rotating young Sun. Both XUV and EUV radiation from the young Sun were therefore higher during the Hadean (4.56–4.0 Ga) but decreased with time, together with x-ray and magnetic activity [455]. However, in what is known as the Faint Young Sun Paradox, the lower luminosity of the young Sun (about 70% lower than today) at the time that it became a main-sequence star meant that during the Hadean/Palaeoarchaean period the Earth must have been at the outer, cold edge of the HZ around the Sun, with frozen water. Nevertheless, evidence exists that this may not have been the case and there are numerous hypotheses concerning the temperature of the early atmosphere, depending on the composition and partial pressures used in the models. While one model indeed suggests that water at the surface of the Earth should have been frozen [456], others suggest various mechanisms for warming up the atmosphere to get a liquid water surface. For example, surface temperatures on the Earth were and still are influenced by absorption of visible and near infrared radiation from the Sun. Absorption of the internal infrared radiation warms the atmosphere and some of this radiation is reflected back down onto the Earth's surface, thus warming it up and contributing to the total radiation energy received from the Sun. On the other hand, the Earth is cooled by emission of thermal infrared radiation, akin to black body radiation [457]. These processes are in thermal equilibrium.

Both CO_2 and H_2O, the so-called greenhouse gases contributing to warming up of the surface of the Earth, are believed to be present in the early atmosphere of the Earth. H_2O is a stronger greenhouse gas than CO_2, absorbing infrared radiation over a wide range of wavelengths [458]. Other greenhouse gases that may have been present include CH_4, which warms the atmosphere far more efficiently than does CO_2. While today most Methane in the atmosphere is of biogenic origin, abiotic production of Methane through serpentinisation reactions of the mafic crust, releasing a small but significant fraction of CH_2 into the atmosphere [459], could have been predominant on the Hadean Earth. The only way CH_4 can be removed from an oxygen-poor atmosphere is through photolysis at wavelengths below 145 nm, thus producing CH_3, CH_2, and CH radicals [460]. This gives an estimated lifetime between 10 000 and 20 000 years for CH_4, much longer than the lifetime of CH_4 in today's oxygenated atmosphere, equal to only 10 years. There is a caveat to high abundances of CH_4 in the atmosphere. If they are too high an organic haze is formed [461], which has a significant albedo effect, reflecting the Sun's radiation back into space and thus cooling the surface.

This albedo, expressing the fraction of solar radiation reflected back into space by the clouds and the surface plays a crucial role in the warming of the Earth. It is determined by the interaction between the variables mentioned earlier and governs the negative feedback of increasing infrared outbound radiation due to increasing temperature that leads to cooler surface temperatures [457]. Moreover, a recent study [462] has added the intriguing hypothesis that tidal heating between the young Moon and the Earth could have contributed to potential heat sources other than the faint young Sun. Ocean condensation from a primordial steam atmosphere may already have occurred around 4.4 Ga [452], but it is difficult to estimate exactly when temperatures became conducive enough to initiate prebiotic processes leading to the emergence of life.

Based on geochemical arguments, the volume of the early oceans is estimated to be between 30% and 50% larger than today, slowly losing mass through H_2O dissociation and hydrogen escape [463, 464]. The larger volume of water probably implied that the early protocontinents were largely submerged. Ocean salinity was likely similar or slightly higher (30%) than modern values [449, 465].

4.6.2.4 Prebiotic chemistry and the emergence of life

The basic initial conditions for prebiotic reactions to occur are the presence of liquid water, organic molecules, and a source (or sources) of energy (figure 4.22). Various catalyzers and other controls must also have been necessary to ensure that the kinetics of the reactions were fast enough and that the reactions went forward, i.e., were irreversible. Kinetic barriers are a requirement to ensure the irreversibility of these reactions. Robert Pascal and Addy Pross have been influential in highlighting the absolute need for such kinetic barriers, and the role of especially high-energy radiation in providing the initial free energy needed to set in motion the prebiotic processes leading to life [466, 467]. Since more than 150 kJ mol^{-1} is necessary to fuel

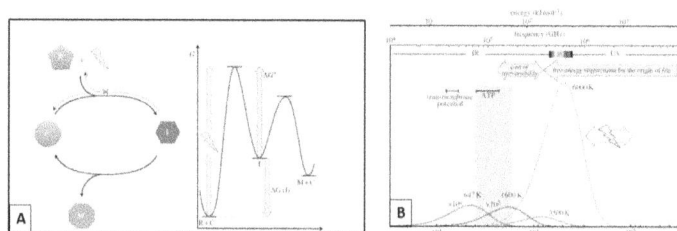

Figure 4.22. Kinetic energy barriers and sources of energy for the initiation of prebiotic reactions. (A) (left) Driving a catalytic cycle (R, reactant; C, catalyst; I, intermediate; M, downstream metabolite) to proceed unidirectionally by coupling with an energy source. (right) Irreversibility requires the waste of an amount of free energy corresponding to the kinetic barrier of the reverse reaction ($\Delta G^{\neq}$). (B) Comparison of different sources of energy available in planetary environments: electromagnetic radiation (correspondence with frequency and wavelength in abscissa), thermal energy (black body radiation curves displaying spectral radiance in ordinate: at 647 K, the critical point of water, blue line; 1600 K, representing typical Hadean magma temperatures red line; and 3500 or 6000 K, dark and light orange lines, surface temperatures of examples of M-stars or G-stars as the Sun, respectively) and lightning ($T \geqslant 10^4$ K). [467]

initial reactions, these authors point out that such high energy could not have been provided by the geothermal sources traditionally proposed as the source of energy for prebiotic reactions. Hadean lavas at 1600 °C would have produced only 70 kJ mol^{-1}. On the other hand, three other processes exist that could produce enough energy: (1) thermal energy at ~3600 Kelvin produced by shock metamorphism caused by impacts; (2) lightning in the atmosphere (a bolt of lightning can reach 29 725 °C); and (3) physical energy produced by photochemical processes at wavelengths of about 800 nm (150 kJ mol^{-1}). The latter is the dominating wavelength of radiation emanating from main-sequence G stars. The amount of 150 kJ mol^{-1} is the minimal energy needed to initiate self-organization of molecules in a manner that could not be reversed.

Photochemical dissociation of CO_2 and CH_4 in the atmosphere fueled by UV and lightning can yield organic molecules, such as HCN and CH_2O [468]. Strecker's synthesis of organic molecules in an experiment reproducing an early, reduced atmosphere of H_2S, CH_4, NH_3 and CO_2 subjected to lightning (spark discharge) was the first major prebiotic experiment undertaken [34]. Smallprim organic molecules were also formed through Fischer–Tropsch processes in the crust during the alteration of the ultramafic (rich in Fe and Mg) crustal rocks by through-flowing hydrothermal fluids that formed serpentine. This produced Methane, longer-chain alkanes, and more complex organic molecules [469, 470]. It has been suggested that organic matter may also have been degassed from the planetesimals forming the Earth [471, 472]. Nevertheless, the most important flux of organic matter to the early Earth must have come from space in the form of the carbonaceous meteorites (carbonaceous chondrites, CC), micrometeorites, and interplanetary dust particles (IDPs) that also contributed volatiles to the Earth's atmosphere and oceans [450, 473]. These materials can contain a huge variety of molecular compounds; for example, reference [474] estimated that more than 14 000 compounds exist in the Murchison meteorite.

Layers of impact spherulites from the early terranes in South Africa and in the Pilbara in Australia (3.5–3.2 Ga) confirm the continuing and significant flux of impactors during the Palaeoarchaean era (3.6–3.2 Ga, [475, 476]), but it is only recently that evidence for the presence of extraterrestrial carbonaceous matter found in these sediments has been documented [477]. This observation is remarkable because it is the first time that extraterrestrial carbonaceous matter has been found in terrestrial sediments and also the oldest evidence (3.33 Ga). Interestingly, micrometeorites have been proposed to be the carriers of this ancient extraterrestrial carbonaceous matter.

The role of high-energy impacts has been invoked by reference [478] as an important source of energy with respect to the possible emergence of life on other planets, such as Mars and even exoplanets. The most important supply of the required high energies came from solar radiation, and this has important implications for the kinds of environments in which life could have emerged, either in the sea or on exposed land. In favor of the latter scenario, the emergence of life on land [479] implies that photochemically formed molecules underwent concentration and complexification in ponds, eventually becoming living cells, before being transported by rivers to the sea, which the cells then colonized. Other suggested favorable

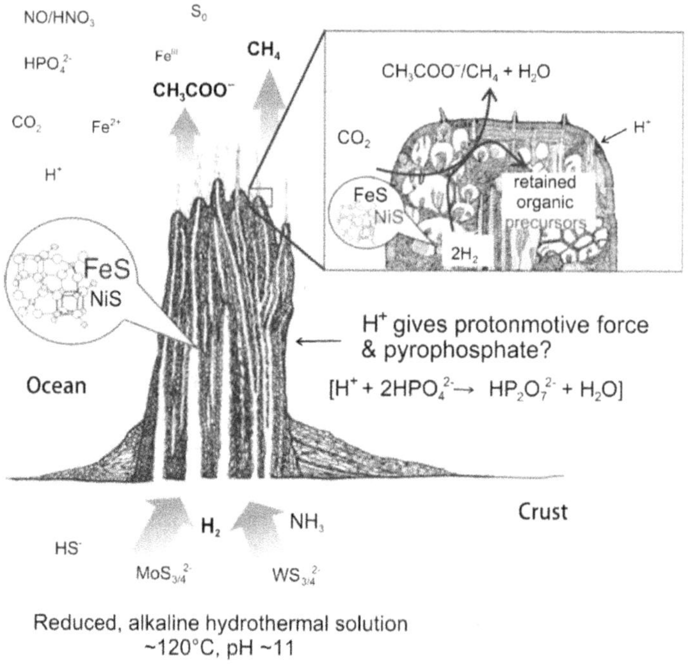

Figure 4.23. The emergence of protolife in a hydrothermal vent [482] John Wiley & Sons (2010).

environments for prebiotic chemistry and the emergence of protocells and of the first living entities include metal-rich, alkaline hydrothermal vents (see figure 4.23) [469, 480–482], or hydrothermal sediments [483].

What is of interest here is the relationship between the proposed locations for the emergence of life and the flux of solar radiation during the Hadean Earth. As was noted above, high-energy radiation is necessary to initiate prebiotic processes to overcome a kinetic barrier that is high enough to prevent reversibility of the chemical reactions. Without free oxygen in the atmosphere and without an ozone layer, radiation could reach the surface of the Earth with few obstacles. Some attenuation could have occurred due to volcanic and impact dust in the atmosphere, but these episodes would have been sporadic and localized, although supervolcano eruptions can eject significant quantities of dust into all levels of the atmosphere, remaining there for many years. We also noted above the possibility of photo-chemistry resulting in an organic haze if the atmospheric CH_4 concentrations were too high. Obviously, energetic radiation, especially at wavelengths below 103 nm and providing energies greater than 100 kJ mol^{-1}, would have illuminated exposed landmasses, and this is one of the reasons why prebiotic processes may have occurred in such environments on land. However, landmasses on the Hadean Earth must have been restricted to sparse emerged volcanic edifices, such as Iceland or Kerguelen today. Indeed, the prevailing hypothesis is that early continents resembled submerged oceanic plateaus, like the Kerguelen Islands today.

Unfortunately, the lack of preservation of Hadean crust excludes a better under-standing of the geological nature of the early Earth as the habitat for the emergence of life. However, if the Palaeoarchaean Earth is, to a certain extent, a representative of the Hadean Earth in this respect, we can make some hypotheses about the environments necessary for the emergence of life, also taking into account the influence of radiation.

As a result of tectonic cycling and impact 'gardening' of the early Earth, its geological record is very patchy and restricted to a few remnants, and even fewer well-preserved remnants), of volcanic and sedimentary rocks that were formed on top of the early protocontinents. There is little evidence for sediment deposition at depths much beyond 100 m or so, while much evidence exists for the deposition in shallow waters, in an offshore to littoral setting. Also, rare deposits of subaerial sedimentation exist in the form of alluvial fans. The question is to what depth the high-energy radiation, necessary for initiating prebiotic chemistry, could penetrate in the early oceans? Radiation penetration in seawater is dependent on wavelength, as well as on the transparency of the waters. During eruption and impacts, large amounts of detritus would decrease transparency, but these events would have been relatively spaced out in time. There would have been periods in between such events that may have been relatively long (perhaps 10^4–10^6 years), when little detritic sedimentation existed in the shallow seas and lagoons, but these were also periods of quieter hydrothermal activity. Nevertheless, a certain fraction of the high-energy radiation required for initial prebiotic processes is strongly attenuated in water. Wavelengths between 200 and 1000 nm (i.e., mid-UV to visible light) are absorbed in the upper 1 cm of the ocean, while higher and lower wavelengths can penetrate to deeper depths. In practice, the longer wavelengths of visible light can penetrate to depths of roughly 50 m in seawater.

Does this mean that locations such as hydrothermal vents and their associated sediments were not conducive to initiating prebiotic chemistry? This is not necessarily the case. Organic molecules could have rained down through the atmosphere to the surface of the Earth, both to ocean and exposed landmasses. At the surface of the ocean, soluble organic matter could have been dissolved and be reprocessed by photochemistry in the upper layers of the water. The products would then sink through the water column to the seafloor and be incorporated into the chemical processing taking place in and around hydrothermal vents. Note that these vents occurred at all depths on the sea floor, including the very shallow near-shore and littoral regions.

While high-energy radiation is apparently required for initiating prebiotic chemistry [467, 479], it is also destructive to organic molecules, especially those that are compositionally and structurally more complex, as was demonstrated by exposure experiments carried out in space and in the laboratory [471, 484]. Thus, at a certain point in the chemical processes leading to the emergence of life, the molecular building bricks of life needed to be protected from radiation. Environments below the penetration depth of UV or in subaqueous sediments or low-temperature hydrothermal vents would be protected. Subaerial environments, on the other hand, would not provide such protection. Since wetting/drying cycles are one of the ways to concentrate prebiotic molecules and promote their incorporation (e.g., RNA inside lipid vesicles), the radiation flux could be a severe hindrance to the survivability of the molecules (see, e.g., [485]). One point of interest regarding the effects of the fainter young Sun is that, while glaciated conditions have been proposed for the Hadean period because of the uncertainties regarding the warming of the early atmospheres, the ancient sedimentary horizons demonstrate absolutely no evidence for glaciated conditions. The formations preserved were largely deposited in shallow waters close to the coastline. These would have been the first areas to be glaciated. On the contrary, there is ample evidence for the presence of liquid water, with waves and storms effortlessly interacting with the underlying sediments [483]. Indeed, volcanic activity accompanied by abundant hydrothermal activity ensured that, at the rock/water interface, temperatures must have been rather warm, up to 70 °C [486].

4.6.2.5 *How does the solar emission vary on different timescales?*

A specific feature of the Sun's internal dynamo and the associated magnetic activity is the modulation of its output. This modulation manifests itself on many different timescales, with the dominant one being the 11-year Schwabe or sunspot cycle. The 400 year long international sunspot number record represents the longest direct time series of solar activity indices available to us and provides an important insight into the existence of some longer term cyclic variations. Indirect methods such as mass spectrometry now also allow us to infer levels of solar activity throughout the duration of the Holocene using the measurement of ^{14}C and ^{10}Be isotopes inside tree rings and ice cores. These records have enabled the identification of longer but more intermittent cyclic variations. These longer cycles include the Gleissberg cycle (80–150 years), the de Vries/Suess cycle (210 years), and the Hallstatt cycle, a quasi-periodic cycle of 2000–2400 years. In general, these datasets indicate that during the last eleven millennia, the Sun has spent most of its time at moderate activity levels, between 15%–20% in so-called Grand minima and around 10%–15% of its time in Grand maxima [487]. The Grand minima appear to be the result of a stochastic and distinct dynamo mode, persisting for both short and longer periods, while the origin and frequency of Grand maxima remains a topic of considerable debate [487]. On shorter timescales a quasi-biennial (2-year) oscillation in various solar activity proxies has also been identified by many authors [488, 489] as well as in helioseismic frequency measurements, that reveal an oscillation with 11 years of periodicity [490].

A direct consequence of these cycles of magnetic activity is the episodic release of stored magnetic energy and its conversion into other forms such as kinetic energy, accelerated particles, electromagnetic radiation, and heat. This energy release manifests itself observationally as solar flares, during which rapid and localized enhancements across the electromagnetic spectrum occur, accompanied by the acceleration of charged particles. The largest enhancements occur in the UV, EUV, and x-ray regions of the spectrum and since these photons travel at the speed of light their impacts are felt rapidly (8 min later) inside the Earth's magnetosphere, ionosphere, and upper atmosphere. The frequency of flares follows well, though not perfectly, the sunspot cycle, and some of the largest flares of previous solar cycles occurred during the declining phases of activity. The reason for this has been an open question for many years, and is still not well understood.

Often associated with some flares, but sometimes occurring with no obvious flares or other signatures at all, are coronal mass ejections (CMEs). CMEs are 'bubbles' of plasma and magnetic fields sporadically expelled from the Sun's atmosphere. In contrast to the flares that often accompany them, their effects on the solar surface are typically global and their speeds range from a few hundred km s^{-1} to over 2000 km s^{-1} for the fastest events. While this is significantly slower than the speed of light, one of the key properties of CMEs is their speed relative to the ambient solar wind. If their speed is larger, they can drive shock waves that accelerate ambient charged particles to high energies, increase their global extent, and the magnetic field inside the bubbles. In particular, the orientation of the CME magnetic field with respect to the Earth's magnetic field is critical in determining the scale of effects within the magnetosphere, the ionosphere, and the upper atmosphere. When a wide and fast CME with Southward directed internal magnetic field arrives at Earth, several conditions may be satisfied for the onset of a geomagnetic storm. The magnetosphere is pushed inward and the position of the magnetopause (the so-called stand-off distance where inward pressure of solar wind and outward magnetic pressure of the Earth compensate) is reduced. The interaction of oppositely directed magnetic fields leads to a process called reconnection, driving large field-aligned currents that open up the magnetic field at the polar cap, thereby allowing energetic particles from the CME and solar wind to enter the atmosphere and to interact with ambient particle species. The combination of all of these effects on the environment near Earth is described by the term 'space weather'. The strength of the ensuing geomagnetic storm following the impact of a CME depends not only on the properties of the CME, but also on the Earth's magnetic field and on the conditions within the magnetosphere. This significantly complicates making a reliable prediction for the level of likely disruption, requiring detailed understanding of the coupling of many scales between the site of reconnection in the solar atmosphere and the Earth. For our digitally reliant and globally connected societies, a large space weather event represents a genuine risk of significant disruption to both ground- and space-based infrastructures, with associated safety, security, and economic implications. For this reason, space weather features on the national risk registers of many nations. Space weather is discussed in more detail in Chapter 5.

In addition to the solar energetic particles which are accelerated during flares and CMEs, the Earth is, and has always been continuously bombarded by cosmic rays that are produced outside our Solar System with typically much higher energy than the particles produced by the Sun. The flux of cosmic rays that reaches the Earth is, however, modulated by the solar magnetic field with the highest cosmic ray fluxes reaching Earth during periods when the solar magnetic field is weakest. The variation in galactic cosmic ray (GCR) flux depends also on the polarity of the global solar field and exhibits a clear lag with respect to sunspot number, which is also polarity dependent.

Stellar irradiance is a key component in almost every step in the formation and evolution of planetary systems, even at the very earliest stages. For example, Gudel [491] describes how the interaction between the circumstellar gas and dust disk during star formation influences the disk structure and subsequent planetary formation, and how strong temperature gradients induced by x-ray heating of the disk can produce the complex chemical conditions that are required for the formation of planets and planetary atmospheres. The current spectral distribution of solar irradiance is dominated by wavelengths in the optical and infrared regions of the electromagnetic spectrum [492]. The total solar irradiance of the early Sun is estimated to be about 75% of the current value [493], but with a significantly higher UV component. The radiation that reaches the Earth's surface also depends strongly on atmospheric attenuation, highlighting the importance of the coupling between external radiation and atmospheric composition and evolution. Today, atmospheric attenuation provides us considerable protection from harmful UV radiation. All UVC (<280 nm), most UVB (280–320 nm), and some UVA (320–400 nm) is blocked by the existence of a stratospheric ozone. However, this was not the case for early Earth, which had a rather different atmospheric chemistry and no Ozone.

Today it is well established that the total solar irradiance (TSI) varies by only about 0.1% over the period of the sunspot cycle, but evidence accumulated over the last 25 years indicates that that 0.1% total variation masks much larger wavelength-dependent changes as large as 10% in the UV [494]. As we have seen from the previous discussion on the emergence of life, variations in UV radiation can have significant impacts on radiative heating, on the production of ozone, and on other chemical reactions in the middle atmosphere.

Our understanding of stellar evolution, coupled with observational evidence from solar mass stars (or solar analogues), such as T-Tauri stars, suggests that in its pre-main sequence and early main-sequence evolutionary phase our Sun must have rotated much more rapidly than it does today. The consequence of that rapid rotation was a much stronger internal magnetic dynamo which would have led to increased magnetic activity. The manifestations of that higher magnetic activity would likely have included larger sunspots covering more of the solar disk, increased frequency of flares and CMEs, and a stronger solar wind. Additional evidence to support the presence of elevated solar activity comes from mineralogical studies, which can only be explained by much higher fluxes of energetic protons bombarding the Earth in its early evolution.

How can we tell from the study of solar analogues what the past evolution of our Sun looked like? Soderblom *et al* [495] describe the factors that control the rotation rate for solar mass stars a few 100 Myr into their evolution. They find that the effect of loss of angular momentum induced by a stellar wind on its rotation is primarily determined by age. The rotation rate in turn controls the internal magnetic dynamo and hence the level of magnetic activity. This feature of stellar evolution then provides us with a means to infer the activity history of our own Sun through the study of solar analogues spanning a range of ages back to early main sequence. Such a study is the goal of the 'Sun in Time' program [491, 496]. Gudel [491] also notes that while it is difficult to determine precisely how much more active the young Sun must have been, if it has followed a typical main-sequence behavior, a rotation period that is one order of magnitude shorter than the current period of 25–35 days is likely. This implies an x-ray flux that would be 100 times larger than the present level.

While such studies tell us that the Sun must have been more active in the past, how that activity will evolve in the future is far less understood. Even on relatively short timescales of stellar evolution, the recent extended minimum phase of solar cycle 23 sparked a fierce debate about the likelihood of an impending Grand minimum. It is clear that convection and rotation, and especially differential rotation, are critical to the generation of the global dipolar field, and helioseismic measurements increasingly indicate the additional importance of shear stress both for field amplification and surface dynamo processes [497]. What remains a challenge for all current dynamo models is to accurately reproduce cyclic effects, and since the cyclic variation of magnetic activity correlates with increased radiation and particle output, this is a key area of future research.

4.6.2.6 How does solar variability affect the Earth's climate and life?
The Sun has provided the dominant external energy source for our planet since its formation. On the basis of observations of flaring activity in T-Tauri stars, Canuto *et al* [498] estimated that the young Sun could have emitted 10^4 times more UV radiation in the Archean period than it does today. Meert *et al* [499] argue that this strong UVB radiation was able to reach the Earth's surface before the development of the ozone layer, and must have been a key selection driver in evolutionary terms, favoring the development of organisms capable of burrowing vertically and to develop exoskeletons. UV radiation also promotes vitamin D synthesis and affords protection against some viruses and bacteria, equally important positive evolutionary drivers. Nitrogen is present in all complex biologically important molecules but molecular nitrogen requires 'nitrogen fixation' for those complex compounds to form. High temperatures are required for this fixation to occur which could be created through lightning, shock heating, as well as by solar UV radiation, as discussed by reference [500]. In particular, this work explores how the increased level of solar activity during the Sun's young and rapidly rotating phase could have provided the conditions necessary for this fixation to occur through the increased frequency for the occurrence of so-called 'superflare ejections' that produced elevated levels of high-energy radiation and energetic particles.

One key way in which solar variability continues to affect the Earth is through climate forcing. Historically the only solar forcing term included in climate simulations was total solar irradiance (TSI). However, the growing recognition of the importance of spectral solar irradiance (SSI) variability, as well as energetic particle precipitation, has led to the inclusion of SSI in stratospheric models [501–504]. Solar forcing has additional effects beyond atmospheric heating and ozone production, also influencing the lower atmosphere and oceans, as demonstrated by [502] and more recently by [505]. Matthes *et al* [466] note that in addition to volcanic activity, solar variability is a central external driver of climate variability, and that the relative stability of the sunspot cycle may provide a useful tool for improving the reliability of climate predictions on timescales as long as decades. While the effect of energetic particles produced during solar flares, CMEs, and galactic cosmic rays (GCRs) on ionization levels in the ionosphere and subsequent changes in chemical composition is well established and relatively well understood [506], a significant controversy remains about the influence of energetic particles on cloud production. The process of aerosol formation seeded by energetic ions has been demonstrated in the lab, but Dunne *et al* [507] find that the connection between GCRs and cloud production is not very strong.

4.6.3 The grand challenges

From the above, it is clear that our understanding of the early environment of Earth, the environment(s) in which life must have emerged, is far from being well-defined. One of the greatest problems is the lack of a well-preserved crust older than 3.5 Ga. The high degree of metamorphic alteration of older crustal fragments from Greenland and Canada complicate deciphering of the environmental record, although new *in situ* techniques and new discoveries of geochemical proxies for environmental signatures are starting to overcome the problem of metamorphic overprint. New areas of very ancient crust may yet be discovered under the Antarctic ice cover, although the scenario of investigating emerging life at places with molten ice is unpopular from the point of view of global warming. Perhaps other deposits hosting reworked ancient zircon crystals dating back to more than 4 Ga may be brought to light in the future. Geochemical signatures in ancient zircons have been used as proxies to infer the presence of relatively low-temperature hydrothermal fluids and hydrated crust, taken as evidence for the existence of oceans at already 4.3 Ga [508, 509], but it was later recognized that the zircons were actually much younger, and date from the Eo-Palaeoarchaean [510].

One important question is, to what extent can we use the well-preserved Palaeoarchaean rocks as a proxy for the Hadean? Did protocontinents exist in the Hadean, as in the Palaeoarchaean? On the basis of the abundant inherited zircon crystals supposedly inherited from the Hadean era [511], it was originally believed that they would have been common, but the redating of the zircons showed them to be Eo-Palaeoarchaean in age and not Hadean. The presence of a felsic crust, i.e., a fractionated crust formed by water interaction with an ultramafic, basaltic so-called protocontinental crust, is testified by geochemical signatures in younger, Eo-Palaeoarchaean rocks [512]. It may not be possible to realistically determine how abundant the Hadean protocontinents were. Only very rare enclaves of such crust have survived even to the Palaeoarchaean. The presence or not of protocontinents and exposed landmasses is of relevance for early prebiotic chemistry and, to a large extent, to the emergence of life. What was the influence of radiation on subaerial environments in the Hadean and how did it affect prebiotic chemistry and the emergence of life?

The cooling of the Earth after the last, giant Moon-forming impact, the timing of the outgassing of the atmosphere, the condensation of the oceans, the heat flux from the mantle, and the early solar emissivity are all closely interlinked. On the early Earth there was a more or less continuous influx of extraterrestrial organic matter, as well as small organic molecules produced in the chemical reaction of hydrothermal fluids with crustal rocks by the chemical Fischer–Tropsch process. Furthermore, organic molecules were formed in the atmosphere by chemical Strecker synthesis. After the Moon-forming impact, and the subsequent ubiquitous volcanic and hydrothermal activity as the Earth was cooling down, the earliest moment when prebiotic chemistry could have been initiated must have been when oceans condensed and when water temperatures became colder than 120 °C. As we have argued above, the temperature must also have been controlled by the composition of the atmosphere. For example, what was the partial pressure of CO_2 during the Hadean, what were the greenhouse gases? Was there an organic haze because of photo-dissociation of CH_4 in the atmosphere? What was the amount of cloud cover and its effects on albedo? This information can be approached through modeling and comparative planetology and hopefully the use of geochemical proxies for mantle temperature and composition. The same holds for the volatile composition of the early Earth's outer envelope.

There is an active discussion going on concerning the origin of the Earth's volatile layer. Were the planetesimals that formed the Earth rich in volatiles that subsequently outgassed, as suggested by [450], or did a majority of them arrive in a late veneer of extraterrestrial origin [513]? Today it is believed that the Earth's volatile inventory is of mixed origin, but there is much room for improvement in our understanding [514].

These considerations are of enormous relevance to the search for extraterrestrial life. Finding traces of life elsewhere, either on Mars, on icy satellites such as Europa or Enceladus, or on exoplanets using biosignature gaseous combinations, will underline the hypothesis that, under certain environmental conditions, chemistry and physics may natuarally lead to biology, that life is a indeed natural consequence of physics and chemistry, 'vital dust' as was proposed by the Nobel Prize laureate Christian deDuve in 1995. The future NASA Europa Clipper and ESA JUICE missions to Jupiter's moon Europa and planned missions to Saturn's moon Enceladus will go a long way in helping us to understand the processes leading to the emergence of life and the importance of solar radiation for prebiotic chemistry, that is provided they find convincing traces of life.

Today the magnetic field of the Earth protects us from radiation originating from the Sun and beyond, but when did the magnetic field appear? The oldest traces of magnetism on Earth are found in rocks as old as 3.42 Ga from the Barberton Greenstone Belt [515, 516]. Previous interpretations of a Hadean magnetic field based on analysis of zircon crystals [517] have been shown to be erroneous [518]. However, there was already a dynamo on Mars that left its imprint in magnetized rocks dating to 4.2 Ga [519]. Mars did not undergo a magma ocean-forming impact as did the Earth and, thus cooled down more rapidly than the Earth. The initiation of a magnetic dynamo requires that the outer core be in a molten state. The magnetic field of Mars was of short duration and rapidly disappeared, with the core cooling rapidly. This contributed largely to the erosion of the volatile envelope of Mars and its climatic degradation. Could a magnetic field have been initiated on Earth much earlier than 3.5 Ga? This question will have to be addressed by modeling and comparative planetology.

It is clear that variations in solar activity have played, and continue to play, a key role in the evolution of the conditions that support life on Earth. While our ability to accurately predict the strength and duration of the solar cycle has made substantial progress in recent decades, dynamo models still struggle to reproduce the observed asymmetry in sunspot number and duration of each solar cycle. A key element of the regeneration of the global magnetic field during every cycle is the polarity reversal, and our vantage point on the Earth–Sun line has severely restricted our ability to obtain good observations of the polar magnetic field throughout the solar cycle. The extended mission phase of Solar Orbiter, launched in February 2020, will provide important breakthroughs when it begins to leave the ecliptic plane. Given that the solar cycle includes two 11-year sunspot cycles and two polarity reversals, it will be important to ensure the continued availability of such measurements beyond the Solar Orbiter's extended mission.

What also still remains beyond our current reach is the reliable identification of the mechanism that triggers flares and CMEs. This is not just needed to improve space weather forecasting tools. It is the fundamental underpinning science that is needed to understand energy release in magnetized plasmas throughout the Universe as well as in the laboratory, including problems as diverse as plasma confinement in tokamaks, the habitability of exoplanets, and the functioning of blazar jets. Progress in this area requires the ability to reliably measure the magnetic field throughout the atmosphere, particularly inside the corona where the energy is released, and to measure corresponding plasma changes that trace where energy is released and transported. Much like the reliable forecasting of space weather events, one of the challenges here is the complex coupling between different scales, which requires observations at both very high spatial and temporal resolution, and over large fields of view. Simulations, particularly data driven, will also be crucial in this respect.

The uncertainties around the solar forcing of climate is another major challenge, both from the perspective of the uncertainty in the forcing term itself, as well as from that in the uncertainty of the simulated response of the climate. Long-term stable and well-calibrated observations in the UV and shorter wavelengths are needed here in order to improve both model validation and simulations. While there is currently significant disagreement over the importance of GCRs in seeding clouds, the effect of energetic particles on climate is an area that needs further exploration, including the impact of variations in the strength of the solar magnetic field that modulates the GCR flux reaching Earth. But the strength of the solar magnetic field may have broader impact. A study by Lockwood [520] found evidence that extreme winter temperatures in the Northern hemisphere may be the result of a solar influence on the occurrence of jet stream 'blocking' events in the Atlantic. While that is both a regional and seasonal effect, it raises the question of how to quantify the importance of longer-term variations in solar magnetic field strength on climate timescales, both regionally and globally. Finally, the greatest uncertainty in our understanding of solar spectral irradiance is in the UV and at even shorter wavelengths where measurements must be made outside the Earth's atmosphere.

References

[1] Crowe M J 2008 The extraterrestrial life debate *Antiquity to 1915: A Source Book* (Notre Dame, IN: University of Notre Dame Press)
[2] Mifflin M 2009 *The Blue Tattoo: The Life of Olive Oatman* (Lincoln: University of Nebraska Press)

[3] Arcichovsky V M 1912 Auf der Suche nach Chlorophyll auf den Planeten *Ann. Inst. Polytechnique DonCesarevitch Alexis a Novotcherkassk* **1** 195–214

[4] Jeans J 1930 *The Universe Around US* (Cambridge: Cambridge University Press)

[5] Lowell P 1895 Evidence of a twilight arc upon the planet Mars *Nature* **52** 401–5

[6] Lowell P 1906 First photographs of the canals of Mars *Proc. R. Soc.* A **77** 132–5

[7] Sinton W M 1957 Spectroscopic evidence for vegetation on Mars *Astrophys. J.* **126** 231

[8] Sinton W M 1959 Further evidence of vegetation on Mars *Science* **130** 1234–7

[9] Rea D G, O'Leary B T and Sinton W M 1965 Mars: the origin of the 3.58- and 3.69-micron minima in the infrared spectra *Science* **147** 1286–8

[10] McCauley J F, Carr M H, Cutts J A, Hartmann W K, Masursky H, Milton D J, Sharp R P and Wilhelms D E 1972 Preliminary Mariner 9 Report on the geology of Mars (A 4. 3) *Icarus* **17** 289–327

[11] Biemann K *et al* 1976 Search for organic and volatile inorganic compounds in two surface samples from the Chryse Planitia region of Mars *Science* **194** 72–6

[12] Glavin D P, Schubert M, Botta O, Kminek G and Bada J L 2001 Detecting pyrolysis products from bacteria on Mars *Earth Planet. Sci. Lett.* **185** 1–5

[13] dos Santos R, Patel M, Cuadros J and Martins Z 2016 Influence of mineralogy on the preservation of amino acids under simulated Mars conditions *Icarus* **277** 342–53

[14] Fornaro T, Boosman A, Brucato J R, ten Kate I L, Siljeström S, Poggiali G, Steele A and Hazen R M 2018 UV irradiation of biomarkers adsorbed on minerals under Martian-like conditions: hints for life detection on Mars *Icarus* **313** 38–60

[15] Garry J R C, ten Kate I L, Martins Z, Nørnberg P and Ehrenfreund P 2006 Analysis and survival of amino acids in Martian regolith analogs *Meteorit. Planet. Sci.* **41** 391–405

[16] Laurent B, Cousins C R, Pereira M F C and Martins Z 2019 Effects of UV-organic interaction and Martian conditions on the survivability of organics *Icarus* **323** 33–9

[17] Poch O, Noblet A, Stalport F, Correia J J, Grand N, Szopa C and Coll P 2013 Chemical evolution of organic molecules under Mars-like UV radiation conditions simulated in the laboratory with the 'Mars organic molecule irradiation and evolution' (MOMIE) setup *Planet. Space Sci.* **85** 188–97

[18] Poch O, Kaci S, Stalport F, Szopa C and Coll P 2014 Laboratory insights into the chemical and kinetic evolution of several organic molecules under simulated Mars surface UV radiation conditions *Icarus* **242** 50–63

[19] Poch O, Jaber M, Stalport F, Nowak S, Georgelin T, Lambert J-F, Szopa C and Coll P 2015 Effect of nontronite smectite clay on the chemical evolution of several organic molecules under simulated Martian surface ultraviolet radiation conditions *Astrobiology* **15** 221–37

[20] Stoker C R and Bullock M A 1997 Organic degradation under simulated Martian conditions *J. Geophys. Res.* **102** 10881–8

[21] ten Kate I L, Garry J R C, Peeters Z, Quinn R, Foing B and Ehrenfreund P 2005 Amino acid photostability on the Martian surface *Meteorit. Planet. Sci.* **40** 1185–93

[22] ten Kate I L, Garry J R C, Peeters Z, Foing B and Ehrenfreund P 2006 The effects of Martian near surface conditions on the photochemistry of amino acids *Planet. Space Sci.* **54** 296–302

[23] Crandall P B, Góbi S, Gillis-Davis J and Kaiser R I 2017 Can perchlorates be transformed to hydrogen peroxide (H_2O_2) products by cosmic rays on the Martian surface? *J. Geophys. Res.: Planets* **122** 1880–92

[24] Pavlov A A, Vasilyev G, Ostryakov V M, Pavlov A K and Mahaffy P 2012 Degradation of the organic molecules in the shallow subsurface of Mars due to irradiation by cosmic rays *Geophys. Res. Lett.* **39** L13202

[25] Benner S A, Devine K G, Matveeva L N and Powell D H 2000 The missing organic molecules on Mars *Proc. Natl Acad. Sci.* **97** 2425–30

[26] Hecht M H *et al* 2009 Detection of perchlorate and the soluble chemistry of Martian soil at the phoenix lander site *Science* **325** 64–7

[27] Kounaves S P *et al* 2010 Wet chemistry experiments on the 2007 Phoenix Mars Scout Lander mission: data analysis and results *J. Geophys. Res.* **115** E00E10-1–E00E10-16

[28] Gaudi B S, Seager S, Mennesson B, Kiessling A, Warfield K and Cahoy K *et al* 2020 The habitable exoplanet observatory (HabEx) mission concept study final report *ArXiv E-Prints* arXiv:2001.06683 (https://jpl.nasa.gov/habex/pdf/HabEx-Final-Report-Public-Release.pdf)

[29] National Academies of Science 2018 *Exoplanet Science Strategy* (Washington, DC: National Academies Press)

[30] Haldane J B S 1929 The origin of life *Ration. Ann.* **148** 3–10

[31] Haldane J B S 1954 The origin of life *New Biol.* **16** 12–27

[32] Oparin A I 1924 Proiskhodenie Zhizni. Moscoksky Rabotichii, Moscow translated by A S Bernal 1967 *The Origin of Life* In: R Carrington *The Origin of Life* (London: Weidenfeld and Nicolson) pp 199–234

[33] Urey H C 1952 *The Planets, Their Origin and Development* (New Haven: Yale University Press)

[34] Miller S L 1953 A production of amino acids under possible primitive Earth conditions *Science* **117** 528–9

[35] Miller S L and Urey H C 1959 Organic compound synthesis on the primitive Earth *Science* **130** 245–51

[36] Kasting J F, Eggler D H and Raeburn S P 1993 Mantle redox evolution and the oxidation state of the Archean atmosphere *J. Geol.* **101** 245–57

[37] Kasting J F and Catling D 2003 Evolution of a habitable planet *Annu. Rev. Astron. Astrophys.* **41** 429–63

[38] Olson J M 2006 Photosynthesis in the Archean era *Photosynth. Res.* **88** 109–17

[39] Walker J C G 1986 Carbon dioxide on the early Earth *Origins Life* **16** 117–27

[40] Zahnle K, Schaefer L and Fegley B 2010 Earth's earliest atmospheres *Cold Spring Harb. Perspect. Biol.* **2** a003467

[41] Cleaves H J, Chalmers J H, Lazcano A, Miller S L and Bada J L 2008 *Orig. Life Evol. Biosph.* **38** 105–15

[42] Plankensteiner K, Reiner H and Schranz B *et al* 2004 Prebiotic formation of amino acids in a neutral atmosphere. by electric discharge *Angew. Chem. Int. Ed. Engl.* **43** 1886–8

[43] Schlesinger G and Miller S L 1983 Prebiotic synthesis in atmospheres containing CH_4, CO and CO_2. II. Hydrogen cyanide, formaldehyde and ammonia. *J. Mol. Evol.* **19** 383–90

[44] Schlesinger G and Miller S L 1983 Prebiotic synthesis in atmospheres containing CH_4, CO, and CO_2. I. Amino acids *J. Mol. Evol.* **19** 376–82

[45] Stribling R and Miller S L 1987 Energy yields for hydrogen cyanide and formaldehyde syntheses: the HCN and amino acid concentrations in the primitive ocean *Orig. Life Evol. Biosph.* **17** 261–73

[46] Furnes H, Banerjee N R, Muehlenbachs K, Staudigel H and de Wit M 2004 Early life recorded in archean pillow lavas *Science* **304** 578–81

[47] Javaux E J, Marshall C P and Bekker A 2010 Organic-walled microfossils in 3.2-billion-year-old shallow-marine siliciclastic deposits *Nature* **463** 934–8

[48] Mojzsis S J, Arrhenius G, McKeegan K D, Harrison T M, Nutman A P and Friend C R L 1996 Evidence for life on Earth before 3,800 million years ago *Nature* **384** 55–9

[49] Rosing M T 1999 ^{13}C-Depleted carbon microparticles in >3700-Ma sea-floor sedimentary rocks from west Greenland *Science* **283** 674–6

[50] Rosing M T and Frei R 2004 U-rich Archaean sea-floor sediments from Greenland—indications of >3700 Ma oxygenic photosynthesis *Earth Planet. Sci. Lett.* **217** 237–44

[51] Schidlowski M 1988 A 3,800-million-year isotopic record of life from carbon in sedimentary rocks *Nature* **333** 313–8

[52] Sugitani K, Grey K and Nagaoka T *et al* 2009 Taxonomy and biogenicity of Archaean spheroidal microfossils (ca. 3.0 Ga) from the Mount Goldsworthy–Mount Grant area in the northeastern Pilbara Craton, Western Australia *Precambrian Res.* **173** 50–9

[53] Wacey D, McLoughlin N and Whitehouse M J *et al* 2010 Two coexisting sulfur metabolisms in a ca. 3400 Ma sandstone *Geology* **38** 1115–8

[54] Westall F *et al* 2006 *Processes on the Early Earth* ed W U Reimold and R L Gibson (Boulder, CO: Geological Society of America) pp 105–31

[55] Westall F *et al* 2011 Implications of in situ calcification for photosynthesis in a ~3.3 Ga-old microbial biofilm from the Barberton greenstone belt, South Africa *Earth Planet. Sci. Lett.* **310** 468–79

[56] Westall F *et al* 2011 Volcaniclastic habitats for early life on Earth and Mars: a case study from ~3.5 Ga-old rocks from the Pilbara, Australia *Planet. Space Sci.* **59** 1093–106

[57] Westall and Matthews, *Grand Challenges 2050* ch 1.3.5

[58] Cleaves H J and Lazcano A 2009 *Chemical Evolution II: From Origins of Life to Modern Society* ed L Zaikowski, J M Friedrich and S R Seidel (New York: Oxford University Press) pp 17–43

[59] Miller S L and Bada J L 1988 Submarine hot springs and the origin of life *Nature* **334** 609–11

[60] Bains W 2004 Many chemistries could be used to build living systems *Astrobiology* **4** 137–67

[61] Morris R V *et al* 2010 Identification of carbonate-rich outcrops on Mars by the Spirit Rover *Science* **329** 421–4

[62] Poulet F, Bibring J-P, Mustard J F, Gendrin A, Mangold N, Langevin Y, Arvidson R E, Gondet B and Gomez C 2005 Phyllosilicates on Mars and implications for early martian climate *Nature* **438** 623–7

[63] Coradini A, Federico C, Forni O and Magni G 1995 Origin and thermal evolution of icy satellites *Surv. Geophys.* **16** 533–91

[64] Néri A, Guyot F, Reynard B and Sotin C 2020 A carbonaceous chondrite and cometary origin for icy moons of Jupiter and Saturn *Earth Planet. Sci. Lett.* **530** 1–10

[65] Roberts J H and Nimmo F 2008 Near-surface heating on Enceladus and the south polar thermal anomaly *Geophys. Res. Lett.* **35** L09201

[66] Sohl F, Choukroun M, Kargel J, Kimura J, Pappalardo R, Vance S and Zolotov M 2010 Subsurface water oceans on icy satellites: chemical composition and exchange processes *Space Sci. Rev.* **153** 485–510

[67] de Marcellus P, Meinert C, Nuevo M, Filippi J J, Danger G, Deboffle D, Nahon L, d'Hendecourt L L S and Meierhenrich U J 2011 Non-racemic amino acid production by

ultraviolet irradiation of achiral interstellar ice analogs with circularly polarized light *Astrophys. J. Lett.* **727** L27

[68] Meinert C, Filippi J-J, Marcellus P D, D'Hendecourt L L S and Meierhenrich U J 2012 *N*-(2-aminoethyl)glycine and amino acids from interstellar ice analogues *ChemPlusChem* **77** 186–91

[69] Modica P, Meinert C, Marcellus P D, Nahon L, Meierhenrich U J and D'Hendecour L L S 2014 Enantiomeric excesses induced in amino acids by ultraviolet circularly polarized light irradiation of extraterrestrial ice analogs: a possible source of asymmetry for prebiotic chemistry *Astrophys. J.* **788** 79

[70] Modica P, Martins Z, Meinert C, Zanda B and D'Hendecourt L L S 2018 The amino acid distribution in laboratory analogs of extraterrestrial organic matter: a comparison to CM chondrites *Astrophys. J.* **865** 41–51

[71] Muñoz Caro G M, Meierhenrich U J, Schutte W A, Barbier B, Arcones Segovia A, Rosenbauer H, Thiemann W H-P, Brack A and Greenberg J M 2002 Amino acids from ultraviolet irradiation of interstellar ice analogues *Nature* **416** 403–6

[72] Myrgorodska I, Meinert C and Martins Z *et al* 2016 Quantitative enantioseparation of amino acids by comprehensive two-dimensional gas chromatography applied to non-terrestrial samples *J. Chromatogr.* A **1433** 131–6

[73] Brinton K L F, Engrand C, Glavin D P, Bada J L and Maurette M 1998 A search for extraterrestrial amino acids in carbonaceous antarctic micrometeorites *Origins Life Evol. Biosphere* **28** 413–24

[74] Clemett S J, Maechling C R, Zare R N, Swan P D and Walker R M 1993 Identification of complex aromatic molecules in individual interplanetary dust particles *Science* **262** 721–5

[75] Clemett S J, Chillier X D F, Gillette S, Zare R N, Maurette M, Engrand C and Kurat G 1998 Observation of indigenous polycyclic aromatic hydrocarbons in 'Giant' carbonaceous antarctic micrometeorites *Origins Life Evol. Biosphere* **28** 425–48

[76] Dartois E *et al* 2013 Ultra carbonaceous antarctic micrometeorites, probing the Solar System beyond the nitrogen snow-line *Icarus* **224** 243–52

[77] Duprat J *et al* 2010 Extreme deuterium excesses in ultracarbonaceous micrometeorites from Central Antarctic Snow *Science* **328** 742–5

[78] Flynn G J, Keller L P, Feser M, Wirick S and Jacobsen C 2003 The origin of organic matter in the Solar System: evidence from the interplanetary dust particles *Geochim. Cosmochim. Acta* **67** 4791–806

[79] Flynn G J, Keller L P, Jacobsen C and Wirick S 2004 An assessment of the amount and types of organic matter contributed to the Earth by interplanetary dust *Adv. Space Res.* **33** 57–66

[80] Glavin D P, Matrajt G and Bada J L 2004 Re-examination of amino acids in Antarctic micrometeorites *Adv. Space Res.* **33** 106–13

[81] Keller L P, Messenger S, Flynn G J, Clemett S, Wirick S and Jacobsen C 2004 The nature of molecular cloud material in interplanetary dust *Geochim. Cosmochim. Acta* **68** 2577–89

[82] Martins Z 2019 Organic molecules in meteorites and their astrobiological significance *Handbook of Astrobiology* (Boca Raton, FL: CRC Press) pp 177–94

[83] Martins Z and Sephton M A 2009 Extraterrestrial amino acids *Amino Acids, Peptides and Proteins in Organic Chemistry* ed A B Hughes (Weinheim: Wiley-VCH Publishers) pp 3–42

[84] Matrajt G, Pizzarello S and Taylor S *et al* 2004 Concentration and variability of the AIB amino acid in polar micrometeorites: implications for the exogenous delivery of amino acids to the primitive Earth *Meteorit. Planet. Sci.* **39** 1849–58

[85] Matrajt G, Muñoz Caro G M and Dartois E *et al* 2005 FTIR analysis of the organics in IDPs: comparison with the IR spectra of the diffuse interstellar medium *Astron. Astrophys.* **433** 979–95

[86] Martins Z 2020 Detection of organic matter and biosignatures in space missions *Astrobiology: Current, Evolving and Emerging Perspectives* ed A Antunes (Poole: Caister Academic Press) pp 53–74

[87] Martins Z, Chan Q H S, Bonal L, King A and Yabuta H 2020 Organic matter in the Solar System—implications for future on-site and sample return missions *Space Sci. Rev.* **216** 54

[88] Chan M A *et al* 2019 Deciphering biosignatures in planetary contexts *Astrobiology* **19** 1075–102

[89] Des Marais D J, Nuth J A III, Allamandola L J, Boss A P, Farmer J D, Hoehler T M, Jakosky B M, Meadows V S, Pohorille A and Spormann A M 2008 The NASA astrobiology roadmap *Astrobiology* **8** 715–30

[90] Wayne R K, Leonard J A and Cooper A 1999 Full of sound and fury: the recent history of ancient DNA *Annual Rev. Ecol. Syst.* **30** 457–77

[91] Glavin D P, Burton A S, Elsila J E, Aponte J C and Dworkin J P 2020 The search for chiral asymmetry as a potential biosignature in our Solar System *Chem. Rev.* **120** 4660–89

[92] Parnell J *et al* 2007 Searching for life on Mars: selection of molecular targets for ESA's Aurora ExoMars Mission *Astrobiology* **7** 578–604

[93] Olson S L, Reinhard C T and Lyons T W 2016 Limited role for methane in the mid-Proterozoic greenhouse *PNAS* **113** 11447–52

[94] Reinhard C T, Olson S L, Schwieterman E W and Lyons T W 2017 False negatives for remote life detection on ocean-bearing planets: lessons from the early Earth *Astrobiology* **17** 287–97

[95] Schwieterman E W, Kiang N Y, Parenteau M N, Harman C E, DasSarma S and Fisher T M *et al* 2018 Exoplanet biosignatures: a review of remotely detectable signs of life *Astrobiology* **18** 663–708

[96] Arney G, Domagal-Goldman S D and Meadows V S 2018 Organic haze as a biosignature in anoxic Earth-like atmospheres *Astrobiology* **18** 311–29

[97] Domagal-Goldman S D, Meadows V S, Claire M W and Kasting J F 2011 Using biogenic sulfur gases as remotely detectable biosignatures on anoxic planets *Astrobiology* **11** 419–41

[98] Seager S, Schrenk M and Bains W 2012 An astrophysical view of Earth-based metabolic biosignature gases *Astrobiology* **12** 61–82

[99] Des Marais D J, Harwit M O, Jucks K W, Kasting J F, Lin D N C, Lunine J I, Schneider J, Seager S, Traub W A and Woolf N J 2002 Remote sensing of planetary properties and biosignatures on extrasolar terrestrial planets *Astrobiology* **2** 153–81

[100] Seager S, Bains W and Hu R 2013 Biosignature Gases in H_2-dominated atmospheres on rocky exoplanets *Astrophys. J.* **777** 95

[101] Segura A, Kasting J F, Meadows V, Cohen M, Scalo J and Crisp D *et al* 2005 Biosignatures from Earth-like planets around M dwarfs *Astrobiology* **5** 706–25

[102] Sousa-Silva C, Seager S, Ranjan S, Petkowski J J, Zhan Z, Hu R and Bains W 2020 Phosphine as a biosignature gas in exoplanet atmospheres *Astrobiology* **20** 235–68

[103] Zhan Z, Seager S, Petkowski J J, Sousa-Silva C, Ranjan S, Huang J and Bains W 2021 Assessment of isoprene as a possible biosignature gas in exoplanets with anoxic atmospheres arXiv:2103.14228 [astro-ph.EP] *Astrobiology*

[104] Huang J, Seager S, Petkowski J, Ranjan S and Zhan Z 2021 Assessment of ammonia as a biosignature gas in exoplanet atmospheres *Astrobiology* **22** 171–91

[105] Seager S, Bains W and Hu R 2013 A biomass-based model to estimate the plausibility of exoplanet biosignature gases *Astrophys. J.* **775** 104

[106] Grenfell J L 2018 Atmospheric biosignatures *Handbook of Exoplanets* ed H J Deeg and J A Belmonte (Cham: Springer International Publishing) pp 1–14

[107] Kaltenegger L 2017 How to characterize habitable worlds and signs of life *Annu. Rev. Astron. Astrophys.* **55** 433–85

[108] Morowitz H and Sagan C 1967 Life in the clouds of Venus? *Nature* **215** 1259

[109] Seager S, Petkowski J J, Gao P, Bains W, Bryan N C, Ranjan S and Greaves J 2021 The venusian lower atmosphere haze as a depot for desiccated microbial life: a proposed life cycle for persistence of the venusian aerial biosphere *Astrobiology* **21** 1206–23

[110] Cavalazzi B *et al* 2019 The dallol geothermal area, northern Afar (Ethiopia)—an exceptional planetary field analog on earth *Astrobiology* **19** 553–78

[111] Kotopoulou E, Delgado Huertas A, Garcia-Ruiz J M, Dominguez-Vera J M, Lopez-Garcia J M, Guerra-Tschuschke I and Rull F 2018 A polyextreme hydrothermal system controlled by iron: the case of Dallol at the Afar Triangle *ACS Earth Sp. Chem.* **3** 90–9

[112] Limaye S S, Mogul R, Smith D J, Ansari A H, Słowik G P and Vaishampayan P 2018 Venus' spectral signatures and the potential for life in the clouds *Astrobiology* **18** 1181–98

[113] Oyama V I *et al* 1980 Pioneer Venus gas chromatography of the lower atmosphere of Venus *J. Geophys. Res. Sp. Phys.* **85** 7891–902

[114] Mukhin L M *et al* 1982 VENERA-13 and VENERA-14 gas chromatography analysis of the Venus atmosphere composition *Sov. Astron. Lett.* **8** 216–8

[115] Rimmer P B, Jordan S, Constantinou T, Woitke P, Shorttle O, Paschodimas A and Hobbs R 2021 Hydroxide salts in the clouds of Venus: their effect on the sulfur cycle and cloud droplet pH *Planet. Sci. J.* arXiv:2101.08582

[116] Greaves J S, Bains W, Petkowski J J, Seager S, Sousa-Silva C and Ranjan S *et al* 2021 On the robustness of phosphine signatures in Venus' clouds *arXiv preprint* arXiv:2012.05844

[117] Greaves J S, Richards A, Bains W, Rimmer P B, Clements D L and Seager S *et al* 2021 Recovery of spectra of phosphine in Venus' clouds *arXiv preprint* arXiv:2104.09285

[118] Greaves J S *et al* 2020 Phosphine gas in the cloud decks of Venus *Nat. Astron.* **5** 655–64

[119] Mogul R, Limaye S S, Way M J and Cordova J A 2021 Venus' mass spectra show signs of disequilibria in the middle clouds *Geophys. Res. Lett.* **48** e2020GL091327

[120] Surkov Y A, Andrejchikov B M and Kalinkina O M 1973 On the content of ammonia in the Venus atmosphere based on data obtained from Venera 8 automatic station *Akad. Nauk SSSR Dokl.* **213** 296–8

[121] Colin L 1980 The pioneer Venus program *J. Geophys. Res.* **85** 7575

[122] Sagdeev R Z *et al* 1986 The VEGA Venus balloon experiment *Science* **231** 1407

[123] Archer P D *et al* 2014 Abundances and implications of volatile-bearing species from evolved gas analysis of the Rocknest aeolian deposit, Gale Crater, Mars *J. Geophys. Res.: Planets* **119** 237–54

[124] Freissinet C *et al* MSL Science Team 2015 Organic molecules in the Sheepbed Mudstone, Gale Crater, Mars *J. Geophys. Res.: Planets* **120** 495–514

[125] Glavin D P *et al* 2013 Evidence for perchlorates and the origin of chlorinated hydrocarbons detected by SAM at the Rocknest aeolian deposit in Gale Crater *J. Geophys. Res.: Planets* **118** 1955–73

[126] Leshin L A *et al* 2013 Volatile, isotope, and organic analysis of Martian fines with the Mars curiosity rover *Science* **341** 1238937-1–1239

[127] Ming D W *et al* 2014 Volatile and organic compositions of sedimentary rocks in Yellowknife Bay, Gale Crater, Mars *Science* **343** 1245267-1–15

[128] Vago J L *et al* 2017 Habitability on early Mars and the search for biosignatures with the ExoMars Rover *Astrobiology* **17** 471–510

[129] Liuzzi G *et al* NOMAD Team 2019 Methane on Mars: new insights into the sensitivity of CH_4 with the NOMAD/ExoMars spectrometer through its first in-flight calibration *Icarus* **321** 671–90

[130] Mumma M J, Villanueva G L, Novak R E, Hewagama T, Bonev B P, DiSanti M A, Mandell A M and Smith M D 2009 Strong release of methane on Mars in northern summer 2003 *Science* **323** 1041–5

[131] Formisano V, Atreya S, Encrenaz T, Ignatiev N and Giuranna M 2004 Detection of methane in the atmosphere of Mars *Science* **306** 1758–61

[132] Geminale A, Formisano V and Sindoni G 2011 Mapping methane in Martian atmosphere with PFS-MEX data *Planet. Space Sci.* **59** 137–48

[133] Fonti S and Marzo G A 2010 Mapping the methane on Mars *Astron. Astrophys.* **512** A51–7

[134] Fonti S, Mancarella F, Liuzzi G, Roush T L, Chizek Frouard M, Murphy J and Blanco A 2015 Revisiting the identification of methane on Mars using TES data *Astron. Astrophys.* **581** A136-1–A136-11

[135] Webster C R *et al* 2015 Mars methane detection and variability at Gale crater *Science* **347** 415–7

[136] Webster C R *et al* 2018 Background levels of methane in Mars' atmosphere show strong seasonal variations *Science* **360** 1093–6

[137] Korablev O, Vandaele A C and Montmessin F *et al* 2019 No detection of methane on Mars from early ExoMars Trace Gas Orbiter observations *Nature* **568** 517–20

[138] Iess L *et al* 2014 The gravity field and interior structure of Enceladus *Science* **344** 78–80

[139] Thomas P C, Tajeddine R, Tiscareno M S, Burns J A, Joseph J, Loredo T J, Helfenstein P and Porco C 2016 Enceladus's measured physical libration requires a global subsurface ocean *Icarus* **264** 37–47

[140] Postberg F *et al* 2018 Macromolecular organic compounds from the depths of Enceladus *Nature* **558** 564–8

[141] Waite J H *et al* 2006 Cassini ion and neutral mass spectrometer: Enceladus plume composition and structure *Science* **311** 1419–22

[142] Grasset O *et al* 2013 JUpiter ICy moons Explorer (JUICE): an ESA mission to orbit Ganymede and to characterise the Jupiter system *Planet. Space Sci.* **78** 1–21

[143] Kissel J and Krueger F R 1987 The organic component in dust from comet Halley as measured by the PUMA mass spectrometer on board Vega 1 *Nature* **326** 755–60

[144] Cody G D *et al* 2008 Quantitative organic and light-element analysis of comet 81P/Wild 2 particles using C-, N-, and O-µ-XANES *Meteorit. Planet. Sci.* **43** 353–65

[145] Sandford S A *et al* 2006 Organics captured from Comet 81P/Wild 2 by the Stardust Spacecraft *Science* **314** 1720–4

[146] De Gregorio B T, Stroud R M, Cody G D, Nittler L R, David Kilcoyne A L and Wirick S 2011 Correlated microanalysis of cometary organic grains returned by Stardust *Meteor. Planet. Sc.* **46** 1376–96

[147] Elsila J E, Glavin D P and Dworkin J P 2009 Cometary glycine detected in samples returned by Stardust *Meteorit. Planet. Sci.* **44** 1323–30

[148] Fray N *et al* 2016 High-molecular-weight organic matter in the particles of comet 67P/ Churyumov–Gerasimenko *Nature* **538** 72–4

[149] Capaccioni F 2015 The organic-rich surface of comet 67P/Churyumov–Gerasimenko as seen by VIRTIS/Rosetta *Science* **347** aaa0628-1–aaa4

[150] Goesmann F *et al* 2015 Organic compounds on comet 67P/Churyumov–Gerasimenko revealed by COSAC mass spectrometry *Science* **349** aab0689-1–aab4

[151] Wright I P, Sheridan S, Barber S J, Morgan G H, Andrews D J and Morse A D 2015 CHO-bearing organic compounds at the surface of 67P/Churyumov–Gerasimenko revealed by Ptolemy *Science* **349** aab0673-1–aab3

[152] Altwegg K *et al* 2016 Prebiotic chemicals—amino acid and phosphorus—in the coma of comet 67P/Churyumov–Gerasimenko *Sci. Adv.* **2** e1600285–e15

[153] Nakamura T *et al* 2011 Itokawa dust particles: a direct link between S-type asteroids and ordinary chondrites *Science* **333** 1113–6

[154] Uesugi M *et al* 2014 Sequential analysis of carbonaceous materials in Hayabusa-returned samples for the determination of their origin *Earth, Planets Space* **66** 102-1–102-11

[155] Wada K *et al* 2018 Asteroid Ryugu before the Hayabusa2 encounter *Prog. Earth Planet. Sci.* **5** 82 1–30

[156] Naraoka H 2023 Soluble organic molecules in samples of the carbonaceous asteroid (162173) Ryugu *Science* **379** abn9033

[157] Yabuta H 2023 Macromolecular organic matter in samples of the asteroid (162173) Ryugu *Science* **379** eabn9057

[158] Lauretta D S *et al* 2017 OSIRIS-REx: sample return from asteroid (101955) Bennu *Space Sci. Rev.* **212** 925–84

[159] Gardner J P, Mather J C, Clampin M, Doyon R, Greenhouse M A, Hammel H B, Hutchings J B, Jakobsen P, Lilly S J and Long K S *et al* 2006 The James Webb space telescope *Space Sci. Rev.* **123** 485–606

[160] Seager S and Sasselov D D 2000 Theoretical transmission spectra during extrasolar giant planet transits *Astrophys. J.* **537** 916

[161] Gillon M, Triaud A H M J, Demory B-O, Jehin E, Agol E and Deck K M *et al* 2017 Seven temperate terrestrial planets around the nearby ultracool dwarf star TRAPPIST-1 *Nature* **542** 456–60

[162] Turbet M, Bolmont E, Bourrier V, Demory B O, Leconte J, Owen J and Wolf E T 2020 A review of possible planetary atmospheres in the TRAPPIST-1 system *Space Sci. Rev.* **216** 100

[163] Gilmozzi R and Spyromilio J 2007 The European extremely large telescope (E-ELT) *The Messenger* **127** 11 https://ui.adsabs.harvard.edu/abs/2007Msngr.127...11G

[164] Sanders G H 2013 The Thirty Meter Telescope (TMT): an international observatory *J. Astrophys. Astron.* **34** 81–6

[165] Bernstein R A, McCarthy P J, Raybould K, Bigelow B C, Bouchez A H and Filgueira J M *et al* 2014 Overview and status of the Giant Magellan Telescope project *Ground-based and Airborne Telescopes* V **vol 9145** (Bellingham, WA: SPIE) p 91451C

[166] Johns M, McCarthy P, Raybould K, Bouchez A, Farahani A, Filgueira J, Jacoby G, Shectman S and Sheehan M 2012 Giant Magellan telescope: overview *Ground-based and Airborne Telescopes IV* (Bellingham, WA: SPIE) vol 8444 p 84441H

[167] Anglada-Escudé G, Amado P J, Barnes J, Berdiñas Z M, Butler R P and Coleman G A L *et al* 2016 A terrestrial planet candidate in a temperate orbit around Proxima Centauri *Nature* **536** 437–40

[168] Spitzer L 1962 The beginnings and future of space astronomy *Am. Sci.* **50** 473–84

[169] Seager S, Turnbull M, Sparks W, Thomson M, Shaklan S B, Roberge A, Kuchner M, Kasdin N J, Domagal-Goldman S and Cash W *et al* 2015 The Exo-S probe class starshade mission *Tech. Instrum. Detect. Exopl. VII* **9605** 96050W

[170] LUVOIR Team 2019 The LUVOIR Mission Concept Study Final Report *ArXiv E-Prints* arXiv:191206219T https://asd.gsfc.nasa.gov/luvoir/reports/LUVOIR_FinalReport_2019-08-26.pdf

[171] Feng F, Tuomi M, Jones H R A, Barnes J, Anglada-Escudé G, Vogt S S and Butler R P 2017 Color difference makes a difference: four planet candidates around τ ceti *Astron. J* **154** 135

[172] Tuomi M, Jones H R A, Jenkins J S, Tinney C G, Butler R P and Vogt S S *et al* 2013 Signals embedded in the radial velocity noise. Periodic variations in the τ Ceti velocities *Astron. Astrophys.* **551** A79

[173] Chan Q H S, Stroud R, Martins Z and Yabuta H 2020 Concerns of organic contamination for sample return space missions *Space Sci. Rev.* **216** 56

[174] Dhingra R D *et al* 2019 Observational evidence for summer rainfall at Titan's north pole *Geophys. Res. Lett.* **46** 1205–12

[175] Poch O, Coll P, Buch A, Ramírez S I and Raulin F 2012 Production yields of organics of astrobiological interest from H_2O–NH_3 hydrolysis of Titan's tholins *Planet. Space Sci.* **61** 114–23

[176] Eshleman V R 1979 Gravitational lens of the Sun: its potential for observations and communications over interstellar distances *Science* **205** 1133–5

[177] Landis G A 2017 Mission to the gravitational focus of the Sun: a critical analysis *55th AIAA Aerospace Sciences Meeting* (American Institute of Aeronautics and Astronautics)

[178] Turyshev S G, Shao M, Toth V T, Friedman L D, Alkalai L and Mawet D *et al* 2020 Direct multipixel imaging and spectroscopy of an exoplanet with a solar gravity lens mission *ArXiv E-Prints* 2020 arXiv:200211871T

[179] Licklider J C R 1960 Man-computer symbiosis *IRE Trans. Hum. Factors Electron.* **HFE-1** 4–11

[180] Schwab K 2021 The Fourth Industrial Revolution: What It Means, How To Respond *Word Economic Forum* (https://weforum.org/agenda/2016/01/the-fourth-industrialrevolution-what-it-means-and-how-to-respond/)

[181] Wiener N 1948 *Cybernetics: or the Control and Communication in the Animal and the Machine* (New York: Wiley)

[182] Ferrucci D *et al* 2010 Building Watson: an overview of the Deepqa project *AI Mag.* **31** 59–79

[183] Boden M 2014 GOFAI *The Cambridge Handbook of Artificial Intelligence* ed F Keith and W M Ramsey (Cambridge: Cambridge University Press)

[184] Newell A and Simon H A 1976 Computer science as empirical inquiry: symbols and search (ACM 1975 Turing award lecture) *Commun. ACM* **19** 113–26

[185] Franklin S 2014 History, motivations, and core themes *The Cambridge Handbook of Artificial Intelligence* ed F Keith and W M Ramsey (Cambridge: Cambridge University Press)

[186] Feldman R A and Ballard D H 1982 Connectionist models and their properties *Cogn. Sci.* **6** 205–54

[187] James W 1890 *The Principles of Psychology* **vol 1** (Cambridge, MA: Harvard University Press)

[188] Little W A 1974 The existence of persistent states in the brain *Math. Biosci.* **19** 101–20

[189] Hopfield J J 1982 Neural neural network and physical systems with emergent collective computational abilities *Proc. Natl Acad. Sci.* **79** 2554–88

[190] Hinton G E and Sejnowski T J 1986 Learning and relearning in Boltzmann machines *Parallel Distributed Processing: Explorations in the Microstructure of Cognition* ed D E Rumelhart, J L McClelland and The PDP Research Group (Cambridge, MA: MIT Press) pp 282–317

[191] Abbas A, Sutter D, Zoufal C, Lucchi A, Figalli A and Woerner S 2021 The power of quantum neural networks *Nat. Comput. Sci.* **1** 403–9

[192] Dunjko V and Briegel H J 2018 Machine learning and artificial intelligence in the quantum domain: a review of recent progress *Rep. Prog. Phys.* **81** 074001

[193] McCulloch W S and Pitts W 1943 A logical calculus of ideas immanent in nervous activity *Bull. Math. Biophys.* **5** 115–33

[194] Rosenblatt F 1958 The perceptron: a probabilistic model for information storage and organization in the brain *Psychol. Rev.* **65** 386–408

[195] Widrow B and Hoff M E 1960 Adaptive switching circuits *IRE WESCON Convention Record (New York)* pp 96–104

[196] Minsky M and Papert S 1969 *Perceptrons: An Introduction to Computational Geometry* (Cambridge, MA: MIT Press)

[197] Rumelhart D E, Hinton G E and Williams R J 1986 Learning representations by back-propagating errors *Nature* **323** 533–6

[198] Werbos P 1974 Beyond regression: new tools for prediction and analysis in the behavioural sciences *Masters Thesis* Harvard University

[199] Medler D A 1998 A brief history of connectionism *Neural Comput. Surv.* **1** 61–101

[200] Hochreiter S and Schmidhuber J 1997 Long short-term memory *Neural Comput.* **9** 1735–80

[201] LeCun Y, Bottou L, Bengio Y and Haffner P 1998 Gradient-based learning applied to document recognition *Proc. IEEE* **86** 2278–324

[202] Krizhevsky A, Sutskever I and Hinton G 2012 Imagenet classification with deep convolutional neural networks *Advances in Neural Information Processing Systems* ed F Pereira, C J C Burges, L Bottou and K Q Weinberger (Cambridge, MA: MIT Press) vol 25

[203] Goodfellow I, Bengio Y and Courville A 2016 *Deep Learning* (Cambridge, MA: MIT Press)

[204] Schmidhuber J 2014 Deep learning in neural networks: an overview *arXiv preprint* (arXiv:1404.7828 v2)

[205] Mao J, Xu W, Huang Y Y and Yuille A 2015 Deep captioning with multimodal recurrent neural networks (m-RNN) *Proc. of the 3rd Int. Conf. on Learning Representations (ICLR)*

[206] Brown T B *et al* 2020 Language models are few-shot learners arXiv:2005.14165

[207] Danks D 2014 Learning *The Cambridge Handbook of Artificial Intelligence* ed F Keith and W M Ramsey (Cambridge: Cambridge University Press)

[208] Silver D, Huang A, Maddison C J, Guez A, Sifre L and Van Den Driessche G *et al* 2016 Mastering the game of Go with deep neural networks and tree search *Nature* **529** 484–9

[209] Hu Z, Tang J, Wang Z, Zhang K, Zhang L and Sun Q 2018 Deep learning for image-based cancer detection and diagnosis—a survey *Pattern Recognit.* **83** 134–49

[210] Holzinger A 2016 Interactive machine learning for health informatics: when do we need the human-in-the-loop? *Brain Inform.* **3** 119–31

[211] Miotto R, Wang F, Wang S, Jiang X and Dudley J T 2018 Deep learning for healthcare: review, opportunities and challenges *Brief. Bioinform* **19** 1236–46

[212] Holzinger A, Langs G, Denk H, Zatloukal K and Müller H 2019 Causability and explainability of artificial intelligence in medicine *Wiley Interdiscip. Rev.: Data Min. Knowl. Discov.* **9** e1312

[213] IEEE Robots 2021 (https://robots.ieee.org/)

[214] Cao H *et al* 2019 Robot-enhanced therapy: development and validation of a supervised autonomous robotic system for autism spectrum disorders therapy *IEEE Rob. Autom Mag.* **26** 49–58

[215] Zhang S, Yao L, Sun A and Tay Y 2019 Deep learning based recommender system: a survey and new perspectives *ACM Comput. Surv. (CSUR)* **52** 1–38

[216] Buolamwini J and Gebru T 2018 Gender shades: intersectional accuracy disparities in commercial gender classification *Conf. on Fairness, Accountability and Transparency Proc. of Machine Learning Research* pp 77–91

[217] Barlow A and Sriskandarajah S 2019 *Artificial Intelligence—Application to the Sports Industry* (https://pwc.com.au/industry/sports/artificial-intelligence-application-to-thesports-industry.pdf)

[218] Hawk-Eye 2021 (https://hawkeyeinnovations.com/)

[219] Shaw D 2017 *How Wimbledon is Using IBM Watson Ai to Power Highlights, Analytics and Enriched Fan Experiences* (https://ibm.com/blogs/watson/2017/07/ibm-watsons-ai-ispower-ing-wimbledon-highlights-analytics-and-a-fan-experiences/)

[220] AlphaGo 2021 (https://deepmind.com/research/case-studies/alphago-the-story-so-far)

[221] Bartneck C, Belpaeme T, Eyssel F, Kanda T, Keijsers M and Šabanović S 2020 *Human–Robot Interaction: An Introduction* (Cambridge: Cambridge University Press)

[222] Vinanzi S, Patacchiola M, Chella A and Cangelosi A 2019 Would a robot trust you? Developmental robotics model of trust and theory of mind *Phil. Trans. R. Soc. B* **374** 20180032

[223] Sandini G, Sciutti A and Vernon D 2021 Cognitive robotics *Encyclopedia of Robotics* ed M Ang, O Khatib and B Siciliano (Berlin: Springer)

[224] Leite I, Martinho C and Paiva A 2013 Social robots for long-term interaction: a survey *Int. J. Soc. Robot.* **5** 291–308

[225] Mou W, Ruocco M, Zanatto D and Cangelosi A 2020 When would you trust a robot? A study on trust and theory of mind in human–robot interactions *Proc. of RO-MAN2020, 29th IEEE Int. Conf. on Robot and Human Interactive Communication (Naples, August 2020)*

[226] Vinanzi S, Cangelosi A and Goerick C 2021 The collaborative mind: intention reading and trust in human–robot interaction *Science* **24** 102130

[227] Balog M, Gaunt A L, Brockschmidt M, Nowozin S and Tarlow D 2016 DeepCoder: learning to write programs *arXiv preprint* arXiv:1611.01989

[228] Hewitt L B, Le T A and Tenenbaum J B 2020 Learning to infer program sketches Proc. of the 36th Conf. on Uncertainty in Artificial Intelligence

[229] Heaven W D 2020 *OpenAI's New Language Generator GPT-3 Is Shockingly Good—and Completely Mindless* (MIT Technology Review)

[230] Cangelosi A and Schlesinger M 2015 *Developmental Robotics: From Babies to Robots* (Cambridge, MA: MIT Press)

[231] Silver D, Schrittwieser J, Simonyan K, Antonoglou I, Huang A, Guez A, Hubert T, Baker L, Lai M and Bolton A *et al* 2017 Mastering the game of go without human knowledge *Nature* **550** 354

[232] OECDAI 2021 National AI policies & strategies https://oecd.ai/dashboards

[233] Alupo C D, Omeiza D and Vernon D 2022 Realizing the potential of AI in Africa ed M I Aldinhas Ferreira and O Tokhi *Towards Trustworthy Artificial Intelligence Systems, Intelligent Systems, Control and Automation: Science and Engineering* (Berlin: Springer)

[234] Novitske L 2018 *The AI Invasion Is Coming to Africa (and It's a Good Thing)* (Stanford Social Innovation Review)

[235] Pillay N and Access Partnership 2018 *Artificial Intelligence for Africa: An Opportunity for Growth, Development, and Democratization* (White Paper, University of Pretoria)

[236] EUAI 2021 *European Commission Ethics Guidelines for Trustworthy AI* (https://ec.europa.eu/futurium/en/ai-alliance-consultation)

[237] Besaw C and Filitz J 2019 AI and global governance: AI in Africa is a double-edged sword *Tech. rep.* (United Nations University) (https://cpr.unu.edu/ai-in-africa-is-a-doubleedged-sword.html)

[238] Wilson B, Hoffman J and Morgenstern J 2019 Predictive inequity in object detection *arXiv preprint* arXiv:1902.11097

[239] Larson J, Mattu S, Kirchner L and Angwin J 2016 How we analyzed the COMPAS recidivism algorithm *ProPublica* May 2016 9

[240] Wolpert D 2011 *The Real Reason for Brains* (https://youtube.com/watch?v=7s0CpRfyYp8)

[241] Maturana H and Varela F 1987 *The Tree of Knowledge—The Biological Roots of Human Understanding* (Boston, MA and London: New Science Library)

[242] Friston K J 2010 The free-energy principle: a unified brain theory? *Nat. Rev. Neurosci.* **11** 127–38

[243] Verschure P F M J 2016 Synthetic consciousness: the distributed adaptive control perspective *Phil. Trans. R. Soc.* B **371** 20150448

[244] Laird J E 2012 *The Soar Cognitive Architecture* (Cambridge, MA: MIT Press)

[245] Anderson J R, Bothell D, Byrne M D, Douglass S, Lebiere C and Qin Y 2004 An integrated theory of the mind *Psychol. Rev.* **111** 1036–60

[246] Sun R 2016 *Anatomy of the Mind: Exploring Psychological Mechanisms and Processes with the Clarion Cognitive Architecture* (Oxford: Oxford University Press)

[247] Langton C G 1997 *Artificial Life: An Overview* (Cambridge, MA: MIT Press)

[248] Adami C 1998 *Introduction to Artificial Life* (New York: Springer)

[249] Aguilar W, Bonfil G S, Froese T and Gershenson C 2014 The past, present, and future of artificial life *Front. Robotics AI* **1** 8

[250] Langton C G 1987 *Artificial Life: The Proc. of an Interdisciplinary Workshop on the Synthesis and Simulation of Living Systems (Los Alamos, NM, Sept. 1987)* (Addison-Wesley)

[251] Bedau M, McCaskill J, Packard P, Rasmussen S, Green D, Ikegami T, Kaneko K and Ray T 2000 Open problems in artificial life *Artif. Life* **6** 363–76

[252] von Neumann J 1966 *The Theory of Self-Reproducing Automata* ed A W Burks (Champaign, IL: University of Illinois Press)

[253] Shelling T C 1971 Dynamic models of segregation *J. Math. Sociol.* **1** 143–86

[254] Berlekamp E R, Conway J H and Guy R K 1982 *Winning Ways for Your Mathematical Plays Games in Particular* **vol 2** (London: Academic)

[255] Wolfram S 1983 Statistical mechanics of cellular automata *Rev. Mod. Phys.* **55** 601–44

[256] Wuensche A and Lesser M J 1992 *The Global Dynamics of Cellular Automata; An Atlas of Basin of Attraction Fields of One-Dimensional Cellular Automata* (Reading, MA: Santa Fe Institute Studies in the Sciences of Complexity, Addison-Wesley)

[257] Mitchell M, Hraber P T and Crutchfield J P 1993 Revisiting the edge of chaos: evolving cellular automata to perform computations *Complex Syst.* **7** 89 130

[258] Epstein J M and Axtell R L 1996 *Growing Artificial Societies: Social Science from the Bottom Up* (Cambridge, MA: Brookings Institution Press MIT Press)

[259] Kauffman S A 1969 Metabolic stability and epigenesis in randomly constructed genetic nets *J. Theor. Biol.* **22** 437 467

[260] Aldana-González M, Coppersmith S and Kadanoff L P 2003 Boolean dynamics with random couplings In E Kaplan, J E Marsden and K R Sreenivasan *Perspectives and Problems in Nonlinear Science. A Celebratory Volume in Honor of Lawrence Sirovich, Applied Mathematical Sciences Series* (Berlin: Springer)

[261] Gershenson C 2004 Introduction to random Boolean networks *Workshop and Tutorial Proc., 9th Int. Conf. on the Simulation and Synthesis of Living Systems (ALife IX) (Boston, MA)* ; M Bedau, P Husbands, T Hutton, S Kumar and H Suzuki pp 160–73

[262] Gershenson C 2012 Guiding the self-organization of random Boolean networks *Theory Biosci.* **131** 181–91

[263] Reynolds C W 1987 Flocks, herds, and schools: a distributed behavioral model *Comput. Graph.* **21** 25–34

[264] Sipper M 1998 Fifty years of research on self-replication: an overview *Artif. Life* **4** 237–57

[265] Ray T S 1994 An evolutionary approach to synthetic biology: zen and the art of creating life *Artif. Life* **1** 195–226

[266] Chris Adami and Titus Brown C 1994 Evolutionary learning in the 2D artificial life system 'Avida' *Proc. Artificial Life IV* ed R Brooks and P Maes (Cambridge, MA: MIT Press) pp 377–81

[267] Lenski R E, Ofria C, Collier T C and Adami C 1999 Genome complexity, robustness and genetic interactions in digital organisms *Nature* **400** 661–4

[268] Lenski R E, Ofria C, Pennock R T and Adami C 2003 The evolutionary origin of complex features *Nature* **423** 139–44

[269] Adami C, Ofria C and Collier T C 2000 Evolution of biological complexity *Proc. Natl Acad. Sci.* **97** 4463–8

[270] Fontana W 1991 Algorithmic chemistry: a model for functional self-organization *Artificial Life II* ed C G Langton, C Taylor, J D Farmer and S Rasmussen (Boston, MA: Addison-Wesley) pp 159–202

[271] Dittrich P, Ziegler J and Banzhaf W 2001 Artificial chemistries—a review *Artif. Life* **7** 225–75 2014/07/08

[272] Hiroki S 2008 Swarm chemistry *Artif. Life* **15** 105–14 2014/07/08

[273] Karl S 1994 Evolving 3D morphology and behavior by competition *Artif. Life* **1** 353–72

[274] Jakobi N 1997 Evolutionary robotics and the radical envelope of noise hypothesis *Adapt. Behav.* **6** 325–68

[275] Lipson H and Pollack J B 2000 Automatic design and manufacture of robotic lifeforms *Nature* **406** 974–8

[276] Bongard J, Zykov V and Lipson H 2006 Resilient machines through continuous self-modeling *Science* **314** 1118–21

[277] Brooks R A 1991 Intelligence without representation *Artif. Intell.* **47** 139 –159

[278] Steels L and Brooks R A 1995 *The Artificial Life Route to Artificial Intelligence: Building Embodied, Situated Agents* (New York City: Lawrence Erlbaum Associates)

[279] Beer R D 2014 Dynamical systems and embedded cognition *The Cambridge Handbook of Artificial Intelligence* ed K Frankish and W Ramsey (Cambridge: University Press)

[280] Maes P 1994 Modeling adaptive autonomous agents *Artif. Life* **1** 135–62

[281] Beer R D 1990 *Intelligence as Adaptive Behavior: An Experiment in Computational Neuroethology* (San Diego, CA: Academic)

[282] Braitenberg V 1986 *Vehicles: Experiments in Synthetic Psychology* (Cambridge, MA: MIT Press)

[283] Gershenson C 2004 Cognitive paradigms: which one is the best? *Cognit. Syst. Research* **5** 135–56

[284] Cliff D, Husbands P and Harvey I 1993 Explorations in evolutionary robotics *Adapt. Behav.* **2** 73–110

[285] Nolfi S and Floreano D 2000 *Evolutionary Robotics: The Biology, Intelligence, and Technology of Self-Organizing Machines* (Cambridge, MA: MIT Press)

[286] Harvey I, Di Paolo E A, Wood R, Quinn M and Tuci E A 2005 Evolutionary robotics: A new scientific tool for studying cognition *Artif. Life* **11** 79–98

[287] Dorigo M *et al* 2004 Evolving self-organizing behaviors for a swarm-bot *Auton. Robots* **17** 223–45

[288] Zykov V, Mytilinaios E, Adams B and Lipson H 2005 Robotics: self-reproducing machines *Nature* **435** 163 164

[289] Halloy J *et al* 2007 Social integration of robots into groups of cockroaches to control self-organized choices *Science* **318** 1155–8

[290] Rubenstein M, Cornejo A and Nagpal R 2014 Programmable self-assembly in a thousand-robot swarm *Science* **345** 795–9

[291] Vásárhelyi G, Virágh C, Somorjai G, Nepusz T, Eiben A E and Vicsek. T 2018 Optimized flocking of autonomous drones in confined environments *Sci. Robot.* **3**

[292] Luisi P L 2002 Toward the engineering of minimal living cells *Anat. Rec.* **268** 208–14

[293] Hanczyc M M, Fujikawa S M and Szostak J W 2003 Experimental models of primitive cellular compartments: encapsulation, growth, and division *Science* **302** 618–22

[294] Rasmussen S, Bedau M A, Chen L, Deamer D, Krakauer D C, Packard N H and Stadler P F 2008 *Protocells: Bridging Nonliving and Living Matter* (Cambridge, MA: MIT Press)

[295] Rasmussen S, Chen L, Deamer D, Krakauer D C, Packard N H, Stadler P F and Bedau M A 2004 Transitions from nonliving to living matter *Science* **303** 963–5

[296] Bachmann P A, Luisi P L and Lang J 1992 Autocatalytic self-replicating micelles as models for prebiotic structures *Nature* **357** 57

[297] Walde P, Wick R, Fresta M, Mangone A and Luisi P L 1994 Autopoietic self-reproduction of fatty acid vesicles *JACS* **116** 11649–54

[298] Čejková J, Novak M, Štěpánek F and Hanczyc M M 2014 Dynamics of chemotactic droplets in salt concentration gradients *Langmuir* **30** 11937–44

[299] Hanczyc M M, Toyota T, Ikegami T, Packard N and Sugawara T 2007 Fatty acid chemistry at the oil- water interface: self-propelled oil droplets *J. Am. Chem. Soc.* **129** 9386–91

[300] Čejková J, Hanczyc M M and Štěpánek F 2018 Multi-armed droplets as shape-changing protocells *Artif. Life* **24** 71–9

[301] Čejkova J, Stepanek F and Hanczyc M M 2016 Evaporation-induced pattern formation of decanol droplets *Langmuir* **32** 4800–5

[302] Jitka Č, Schwarzenberger K, Eckert K and Tanaka S 2019 Dancing performance of organic droplets in aqueous surfactant solutions *Colloids Surf., A* **566** 141–7

[303] Qiao Y, Li M, Booth R and Mann S 2017 Predatory behaviour in synthetic protocell communities *Nat. Chem.* **9** 110–9

[304] Čejková J, Banno T, Hanczyc M M and Štěpánek F 2017 Droplets as liquid robots *Artif. Life* **23** 528–49

[305] Kriegman S, Blackiston D, Levin M and Bongard J 2020 A scalable pipeline for designing reconfigurable organisms *Proc. Natl Acad. Sci.* **117** 1853–9

[306] Blackiston D, Lederer E, Kriegman S, Garnier S, Bongard J and Levin M 2021 A cellular platform for the development of synthetic living machines *Sci. Robot.* **6**

[307] Taylor T and Dorin A 2020 *Rise of the Self-Replicators* (Berlin: Springer)

[308] Bedau M A, McCaskill J S, Packard N H and Rasmussen S 2009 Living technology: exploiting life's principles in technology *Artif. Life* **16** 89–97 2011/11/03

[309] Bedau M A, McCaskill J S, Packard N H, Parke E C, S R and Rasmussen 2013 Introduction to recent developments in living technology *Artif. Life* **19** 291–8 2014/07/08

[310] Standish R K 2003 Open-ended artificial evolution *Int. J. Comput. Intell. Appl.* **3** 167–75

[311] Moreno A and Ruiz-Mirazo K 2006 The maintenance and open-ended growth of complexity in nature: information as a decoupling mechanism in the origins of life *Reframing Complexity: Perspectives from the North and South* ed F Capra, A Juarrero, P Sotolongo and J van Uden (ISCE Publishing)

[312] Taylor T, Bedau M, Channon A, Ackley D, Banzhaf W, Beslon G, Dolson E, Froese T, Hickinbotham S and Ikegami T *et al* 2016 Open-ended evolution: perspectives from the oee workshop in york *Artif. Life* **22** 408–23

[313] Roli A and Kauffman S A 2020 Emergence of organisms *Entropy* **22**

[314] Hernández-Orozco S, Hernández-Quiroz F and Zenil H 2018 Undecidability and irreducibility conditions for open-ended evolution and emergence *Artif. Life* **24** 56–70

[315] Kurt G 1931 Über formal unentscheidbare sätze der principia mathematica und verwandter systeme i *Monatsh. Math. Phys.* **38** 173–98

[316] Turing A M 1936 On computable numbers, with an application to the Entscheidungsproblem *Proc. London Math. Soc., Ser. 2* 230–65

[317] Chaitin G J 1975 Randomness and mathematical proof *Sci. Am.* **232** 47–52

[318] Rahwan I *et al* 2019 Machine behaviour *Nature* **568** 477–86

[319] Nicolis G and Prigogine I 1977 *Self-Organization in Non-Equilibrium Systems: From Dissipative Structures to Order Through Fluctuations* (Chichester: Wiley)

[320] Martyushev L M and Seleznev V D 2006 Maximum entropy production principle in physics, chemistry and biology *Phys. Rep.* **426** 1–45

[321] Ramaswamy S 2017 Active matter *J. Stat. Mech: Theory Exp.* **2017** 054002

[322] van Leeuwenhoek A 1700 IV. Part of a letter from Mr Antony van Leeuwenhoek, concerning the worms in sheeps livers, gnats, and animalcula in the excrements of frogs *Phil. Trans. R. Soc.* **22** 509–18

[323] Corti B 1774 *Osservazioni Microscopische sulla Tremella e sulla Circolazione del Fluido in Una Pianta Acquajuola* (Italy: Appresso Giuseppe Rocchi, Lucca)

[324] Brown R 1828 A brief account of microscopical observations made in the months of June, July, and August, 1827, on the particles contained in the pollen of plants; and on the general existence of active molecules in organic and inorganic bodies *Phil. Mag.* **4** 161–73

[325] Engelmann R W 1882 Ueber sauerstoffausscheidung van pfalnzenzellen im mikrospektrum *Pflügers Arch. Gesamte Physiol. Menschen Tiere* **27** 485–73

[326] Secchi E, Rusconi R, Buzzaccaro S, Salek M M, Smriga S, Piazza R and Stocker R 2016 Intermittent turbulence in flowing bacterial suspensions *J. R. Soc. Interface* **13** 20160175

[327] Haas P A, Höhn S, Honerkamp-Smith A R, Kirkegaard J B and Goldstein R E 2018 The noisy basis of morphogenesis: mechanisms and mechanics of cells sheet folding inferred from developmental variability *PLoS Biol.* **16** e2005536

[328] Besson S and Dumais J 2011 Universal rule for the symmetric division of plant cells *Proc. Natl. Acad. Sci. USA* **108** 6294–9

[329] Ratcliff W C, Denison R F, Borrello M and Travisano M 2012 Experimental evolution of multicellularity *Proc. Natl Acad. Sci. USA* **109** 1595–600

[330] Brunet T, Larsen B T, Linden T A, Vermeij M J A, McDonald K and King N 2019 Light-regulated collective contractility in a multicellular choanoflagellate *Science* **366** 326–34

[331] ESA 2020 Space for the oceans http://www.esa.int/Enabling_Support/Preparing_for_the_Future/Space_for_Earth/Space_for_the_oceans

[332] Hodgkin A L and Huxley A F 1952 A quantitative description of membrane current and its application to conductance and excitation in nerve *J. Physiol.* **117** 500–44

[333] Franklin R E and Goslin R G 1953 Molecular configuration in sodium thymonucleate *Nature* **171** 740–1

[334] Watson J D and Crick F H 1953 Molecular structure of nucleic acids: a structure for deoxyribose nucleic acid *Nature* **171** 737–8

[335] Cole K S 1949 Dynamic electrical characteristics of the squid axon membrane *Arch. Sci. Physiol.* **3** 253–8

[336] Meron E 1992 Pattern formation in excitable media *Phys. Rep.* **218** 1–66

[337] Bragg W L 1913 The diffraction of short electromagnetic waves by a crystal *Proc. Camb. Phil. Soc.* **17** 43–57

[338] Murray J D 1989 *Mathematical Biology* (Berlin: Springer)

[339] Rayleigh L 1916 On convection currents in a horizontal layer of fluid, when the higher temperature is on the under side *Phil. Mag.* **32** 529–46

[340] Keller E F 2003 *Making Sense of Life. Explaining Biological Development with Models, Metaphors, and Machines* (Cambridge, MA: Harvard University Press)

[341] Castets V, Dulos E, Boissonade J and De Kepper P 1990 Experimental evidence of a sustained standing turing-type nonequilibrium chemical pattern *Phys. Rev. Lett.* **64** 2953–6

[342] De Kepper P, Castets V, Dulos E and Boissonade J 1991 Turing-type chemical patterns in the chlorite-iodide-malonic acid reaction *Physica* D **49** 161–9

[343] Ouyang Q and Swinney H L 1991 Transition from a uniform state to hexagonal and striped turing patterns *Nature* **352** 610–2

[344] Goldstein R E 2018 Coffee stains, cell receptors, and time crystals: lessons from the old literature *Phys. Today* **71** 32–8

[345] Keller E F and Segel L A 1971 Model for chemotaxis *J. Theor. Biol.* **30** 225–34

[346] Lee K J, Cox E C and Goldstein R E 1996 Competing patterns of signaling activity in *dictyostelium discoideum Phys. Rev. Lett.* **76** 1174–7

[347] Taylor G I 1951 Analysis of the swimming of microscopic organism. *Proc. R. Soc.* A **209** 447–61

[348] Lighthill M J 1952 On the squirming motion of nearly spherical deformable bodies through liquids at very small Reynolds numbers *Commun. Pure Appl. Math.* **5** 109–18

[349] Purcell E M 1977 Life at low Reynolds number *Am. J. Phys.* **45**–11

[350] Berg H C and Purcell E M 1977 Physics of chemoreception *Biophys. J.* **20** 193–219

[351] Steinberg M S 1962 On the mechanism of tissue reconstruction by dissociated cells, I. Population kinetics, differential adhesiveness, and the absence of directed migration *Proc. Natl Acad. Sci. USA* **48** 1577–82

[352] Foty R A, Forgacs G, Pfleger C M and Steinberg M S 1994 Liquid properties of embryonic tissues: measurement of interfacial tensions *Phys. Rev. Lett.* **72** 2298–301

[353] de Gennes P G 1979 *Scaling Concepts in Polymer Physics* (Ithaca, NY: Cornell University Press)

[354] Flory P J 1953 *Principles of Polymer Chemistry* (Ithaca, NY: Cornell University Press)

[355] Langer J S 1980 Instabilities and pattern formation in crystal growth *Rev. Mod. Phys.* **52** 1–28

[356] Helfrich W 1973 Elastic properties of lipid bilayers: theory and possible experiments *Z. Naturforsch.* C **28** 693–703

[357] Smith S B, Finzi L and Bustamante C 1992 Direct mechanical measurements of the elasticity of single DNA molecules by using magnetic beads *Science* **258** 1122–6

[358] Marko J F and Siggia E D 1995 Stretching DNA *Macromolecules* **28** 8759–70

[359] Svoboda K, Schmidt C F, Schnapp B J and Block S M 1993 Direct observation of kinesin stepping by optical trapping interferometry *Nature* **365** 721–7

[360] Magnasco M O 1993 Forced thermal ratchets *Phys. Rev. Lett.* **71** 1477–80

[361] Rousselet J, Salome L, Ajdari A and Prost J 1994 Directional motion of Brownian particles induced by a periodic asymmetric potential *Nature* **370** 446–7

[362] Jülicher F, Ajdari A and Prost J 1997 Modeling molecular motors *Rev. Mod. Phys.* **69** 1269–81

[363] Bustamante C, Keller D and Oster G 2001 The physics of molecular motors *Acc. Chem. Res.* **34** 412–20

[364] Lubkins S and Rand R 1994 Oscillatory reaction-diffusion equations on rings *J. Math. Biol.* **32** 617–32

[365] Hänggi P and Marchesoni F 2009 Artificial Brownian motors: controlling transport on the nanoscale *Rev. Mod. Phys.* **81** 387–442

[366] Guerin T, Prost J, Martin P and Joanny J F 2010 Coordination and collective properties of molecular motors: theory *Curr. Opin. Cell Biol.* **22** 14–20

[367] Nadrowski B, Martin P and Jülicher F 2004 Active hair-bundle motility harnesses noise to operate near an optimum of mechanosensitivity *Proc. Natl Acad. Sci. USA* **101** 12195–200

[368] Crocker J C and Grier D G 1996 Methods of digital video microscopy for colloidal studies *J. Coll. Int. Sci.* **179** 298–310

[369] Polin M, Tuval I, Drescher K, Gollub J P and Goldstein R E 2009 *Chlamydomonas* swis with two 'gears' in a eukaryotic version of run-and-tumble locomotion *Science* **325** 487–90

[370] Wan K Y and Goldstein R E 2018 Time-irreversibility and criticality in the motility of a flagellate microorganism *Phys. Rev. Lett.* **121** 058103

[371] Squires T M and Quake S R 2005 Microfluidics: fluid physics at the nanoliter scale *Rev. Mod. Phys.* **77** 977–1026

[372] Huebner J S, Jakobsson E, Dam H G, Parsegian V A and Austin R H 1997 How should physicists, biologists work together? The 'Harness the hubris' debate continues *Phys. Today* **50** 11–95

[373] Parsegian V A 1997 Harness the hubris: useful things physicists could do in biology *Phys. Today* **50** 23–7

[374] Finlayson B A and Scriven L E 1969 Convective instability by active stress *Proc. R. Soc.* A **310** 183–219

[375] Marchetti M C, Joanny J F, Ramaswamy S, Liverpool T B, Prost J, Rao M and Simha R A 2013 Hydrodynamics of soft active matter *Rev. Mod. Phys.* **85** 1143–89

[376] Ramaswamy S 2019 Active fluids *Nat. Rev. Phys.* **1** 640–64

[377] Gompper G, Winkler R G, Speck T, Solon A, Nardini C, Peruani F, Löwen H, Golestanian R, Kaupp U B and Alvarez L 2020 The 2020 motile active matter roadmap *J. Phys. Condens. Matter* **32** 193001

[378] Shaebani M R, Wysocki A, Winkler R G, Gompper G and Rieger H 2020 Computational models for active matter *Nat. Rev. Phys.* **2** 181–99

[379] Prost J and Bruinsma R 1996 Shape fluctuations of active membranes *Europhys. Lett.* **33** 321–6

[380] Ramaswamy S, Prost J and Toner J 2000 Nonequilibrium fluctuations, travelling waves, and instabilities in active membranes *Phys. Rev. Lett.* **84** 3494–7

[381] Nédélec F J, Surrey T, Maggs A C and Leibler S 1997 Self-organization of microtubules and motors *Nature* **389** 305–8

[382] Toner J and Tu Y 1995 Long-range order in a two-dimensional dynamical XY model: how birds fly together *Phys. Rev. Lett.* **75** 4326–9

[383] Simha R A and Ramaswamy S 2002 Hydrodynamic fluctuations and instabilities in ordered suspensions of self-propelled particles *Phys. Rev. Lett.* **89** 058101

[384] Saintillan D and Shelley M J 2008 Instabilities and pattern formation in active particle suspensions: kinetic theory and continuum simulations *Phys. Rev. Lett.* **100** 178103

[385] Dombrowski C, Cisneros L, Chatkaew S, Kessler J O and Goldstein R E 2004 Self-concentration and large-scale coherence in bacterial dynamics *Phys. Rev. Lett.* **93** 098103

[386] Lacoste D and Bassereau P 2014 An update on active membranes ed G Pabst, N KuÄerka, M-P Nieh and J Katsaras *Liposomes, Lipid Bilayers and Model Membranes: From Basic Research to Application* pp 271–87 (Boca Raton, FL: CRC Press, Taylor and Francis Group)

[387] Turlier H and Betz T 2019 Unveiling the active nature of living-membrane fluctuations and mechanics *Annu. Rev. Condens. Matter Phys.* **10** 213–32

[388] Turlier H, Fedosov D A, Audoly B, Auth T, Gov N S, Sykes C, Joanny J-F, Gompper G and Betz T 2016 Equilibrium physics breakdown reveals the active nature of red blood cell flickering *Nat. Phys.* **12** 513–9

[389] Kruse K, Joanny J F, Jülicher F, Prost J and Sekimoto K 2005 Generic theory of active polar gels: a paradigm for cytoskeletal dynamics *Eur. Phys. J.* E **16** 5–16

[390] Prost J, Jülicher F and Joanny J F 2015 Active gel physics *Nat. Phys.* **11** 111–7

[391] Doostmohammadi A, Ignés-Mullol J, Yeomans J M and Sagués F 2018 Active nematics *Nat. Commun.* **9** 3246

[392] Stenhammar J, Wittkowski R, Marenduzzo D and Cates M E 2015 Activity-induced phase separation and self-assembly in mixtures of active and passive particles *Phys. Rev. Lett.* **114** 018301

[393] Banani S F, Lee H O, Hyman A A and Rosen M K 2017 Biomolecular condensates: organizers of cellular biochemistry *Nat. Rev. Mol. Cell Biol.* **18** 285–98

[394] Shin Y, Chang Y-C, Lee D S W, Berry J, Sanders D W, Ronceray P, Wingreen N S, Haataja M and Brangwynne C P 2018 Liquid nuclear condensates mechanically sense and restructure the genome *Cell* **175** 1481–91

[395] Zwicker D, Hyman A A and Jülicher F 2015 Suppression of Ostwald ripening in active emulsions *Phys. Rev.* E **92** 012317

[396] Guillot C and Lecuit T 2013 Mechanics of epithelial tissue homeostasis and morphogenesis *Science* **340** 1185–9

[397] Collinet C, Rauzi M, Lenne P-F and Lecuit T 2015 Local and tissue-scale forces drive oriented junction growth during tissue extension *Nat. Cell Biol.* **17** 1247–58

[398] Ranft J, Basan M, Elgeti J, Joanny J-F, Prost J and Jülicher F 2010 Fluidization of tissues by cell division and apoptosis *Proc. Natl Acad. Sci. USA* **107** 20863–8

[399] Popović M, Nandi A, Merkel M, Etournay R, Eaton S, Jülicher F and Salbreux G 2017 Active dynamics of tissue shear flow *New J. Phys.* **19** 033006

[400] Girkin J M and Carvalho M T 2018 The light-sheet microscopy revolution *J. Opt.* **20** 053002

[401] Takebe T and Wells J M 2019 Organoids by design *Science* **364** 956–9

[402] Haas P A and Goldstein R E 2019 Nonlinear and nonlocal elasticity in coarse-grained differential-tension models of epithelia *Phys. Rev.* E **99** 022411

[403] Vagne Q, Vrel J-P and Sens P 2020 A minimal self-organisation model of the Golgi apparatus *eLife* **9** e47318

[404] Thompson D W 1942 *On Growth and Form* 2nd edn (Cambridge: Cambridge University Press)

[405] Discher D E, Janmey P and Wang Y-L 2005 Tissue cells feel and respond to the stiffness of their substrate *Science* **310** 1139–43

[406] Harris A R, Jreij P and Fletcher D A 2016 Mechanotransduction by the actin cytoskeleton: converting mechanical stimuli into biochemical signals *Annu. Rev. Biophys.* **47** 617–31

[407] Engler A J, Sen S, Sweeney H L and Discher D E 2006 Matrix elasticity directs stem cell lineage specification *Cell* **126** 677–89

[408] Fernandez-Sanchez M-E, Brunet T, Röper J-C and Farge E 2015 Mechanotransduction's impact on animal development, evolution, and tumorigenesis *Annu. Rev. Cell Dev. Biol.* **31** 373–97

[409] Broders-Bondon F, Nguyen Ho-Bouldoires T H, Fernandez-Sanchez M-E and Farge E 2018 Mechanotransduction in tumor progression: the dark side of the force *J. Cell Biol.* **217** 1571–87

[410] Roca-Cusachs P, Conte V and Trepat X 2017 Quantifying forces in cell biology *Nat. Cell Biol.* **19** 742–51

[411] Geiger B, Bershadsky A, Pankov R and Yamada K M 2001 Transmembrane crosstalk between the extracellular matrix and the cytoskeleton *Nat. Rev. Mol. Cell Biol.* **2** 793–805

[412] Martino F, Perestrelo A R, Vinarský V, Pagliari S and Forte G 2018 Cellular mechano-transduction: from tension to function *Front. Physiol.* **9** 824

[413] Charras G and Yap A S 2018 Tensile forces and mechanotransduction at cell–cell junctions *Curr. Biol.* **28** R445–57

[414] Gottlieb P A (ed) 2017 *Piezo Channels, Volume 79 of Current Topics in Membranes* (New York: Academic)

[415] Eder D, Aegerter C and Basler K 2017 Forces controlling organ growth and size *Mech. Develop.* **144** 53–61

[416] Van Essen D C, Smith S M, Barch D M, Behrens T E J, Yacoub E and Ugurbil K 2013 The WU-Minn human connectome project: an overview *NeuroImage* **80** 62–79

[417] Zecchi K A and Heimburg T 2021 Non-linear conductance, rectification, and mechano-sensitive channel formation of lipid membranes *Front. Cell Dev. Biol.* **8** 1751

[418] Deubner J, Coulon P and Diester I 2019 Optogenetic approaches to study the mammalian brain *Curr. Opin. Struct. Biol.* **57** 157–63

[419] Calarco J A and Samuel A D T 2019 Imaging whole nervous systems: insights into behavior from brains inside bodies *Nat. Meth.* **16** 14–5

[420] Dragomir E I, Åtih V and Portugues R 2020 Evidence accumulation during a sensorimotor decision task revealed by whole-brain imaging *Nat. Neurosci.* **23** 85–93

[421] Gao R *et al* 2019 Cortical column and whole-brain imaging with molecular contrast and nanoscale resolution *Science* **363** 543–8

[422] Richards B A *et al* 2019 A deep learning framework for neuroscience *Nat. Neurosci.* **22** 1761–70

[423] Bassett D S and Sporns O 2017 Network neuroscience *Nat. Neurosci.* **20** 353–64

[424] Schrödinger E 1944 *What is Life? The Physical Aspect of the Living Cell* (Cambridge: Cambridge University Press)

[425] Xu C, Hu S and Chen X 2016 Artificial cells: from basic science to applications *Mat. Today* **19** 516–32

[426] https://nature.com/collections/ngmpdftlqt/

[427] Zwicker D, Seyboldt R, Weber C A, Hyman A A and Jülicher F 2017 Growth and division of active droplets provides a model for protocells *Nat. Phys.* **13** 408–13

[428] Jia H and Schwille P 2019 Bottom-up synthetic biology: reconstitution in space and time *Curr. Opin. Biotech.* **60** 179–87

[429] Gibson D G *et al* 2010 Creation of a bacterial cell controlled by a chemically synthesized genome *Science* **329** 52–6

[430] Breuer M *et al* 2019 Essential metabolism for a minimal cell *eLife* **8** e36842

[431] Braun D, Goddard N L and Libchaber A 2003 Exponential DNA replication by laminar convection *Phys. Rev. Lett.* **91** 158103

[432] Kreysing M, Keil L, Lanzmich S and Braun D 2015 Heat flux across an open pore enables the continuous replication and selection of oligonucleotides towards increasing length *Nat. Chem.* **7** 203–8

[433] Bell G and Mooers A O 1997 Size and complexity among multicellular organisms *Biol. J. Linn. Soc.* **60** 345–63

[434] Weismann A 1892 *Essays on Heredity and Kindred Biological Problems* (Oxford: Oxford University Press)

[435] Huxley J S 1912 *The Individual in the Animal Kingdom* (Cambridge: Cambridge University Press)

[436] Solari C A, Kessler J O and Michod R E 2006 A hydrodynamics approach to the evolution of multicellularity: flagellar motility and germ-soma differentiation in volvocalean green algae *Am. Nat.* **167** 537–54

[437] Goldstein R E 2015 Green algae as model organisms for biological fluid dynamics *Annu. Rev. Fluid Mech.* **47** 343–75

[438] Brunet T and King N 2017 The origin of animal multicellularity and cell differentiation *Dev. Cell* **43** 124–40

[439] Sella G and Hirsh A E 2005 The application of statistical physics to evolutionary biology *Proc. Natl Acad. Sci. USA* **102** 9541

[440] Drossel B 2001 Biological evolution and statistical physics *Adv. Phys.* **50** 209–95

[441] Wolf Y I, Katsnelson M I and Koonin E V 2018 Physical foundations of biological complexity *Proc. Natl Acad. Sci. USA* **115** E8678

[442] Kasting J F 1993 Earth's early atmosphere *Science* **259** 920–6

[443] Kasting J F, Toon O B and Pollack J B 1988 How climate evolved on the terrestrial planets *Sci. Am.* **258** 90–7

[444] Bottke W F, Vokrouhlický D, Marchi S, Swindle T, Scott E R D, Weirich J R and Levison H 2015 Dating the Moon-forming impact event with asteroidal meteorites *Science* **348** 321–3

[445] Schlichting H E and Mukhopadhyay S 2018 Atmosphere impact losses *Space Sci. Rev.* **214** 34

[446] Lebrun T, Massol H, ChassefièRe E, Davaille A, Marcq E, Sarda P, Leblanc F and Brandeis G 2013 Thermal evolution of an early magma ocean in interaction with the atmosphere *J. Geophys. Res. (Planets)* **118** 1155–76

[447] Stüeken E E, Som S M and Claire M *et al* 2020 Mission to planet Earth: the first two billion years *Space Sci. Rev.* **216** 31

[448] Pahlevan K, Schaefer L and Hirschmann M M 2019 Hydrogen isotopic evidence for early oxidation of silicate Earth *Earth Planet. Sci. Lett.* **526** 115770

[449] Sleep N H 2010 The Hadean-Archaean environment *Cold Spring Harb. Perspect. Biol.* **2** a002527

[450] Bekaert D V, Broadley M W and Marty B 2020 The origin and fate of volatile elements on Earth revisited in light of noble gas data obtained from comet 67P/Churyumov-Gerasimenko *Sci. Rep.* **10** 5796

[451] Mezger K, Maltese A and Vollstaedt H 2021 Accretion and differentiation of early planetary bodies as recorded in the composition of the silicate earth *Icarus* **365** 114497

[452] Mojzsis S J, Brasser R, Kelly N M, Abramov O and Werner S C 2019 Onset of Giant planet migration before 4480 million years ago *Astrophys. J.* **881** 44

[453] Fuller-Rowell T, Solomon S, Roble R and Viereck R 2004 Impact of solar EUV, XUV, and x-ray variations on Earth's atmosphere *Solar Variability and its Effects on Climate. Geophysical Monograph 141* (AGU Publications) vol 141 p 341

[454] Lammer H, Leitzinger M and Scherf M *et al* 2020 Constraining the early evolution of Venus and Earth through atmospheric Ar, Ne isotope and bulk K/U ratios *Icarus* **339** 113551

[455] Ribas I, Guinan E F, Güdel M and Audard M 2005 Evolution of the solar activity over time and effects on planetary atmospheres. I. High-energy irradiances (1–1700 Å) *Astrophys. J.* **622** 680–94

[456] Sagan C and Mullen G 1972 Earth and Mars: evolution of atmospheres and surface temperatures *Science* **177** 52–6

[457] Catling D C and Kasting J F 2007 *Planetary Atmospheres and Life* (London: Cambridge University Press)

[458] Elkins-Tanton L T 2008 Linked magma ocean solidification and atmospheric growth for Earth and Mars *Earth Planet. Sci. Lett.* **271** 181–91

[459] Guzmán-Marmolejo A, Segura A and Escobar-Briones E 2013 Abiotic production of methane in terrestrial planets full access *Astrobiology* **13** 550–9

[460] Atreya S K, Adams E Y, Niemann H B, Demick-Montelara J E, Owen T C, Fulchignoni M, Ferri F and Wilson E H 2006 Titan's methane cycle *Planet. Space Sci.* **54** 1177–87

[461] Pavlov A A, Brown L L and Kasting J F 2001 UV shielding of NH_3 and O_2 by organic hazes in the Archean atmosphere *J. Geophys. Res.* **106** 23267–88

[462] Heller R, Duda J-P, Winckler M, Reitner J and Gizon L 2021 Habitability of the early Earth: liquid water under a faint young Sun facilitated by strong tidal heating due to a nearby Moon *Palaontol. Z.* **95** 563–75

[463] Pope E C, Bird D K and Rosing M T 2012 Isotope composition and volume of Earth's early oceans *Proc. Natl Acad. Sci.* **109** 4371–6

[464] Sharp Z D, McCubbin F M and Shearer C K 2013 A hydrogen-based oxidation mechanism relevant to planetary formation *Earth Planet. Sci. Lett.* **380** 88–97

[465] Marty B, Avice G, Bekaert D V and Broadley M W 2018 Salinity of the Archaean oceans from analysis of fluid inclusions in quartz *C. R. Geosci.* **350** 154–63

[466] Matthes K, Funke B and Andersson M E *et al* 2017 Solar forcing for CMIP6 (v3.2) *Geosci. Model Dev.* **10** 2247–302

[467] Pascal R, Pross A and Sutherland J D 2013 Towards an evolutionary theory of the origin of life based on kinetics and thermodynamics *Open Biol* **3** 130156

[468] James Cleaves H, Chalmers J H, Lazcano A, Miller S L and Bada J L 2008 A reassessment of prebiotic organic synthesis in neutral planetary atmospheres *Origins Life Evol. Biosphere* **38** 105–15

[469] Martin W F, Baross J, Kelley D and Russell M J 2008 Hydrothermal vents and the origin of life *Nat. Rev. Microbiol.* **6** 805–14

[470] Shock E L and Schulte M D 1998 Organic synthesis during fluid mixing in hydrothermal systems *J. Geophys. Res.* **103** 28513–28

[471] Bertrand M, Chabin A, Brack A, Cottin H, Chaput D and Westall F 2012 The PROCESS experiment: exposure of amino acids in the EXPOSE—experiment on the International Space Station and in laboratory simulations *Astrobiology* **12** 426–35

[472] Marty B, Alexander C M O and Raymond S N 2013 Primordial Origins of Earth's Carbon *Rev. Mineral. Geochem.* **75** 149–181

[473] Duprat J, Dobrică E and Engrand C *et al* 2010 Extreme deuterium excesses in ultra-carbonaceous micrometeorites from central antarctic snow *Science* **328** 742

[474] Schmitt-Kopplin P, Gabelica Z and Gougeon R D *et al* 2010 High molecular diversity of extraterrestrial organic matter in Murchison meteorite revealed 40 years after its fall *Proc. Natl Acad. Sci.* **107** 2763–8

[475] Lowe D R, Byerly G R and Kyte F T *et al* 2003 Spherule beds 3.47–3.24 billion years old in the Barberton Greenstone Belt, South Africa: a record of large meteorite impacts and their influence on early crustal and biological evolution *Astrobiology* **3** 7–48

[476] Lowe D R, Byerly G R and Kyte F T 2014 Recently discovered 3.42–3.23 Ga impact layers, Barberton Belt, South Africa: 3.8 Ga detrital zircons, Archean impact history, and tectonic implications *Geology* **42** 747–50

[477] Gourier D, Binet L and Calligaro T *et al* 2019 Extraterrestrial organic matter preserved in 3.33 Ga sediments from Barberton, South Africa *Geochim. Cosmochim. Acta* **258** 207–25

[478] Sasselov D D, Grotzinger J P and Sutherland J D 2020 The origin of life as a planetary phenomenon *Sci. Adv.* **6** eaax3419

[479] Pascal R 2012 Suitable energetic conditions for dynamic chemical complexity and the living state *J. Syst. Chem.* **3**

[480] Baross J A and Hoffman S E 1985 Submarine hydrothermal vents and associated gradient environments as sites for the origin and evolution of life *Orig. Life* **15** 327–45 :

[481] Russell M J and Hall A J 1997 The emergence of life from iron monosulphide bubbles at a submarine hydrothermal redox and pH front *J. Geol. Soc.* **154** 377–402

[482] Russell M J, Hall A J and Martin W 2010 Serpentinization as a source of energy at the origin of life *Geobiology* **8** 355–71

[483] Westall F, Hickman-Lewis K and Hinman N *et al* 2018 A hydrothermal-sedimentary context for the origin of life *Astrobiology* **18** 259–93

[484] Cottin H, Guan Y Y and Noblet A *et al* 2012 The PROCESS experiment: an astrochemistry laboratory for solid and gaseous organic samples in low-earth orbit *Astrobiology* **12** 412–25

[485] Deamer D W, Singaram S, Rajamani S, Kompanichenko V and Guggenheim S 2006 Self-assembly processes in the prebiotic environment *Phil. Trans. R. Soc.* B **361** 1809–18

[486] Tartese R, Chaussidon M, Gurenko A, Delarue F and Robert F 2017 Archean oceans reconstructed from oxygen isotope composition of early-life remnants *Geochem. Persp. Lett.* **3** 55 65

[487] Usoskin I G 2017 A history of solar activity over millennia *Living Rev. Sol. Phys.* **14** 3

[488] Badalyan O G and Obridko V N 2011 North-south asymmetry of the sunspot indices and its quasi-biennial oscillations *Nature* **16** 357–65

[489] Kane R P 2005 Differences in the quasi-biennial oscillation and quasi-triennial oscillation characteristics of the solar, interplanetary, and terrestrial parameters *J. Geophys. Res. (Space Phys.)* **110** A01108

[490] Broomhall A-M, Fletcher S T and Salabert D *et al* 2011 Are short-term variations in solar oscillation frequencies the signature of a second solar dynamo? *J. Phys.: Conf. Ser.* **271** 012025 (GONG-SoHO 24: A New Era of Seismology of the Sun and Solar-Like Stars)

[491] Manuel G 2007 The Sun in time: activity and environment *Living Rev. Sol. Phys.* **4** 3

[492] Claire M W, Sheets J, Cohen M, Ribas I, Meadows V S and Catling D C 2012 The evolution of solar flux from 0.1 nm to 160 μm: quantitative estimates for planetary studies *Astrophys. J.* **757** 95

[493] Bahcall J N, Pinsonneault M H and Basu S 2001 Solar models: current epoch and time dependences, neutrinos, and helioseismological properties *Astrophys. J.* **555** 990–1012

[494] Lean J L, Rottman G J, Lee Kyle H, Woods T N, Hickey J R and Puga L C 1997 Detection and parameterization of variations in solar mid- and near- ultraviolet radiation (200–400 nm) *J. Geophys. Res.* **102** 29939–56

[495] Soderblom D R, Stauffer J R, MacGregor K B and Jones B F 1993 The evolution of angular momentum among zero-age main-sequence solar-type stars *Astrophys. J.* **409** 624

[496] David Dorren J and Guinan E F 1994 HD 129333: the Sun in its infancy *Astrophys. J.* **428** 805

[497] Brun A S and Browning M K 2017 Magnetism, dynamo action and the solar-stellar connection *Living Rev. Sol. Phys.* **14** 4

[498] Canuto V M, Levine J S, Augustsson T R and Imhoff C L 1982 UV radiation from the young Sun and oxygen and ozone levels in the prebiological palaeoatmosphere *Nature* **296** 816–20

[499] Meert J G, Levashova N M, Bazhenov M L and Landing E 2016 Rapid changes of magnetic field polarity in the late Ediacaran: linking the Cambrian evolutionary radiation and increased UV-B radiation *Gondwana Res.* **34** 149–57

[500] Airapetian V S, Glocer A, Gronoff G, Hébrard E and Danchi W 2016 Prebiotic chemistry and atmospheric warming of early Earth by an active young Sun *Nat. Geosci.* **9** 452–5

[501] Austin J, Tourpali K and Rozanov E *et al* 2008 Coupled chemistry climate model simulations of the solar cycle in ozone and temperature *J. Geophys. Res. (Atmos.)* **113** D11306

[502] Gray L J, Beer J, Geller M and Haigh J D *et al* 2010 Solar influences on climate *Rev. Geophys.* **48** RG400i

[503] Haigh J D 1996 The impact of solar variability on climate *Science* **272** 981–4

[504] Matthes K, Kuroda Y, Kodera K and Langematz U 2006 Transfer of the solar signal from the stratosphere to the troposphere: northern winter *J. Geophys. Res.* **111** D06108

[505] Matthes K, Biastoch A and Wahl S *et al* 2020 The flexible ocean and climate infrastructure version 1 (FOCI1): mean state and variability *Geosci. Model Dev.* **13** 2533–68

[506] Sinnhuber M, Nieder H and Wieters N 2012 Energetic particle precipitation and the chemistry of the mesosphere/lower thermosphere *Surv. Geophys.* **33** 1281–334

[507] Dunne E M, Gordon H and Kürten A *et al* 2016 Global atmospheric particle formation from CERN CLOUD measurements *Science* **354** 1119–24

[508] Mojzsis S J, Harrison T M and Pidgeon R T 2001 Oxygen-isotope evidence from ancient zircons for liquid water at the Earth's surface 4,300 Myr ago *Nature* **409** 178–81

[509] Wilde S A, Valley J W, Peck W H and Graham C M 2001 Evidence from detrital zircons for the existence of continental crust and oceans on the Earth 4.4 Gyr ago *Nature* **409** 175–8

[510] Whitehouse M J, Nemchin A A and Pidgeon R T 2017 What can Hadean detrital zircon really tell us? A critical evaluation of their geochronology with implications for the interpretation of oxygen and hafnium isotopes *Gondwana Res.* **51** 78 91

[511] Arndt N T and Nisbet E G 2012 Processes on the young earth and the habitats of early life *Annu. Rev. Earth Planet. Sci.* **40** 521–49

[512] Kamber B S 2015 The evolving nature of terrestrial crust from the Hadean, through the Archaean, into the Proterozoic *Precambrian Res.* **258** 48 82

[513] Wang Z and Becker H 2013 Ratios of S, Se and Te in the silicate Earth require a volatile-rich late veneer *Nature* **499** 328–31

[514] Morbidelli A and Wood B J 2015 *Late Accretion and the Late Veneer* (Washington, DC: American Geophysical Union (AGU))

[515] Biggin A J, de Wit M J, Langereis C G, Zegers T E, Voûte S, Dekkers M J and Drost K 2011 Palaeomagnetism of Archaean rocks of the Onverwacht Group, Barberton Greenstone Belt (southern Africa): evidence for a stable and potentially reversing geo-magnetic field at ca. 3.5 Ga *Earth Planet. Sci. Lett.* **302** 314–28

[516] Usui Y, Tarduno J A, Watkeys M, Hofmann A and Cottrell R D 2009 Evidence for a 3.45-billion-year-old magnetic remanence: hints of an ancient geodynamo from conglomerates of South Africa *Geochem. Geophys. Geosyst.* **10** Q09Z07

[517] Tarduno J A, Cottrell R D and Bono R K *et al* 2020 Paleomagnetism indicates that primary magnetite in zircon records a strong Hadean geodynamo *Proc. Natl Acad. Sci.* **117** 2309–18

[518] Borlina C S, Weiss B P and Lima E A *et al* 2020 Reevaluating the evidence for a Hadean-Eoarchean dynamo *Sci. Adv.* **6** eaav9634

[519] Stevenson D J 2001 Mars' core and magnetism *Nature* **412** 214–9

[520] Lockwood M, Harrison R G, Woollings T and Solanki S K 2010 Are cold winters in Europe associated with low solar activity? *Environ. Res. Lett.* **5** 024001

Part II

Physics developments to tackling major issues affecting the lives of citizens

IOP Publishing

EPS Grand Challenges

Physics for Society in the Horizon 2050

Mairi Sakellariadou, Claudia-Elisabeth Wulz, Kees van Der Beek, Felix Ritort, Bart van Tiggelen, Ralph Assmann, Giulio Cerullo, Luisa Cifarelli, Carlos Hidalgo, Felicia Barbato, Christian Beck, Christophe Rossel and Luc van Dyck

Chapter 5

Physics for health

Ralph Assmann, Darwin Caldwell, Giulio Cerullo, Henry Chapman, Dana Cialla-May, Marco Durante, Angeles Faus-Golfe, Jens Hellwage, Susanne Hellwage, Christoph Krafft, Timo Mappes, Thomas G Mayerhöfer, Thierry Mora, Andreas Peters, Chiara Poletto, Juergen Popp, Felix Ritort, Petra Rösch, Lucio Rossi, Marta Sales-Pardo, Iwan Schie, Michael Schmitt, Friedrich Simmel and Aleksandra Walczak

5.1 Introduction

Ralph W Assmann[1,4], Giulio Cerullo[2] and Felix Ritort[3]
[1]Deutsches Elektronen-Synchrotron (DESY), Hamburg, Germany
[2]Politecnico di Milano, Milan, Italy
[3]Universitat de Barcelona, Barcelona, Spain
[4]Present affiliation: GSI Helmholtzzentrum für Schwerionenforschung, Darmstadt, Germany

The fundamental research on the physics of elementary particles and nature's fundamental forces led to numerous spin-offs and has tremendously helped human well-being and health. Prime examples include the electron-based generation of x-rays for medical imaging, the use of electrical shocks for treatment of heart arrhythmia, the exploitation of particle's spin momenta for spin tomography (NMR) of patients, and the application of particle beams for cancer treatment. Tens of thousands of lives are saved every year from the use of those and other physical principles. A strong industry has developed in many countries, employing hundreds of thousands of physicists, engineers, and technicians. Industry is designing, producing, and deploying the technology that is based on advances in fundamental physics.

Major research centers have been established and provide cutting-edge beams of particles and photons for medical and biological research, enabling major advances in

© 2024 The Authors.

the understanding of structural biology, medical processes, viruses, bacteria, and possible therapies. Those research infrastructures serve tens of thousands of users every year and help them in their research. Modern hospitals are equipped with a large range of high technology machines that employ physics principles for performing high-resolution medical imaging and powerful patient treatment. Professors and students at universities use even more powerful machines for conducting basic research in increasingly interdisciplinary fields like biophysics and robotics. New professions have developed involving physicists and reaching out to other domains. We mention the rapidly growing professions of radiologists, health physicists, and biophysicists.

While physics spin-offs for health are being heavily exploited, physicists in fundamental research keep advancing their knowledge and insights on the bio-chemical mechanisms at the origin of diseases. New possibilities and ideas keep constantly emerging, creating unique added value for society from fundamental physics research. This chapter does not aim to provide a full overview of the benefits of physics for health. Instead, the authors concentrate on some of the hot topics in physics- and health-related research. The focus is put on new developments, possible new opportunities, and the path to new applications in health.

Angeles Faus-Golfe and Andreas Peters describe the role of particle accelerators and the use of their beams for irradiating and destroying cancer cells. State-of-the-art machines and possibilities for new irradiation principles (i.e., the FLASH effect) are introduced. As physics knowledge and technology advance, tumors can be irradiated more and more precisely, damage to neighboring tissue can be reduced, and irradiation times can be shortened.

Darwin Caldwell looks at the promise and physics-based development of robotic systems in the macroscopic world, where they are complementing human activities in a number of tasks from diagnosis to therapy. Friedrich Simmel looks at the molecular and cell-scale world and explains how nanorobotics, biomolecular robotics, and synthetic biology are emerging as additional tools for human health (e.g., as nano-carriers of medication that is delivered precisely).

Henry Chapman and Jürgen Popp describe the benefits of light for health. Jürgen Popp is considering the use of lasers that have advanced tremendously in recent years in terms of power stability and wavelength tunability. Modern lasers are used in several crucial roles in cell imaging, disease diagnosis, and precision surgery. Henry Chapman considers the use of free-electron lasers for understanding features and processes in structural biology. He shows that the advance of those electron accelerator-based machines has allowed tremendous progress in the determination of the structures of biomolecules and the understanding of their function.

Aleksandra Walczak, Chiara Poletto, Thierry Mora, and Marta Sales describe physics research against pandemics, a multi-disciplinary problem at the crossing of immunology, evolutionary biology, and networks science. Pandemics are also multi-scale problems at the spatial and temporal levels: from the small pathogen to the large organism; and from the infective process at cellular scale (hours) to its propagation community-wide (months). Simple mathematical models such as SIR (susceptible-infected-recovered) have been a source of inspiration for physicists who model key quantities at an epidemic outbreak, such as the effective reproductive number R, in situations where a disease has already spread. A prominent example is the recent COVID-19 pandemic that has been more than a health and economic crisis. It illustrates our vulnerability where interdisciplinary and multilateral science play a crucial role addressing a global challenge such as this.

Promise and progress in further diagnostics and therapies are also considered. Lucio Rossi explains the progress in magnetic field strength as it can be achieved with superconducting magnets, while Marco Durante discusses the progress in charged particle therapy for medical physics.

5.2 Accelerators for health

Angeles Faus-Golfe[1] and Andreas Peters[2]
[1]Laboratoire de Physique des 2 Infinis Irene Joliot-Curie,IN2P3-CNRS—Université Paris-Saclay, Orsay, France
[2]HIT GmbH at University Hospital Heidelberg, Germany

Energetic particles, high-energy photons (x-rays and gamma rays), electrons, protons, neutrons, and various atomic nuclei and more exotic species are indispensable tools in improving human health.

The potential of accelerator-reliant therapy and diagnostic techniques has increased considerably over the past decades, playing an increasingly important role in identifying and curing affections, such as cancer, that otherwise are difficult to treat; they also help to understand how major organs such as the brain function and thus to determine the underlying causes of diseases of growing societal significance such as dementia.

5.2.1 Motivation to use and expand x-rays and particle therapy

The use of **x-rays** in radiotherapy (RT) is now the most common method of RT for cancer treatment. While x-ray therapy is a mature technology there is room for improvement. The current challenges are related to the **accurate delivery of x-rays to tumours** involving sophisticated techniques to **combine imaging and therapy**. In particular, the ability to achieve better definition and efficiency in 4D reconstruction (3D over time) distinguishing volumes of functional biological significance. Further technical improvements to reduce the risk of a treatment differs from the prescription and moving towards 'personalised treatment planing' are being made. Some of these techniques such as: image-guided radiation therapy (IGRT), control of the dose administered to the patient (*in vivo* dosimetry) or adaptive RT to take into account the morphology changes in the patient, are the state of the art and are being implemented in the routine operation of these types of facilities. An example is the so-called MR linac, which provides magnetic resonance (MR) and RT treatment at the same time. Finally, the reduction of the accelerator costs and the increase of reliability/availability in challenging environments are also important research challenges to expand this kind of RT in low- and middle-income countries (LMICs).

Low-energy electrons have historically been used to treat cancer for more than five decades, but mostly for the treatment of superficial tumours given their very limited penetration depth. However, this limitation can be overcome if the electron energy is increased between 50 and 200 MeV (i.e., **very high-energy electrons, VHEE;** figure 5.1). With the recent developments of high-gradient normal conducting (NC) radio frequency **(RF) linac technology** (figure 5.2) (CLIC Project n.d.) or even the novel acceleration techniques such as the **laser-plasma accelerator** (LPA) (figure 5.3), VHEE offer a very promising option for anticancer RT. Theoretically, VHEE beams offers several benefits. The ballistic and dosimetry properties of VHEE provide small-diameter beams that could be scanned and focused easily, enabling

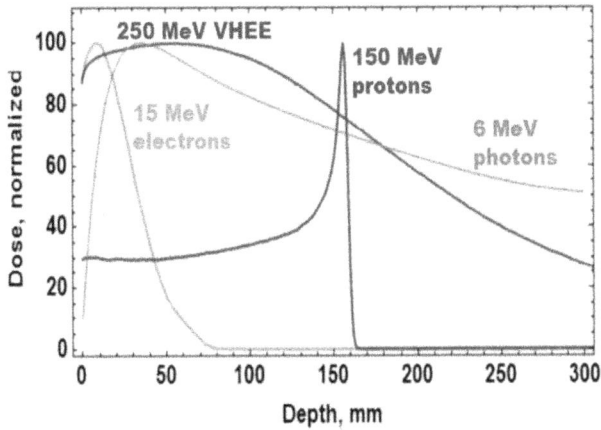

Figure 5.1. Dose profile for various particle beams in water (beam widths $r = 0.5$ cm).

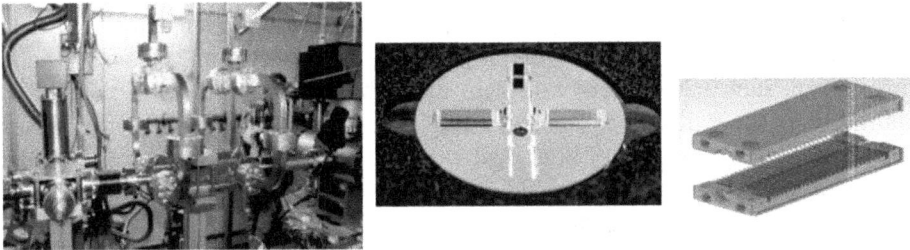

Figure 5.2. CLIC RF X-band cavity prototype (12 GHz, 100 MV m^{-1}).

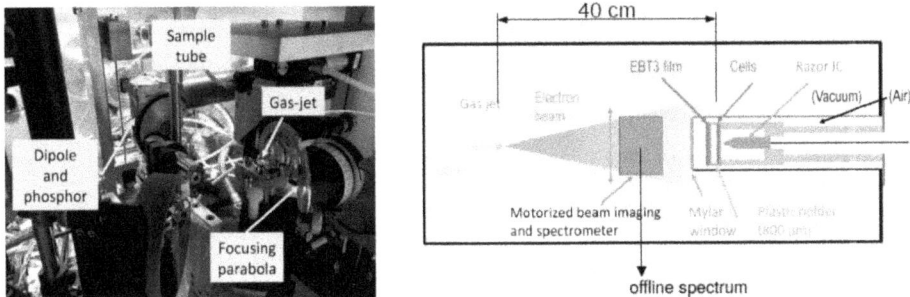

Figure 5.3. Setup of Salle Noire for cell irradiation at LOA-IPP laser-driven wakefield electron accelerator.

finer resolution for intensity-modulated treatments than is possible with photons beams. Electron accelerators are more compact and cheaper than proton therapy accelerators. Finally, VHEE beams can be operated at very high-dose rates and fast electromagnetic scanning providing uniform dose distribution throughout the target and allowing for unforeseen RT modalities in particular the FLASH-RT.

FLASH-RT is a paradigm-shifting method for delivering ultra-high doses within an extremely short irradiation time (tenths of a second). The technique has recently been shown to preserve normal tissue in various species and organs while still maintaining anti-tumour efficacy equivalent to conventional RT at the same dose level, in part due to decreased production of toxic reactive oxygen species. The 'FLASH effect' has been shown to take place with electron, photon, and more recently for proton beams. However, the potential advantage of using electron beams lies in the intrinsically higher dose that can potentially be reached compared to protons and photons, especially over large areas as would be needed for large tumours. Most of the preclinical data demonstrating the increased therapeutic index of FLASH has used a single fraction and hypo-fractionated regimen of RT and using 4–6 MeV electron beams, which do not allow treatments of deep-seated tumours and trigger large lateral penumbra (figure 5.4). This problem can be solved by increasing the electron energy to values higher than 50 MeV (VHEE), where the penetration depth is larger.

Many challenges, both technological and biological, have to be addressed and overcome for the ultimate goal of using VHEE and VHEE-FLASH as an innovative modality for effective cancer treatment with minimal damage to healthy tissues.

From the accelerator technology point of view the major challenge for VHEE-RT is the demonstration of a suitable high-gradient acceleration system, whether conventional, such a X-band or not, with the stability, reliability, and repeatability required to be operated in a medical environment. In particular, for the VHEE-FLASH is the delivery of very high dose rate, possibly over a large area, providing uniform dose distribution throughout the target (Faus-Golfe 2020).

All this asks for a large beam test activity in order to experimentally characterize VHEE beams and their ability to produce the FLASH effect and provide a test bed for the associated technologies. It is also important to compare the properties of the electron beams depending on the way they are produced (RF linac or LPA technologies). Preliminary VHEE experimental studies have been realized using NC RF accelerator facilities in NLTCA at SLAC and CLEAR at CERN, giving very promising results for the use of VHEE electron beams. Furthermore, some

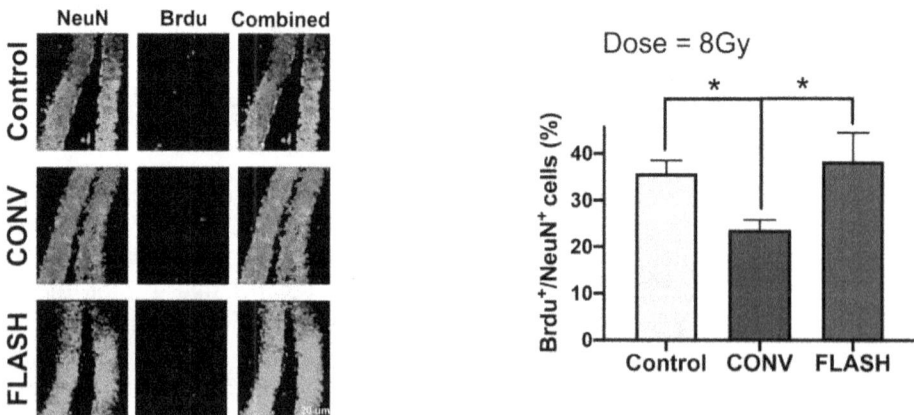

Figure 5.4. FLASH preservation of the neurogenic niche in juvenile mice (courtesy of C Limoli).

experimental tests have been carried out with laser-plasma sources at ELBE-DRACO in HZDR, at LOA in IPP, and at the SCAPA facility at the University of Strathclyde. In particular for the FLASH, experimental studies have been realized at low energies with the Kinetron linac at the Institute Curie at Orsay and the eRt6-Oriatron linac at CHUV and at very high energies at CLEAR at CERN.

Proton and ion beam therapy has growing potential in dealing with difficult-to-treat tumours, for example, because of the risk of damaging neighbouring sensitive tissues such as the brainstem or visual nerves in the case of head tumour treatments. Also, some treatments may benefit from the use of particles that deliver doses with greater radiobiological effectiveness (RBE) and higher local precision, notably carbon, and in the near future also helium ions.

Recent investigations using ultra-short and ultra-high dose rates (called FLASH) of electron beams showed growth retardation of tumours with the same effect as in conventional therapy, but with minimized impact to the surrounding tissue. FLASH with proton and ion beams is expected to offer additional healthy tissue sparing from beam stopping in the tumour—but the research on this topic is still not completed, and the experiments and evaluations are ongoing. Healthy tissue sparing with FLASH would enable a dose increase as well as a significant reduction of treatment time without additional aggravations. These new key findings may influence the accelerator development for particle therapy considerably in the next future.

In x-ray radiotherapy the integration of imaging devices for image-guided radiation therapy is standard. In-room CTs and now the introduction of magnetic resonance imaging (MRI) in the form of the recent clinical adoption of the MR linac show these necessary advances for better positioning of the patients and online observation of the tumours. Whereas CTs are also introduced in particle therapy facilities, the combination of MRI and hadron beams is delicate due to the high magnetic fields of these diagnostic devices diverting the proton/ion beam. In

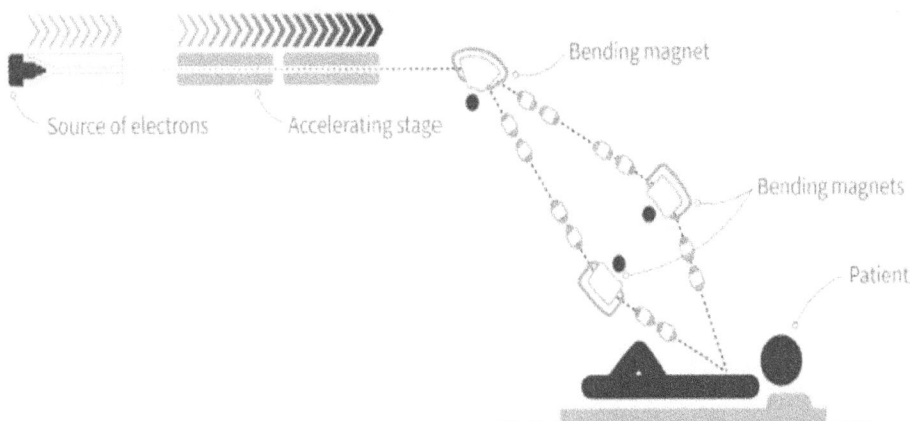

Figure 5.5. FLASH facility cartoon for CHUV—Lausanne.

addition, also the stray fields may lead to effects on the treatment systems in the near future. The investigations on this topic have started (see below).

In the last three decades about 25 proton and 4 hadron therapy facilities were built in Europe (figure 5.6). But only a few more are actually under planning or construction, among them only one hadron facility project (SEEIIST[1]). Thus, new

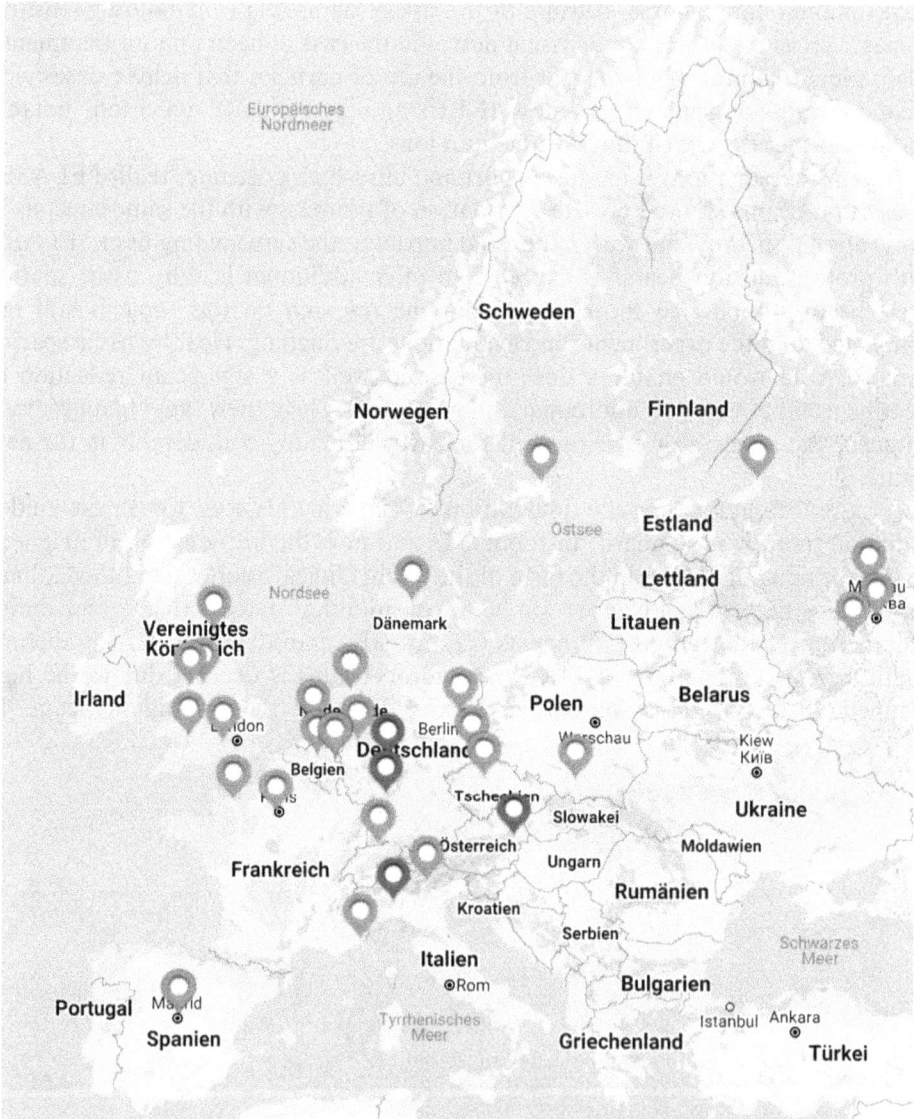

Figure 5.6. Particle therapy facilities in Europe; see https://www.ptcog.ch/index.php/facilities-world-map. SEEIIST—The South East European International Institute for Sustainable Technologies http://seeiist.eu/.

---

[1] https://seeiist.eu/; SEEIIST—The South East European International Institute for Sustainable Technologies.

efforts are required to make these techniques smaller, cheaper (in investment and operating costs), and easier to maintain, which will be discussed in a separate chapter below showing potentials for the next three decades.

5.2.2 Further developments in the next decades

5.2.2.1 Introduction of helium for regular treatment

For protons and 3He/4He similar radio-biological properties have been determined, but the lateral scattering is reduced by nearly 50% in the case of helium ions versus protons (figure 5.7). In recent years, helium ions again became of interest for clinical cases where neither protons nor carbon ions are ideally suited, especially for treating paediatric tumours. Currently, patient irradiations with scanned ^{4}He ions at the Heidelberg Ion beam Therapy Center (HIT) in Germany are used only in 'treatment attempts' ('individuelle Heilversuche') and will go into regular operation by mid 2024. Other ion therapy facilities in Europe (e.g., CNAO in Italy and MedAustron in Austria) have also started technical upgrades to produce helium ion beams in the near future.

Recent studies have shown that 3He ions can be a viable alternative to ^{4}He, as they can produce comparable dose profiles, demanding slightly higher kinetic energy per nucleon, but less total kinetic energy. This results in 20% less magnetic rigidity needed for the same penetration depth which may be of importance for the design of future compact therapy accelerators like superconducting synchrotrons or energy-variable cyclotrons.

5.2.2.2 Image-guided hadron therapy using MRI

The observation of the patient's position during treatment has become more and more a standard procedure. 3D camera systems make it possible to control the location of the patient within tenths of millimetres observing the exterior of the body. But organs struck by tumours can move dramatically within the abdomen,

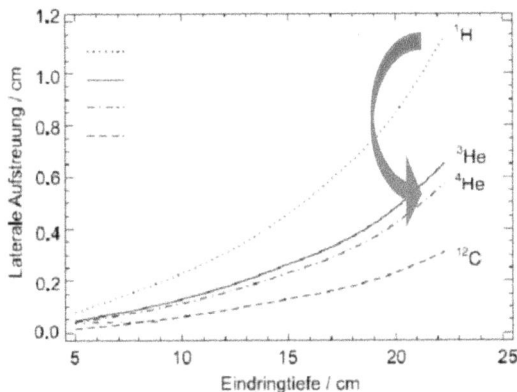

Figure 5.7. Lateral scattering of different light ions, adapted from Fiedler (2008).

which is not visible from outside. To observe the soft tissues an MRI scanner is the preferred diagnostic tool. But due to the moderate to high fields used the charged particle beam is affected. In addition, conventional MRI scanners are not constructed to have an inlet for radiation to be applied in parallel to the diagnostic procedure. Nevertheless, studies like ARTEMIS at HIT in Heidelberg have been started to investigate possible arrangements of open, low-field MRIs (figure 5.8). Still the deflection and/or distortion of the beam can be observed, but algorithms will be developed to cancel out this influence of the MRI's magnetic field. A second goal of such studies is to optimize the design of the measurement coils to spare out space for the beam entrance fields. A further aspect should not be underestimated—the impact of the magnetic (stray) fields on the QA devices and the online monitoring detectors used during treatment. All these have to be examined and possibly adapted to be magnetic field compatible; at worst new or alternative detector principles have to be applied. A lot of technical solutions have to be found in the next years before MRI scanners can be introduced in regular particle beam therapy.

5.2.2.3 The future of compact accelerator concepts

To enhance the coverage of particle therapy in Europe and worldwide and to enlarge the number of patients that can profit from this special treatment, the investment costs of such facilities should be reduced as much as possible, which requires smaller and simpler machines to reduce manufacturing and operating effort. Especially the size of the accelerator has an important influence on the building costs. And the amount of beam losses demands more or less concrete for radiation shielding, and

Figure 5.8. Setup of an MRI scanner from Esaote at HIT's experimental place.

should be minimized by design. In addition, the operating and maintenance team needed should be small, but adequate and well-trained, and sustained by a modern control system, which predicts pre-emptive maintenance measures through AI algorithms and thus guarantees highest availability.

Energy-variable cyclotrons

Most proton therapy facilities use cyclotrons to produce the beam, as these accelerators are compact, have only a few tuning parameters, and are thus simple to operate. Over the last three decades the size and weight have dramatically shrunk by factors 3–10 while the magnetic fields using superconductivity were increased up to a factor 4 (figure 5.9). These improvements still show the potential of cyclotrons while parallel developments like proton linacs are still larger and have a much more complex technique. But the actual cyclotron generation consists still of fixed energy machines, which demands a degrader for energy variation causing two main drawbacks: (a) high local beam losses and (b) relatively low currents for mid-depth and skin-deep tumours, which may be a problem to use the FLASH-based treatment procedure. To overcome these disadvantages first studies of energy-variable cyclotrons were undertaken and published. These show the possibilities to consequently advance the cyclotron systems to lighter, high-field superconducting arrangements, but now iron-free and thus capable to vary the magnetic field and energy in reasonable times. In addition, the produced currents of such a setting do not depend on the energy because no loss mechanism is involved anymore. The resulting machines would be 'FLASH'-ready and have the additional possibility to be enhanced to ^{3}He/^{4}He beam combined with proton therapy in one cyclotron setup. More simulation studies and prototype constructions are needed to reach these goals in the next 5–10 years.

Superconducting fast-ramped synchrotrons

Wherever the flexibility of different ions from protons, helium, carbon to oxygen ions is asked to have the ability to study different dose distributions and linear energy transfer (LETs), a flexible accelerator concept with a high bandwidth of magnetic rigidities is needed. Today small synchrotrons with circumferences of 60–80 m and iron-dominated normal-conducting dipoles and quadrupoles are built for such purposes. Together with the injector linac and a variety of ion sources on one

Table 1 Comparison of Main Parameters of Several Commercial Cyclotrons

	Mevion S250	IBA S2C2	Varian Proscan	IBA C230
Magnet Type	Superconducting	Superconducting	Superconducting	Copper
R pole (m)	0.34	0.50	0.80	1.05
D Yoke (m)	1.80	2.50	3.10	4.30
Height (m)	1.20	1.50	1.60	2.10
B_0 (T)	8.9	5.7	2.4	2.2
B_r (T)	8.2	5.0	3.1	2.9
Mass (mt)	25	50	100	250
T_f (MeV)	254	230/250	250	235

Figure 5.9. Comparison of main parameters of several commercial cyclotrons; see 'Compact, low-cost, lightweight, superconducting, ironless cyclotrons for hadron radiotherapy', PSFC/RR-19-5.

side and all the high-energy beam transport lines on the other side such a facility has a large required space in contrast to the treatment rooms. As a result, the building costs are high and such facilities were set up only in combination with large university hospitals. To shrink the size of these accelerator arrangements the main idea is to equip the synchrotron with superconducting magnets. This is a big challenge, because curved 3–4 T magnets ramped at 1–2 T s^{-1} are needed for an efficient operation. Studies at Toshiba in Japan and at CERN (NIMMS, HITRIplus[2]) are underway to prepare prototype magnets.

In addition, modern control systems providing the multiple energy extraction method—using several post-accelerations in the same synchrtotron cycle—have to be developed to enhance the duty cycle of such machines. A new approach using time-sensitive networking (TSN), the next generation of (real-time) ethernet in industry, will be implemented at HIT (partly within HITRIplus) in the next years.

However, an open question still exists: Can synchrotrons be used later on for FLASH therapy, as the particle filling is limited by space charge restrictions? In addition, the necessary high dose rates would demand short extraction times (or fast extraction methods) with rapid refilling of the synchrotron and short cycle times (see above). These are all big challenges.

The first proton and ion linac-based facilities

After several years of R&D and developments in research at international laboratories (CERN, TERA, ENEA, INFN, ANL) the first linac-based proton therapy facilities are under construction and commissioning in UK (STFC Daresbury) and in Italy. With the use of high-frequency copper structures, designed to achieve relatively compact solution and high repetition rate operation, linacs will allow the production of beams with fast energy variation (without the need for mechanically moved beam energy degraders), as well as small emittance beams that are potentially suited for the further development of mini-beam dose-delivery techniques. The shielding around the linac can also be reduced compared to other installations allowing a more flexible solution that could be installed in existing buildings and other restricted space settings. Also, dedicated designs for He and C ions are being studied, which would require less power compared to presently operating synchrotrons and allow for flexible pulsed operation. HG RF technology as well as high-efficiency klystrons are key developments for the future and further spread of this approach. Furthermore, the linac technology is attractive as a booster option to increase the output energy of cyclotron-based existing facilities.

Towards a VHEE RT facility

NC RF linac is the technology being used for most of the VHEE research. The main advantages of the linacs are the flexibility and the compactness. Regarding the linac design in the energy range of interest for VHEE applications there are different possibilities offering the desired performances and compactness with different

---

[2] **NIMMS** (Next Ion Medical Machine Study). https://www.hitriplus.eu/ Heavy Ion Therapy Research Integration plus.

degrees of technology maturity. The S-band technology is the most mature one; HG compact linacs of this type are already available from various industrial partners. The C-band and X-band RF linacs are still less mature and are mainly constructed in labs with the help of industries for machining. Recently a considerable effort is being made from the industrialization point of view. The current and next future available machines for VHEE are the eRT6-Oriatron at Centre Hospitalier Universitaire Vaudois in Laussane; ElectronFlash at IC in Orsay; CLARA at Daresbury; AWA at ANL; and CLEAR at CERN (all based on NC RF linacs). A VHEE-FLASH facility based on a CLIC X-band 100 MeV linac is being designed in collaboration with CHUV to treat large, deep-seated tumours in FLASH conditions. The facility is compact enough to fit on a typical hospital campus (figure 5.5). Another proposal in this sense is the upgraded PHASER proposal at SLAC. Finally, ELBE at HZDR and the next future PITZ at DESY are based on SCRF linac technology.

Recent advances in the high-gradient RF structures where more than 100 MeV m^{-1} are now achievable in the lab environment are transforming the landscape for VHEE RT. VHEE RT requires beam energies between 50 and 200 MeV, an improved dose conformity and scale to higher doses rates, in the case of the FLASH-RT until 50 Gy s^{-1} are needed. Novel high-gradient technologies could enable ultra-compact structures, with higher repetition rates and higher currents. An international R&D global effort is being made by major accelerator laboratories and industry partners and is focused on two aspects: material origin and purity, surface treatments, and manufacturing technology on the one hand and the consistency and reproducibility of the test results on the other. Some promising R&D in the next decade are the distributed coupling accelerator developed at SLAC and the use of cryogenic copper that is transforming the linac design offering a new frontier from beam brightness, efficiency, and cost capability. Another approach for the next generation of compact, efficient, and high-performance VHEE accelerator is the use of higher-frequency millimetric waves (~100 GHz) and higher repetition rates using THz sources.

An important R&D effort to apply these technologies in the medical industry has to be made in the next decade, if successful, this could be a step further in the quest for compact and efficient VHEE RT in the range of hundreds of MeV. For achieving these aims, a synergistic and multidisciplinary research effort based on accelerator technology as well as physical and radiobiological comparisons to see how well VHEE can meet the current assumptions and become a clinical reality is needed (Very High Energy 2020).

Therapy facilities based on laser plasma acceleration

As high-performance lasers have increased greatly in recent years in terms of power and repetition rate their use for particle therapy may be possible in the future. The actual limit of about 100 MeV achieved for the highest proton energies driven by ultra-intense lasers using Target Normal Sheath Acceleration (TNSA) depicts a major milestone on the way to the needed energies. But still the broad energy spread of the accelerated protons is not feasible for treatment modalities. The reached energies for laser-accelerated ions is still a magnitude lower and thus far from the necessary values. The very short dose peaks may be attractive for FLASH therapy, but then the repetition rate of the Petawatt lasers should reach 100 Hz and more, which is not the case today. In addition, the target configuration has to resist this high load on a long-term basis—a therapy facility runs several 1000 h a year. Furthermore, the reliability of a laser-based proton or ion accelerator must reach 98% or more to be of practical use in a medical facility. But this technique should be explored with high effort in the next decade to identify the long-term potential.

Concerning the VHEE there is an intense R&D effort in LPA to be applied in the next generation of VHHE-RT facilities. The major challenge for the LPA technique is the beam quality, reproducibility, and reliability needed for RT applications. This R&D is being carried out in facilities such as the DRACO at ELBE at Helmholtz-Zentrum Dresden-Rossendorf (HZDR) and in the Laboratoire d'Optique Appliquée (LOA) at Institute Polytechnique de Paris (IPP) where a new beamline dedicated to VHEE medical applications known as IDRA is being constructed. The new beamline will provide stable experimental conditions for radiobiology and dosimetry R&D (Very High Energy 2020).

A wide international R&D programme, in particular we highlight the role of the EU network EUPRAXIA (http://www.eupraxia-project.eu/), will be needed in the next decade in order to convert these 'dream' facilities into reality.

5.3 Bionics and robotics

5.3.1 Bioinspired micro- and nanorobotics

Friedrich Simmel[1]
[1]Technische Universität München, Garching, Germany

5.3.1.1 General overview

Robotic systems transform the way we work and live, and will continue to do so in the future. At the macroscopic scale, robots greatly speed up and enhance manufacturing processes, provide assistance in diverse areas such as healthcare or environmental remediation, and perform robustly in environments that are inaccessible, too harsh, or too dangerous for humans. In the lab they can, in principle, perform large numbers of experiments in parallel, reproducibly and without getting tired. This supports, for example, the search for new chemical compounds in combinatorial approaches, and can also generate the large datasets required by data-hungry machine learning techniques.

Robotic systems are 'reprogrammable multifunctional manipulators' and typically comprise sensors and actuators connected to and coordinated by an information-processing unit. Sensors provide information about the environment, which is evaluated by a computer and then used to decide on the necessary actions—which often means mechanical motion of some sort. At a macroscopic scale, a wide range of sensors, electromechanical components, and powerful—potentially networked—electronic computers are available to realize robotic systems, which are essentially all powered by electricity. Is it possible to realize robotic functions also at the micro- or even nanoscale, where we have to work with molecular components, and on-board electronics is not available?

In fact, over the past years researchers have begun to work on the development of 'molecular robotic systems', in which sensors, computers, and actuators are integrated within molecular-scale systems. Among the many possible applications for molecular-scale machinery and robots are, most prominently, the generation of nanomedical robots that autonomously detect and cure diseases at the earliest stages, and the generation of molecular assembly lines that will enable the programmable synthesis of chemical compounds.

Biology as a guide for molecular and cell-scale robotics

Biology has inspired the development of robotic systems at the macroscale in manifold ways—many robot body plans are derived from those of animals (humanoid, dog, insect-like robots, etc), and the movements and actions of these robots resemble those of their living counterparts. Roboticists are concerned with 'motion planning', 'robotic cognition', etc, and therefore ask similar questions as neuroscientists. The field of swarm robotics is inspired by the observation of social behavior in biology.

But also at the cellular and molecular scale we can find inspiration for robotics—cells, like robots, have sensors and actuators; they store and process information, they move, manufacture, interact with other cells, they can even self-replicate (which robots cannot, so far). To name but a few, specific examples for biological functions that are of direct relevance for molecular robotics are protein expression, molecular motors, bacterial swimming and swarming, chemotaxis, cell shape changes and the cytoskeleton, cell-cell communication, the immune system, muscle function, etc.

What's important is that biology is a very different 'technology' than electro-mechanics and electronic computers—biological systems are self-organized chemical systems far from thermal equilibrium. If we want to build bioinspired robots at this small scale, we will have to apply other principles than those developed for macroscale robotics.

From molecular machines to robots

Over the past decades, one of the major topics in biophysics (and in supramolecular chemistry as well) was the study of molecular machines and motors. Research in this field has clarified how machines operate at the nanoscale and how they differ from macroscopic machines. The small size of the machines changes everything—these machines operate in a storm of Brownian motion, in which the forces they are able to generate are small compared to the thermal forces. Motion sometimes is achieved via a ratchet mechanism that utilizes and rectifies thermal motion (which is only possible out of equilibrium); in some cases it also involves a power stroke, in which a chemical reaction (ATP hydrolysis) effects a conformational change in the machine or motor, biasing its movement in one direction.

Research on biological molecular machines informs molecular robotics in several ways. First, these machines have taught us how they work (at least in principle), and have given examples for what they can achieve (e.g., transport molecules from point A to B, synthesize molecules, exert forces, pump ions/molecules across membranes, etc). They also have indicated speeds ($\mu m\ s^{-1}$) and forces (piconewton) that can be achieved with molecular machinery. Their spatial organization often plays a role in their function (e.g., huge numbers of myosin molecules acting together in muscle, ATP synthase or flagellar motors embedded in membranes, etc) and thus needs to be controlled.

Even if the construction of powerful synthetic molecular machines will not succeed in the near future, experimental work in biology and biophysics has provided protocols that allow the extraction, purification, and chemical or genetic modification of biological machines and their operation in a non-biological context. Many researchers have already started to harness biological motors such as kinesin and myosin in an artificial context (e.g., for molecular transport, biosensing, agent-based computing, as active components of synthetic cell-like structures or synthetic muscles). Other molecular machines such as ATP synthase have been used to power biochemical reactions within synthetic cells.

But then, not every machine or motor should be called a robot. What we expect from a molecular robot is the programmable execution of multiple and more complex tasks (as opposed to non-programmable, repeated execution of always the same task), potentially with some sort of decision-making or context-dependence. This likely will require the combination of multiple molecular components into a consistent system that can be continuously operated or driven out of equilibrium. This requirement also defines one of the major challenges for the field, namely systems integration of molecular machines and other components to perform useful tasks, which also comes with challenges for energy supply and interfacing with the environment or non-biological components.

DNA-based robots

DNA molecules turn out to be ideal to experimentally explore ideas in nanoscale biomolecular robotics. DNA intrinsically is an information-encoding molecule, based on which a wide variety of schemes for DNA-based molecular computing have already been developed. Further, DNA nanotechnology—notably the so-called 'DNA origami' technique—has enabled the sequence-programmable self-assembly of almost arbitrarily shaped molecular objects. Various chemical and physical mechanisms have been employed to switch DNA-based molecular objects between different conformations, and to realize linear and rotary molecular motors. Thus, in principle, all the major functional components of a robotic system—sensors, actuators, and computers—can be realized with DNA alone.

As mentioned, in order to realize robot-like systems, these separate functions have to be integrated into consistent multifunctional systems. However, only few experimental examples have convincingly demonstrated such integration so far. In one example by Nadrian Seeman and coworkers, a 'molecular assembly line' was shown to be capable of programmable assembly of metallic nanoparticles by a molecular walker. The walker could collect nanoparticles from three assembly stations, which were controlled to either present a nanoparticle or not. This resulted in a total of $2^3 = 8$ different assemblies that could be 'programmably' realized with the system. In the context of nanomedicine, origami-based molecular containers have been realized that open up only when certain conditions—e.g., the presence of certain molecules on the surface of cells—are met. The containers can then present previously hidden molecules that trigger signaling cascades in the cells, or release drugs. This also exemplifies the two main fields of application that have been envisioned for DNA robots: programmable molecular synthesis with molecular 'assembly lines', and the delivery of drugs by nanomedical robots.

While these prototypes are extremely promising, many challenges remain—and some of these probably pertain also to nanorobots realized with other molecules than DNA. First, information-processing capabilities of individual molecular structures are quite limited—essentially, they are based on switching between a few distinct states (i.e., of similar energy), but separated by a 'high enough' activation barrier), which means that their computational power should be similar to that of finite state machines. Second, current instantiations of DNA robots are quite slow (movements with speeds of nm s^{-1} rather than µm s^{-1}), and do not allow fast operation or response to changes in the environment. Thirdly, molecular robots are, of course, small, which poses a problem in many instances, where we might want to integrate them into larger systems, let them move across larger length scales, and operate many of them in parallel.

Active systems

One of the major visions in bioinspired nano- and microrobotics is the realization of autonomous behaviors, which, as a subtask, involves autonomous motion. In this context, over the past decade there has been huge interest in the realization and study of active matter systems, which include self-propelling colloidal particles, and active biopolymer gels actuated by ATP-consuming molecular motors. In contrast to other, more conventional approaches based on manipulation by external magnetic or electric fields, such systems promise to move 'by themselves' and also display interesting behaviors such as chemotaxis or swarming.

As before, autonomously moving particles or compartments *alone* will not make a robot, and again the challenge will be to integrate such active behavior with other functions. For instance, it would be desirable to find ways to control and program active behavior—the output of a sensor module could be used to control a physicochemical parameter that is important for movement. Active particles that move in chemical gradients need to be asymmetric, and potentially this asymmetry (or some symmetry-breaking event) could be influenced by a decision-making molecular circuit. Another challenge will be to find the 'right chemistry' that allows active processes and other robotic modules to operate under realistic environmental conditions—(e.g., inside a living organism).

Collective dynamics and swarms

If single nanorobots are unavoidably slow and rather dumb, maybe a large collection of such units can do better? In order to overcome their limitations, a conceivable strategy is to couple large numbers of robots via some physical or chemical interaction and let them move, compute, operate collectively. First steps in this direction have been taken by emulating population-based decision-making processes such as the 'quorum sensing' phenomenon known from bacteria, or swarm-like movement of microswimmers. Ideally, for a robotic system one would like to be able to couple such collective behaviors to well-defined environmental inputs and functional outputs—(e.g., if there is light, self-organize into a swarm, move towards the light source and release (or collect) molecules, etc).

There are various challenges associated with these ideas. How can one program the behavior of a swarm? As swarming depends on the interactions between the constituting particles/robots and their density, one could think of changing these control parameters in response to an external input. 'Programming' the behavior of such dynamical systems would mean to choose between different types of behaviors that are realized in different regions of their phase space. Potentially, the behavior of a whole collection of particles could be influenced by a single or a few particles (leader particles) and programming the swarm would amount to programming or selecting these leaders.

Cells as robots

As mentioned above, biological cells really behave a little like microscale robots. Biology has tackled the 'systems integration' challenge and realized out-of-equilibrium systems, in which various functionalities play together, behave in a context-dependent manner, and which are controlled by genetic programs. From a robotics perspective we can therefore ask whether we can (i) build synthetic systems that imitate cells but perform novel functions and thus act as cell-scale soft robots, or (ii) engineer extant cells to become more like robots?

Essentially the same approaches to engineer biological systems are pursued in synthetic biology, and they come with the same challenges. Regarding the first ('bottom-up') approach—putting together all the necessary parts to generate a synthetic living system is yet another systems engineering challenge. In order to realize synthetic cells, metabolic processes need to be compartmentalized and coupled to information processing, potentially growth, movement, division, etc, which has not succeeded so far (cf the separate EPS challenge section 4.4). The second approach circumvents the challenge of realizing a consistent multifunctional molecular system, but engineering of extant cells is difficult due to the sheer complexity of these systems. Engineered modules put additional load on a cell (whose exclusive goal is to self-sustain, maybe grow and divide), which compromises their fitness, and they also often suffer from unexpected interactions with other cellular components (also known as the 'circuit-chassis problem' in synthetic biology).

Power supply

A major issue that has to be tackled for all of the approaches described above is power supply. How are we going to drive the systems continuously to generate robotic behaviors? Cells come with their own metabolism, which means that cell-based robots would simply have to be fed in the same way as cell or tissue cultures. In the absence of a metabolism, however, molecular robotic systems driven by more complex chemical fuels such as ATP or nucleic acids probably will need to be supplied with these fuels using fluidics. When operation inside of a biological organism is desired, biologically available high-energy molecules might be used as fuels.

In the context of nano- and microrobotics, external driving with magnetic or electric fields via light irradation or heating is heavily investigated. Here one challenge is to convert these globally applied inputs into local actions—potentially by some local amplification mechanism (e.g., plasmonic field enhancement) and/or by combining actuation with molecular recognition or computation that lead to action only when certain conditions are met. So far, in most cases externally supplied energy was used for mechanical actuation (e.g., motion or opening/closing of containers), but only rarely to power complex behaviors or information processing.

In some cases it is not yet clear whether the energy balance will work out— depending on the mechanism employed and the efficiency of the robots, it may not be possible to deliver enough power to small systems to enable movement or more complex behaviors in the presence of overwhelming Brownian motion and other small-size effects. For instance, rotational diffusion of nanoparticles can be too fast to allow for directional movement, energy dissipation in aqueous environments may be too fast to heat up nanoscale volumes, etc.

Hybrid systems

In light of the limitations in our ability to generate autonomous biomolecular robots with similar capabilities as macroscopic robots, a realistic and potentially powerful approach will be to focus on hybrid systems, in which non-autonomous molecular systems are combined with already established robotic (electromechanical) systems. Such an approach would combine the advantages of the different technologies involved—bionanotechnology and synthetic biology on the one hand, and electronic computing and electromechanical actuation on the other. For instance, it will be very hard if not impossible to outcompete electronic computers using the limited capabilities of molecular systems alone. On the other hand, biomolecular systems and/or cells are at the right scale and 'speak the right language' to interact with other molecular/biological systems.

It is obvious that one of the major challenges for this approach is interfacing—finding effective ways to transduce biological into electronic signals, and vice versa. In one direction this coincides with the well-known challenges involved in biosensing, for which a biological signal (the presence of biomolecules) needs to be converted into electrons or photons (which also can be further converted into electrons). In the robotics context, an additional requirement will be the speed of the sensing event, which has to be quick enough to allow responding to changes in the environment in which the robot operates.

In the other signaling direction, efforts have been made to control molecular machines with light, electric, or magnetic fields, which enables potentially fast and computer-controlled external manipulation of these systems. Also various attempts have been made to influence the behavior of cells using external stimuli, notably in areas such as neuroelectronics or optogenetics. Similar approaches might be adopted to control bio-based micro- and nanorobots.

In the case a bidirectional biointerface has been successfully established, one can imagine hybrid robotic systems, in which computer-controlled signals direct the behavior of the biomolecular part of the robot, and sensory information is fed back to the computer, enabling the implementation of feedback or more complex control mechanisms—in this approach, micro- and nanorobots will be all sensors and actuators, with a brain outsourced to an external electronic computer.

Applications envisioned for micro- and nanorobotics

Micro- and nanorobots will be used when a direct physical interaction with the molecular or cellular world is required. As already mentioned above, the main application envisioned for such robotic systems will be in nanomedicine. One instantiation of nanomedical robots are advanced delivery vehicles that can sense their environment, release drugs on demand, or stimulate cell-signaling events. They may potentially be equipped with simple information-processing capabilities that can integrate more complex sensory information (e.g., to evaluate the presence of a certain tissue, cell type, and thus location in the body), and which may also be used to evaluate diagnostic rules based on this information (such as 'if condition X is met, bind to receptor Y, release compound Z', etc). Given the limited capabilities of small-scale systems, it is not clear how programmable such robotic devices will be. It is well conceivable, however, to come up with modular approaches, in which the same basic chassis is modified with different sensors and actuators, depending on the specific application. Autonomous robots will have to find their location by themselves, which for some applications may be achieved by circulation and targeted localization in the organism. Alternatively, hybrid approaches are conceivable, and allow for active control from the outside (e.g., by magnetic or laser manipulation, depending, of course, on the penetration depth of these stimuli in living tissue). There are many additional challenges for such devices, which are similar to those for conventional drugs (e.g., degradation, allergenicity, dose, circulation time, etc).

Apart from nanomedical robots, for which the first examples are already emerging, a wide range of applications can be envisioned in biomaterials and hybrid robotics. Hybrid robots could use a biomolecular front end that acts as sensor for (bio)chemicals and actuator that allows the release or presentation of molecules (here the overall robot would not be microscale). Potentially, surfaces or soft bulk materials (such as gels) could be modified with robotic devices, resulting in novel materials that can be programmed and change their properties in response to environmental signals in manifold ways (resulting in materials that are smarter than 'smart materials').

One of the most fascinating applications would be programmable, molecular assembly lines. Rather than aiming for a universal assembler, a more modest and achievable goal is the programmable assembly of a finite number of possible assembly outcomes starting from a defined set of components—similar to the DNA-based molecular assembly line mentioned above. This is not unlike a macroscopic assembly line, which is optimized for the production of one defined product with optional variations. Notably, biological processes such as RNA or protein synthesis already look a little like programmable assembly: RNA polymerase and ribosomes read off instructions from a molecular tape (DNA and mRNA, respectively), and use them to assemble other macromolecules. Maybe similar systems can be conceived that allow sequence-programmable synthesis of non-biological products.

In any case, molecular assembly means control over chemical reactions, which also means that we cannot ignore basic chemical rules and simply synthesize anything we want. Another issue is the scale of the process: in order to synthesize appreciable quantities of molecules or molecular assemblies, large numbers of assembly robots will have to be embedded into the active medium of a synthesis machine (similar, maybe, to a DNA or peptide synthesizer), where they can synthesize larger quantities in parallel.

5.3.1.2 *Challenges and opportunities*
In the sections above, a wide range of challenges and opportunities associated with the development of future bioinspired micro- and nanorobots were mentioned, which are summarized more concisely here. A variety of challenges relate to the way robotic functions will be implemented in the first place:

- Some robotic functions will be realizable with nanoscale systems (composed of supramolecular assemblies, DNA structures, biological motors); others will require larger cell-like systems into which multiple functions are integrated. It will be important to understand the size dependence of these functions and clarify what can be achieved at which length scale, and with which components. Systems integration—combining the components into consistent and functioning systems—will be *the* major challenge for bioinspired small-scale robotics.

- Autonomous behavior versus external control. How complex do robotic systems need to be to act autonomously—and is autonomy required? For many applications, external control will be a more feasible and powerful approach, which also benefits from established macroscale technologies.

- Hybrid robots that combine the advantages of biological systems and traditional robotics seem promising—a major challenge is the realization of efficient interfaces for signal transduction and actuation between biological and conventional robotics.

- Robotic systems should perform useful tasks, which means they need to be operated under realistic conditions. Depending on the field of application, various practical issues will have to be tackled—operation inside a living organism or on the surface of a sensor chip come with very different requirements.

Many physical challenges arise from the question of whether we can create something like a robot at such small scales at all:

- Can small robots sense and respond to small numbers of molecules or photons? There are biological examples of extremely sensitive sensors that require special molecular architectures and amplification processes, which may guide the design of microrobotic sensors. A related question that has been studied extensively in biology is chemotaxis, where cells sense the presence of a concentration gradient.

- What forces can be usefully applied by small robots? Biological molecular motors are known to generate piconewton forces, and the same magnitude is also expected from artificial motors. However, synthetic molecular motors have not yet been used to perform any useful task.

- How fast can a nano- or microrobot move or act? Do we need to achieve directional movement or will diffusion be sufficient? Of course, this again will depend on the application. Interestingly, in biology we do not find molecular motors for transport of molecules in bacteria (diffusion is sufficient), but in the much larger eukaryotic cells. Further, the smallest bacteria tend to be non-motile—apparently motility only pays off above a certain size. There are various further issues such as the dominance of Brownian effects at the nanoscale.

- How much energy is required to perform robotic functions—and how will it be supplied? Should nano- or microrobots be autonomously driven by chemical reactions, or actuated externally via physical stimuli?

- What is the computational power of small robots? How much information can be stored, how fast can it be processed? As the computing power of small systems necessarily is very limited, nanorobotics means 'robotics without a brain'. Nanorobots will tend have 'embodied intelligence', in which input-output relationships between sensor and actuator functions are custom-made and hardwired rather than freely programmable.
- Can one realize collective behaviors of large numbers of nanorobots—and how can one 'program' such systems?

Ultimately, the development of micro- and nanorobotics will lead to advancements in many application areas:

- Nanomedical robots operating in living organisms will be able to deliver potent drugs at the right spot, and in a context-dependent manner. They can also act as sentinels, permanently monitoring, recording, and reporting the presence of disease indicators.
- Small robots will extend the capabilities and enhance the sensory spectrum of larger robots. They will constitute the molecular interface of hybrid robotic systems, which enables bidirectional communication with biological systems. Such systems can be integrated (e.g., with mobile robotic units that perform environmental sensing and monitoring).
- Robotic components extend the functionality of 'smart' materials and surfaces. Embedding robotic functions into materials will enhance their ability to adjust to their environment and change shape and mechanical properties in response to environmental cues. Conceivably, such materials will have something akin to a metabolism that supports these functions and enables continuous operation. Surfaces coated with nanorobotic components will have enhanced sensor properties; they may be able to actively transport matter, take up or release molecules to the surroundings, and change their structure and physicochemical properties.
- Not the least, work on bioinspired small robotic systems will elucidate physical limits of sensors and actuators, and will clarify what it takes to display intelligent (or seemingly intelligent) behavior; it will also result in methodologies that allow programming the behavior of dynamical, out-of-equilibrium systems. Ultimately, the realization of bioinspired robots may also contribute to a better understanding of complex biological phenomena and behaviors.

5.3.2 Robotics in healthcare

Darwin Caldwell[1]

[1]Instituto Italiano di Tecnologia, Genoa, Italy

5.3.2.1 Increasing demands for healthcare

The provision of appropriate healthcare is unquestionably a major worldwide societal challenge that impacts all nations and peoples. The growth in medical robots driven by advances in technologies such as actuation, sensors, control theory, materials, AI, computation, medical imaging, and of course supported by increased doctor/patient acceptance may be a solution to a number of these convergent problems that include:

- Aging society and the increasing burden of dementia. This has been recognized in developed countries for many years, but is now becoming increasingly common globally. According to the OECD around 2% of the global population is currently over 80, and this is expected to reach 4% by 2050. In Europe, people over 80 already represent around 5% of the population and are expected to reach 11% by 2050 (OECD 2020).

- Increased healthcare spending and costs. According to the OECD in almost every country spending on healthcare consistently outstrips GDP growth. It has risen from 6% in 2010 to a predicted 9% of GDP by in 2030 and will continue to increase to 14% by 2060 (https://www.oecd.org/health/health-carecostsunsustainableinadvancedeconomieswithoutreform.htm) (Moses *et al* 2013). The per capita health spending across OECD countries grew in real terms, by an average of 4.1% annually over the 10-year period between 1997 and 2007. By comparison, average economic growth over this period was 2.6%, resulting in an increasing share of the economy being devoted to health in most countries (OECD 2009).

- Shortage of workers in medical and social care professions. A relative decrease in the proportion of healthcare workers is increasing the demands on the active workforce, which is itself ageing in many countries. Data from the World Health Organization (WHO) indicates that the average age of nurses is over 40 in several European countries. Furthermore, the WHO states that the global health worker shortfall is already over 4 million (WHO 2008).

- The rapidly growing population and the need to provide basic medical needs in developing countries.

- Changing family structures. Family sizes are much smaller than in the past and the percentage of elderly people living alone is rapidly increasing.

- Increased acceptance of technology (Abou Allaban *et al* 2020a)—Although there is among many groups (particularly elderly users) an ambivalence about the use of robot technology and a preference for human touch, there is little hostility, and among younger people who are more familiar with computers, AI, smart technology, and advanced communications, there is a growing

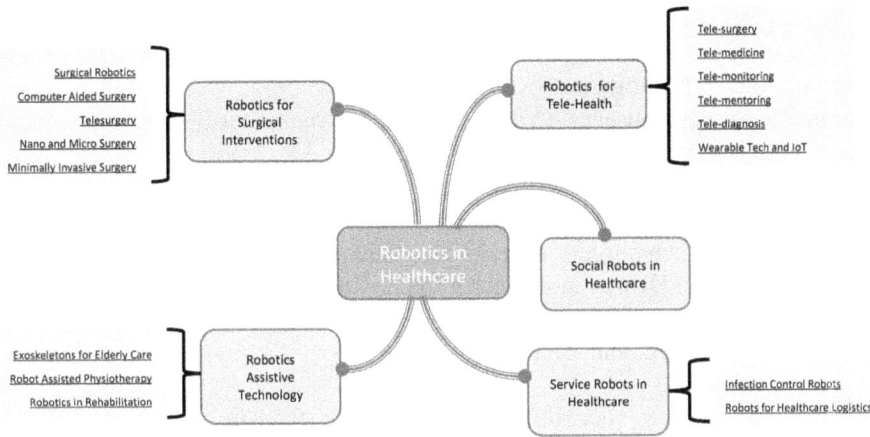

Figure 5.10. Uses of robotics in healthcare.

acceptance of the benefits of robotic and related technologies (Wu *et al* 2014). This increasing acceptance, and indeed in some instance reliance, on robots has been accelerated by COVID-19 (Zemmar *et al* 2020).

Against all of these demands and counter demands there is a widespread belief that medical and healthcare robotics is essential to transform all aspects of medicine— from surgical intervention to targeted therapy, rehabilitation, and hospital automation (figure 5.10).

5.3.2.2 *Robotics and sustainable healthcare (Yang* et al *2018)*

As the demands on health systems grow, it is perhaps inevitable that we should turn to technology and particularly robotics, both to provide the extra capacity/ productivity that will be needed by aging and rapidly increasing populations, and also to continue to enhance and improve the quality of life provided by the healthcare systems. Although robots and robotic systems represent a significant investment cost, experience in others sectors, such as manufacturing, has demonstrated that the use of robotic technology can also offer significant savings and increases in efficiency/productivity, while also contributing to the establishment of high-quality, sustainable, and affordable healthcare systems. Important application domains that could benefit include medical training, rehabilitation, prosthetics, surgery, diagnosis, and physical and social assistance to disabled and elderly people (Stahl *et al* 2016, Wang *et al* 2021).

5.3.2.3 *Robotics for medical interventions (Mattos* et al *2016)*

Surgical robotics
Robotic surgery involves using a computer-controlled motorised manipulator/arm that has small instruments attached to this assembly. Using artificial sensing this arm can be programmed to move and position tools to carry out surgical tasks. The

surgeon may or may not play a direct role during the procedure. The history of surgical robotics dates back almost 40 years and arose from a convergence of several key advances in technology (Satava 2002).

Minimally invasive surgery (MIS): Driven by the potential to create smaller incisions, with lower risks of infection, reduced pain, less blood loss, shorter hospital stays, faster recovery, and better cosmesis, the first laparoscopic cholecystectomy was performed in the mid-1980s (Antoniou *et al* 2015). The benefits of this approach over conventional open surgery quickly became apparent to surgeons, patients, and healthcare providers. However, although the potential benefits were clear there were also several technical and human factors problems. These included:

- Poor visual access and depth perception, due to the quality of the 2D cameras and displays.
- Difficult hand-eye coordination due to the fulcrum effect of using long instruments inserted through a cannula. This produced motion reversals, and a scaled and limited range of motion that was dependent on the insertion depth of the tooling.
- Little or no haptic feedback and a reduced number of degrees of freedom and dexterity.
- Camera instability and loss of spatial awareness within the body cavity (the which way is up problem).
- Increased transmission of physiologic tremors from the surgeon through the long rigid instruments.

As a consequence, when performing MIS a surgeon must master a new and different set of technical and surgical skills compared to performing a conventional procedure.

VR, telepresence, and telesurgery: Around the time of the first developments in MIS, NASA was studying options for providing medical care to astronauts. Their research teams were particularly interested in using emerging concepts in virtual reality (VR), haptics, and telepresence. Teams within the US Army (Satava 2002) became aware of this work and were interested in the possibility of decreasing battlefield mortality by bringing the surgeon and operating theatre closer to the wounded soldier (figure 5.11).

Robots in surgery: Although the first MIS was only performed in 1987, the history of surgical robotics does in fact slightly pre-date this, with Kwoh *et al* performing neurosurgical biopsies in 1985 (Kwoh *et al* 1988), and in 1988 Davies *et al* performed a transurethral resection of the prostate (Davies 2000). While these and other surgical robots were being developed, clinicians working with the various robotics teams realised that surgical robots and concepts in telepresence/telesurgery had the potential to overcome limitations inherent in MIS by:

- Using software to eliminate the fulcrum effect and restore proper hand-eye coordination. At the same time the software made movement and force scaling possible so that large movements or grasp forces at the surgeon's console could be transformed into micro motions and delicate actions inside the patient.

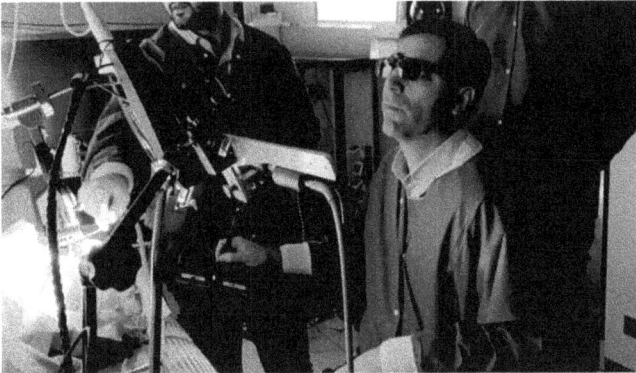

Figure 5.11. VR/AR, advanced user interfaces and telecommunications, and robot technology combine to create an enhanced surgical experience.

- Increasing dexterity using instruments with flexible wrists (designed to at least partially mimic human wrist action) to give increased degrees of freedom which greatly improves the ability to manipulate tissues.
- Using software filtering to compensate for any surgeon induced tremor, making increasingly more delicate operations possible.
- Improving surgeon comfort by designing dedicated ergonomic consoles/ workstations that eliminate the need to twist, turn, or maintain awkward positions for extended periods.

Computer-assisted surgery (Buettner et al 2020)

Computer assisted surgery (CAS) is also on occasion known as image-guided surgery (IGS), computer-aided surgery, computer-assisted intervention, 3D computer surgery, or surgical navigation. It is a broad term used to indicate a surgical concept and set of methods whereby the motions of the surgical instruments being manipulated by the clinician are tracked and subsequently integrated with intra-operative and/or preoperative images of the patient. The pre/interoperative images can be produced by a combination of x-rays, computers, and/or other equipment. (e.g., medical ultrasound, ionizing techniques such as fluoroscopy CT, x-ray, and tomography, fixed C-Arms, CT, or MRI scanners). This information is used either directly or indirectly to safely and precisely navigate to and treat a condition: tumour, vascular malformation, lesion, etc, as demanded by the treatment. This increases the efficiency and accuracy of the procedure, and reduces the risk to nearby critical tissues and organs.

Although CAS can be used in traditional open surgery the need for and use of CAS, as with robotic surgery, has been driven by advances in MIS, where the minimal access needed to provide medical benefits results in restricted visualisation of the site and can create difficulties in understanding of the exact spatial location of the tooling within the body. The key difference between CAS and robotic surgery is the lack of a robot!

CAS and IGS systems can use different tracking techniques including mechanical, optical (cameras), ultrasonic, and electromagnetic or some combination of these systems to capture, register, and relay the patient's position/anatomy, the surgeon's precise movements relative to the patient, and the motion of the tooling/instruments, to a computer which displays images of the instruments' exact position inside the body on a monitor (2D or 3D). It is also possible to relay these images to virtual and augmented reality (VR/AR) headsets. This imaging and display are usually (and ideally) performed in real-time, although there can be delays of seconds depending on the modality and application. The tracking technology and the ability to track the location of the instruments within the body is often likened to GPS.

CAS and IGS have become the standard of care in providing navigational assistance during many medical procedures involving the brain, spine, pelvis/hip, knee, lung, breast, liver, prostate, and in otorhinolaryngology, orthopaedic, and cardiovascular systems. The clinical advantages include decreased intraoperative complications, increased surgeon confidence, improved preoperative planning, more complete surgical dissections, and safer junior physician training/mentoring.

Tele and remote surgery (Choi et al 2018)

Telesurgery allows a surgeon to operate on (or be virtually present with) a remote patient, by using a robotic surgical system at the patient's site. This involves the use of teleoperation technologies that use real-time, bidirectional information flow: surgical commands must be sent (in real time) to the remote robot, and everything that is happening in the surgical/patient site must be immediately perceived by the surgeon, through visual, auditory, and on occasion haptic feedback. This separation of the surgeon and the patient is already common with most current surgical robots (e.g., the Da Vinci Surgical system), with the surgeon sitting at an operating console that is a few metres away from the robot and patient, but connected with a dedicated wired connection. Telesurgery, on the other hand, refers to medium- and long-distance teleoperation with distances measured in kilometers or even thousands of kilometers between the surgeon and the robot/patient. The connection may use wired as well as 'conventional' broadband, a dedicated connection, wireless, or some combination of wired and wireless.

Potential benefits of telesurgery include:

- Eliminating the need for long-distance travel, along with associated costs and risks. This will provide for urgent emergency interventions and expert surgical care in underserved regions such as remote/rural areas, underdeveloped countries, in space, at sea, and on the battlefield.
- Surgical training: Telementoring can enhance the training of novice surgeons and bring new skills to more experienced medics. This could revolutionize surgical education.
- Surgical collaboration: Real-time collaboration between surgeons at different medical centers using shared, simultaneous perceived, high-definition visual feedback.
- Surgical data: Telesurgery focuses on moving data instead of the patient or the surgeon. This data is extremely rich, and includes sensory information,

records of the surgical workflow, actions, and decision-making processes. This data can be used to assess surgical quality, or develop AI to enable autonomous surgical supervision, assistance, or even fully robotic surgery.

- Robotic precision: The use of robotic systems increases surgical precision and quality, removing physiologic tremor, reducing adjacent tissue damage, and permitting previously impossibly delicate surgery.

This large separation between the surgeon and surgical robot brings significant medical advantages but also creates many technical challenges, most of which are directly linked to the need to instantaneously transfer, in a safe, secure, and reliable way, massive amounts of data between both ends of the system. Fortunately, advances in data communication, fibre optic broadband, and particularly the increasingly widespread availability of 5G mean that the long-dreamed goal of truly remote telesurgery is now increasingly possible (Acemoglu *et al* 2020a). This will be explored in the later section on general requirements in Robotics in Tele-Healthcare.

Micro- and nanomedical robots (Nelson et al 2010, Li et al 2017, Soto et al 2020)

Current medical and surgical robots draw most of their design inspiration from conventional industrial robotics, with modifications to suit the particular requirements of operations in, on, and near tissue. Miniaturization of these robotic platforms, and indeed creation of completely novel, versatile micro- and nanoscale robots, will allow access to remote and hard to reach parts of the body, with the potential to advance medical treatment and diagnosis of patients, through cellular level procedures, and localized diagnosis and treatment. This will result in increased precision and efficiency. These advances will have benefits across domains such as therapy, surgery, diagnosis, and medical imaging.

Medical micro/nanorobots are untethered, small-scale structures capable of performing a pre-programmed task using conventional (electric) and unconventional (chemical, biological) energy sources and actuators, to create mechanical actions. Due to their size (sub mm) they face distinct challenges when compared to large (macroscale) robots: with viscous forces dominating over inertial forces, and motion and locomotion being governed by low Reynolds numbers and Brownian motion. Thus, the design requirements, parameters, and operation of these robots is almost unique in the mechanical world, although these features are common in biological systems. Many micro/nanorobots are made of biocompatible materials that can degrade and even disappear upon the completion of their mission.

Despite the challenges caused by the miniature scale, the potential benefits are significant in areas such as targeted delivery, precision surgery, sensing of biological targets, and detoxification.

Targeted drug delivery—Motile micro/nanorobots have the potential to 'swim' directly to a very specific target site within any part of the body and deliver a precise dosage of a therapeutic payload deep into diseased tissue. This highly directable approach will retain the therapeutic efficacy of the drug/payload while reducing side effects and damage to other tissue.

Surgery—micro/nanorobots could navigate through complex biological media or narrow capillaries to reach regions of the body not accessible by catheters or invasive surgery. At these sites they can be programmed or teleoperated to perform the required intervention (e.g., take biopsy samples or perform simple surgery). The use of tiny robotic surgeons could help reduce invasive surgical procedures, thus reducing patient discomfort and post-operative recovery time.

Medical diagnosis—isolating pathogens or measuring physical properties of tissue in real-time could be a further function performed by micro/nanorobots. This would provide a much more precise diagnosis of disease and/or vital signals. The integration of micro/nanorobots with medical imaging modalities would provide accurate positioning inside the body.

The use of micro- and nanorobots in precision medicine still faces technical, regulatory, and market challenges before their widespread use in clinical settings. Nevertheless, recent translations from proof of concept to *in vivo* studies demonstrate their potential in the medium to longer term.

Medical capsule robots (MCRs) are smart medical systems that enter the human body through natural orifices (often they are swallowed) or small incisions. They are an already operational form of micro/nanorobot that use task specific sensors, data processing, actuation, and wireless communication to perform imaging and drug delivery operations by interacting with the surrounding biological environment. MCRs face a number of challenges such as size (for non-invasive entry they must be less than 1cm in diameter), power consumption (they must carry on-board batteries), and fail safe operation (they operate deep within the body). MCR design and development focuses on miniaturization of the electronics and mechanical structure, packaging, and software (Beccani *et al* 2016).

5.3.2.4 *Robotics in tele-health (De Michieli* et al *2020)*

Tele-healthcare involves the remote (tens of km to thousands of km) connection of different subjects involved in the healthcare process. This includes the patients, the clinicians, and others. Tele-health systems allow different kind of interactions: (i) clinician to clinician, (ii) clinician to patient, and (iii) patient to mobile health technology. Each interaction has a different purpose and requirements.

A common scenario where tele-health can provide an important benefit is when the patient or medic is not able to travel to the clinical setup, due to distance limitations, cost, poor transportation links, or other external factors. The COVID-19 pandemic has made this scenario even more real (Chang *et al* 2020, Leochico 2020, Prvu Bettger 2020), yet through the technology and application of tele-medicine, medical treatment, assessment, monitoring, or rehabilitation can continue to be provided without the need to come in person to the clinic. This offers important advantages in terms of infection prevention.

A second application of tele-health is continuous monitoring. Classical tele-health approaches make use of simple technologies such as phones, email, or video-chats, but the latest IoT (Internet of Things) technologies allow continuous monitoring and feedback of many different medical data and processes (Dabiri *et al* 2009), This makes use of developments in ubiquitous computing, cloud storage, and intuitive human-machine interaction. Tele-Health is also exploring the emerging possibility of using the sensory capabilities of wearable devices that offer a unique opportunity to enhance the monitoring and intervention abilities of medical staff and relatives (De Marchi *et al* 2021). This data can be analysed and sent to nurses, doctors, and public health agencies for monitoring purposes. In addition to providing a greater range of data, wearables can also reduce the demand for staff to make manual observations.

Wearable technologies for health monitoring (Best 2021)

Wearable technologies are patient portable devices that can range from simple systems with a single sensor that measures only one variable to complex devices that can gather a wide range of vital health and lifestyle parameters. They are typically worn on the arm/wrist, or around the chest/waist, but the positioning can be dependent on the variables being measured and each generation of new wearable devices adds greater functionality in increasingly compact packages that have better and better battery life. The information collected by the wearable device is typically stored locally before being relayed directly to the clinicians for monitoring or analysis, but this feed can also on occasion be continuous when the clinical need and device support this functionality.

While many wearable devices were historically custom-made medical devices, the past 10–15 years have seen a vast growth in consumer-grade wearables, such as smart watches and fitness trackers that have increasingly sophisticated and accurate sensing technology (exercise level, sleep patterns, SpO2, heart rate, and even single lead electrocardiographs). These wearables (which generally link to mobile phones) can measure data continuously and can be programmed to alert the wearer and the clinician if values deviate significantly from the norm.

The already existing widespread use and acceptance of wearables means that this technology is likely to have an increasingly important impact on healthcare, providing a much more accurate profile of patient health over a prolonged period.

5G technology and tele-health (Acemoglu et al 2020b)

Improvements in internet speeds and connectivity have meant that even with the ever- increasing data levels demanded by high-resolution systems, wired tele-health connections have been possible for several years. However, there were always constraints when the data had to be transferred wirelessly, and this meant that there were limits on the scope of telemedicine and where this service could be provided. To provide truly flexible operation for remote and tele-health applications away from fixed communication points there needed to be a step forward in wireless commu- nications. This was provided by 5G mobile networks. 5G technology is not just the evolution of 4G. It is a completely new platform capable of enabling new services. 5G provides the following major benefits compared to previous technologies:

- Ultra-high speed broadband up to 10 Gbps.
- Ultra reliability and ultra-low latency down to 1 ms.
- Massive connectivity, able to handle one million devices per square km.

- Quality of service, ensured by multiple input multiple output (MIMO) antennas with beamforming capabilities (i.e., able to direct their power to where it is requested by a user).
- 5G virtual network (VN) element infrastructure, which increases the network's flexibility and resilience, allowing network slicing to create independent virtual networks on the same physical infrastructure. Each VN has its own characteristics in terms of security, traffic routing, and quality of service. Network slicing enables the customization of services for users or applications with specific requirements, making it possible to allocate and reserve resources for mission-critical services such as a telesurgery.

Tele-diagnosis and monitoring (Ding et al 2020)
Tele-diagnosis refers to the ability to make a remote diagnosis using platforms designed to transmit the physical examination records and medical reports to the examining specialist. The tele-diagnosis system should ensure that images and videos preserve the diagnostic quality even after compression for transmission. There can be limitations due to compression, bandwidth issues, and lag, but improvements in internet bandwidths, video imaging, the widespread roll-out of 5G and demands of COVID-19 which has increased user (both medic and patient) acceptance of the technologies, mean that tele-diagnosis and all aspects of tele-health have received a significant boost, and it is expected that this will continue, becoming more common in forthcoming years. This will include fewer network-critical services such as tele-mentoring and tele-assistance.

5.3.2.5 Robotics assistive technology
The elderly population is projected to quadruple between 2000 and 2050, with many experiencing mobility impairments due to physiological muscular decay or associated health conditions, such as stroke, which shows an increasing rate of incidence with the age of the subject. In its '*2030 Agenda for Sustainable Development*' the UN studied future housing, healthcare, employment, and social protection needs (DoE 2017), and identified mobility as being of paramount importance to quality of life, social inclusion, and independence (Richardson *et al* 2015). While walking sticks, wheelchairs, or walking frames can provide reasonable assistance, many elderly/disabled still require additional assistance when walking (Charron *et al* 1995).

Stroke is the leading cause of disability in industrialised countries. Fortunately, over 65% of patients survive, but the majority have residual disabilities, with up to a third having severe disabilities. Hemiplegia, the most common impairment resulting from stroke, leaves the survivor with a stronger unimpaired side and a weaker impaired one (hemiparesis). In addition to stroke, traumatic injuries as well as conditions such as muscular dystrophy, arthritis, and regional pain syndromes also add to the major causes of disability and functional dependence. Deficits in motor control and coordination synergy patterns, spasticity, and pain are some of the most common symptoms of these conditions (Parker *et al* 1986).

Evidence has shown that intensive and repetitive physiotherapy can have substantial benefits (Carr *et al* 1987) including regaining motor control and muscle strength as well as in restoring/retaining the joints' range of motion. Despite the benefits of intensive physiotherapy, the associated disabilities are seldom considered life-threatening; therefore, they rate relatively low on the priority list for urgent medical assistance. In addition, manipulative physiotherapy is labour-intensive and therefore fatiguing for the therapists as well as the patient; it requires high levels of one-to-one attention in an environment with an international shortage of physiotherapists, and patients must receive individualised treatment.

Against this background, wearable technologies, robots, exoskeletons, and power- assistive techniques seem to offer a promising solution and are increasingly viewed as a potential replacement for the physical labour. This will leave therapists with greater time to develop treatment plans and enhance therapeutic outcomes.

Rehabilitation robots

The goal of rehabilitation is the reintegration of an individual with a disability into work/society. This can be achieved by:

- enhancing existing capabilities (achievable through therapy and training) or by
- providing alternative means to perform various functions or to substitute for specific sensations (achievable using assistive technologies) (Robinson 1995).

Rehabilitation robotics is the application of robotic methods to train or assist an individual with a disability and support their reintegration to work/society. This broad definition covers a variety of different mechatronic machines that support both gait and arm therapy in a clinical setting. It can also include powered orthotics for use in daily life environments, actuated prosthetics, and can even include intelligent wheelchairs (Priplata et al 2003).

Many rehabilitation robots are based on traditional robot designs (often collaborative robots used in the industrial sector that have been modified to provide the levels of safety needed for close interaction with patients). They usually connect to a distal segment (wrist, hand, ankle, foot) of the patient and can be used to support rehabilitation of either upper or lower limbs (e.g., Burt [Barrett Technology], ROBERT [Life Science Robotics]). Although traditional arm-inspired designs have been the most common approach in rehabilitation, the specialised/ individual nature of the requirements means custom designs are also common (e.g., InMotion ARM [Bionik Laboratories], Hunova [Movendo Tech]; Saglia et al 2013, 2019). For both the arm-inspired and custom designs, the approach taken is often similar involving initial assessment and subsequent retraining of the range of motion of a limb (arm, leg) and increasing power functions.

Gait rehabilitation robots

Strokes often result in loss of mobility, which can easily lead to patients becoming house bound and completely dependent on others for many aspects of daily life. Hence, restoration of walking is a vital goal. Robots can play an important part in locomotion therapy and gait rehabilitation. Systems used in gait rehabilitation typically must provide support for the whole body weight of the patient. This can be achieved using multidirectional overhead body weight support systems (e.g., ZeroG or FLOAT) that require the user to wear a harness that can provide full or partial gravity compensation. Wearing the support, the patients can start training earlier, while muscle control/coordination is poor and muscle strength is low, without the risk of falling. The Andago system provides a similar form of body weight support, but the unit can move freely without the need for ceiling-mounted rails. To take full advantage of the potential of these body weight support systems they can/should be linked to a mobile robot (e.g., a lower limb exoskeleton such as ReWalk or Twin; Laffranchi *et al* 2021) or as part of a stationary gait trainer (e.g., Lokomat, Walk Training Assist; Baronchelli *et al* 2021).

Systems of this type permit patients to relearn walking and safely train after stroke, spinal cord or brain injuries, incomplete paraplegia, or orthopedic patients. For the exoskeleton-based systems this training can in addition be in an unrestricted three-dimensional space. Although free movement is possible and will become more common in the future, the leading robot gait rehabilitation systems such as the Lokomat remain high in cost, and are large, heavy, static structures found in specialised rehabilitation centres.

Balance trainers such as Toyota's Balance Training Assist or the Balance Tutor can be considered an adjunct or subset of gait trainers and can be used for balance training. They often include a platform (fixed, mobile, or in the form of a treadmill) and a weight support system, which can be programmed to disturb the patient's balance allowing them to safely relearn, regain, and enhance their stability.

Exoskeletons for rehabilitation (Pons 2008)

An exoskeleton is a wearable robot that assists with the execution of physical activities by delivering forces/torques at the human joints (Toxiri *et al* 2019). One important and very specific feature of exoskeletons is the physical connection between the human and the robot. This human–robot interaction (HRI) is twofold: firstly, cognitive (the human controls the exoskeleton); secondly, physical (the human and robot are bidirectionally coupled). This physical coupling is one of the most exciting aspects, however, it does bring challenges since it involves the cooperation of two dynamic systems (i.e., human motor control and robot control). Therefore, human-activity recognition (HAR) is required to understand what kind of action the user is performing or wants to perform (Poliero *et al* 2019).

Exoskeletons can be categorized in several ways, but among the most common are:

- *Rigid* or *soft*. Rigid exoskeletons use structural frames and links to support part of the exoskeleton weight and transmit forces and torques. Soft exoskeletons, also known as *exosuits*, are lightweight and use soft/softer materials (fabrics and plastic) that can more easily be integrated with clothing to increase user acceptance. This softness can also be extended to the actuators which may be 'hard' but behave in a soft way such as variable stiffness actuators, series elastic actuators, or pneumatic actuators (Vanderborght *et al* 2013), or can truly follow the soft material paradigm and actually be soft actuators (Di Natali *et al* 2019).

- The assisted joint (e.g., shoulder, neck, wrist, back, hip, knee, ankle, or any combination of these).
- Actuation (passive, active or *quasi-passive*). *Passive* exoskeletons use elastic components such as springs to provide a fixed level of assistance (Huysamen *et al* 2018). Passive actuation is not suited for many impaired users since they have little residual strength to 'charge' the mechanical elements. *Active* exoskeletons represent the optimal fit, exploiting controllable elements like electrical motors or pneumatic actuators (Toxiri *et al* 2018). *Quasi-passive* exoskeletons take advantage of controllable elements not to directly provide assistance to the user but, rather, control the engagement of the connected passive elements (Di Natali *et al* 2020).

Currently most exoskeletons for healthcare are fairly bulky, stationary, systems that target improved rehabilitation. Fewer fully wearable systems are being used due to their complexity, weight, and performance. However, research within industrial and military domains is spilling over into healthcare, and there is now increasing potential for portable, home-based systems both for rehabilitation and as a mobility assist.

5.3.2.6 Social robotics for healthcare (Henschel et al 2021)

Social robots are often characterised as providing support or training to a person in a caring/social interaction scenario. This may be interpreted by external observers as appearing social. They are physically embodied, often in human form, although animal- and cartoon-like appearances are not uncommon. Although many social robots emulate human form, features, behaviours, and expressions, care must be taken to avoid imitating human appearance or motion too closely to avoid falling into the Uncanny Valley (Pandey *et al* 2018), where behaviours and appearances can become intimidating or disturbing. Social robots have some level of autonomy and can directly interact, communicate, and cooperate with people following social accepted behaviours and rules. They do not typically perform a mechanical task, although they may be capable of movement including on occasion locomotion.

For healthcare-based scenarios many see social robots as playing a part in addressing mental health issues, with the robots acting as social partners in applications such as robot companions, robots as educators for children or children/adults on the autistic spectrum (Cao *et al* 2019, Stower 2019), robots as assistants/companions for the elderly/disabled, and those studying affect, personality, and adaptation (Čaić *et al* 2019, Dawe *et al* 2019, Johanson *et al* 2020). The development of social robots can also create a bidirectional paradigm with the human learning from the robot and the robot learning from the human. This has seen particular relevance in cognitive neuroscience.

A early example of a social robot developed to support developmental psychology, cognitive neuroscience, and embodied cognition was the iCub humanoid. iCub was a child- sized robot designed to manipulate its surroundings, imitate its human partners, and communicate with them (Tsagarakis *et al* 2007). Over the past 20 years iCub and social robots in general have investigated human social concepts such as decision making, intent, trust, attention, perception attachment empathy acceptance, and disclosure.

Although there is a substantial body of research in social robotics with some very positive progress, there is nonetheless a substantial gap between the capabilities of social robots and the expectations of the general public, with potential users expecting s social robot to be 'like a friend' but being disappointed when the robot is not able to go beyond the smart-speaker like single-turn structure of conversation. Nonetheless the ongoing research continues to make important progress and this will only increase, resulting in robots that are more 'social' and personal over time (de Graaf *et al* 2015).

5.3.2.7 Service robots in healthcare

Service robots have expanded rapidly into many areas in the past 20 years and are increasingly common in industrial and even home settings. The healthcare sector has also been seen as an area of potential growth, although pre-COVID-19 many of these opportunities were seen as potential rather than realised goals. This was perhaps due to the challenges associated with the very nature of the services needed in healthcare. However, there is now no doubt that the COVID-19 pandemic has created an environment that has accelerated deployment and use of service robots within hospitals and across the wider healthcare system. Within this context robots are now being used either semi- or fully autonomously to provide services such as effective cleaning in hospitals/care homes and logistics of patients and supplies. This has benefits for patients, healthcare workers, and healthcare systems.

Infection control robots

Infection control and sterilisation has always been a problem within hospitals and healthcare settings (Carling *et al* 2013), and robotic solutions for sterilisation had already been developed prior to the pandemic. However, with the onset of the pandemic several robotics companies have introduced mobile robotic platforms providing disinfection solutions. These robots often have the capacity to navigate autonomously to the desired sterilisation site using lifts, passing through doorways and along corridors. At the target site they can use a variety of sterilisation/disinfection techniques including:

- UVD (ultraviolet disinfection)—disinfection using high-intensity continuous or pulsed UVC light (Beal *et al* 2016, Omron 2022).
- Hydrogen peroxide—surface decontamination using a dry aerosol of hydrogen peroxide disinfectant (Anderson *et al* 2006).
- Physical disinfection of contact surfaces—The human support robot uses a fully autonomous arm manipulator to spray disinfecting liquids on to commonly touched surfaces such as door handles, lift call buttons, handrails, tables, walls, etc The arm and a cleaning brush are subsequently used to clean the region much as would be done by a human cleaner (Ramalingam *et al* 2020).

Robots for healthcare logistics

As with any large and multifaceted organisation, hospitals and care homes have complex logistical requirements ranging from those that are medical centred such as the transport of medical samples, medicine, and medical supplies, laboratory samples, to the delivery and coordination of daily necessities such as food and linen. Robots with autonomous navigation capabilities that are used in industry and housing/offices offer a pre-existing entry point for this operation, particularly when considering that it is estimated that nurses spend only 70% of their time dealing with their patients with the remaining time being used to track down records, supplies, test results, or in other 'logistical' tasks not directly related to patient care. If robots could perform some or all of these search tasks this would increase overall efficiency and accuracy, and allow the staff to concentrate on their key roles (Bloss 2011).

A number of robots and robotics solutions are already available for logistical operations such as delivery of patient records, pharmacy supplies, diagnostic bloods, and fluid samples and test results.

The use of service robot across the healthcare sector, can and is, having important benefits in terms of removing human cleaning staff from potential disease exposure, preventing the spread of infection, improving levels of cleaning by reducing human error, ensuring 24 h availability, and allowing front-line staff to reduce direct contact, focusing their attention on higher priority tasks and creating separation from direct exposure to infection.

5.3.2.8 Future challenges and trends

Healthcare is a major societal challenge with daunting forecasts for the future, especially given the global population aging trend, the increasing number of patients, the decreasing proportion of healthcare workers, and the increasing costs of care. Fortunately, technological progress and the rise of healthcare robotics offer hope for the establishment of sustainable, affordable, and high-quality healthcare systems.

Robotics can offer significant contributions to progress in disease detection, diagnosis, and treatment at early stages; patient care; earlier more intense personalised rehabilitation; remote treatment; and the training of medical personnel, improving the skill levels of trainees and lengthening the effective career of experienced surgeons/clinicians. All of this can be achieved while enhancing the safety, precision, quality, and efficiency with which care is provided. But these opportunities do not come without challenges and barriers. There needs to be a recognition that while individual technologies can and will bring important but still incremental levels of improvement, major impacts are expected to arise from their full integration into a complete robotic healthcare system.

As noted throughout this chapter the possible areas of application of robotic technology are vast and diverse and therefore identifying generic issues and barriers to implementation is not easy. However, it is clear that there are challenges that encompass economic, social/societal, clinical, and research-related domains.

Technical

Technical challenges are the first and most obvious barriers to the growth, development, and successful use of any and all robotic technologies in healthcare. If the technical issues cannot be resolved subsequent challenges would in all likelihood not arise. Foremost among the technical challenges are the mechatronics: the hardware, software, and control systems that form the foundation of any robot. With respect to healthcare robotics (as is also true for other robotics sectors) many of these challenges focus on mechanical design (e.g., miniaturisation, use of novel materials, new construction/fabrication techniques, rapid, and customised/personalised prototyping, dexterity, and flexibility to reach difficult parts of the anatomy); actuation (smaller, faster, more precise, and more efficient motors or even other actuation technologies etc); more and better sensors (e.g., smaller, detection of a wider range of parameters); multifunctional sensors; real-time detection of conditions such as cancer; wireless operation, MEMS, haptics, etc; materials (e.g., biocompatible, biodegradable, lighter, MRI friendly, plastics versus metal versus ceramics, 3D printing, biological materials, etc); imaging and display (e.g., VR/AR, 3D visualisation, improved resolution, new imaging techniques and technologies, etc); new robot anatomies and structures (e.g., bioinspired, continuum robots, soft robots, robots capable of controlled access and operation in confined body spaces) and miniaturisation (e.g., chip-on-the-tip stereoscopic imaging); and integration of the most effective imaging technologies (e.g., NBI, fluorescence, autofluorescence, OCT, and MRI) into surgical robotic systems to fully exploit the benefits of real-time disease detection.

There will also, of course, be many challenges for the software. This will impact algorithmic development driven by advances in machine learning, deep learning, and AI; the use and development of assistive and augmentation algorithms to help and guide the user in surgery/rehabilitation/training, etc; virtual bodies (e.g., robust real-time algorithms to perceive the 3D structure of the body/tissue/limb, and realistically model tissue motion/deformation with lifelike visual and haptic sensations); and analysis of vast amounts of wearable data covering ever-evolving aspects of daily life. Alongside these technical/operational challenges of software development will be growing cyber-security concerns that will be further compounded by the growth of tele-health, wearable technology, and issues arising from remote access. These areas, as indeed, many aspects of tele-health, will bring to the forefront previously lower impact problems such as the risks of intrusions, privacy, and security with personal data and even malicious attacks, which must all be addressed by considering legal and ethical perspectives, in addition to technical obstacles.

A further aspect of the debate on software in healthcare robotics that is not so common in other robotics sectors is: should robots be autonomous? In most other areas of robotics full robotic autonomy is the goal, but in healthcare there remain many unresolved issues around the level of autonomy (tele-operated versus semi-autonomous versus full autonomy) that again go beyond the technical to impact ethical, legal, societal, and of course employment issues.

Finally, more than in any other sector there is a very high likelihood that healthcare robots (e.g., surgical systems, wearable devices, exoskeletons, or logistics robots) will on many occasions come close to or into physical contact with a person, and often this person will be unwell. In any such interactions safety will always be paramount. While guidelines are in place for collaborative robots this may not be enough for the health sector and it may also be valuable to take guidance from the procedures and processes within aerospace or other sectors where there are critical life-threatening interactions between machines/people. Of course, on the positive side robots will also have an increasingly important part to play providing levels of safety and error prevention beyond what is possible with direct human supervision.

Acceptance of healthcare robotics

A major challenge in all aspects of robotics for healthcare is acceptance: by patients, by medical personnel, by healthcare systems, and by society. This acceptance will be influenced by many factors (e.g., age profile, prior experience with and of technology, national/cultural, application area, effectiveness of the technology, attitudes of the physicians, cost, etc). But what is universally clear is that robots will only be accepted if the science and user experience prove the benefits. To achieve this it is important that all users are fully informed with proactive, positive public engagement campaigns that introduce the technology and benefits, effective training of healthcare staff on how to operate these robots, and education that the robots should not be seen as a professional or job-related threat. Furthermore, by giving robots more exposure—making them visible in everyday environments—they are more likely to be accepted.

At the same time the robotics community must recognise that anxieties exist, and understand that the user's (patient or medic) willingness to adopt a system is influenced by the perceived usefulness, the ease of use, avoidance of frustration and embarrassment, managing expectations on the device performance and the delivery against those expectations, trust in reliability, and social/professional pressure. A potential user's intent to use can be further affected by their self-efficacy (i.e., the user's confidence in their ability to face a certain situation). This is critical in many cases of robot acceptance.

It is also critical to note that often this technology will be targeted at older users with reduced motor skills, coordination, muscle strength and/or control, and sensory losses. Designers should pro-actively engage, listening to the user's individual needs, to create age friendly geriatric (Gero)-technologies with recognition that older generations are less comfortable with technology and that older patients might need more time to adapt, while wanting to avoid being stigmatised by using devices that highlight their deteriorating health. Once again there are likely to be ethical considerations for all users, but particularly elderly users who feel that the monitoring (potentially by family members or relatives) is intrusive or dehumanising.

Healthcare robotics and regulatory approval

To ensure patient and clinician safety, robotic devices must obtain regulatory approval. A manufacturer must demonstrate that their device is safe; however, safety validation is complex, time-consuming, and costly. In part this is due to the newness of the area, the lack of examples of best practices, and a lack of clear guidance on applicable safety standards and protocols. Obtaining the required expertise against this background is challenging, especially for start-ups and small-to-medium enterprises. This can and does add to the long lead time for development from laboratory prototypes to clinical systems and often precipitates the failure of potentially life-enhancing developments. The approach to the development, testing, and approval of vaccines and drugs seen during the COVID-19 pandemic may offer some clues as to how these processes could be improved and streamlined for the benefit of all, without risking safety.

Jobs and healthcare robotics

As in other sectors, those working in healthcare fear that if robots develop the same, similar, or better capabilities to humans, they will replace their human counterparts, resulting in job losses. Concerns around the replacement of people by machines is even more pertinent in this sector because human interpersonal interaction is at the centre of healthcare professions and again ethical and societal issues could be as important and maybe even more important than technical concerns, and each of these topics will need careful consideration.

At the same time there is little doubt that the use of robots will indeed cause changes in the nature of work within the healthcare profession, and there will certainly be a loss of typically low-skilled jobs, as in other sectors. But equally this reduction in more menially and potentially less fulfilling jobs could be seen as a benefit. Further, against a background of increasing demand for healthcare services and reducing numbers of people entering the professions (and willing to undertake these menial jobs) it is certain that there must be a displacement of work profiles. This will create both personal fears and ethical risks that must be addressed by society.

Conclusions

Although the road for healthcare robotic seems long, it is clear the technologies have the potential to bring many benefits to patients, medics, hospitals, and the public healthcare systems in general, bringing better treatment and outcomes, fewer infections, complications, and errors, increasing customer satisfaction and, at the same time, contributing to a more productive healthcare sytem. These are reassuring results offering hope for a brighter healthcare future.

5.4 Physics for health science

5.4.1 Light for health

Thomas G Mayerhöfer[1], Susanne Hellwage[1], Jens Hellwage[1,2], Dana Cialla-May[1], Christoph Krafft[1], Iwan Schie[1,3], Petra Rösch[4], Timo Mappes[4,5], Michael Schmitt[4] and Juergen Popp[1,2,4]

[1]Leibniz Institute of Photonic Technology, Member of Leibniz Health Technologies, Albert-Einstein-Str. 9, 07745 Jena, Germany
[2]InfectoGnostics Research Campus Jena, Centre for Applied Research, Philosophenweg 7, 07743 Jena, Germany
[3]Department for Medical Engineering and Biotechnology, University of Applied Sciences Jena, 07745 Jena, Germany
[4]Institute of Physical Chemistry and Abbe Center of Photonics, Friedrich Schiller University Jena, Helmholtzweg 4, 07743 Jena, Germany
[5]Stiftung Deutsches Optisches Museum, Carl-Zeiss-Platz 12, 07743 Jena, Germany

5.4.1.1 Introduction (Popp and Strehle 2006, Popp et al 2011)

Light for Health—what is the meaning of this title? Before we discuss this in detail, we want to introduce another term which is closely connected, namely Biophotonics. Biophotonics derives from two Greek words, namely, βιoσ life, and φωσ, light. Correspondingly, under the term Biophotonics we understand the application of optical technologies or methods like microscopy to solve problems in medicine and life sciences. In this sense, the use of light for health applications is nothing new, and has already established for a very long time. One cannot imagine health-related science without light, all the more, because one may extend the definition of light beyond the invisible parts of the electromagnetic spectrum, namely, into the infrared spectral range on the one side and down to the ultraviolet, and even reaching x-rays, on the other side.

Light may be used to diagnose diseases early, if possible before the outbreak, and to prevent or treat diseases (e.g., by identifying infectious bacteria that are resistant to antibiotics). But even more, light helps to understand fundamental processes in cells (e.g., by using conventional or high-resolution microscopes), and thus enables us to elucidate the causes of illnesses. Thereby it can help to fight diseases, for example, by finding the right dose of a pharmaceutical substance or finding the right drug at all. In addition, it can also be employed to track and destroy cancerous cells and tissue, and all that with minimal side effects when compared to systemic drug therapy (figure 5.12).

What is it that renders light unique? Light is able to measure touch-free. In a strict sense, this is not completely accurate—light always changes matter, even if it is not absorbed, but usually such interactions are transient and reversible, so that light does not disturb processes in living matter it illuminates. Light interactions are instant. Instead of waiting a couple of days (e.g., for a hemogram) a result can be gained much faster within hours or, in some cases, in real time. This means it would be possible for a medical doctor to find the cause of a disease immediately and react

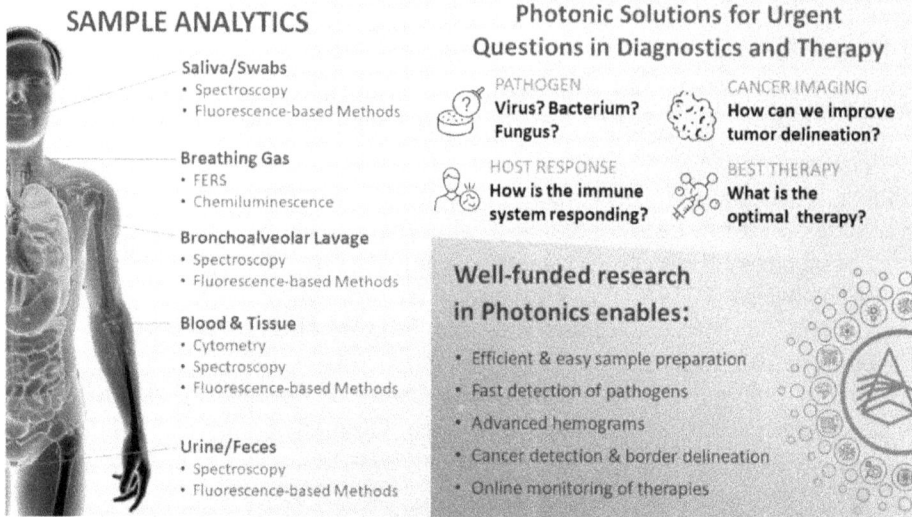

Figure 5.12. Overview of the application of Light for Health (with permission from 'Leibniz Gesundheitstechnologien', Christian Döring). Reproduced from Osibona *et al* (2021) CC BY 4.0.

instantly. In addition, it is the ideal tool to illuminate the world of microbes and cells, down to even sub-cellular structures.

The goals of using light-based technologies and methods for health applications are manifold. First, light-based technologies can solve the urgent problems our civilization is facing be it either strongly aging societies like (e.g., in Europe, North America, or China) or poorer societies such as in parts of Africa. This might be surprising at first glance. However, the focus is not on high-end instrumentation only, but also on portable and wearable instruments that sample data more often or even continuously. These data can be analyzed in rural areas or at the point-of-care or be remotely collected and evaluated. Such technologies can help to reduce costs and enable medical diagnosis and treatments without a medical doctor present. In any case, the direct benefit for the patient is paramount. In addition to this benefit, health-based technologies help to secure existing jobs and to create new ones. And, last but not least, light-based techniques and technologies allow us to satisfy our curiosity to understand life processes at the cellular level, to understand the cause of diseases for the ultimate goal to enable the early diagnosis and treatment of diseases.

Since there are several possibilities to subdivide the vast field of biophotonics and because it is not possible to even cover the most important methods and techniques (this would also go far beyond the purpose of this contribution), we selected examples of what we consider important technological achievements of light for health and offer an outlook of what may be possible in the future. In doing so we focus on examples of photonics approaches which are already used in clinical routines for therapy and diagnosis. Furthermore, we use the example of optical coherence tomography (OCT) to show the challenges one faces when introducing a new method into the clinical workflow. We also present an example of a very promising point-of-care approach that is close to being ready for clinical use. The latter highlights the potential of photonic approaches to address currently unmet medical needs. Finally, we close this subchapter by giving an outlook on future perspectives. However, we start with a brief introduction into the history of microscopy because modern microscopy, or the achievements in light microscopy, can be seen as one of the drivers of Biophotonics, and many biophotonic approaches involve microscopes or innovative microscopic/imaging concepts.

5.4.1.2 Microscopy (Popp et al 2011, Popp 2014, Meyer et al 2019)
Understanding the world in ever smaller dimensions has spurred scientists to develop ever better optical instruments. Our current view of the microscopic world is based to a large extent on discoveries made with the help of the microscope (figure 5.13). This applies to both the animate and inanimate part of nature. Over time, the light microscope has become an outstanding instrument for medicine and the natural sciences. The time of the invention of the microscope cannot be clearly pinpointed. For more than two centuries, starting from the beginning of the 17th century, scientists and craftsmen endeavored to develop the device into the instrument we all know today.

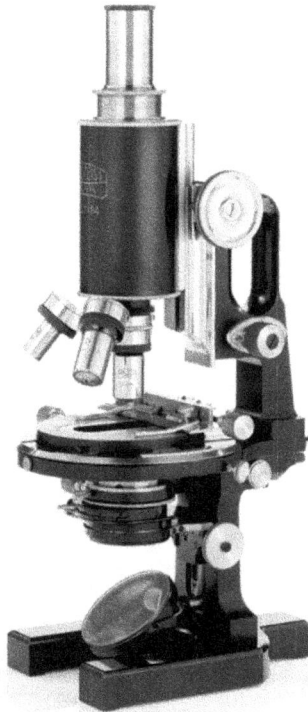

Figure 5.13. A research microscope of 1912 (Stiftung Deutsches Optisches Museum (D.O.M.), License CC0).

While in antique times the use of lenses for the cauterization of wounds is documented, it looks like as if its employment for magnification purposes cannot be attested. After the fall of the Roman empire, Arab scholars took charge of increasing the knowledge in the Middle Ages. As a first milestone, at around 1000 A.D., the invention of a reading stone took place, most likely by Ibn Al Haitham.

It is not clear who first invented the microscope. But for clarity, let's first define the term. If we accept any magnifying instrument as a microscope, then also devices with only one lens fall into this category. A stricter definition of the term requires that a microscope consists of at least two lenses, an objective lens to magnify the object and an eyepiece lens to further enlarge this magnified image. The idea of employing two lenses most likely originated in the beginning of the 17th century, but it is not clear when the first device was actually built. Instruments with two lenses were originally of inferior quality compared to instruments using only one lens, when it comes to high-power magnification. First of all, one problem was the fabrication of glasses without streaks and other defects which increased if a device with two lenses was to be constructed. A further challenge was chromatic aberration (figure 5.14). Since the refraction of light is increasing with decreasing wavelengths (dispersion), a collecting lens will have the focal spots of blue light closer to the glass than for red light. This was a common problem in early microscopes and telescopes until compound optics using differently refracting glasses were engineered. This was a big hurdle for Robert Hooke (1635–1703), who was the first to publish an extensive scientific work about microscopy called 'Micrographia'. His instruments achieved approximately 50× magnification with an incident light setting, using a cobbler's ball for illumination.

While Isaac Newton (1643–1727) had erroneously claimed in 1666 that the problem of achromatism could not be solved, it was Antoni van Leeuwenhoek (1632–1723) who focused his developments on instruments with only one lens, with which he achieved magnifications of up to 260×. One of Leeuwenhoek's secrets was to fabricate lenses virtually free from glass defects, allowing him to reach such high magnifications that he was the first to publish sketches of living bacteria and

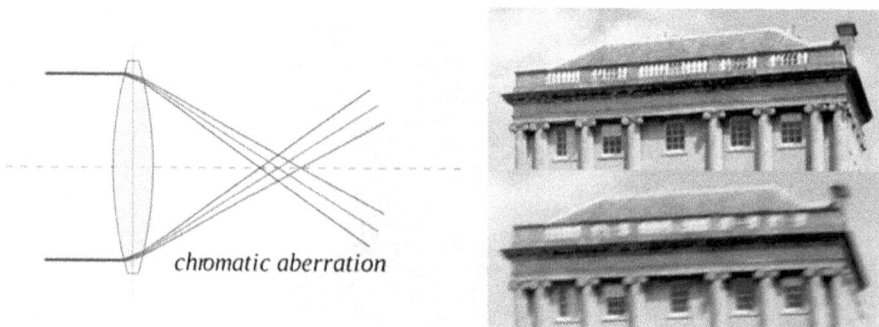

chromatic aberration

Figure 5.14. Left panel: Chromatic aberration of a single lens causes different wavelengths of light to have differing focal lengths. Right panel: Photographic example showing high-quality lens (top) compared to lower-quality model exhibiting transverse chromatic aberration (seen as a blur and a rainbow edge in areas of contrast). Taken and adapted from https://en.wikipedia.org/wiki/Chromatic_aberration.

spermatozoa. While van Leeuwenhoek did not use today's definition of what he documented, this still can be seen as the beginning of modern life sciences research.

Despite this success, the research for better microscopes lacked systematic efforts, since too many disciplines from glass fabrication to physics and engineering were involved. This changed dramatically thanks to Carl Zeiss (1816–1888), an engineer who started a collaboration with the physicist Ernst Abbe (1840–1905) in Jena, Germany. By intense theoretical studies and systematic practical investigation of the optical properties of lenses Abbe formulated the theory of (microscopic) image formation and revolutionized the entire field. Consequently, Abbe addressed optical aberrations in a holistic manner, asking for optical glass with new refractive indices and dispersion. Therefore, Zeiss and Abbe motivated the chemist Otto Schott (1851–1935) to develop optical glasses in Jena. While achromatic lenses were introduced in the 18th century by merging the images of two wavelengths (red and blue) into one plane, Abbe created the apochromatic lens by using the glasses of Schott. Here, three colors of the spectrum (red, green, and blue) are brought into focus in the same plane, while the spherical aberration is corrected for two wavelengths as well. In 1873 Abbe defined the resolution limit in wide-field microscopy as the wavelength of the light used divided by the sum of the numerical aperture of illumination and observation.

With the collaboration of Zeiss, Abbe and Schott a new age dawned, allowing the reliable fabrication of a comparably large number of high-quality optics. This started the boom of the use of the microscope in medicine and the life sciences. The differences in morphology (e.g., of a bacterium) are optically defined by differences in the refractive index only. These differences are, however, usually comparably small, which renders a lot of features hard to identify and observe. In the 19th century this issue was already solved by chemical staining of the samples. Until today, for bacteria it is common practice to color the samples by crystal violet. This staining technique is called Gram and divides bacteria into two groups: Gram-positive bacteria that become purple by Gram staining, and Gram-negative bacteria that appear pink afterward. Another example is Haematoxylin and eosin (H&E) staining, which is a standard staining technique in pathology.

In order to avoid the chemical treatment of the samples and to observe living cells phase contrast microscopy was invented in 1932 by Frits Zernike (1888–1966) and commercialized in 1941. The Nobel Prize in Physics was awarded to him for this in 1953. A few years later Georges Nomarski (1919–97) invented differential interference microscopy, a method visualizing the local differences of the refractive index of a sample.

The contrast of these methods is based on differences of the refractive index, ranging from roughly 1.35–1.5, depending on the water content, and causing optical inhomogeneities. However, these inhomogeneities often arise from differences in the chemical composition. A different kind of contrast is used by methods that are chemically specific. The most well-known property in relation to this is fluorescence. Here, the sample is illuminated by light with a short wavelength. Chemicals called fluorophores are exited by this light and re-emit the absorbed energy in a lossy process at defined higher wavelengths. This emitted light is then easily detected. The most famous fluorophore is the green fluorescing protein (GFP). It is naturally expressed in the jellyfish Aequorea Victoria as described by Osamu Shimomura in 1961, but can be prepared also outside the Jellyfish, which was accomplished first by Douglas Prasher. Martin Chalfie and Roger Tsien managed to express the cloned GFP in bacteria in the 1990s for which they received the Nobel Prize in Chemistry together with Osamu Shimomura in 2008.

Further chemical contrast methods are based on the ability of molecules to vibrate. These vibrations can be excited either by visible light (strictly speaking, molecules vibrate also without being excited, and the excitation just increases the amplitude of the vibration), where a small part of the light interacts with the molecules in this way and has a lower frequency afterwards (Raman spectroscopy), or by infrared light the frequency of which is the same as the vibrations (infrared spectroscopy). The spectral signatures allow on the one hand an identification of the molecules as a spectrum can be seen as a fingerprint. On the other hand, changes of the chemical structure can be followed. More importantly, these techniques do not require labeling as in the case of fluorescence, which avoids an alteration of the molecule and a disturbance of cell processes.

To obtain as much information about morphology, structure, and chemical content of a sample as possible, it is of advantage to combine a number of different techniques or modalities within one instrument. Such devices allow multimodal microscopy and imaging and are of particular use for medical applications (e.g., in pathology) where they can complement the classical light microscope or serve as tools for surgical procedures, where they can reveal the borders of cancer and healthy tissue.

In recent years a fundamental limit has been circumvented, which we introduced already, namely the Abbe limit. A resolution limit of about 200 nm in wide-field microscopy often does not allow to identify relevant structures (e.g., to resolve distinct distributions of actin inside dendrites and spines in the brain). One prominent method to go beyond the Abbe limit is called STED, which stands for stimulated emission depletion in confocal microscopy. Put in simple terms, by this technique the fluorescence is suppressed except in the center of the focal spot, which usually leads to a resolution of 30–80 nm (figure 5.15), but values below 4 nm have been reported in literature. For using a stochastic method with significantly less phototoxicity than STED, stochastic optical reconstruction microscopy (STORM) was introduced. For 'the development of super-resolved fluorescence microscopy' Stefan Hell, Eric Betzig, and William E Moerner were awarded the Nobel Prize for Chemistry in 2014. In the meantime, a cornucopia of different (scanning) techniques became available like photo-activated localization microscopy (PALM), structured illumination microscopy (SIM), or total internal reflection fluorescence microscopy (TIRFM), the latter making use of the evanescent field of light.

The development of the microscope is still not at its end and further improvements can be expected to allow 3D imaging with high resolution from the components of cells up to whole organ and body imaging in video rate and beyond. It will still improve and with it its value for medicine and the life sciences.

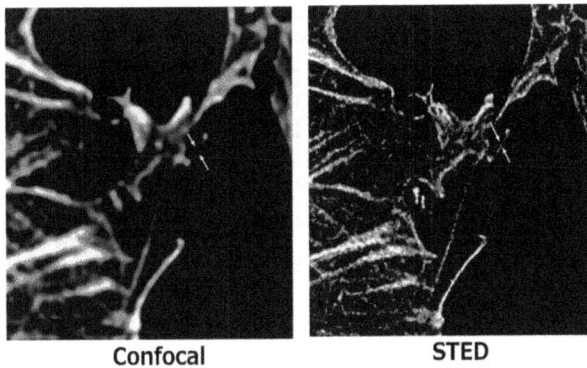

Confocal   STED

Figure 5.15. Resolution improvement between traditional confocal microscopy and stimulated emission depletion (STED) microscopy (adapted from wikipedia commons, from the page https://en.wikipedia.org/wiki/ STED_microscopy under the license CC BY-SA 4.0).

5.4.1.3 Photodynamic therapy (Agostinis et al 2011, Popp et al 2011)

Photodynamic therapy (PDT) involves light and a photosensitizing chemical substance used in conjunction with molecular oxygen to induce phototoxicity and elicit cell death in diseased tissue. It is recognized as a treatment strategy that is both minimally invasive and minimally toxic to healthy tissue. PDT's advantages lessen the need for delicate surgery, lengthy recuperation, and formation of scar tissue and disfigurement. PDT is applied to treat a wide range of medical conditions in skin (e.g., acne, psoriasis), macular degeneration in eye retina, atherosclerosis, and malignant cancers (e.g., skin, bladder, lung, breast, head and neck, gastrointestinal), and has also shown some efficacy in anti-viral and anti-microbial treatments.

As depicted in figure 5.16, the workflow of PDT applications involves three components and is a multi-stage process. First, a photosensitizer (PS) with negligible dark toxicity is administered, either systemically or topically, in the absence of light. Second, light activates the PS when a sufficient amount appears in the diseased tissue. The wavelength of the light source needs to be appropriate for exciting the photosensitizer. Third, the PS produces radicals and/or reactive oxygen species (ROS) that are highly cytotoxic and kills the target cells. The light dose is controlled to supply enough energy for stimulation, but not enough to damage neighboring healthy tissue.

The key characteristic of a PS is the ability to preferentially accumulate in diseased tissue and induce a desired biological effect via generation of ROS. Among the specific criteria are also strong absorption with a high-extinction coefficient in the red/near infrared region of the electromagnetic spectrum (600–850 nm) allowing deeper tissue penetration. Whereas some PDT protocols required rapid clearance of PS within 1–2 days to minimize patient photosensitivity, prolonged production of photoreaction was reported for up to 20 days post-administration. Tailored PSs are key to the development of PDT. They are categorized as porphyrins, chlorins, and dyes. The major difference between PSs is the parts of the cells they target. 5-amino-levulinic acid (5-ALA) as member of the porphyrin family localizes in the mitochondria, m-tetrahydroxyphenylchlorin (mTHPC) as member of the chlorin

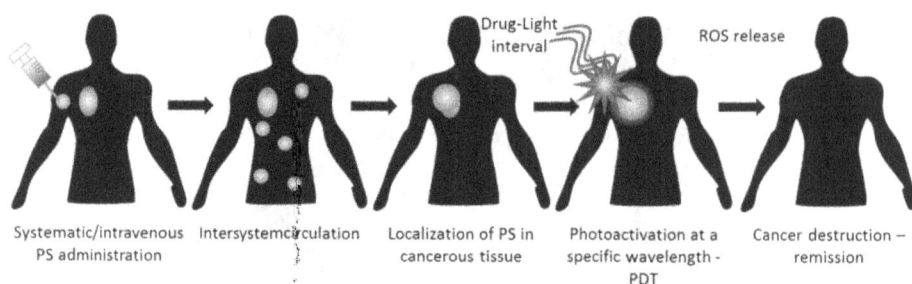

Figure 5.16. Workflow of photodynamic therapy (PDT) after photosensitizer (PS) administration.

family in the nuclear envelope, and the dye methylene blue in the lysosomes. PSs have also been attached to polyethylene glycol (PEG) copolymers that have a hydro-dynamic size of 30–40 nm and provide combinatorial phototherapy with photo-thermal and photodynamic therapeutic mechanisms. To overcome the poor solubility of many PSs in aqueous media, particularly at physiological pH, alternate delivery strategies are used in new generations such as in liposomes and nanoparticles. Antibody-directed PSs have been developed to further enhance targetability.

Illumination systems are another key to the application of PDT. To achieve selected target cell destruction, while protecting normal tissues, the PS can be applied locally to the target area and globally illuminated (e.g., by exposure to daylight) or an external broadband (halogen) or narrow band (halogen with filters, laser, LED) light source. Local illumination offers another way for selected target cell destruction. For internal tissues and cancers, intravenously administered PSs can be illuminated using endoscopes and fiberoptic catheters.

5.4.1.4 Augmented surgery (Gorpas et al 2011, Popp et al 2011, Krafft et al 2018)

Augmented reality (AR) is an interactive experience of the real-world environment where the objects that reside in the real world are enhanced by computer-generated perceptual information, including photonic modalities. AR can be defined as a system that incorporates three basic features: a combination of real and virtual worlds, real-time interaction, and accurate 3D registration of virtual and real objects. Whereas first AR experiences were introduced in entertainment and gaming business, subsequent AR applications have spanned commercial industries such as education, communications, and medicine. A potential medical AR application is to project photonic-generated data about pathological tissue margins into the surgical field of view which assists the surgeon during tumor resection. Neuronavigation is a related technique. Here, the advent of modern neuro-imaging technologies such as computed tomography and magnetic resonance imaging along with the capabilities of digitalization enabled the real-time quantitative spatial fusion of images of the patient's brain. The purpose of this technique was to guide the surgeon's instrument or probe to a selected target during neurosurgery. However, the accuracy of pre-operative recorded images was limited due to the shift of soft brain tissue, in particular after removal of larger tumor masses.

Hardware components needed for AR are input devices, sensors, a processor, and a display. Modern mobile computing devices like smartphones and tablet computers contain these elements which often include a camera and microelectromechanical motion tracking sensors such as an accelerometer, GPS, and solid-state compass making them suitable AR platforms. Input device techniques include speech recognition systems that translate user's spoken words into computer instructions, and gesture recognition systems that interpret user's body movements by visual detection or from sensors embedded in a peripheral device such as a pointer, glove, or other body wear. Various display technologies are used in AR including optical projection systems, monitors, handheld devices, and systems that are worn on the human body such as head-mounted displays and eyeglasses; even contact lenses and virtual retinal displays are in development. AR applications often rely on computationally intensive computer vision algorithms. To compensate for the lack of computing power, offloading data processing via networks to a distant machine is often desired.

First research towards AR with biophotonic tools has been reported for fiber probe-coupled autofluorescence intensity and lifetime which enables label-free and real-time identification of cancerous tissue. Figure 5.17 illustrates how the

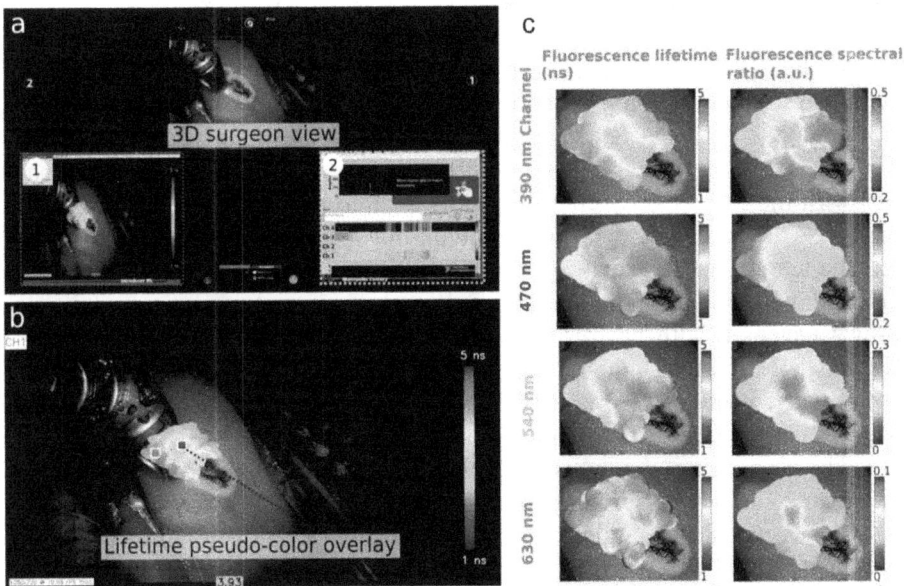

Figure 5.17. Augmentation of multispectral time-resolved fluorescence spectroscopy (ms-TRFS) derived data on the surgeon console during oral cavity surgery. The 445 nm aiming beam used to identify the location probed by ms-TRFS is visible at the distal end of the tool. The panels labeled with (1) and (2) correspond to the lifetime maps that can be visualized and the linear representation of lifetime values, respectively. (b) The white-light image augmented with lifetime values from channel 1 of the instrument. (c) Matrix of the distribution maps for the autofluorescence parameters. Note that each of these maps can be displayed/augmented in real-time during the scanning procedure if needed. Adapted from Gorpas *et al* (2011).

information acquired by a probe measurement technique is transformed into a two-dimensional image resulting in detailed visualization of the fluorescence-based contrast. For each location it takes 40 ms to acquire, process/analyze, and display the AR parameters on the surgeon's console.

Such an AR approach is particularly interesting during robotic surgery which can provide high surgical precision and improve recovery time for the patient and is becoming a preferred means for numerous cancer and disease treatment. Here, due to the loss of tactile feedback, the assessment of tumor margins is based only on visual inspection which is neither significantly sensitive nor specific. The presented augmented surgery can serve as a framework for the clinical translation of other point-scanning label-free optical techniques such as elastic scattering spectroscopy, diffuse reflectance spectroscopy, or Raman spectroscopy.

5.4.1.5 OCT—the challenge to translate an idea into an instrument ready for clinical use (Popp et al 2011, Fujimoto and Swanson 2016, de Boer et al 2017)

Optical coherence tomography (OCT) provides structural information of specimens in depth based on the change in the refractive index between microstructures. From a layman's perspective it is akin to ultrasound imaging but using light waves instead of mechanical waves. And while in ultrasound it is possible to electronically measure the time of flight between the excitation and the reflection, to the speed of light a low-coherence interferometry approach is used. The signal is generated through light scattering in structured samples with transition zones of chemically different composed tissue locations, resulting in changes in the refractive index and imaging contrast.

To record a signal, the specimen is illuminated with a broadband illumination source with the central wavelength typically located in the region between 400 and 1500 nm. The bandwidth of the illumination spectrum is proportional to the coherence length, and as such, related to the axial resolution (i.e., shorter central wavelength with broader bandwidth), results in higher axial resolution, while longer central wavelength with narrower bandwidth results in lower axial resolution. In either case the light is considered to have a low coherence, meaning if an interferometric measurement is performed (e.g., using a Michelson interferometer) only photons that travel the same distance will interfere, resulting in a measurable signal. The photons will be ballistically and elastically scattered by microstructures in the sample and detected in the same wavelength range. Typically used sources are broadband lasers, superluminescent diodes (SLED), or halogen lamps. The excitation light is guided into an interferometric detection scheme with a fixed reference arm and, in the case of time-domain OCT, a moving sample arm. Based on the interaction with the sample and the relative length-difference between sample and reference arm interference pattern can be measured or not. The detector is recording the scattered photons from different depths of the tissue and based on a measurable interference a depth profile can be recorded. Depending on the central wavelength and the bandwidth a spatial resolution of typically 2–10 μm can be reached.

Today, time-domain OCT systems are rarely implemented and most work is performed using Fourier-domain OCT, where two possible implementations are available: spectral-domain and swept-source OCT. In the spectral domain a broadband excitation source is used in combination with a Michelson interferometer scheme. However, instead of a scanning arm spectrometer-based detection is used and a spectral interferogram, which contains the information about the depth profile of the sample, is used. This implementation significantly increases the acquisition speed, as no mechanical sample-arm scanning is required. The other method to acquire an interferogram is based on swept-source excitation, which uses a wavelength-tunable illumination source to generate the interferogram. Here, cheaper detectors can be used and problems, such as signal-roll-off in spectrometer-based implementations, does not occur, resulting in higher penetration depths.

OCT is a great example for the translation of a modality to clinical applications. As a matter of fact, OCT is one of the most widely used modalities for the diagnosis of eye-related disease, including glaucoma, age-related macular degeneration, diabetic retinopathy, diabetic retinopathy, and others. The human retina, which is a light-sensitive, multi-layered tissue located at the back of the eye, provides ideal conditions for the application of OCT. Due to the layered structure with very well-defined bands the retina is perfectly suited for analysis with OCT, and the application of the established method in ophthalmology and routine clinical diagnostics in eye diseases are performed on a daily basis. Thereby the retina is visualized as a three-dimensional, micrometer-resolved cross-sectional image with mm depth. Trained personnel can identify anatomical changes in the retinal layers and diagnose diseases, such as macular degeneration or glaucoma.

There are significant efforts to translate OCT to other clinical applications and different pathologies, such as intravascular imaging of atherosclerotic plaque, fibrotic diseases, and cancer diagnostics. While the premise of high spatial resolution with depth information is very intriguing there are methodical and technical challenges that need to be addressed first. For example, one of the problems is the access to the regions of interest where a measurement has to be performed. These locations are, except for skin, not directly and easily assessable with the typical systems and additional developments for *in vivo* endoscopy probes are required. There have readily been multiple implementations for *in vivo* endoscopy probes, which are typically based on four types of scanning geometries (i.e., micro-electro-mechanical-systems (MEMS), piezo-tube scanner, micro-motors, and optical rotary joints). The systems are technologically quite challenging, time-consuming, and usually manufactured individually. From the methodical point of view a significant challenge is the low penetration depth and the often heterogeneous, low-contrast data. The typically reached penetration depth of 1 mm is very sufficient to extract the relevant information about the layer structure. In many clinical settings the relevant features occur way below the depth of 1 mm, resulting in hard to interpret results. Moreover, tissue is not as well structured as the well-defined layers and the resulting image information is very homogeneous with very few features, making an analysis challenging. The homogeneity, at least at a depth of 1 mm, also appears when underlying changes in the physiological condition occur. An additional challenge is the interpretation of the data from *in vivo* studies and comparison to pathological reference information. Currently, in research it is standard to compare

OCT data acquired from biopsies with pathological hematoxylin and eosin (H&E) staining of thin section slides and a one-to-one correlation is often hard to achieve. This correlation of the visual information from OCT data becomes even more difficult in *in vivo* applications, as correlation between measurement sites and a removed sample is not easily achieved and has to be addressed, for example, through combination with other optical reference modalities.

In summary, OCT is a good example of the translation of an optical modality to real applications, particularly in ophthalmology. Moreover, there are multiple very promising clinical applications where the method offers great potential. Nevertheless, the translation also has to address methodical and technological challenges. Specifically, the development of novel and high-resolution ultrasound transducers in combination with image processing approaches readily reaching resolutions of sub 50 μm, while providing depths of up to 10 mm. The next few years will be very exciting in terms of technological development and additional translation to clinical applications.

5.4.1.6 Point-of-care, particularly with respect to the use of light-based techniques for infectious diseases (Popp and Bauer 2015, Tannert et al 2019, Matanfack Azemtsop et al 2020)

As mentioned above the first classification of bacteria was made possible by the advent of the earliest microscopes using different morphologies of the cells. Nevertheless, this information is limited, since only a few basic morphologies are known for bacteria. In a further step, bacteria with different cell-wall properties could be distinguished on the basis of Gram staining. Morphology and Gram staining is still used today in the routine identification of bacteria. However, with Robert Koch, the cultivation of bacteria became possible leading to the discovery of Anthrax and tuberculosis-causing bacteria. Koch also showed for the first time that the differentiation of bacteria (i.e., pathogens) based on biochemical tests was possible. This method is based on the fact that bacteria can metabolize different substances. Therefore, selective agars or enrichment agars are used to make different properties of the bacteria visible by means of indicator substances used during cultivation. The characteristic metabolic profile can then be used to distinguish the bacterial species. Today, the actual identification takes place with the help of automated systems such as the API (analytical profile index) system. The cultivation of bacteria in the presence of antibiotics with Etests or disk diffusion then provides information about potential (multi) resistances of pathogens.

In addition to cultivation-based methods, also nucleic acid-based techniques, such as polymerase chain reaction (PCR), are used to identify pathogens. Here, after biomass enrichment, first the DNA of the pathogens is isolated and then selected discriminative targets of this DNA are amplified in order to identify the species.

Cultivation-based tests are still the gold standard of bacteria identification; nevertheless, they require a lengthy cultivation time for biomass production. To shorten the complete identification process light-based methods are available. Using a short initial cultivation of only several hours, microcolonies (figure 5.18) can be used. Here the biomass of the microcolonies are dried as a homogeneous film on

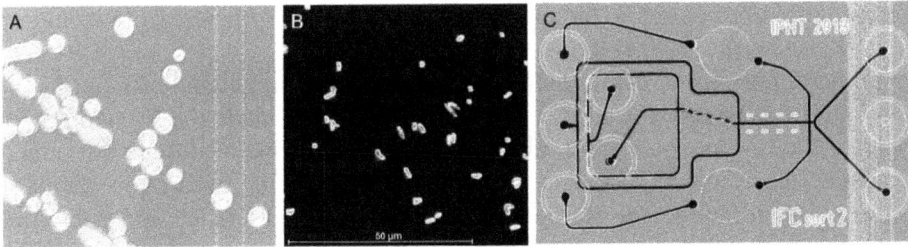

Figure 5.18. (A) Microcolonies on an agar plate; (B) microscopic image of single bacteria cells; and (C) microfluidic chip to sort bacteria.

suitable sample holders and the pathogens can then be identified by means of the molecular fingerprint methods Infrared or Raman spectroscopy (see section 5.4.1.2). Applying a Raman fiber probe even getting measurements directly on an undisturbed microcolony on an agar plate is possible.

When Raman spectroscopy is combined with a microscope, spatial resolutions on the order of bacteria size can be achieved. In figure 5.18(B) a dark field microscopic image of bacteria is presented. Using the excitation laser of the Raman microscope, single bacterial cells can consequently be measured. It is therefore possible to identify bacteria directly by analyzing the fingerprint Raman spectrum, without prior cultivation. Nevertheless, the bacteria cannot always be found as a pure sample. Therefore, to apply Raman microscopy it is necessary to isolate bacteria from their natural habitat. For pathogens this means destruction-free isolation methods need to be acquired for medical samples (e.g., blood, urine, or ascites) but also from environmental samples (e.g., water or soil). Despite the isolation time, identification based on Raman microscopy can be carried out in under two hours. After the identification of the pathogen species discrimination of sensitive and resistant isolates is also possible with this method. However, it is not possible to determine to which antibiotic the isolates are resistant or whether multi-resistant pathogens are present.

Microfluidic chips can be used to obtain further information on the resistance profile of individual pathogens. By combining Raman microscopy with different microfluidic chips, the changes of pathogens in the presence of different antibiotics can be analyzed. In this way, a resistance profile of the pathogens is obtained within approximately 90 min.

Besides the identification of species or the determination of resistances, microfluidic chips in combination with Raman microscopy can also be used for other applications. Since the Raman spectrum of a cell reflects normally the whole biochemical profile of this cell, the different contents in nucleic acids, protein, carbohydrates, and lipids may lead to an overall differentiation of phenotypes, growth states, or physiological states. Using this Raman spectroscopic information inside a microfluidic sorting chip (figure 5.18(C)) the cells can be then sorted into the different classes. The method is especially suitable for bacterial consortia which are heterogenous and/or cannot be cultivated. One application is the differentiation of cells with or without specific biomarkers such as special pigments, storage material like PHB, or secondary metabolites.

Another application uses stable isotopes which are metabolized during cell growth and can be recognized by characteristic changes inside the Raman spectra. Applying the isotope incorporation technique, it is possible to differentiate microorganisms according to, for example, metabolic activity or carbon metabolic pathways. In addition, the Raman-based cell sorting allows to add subsequent other analytical techniques such as PCR or whole genome analysis to the different sorted sample portions and to gain even more information about the sample.

Overall, the combination of Raman spectroscopy, chip-based sampling strategies as well as chemometric spectroscopic data analysis methods represents a powerful point-of-care approach comprising the entire process chain (i.e., from sampling to the final diagnostic result). By doing so, culture-free isolation and identification of pathogens, their host response, and their antibiotic resistance can be achieved. Thus, this promising approach offers the potential to overcome an unmet medical need by reducing the critical parameter 'time' to initiate a personalized lifesaving therapy as compared to the gold standard microbiology.

5.4.1.7 Outlook and future challenges

Big data—artificial intelligence
The most dramatic leap in the use of light-based concepts for medical diagnostics and therapy will come from advances in digitalization. For a broad/interdisciplinary use of optical/photonic methods for medical diagnostics and therapy, automated evaluation approaches are indispensable. They only allow the researched photonic approaches to be used in the sense of a qualitative or quantitative evaluation. The mere observation of the generated photonic data such as spectrum or image is very often not sufficient). In this context, artificial intelligence (AI) concepts utilizing deep learning approaches in particular are playing an increasingly important role in the successful implementation of light-based methods. Intensive cross-fertilization between optics/photonics and AI will enable experiments and photonic technological concepts to be advanced to such an extent that routine use far beyond specialized optics laboratories becomes conceivable. Because of the fact that the measurements of spectra, images, and time traces becoming easier and faster the speed of the analysis needs to be accelerated as well. This requires fully automatic data analysis pipelines to minimize the user intervention and to speed up the analysis by optimized algorithms. Further leaps in innovation can be expected in particular if research into new AI-based light analysis tools is carried out in line with research into new light-based processes. In this context, multimodal methods are particularly worth mentioning. As mentioned above, it has proven to be very advantageous to combine several optical methods in order to significantly increase the information content or to expand the application range of optical methods. Especially in such multiscale spectroscopy and correlative multimodal imaging, in which two or more optical modalities are used in combination, AI-based concepts can on the one hand guide the design of the light-based measurement methods in (e.g., the sense of

compressed sensing), and on the other hand AI concepts are the key to the application of such new multimodal photonic technologies, in that extreme amounts of data (big data) can be managed and visualized and evaluated in a user-centered manner.

Overall, there are great synergies between light/optics/photonics and AI, which are just being tapped in the context of light-based health technologies. Light or optics/photonics form an ideal platform for the application of AI concepts. In the first place, the automated evaluation of large amounts of data (big data) should be mentioned here. In addition, AI opens up great potential in the derivation of secondary data and conclusions from the primary light-based information, which in turn enables new design possibilities for optics and sensor technology. In doing so the optical process can be revolutionized: if, for example, the core purpose of an imaging system is to quantify a certain application-specific pattern, then it must be optimized above all to optimally illuminate, transmit, image, and recognize this type of correlation. The requirements for pattern recognition must therefore already be represented at all levels in the design process. This is only possible through digital design methods. AI-based algorithms will also play a major role in the recording of image data in the future. In this context, AI-based compressed sensing and AI-based compressed acquisition as well as AI-generated inverse modelling of the measurement processes should be mentioned.

New light applications—smart photonic materials

As explained above, with the help of STED and TIRFM microscopy, it is already possible today to obtain a high-resolution 'live image' of individual body cells. STED and STORM microscopy build on light microscopy and allow an enormous leap in the resolution of images. Until then, the limits of light microscopes were defined by wide-field imaging. STED microscopy is based on findings and advances in quantum physics/optics. Overall, quantum physics has given us a new understanding of the connection between light and information as well as access to new light sources and measuring methods or contrast phenomena for medical diagnostics and therapy in recent decades.

The so-called second quantum revolution will usher in another new era of optical imaging methods for biomedical issues through the use of non-classical light sources in the coming years. By using quantum effects, it may be possible in the near future to create images beyond our current state. The measurability of new characteristics of a light field, beyond the intensity distribution, opens up new conceptual and technological possibilities for optical health technologies. Here, light sources that can generate noise-free quantum states of light, imaging methods that rely on the correlations between different photons, and detector systems going far beyond the concept and capability of classical cameras to measure correlations in light should be mentioned. Furthermore, progress in the generation of extreme light (i.e., ultrashort pulses in the XUV or x-ray) allowing intensities higher than currently achievable will open up completely new possibilities for medical imaging and diagnostics.

The ability to structure materials on the nanoscale range creates the prerequisite for a completely new control of the propagation of light, which does not occur in natural materials. Particularly outstanding is the ability of nanosystems to concentrate light to regions well below the wavelength. The combination of these nano-photonic materials with elements of classical and quantum optics opens up imaging beyond classical limits (e.g., the Abbe Limit in wide-field imaging).

In summary optical medical research in the future must and will be researched along the entire chain, starting from the light source, via the light–matter interaction and detection to information evaluation, as a holistic and interdisciplinary problem in order to rethink it and design it for the future. By converging photonics, electronics, material design, and big data new optical health technologies will emerge to detect and fight diseases far before their onset. This is particularly important not to lose the fight against rising healthcare costs, requiring a massive restructuring of healthcare. Instead of the mere treatment of acute diseases, the prevention of diseases, the shortening of the time to convalescence, and the improvement of the quality of life must be given more and more priority in the future. This requires, in particular, personalized and precise treatment and an alignment of principles along the 'continuum of integrated care'. Here, new photonics concepts utilizing the above-mentioned approaches will play an important role.

Translational hurdles

In addition to the continuous further development of photonics technology research in the area of developing new, powerful hardware and software components, a particularly great challenge in the coming years will be simplifying and shortening the path of technological innovations into clinical routine. Translating biophotonic research results faces major challenges, especially with regard to, for example, the EU Medical Device Regulation (MDR) for translational research. Currently, the regulation makes it significantly more difficult or impossible to test biophotonic approaches on patients in the form of preclinical or clinical studies. Numerous light-based technologies have proven their potential for certain diagnostic and therapeutic questions in proof-of-principle studies, but the actual performance has not yet been demonstrated under routine clinical conditions in the form of comparative studies on a large cohort of patients. Here, funding for such validation studies is urgently needed to generate a marketable product. It only makes economic sense for industry to support a biophotonic proof-of-concept approach if there is a regulation-compliant study on a large cohort of patients that clearly demonstrates the added value.

What such translational research could look like in concrete terms is demonstrated by the 'Leibniz Centre for Photonics in Infection Research' (LPI) (see https://lpi-jena.de/en/), which was recently included in the national roadmap by the German government. Within the framework of the LPI, diagnostic and targeted therapeutic light-based approaches will emerge through the combination of photonic methods with infection research, which, after appropriate approval, will be transferred directly to industrial production and clinical application. A central element within the framework of the LPI are technology scouts who, at the beginning of the value chain, work together with clinicians to identify precisely fitting photonic solution approaches that have already successfully proven their potential in proof-of-concept studies for specific medical issues. LPI can thus serve as a blueprint for other medical problems, such as cancer or neurodegenerative diseases, to overcome the Valley of Death of clinical translation in these areas as well.

5.4.2 Synchrotrons and free-electron lasers for structural biology

Henry N. Chapman[1,2,3]
[1]Center for Free-Electron Laser Science (CFEL), Deutsches Elektronen-Synchrotron (DESY), Hamburg, Germany
[2]Department of Physics, Universität Hamburg, Hamburg, Germany
[3]Centre for Ultrafast Imaging, Hamburg, Germany.

The field of structural biology studies the forms of macromolecules such as proteins, RNA, and DNA, and the larger complexes that they can produce. Proteins are the nanomachines that drive the processes of life, and DNA and RNA carry the information to produce those machines.

Mammalian cells, for example, usually contain over 40 million macromolecules to catalyse reactions, regulate the flow of molecules for energy or signalling, sense and respond to other stimuli (such as forces), exert forces, repel pathogens, repair damage, or replicate and divide. By revealing the structures of proteins at the atomic scale, the mechanisms and functions of proteins can be uncovered, giving insights into disease, infection, the action of medicines, energy conversion, and food production. More recently, it has given the ability to design and construct new functioning materials. Accelerator-based x-ray light sources, and in particular synchrotron radiation facilities, have been largely responsible for creating and propelling the field of structural biology over the last thirty years by revealing protein structures through the method of macromolecular crystallography, as well as the study of the interactions between proteins via solution scattering techniques. There are presently several large developments in the field that promise profound changes and a more complete understanding of biological processes. The first is the ability to predict the 3D structure of a protein from its genetic code, using artificial intelligence methods based on examining the huge number of structures determined so far. Another is the effort to map out every protein molecule in the cell to understand the entire system, gained primarily through improvements in electron microscopy. Recent breakthroughs in electron microscopy also provide structures of proteins from many images of single uncrystallised molecules, which can be sorted and classified to observe conformational variabilities of proteins. Finally, large-scale, x-ray free-electron laser facilities have come online and are now providing movies of proteins undergoing reactions and responding to stimuli, and providing structures over relevant ranges of physiological environmental conditions that are not accessible in electron microscopy (Schlichting 2015, Spence 2017, Chapman 2019). Together, these developments are ushering in a new technological age of protein design using the raw materials of amino acids to create molecular machines to achieve specific functions (Huang *et al* 2016).

Proteins are linear polymers made out of 20 different amino acid building blocks. These are arranged in the sequence prescribed by the genetic code. Each amino acid type has different physical–chemical properties, such electronegativity and hydrophobicity, which together influence how the polymer chain folds and coils into its unique complex three-dimensional structure that confers the protein with particular structural properties and functions. Such functions range from catalysing specific reactions, regulating the flow of signalling molecules, sensing and responding to those signals as well as other stimuli (such as forces). By actually imaging the molecular structure, an understanding of the mechanisms of proteins can often be revealed. This necessarily requires radiation of short enough wavelength to resolve interatomic spacings. For x-rays there are no atomic-resolution lenses to directly form this image, so the molecular structure must be inferred from the scattered radiation that interferes to produce a diffraction pattern. One of the earliest and most iconic examples is Rosalind Franklin's Photograph 51 of DNA fibres, taken in 1952 using a laboratory x-ray tube and which led to the proposal of the double- helix structure by Francis Crick and James Watson.

The first protein structures soon followed, applying an earlier discovery that purified proteins can be crystallised. Crystals give strong, measurable diffraction peaks, called Bragg reflections, which can be mapped out in three dimensions as the crystal is rotated. The strength of each reflection is proportional to a particular spatial modulation of the electron density of the molecule. A complete set of these spatial frequencies can be combined to create an image of the molecule, except for the fact that the phase of the diffracting wavefields cannot be recorded (which give the relative alignments of the spatial frequencies). This was first solved by an approach akin to holography, by recording diffraction from crystals with and without a heavy reference atom. Those first structures, hemoglobin and myoglobin, earned Nobel Prizes to Max Perutz and John Bernal. The molecular structure of hemoglobin immediately revealed how this molecule has a high affinity to oxygen when blood cells are in the lung and yet release it in other parts of the body to power many other cellular functions. The protein binds four oxygen molecules but undergoes a conformational change upon binding the first, making affinity dependent on oxygen partial pressure. Now, the protein data bank, an on-line repository of protein structures, has accumulated over 200 000 structures, indicating the enormity of the global efforts of structural biology. The vast majority of those structure were generated in the last two decades using synchrotron radiation facilities with specialised beamlines and instruments dedicated to macro-molecular crystallography.

One reason for the success of crystallography is that by crystallising a protein, the molecules are brought into an alignment and regularity which provides a cooper-ative increase of the diffraction signal, proportional to the number of molecules or the volume of the crystal. This amplification of the signal enabled large enough crystals to be measured with weak laboratory x-ray sources (sometimes over days) and which now can be accomplished even with microscopic crystals in seconds at a synchrotron radiation facility. Accelerating charged particles radiate, and when those particles are moving near the speed of light (as they are in accelerators such as storage rings and synchrotrons that were built for high-energy physics experiments) they emit very energetic x-ray beams that are directed into a narrow cone in the forward direction due to the relativistic motion of the charges. Although a nuisance for the experiments the machines were built for, this directed radiation was a boon for crystallography which requires a collimated or laser-like beam to impinge on the crystal. The synchrotron radiation from electrons in storage rings was initially exploited parasitically from the early 1970s, and some of the first uses were to measure the weak scattering from biological fibres and macromolecular crystals. Particle physics laboratories such as DESY in Germany, the SLAC National Accelerator Laboratory and Brookhaven National Lab in the USA, Daresbury in the UK, and KEK in Japan were centers of innovation which designed and constructed dedicated machines, optimised to create bright beams of x-rays, tunable over wide wavelength ranges. A particular innovation was the development of the wiggler and undulator devices—periodic magnetic structures that oscillate electrons side to side as they travel—to provide further amplification factors proportional to the square of the number of periods in the case of undulators. Designs of the storage ring electron optical layout confined the electron beam to ever smaller dimensions to increase the brightness of the x-ray beams they produced. There has been an unabated exponential increase in the brightness output of new or upgraded facilities as a function of time that continues even today, so that we now have an increase of over 10^{22} in average brightness at a so-called fourth-generation facility as compared with the first storage rings. Today, there are more than 50 facilities all over the world, with each facility typically housing 20 or more beamlines, each with specific instrumentation for various analysis techniques for the study of all phases of matter. The instruments serve broad disciplines such as materials science, medical imaging, atomic and plasma physics, and structural biology (Jaeschke *et al* 2020).

Synchrotron radiation facilities began to impact and define the field of structural biology in the 1990s with the construction of dedicated macromolecular crystallography beamlines consisting of a precision goniometer to orient and rotate the crystal and a large-area detector to digitally capture the diffraction patterns that could be directly processed to obtain an electron density map. Unlike laboratory sources which primarily emit at discrete fluorescence photon energies, the wavelength of synchrotron radiation sources can be continuously tuned. This provides a powerful approach to solve the 'phase problem' mentioned above, by exploiting changes in the scattering properties of specific atoms at energies near resonances in their absorption. This method of multiple-wavelength anomalous diffraction (MAD) phasing allowed the structures of native metalloproteins to be obtained, as well as proteins containing the amino acid methionine (which contains a sulphur atom and which can be replaced with selenomethionine containing a more suitable selenium atom). Other proteins not amenable to this method could be solved by a method called molecular replacement, where a structure of a similar protein is used as a starting point to refine the atomic coordinates, subject to many constraints such as bond lengths and angles. As the x-ray brightness of facilities increased, robotic handling of crystals and the inexorable improvements in computing capabilities removed bottlenecks and opened up new possibilities for large-scale screening experiments. Today, as long as the quality of the crystal is sufficient, the protein structure can be obtained in minutes—a task that was worthy of a PhD dissertation at the turn of the twenty-first century.

The short wavelength of x-rays needed to resolve molecules at the atomic scale unavoidably means that the radiation is energetic enough to ionise those atoms. Biological materials cannot withstand too much x-ray exposure, as determined by a maximum tolerable dose (energy deposited per atom or per unit mass) at which the very structure under investigation is degraded. Obtaining large crystals can be difficult, and while bright x-ray sources can make up for the loss of diffracting strength (proportional to crystal volume) this comes at the cost of increased dose and damage through the larger exposure. The radiation sensitivity of protein crystals can be extended by cryogenically cooling them, to immobilise the products of radiolysis, such as free radicals, and prevent them from reacting with protein molecules.

Without the practice of cryo-crystallography it would not have been possible to obtain the vast majority of the 200 000 structures in the protein data bank. The observations and analyses of the molecular structures revealed using synchrotron radiation have brought fundamental new understanding and knowledge about biology. The structure of the green fluorescent protein (GFP) from a jellyfish showed how a barrel-like arrangement of the protein chain that houses a chromophore confers bright green fluorescence and showed how to mutate the protein to change the emission wavelength, thereby providing a visible marker that can be genetically targeted to other proteins for studying cancer and other diseases, and revolutionising microscopy of the cell. Synchrotron radiation also revealed the structure and mechanism of adenosine triphosphate (ATP) synthase, the remarkable macromolecular machine that produces the molecule ATP that is used as a fuel to drive most protein reactions. The ATP synthase is powered using a potential gradient across the cell membrane. Many proteins are located in the lipid membranes of cells, which are notoriously hard to crystallise, but cryo-cooling gave the possibility to measure some small crystals of ion-channel proteins to reveal how the channel can selectively pass potassium ions and not sodium ions, at the extremely high rates as needed for electrical signalling in the nervous system. And the technique was crucial to obtain the structure of the ribosome—the large complex machine consisting of protein and RNA which assembles new proteins amino acid by amino acid, by reading the genetic code provided by a strand of messenger RNA (mRNA). All these examples garnered separate Nobel Prizes (Jaskolski *et al* 2014). There were two more for the structure of a G-protein coupled receptor and the structure of the protein that transcribes a DNA sequence into the mRNA strand.

Despite all these successes, cryogenic cooling may lead to subtle changes in the molecular structure which may mislead the interpretation of how a protein catalyses a reaction, for example, or how a drug molecule can bind to and inactivate a viral protein. Cooling a protein to liquid nitrogen temperatures also prevents us from studying the evolution of a reaction or studying protein dynamics. In 2009, a revolutionary new machine was brought online that can address this problem: the x-ray free-electron laser, or XFEL. Unlike circular storage rings, an XFEL employs a linear accelerator coupled with a 100 m long undulator device to create x-ray pulses with a billion times higher peak brightness (outpacing the previous trend of synchrotron radiation improvements). Each pulse has a duration of tens of femtoseconds, packed with as many photons as used to generate a protein crystal dataset at a synchrotron. The x-rays are created from accelerated electrons via an amplification process in the undulator. With energies of several giga-electron-volts, the electrons travel at close to the speed of light, and therefore they keep up with the light field that they generate in the undulator. That light field in turn influences the trajectories of the electrons and forces them to separate into bunches that are a wavelength apart. The bunched electrons radiate in phase, giving an amplification proportional to the square of the number of electrons. The process grows exponentially along the length of the undulator due to this positive feedback. As with synchrotron sources, the machine addresses a plethora of scientific needs, but emphasises ultrafast processes on the timescale of atomic motions and chemical processes. Today there are five operating x-ray FEL facilities in the world, with more in construction. The largest is the European XFEL in Germany, a facility that is three kilometers long with the potential to operate five undulators simultaneously (figures 5.19 and 5.20).

The short intense pulses of XFELs break the dependence between exposure, dose, and crystal size. When focused to a small spot, a single XFEL pulse vaporises any material in its path by stripping electrons from all atoms, creating an expanding plasma. However, it takes time for atoms to begin moving since they have inertia, and by the time atoms have moved a fraction of a bond length the pulse has already

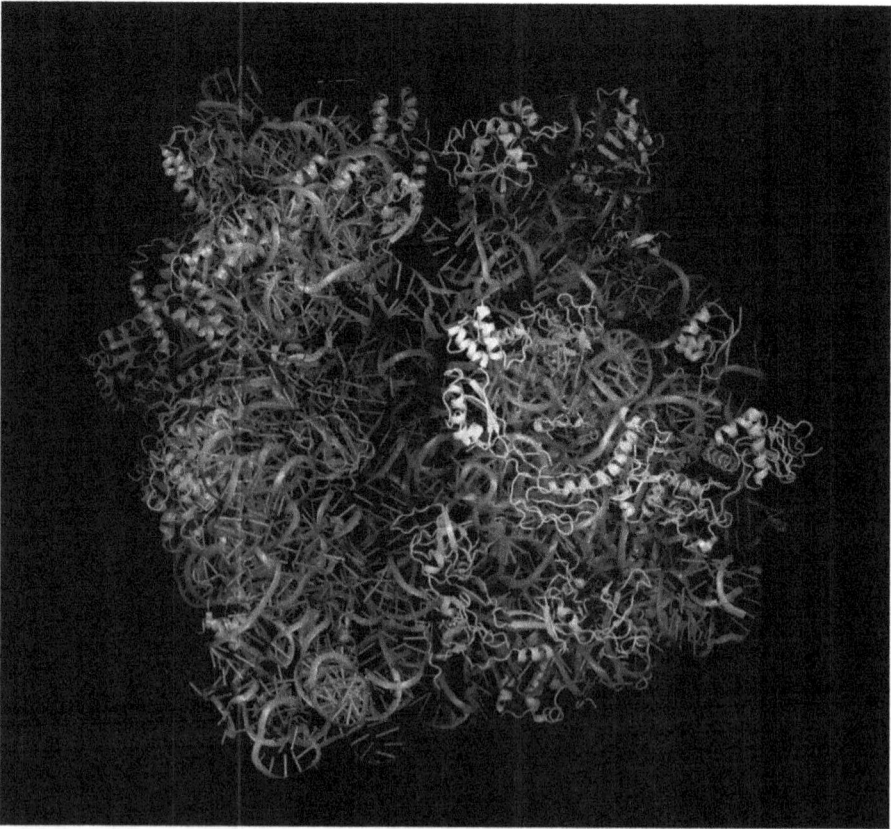

Figure 5.19. The molecular structure of the ribosome, as determined by x-ray crystallography at DESY in Hamburg, showed researchers how this complicated nano-machine synthesises new proteins by reading genetic information encoded in messenger RNA molecules (credit: Joerg Harms, MPSD).

Figure 5.20. A beamline at the Linac Coherent Light Source of the SLAC National Accelerator Center in USA, used for serial femtosecond crystallography measurements (credit: SLAC National Accelerator Laboratory).

passed (Neutze *et al* 2000). Thus, the exposure to biological materials and protein crystals can far exceed the previous dose limits, giving much stronger diffraction. This concept of 'diffraction before destruction' opened up the method of protein nano-crystallography (Chapman 2019). Since a crystal lasts only one pulse, the full three-dimensional diffraction dataset requires a stream of nanocrystals to be rapidly fed into the beam one by one: at the European XFEL located in Hamburg, Germany, pulses arrive with as little as 220 ns between them. One method is to deliver crystals across the path of the x-ray beam in a high-speed liquid microjet. Crystals are exposed in random orientations, which must be inferred from the diffraction pattern itself. This approach, called serial femtosecond crystallography, measures many tens of thousands of individual crystallites in minutes, at the appropriate physiological temperature, and gives structures that are completely free of radiation damage. Most importantly, the short exposure time of each pulse freezes out all motion and can capture various stages in chemical reactions at femtosecond precision. That is most readily achieved in photoactive proteins (such as those involved in vision or photosynthesis) by triggering a reaction with a short optical pulse from a laser that is precisely timed to arrive at a particular time before the x-ray pulse. Datasets can be collected at different delay times and the resulting structures assembled into a 3D movie. The first conformational changes of the chromophore responding to a visible light photon have now been captured in exquisite detail and several groups are racing to understand how the large photo-system II complex breaks down water molecules and creates a separation of charges (to power ATP synthase), releasing molecular oxygen as a by-product (Brändén and Neutze 2021). Such knowledge will help us to understand how to more efficiently (and cleanly) capture the energy of sunlight.

5.4.2.1 Futures

Nature utilises 20 different amino acids to make proteins and so for an average sized-protein of about 300 amino acid residues there are $20^{300} = 2 \times 10^{390}$ different possibilities of protein structures. There is not enough matter in the Universe to make one copy of each of these, which of course dwarf the number of unique proteins thought to exist in nature (less than 10^{10}), let alone the number of proteins whose structures are solved (200 000 in the protein data bank). While not every random sequence of amino acids would create a protein to do something useful (just as not every random sequence of letters produces interesting literature) it is clear that nature has only sampled a tiny fraction of the possible useful protein structures that can be made. There is therefore a goal to design new proteins to carry out functions of nanomachines or form new materials, and specifically to design new proteins so far from nature that they cannot be thought to have evolved from existing sequences. Using our literature analogy, we can aspire to form completely new genres. This requires the long-sought ability to predict the three-dimensional fold of the protein —the way the polymer chain coils and bends on itself—from the sequence of amino acids that make up that chain.

In the year 2021, that ability was demonstrated to remarkable precision by the computer program AlphaFold 2, developed by Google labs (Jumper *et al* 2021). The program uses a deep neural network that was trained on the structures in the protein data bank—most of which were determined using x-ray crystallography at synchrotron radiation sources (Burley *et al* 2022) Are those sources now obsolete? The answer seems to be: no, and further experimental capabilities are needed. Proteins are not static objects, and as seen from the very first structure of hemoglobin, their conformations change as they carry out their functions, respond to their environment, and interact with other proteins. The macromolecular machines that produce ATP or transcribe DNA are exceedingly complex and intricate in their mechanisms, which are only currently understood at a superficial level. Many such machines are driven by chemical or voltage gradients, Brownian motion, or entropy, and it is not yet possible to predict how a such a system operates or how it responds in changing environments. Even the act of breaking such a mechanism—to inhibit the action of a virus or other pathogen—cannot be predicted from a static structure. This knowledge requires mapping out the conformational energy landscapes of proteins in full detail and under a range of environmental conditions or subjected to stimuli. That is, structures must be measured as they fluctuate or are directed to carry out reactions. As seen above, XFELs are answering this call, using intense femtosecond pulses to collect diffraction snapshots of small crystals, but more is required to expand the scope of these measurements and to extract structures and dynamics from large datasets.

Serial crystallography at XFELs must be expanded to achieve high-throughput multidimensional measurements. The kinetics of drug binding, for example, could be mapped by measuring diffraction at various times after mixing crystals with those molecules, over a range of temperatures. Currently the time needed per measurement point in this matrix is limited by the repetition rate of the x-ray source (which can potentially reach millions per second) and the matching rate of the detector. It may be feasible to probe tens of conditions per second. Since serial crystallography can be carried out by ejecting crystals across the x-ray beam in liquid jets, the experiment can be controlled and automated using microfluidic technologies. In this way, entire libraries of potential drug molecules or compounds could be studied, greatly expanding current activities of crystallographic screening. During the current COVID-19 pandemic, crystallography-based searches for compounds that inhibit certain viral proteins have taken place, measuring thousands of conditions over month-long campaigns, or several minutes per compound. This potentially could be accomplished in minutes. With automation comes the ability to control and use the analysis of resulting structures to guide measurements and to optimise searches. Development must be made to integrate production, purification, and chemical synthesis with the analytical capabilities of serial crystallography.

Today, molecular movies have been made for proteins undergoing reactions triggered by short light pulses, precisely timed to x-ray pulses. While such studies have given some general insights of the actions of proteins, these photoactive proteins represent but a small fraction of all proteins. The means is required to precisely trigger reactions in systems that are not sensitive to light. One promising method is to engineer caged compounds, which are released by a flash of light to begin the reaction. More generally, there are efforts to use advanced data processing and pattern matching techniques to precisely order a set of snapshots recorded at fluctuating times from experiments where the temporal evolution of a process is only loosely imposed. This relies upon machine learning to place those snapshots on a manifold, in a high-dimensional phase space, which must be discovered in the data itself. Recent tests have shown a remarkable ability to achieve a time resolution much better than the duration of the x-ray pulses, perhaps aided by temporal fluctuations in those pulses (Hosseinizadeh *et al* 2021). Such machine learning methods could be used to feed back into the generation of the temporal profiles of x-ray pulses, perhaps by modulating the electron beam in the accelerator, to optimise the ability to place events according to their sequence in time.

The crystallisation of proteins has been key to obtaining structures of macro-molecules using x-rays from the earliest days. An alternative approach is to use a powerful electron microscope to image the molecules directly, but this suffers from a similar dependence of damage on dose as for x-rays. It is now possible to assemble 3D molecular structures at atomic resolution from very noisy low-exposure images, but this too requires cooling them to cryogenic temperatures. Nevertheless, cryo-electron microscopy is revolutionising structural biology by obtaining structures of molecules that cannot crystallise (Kühlbrandt 2014). And while structures may be distorted by the low temperatures, it is possible to sort the images into different structures to observe distinct conformations (such as a ribosome in different stages of constructing a polypeptide). X-ray free-electron lasers promise similar capabilities, but without the need to cool the molecules. This is being pursued by reducing the size of crystals in diffraction experiments down to the smallest they can go—to single molecules. The diffraction patterns of single molecules are certainly weak without the amplification provided by the crystalline lattice but, by using similar algorithms as those used to sort and assemble data in cryo-electron microscopy, it may be possible to build up diffraction datasets of single molecules that are streamed across the x-ray FEL beam as an aerosol or in a thin liquid jet. As is the case for crystals, it should be possible to trigger reactions and obtain structures as a function of time, but now molecules that follow different reaction pathways could be sorted and examined independently. From large ensembles of single-molecule data it will be possible to capture rare events that happen spontaneously, such as the precise moment a molecule binds.

High-power x-ray FEL pulses are needed for single-molecule diffractive imaging, which might be achieved by adapting chirp-pulse amplification schemes to the x-ray regime. Stronger diffraction signals will require even shorter pulses—at the scale of attoseconds—to ensure 'diffraction before destruction' of single molecules. The total pulse energy may not necessarily be much larger than what is generated today using a kilometer-long linear accelerator. With high enough control of electron beams with the next generation of particle accelerators it may be possible to shrink the facility to a size that can fit into a laboratory. Together with electron microscopes, large-scale x-ray facilities, and computational prediction, these will help to vastly expand the study of the machinery of life and usher in the new technical age of protein design—the amino age.

5.5 Physics research against pandemics

Thierry Mora[1], Chiara Poletto[2], Marta Sales-Pardo[3] and Aleksandra M Walczak[1]

[1]Laboratoire de physique de l'école normale supérieure, CNRS, PSL University, Sorbonne Université, and Université de Paris, 75005 Paris, France

[2]INSERM, Sorbonne Université, Pierre Louis Institute of Epidemiology and Public Health, Paris, France

[3]Department of Chemical Engineering, Universitat Iovira i Virgili, 43007 Tarragona, Catalonia, Spain

5.5.1 Introduction

Diseases and epidemics have always accompanied humanity. Viruses need hosts to survive and both humans and livestock fit the role. Science has contributed significantly to our current ability to fight infections through vaccination, medicine, and therapy.

Epidemics are emerging collective phenomena. In 2019 the SARS-CoV2 pandemic hit the world. The curve of new cases we were checking every day is the result of interacting phenomena occurring at completely different spatial and temporal scales. More specifically it is the result of interacting systems that co-evolve. Humans react to the epidemic and change their behavior in the attempt to contain it. Viruses, on the other hand, continuously mutate and diversify. Can quantitative science in general, and physics in particular, help us grasp this complexity and better predict and control? We can schematically break down this complex interdependence in four main steps, from the micro to the global scale (figure 5.21): (1) predicting the emergence of a new mutant variant strain; (2) predicting its ability to escape the hosts' immune system; (3) predicting its spread through the population; and (4) predicting the behavior of the population in response to the epidemic. Each one of these scales contributes to whether an outbreak results in an epidemic or not, and how this epidemic will progress and evolve. Physicists have been working on all of these scales, bringing about both important methodological and practical insights that have had a deep impact on biology, medicine, and public health. Here, we walk through these scales, showing how they interact.

5.5.2 The physics of host–pathogen evolution

5.5.2.1 The diversity of immune responses

At any time, each person is surrounded by many pathogens. Our immune systems manage to recognize most of them in time and stop us from getting sick. Our immune systems protect us by producing a wide range of defense mechanisms ranging from mechanical and anatomical barriers, such as our skin, physical reactions such as scratching or crying, to the action of cells of the innate and adaptive immune systems. The innate immune system provides a first line of defense by recognizing stereotyped properties of pathogens. Cells recognize general properties of foreign, potentially harmful organisms such as molecules characteristic of

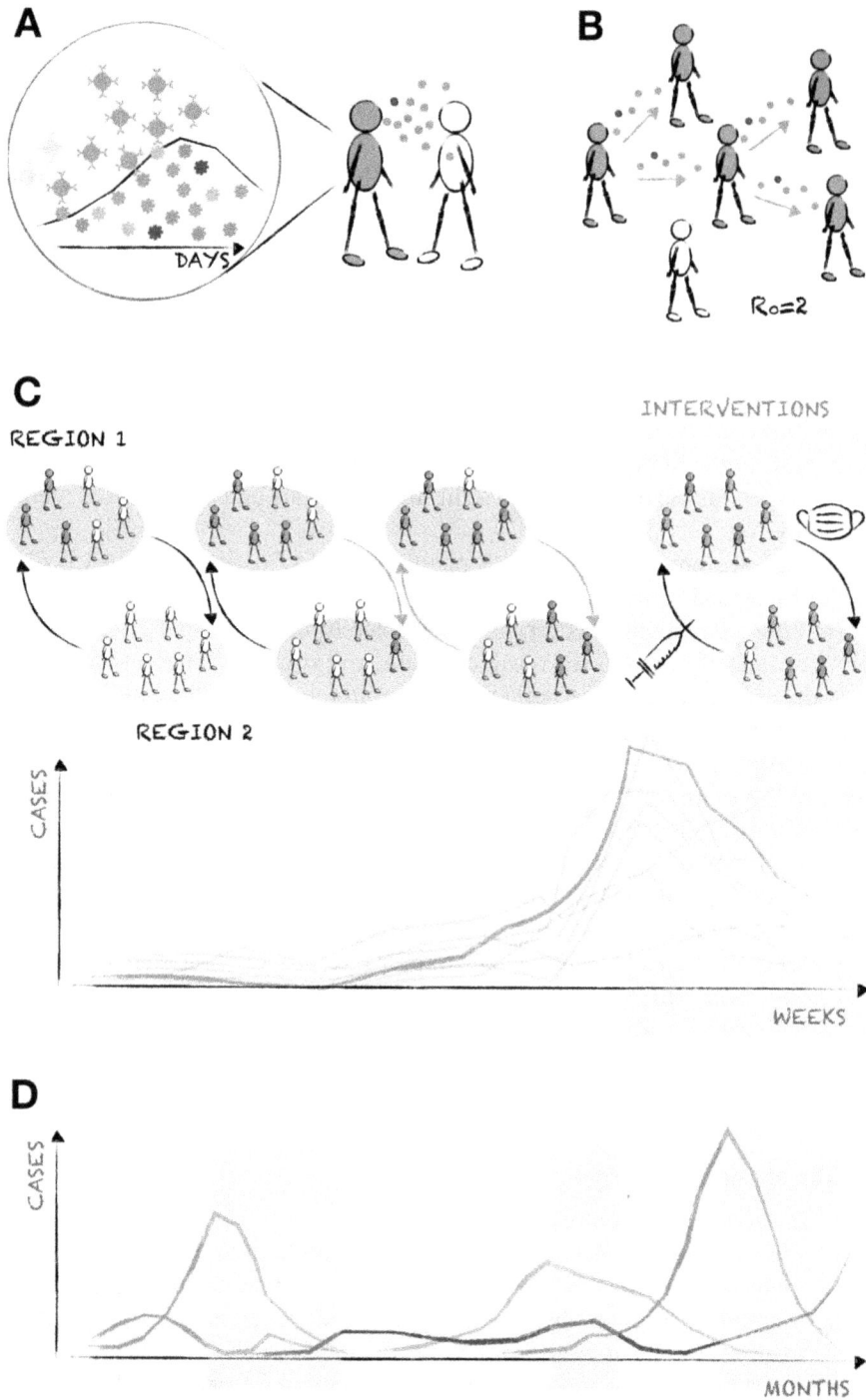

Figure 5.21. The many scales of host–pathogen interactions. (A) Within an infected host the cells of the immune system target existing viruses (and pathogens in general) by binding and neutralizing them. Each host has a vast repertoire of immune cells (denoted by different colors) from which those that best target the

infection are chosen (same colour as the viral strains). In some cases, these cells can further somatically evolve to increase their recognition power. As the virus replicates within the host, it can also mutate within the host such that the strain that infects the next susceptible individual can be different than the one that infects the infecting host. The color of each infected host denotes the most common viral strain within the host. **(B)** Infected hosts infect on average R_0 other individuals. The immune systems of the infected hosts exert selection pressure on the virus to mutate and diversify so that it can continue infecting. After an infection, hosts develop immunity to the infecting strain (and often similar strains). Through mutations, some new mutants can escape host immunity and re-infect previously recovered individuals. **(C)** The rate at which the disease spreads within the population depends on the network of interactions between individuals and their behaviour. The disease will first spread within a region of close-knitted interactions and slowly spread to other regions via mobility. Interventions such as vaccinations, face masks, or physical isolation decrease the rate of spread. These dynamics result in the increase and slow down of the number of new cases. An epidemic is one stochastic realisation among many possible ones. Factors such as social-contact structure of the population, immunity profile, strength, kind, and adherence to physical distancing measures, among others, drive and (constrain) the space-phase of possible trajectories. **(D)** On longer timescales, we see repeated patterns of increase and decrease, governed by events on the smaller scales: interventions, population structure, and emergence of new strains.

bacterial membranes, amoeba, and other parasites. The adaptive immune system is more specific, and uses a large ensemble of cells carrying diverse receptors that are specific for pathogenic molecules. This group of B- and T-cells is called a repertoire. Importantly, the adaptive immune repertoire updates its composition based on the pathogens it encounters, meaning it is a constantly evolving and changing system. The innate system informs the adaptive system of newly detected threats, acts as a messenger between its cells, and the two systems work closely to control all infections.

There are about a billion different B-cells and T-cells, each coming in clones of different sizes, making up together about a trillion different cells that patrol our body. This diversity of cells protects us against pathogens, including those that did not exist when we were born, while at the same time being able to discriminate between molecules natural to our body (good) and those that are foreign (bad). This dynamic ensemble of ever-changing cells is self-organised and distributed, meaning there is no leader cell. Cells communicate between each other to control their numbers and proliferate in case of an infection through signaling molecules, but there is no top-down chain of command. For a few decades now, biological physicists have been studying the equations that describe the interactions between these cells and lead to the composition of the repertoire that protects us against pathogenic threats. These equations are both stochastic, meaning that the forces that drive the repertoire are random and can take different values at different times, but also the interactions are nonlinear, meaning that the existing cells influence the frequency of each other in complex ways. Together these elements make predicting the future frequencies of cells in the repertoire difficult. Physicists have a lot of experience in solving this class of equations and finding the parameters they are sensitive to and those that matter less. These skills are important for predicting future states of repertoires and correctly modelling these systems (Perelson and Weisbuch 1997).

Lastly, pathogen-recognizing receptors of the adaptive immune system are not hard coded in our DNA, like the innate system receptors. They are constantly generated by a highly stochastic process, meaning that each one of us has a set of cells that is unique to us. In fact, it is so unique it can be used to identify us. But random does not mean it can be anything. There are rules that this randomness obeys and physicists have managed to learn these stochastic rules thanks to which the functioning immune repertoire is generated. This uniqueness and randomness imply that we can all have different cells protecting us against the same viral challenge, and that they will protect us equally well. On the one hand, this means there are many good solutions to the same problem; on the other hand it means that we cannot just hope to engineer the exact same solution for everyone. The best protection helps our immune systems find its own solution, which is why both vaccines and immunotherapy work. In summary, this large but finite and constantly changing army of cells controls most of the pathogens we encounter, most of the time.

5.5.2.2 *The diversity of viruses*

Where do these new mutant viral strains come from? From the point of view of the human population, they either are mutations from existing strains of human viruses, as in the case of seasonal influenza, or come from spillover of a virus from animals (called *zoonosis*) through adaptive mutations, as in the recent H1N1 influenza pandemic or SARS and MERS betacoronavirus outbreaks. Yet, even the strains that come from animal reservoirs come from mutations: before it mutates, the strain that infects animals is not well adapted to humans, and specific mutations allow it to infect and spread through the human population.

The rules underlying evolution are simple. Mutations (small local changes to the organism's DNA) and recombination (rearrangements of different fragments of DNA) generate diversity. For many microbes such as viruses, mutations play a dominant role. These changes are selected according to the survival and reproductive success of organisms that carry them. Mutation first appears in one organism in the population and then spreads with the rate this individual reproduces and has offspring. The first couple of reproductive steps are very perilous since even a very beneficial mutant may not manage to reproduce, simply due to bad luck. Once the individual manages to produce enough of its offspring—and we know how to calculate the chances of that—selection takes over and its final frequency in the population depends on whether it's a beneficial, deleterious, or neutral mutant. The life of an evolving mutant virus, as of any organism, is driven by these initial random events. Due to this initial randomness, even a mutation that is quite deleterious for the organism can rise to prominence and finally be present in every individual in the population.

Mutations and recombination events are very rare. Most of the time DNA gets replicated faithfully, with only occasional errors. For example, the mutation rate when making copies of viral DNA is one every 100 000 base pairs. That number seems small, but if there are 10 000 potentially mutating positions in each virus, and if there are 25 million copies of the virus (a fair estimate for the flu), millions of sites on the viral genome will mutate in each viral replication in each infected person.

Importantly, the number of currently infected hosts increases the probability of a mutation happening. The more mutations that happen, the larger the probability that one of these mutations is beneficial and helps the virus infect new hosts and spread. As a result, although viral evolution is driven by rare and random events, we can see and feel its outcome on large scales.

Ultimately, the fate of every single mutation is to either take over the whole population, or disappear. However, before that happens the organism carrying this mutation can acquire a new mutation but in a different place in its genome. Then the fate of our initial mutation depends on all the other mutations in the organism. For this reason, a viral population carrying many strains with different sets of mutations may co-evolve with the hosts for a very long time, during which the fate of each mutation individually is settled.

One outcome of the evolutionary process is that a single individual and its descendants take over the population—every virus in the population is the offspring of the same ancestor in which the mutation appeared. While this seems very counterintuitive it is an example of the rich-get-richer law and a consequence that individuals with a selective advantage will have more offspring than other individuals. In many practical cases, especially at short times, mutants with similar reproductive success can appear and none of them will completely take over the population, especially if this population is large and not well mixed. This co-existence of several strains often happens in viral populations. While some strains may have a selective advantage, new, fitter mutants are likely to appear on the background of less fit strains, reshuffling the hierarchy of strains and completely changing the race. Of course, this does not mean the new mutant will take over, since it can also share the same fate. Physicists in the last two decades have been instrumental in highlighting and understanding this regime, which is formally analogous to thermodynamic systems driven out of equilibrium (Tsimring *et al* 1996, Desai and Fisher 2007). This work has shifted the focus of evolution from a winner-takes-all to a more complex picture of viral evolution.

5.5.2.3 Viral-immune interactions

Unlike other organisms, viruses need host organisms to reproduce: they do not have their own machinery to replicate their DNA, they need to 'borrow' it from the host. After they infect a host, its immune system attempts to get rid of them, so they need to move on and infect new hosts to survive. This means that one of the ways in which viral mutations are beneficial is by making it easier to infect new hosts, or even new types of hosts as it happens when a virus jumps from one species to another. Other intrinsic properties of the virus can also be increased, such as structural stability of its proteins and capsid or how fast it reproduces. Another type of beneficial mutations, called antigenic mutations, allow the virus to escape the immune response. Beneficial antigenic mutations will change viral proteins in such a way that immune systems of hosts that were previously immunized against the virus will no longer recognize them, because their T-cells, B-cells, or antibodies can no longer bind to them. This gives the virus time to increase the in-host population size, infect new hosts, and in the long run maybe even find better mutations (figure 5.21(A)).

Ultimately, it guarantees the survival of the viral population. In response, individuals infected with this new escape strain will update their immune system, exerting further pressure on the virus to find new antigenic mutations, fueling a continual arms-race (figure 5.21(B)). Biophysicists in recent years have been working on both identifying regions in viral proteins that are prone to antigenic and intrinsic mutations and using stochastic evolutionary models to understand how mutations in different regions influence viral evolution (Nourmohammad *et al* 2016, Chakraborty and Barton 2017). This problem bridges the molecular scale, where the biophysical properties of the amino acid residues that make up viral proteins and underlie their binding properties matter, and the population scale, where a fine understanding of nonlinear stochastic equations is needed. Of course, not every detail matters, and physicists have been working on figuring out which details are important, both through theoretical analysis and quantitative experiments.

We have been focusing on viruses, but the same evolutionary dynamics and co-evolution is valid for antibiotic resistance. Antibiotics are specialised molecules that target and lead to the death of bacteria that infect their hosts. However, bacteria, like all other organisms evolve and beneficial mutations are such that make the bacteria reproduce even in the presence of antibiotics. To overcome this newly evolved resistance, host organisms are either given new types of antibiotics, or higher doses. Physicists are also actively studying this problem using similar approaches as for viral-host co-evolution and have set up fascinating experiments where they identified the mutations that lead to antibiotic resistance even at high doses (Bush *et al* 2011). This kind of quantification is useful for guiding the design of efficient antibiotic regimes to suppress the evolution of antibiotic resistance. Interestingly, often mutations in different places on the DNA work together to produce stronger beneficial effects. This can help us make predictions, as we will see below.

5.5.2.4 *Predicting viral-host co-evolution*

Since the main events that drive the evolution of pathogens are random, is there any hope in predicting its outcome? In the last couple of years physicists have taken the idea of predicting influenza strains seriously (Łuksza and Lässig 2014, Neher *et al* 2014). They have shown that by taking a probabilistic approach combined with biophysical knowledge at the molecular scale and stochastic equations of evolution, one can predict this year's dominant flu strains, and one can even predict the frequency of these strains compared to other strains. Although viral evolution is driven by random rules, this does not mean everything is equally likely to happen. There are stochastic rules that make certain outcomes more likely than others. We can explore our knowledge of molecular biophysics and stochastic processes to assign probabilities to the outcomes of viral evolution. Specifically, starting with flu strains that have been measured in February of a given year, they use stochastic evolutionary models to calculate the frequency of each strain in the autumn of that year, when the flu starts spreading in the Northern Hemisphere (and analogously in August for the southern hemisphere). These equations account for the competitive interaction of the different strains but also for the interaction with the immune system. Since these are not exactly known, different scenarios must be considered.

The basic idea is similar to the one used in good portfolio design or weather prediction: we do not put all our eggs in one basket but sum over all possible outcomes, weighted by how likely they are. Taken together, the stochastic rules work well and these methods are now used by the WHO to decide which flu strains should be targeted in the flu vaccine for a given year.

Another example of quantitative random rules learned by physicists (Murugan *et al* 2012) describe the stochastic process that drives the huge diversity of B- and T-cell receptors and antibodies that the body can produce. These receptor proteins are encoded in the DNA. However, if we wanted to encode a billion different receptors in the DNA, the DNA would physically not fit into the cell. Instead, the DNA is edited in each B- and T-cell in a way that pre-existing gene building blocks are assembled like legos to produce hundreds or thousands of different receptor combinations. Additional random insertions and deletions of base pairs further increase diversity, resulting in the billion different receptors in a given person. Again, although the rules are random, not every outcome is equally likely. Physicists have quantified these random rules and can calculate the probability of any of us producing a concrete B- and T-cell receptor. Since the process is random, different people have different sets of billions of receptors. But knowing these probability rules allows us to predict that some receptors are more likely to be shared between different people, simply because they are more likely to be produced. These widely shared receptors and antibodies are called 'public'. In general, the machinery can produce much greater diversity than is realised in one person. Even the population of the whole world is not enough to exhaust all the possible receptors that can be generated. However, this is yet another example that random does not mean unpredictable. We can predict exactly how many receptors will be shared between different sets of people, and even which ones (Elhanati *et al* 2018).

In predicting future flu strains we solve one part of the prediction problem: knowing existing viral strains, can we predict their future frequencies. However, can we also predict new mutations that will appear and estimate how likely they are to survive the competition with other mutations and the challenges of the immune system? This is an important question for predicting escape mutations: mutations that are not recognised by the immune system. One approach that has been pioneered by physicists is making quantitative measurements of binding between viruses and B-cell receptors simultaneously in large libraries of immune cell mutants (Adams *et al* 2016). These methods have been extended to measure the binding properties of all possible single mutants in a key SARS-CoV-2 protein (the receptor binding domain of the spike protein) (Starr *et al* 2020). These measurements allow us to predict the ability of the virus to enter human cells, which is linked to disease severity and transmissibility, including for mutant strains that haven't emerged yet. These types of quantitative measurements allow us to explore the evolutionary trajectories of viral evolution and show which mutations close off paths for future mutants. Such knowledge allows us to propose vaccines that can guide the immune system. Physicists have shown that having such knowledge is important to design vaccine schedules that elicit protective immune cells (broadly neutralizing antibodies) that protect us against many viral mutants (Wang *et al* 2015).

5.5.2.5 Quantitative measurements and prediction scales

More generally, an important challenge is to measure or predict the binding affinity of immune receptors and antibodies to their target on pathogens, which in turn determines the efficacy of these receptors to fight the disease. Since both the number of receptors and antigens are extremely large, it is impractical to measure all pairs one by one. Massively parallel methods need to be designed, and biophysics-inspired computational methods need to be developed to fill in the 'gaps' of receptor-antigen pairs that we will not be able to measure. Promising methods that combine droplet microfluidics (Gérard *et al* 2020) and binding assays (Zhang *et al* 2018) are paving the way to that goal. The idea is to encapsulate single cells in small droplets of water suspended in oil, and then manipulate these droplets in a microfluidic chip to perform biochemical reactions in each cell in a compartmentalized way. By combining phenotyping and binding assays in each droplet in this way, and by introducing unique molecular barcodes prior to sequencing, one can associate the full sequence of the receptors to its affinity for a set of antigenic targets, which can themselves be sequenced. To complement these experiments and extrapolate their results for sequences that were not directly tested, we expect machine learning methods to be useful. As they are improved and scaled up in the future, these techniques will help to build a complete specificity map between receptors and pathogens, which will serve as a basis for rational drug and vaccine design.

As we see, the immune system and viruses are constantly co-evolving: the immune systems of the world's hosts exert pressure on viruses and viruses force our immune systems to constantly update itself. Vaccines prepare our immune systems for challenges that we have not yet seen, such that if we encounter the virus we have pre-existing protection. Additionally, the immune system is flexible and protection against one virus can also protect us against similar viruses, as was originally shown by Edward Jerne using the cowpox virus to immunize people against smallpox in the 18th century. Similarly, using sequencing measurements and analysis, physicists have shown that T-cells reactive to the common cold, which is also a virus from the coronavirus family, can recognize SARS-CoV2 proteins (Minervina *et al* 2021). Using quantitative experiments to understand the range of this cross-reactivity is an important challenge that can help us predict and trigger immune responses. This may also allow us to extend the prediction of viral strains to longer timescales.

There are many different molecular solutions to the same challenge, yet we show that we can statistically predict the immune pressures exerted on evolving viruses. How can we reconcile this large molecular diversity with predictability? One answer is that the phenotype is more constrained and reproducible between experiments than the molecular implementation in terms of actual mutations in the DNA (Lässig *et al* 2017). In short, since the selection pressure acts on the phenotype, it is constrained, and thanks to that, predictable. As we showed in the example of the immune system, there are many ways to obtain the same results in terms of mutations. The same results have been observed for antibiotic resistance and *E. coli* evolution under the pressure of the innate immune system in mouse guts.

5.5.2.6 Sub-summary

In summary, host-pathogen co-evolution occurs on many different scales: from the molecular, cellular, organismal to the worldwide population scale. On most of these scales we are dealing with random interactions between many elements. However, thanks to experimental and theoretical work we now have a command of these stochastic rules which allows us to make statistical predictions. Since this co-evolutionary race is constantly happening, why do we not see pandemic outbreaks all the time? We explore this question in the next section.

5.5.3 The physics of epidemics

With COVID-19 the whole world has experienced first hand the exponential growth of a serious infectious disease. This marked feature of many infectious disease epidemics challenges our intuition and leaves us unprepared.

The exponential growth in the number of infected individuals can be explained by simple contagion rules as shown in the influential work of Kermack and McKendrick in 1927 (Kermack *et al* 1927). This work laid down the equations of the SIR (Susceptible-Infected-Recovered) model which has become the core of modern epidemiology because it is able to reproduce fundamental characteristics of epidemics. In particular, the model allows for the clear framing of a key concept in epidemiology, that of the reproductive number R_0. This is the average number of individuals an infected individual infects before recovering when the population is fully susceptible (figure 5.21(B)), and quantifies the transmission potential of an epidemic (see box 5.1).

Box 5.1

Imagine that a few individuals bring a new pathogen into a susceptible population (S). If we assume two people interact at random, the probability for each susceptible individual to become infected at a given time is proportional to the number of infectious individuals in the population (I). The proportionality coefficient depends on two factors: the average number of contacts a person has at each time, k, and the rate of transmission per contact, β. Infected individuals can also recover with rate μ, thus ceasing to be a source of infection (R). It turns out that a simple model that allows transitions between the different states (S, I, and R) results in an exponential growth of the number of infected individuals in the early stages of an epidemic outbreak. The relationship between the generation of new infectious contacts $k\,\beta$ and the recovery rate μ determines the growth rate and the size of the epidemic. Intuitively, the more susceptible individuals an individual can infect before recovering, the larger the epidemic wave will be. The ratio $k\,\beta/\mu$, which can be interpreted as the average number of individuals an infected individual infects before recovering when the population is fully susceptible, is what is called the basic reproductive number R_0. It controls the final size of the outbreak and, together with the rate of recovery μ, the growth rate of the epidemic, $\mu(R_0 - 1) = k\,\beta - \mu$. If R_0 is greater than one, the epidemic grows exponentially; if it is instead smaller than one, it goes extinct because each infected person on average will infect less than one person. Importantly, a change in number of contacts (e.g., due to implementation or relaxation of social distancing

---

measures), in transmission per contact (e.g., increased hygiene or face masks), or a reduction in infectious period (e.g., identification and isolation of a case) can change R_0 and as a result have an exponential impact on the unfolding of the epidemic. This is the reason R_0 has important public health implications. It tells us how far we are from suppressing an outbreak, what will be the outbreak impact if left uncontrolled, and how strong is the effort needed to curb it.

The Kermack and McKendrick model relies on the homogeneous mixing assumption (i.e., individuals enter in contact at random). This is very convenient and widely adopted. However, we know that assuming a homogeneous number of contacts per individual is far from reality as the number of connections an individual has is rather heterogeneous. The question is then: How does this heterogeneity affect the spread of an epidemic? It turns out that the precise topology of the network of human interactions can have a dramatic effect on the way an epidemic spreads through the population (Pastor-Satorras *et al* 2015, Kiss *et al* 2017). For instance, in populations in which most individuals have a few connections but there are few with an (outrageously) large number of contacts, the epidemic will almost surely be able to spread (if no measures are put into place) no matter the value of epidemiological parameters.

However, we know that networks of interactions between individuals have more complex structures. Networks of relevant interactions for disease contagion can be quite different: a network of sexually transmitted diseases is not the same as a network of contagion of airborne transmitted diseases (such as influenza, MERS, or SARS viruses). For the former case, only a fraction of the population is at risk which makes it possible to design containment measures that target specific parts of the population. For the latter case in which the majority of the population is at risk of becoming infected, we know that the network of contacts between individuals has a large-scale structure (often called communities) that shapes the spatiotemporal pattern of spread (Guimerà *et al* 2005). For instance, we can visualize this structure as city or country boundaries that are overcome through mobility. Nonetheless, even within the city population we can define some coarse-grained structure of an otherwise very complex network of interactions.

The structure of the network Is extremely inflrmative when it comes to understanding contagion and designing containment and vaccination strategies (Danon *et al* 2011). Containment strategies (especially in the case of airborne transmitted diseases such as SARS or pandemic influenza) aim precisely at preventing contact between different regions (communities/components) of the network and therefore effectively isolating different foci of the epidemic (figure 5.21(C)). An obvious first choice for containment is thus to reduce mobility, including inter-country and inter-city mobility, since these are *de facto* transmission highways within the population. However, quantifying the impact of travel restrictions is not as simple as it may seem, as it requires a deep understanding of the mechanisms driving the epidemic invasion. A convenient framework to tackle this issue is provided by multiscale models rooted on the metapopulation approach that are able to integrate informa-tion on the human mobility network, such as the flight network (Colizza *et al* 2006).

However, there are other aspects that we have to take into account such as the fact that interactions between individuals are neither static (Masuda and Holme 2013) nor happen in the same environment. For instance, the proximity interaction among people who live under the same roof is not the same as that among people who work desk-to-desk in the same office every day, share a bus ride during 15 min, or go for a walk outdoors once a week. In reality, interactions between individuals take place on multiple layers which have different levels of infectiousness associated (or equivalently the networks in different layers have different weights) and are not active at the same time (Kivelä *et al* 2014). Spread is typically driven by the layers with a larger density of connections with high strength. The epidemic then cascades onto other layers, effectively resulting in an increase in the number of infected individuals.

Precise and real-time information on real network structures is impossible to obtain. Still, thanks to extensive data becoming available at all temporal and spatial scales, structural and dynamical properties of these networks are described with increasing levels of detail (Barrat *et al* 2013, Barbosa *et al* 2018, Eames *et al* 2015, Masuda and Holme 2013). This information enables the design of idealised networks of contacts that reproduce these properties for numerical and theoretical exploration. Most of these findings are valuable to understand the effect of social distancing measures that reduce interaction strength. They also highlight how network properties, and their change in adaptation to the epidemic, can qualitatively alter the epidemic progression, by both speeding it up or slowing it down to a polynomial, rather than exponential, growth in the number of cases. Theoretical results are also informative about how to design effective vaccination strategies that reduce epidemiological parameters and can aid in estimating the percentage of the population that needs to be vaccinated to achieve herd immunity (i.e., the number of immune people necessary to prevent the epidemic from spreading throughout the population) as well as which part of the population should be targeted first in vaccination efforts.

5.5.4 Facing outbreaks

Models are able to encode the driving forces of epidemics and as a result provide a coherent framework to make sense of epidemiological data. As such, they are critical tools to face an outbreak. When a new pathogen emerges in the human population, the lack of knowledge of its epidemiological characteristics makes identifying the most effective interventions very hard. In physics terms, the propagation of an infection is all about the interplay between dynamical processes unfolding at different timescales—dynamics of symptoms, infectiousness, and human interac-tions. By comparing models to data, it is possible to estimate key parameters ruling these processes. The COVID-19 emergency has showcased the crucial role models play in providing epidemiological understanding. Models have allowed for answer-ing critical questions (Ferretti *et al* 2020, Pullano *et al* 2020, Wu *et al* 2020): How fast does the epidemic spread? To what extent can cases be detected? Can asymptomatic cases transmit the infection? Critical unknowns also stem from the heterogeneous properties of individuals and populations, and their impact on the propagation of the infection (Althouse *et al* 2020, Arenas *et al* 2020, Chinazzi *et al* 2020, Davies *et al* 2020, Gatto *et al* 2020, Schlosser *et al* 2020): What is the role of children in the epidemic? What are the main pathways of epidemic spatial propagation? What is the contribution of superspreaders and superspreading events in the propagation of the outbreak? Answering these questions deepens our fundamental understanding of the epidemic dynamics. More importantly, it has immediate practical consequences. For instance, knowing how fast the epidemic spreads tells us how fast we need to react to avoid peaks in hospitalisations, or excess deaths (Wu *et al* 2020). Estimating the number of unseen cases is essential to identifying the critical gaps in the epidemic surveillance and the best strategies to improve it (Pullano *et al* 2020). Knowing the role of children versus adults in disease transmission tells us up to what extent school closure is effective in preventing transmission (Davies *et al* 2020). More in general the quantification of age-variation in the infection and severe form of the disease inform the design of prioritised vaccination strategies (Bubar *et al* 2021). Understanding the role of asymptomatic versus symptomatic transmission and the heterogeneities in the transmission risk across individuals and settings is essential for planning contact tracing strategies (Aleta *et al* 2020, Ferretti *et al* 2020, López *et al* 2021).

Fitting models to data is a critical step that can become quite daunting. As an example, consider the measurement of the basic reproductive number ($\mathbf{R_0}$) which is often (albeit not exclusively) obtained by fitting an exponential growth model to the initial stages of the spread of an epidemic. However, data from the early stages of spreading are noisy. Moreover, we must account for the fact that data of the same disease in different places can be affected by a number of socio-demographic factors which can result in the variability of the estimates of some epidemiological parameters, including $\mathbf{R_0}$. Thus, relying on a single estimate is problematic. Handling the uncertainty of the estimate is a question which has been tackled reasonably well. However, understanding the context-to-context variability and how this uncertainty propagates within the model is another source of error that can seriously bias *in silico* predictions. Most importantly, this uncertainty should be taken into account when calibrating models to make forecasts (Chowell 2017).

Models can also help provide near real-time information of the current epidemic situation. Models have facilitated the creation of tools for anticipating the epidemic trajectory for both short-term epidemic forecasts (figure 5.21(C)) and long-term scenario analyses (figure 5.21(D)). However, this does not come without challenges. Medical and surveillance data are not collected by a controlled experiment. These data are limited and biased, an issue which is exacerbated during a public health emergency, where testing availability may be limited and surveillance protocols are continuously changing. These issues can seriously hinder our ability to even diagnose the current state of a pandemic. A central quantity to diagnose the real-time status of an epidemic outbreak is the effective reproductive number $\mathbf{R_t}$, which tries to capture which is the number of infected people per infected person in a stage in which the disease has already spread community-wide. In the literature, different ways for measuring this number have been proposed (Gostic *et al* 2020), which is a *de facto* indicator for authorities as to whether certain epidemic containment measures are being effective, and whether existing measures can be relaxed. Methods to estimate R_t rely on a very precise estimation of epidemiological factors and on the availability of high-quality, instantaneous data on the number of infected people. For diseases like COVID-19, causing a wide spectrum of symptoms, including asymptomatic and mild symptoms, we know that known cases are an underestimation of real cases. The lack of details on how data have been collected and its possible biases can severely undermine the ability of current approaches to accurately compute R_t, and makes it hard in general to calibrate models with which to build reliable scenarios. Therefore, for models to be of use, it is crucial to know the details about how these data are collected, to understand their biases and limitations. To be helpful, physics approaches need not only to embrace data in all its complexity but also to establish a strong interdisciplinary dialogue with medical, surveillance, and case management experts.

Despite the issues with data, scientists have been able to circumvent some data-related issues by using Bayesian approaches that exploit inherent uncertainty in the parameters to calibrate models using past data on reported number of infected individuals and subsequent deaths (with an appropriate time lag). Indeed, there are many models that can fit past data, but that does not necessarily mean that all such models are able to make reliable future predictions in an appropriate time horizon (usually two weeks in an epidemiological context). This situation has put the stress on the need to develop tools for evaluating forecasts; even for the early epidemic stages, forecast evaluation can help uncover reliable models for a given outbreak and help build meaningful scenarios (Chowell 2018).

One of the cruxes in epidemiological research is precisely forecasting. The inherent uncertainty in parameter estimation makes it impossible to make reliable forecasts of the future evolution of the pandemic in a long time horizon (Castro *et al* 2020). These predictions are uncertain in the same way weather forecasts are, and therefore we need to be able to assign uncertainty to any forecast we make. The recent pandemic has brought together a large community of scientists in a joint effort to use available knowledge to make reliable predictions. Many years of research have produced a wealth of models that can be used to make predictions of the evolution of the pandemic, many of which are quite different in their inner details but whose predictions can be pooled together to make better forecasts (see, e.g., https://covid19forecasthub.org/ and https://covid19forecasthub.eu/). These efforts have shown the power of making ensemble forecasts, but at the same time have put in the spotlight how hard it is to make and interpret forecasts in a constantly changing situation. These difficulties go beyond model uncertainty because they arise from the need to incorporate sudden and profound changes in the epidemic conditions, as the ones caused by the emergence of variants and human response to the epidemic.

One year after COVID-19 emerged, the widespread dissemination of the mutant virus Alpha was like a new pandemic sweeping the world. The variant differed from the wildtype in several aspects, including transmissibility and severity, generating a rapid change in the trajectory of cases, hospitalizations, and deaths (Volz *et al* 2021). This posed a great obstacle to the epidemic forecast and the planning of mitigation measures and required a prompt effort to re-assess the epidemic situation and re-evaluate key epidemiological parameters. Following Alpha other viral variants of concern were detected (Beta, Gamma, Delta). Several variants were simultaneously circulating, showing complex dynamical patterns often difficult to interpret. The advantage conferred by a mutation critically depends not only on its molecular properties, such as its enhanced ability to infect cells, or to escape existing immunity (resulting from previous infections or vaccination), but also on the characteristics of the environment and the host population. For instance, theoretical studies on multi-strain spread on mobility networks show that the phase space of strain (co-)dominance varies according to the mobility level (Poletto *et al* 2013). Still, theoretical research on the complex interplay between environment, human social behavior, and multi-strain/multi-pathogen interaction is still behind work on single-pathogen/single strain epidemics. The paucity of resolved data to monitor the spread of distinct variants at the population scale was initially a key limitation in studying multi-strain dynamics. This is rapidly changing, thanks to the increasing availability and rapid share of genetic data that is enabling the reconstruction and visualization of the spatiotemporal pattern of strains' co-circulation (e.g., https://nextstrain.org/sars-cov-2, https://covariants.org/, https://cov-lineages.org/, https://www.cogconsortium.uk/).

A second challenge in forecasting eIidemic outbreaks is represenIed by the need to incorporate not only changes in containment measures but also the adoption of those measures by individuals. Public health emergencies trigger a change in human behavior and create a feedback loop between infection dynamics and behavior. Governments may take measures to restrict human interactions and reduce the chance of contagion. For instance, cancellation of events and mass gatherings, travel bans, school closures, and curfews were adopted to curb recent epidemics of Ebola, H1N1, or SARS. To curb the COVID-19 pandemic, these kinds of interventions have been of unprecedented strength and extent, including lockdowns and complete travel bans.

The effect these measures have on human behavior is extremely complex. Adherence to measures varies greatly both geographically and in time and depends on socio-demographic factors. Upon the announcement of a measure people may decide to adopt protective measures beyond what is required by the imposed regulations. In fact, even in the absence of interventions, people may change their behavior to reduce as much as possible the risk of contagion. However, if restrictions last too long, people may find it increasingly difficult to respect both self-imposed and external measures. Importantly, all of these behaviors depend not only on the epidemic situation but also on mass media and social network communication around the epidemic.

The feedback between behIvior and contagion is one of the central problems of the physics of epidemics. A large body of work shows that behavioral change may alter the epidemic spreading profoundly (Funk *et al* 2010). Awareness may suppress spreading. However, changes in behavior (e.g., by relaxing social distancing) may also generate interesting dynamical features, such as multiple peaks or oscillations. Nonetheless, the great majority of studies from before the COVID-19 epidemic were purely theoretical. This was because very limited data were available on the behavior change of a population in response to an outbreak, thus hindering the validation of such theories with data.

The COVID-19 pandemic has completely changed our outlook on epidemics (Perra 2021). The pandemic has brought about abrupt societal changes which have generated a desperate need to monitor human behavior. Massive efforts have been dedicated to track individual mobility, physical interactions, and attitudes. Large-scale surveys have quantified compliance to recommendations (e.g., face masks and hygiene measures), and changes in social encounters. Through data-for-good programs, tech and communication giants such as Google, Apple, Facebook, Telefonica, Vodafone, and Orange, among others, have put massive amounts of data at the service of the scientific community. These data are extremely valuable because they provide proxy information on human-to-human interactions. As such, they facilitate the modelling of the underlying network of interactions on which a disease propagates and assess adhesion to measures (Perra 2021). Still, while these data are useful, they can only partially address such a multifaceted problem. As always, a deep comprehension of the phenomenon must be based on the combination of diverse and complementary information sources (e.g., digital proxies, surveys). This may require the development of new approaches. The amount of data generated in one year of an epidemic will require a long time to be digested.

5.5.5 Conclusions

Adapting to questions of current interest with flexibility and openness is one of the cornerstones of the scientific endeavour. Through interactions with scientists in other fields, physicists studying random processes and network science are in the position to contribute to important societal questions in the face of a pandemic. Statistical physics offers tools to describe rare random processes and complex nonlinear patterns of interaction. Importantly, in the same way that machine learning tools do, physics tools can learn from data with the added bonus of providing interpretable, testable generative models that enable us to build theories that bridge scales and incorporate knowledge from other disciplines. We have argued that effects on the microscopic scale, such as single molecular mutations, can influence observable outcomes, such as the spread of a deadly epidemic at the global population scale. Clearly, a challenge for everyone from scientists to policymakers is first to prevent pandemics from happening and, if they happen, to contain them. Physicists can and do contribute meaningfully to this cause. One of the most remarkable examples is physicists' contribution to developing the three-drug combination, commonly known as a triple cocktail against HIV (Perelson *et al* 1996), which consists of administering three different drugs at carefully timed intervals. Through understanding the interplay between the timescales of mutation of the rapidly evolving HIV viruses and those of the immune response, they contributed to quenching the spread of the AIDS epidemic, at least in countries where the triple cocktail was economically available.

Working closely with specialists in other fields, physicists are continuing to offer solutions to different pathogenic threats, with different molecular and mutational properties. The major long-term challenge is being able to predict the emergence of a major epidemic, such as HIV or Sars-CoV-2 even before the virus emerges. This involves knowing the physics of processes occurring on multiple inter-related scales: estimating the probability for a variant to mutate and cross species barriers, the probability that it has high enough transmission rates to start spreading, and what physical contagion mechanisms and social behaviours will help it spread, which in turn depend on the molecular properties of the pathogen. It is extremely important to emphasise that model predictions are statistical and have associated uncertainties. Physics approaches describe viral evolution, immune response, and disease spread by associating probabilities to plausible events in the same way weather forecasts do. It is therefore critical to understand the properties of the probability distributions, and especially to understand rare, disruptive events in the tail of the distributions because these can have a dramatic, systems-wide effect that is not well-represented by the typical or average events.

A prominent example of the importance of the tail of the distribution are superspreading events. Many of them were documented during the COVID-19 pandemic (e.g., the epidemic cluster of ~700 people in the Diamond Princess cruise ship that quickly rose from a Hong Kong resident visiting the ship). In South Korea a single infected person was able to generate a cluster of more than 5000 cases (Althouse *et al* 2020). The central role of events of this kind in the epidemic trajectory has urged a paradigm shift in predictive modelling from the concept of $\mathbf{R_0}$, the average number of secondary infections (i.e., infections generated by an infectious case), to the concept of secondary infection distributions. This calls for new methods to better reconstruct this distribution, identify transmission hotspots, and understand how these can be targeted by interventions to rapidly control the epidemic (Althouse *et al* 2020).

At the molecular scale, rare events in the tail of the distribution also drive dramatic changes in the epidemic. For instance, the emergence of particular variants and substrains require the occurrence of a distinct set of beneficial mutations, each occurring randomly as the result of a replication error of a single nucleotide in the genome of a single virus. This mutation is then amplified to affect the entire population through selection or genetic drift. Which mutations occur and when is largely determined by these rare, partly unpredictable events.

Prediction involves not just predicting the most likely outcome at a given time but understanding limits and best/worst case scenarios. Since we are dealing with stochastic dynamical processes, small differences at one time or one scale can lead to big differences in the global outcome. Current research efforts are being dedicated to improving forecasting by better considering the whole spectrum of possibilities and encompassing a wide range of plausible scenarios. This could be done by incorporating not only bottom-up approaches, but also top-down approaches that look blindly at data and identify important variables that are predictive. At the same time, improvements in forecast evaluation tools will increase our capacities to compare models and learn which assumptions work and which might fail in our models.

Most importantly, ensemble predictions and scenario analysis may be hard to interpret and integrate in the decision-making process. Policymakers must interpret model results in light of their assumptions and limitations. Long-lasting interdisciplinary collaborations between researchers and policymakers are needed to create a culture of outbreak analysis and predictive modelling. Finally, a critical challenge is represented by the communication with the general public and journalists. This is essential to build trust in science and contrast misinformation (Gallotti *et al* 2020). At the same time, communicating epidemic assessment and predictions as well as the uncertainty around them is difficult. Science does not follow a straight line; early hypotheses may be revised as new knowledge accumulates. This may confuse society and lead them to think that changes in containment measures or vaccination strategies happen on a whim as opposed to being the result of careful processing of new knowledge.

There are many other challenges that affect specific scales. At the level of host–pathogen interactions, a big current challenge is combining theory and quantitative experiments to elucidate the so-called genotype–phenotype map: can we find statistical rules that translate the molecular sequence to relevant phenotypes such as immunogenicity? This is a critical step for drug design and for predicting immune escape in disease spread.

At the level of human populations, the challenge is to fully exploit the potential of epidemiological and behavioral data. Physics models would not be useful without the availability of high-quality and rich data. Data should be shared rapidly and widely and made openly available to researchers, accompanied by standardized protocols about data collection and processing and privacy requirements. During the COVID-19 pandemic a wealth of data was made available by government and public health agencies. In addition, initiatives of data gathering, aggregating, and sharing have made these data ready to use by the scientific community—see, e.g., https://coronavirus.jhu.edu/data and https://global.health/. At the same time data for good initiatives have provided anonymised and aggregated information on human behavior. All of this has enabled critical discoveries. Still a strong coordination effort is needed (Oliver *et al* 2020), to get modellers and data scientists involved in the data collection process—thus ensuring appropriate data are collected —and to ensure that researchers working with data know how the data were obtained and what the data actually represent. With this data in hand we can minimize and correct data biases and make more realistic forecasts and scenarios.

As a final point, a ubiquitous problem that we face at all scales is that of heterogeneity: in immunological response and in human reaction, which include gender, age, social status, and geographical diversity. Heterogeneity is indeed an inherent element of random processes. Quantifying the role of heterogeneity at all scales is an open problem important for treatment, vaccination, and response. Again, physics has the tools for addressing and linking heterogeneity at the molecular, cellular, and organismal scales to population-level processes. Our challenge is to make it happen.

The utilitarian goal of predicting evolution and disease spread is within our reach. As the positive example of dominant flu strain prediction shows, combining the processes at different scales, theory, and data results in successful predictions that can be used by policymakers such as the WHO.

5.6 Further diagnostics and therapies

Marco Durante[1]
[1]GSI Helmholtzentrum für Schwerionenforschung and Technische Universität, Darmstadt, Germany

5.6.1 High-resolution imaging

Marco Durante[1]
[1]GSI Helmholtzentrum für Schwerionenforschung and Technische Universität, Darmstadt, Germany

5.6.1.1 General overview

Imaging was the first application of x-rays, with the famous pictures of Mrs Röntgen's hand published on *The New York Times* on January 16, 1896. In over a century of progress, figure 5.22 shows the progress in imaging from the planar x-ray image produced with the orthovoltage tube of Röntgen to the modern CT angiography.

Imaging has therefore changed medicine maybe more than any other technical advances. Therapies like surgery, radiotherapy, and any other medical and pharmacological intervention strongly depend on the imaging obtained with physics techniques. Imaging today follows the pathway described in figure 5.23 and uses many different techniques, as described in table 5.1.

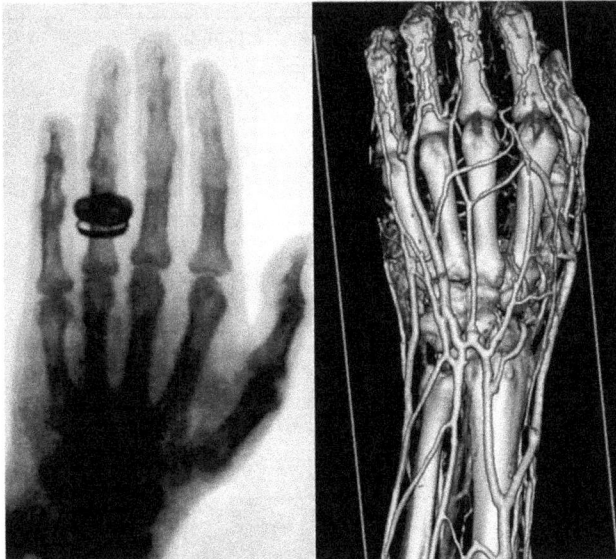

Figure 5.22. X-ray image of a hand obtained by Wilhelm Conrad Röntgen in December 1895 (left), showing the bones and the wedding ring of the wife. A modern CT angiography of a hand (right), showing high-resolution details of the joints and vessels.

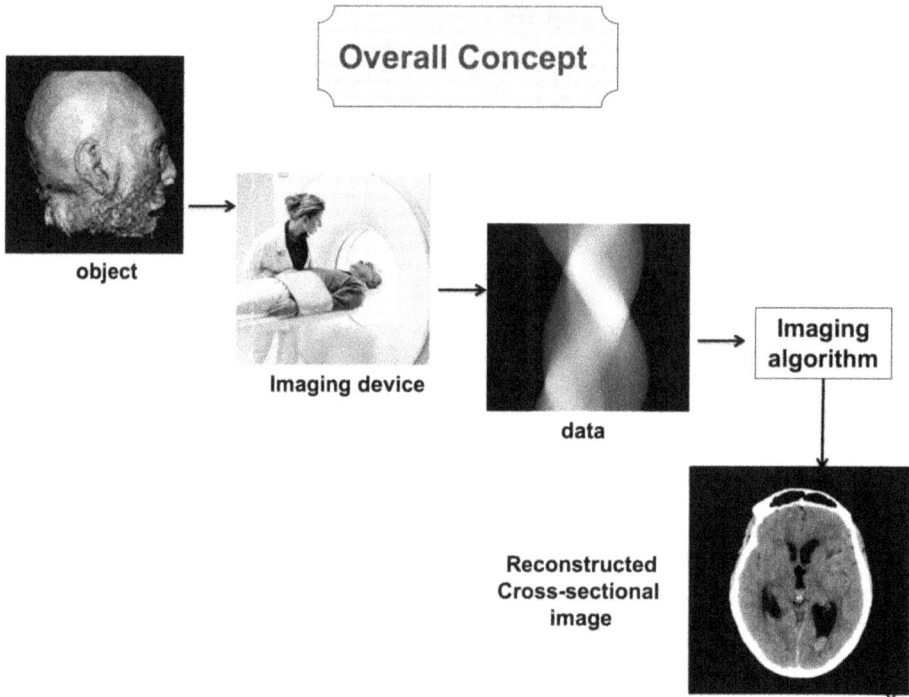

Figure 5.23. Imaging workflow in medicine.

Table 5.1. Main medical imaging techniques in use today.

	Radiation	Techniques
Radiology with external sources	X-rays	Radiography
		Computer tomography (CT)
		Mammography
		Angiography
		Fluoroscopy
	Magnetic field and radiofrequency	Magnetic resonance imaging (MRI)
	Ultrasounds	Echography
Endoradiology	Gamma-rays	Scintigraphy
		Single-photon emission computed tomography (SPECT)
	Positron annihilation	Positron emission tomography (PET)

Different methods are used in different branches of medicine. Mammography is, for instance, specialized for the detection of breast cancer, and angiography for cardiovascular diseases analyses. Other methods are used for many different methods. Echography is for instance popular during pregnancy, but is also widely used to detect abdominal diseases. SPECT and PET are largely used both in oncology (often associated to CT) and cardiology in myocardial perfusion tests. CT and MRI are largely used in oncology, with complementary targets: while CT has a superior resolution of bones, MRI shows excellent images of the soft tissues (figure 5.24).

When using endoradiology, image fusion is generally necessary to combine the functional and anatomical images. Figure 5.25 shows a fusion of CT and PET imaging in oncology. A whole-body PET-CT scanner has been recently built by the EXPLORER consortium, led by the University of California at Davis (figure 5.26).

5.6.1.2 Challenges and opportunities

The new frontiers of imaging is the improvement of both *hardware* and *imaging* parts. The main hardware challenge is the increase of the image resolution. The new ultra-high resolution CT scanners reached resolution of 150 μm, starting to touch the area of microscopy (figure 5.27).

Similarly, for MRI there is a rush to increase the intensity of the magnetic field, which immediately reflects into increased resolution (figure 5.28).

The cost of MR scanners largely increasing with the field intensity, ranging from less than 500 k€ for a conventional 1.5 T scanner to over 5000 k€ for a 7 T scanner. But for preclinical studies at CEA in Saclay researchers have already installed a 11.7 T MRI magnet. Whether these high magnetic fields are really needed in clinics is a big question for the coming years, where it will be defined what the 'optimal' scanner will be.

The perspectives are even more exciting for software, where we are witnessing the invasion of AI in medicine and in diagnostics. AI systems have demonstrated an

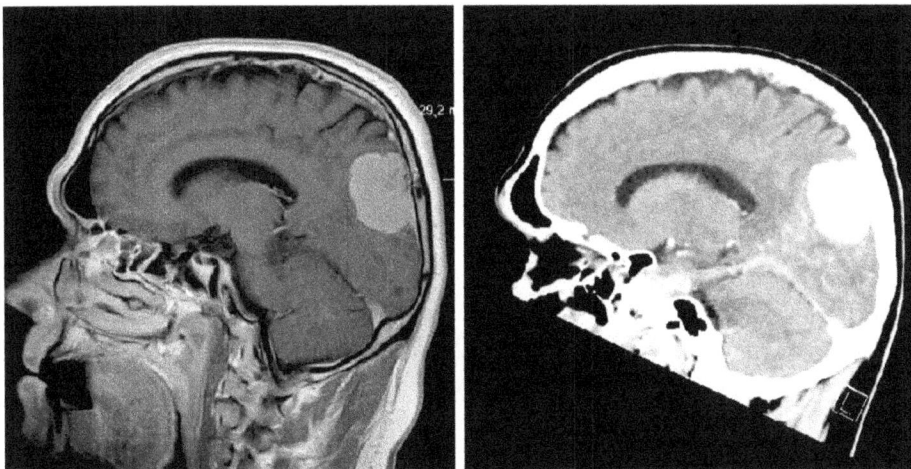

Figure 5.24. MRI (left) and CT (right) images of the same brain tumour.

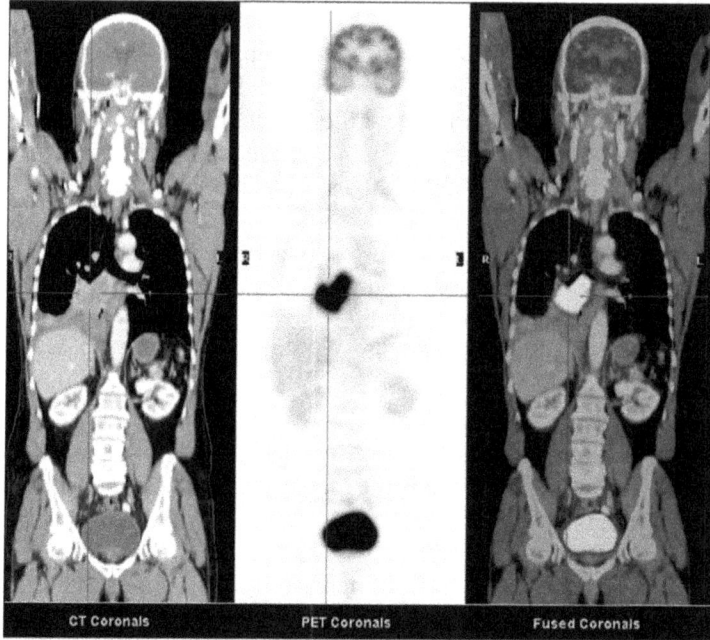

Figure 5.25. Fusion image of a CT and a 18FDG PET clearly identifying a tumor with high aerobic glycolysis.

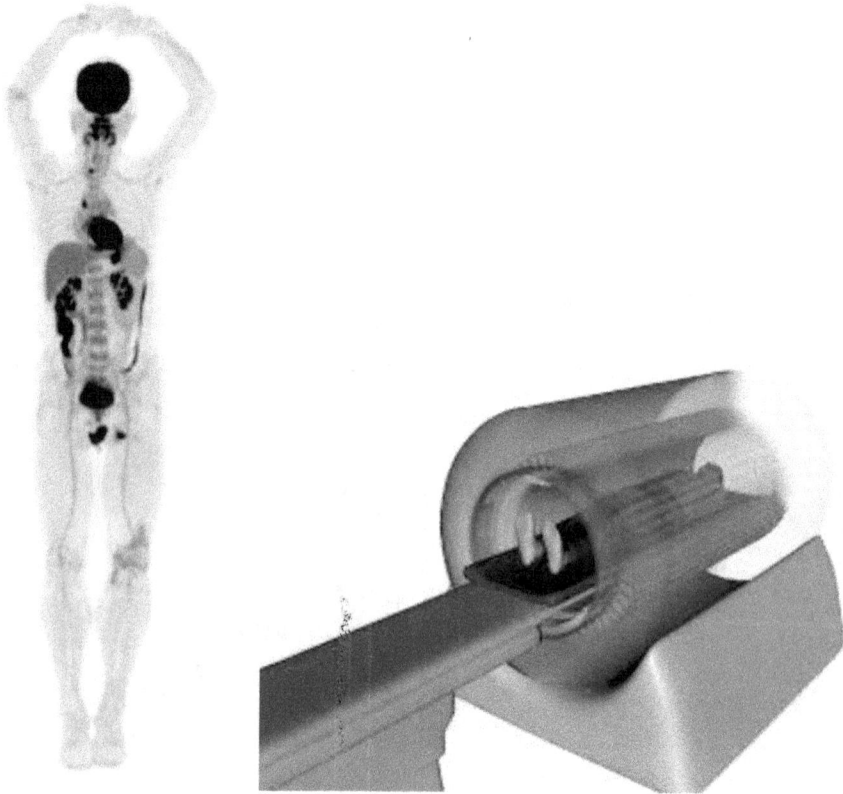

Figure 5.26. The EXPLORERE whole-body PET-CT and a whole-body image produced by the new scanner.

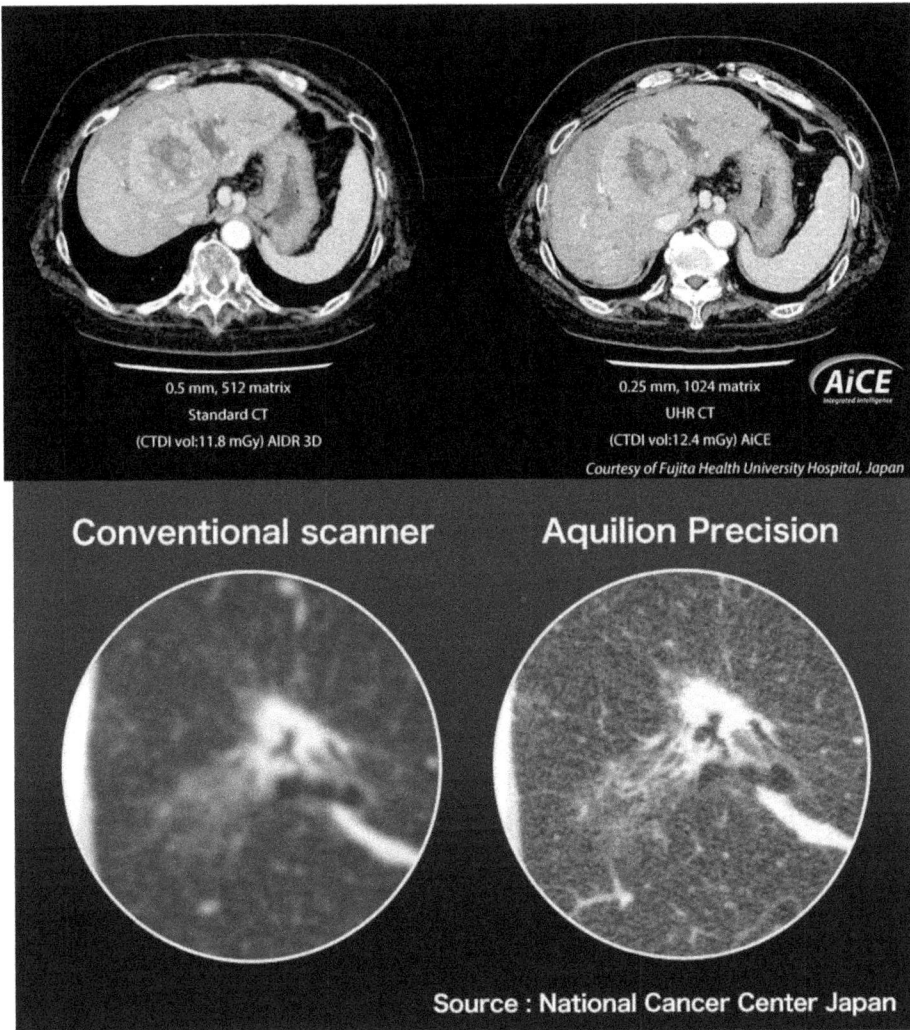

Figure 5.27. Differences between a conventional scanner and ultra-high resolution CT.

extraordinary ability to capture diseases in images notoriously affected by many artifacts such as mammography. Nevertheless, the initial concerns that AI will replace radiologists have gradually vanished, considering that the answer to a single question is not sufficient to draw a solid diagnosis. However, it is clear that future radiologists will all use AI as a sharp-eyed partner to assist physicians in diagnostic procedures.

Another large impact of AI in diagnostics is in the field of radiomics (i.e., the method to extract molecular and histological features from radiographic medical images using data-characterisation algorithms) (figure 5.29). The procedure is very intriguing, because in principle it can overrule the invasive and slow biopsies replacing them with deep learning from conventional radiology. Shape, intensity,

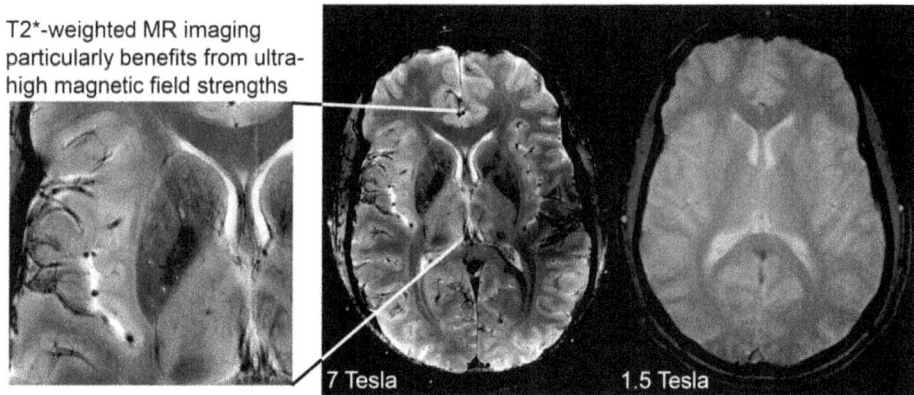

T2*-weighted MR imaging particularly benefits from ultra-high magnetic field strengths

7 Tesla

1.5 Tesla

Figure 5.28. Comparison of brain images (T2-weighted) using a conventional 1.5 T and a new 7 T scanner.

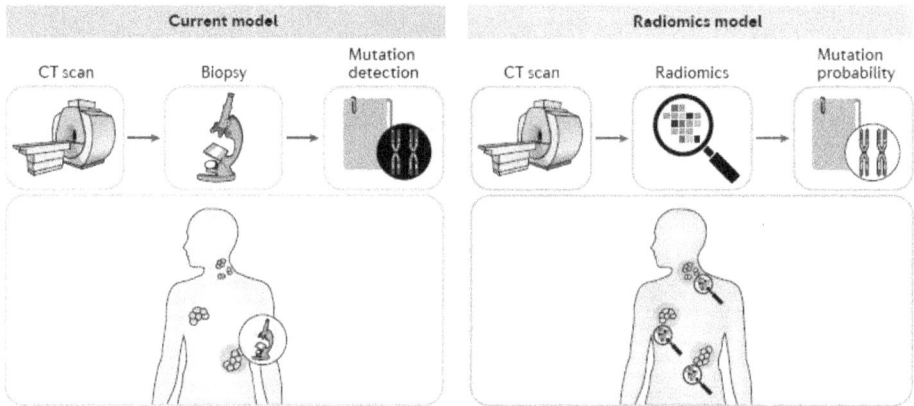

Current model

Radiomics model

CT scan   Biopsy   Mutation detection

CT scan   Radiomics   Mutation probability

Figure 5.29. An image of how radiomics can change the diagnostic procedure, currently needing biopsies and molecular tests to reach diagnosis.

and texture are typical image features that can be analysed to extract biomarkers using radiomics.

The field of functional imaging is also the one with the highest potential. Functional imaging in oncology generally uses the cancer metabolisms, mostly using 18FDG. However, recent studies are targeting cancer-associated fibroblasts, using fibroblast activation protein inhibitor (FAPI) labelled with ^{68}Ga. The shift from tumour metabolism to the tumour microenvironment may provide new, important information for appropriate treatment. Other compounds (e.g., 18F-FMISO or Cu-ATMS) target hypoxic regions in tumors that are treatment-resistant.

The other popular functional imaging method is fMRI that measures changes in blood oxygenation levels, which are correlated with the activity of a specific part of the brain. Apart from initial studies on brain anatomy, fMRI is now becoming a powerful tool in diagnostics and psychiatry. SPECT can also be used for cerebral

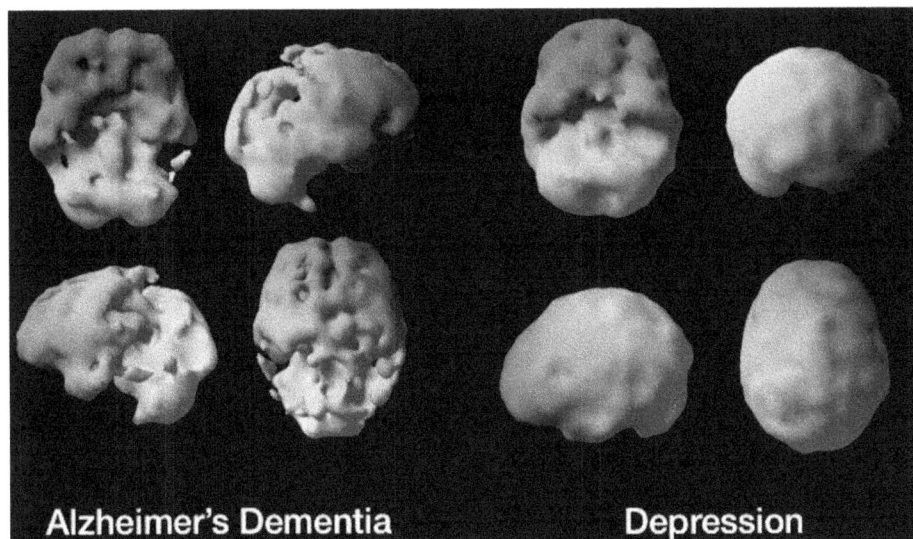

Figure 5.30. 99mTc-HMPAO for cerebral blood flow measurements by SPECT can reveal diseases such as Alzheimer's and depression.

blood flow measurements using compounds such as HMPAO that converts rapidly from the lipophilic form, which passes the blood-brain barrier (BBB), to the hydrophilic form, which is unable to pass the BBB, and is trapped in the brain (figure 5.30).

5.6.2 Innovative therapy

Marco Durante[1]

[1]GSI Helmholtzentrum für Schwerionenforschung and Technische Universität, Darmstadt, Germany

5.6.2.1 General overview

Soon after the discovery of x-rays it became clear that radiation could have been used to destroy tumors, thus starting what we call radiotherapy. The goal of radiotherapy is very simple: to provide a dose as high as possible to the tumour, in order to sterilize it, but simultaneously to minimize the dose to the normal tissue, so that side effects are acceptable. The curves representing the tumour control probability (TCP) and the normal tissue complication probability (NTCP) as a function of the dose must stay separated (figure 5.31). The role of the physics is to enlarge their distance (therapeutic window), thus making the treatment effective and safe.

Figure 5.32 gives a schematic view of the progress in radiotherapy in the past century focussing on the most common male tumour treated in the clinics: prostate cancer. Radiotherapy starts with orthovoltage x-rays that cross-fire the tumour target in a box. In fact, the depth–dose-distribution of x-rays is unfortunate: the entrance dose is always higher than the dose deep in the body, where the tumour is situated (figure 5.33).

Cross-firing from different angles is therefore necessary to enhance the target avoiding unacceptable toxicity in the entrance.

To make the depth–dose distribution more 'flat' it is necessary to increase the photon energy, and therefore the dose distribution improves with the introduction of ^{60}Co γ-rays (about 1 MeV energy) and later with linear electron accelerators (linacs),

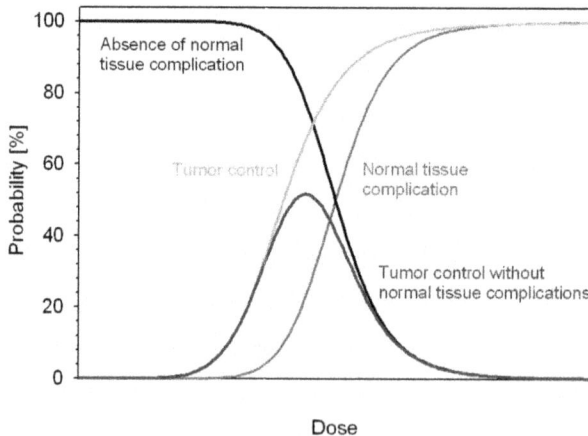

Figure 5.31. The therapeutic window in radiotherapy. The green and red curve should be as separated as possible, so that the blue curve can have a high peak.

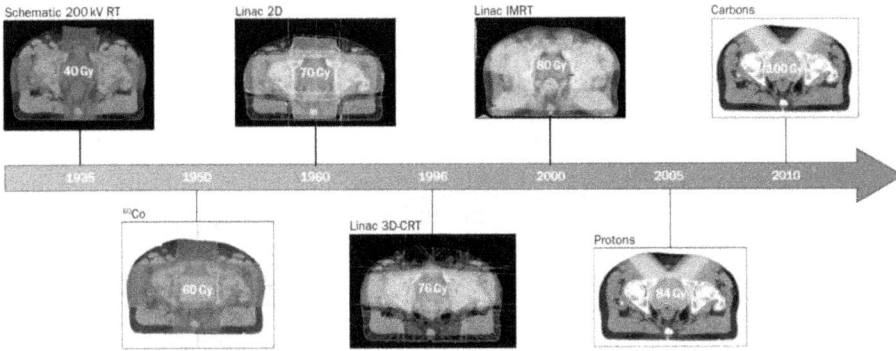

Figure 5.32. The evolution of prostate cancer radiotherapy in the past century. Abbreviations: RT = radiotherapy, kV = kilovolt, 60Co = cobalt-60 γ-rays, linac = linear electron accelerator, CRT = conformal RT, IMRT = intensity-modulated RT. The dose is measured in Gy (1 Gy = 1 J kg^{-1}), and should be as high as possible in the target.

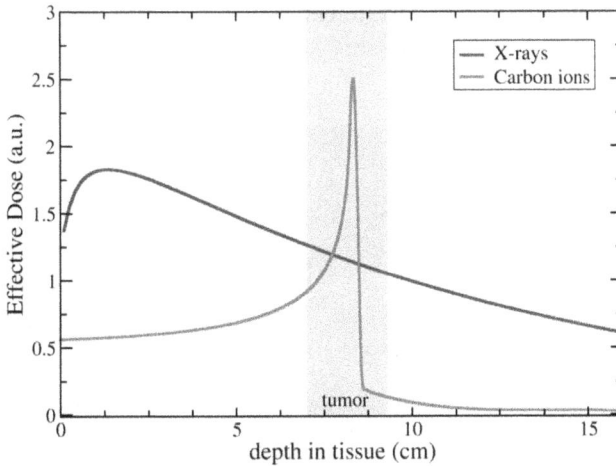

Figure 5.33. Depth–dose distribution of x-rays and charged partciles (here ^{12}C-ions) in human tissue.

which can reach electron energies around 25 MeV, even if the current linacs are normally operated at 6 MeV. However, the cutting-edge technology is the use of protons and heavier ions, because the depth–dose distribution is favourable (figure 5.33). Charged particles deposit indeed most of their energy at the end of the range (the Bragg peak), where the tumour is situated. The exact position of the Bragg peak can be modulated by changing the accelerator energy, and the Bragg peak can be widened (spread-out-Bragg-peak, SOBP) to cover the whole tumour size (generally several cm) by superimposition of beams of different energy.

In addition to the physical advantages, slow ions in the Bragg peak are biologically more effective than fast ions in the entrance channels, whose radio-biology is more similar to x-rays. For this reason, heavy ions are particularly indicated for radioresistant, hypoxic tumors.

As a consequence of their physical and biological advantages, particle therapy is growing worldwide. According to the statistics of the Particle Therapy Co-operative group (PTCOG), in January 2021 there were 111 particle therapy centers actively treating cancer patients, but only 12 using ions heavier than protons (C-ions). Despite the favorable dose distribution shown in figure 5.32, only about 1% of prostate cancer patients are treated with particles. The main problem remains the high cost. Particle therapy centers require indeed large circular accelerators (cyclotrons or synchrotrons) and heavy gantries for beam delivery at different angles (figure 5.34). As a result, the cost of a heavy ion therapy can be as high as 200 M€, versus only 5 M€ for a conventional linac.

In addition to external beam therapy (teletherapy) there has been strong progress in target radionuclide therapy, an internal therapy based on the idea of delivering molecules labeled with radioisotopes to the tumor. The radiopharmaceutical is comprised of a radioisotope with appropriate half-life, a linker, and a small molecule (peptide, antibody, etc) able to target cancer cells. A gold standard for targeted β-therapy is ^{177}Lu, which has been used with success in many tumors including neuroendocrine and advanced prostate cancer.

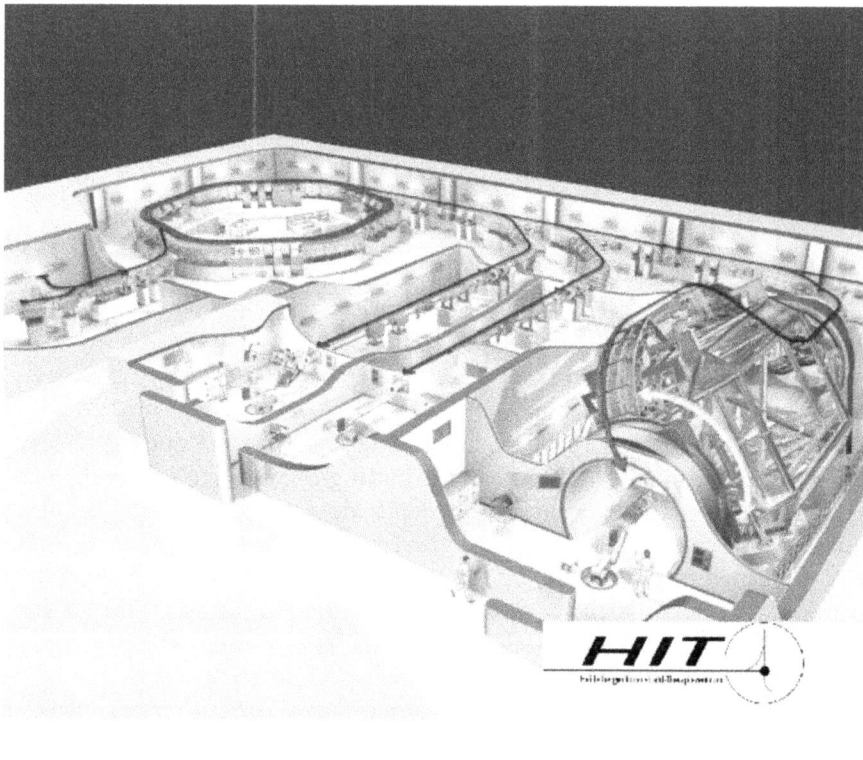

Figure 5.34. Layout of the HIT facility in Heidelberg, where patients can be treated with protons or C-ions. Two treatment rooms have fixed horizontal lines, while one is equipped with a 670-ton rotating gantry.

5.6.2.2 Challenges and opportunities

It is generally acknowledged that the number of cancer patients receiving radio-therapy (slightly more than 50%) is too low and should be increased in the coming years. While most patients will be treated with IMRT, particle therapy is at a tipping point. There are many efforts to reduce the cost of using superconducting magnets for synchrotrons with reduced footprints. Laser-driven particle accelerators are still immature for medical applications, but they bear the potential of having table-top accelerators for particle therapy, even if this goal goes probably beyond 2050. The same seems to be true for other innovative accelerators such as dielectric wall or fixed-field alternating gradient (FFAG) accelerators.

Beyond the cost, particle therapy has to meet the challenges posed by innovative methods in conventional radiotherapy, which is rapidly improving thanks to image guidance. MRI-linacs are already in use in clinics, and allow treatment of patients with full control of organ movements using MRI imaging. Particle therapy is still lagging behind in image guidance, even if the nuclear interaction of charged particle offer beam monitoring and range verification opportunities that are not possible with photons, such as secondary radiation, proton radiography, and PET.

This includes prompt γ-ray detection to assess the beam range; dedicated monitors are already commercially available. Particle radiography is another attractive possibility, both for imaging and for the possibility of reducing the uncertainty in the conversion between the Hounsfield units coming from the CT and the water-equivalent path length needed for particle treatment planning. PET has been used already for C-ion therapy and would greatly benefit if therapy could be done using radioactive ion beams (e.g., ^{11}C or ^{10}C), positron emitters that can be produced by fragmentation of stable ^{12}C. The results would be an enormous increase in the signal-to-noise ratio and the possibility of monitoring online the beam delivery in the tumour (figure 5.35). The use of radioactive beams in therapy had already been proposed in the 70s at the Lawrence Berkeley Laboratory, but has always been hampered by the low intensity of the radioactive beams. The recent intensity upgrades in the accelerators and the push for improving precision of particle therapy has raised interest in this project. CERN has a project for a cyclotron producing radioisotopes to be injected in conventional medical synchrotrons, whilst GSI, exploiting the intensity upgrade for the construction of the new FAIR accelerator, is actively measuring these beams in the framework of an EU ERC Advanced Grant.

In targeted radionuclide therapy, the new challenge is probably the use of α-emitters. Being short range and with high biological effectiveness, α-particles are indeed ideal for effective tumour cell killing with reduced damage to the surrounding normal tissue compared to the long-range β-particles. Some of the clinical results obtained with radioisotopes such as ^{211}At, ^{213}Bi, ^{223}Ra, ^{212}Bi, ^{225}Ac in metastatic patients have been stunning and have attracted great interest in this target therapy modality. The main challenge for this therapy is actually the radioisotope produc-tion. Availability of radioisotopes is a general problem for both diagnostic and therapy because many of them are produced in nuclear reactors that are in the process of being shut down in many countries. Even ^{99m}Tc, the most commonly used radioisotope in the world for radiodiagnostics, is insufficient. The problem is even

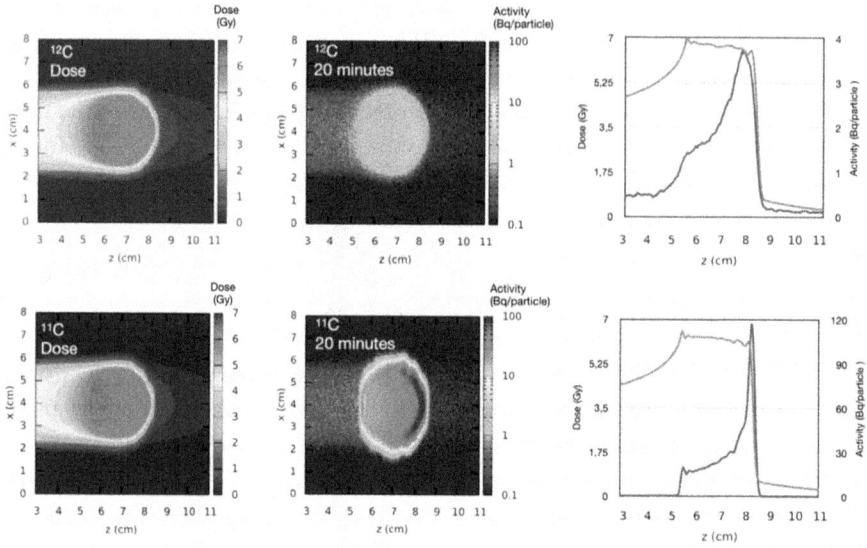

Figure 5.35. Monte Carlo simulations of PET images of stable and radioactive carbon beams. The two columns refer to two different signal acquisition times: 20 s (left) and 20 min (right).

more grave for a-emitters: for instance, [211]At is only produced at ARRONAX, the radioisotope factory in Nantes (France). Radioisotope production will certainly be a major challenge in the coming years.

5.6.3 Superconducting magnets from high-energy physics hadron colliders to hadron therapy

Lucio Rossi[1]

[1]Università di Milano—Dipartimento di Fisica and INFN-Department of Milano, LASA Laboratory, via Fratelli Cervi 201, 20056 Segrate, Milano, Italy

Superconducting magnet development has accompanied the energy and luminosity increase of the hadron colliders in the last 50 years. The first-generation superconducting magnets, in particular the study and R&D for the pioneering Tevatron at Fermilab, were instrumental for the development of the MRI technology. The present accelerator magnet technology, following the industrially built HERA, RHIC, and LHC colliders, is mature for direct use in medical environments and in particular for heavy particle (hadrons) therapy. The first application in prototyping phase is a gantry (i.e., the beam delivery system allowing irradiation of the patient from multiple directions). In this contribution, after a review of the achievement of superconducting magnet technology and of the development for next- generation colliders, we discuss the main development lines that should enable a superconducting ion gantry to be commissioned in this decade.

5.6.3.1 Introduction

Magnets have been the most important application of superconductivity, and accelerators can certainly be credited to be among the key drivers of the development of this technology.

Accelerator technologies, and thus accelerator magnets, need long R&D with programs spanning over decades, which allows investigation and R&D to be done in a coherent way and thus to be fruitful. The latest example, the LHC, is the summit of over 30 years of development of superconducting magnets (SCM). Its enormous size and its long-awaited goal, the Higgs particle, whose discovery about ten years ago at the Large Hadron Collider (LHC) at CERN (Last update in the search for the Higgs boson 2012) have been heralded worldwide, and the possible unveiling of the new world beyond the Standard Model make the LHC (LHC DESIGN Report 2004, The Large Hadron Collider 2018) the crossroad between past and future of high-energy physics (HEP). But LHC may be the crossroad also for medical sector superconducting technology. Thanks to the experience gained with superconductivity for LHC and its predecessors, now the accelerator magnets community is exploring the possibility of using superconducting accelerator magnet technology for particle irradiation-based cancer therapy.

5.6.3.2 Accelerator magnets and hadron therapy

Since the pioneering time of Ernest O Lawrence in Berkeley, where he designed, built, and put in operation the first cyclotrons in the 1930s, the demand for higher energy accelerators has driven the quest for larger size rings and higher field magnets. Not surprisingly superconductivity has become a key technology for modern accelerators and, conversely, particle accelerators have promoted super-conducting (SC) magnet technology (Wilson 1999). The timeline of record accel-erator energy is shown in the compilation reported in figure 5.36, where we have collected the main accelerators for nuclear and particle physics, classifying them according to the center-of-mass energy and highlighting the projects that are based on superconductivity. Many of them, the most powerful ones actually, feature SC magnets, or superconducting RF cavities, as the main technology and performance driver.

HEP has been pursuing superconductivity for many large projects, the most recent one being the already cited LHC at CERN, and plans are being made for further stronger and larger colliders, as shown on the right side of the timeline of figure 5.36. We wish in particular to refer to the some of the monographs (Wilson 1983, Mess *et al* 1996, Ašner 1999) as well as to the more recent reviews on SC magnets and conductors for accelerators (Tollestrup and Todesco 2008, Rossi and Todesco 2011, Rossi and Bottura 2012, Bottura and Rossi 2015).

As far as the medical sector is concerned, superconductivity has been limited to a few types of low-energy cyclotrons or synchrocyclotrons, mostly for hadron (protons or ions) therapy. The use of particles accelerated at a few hundreds of MeV per mass unit for cancer therapy is a well-established technique (Degiovanni and Amaldi 2015). Particle therapy can provide a powerful tool for the treatment of tumours that are not curable with conventional x-rays or that are better treated by particle beams. There are 24 particle therapy centers in Europe and 105 in all the world. However, only four of them in Europe and only 12 in the world employ heavy ions (Carbon ions), the rest employing only protons, despite the fact that heavy ions, like carbon, are effective for some tumours resistant to x-rays and protons, and are superior in preserving the healthy tissue surrounding the cancerous zone. This is due to the

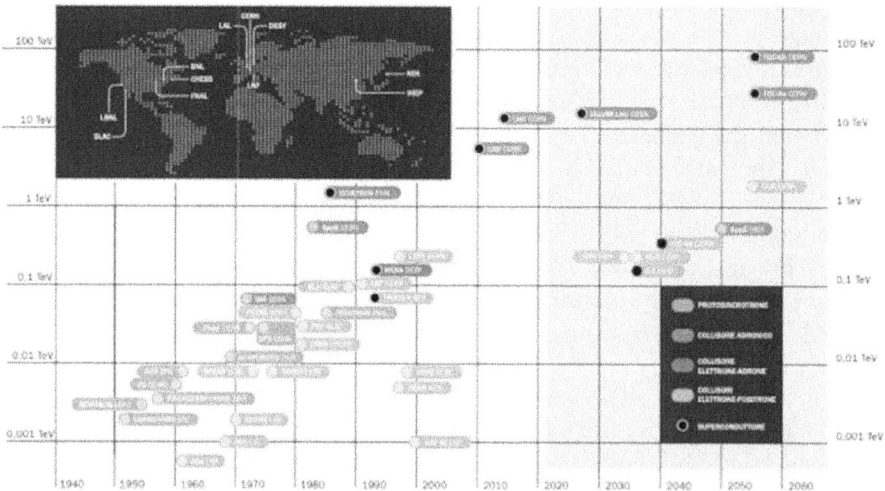

Figure 5.36. The history of accelerators and colliders with center of mass energy versus time. Black dots indicate the superconducting machines. Courtesy of INFN-Communication Office.

much larger size and cost of the required infrastructure for ions than for protons (not to mention x-ray).

As a further consideration, almost all proton centers are equipped with a gantry, allowing beam delivery from multiple directions, boosting the treatment effectiveness. On the contrary, very few ion centers are equipped with a gantry. The first, and so far unique in Europe, has been the center of HIT (Heidelberg Ion Therapy center) (Haberer *et al* 2004) that has been in operation since 2012 and employs classical resistive magnets, which makes it have a length of 26 m and a weight of about 600 tons (rotatable part). A step forward has been carried out by the HIMAC center that from 2018 has been routinely using a more compact ion gantry based on SC magnets, designed and manufactured by Toshiba (Iwata *et al* 2012). This has allowed the Japanese center to reduce the size and weight to about 13 m and 300 tons.

A further reduction of size and weight, which eventually would reflect in reduction of cost, would certainly enable a wider diffusion of ion gantries, thus contributing to better coverage of the territory by ion therapy centers. A large collaboration in Europe is looking to leverage the experience gained in the HEP laboratories for collider SC magnets to try to make a bold step and design a superconducting gantry for ion less than 100 tons in weight. This would allow to reduce the direct cost of the gantry and footprint of the infrastructure. This application would be the second, more direct, beneficial spin-off from HEP accelerators into the medical sector, after the development of the superconductor for the Tevatron that was essential for enabling superconducting MRI magnets. Today MRI is a 5 billion dollars business, and thousands of the large magnets fabricated each year for large size imaging are superconducting ones, thanks to the high-quality, multistrand, high current density superconductor developed first for Tevatron.

5.6.3.3 *Accelerator magnets*

HEP synchrotrons and especially HEP collider rings are the most challenging applications for SC magnets, as field quality, stability, and cost are very demanding and by far superior to other systems (like Fusion, MRI, high-field solenoids, etc). They are superconducting, with a few exceptions such as fast-cycling synchrotrons like the J-PARC complex; however, the fast-cycling FAIR-SIS100, for example, is superconducting as well. Magnets for superconducting linear colliders such as ILC, XFEL, and ESS are superconducting, too, despite the very low-field/gradient, because of the big gain in compactness and integration within the cryo-module (cryostat) hosting the superconducting radio-frequency cavities. There has been considerable effort in SC magnets for low-energy accelerators for nuclear physics, like cyclotrons or synchrocyclotrons, an effort that continues mainly driven today by medical applications. These SC magnets are similar in design and layout to magnets for MRI or particle detectors for momentum spectrometry, having the shape of solenoid or circular split coils. Here we will limit our consideration only to SC magnets for HEP colliders and their implication for the medical sector.

Superconducting magnets become of interest when the required field is above the iron saturation limit (i.e., approximately 2 T). In some instances, however, SC magnets are becoming a choice of preference for their compactness and their low-energy consumption also at low field, a topic that is increasingly important in the design of new accelerators.

Dipoles

The first function of a magnet is to guide and steer charges particles (i.e., to provide an adequate centripetal force to keep them in orbit in a circular accelerator or just to bend them in a transfer line; see figure 5.37, left). The adequate force can be given only by a transverse magnetic field, except for very low-energy beams, where solenoids and sometimes electrostatic deflectors are used. Transverse field means a magnetic field whose main component is perpendicular to the particle trajectory. For high-energy accelerators, the beam region is a cylinder that follows the beam path and with the smallest practical dimension, as shown in figure 5.37, right. As discussed later, the cost and technical complexity of the magnetic system are proportional to the energy stored in the magnetic field, which explains why it is important to minimize the size of the magnet bore.

Even though static magnetic fields do not accelerate, in circular accelerators the bending (dipole) field eventually determines the final energy reach. In relativistic conditions the relation between the beam energy, E_{beam} in TeV, the dipole field B in T, and the radius of the beam trajectory inside the bending field R in km takes a very simple form:

$$E_{\text{beam}} \cong 0.3BR$$

Since the dipole field typically covers 2/3 of the accelerator, R is about 2/3 of the average radius of the ring. The above relation clearly shows the interest for the highest possible field for a given tunnel.

Figure 5.38 shows an artistic sketch of the LHC dipole, which is called the Twin-Dipole since it hosts two opposite dipole fields in the same magnet, to bend the two counter circulating proton beams of the colliders. There are 1232 main dipoles installed in the LHC ring.

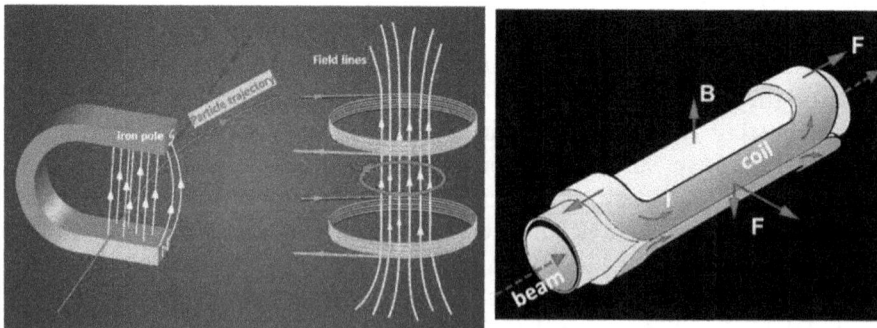

Figure 5.37. Effect of dipole field on charged particle (left, CERN/AC archives) and schematic of an accelerator magnet (right).

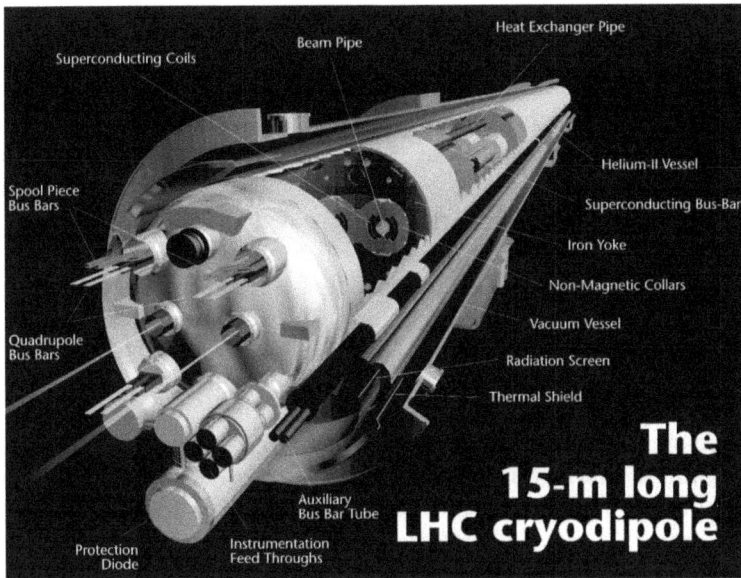

Figure 5.38. The LHC main dipole (CERN/AC archives).

Figure 5.39. Left: Effect of a triplet array of quadrupoles (strong focusing principle). Right: Cross-section of an LHC main quadrupole (CERN archives).

Quadrupoles and high-order magnets

The other key function of magnets is to provide the stability of the beam in the plane transverse to the particle motion. This is realized by the use of quadrupoles, sextupoles, octupoles, and in certain cases of higher-order harmonic magnets; in the LHC, magnets up to the dodecapole order are employed to fulfil the request of the sophisticated beam dynamics. For example, the action of a quadrupole on a particle beam is schematized in figure 5.39, left, showing the effect of the so-called strong focusing on a particle beam. In figure 5.39, right, a picture of the cross-section of one of the 400 main quadrupoles is shown.

While the main quadrupoles are usually of similar—although reduced—size and complexity with respect to the main dipoles, sextupoles and octupoles are of smaller size, peak field, and stored energy, thus featuring less complexity than the main dipoles. It is worth noticing that the quadrupoles that are in the so-called insertion regions, just before the collision points, usually called the low-β section, are necessary for the optics manipulation to focus the beam to the smallest attainable size (i.e., the minimum β-function) at the collision point. This set of quadrupoles must have an aperture significantly larger than the lattice quadrupoles and requires an integrated field gradient much larger than the regular lattice quadrupoles. Larger aperture and higher gradient result in a peak field that could be close to that of the main dipoles. In exceptional cases, like the High Luminosity LHC project, actually the low-β insertions around the ATLAS and CMS experiment would be equipped with quadrupoles of much larger peak field than the lattice LHC dipoles.

Accelerator dipoles and quadrupoles: design and field evolution

Accelerator magnets are very demanding, as mentioned above. Before examining in detail these unique characteristics, we want to underline that such characteristics must be obtained in hundreds or thousands magnets. The solutions to the required specifications must have a cost within reasonable limits. The HEP community cannot afford the luxury of very expensive perfect magnets. For example, each one of the 1232 lHC dipoles cost about 1 MCHF (value of 2001). This explains why one single component, the main LHC dipole, accounted for 70% of the cost of the whole magnetic system (including all other magnets, protection, powering, cryostat, interconnections, etc) and, alone, for 40% of the *total LHC collider project cost* (significantly the whole magnet system accounts for 1700 MCHF of the 3400 MCHF total LHC cost, 2008 price). However, 1 MCHF is surprisingly small for a 15 m long, 30 tons magnets with 7 MJ stored energy and rated for 9 T, 1.9 K, 13 kA operative conditions. Let's now consider the main design topics of accelerator magnets (Rossi and Todesco 2007): superconductor, critical current density, electro-magnetic design, and quench-protection. For brevity we do not discuss structural design and stability design, which are also quite important.

Superconductor

There are tens of thousands of superconducting materials. However, only a few of them have characteristics that make them usable in real devices (i.e., with a critical temperature well above 4.2 K, LHe normal boiling temperature), critical fields above 10 T, critical current density above 1000 A mm^{-2} at the operating temperature and field [e.g., for LHC is 1.9 K and nearly 9 T]). In addition to these characteristics, the superconductor for accelerator magnets needs: (i) 3–20 μm effective filament diameters, to maintain a good field quality during the ramp from injection to flat top; (ii) a geometry and a ductility such as to be assembled in compact cables; and (iii) to be robust enough to be wound in a coil and to withstand mechanical stresses largely exceeding 100 MPa. In figure 5.40 the engineering critical current density J_e of the few available practical superconductor is plotted versus magnetic field. J_e is defined as $J_e = J_c \cdot ff$ where J_c is the critical current density of the superconducting material and $ff = A_{SC}/A_{tot}$ is the volumetric fraction of super-conductor in the wire (that contains also stabilizing copper and other passive

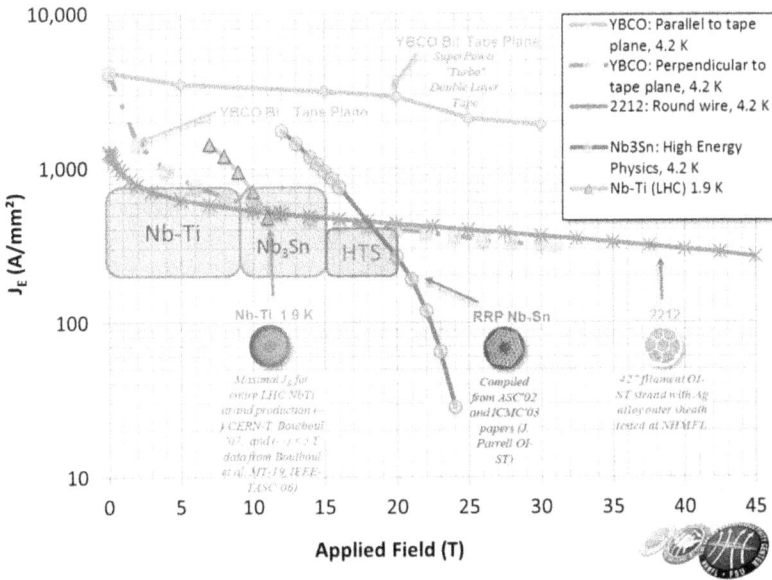

Figure 5.40. Engineering current density of practical superconductor versus field. The useful range for accelerator magnets is 200–800 A mm^{-2}. The data and curves are courtesy of Peter Lee (NHHMFL-FSU) and refer to year 2013 when the 16 T FCC baseline was decided. The working range is indicated by coloured boxes for Nb–Ti, for the Nb$_3$Sn developed for HL-LHC, and for HTS that is under development.

materials, like barriers and alloys). As is known, all superconducting magnets in existing accelerators are wound with Nb–Ti superconductor, High Luminosity LHC being the first project breaking the 10 T-wall (i.e., attempting to use Nb$_3$Sn for collider magnets rated for 11.5 T, and MgB$_2$ for the powering line; Bottura *et al* (2012), Rossi and Bruning (2016)).

In figure 5.41 a cross-section of the Nb–Ti/copper wires used in the LHC magnet is shown, together with the cable grouping many wires to reach at least 13 kA at 11 T, 1.9 K. The flat cable shown in figure 5.41 whose aspect ratio is ~10 and has 20–40 wires has a slightly trapezoidal cross-section with the two large faces making a 0.5°– 1.5° angle, which is called the 'Rutherford cable'.

High critical current density

The average current density in the whole coil section, J, is the basic parameter since the field $B \propto J \cdot \cos\vartheta \cdot t$, with t being the thickness of the coil. Since the coil thickness, differently from other applications like solenoids, detector magnets, etc, cannot be made very large, the only way to produce high field is to use high J. This implies both high critical current density, J_c, the most relevant material property and obliges using at least 75%–80% of the available J_c, with the consequence that J_c has to be high and also very uniform. The LHC dipoles work at 86% of the crossing of the load line with the critical current curve, and they have been designed and tested to 93% of J_c, so a variation larger than 5% directly affects the performance of the entire accelerators, while it is negligible in other systems (Hull *et al* 2015).

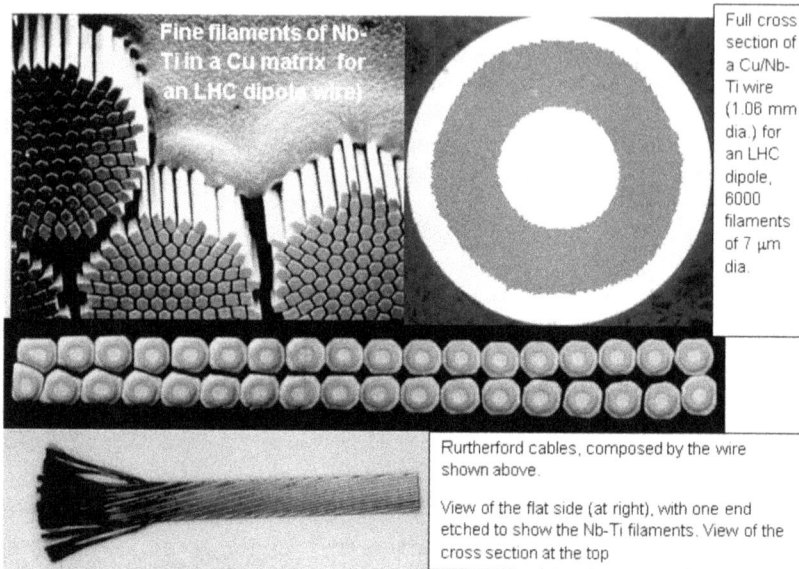

Figure 5.41. LHC superconducting wire and cable (CERN/AC archives).

The high J_c must be preserved by using low stabilizing content (i.e., the fraction of copper in the superconducting wire; see figure 5.41). A minimum amount of copper is necessary, but we need to keep it at minimum compatible with the stabilization and protection requirement, keeping high the filling factor *ff*, to avoid excessive lowering of *J*. Lack of sufficient stabilizing copper can generate big damage as in the case of the well-known LHC incident (Rossi 2010). Usually Cu/SC (or Cu/non-Cu) ranges between 1.5 and 2 for Nb–Ti-based coils and 0.9–1.5 for Nb_3Sn magnets, just the minimum for stabilization and dangerously near to the minimum for protection: we need to protect magnets with such low copper content despite strong force density, high stresses, and a stored energy in the 1–10 MJ ranges. Finally, the filling factor of the cables also needs to be high, keeping voids at 10%–12%: only Rutherford cable can achieve this compaction level. Insulation also must be minimized in volume, despite the fact that voltage could be as high as a hundred volts between coil sections, and a few kV to ground. Usually, the insulation fraction is 10%–15% of the total of that for thin cables, a rather small value.

Where all the factors are folded together, one realizes that the effective (overall) current density is $J_{overall} \sim 0.3\,J_c$. This allows our magnets to work at $J_{overall}$ of 300–600 A mm^{-2} (i.e., *3–10 times higher than any other magnet applications*).

Electromagnetic design and field quality

The coil cross-section is usually a circular sector filled with Rutherford cable arranged with spacers to mimic a distribution of current $J_0\cos\vartheta$ (i.e., a current density that is maximum at the midplane, J_0, decreasing like cosine function when increasing azimuth, and becoming zero at the pole [vertical axis]; Rossi and Todesco 2007, 2006). This is the most efficient distribution to generate transverse field in an

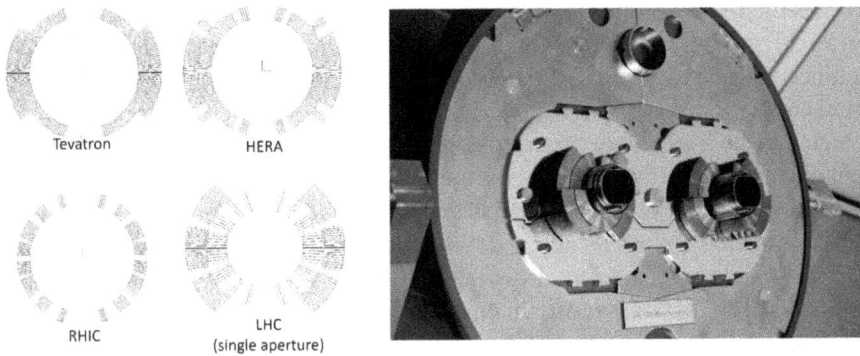

Figure 5.42. Cross-section of the main dipoles of the most important Hadron Colliders (left) and picture of a mock-up of the cross-section of a real LHC main dipole (Twin-Dipole).

infinitely long cylinder. The coil aperture being 50–100 mm and the magnet length being 5–15 m, the approximation is very good, indeed!

In figure 5.42 we show the cross-section of the main dipoles for the most important hadron colliders and a figure of a real LHC main dipole cross-section. A perfect $J_0\cos\vartheta$ would yield a pure dipole field. A perfect $J_0\cos2\vartheta$ (i.e., with maximum at 0°, 90°, etc) would yield a perfect quadrupolar field, a $J_0\cos3\vartheta$ current distribution would yield a perfect sextupole field, etc. The winding cannot be perfect, of course: however, the requested relative field accuracy is of the order of 10^{-5}. This calls for very accurate design and tight control of the tolerances. In the LHC all components have very strict tolerances from a few μm (superconductor) to 15–30 μm for the biggest mechanical components.

In addition, the bending strength of each dipole must be equal to nearly 0.01%, because a full sector (154 dipoles in the LHC) is powered in series, This is actually one of the main constraints of the use of SC magnets in accelerators (e.g., the RF cavities in a linac may differ from each other in integrated gradient by 5% without serious drawbacks). In hadron collider the single weakest dipole determines the energy of the beam.

Protection and quench

The main issues for HEP magnets are the large current density in the conductor and the large magnetic energy stored per unit of coil mass. When a magnet undergoes a sudden irreversible transition from superconducting to normal conducting state, which is called 'quench', the magnetic stored energy is turned into heat via Joule effect. SC accelerator magnets can tolerate temperature increases up to 300 K, if well-conceived and manufactured. The hot-spot temperature $T_{\text{hot spot}}$ can be estimated using an adiabatic heat balance, equating the joule heat produced during the discharge to the enthalpy rise of the conductor:

$$\int_{T_{\text{op}}}^{T_{\text{hot spot}}} \frac{C}{\rho}\,dT = \frac{1}{A_{\text{Cu}}A_{\text{tot}}} \int_{t_{\text{quench}}}^{\infty} I^2\,dt$$

where C is the total volumetric heat capacity of the superconductor composite, and ρ is the resistivity. The speed at which the energy is dumped depends on the magnet inductance and the resistance of the circuit formed by the quenching magnet and the external circuit $R = R_{\text{quench}} + R_{\text{ext}}$. The characteristic time of the dump is then $\tau \approx L/R$, which can be made short by decreasing the magnet inductance by use of large current cables. Protection (i.e., fast dump to avoid excessive T_{hotspot}) is then achieved by triggering heaters embedded in the winding pack, and fired at the moment a quench is detected to spread the normal zone over the whole magnet mass. In the LHC the typical time to detect a quench is 10–20 ms and the heaters must activate the quench in 100 ms or less. A further complication in collider is due to the fact that the main magnets are powered in series, and the stored energy in a single circuit is much larger than the few megajoule energy stored in a single magnet. In the case of each of the eight circuits formed by the series of 154 LHC main dipoles, the stored magnetic energy attains the gigajoule level, which was the reason for the gravity of the above-mentioned LHC incident of 2008 (Rossi 2010).

5.6.3.4 Superconducting accelerator magnets: outlook to future

As previously mentioned, LHC has been the summit of 30 years of development of accelerator magnets, all wound with Nb–Ti superconductor. For more than ten years a considerable amount of effort has been occuring in the US laboratories, BNL, FNAL, and LBNL, as well as in Europe at CERN, CEA (France), and—more recently—at CIEMAT (Es), INFN (It), PSI (Ch), to develop accelerator magnets wound with Nb_3Sn. As can be seen in figure 5.40, Nb_3Sn extends the attainable region from 8–9 to 15–16 T. The High Luminosity LHC project at CERN (Brüning and Rossi 2019, Rossi and Brüning 2019) has the main technological scope of design and manufacture, and operated the first set of Nb_3Sn magnets in a collider, ever. To go beyond this field level in the 20 T region use of HTS (high-temperature superconductor) is necessary (Rossi and Senatore 2021), as can be seen in figures 5.40 and 5.43, which summarizes the past and the possible future of HEP accelerator magnets.

However, these technologies, Nb_3Sn and HTS, are not yet mature enough for societal applications, and in the next chapter we will deal with how the technology used for present colliders can spill into the medical sector.

5.6.3.5 From HEP to hadron therapy: the new journey of accelerator magnets

As mentioned in the first section of this contribution, the use of the slim, tubular shape accelerator magnets can open the way to ion gantries of limited size and weight. Two points need to be specifically addressed for such a step into medical applications: (1) the need to ramp the field in coils that must be indirectly cooled (no use of liquid because gantry is rotating), with a ramp rate of 0.1–1 T s^{-1} which is from 10 to 100 times faster than the sweeping rate of HEP colliders; (2) the fact that the relatively low magnetic rigidity of the beam, together with fields of 3 T or more, implies the necessity to build strongly curved magnets, which has never been attempted for accelerator magnets.

Magnetic field evolution for Hadron Collider

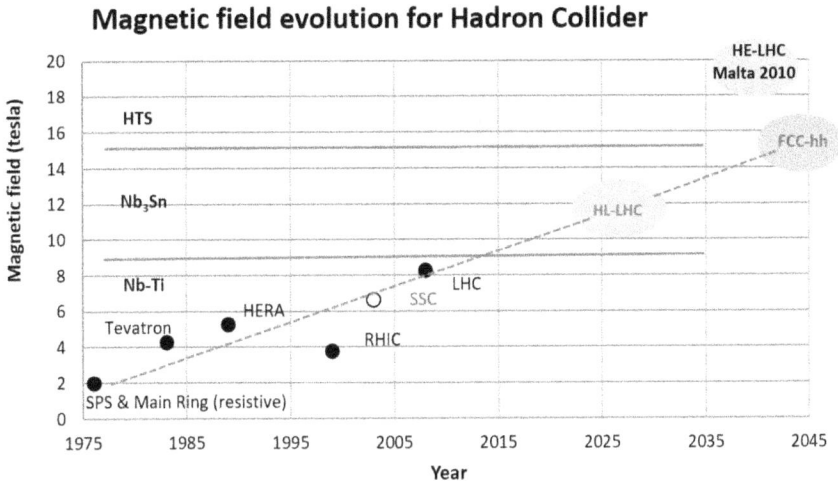

Figure 5.43. Evolution of the field in the SC magnets of the main past and future hadron colliders.

The recent rush for superconducting gantry

The TERA foundation (http://www.tera.it/) with the support of CERN in the last three years worked out a preliminary design for a light (~50 tons) rotatable gantry for ions. The gantry features SC magnets of cosϑ layout, with ramping operation, and has been named SIGRUM (Amaldi *et al* 2021). In 2020, following the phase of preliminary study, a larger collaboration led by CERN and extended to CNAO (CNAO n.d.), MedAustron (MedAustron n.d.), and INFN (departments of Milano-LASA and Genova) took over the SIGRUM design study and called an international review to assess the design. The outcome of the review was very positive, endorsing the basic design of SIGRUM. However, the review panel questioned the choice of the 3 T dipole field on which SIGRUM is based (Karppinen *et al* 2022) and has recommended, besides the construction of a demonstrator magnet, to make an effort toward increasing the dipole field and also to explore the feasibility of switching to a CCT (Canted Cosine Theta) dipole design.

The CCT design

The CCT is a new magnet type that has been pursued at LBNL for HEP and for a superconducting gantry for proton beams (Brouwer *et al* 2020). The schematic of a CCT dipole, basically composed of two slanted solenoids, is shown in figure 5.44, together with a picture of the construction at CERN of 2 m-long prototype (Kirby *et al* 2020), designed and built at CERN for the High Luminosity Project (Brüning and Rossi 2015). The CCT for High Luminosity LHC is the first application of CCT magnets in a collider. Superconducting CCT magnets as possible backbone of SC gantry for ions is the investigation topic of a large European collaboration, formed by CERN, CEA (Saclay, Fr), CIEMAT (Madrid), INFN (Milano and Genova), PSI (Ch), Uppsala University (Se), and Wigner Research Center for Physics (Hu). This collaboration has received two grants, one as part of the H2020-HITRI*plus* program (HITRIplus n.d.) and the other as part of the H2020-I.FAST program (I.

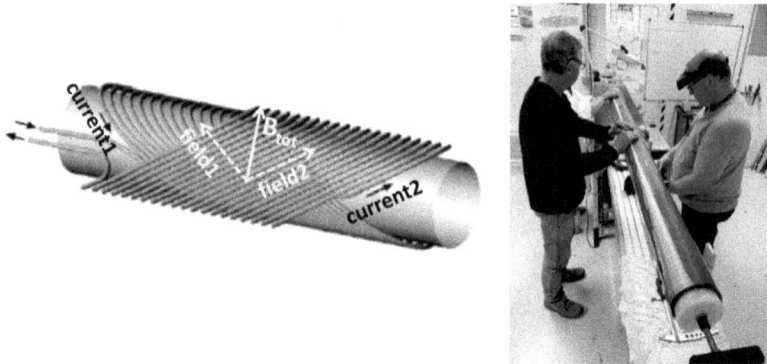

Figure 5.44. Left: CCT principle where oppositely slanted solenoids generate pure dipole field. Right: The High Luminosity CCT prototype during winding at CERN (Courtesy of Glyn Kirby, CERN).

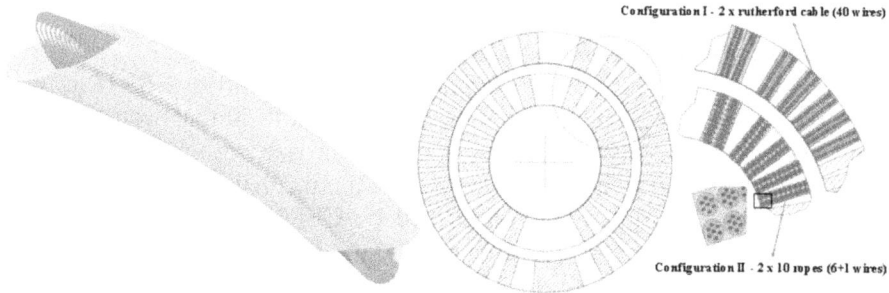

Configuration I - 2 x rutherford cable (40 wires)

Configuration II - 2 x 10 ropes (6+1 wires)

Figure 5.45. Sketch of the 4 T CCT dipole for HITRIplus. Left: 3D view along the length (1 m). Right: Cross-section showing the positioning of the conductor in the former grooves.

FAST n.d.). Both programs started in spring 2021. HITRI*plus* aims at building a 4 T demonstrator of a CCT SC dipole for ion gantry, with the necessary curved shape, with Nb–Ti conductor. I.FAST comprises also three companies in the consortium, in addition to the above-mentioned laboratories: BNG (De), Elytt (Es), and Scanditronics (Se). I.FAST plans to test a straight CCT wound with HTS in the 4–5 T range. These programs are extensively reported (Rossi *et al* 2022a). It is worth noticing that in the frame of this collaboration a working group is considering how to adapt the field description used both by magnet designers and by beam optics designers, since the usual 2D field multipole decomposition is no longer valid for strongly curved magnets. In figure 5.45 a 3D view of the CCT magnet considered for HITRI*plus* is shown together with a cross-section of the coil-former package. In figure 5.45 it is shown that conductors lay in the grooves of the formers (the dashed sectors in the two annular formers) and that two solutions are being considered for the conductor shape: a Rutherford cable, similar to the one used to wind HEP magnets, and a rope cable, of the type 6 strands around a central core.

HITRIPlus and IFAST should be able to assess in three years if CCT is a viable design to compete and maybe replace the classical cosϑ layout magnets.br

The cosϑ design—INFN SIG project
The above-mentioned collaboration among CERN, CNAO, INFN, and MedAustron is moving along two lines below described:

(1) CERN has refined the design of the 3 T dipole SIGRUM gantry and in (Karppinen *et al* 2022) a fairly detailed design is presented for a 70 mm aperture combined function 3 T dipole with superimposed a 5 T m^{-1} gradient, capable of withstanding a ramp rate of 0.1 T s^{-1}. Figure 5.46, left side, shows a sketch of the magnet.

(2) INFN is investigating the possibility to improve the magnet performance, as recommended by the review panel. The study is supported by a 4-year grant of 1 M€ value that has been awarded by INFN management to study various technologies for the ion gantry, and had an effective start on the 1st of January 2022. The grant is complemented by contributions by CERN (250 k€) and CNAO (350 k€). The grant, called *SIG,* includes various items:

    a. Design, manufacture, and testing of a cosi dipole short prototype.
    b. Design of a scanning magnet system.
    c. Design and prototyping of the main components of a new dose-delivery system (DDS), integrated with a new range verification system (RVS) for ions.

The superconducting demonstrator is the most critical item of the SIG program, taking 600 k€ of the INFN budget, matched by the 600 k€ contribution of CNAO and CERN.

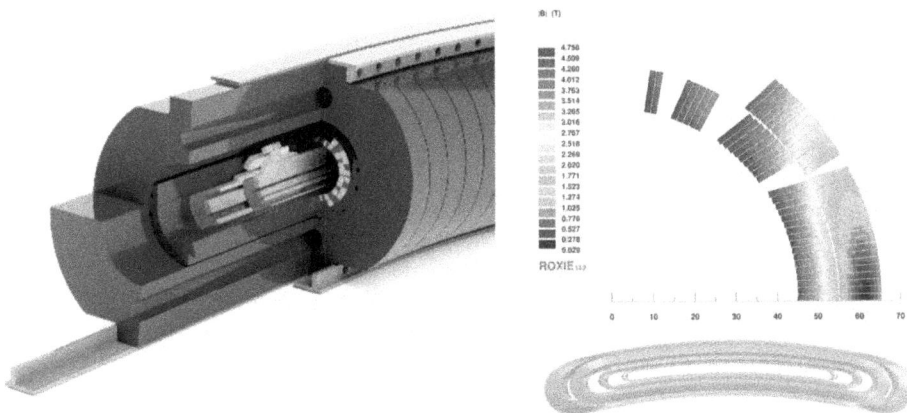

Figure 5.46. Sketch of the SIGRUM magnets (left, courtesy of M Karppinen, CERN) and of the SIG demonstrator coil cross-section with field map and coil top view (right).

For SIG we are investigating the following parameters range (Rossi *et al* 2022b):

- 4–5 T dipole field, which entails a bending radius of 1.35–1.65 m.
- 70–90 mm coil diameter (aperture).
- 0.4 T s^{-1} field continuous ramp rate up-down, with indirect cooling of the coils.

In figure 5.46, on the right side, a coil cross-section with field map for a possible SIG design is shown, together with a top view of the coil, well showing the 'banana shape'.

The Tevatron, HERA, and LHC (at 4.2 K) dipoles feature similar field and aperture (only RHIC has a lower field) (Rossi and Bottura 2012, Bottura and Rossi 2015), but these colliders have a very low-field ramp (0.01–0.03 T s^{-1}). Only the INFN-Discorap prototype dipole (Fabbricatore *et al* 2011, Sorbi *et al* 2013), a built-in collaboration with GSI for FAIR-SIS300, has a faster ramp rate, at about 0.6–1 T s^{-1}, but requires direct cooling, with forced flow supercritical Helium.

The main challenge of the SIG prototype is the strong curvature, of 1.65 m. For comparison, DISCORAP has a 65 m bending radius, with the other projects featuring almost straight magnets.

The HIMAC gantry by Toshiba (Iwata *et al* 2012) is based on dipoles of 2.9 and 2.3 T, with a particular shape (curved race-track coils), with curvature radius of about 3 m. To go beyond this achievement of HIMAC the route of accelerator magnets, of tubular shape and rather low weight, seems the most promising one: with SIG 4 T dipoles, and a new mechanical structure under study at CERN, the weight of the gantry can be of 50–80 tons only.

To conclude this overview, we refer to the plot of figure 5.47, where magnets of different projects are reported in a chart where the squared product of field times the magnet aperture (this parameter gives the class of the magnet since it is proportional

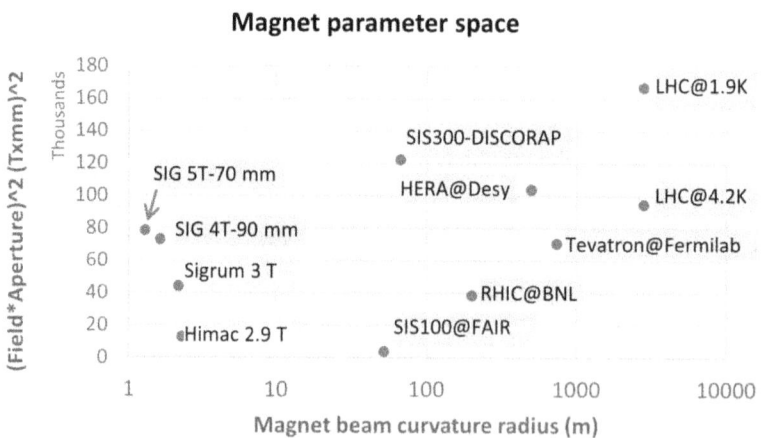

Figure 5.47. Collections of various past and future (SIG and Sigrum) magnets in a special parameter space (see text for details).

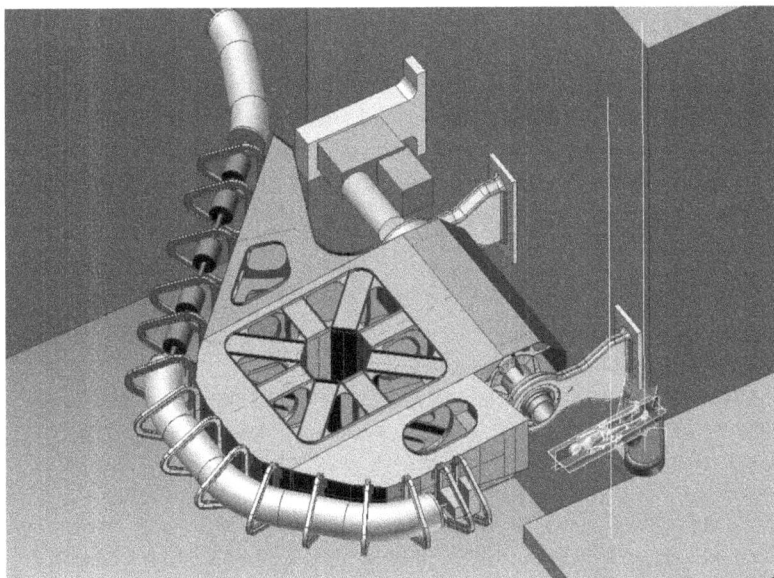

Figure 5.48. Artistic view of SIGRUM in operation (courtesy of M Pullia, CNAO foundation).

to the stored energy) is plotted versus the bending radius. Here the breakthrough made by the HIMAC magnets toward smaller bending radius, and the further step required by the SIG project both in bending radius and in stored energy, is rather evident in this magnet panorama.

The SIG project, as part of the global SIGRUM collaboration, aims at testing the first prototype by 2025 and then to be able to proceed to the gantry construction. A possible sketch of the SIGRUM gantry installed in the CNAO facility is shown in figure 5.48.

5.6.3.6 Conclusion

Superconducting magnets have accompanied the progress of HEP colliders. The community is pursuing the R&D on new more advanced technology for next-generation magnets for 15 T or more, necessary for the post-High Luminosity LHC collider. However, the same community is exploring if the presently available accelerator magnet technology can make a key contribution to cancer therapy, by enabling light and less expensive ion gantries. If successful, this development would give a significant boost to the performance of present hadron therapy centers based on heavy ions.

Acknowledgements

The author is grateful to the many collaborators in the LHC and High luminosity LHC times and, for the discussion on magnet design to, Luca Bottura, Mikko Karppinen, Glyn Kirby, Diego Perini, Davide Tommasini, and Ezio Todesco of the

CERN Technology department. He wants to thank also Gabriele Ceruti, Ernesto De Matteis, Samuele Mariotto, Marco Prioli, Massimo Sorbi, Marco Statera, and Riccardo Valente of the LASA Laboratory (University of Milano and INFN-Milano) for their contributions to the gantry magnet design. This work is partly supported by the European Commission via the H2020-HITRIPlus grant no. 101008548 and H2020-I.FAST grant no. 101004730, and by INFN via the grant CSN5-Call2021-SIG.

References

Abou Allaban A, Wang W and Padr T 2020a A systematic review of robotics research in support of in-home care for older adults *Information* **11** 75

Acemoglu A, Kriegstein J and Caldwell D G *et al* 2020b 5G robotic telesurgery: remote transoral laser microsurgeries on a Cadaver *IEEE Trans. Med. Robot. Bionics* **2** 511–8

Acemoglu A, Peretti G, Hysenbelli J, Trimarchi M and Caldwell D G *et al* 2020a Operating from a distance: the first 5G telesurgery experience *Ann. Intern. Med.* **173** 940–1

Adams R M, Mora T, Walczak A M and Kinney J B 2016 Measuring the sequence-affinity landscape of antibodies with massively parallel titration curves *eLife* **5** e23156

Agostinis P *et al* 2011 Photodynamic therapy of cancer: an update *CA. A Cancer J. Clin.* 250–81

Aleta A *et al* 2020 Modelling the impact of testing, contact tracing and household quarantine on second waves of COVID-19 *Nat. Human Behav.* **4** 964–71

Althouse B M *et al* 2020 Superspreading events in the transmission dynamics of SARS-CoV-2: opportunities for interventions and control *PLoS Biol.* **18** e3000897

Amaldi U *et al* 2021 Sigrum-A Superconducting Ion Gantry with Riboni's Unconventional Mechanics. No. CERN-ACC-NOTE-2021-0014

Anderson B, Rasch M, Hochlin K, Jensen F H, Wismar P and Fredriksen J E 2006 Decontamination of rooms, medical equipment and ambulances using an aerosol of hydrogen peroxide disinfectant *J. Hosp. Infect.* **62** 149–55

Antoniou S A, Antoniou G A, Antoniou A I and Granderath F A 2015 Past, present, and future of minimally invasive abdominal surgery *JSLS* **19** e2015.00052

Arenas A *et al* 2020 Modeling the spatiotemporal epidemic spreading of COVID-19 and the impact of mobility and social distancing interventions *Phys. Rev. X* **10** 041055

Ašner F M 1999 *High Field Superconducting Magnets* (Oxford: Oxford University Press)

Barbosa H *et al* 2018 Human mobility: models and applications *Phys. Rep.* **734** 1–74

Baronchelli F, Zucchella C, Serrao I D and Bartolo M 2021 The effect of robotic assisted gait training with Lokomat® on balance control after stroke: systematic review and meta-analysis *Front. Neurol.*

Barrat A *et al* 2013 Empirical temporal networks of face-to-face human interactions *Eur. Phys. J. Spec. Top.* **222** 1295–309

Beal A, Mahida N, Staniforth K, Vaughan N, Clarke M and Boswell T 2016 First UK trial of Xenex PX-UV, an automated ultraviolet room decontamination device in a clinical haematology and bone marrow transplantation unit *J. Hosp. Infect.* **93** 164–8

Beccani M, Aiello G, Gkotsis N and Tunc H *et al* 2016 Component based design of a drug delivery capsule robot *Sens. Actuators* A **245** 180–8

Best J 2021 Wearable technology: COVID-19 and the rise of remote clinical monitoring *Brit. Med. J.* **372** n413

Bloss R 2011 Mobile hospital robots cure numerous logistic needs *Ind. Robot* **38** 567–71

Bottura L, de Rijk G, Rossi L and Todesco E 2012 Advanced accelerator magnets for upgrading the LHC *IEEE Trans. Appl. Supercond.* **22** 8

Bottura L and Rossi L 2015 Magnets for particle accelerators *Applied Superconductivity—Handbook on Devices and Applications* ed P Seidel (New York: Wiley-VCH) pp 448–86

Brändén G and Neutze R 2021 Advances and challenges in time-resolved macromolecular crystallography *Science* **373** eaba0954

Brouwer L *et al* 2020 Design and test of a curved superconducting dipole magnet for proton therapy *Nucl. Instrum. Meth. Phys. Res. Sect. A, Accel., Spect., Detect. Assoc. Equip.* **957** 163414

Brüning O and Rossi L 2015 *The High Luminosity Large Hadron Collider Advanced Series on Direction in High Energy Physics* (Singapore: World Scientific) vol 24

Brüning O and Rossi L 2019 The high-luminosity large hadron collider *Nat. Rev.—Phys.* **1** 241–3

Bubar K M *et al* 2021 Model-informed COVID-19 vaccine prioritization strategies by age and serostatus *Science* **371** 916–21

Buettner R, Wannenwetsch K and Loskan D 2020 A systematic literature review of computer support for surgical interventions *2020 IEEE 44th Annual Computers, Software, and Applications Conf. (COMPSAC)* pp 729–34

Burley S K *et al* 2022 A comprehensive review of 3D structure holdings and worldwide utilization by researchers, educators, and students *Biomolecules* **12** 1425

Bush K *et al* 2011 Tackling antibiotic resistance *Nat. Rev. Microbiol.* **9** 894–6

Čaić M, Mahr D and Oderkerken-Schröder G 2019 Value of social robots in services: social cognition perspective *J. Serv. Mark.* **33** 463–78

Cao W, Song W, Li X, Zheng S, Zhang G and Wu Y *et al* 2019 Interaction with social robots: improving gaze toward face but not necessarily joint attention in children with autism spectrum disorder *Front. Psychol.* **10** 1503

Carling P C and Huang S S 2013 Improving healthcare environmental cleaning and disinfection current and evolving issues *Infect. Control Hosp. Epidemiol.* **34** 507–13

Carr J H and Shepherd R B 1987 *A Motor Relearning Programme for Stroke* (Oxford: Butterworth Heinemann)

Castro M, Ares S, Cuesta J A and Manrubia S 2020 The turning point and end of an expanding epidemic cannot be precisely forecast *Proc. Natl Acad. Sci.* **117** 26190–6

Chakraborty A K and Barton J P 2017 Rational design of vaccine targets and strategies for HIV: a crossroad of statistical physics, biology, and medicine *Rep. Prog. Phys.* **80** 032601

Chang M C and Boudier-Revéret M 2020 Usefulness of telerehabilitation for stroke patients during the COVID-19 pandemic *Am. J. Phys. Med. Rehabil.* **99** 582

Chapman H N 2019 X-ray free-electron lasers for the structure and dynamics of macromolecules *Annu. Rev. Biochem.* **88** 35–58

Charron P M, Kirby R L and MacLeod D A 1995 Epidemiology of walker-related injuries and deaths in the United States *Am. J. Phys. Med. Rehabil.* **74** 237–9

Chinazzi M *et al* 2020 The effect of travel restrictions on the spread of the 2019 novel coronavirus (COVID-19) *Science* **368** 395–400

Choi P J, Oskouian R J and Tubbs R S 2018 Telesurgery: past, present, and future *Cureus* **10** e2716

Chowell G 2017 Fitting dynamic models to epidemic outbreaks with quantified uncertainty: a primer for parameter uncertainty, identifiability, and forecasts *Infect. Dis. Model.* **2** 379–98

Chowell G, Vespignani A, Viboud C and Simonsen L 2018 The RAPIDD Ebola Forecasting Challenge, Special Issue *Epidemics* **22** 1

CLIC Project n.d. (http://clic-study.web.cern.ch/clic-study/)

CNAO n.d. Fondazione CNAO, Pavia, Italy (https://fondazionecnao.it/)

Colizza V, Barrat A, Barthélemy M and Vespignani A 2006 The role of the airline transportation network in the prediction and predictability of global epidemics *Proc. Natl Acad. Sci.* **103** 2015–20

Dabiri F, Massey T, Noshadi H and Hagopian H *et al* 2009 A telehealth architecture for networked embedded systems: a case study in *in vivo* health monitoring *IEEE Trans. Inf. Technol. Biomed.* **13** 351–9

Danon L *et al* 2011 Networks and the epidemiology of infectious disease *Interdisciplinary Perspectives on Infectious Diseases* **2011** e284909

Davies B 2000 A review of robotics in surgery *Proc. Inst. Mech. Eng.* **214** 129–40

Davies N G *et al* 2020 Age-dependent effects in the transmission and control of COVID-19 epidemics *Nat. Med.* **26** 1205–11

Dawe J, Sutherland C, Barco A and Broadbent E 2019 Can social robots help children in healthcare contexts? A scoping review *BMJ Paediatr. Open.* **3** e000371

de Boer J F, Leitgeb R and Wojtkowski M 2017 Twenty-five years of optical coherence tomography: the paradigm shift in sensitivity and speed provided by Fourier domai—OCT *Biomed. Opt. Express* **8** 3248–80

de Graaf M M A, Ben Allouch S and van Dijk J A G M 2015 What makes robots social? A user's perspective on characteristics for social human–robot interaction *Social Robot* ed A Tapus, E André, J-C Martin, F Ferland and M Ammi (Cham: Springer International Publishing)

De Marchi F, Contaldi E, Magistrelli L and Cantello R *et al* 2021 Telehealth in neurodegenerative diseases: opportunities and challenges for patients and physicians *Brain Sci.* **11** 237

De Michieli L, Mattos L S, Caldwell D G, Metta G and Cingolani R 2020 Technology and telemedicine *Encyclopedia of Gerontology and Population Aging* ed D Gu and M E Dupre (Cham: Springer International Publishing)

Degiovanni O A and Amaldi U 2015 History of Hadron therapy accelerators *Physica Med.* **31** 322–32

Desai M M and Fisher D S 2007 Beneficial mutation–selection balance and the effect of linkage on positive selection *Genetics* **176** 1759–98

Di Natali C and Poliero *et al* 2019 Design and evaluation of a soft assistive lower limb exoskeleton *Robotica* **37** 2014–34

Di Natali C, Sadeghi A and Mondini A *et al* 2020 Pneumatic quasi-passive actuation for soft assistive lower limbs exoskeleton *Front. Neurorobot.* **14** 31

Ding X and Clifton D *et al* 2020 Wearable sensing and telehealth technology with potential applications in the coronavirus pandemic *IEEE Rev. Biomed. Eng.* **14** 48–70

D. O. E. United NationsAffairs S 2017 World population ageing 2017 highlights (Accessed 11 March 2021)

Eames K, Bansal S, Frost S and Riley S 2015 Six challenges in measuring contact networks for use in modelling *Epidemics* **10** 72–7

Elhanati Y, Sethna Z, Callan C G, Mora T and Walczak A M 2018 Predicting the spectrum of TCR repertoire sharing with a data-driven model of recombination *Immunol. Rev.* **284** 167–79

Evans L 2018 *The Large Hadron Collider: A Marvel of Technology* (EPFL Press) 2nd edn

Fabbricatore P *et al* 2011 The construction of the model of the curved fast ramped superconducting dipole for FAIR SIS300 synchrotron *IEEE Trans. Appl. Supercond.* **21** 1863–7 Part: 2 Published: Jun

Faus-Golfe A 2020 Design of a compact 140 MeV electron linear accelerator *ARIES D3.4* (https://edms.cern.ch/ui/file/1817163/1.0/ARIES-Del-D3.4-Final.pdf)

Ferretti L *et al* 2020 Quantifying SARS-CoV-2 transmission suggests epidemic control with digital contact tracing *Science*

Fiedler F 2008 *PhD Thesis* FZD Rossendorf/TU Dresden

Fujimoto J and Swanson E 2016 The development, commercialization, and impact of optical coherence tomography *Invest—Ophthalmol. Vis. Sci.* **57** OCT1–OCT13

Funk S, Salathé M and Jansen V A A 2010 Modelling the influence of human behaviour on the spread of infectious diseases: a review *J. R. Soc. Interface* **7** 1247–56

Gallotti R, Valle F, Castaldo N, Sacco P and De Domenico M 2020 Assessing the risks of 'infodemics' in response to COVID-19 epidemics *Nat. Hum. Behav.* **4** 1285–93

Gatto M *et al* 2020 Spread and dynamics of the COVID-19 epidemic in Italy: effects of emergency containment measures *Proc. Natl Acad. Sci.* **117** 10484–91

Gérard A *et al* 2020 High-throughput single-cell activity-based screening and sequencing of antibodies using droplet microfluidics *Nat. Biotechnol.* **38** 715–21

Gorpas D *et al* 2011 Autofluorescence lifetime augmented reality as a means for real-time robotic surgery guidance in human patients *Sci. Rep.* **9** 1187

Gostic K M *et al* 2020 Practical considerations for measuring the effective reproductive number, Rt *PLoS Comput. Biol.* **16** e1008409

Guimerà R, Mossa S, Turtschi A and Amaral L a N 2005 The worldwide air transportation network: anomalous centrality, community structure, and cities' global roles *Proc. Natl Acad. Sci.* **102** 7794–9

Haberer T *et al* 2004 The Heidelberg ion therapy center *Radiother. Oncol.* **73** S186–90

Henschel A, Laban G and Cross E S 2021 What makes a robot social? A review of social robots from science fiction to a home or hospital near you *Curr Robot Rep.* **2** 9–19

HITRIplus n.d. *Heavy Ion Therapy Research Integration* (https://hitriplus.eu/)

Hosseinizadeh O A, Breckwoldt N, Fung R, Sepehr R, Schmidt M, Schwander P, Santra R and Ourmazd A 2021 Few-fs resolution of a photoactive protein traversing a conical intersection *Nature* **599** 697–701

Huang P-S, Boyken S E and Baker D 2016 The coming of age of de novo protein design *Nature* **537** 320–7

Hull J R, Wilson M N, Bottura L, Rossi L, Green M A, Iwasa Y, Hahn S, Duchateau J and Kalsi S S 2015 *Superconducting Magnets Applied Superconductivity—Handbook on Devices and Applications* (Germany: Wiley- VCH Verlag Gmbh & Co. KGa) ch 4

Huysamen K, Bosch T, de Looze M, Stadler K S, Graf E and O'Sullivan L W 2018 Evaluation of a passive exoskeleton for static upper limb activities *Appl. Ergon.* **70** 148–55

I.FAST n.d. Innovation Fostering in Accelerator Science and Technology (https://ifast-project.eu/home)

Iwata Y *et al* 2012 Design of a superconducting rotating gantry for heavy-ion therapy *Phys. Rev. Spec. Top. PRST-AB. Accel. Beams.* **15** 044701

Jaeschke E, Khan S, Schneider J R and Hastings J B 2020 *Synchrotron Light Sources and Free-Electron Lasers* (Berlin: Springer)

Jaskolski M, Dauter Z and Wlodawer A 2014 A brief history of macromolecular crystallography, illustrated by a family tree and its Nobel fruits *FEBS J.* **281** 3985–4009

Johanson D L, Ho S A, Sutherland C J, Brown B, MacDonald B A and Jong Y L *et al* 2020 Smiling and use of first-name by a healthcare receptionist robot: effects on user perceptions, attitudes, and behaviours *Paladyn J. Behav. Robot.* **11** 40–51

Jumper J *et al* 2021 Highly accurate protein structure prediction with alphafold *Nature* **596** 583–9

Karppinen M, Ferrantino V, Kokkinos C and Ravaioli E 2022 Design of a curved superconducting combined function bending magnet demonstrator for hadron therapy *IEEE Trans. Appl. Supercond.* **32** 1–5

Kermack W O, McKendrick A G and Walker G T 1927 A contribution to the mathematical theory of epidemics *Proc. R. Soc.* A **115** 700–21

Kirby G *et al* 2020 Assembly and test of the HL-LHC twin aperture orbit corrector based on canted cos-theta design *J. Phys.: Conf. Ser.* **1559** 012070

Kiss I Z, Miller J and Simon P L 2017 *Mathematics of Epidemics on Networks: From Exact to Approximate Models* (Cham: Springer International Publishing)

Kivelä M *et al* 2014 Multilayer networks *J. Complex Netw.* **2** 203–71

Krafft C, von Eggeling F, Guntinas-Lichius O, Hartmann A, Waldner M J, Neurath M F and Popp J 2018 Perspectives, potentials and trends of *ex vivo* and *in vivo* optical molecular pathology *J. Biophotonics* **11** e201700236

Kühlbrandt W 2014 Cryo-EM enters a new era *eLife* **3** e03678

Kwoh Y S, Hou J and Jonckheere E A *et al* 1988 A robot with improved absolute positioning accuracy for CT guided stereotactic brain surgery *IEEE Trans. Biomed. Eng.* **35** 53–161

Laffranchi M and D'Angella S *et al* 2021 User-centered design and development of the modular TWIN lower limb exoskeleton *Front. Neurorobot.* **15**

Lässig M, Mustonen V and Walczak A M 2017 Predicting evolution *Nat Ecol Evol* **1** 1–9

Last update in the search for the Higgs boson 2012 Seminar held at CERN, 4 July (https://indico.cern.ch/conferenceDisplay.py?confId=197461)

Leochico C F D 2020 Adoption of telerehabilitation in a developing country before and during the COVID-19 pandemic *Ann. Phys. Rehabil. Med.* **63** 563–4

LHC DESIGN Report 2004 Vol. I The LHC Main Ring Design Report, CERN Report, CERN-2004-003, 4 June

Li J, Esteban-Fernández de Ávila B, Gao W, Zhang L and Wang J 2017 Micro/nanorobots for biomedicine: delivery, surgery, sensing, and detoxification *Sci. Robot.* **2** eaam6431

López J A M *et al* 2021 Anatomy of digital contact tracing: role of age, transmission setting, adoption and case detection *Sci. Adv.* **7** eabd8750

Łuksza M and Lässig M 2014 A predictive fitness model for influenza *Nature* **507** 57–61

Masuda N and Holme P 2013 Predicting and controlling infectious disease epidemics using temporal networks *F1000Prime Reports* **5** 6

Matanfack Azemtsop G, Rüger J, Stiebing C, Schmitt M and Popp J 2020 Imaging the invisible—bioorthogonal Raman probes for imaging of cells and tissues *J. Biophotonics* **13** e202000129

Mattos L S, Caldwell D G, Peretti G, Mora F, Guastini L and Cingolani R 2016 Microsurgery robots: addressing the needs of high-precision surgical interventions *Swiss Med. Wkly.* **146** w14375

MedAustron n.d. *The Center for Ion Therapy and Research, Wiener Neustadt, Austry* (https://medaustron.at/en)

Mess K-H, Schmüser P and Wolff S 1996 *Superconducting Accelerator Magnets* (Singapore: World Scientific)

Meyer T, Schmitt M, Guntinas-Lichius O and Popp J 2019 Toward an all-optical bi—sy *Opt. Photonics News* **30** 26–33

Minervina A A *et al* 2021 Longitudinal high-throughput TCR repertoire profiling reveals the dynamics of T-cell memory formation after mild COVID-19 infection *eLife* **10** e63502

Moses H, Matheson D H, Dorsey E R, George E R, Sadoff D and Yoshimura S 2013 The anatomy of health care in the United States *JAMA* **310** 1947–63

Murugan A, Mora T, Walczak A M and Callan C G 2012 Statistical inference of the generation probability of T-cell receptors from sequence repertoires *Proc. Natl Acad. Sci.* **109** 16161–6

Neher R A, Russell C A and Shraiman B I 2014 Predicting evolution from the shape of genealogical trees *eLife* **3** e03568

Nelson B J, Kaliakatsos I K and Abbot J J 2010 Microrobots for minimally invasive medicine *Annu. Rev. Biomed. Eng.* **12** 55–85

Neutze R, Wouts R, van der Spoel D, Weckert E and Hajdu J 2000 Potential for biomolecular imaging with femtosecond x-ray pulses *Nature* **406** 753–7

Nourmohammad A, Otwinowski J and Plotkin J B 2016 Host-pathogen coevolution and the emergence of broadly neutralizing antibodies in chronic infections *PLoS Genet.* **12** e1006171

Oliver N *et al* 2020 Mobile phone data for informing public health actions across the COVID-19 pandemic life cycle *Sci. Adv.* **6** eabc0764

OMRON 2022 Automating UV Disinfection Process Safely and Wisely (https://web.omron-ap.com/th/ld-uvc) Accessed 10 January 2022

Organisation for Economic Co-operation and Development (OECD) 2009 *Health at a Glance 2009: OECD Indicators* (Paris: OECD Publishing)

Organisation for Economic Co-operation and Development (OECD) 2020 *Historical Population Data and Projections (1950–2050)* (https://stats.oecd.org/) (Accessed 1 Aug 2020)

Osibona O, Solomon B D and Fecht D 2021 Lighting in the home and health: a systematic review *Int. J. Environ. Res. Public Health* **18** 609

Pandey A K and Gelin R 2018 A mass-produced sociable humanoid robot: pepper: the first machine of its kind *IEEE Robot. Autom. Mag.* **25** 40–8

Parker V M, Wade D T and Langton H R 1986 Loss of arm function after stroke: measurement, frequency, and recovery *Int. Rehab. Med.* **8** 69–73

Pastor-Satorras R, Castellano C, Van Mieghem P and Vespignani A 2015 Epidemic processes in complex networks *Rev. Mod. Phys.* **87** 925–79

Perelson A S, Neumann A U, Markowitz M, Leonard J M and Ho D D 1996 HIV-1 dynamics *in vivo*: virion clearance rate, infected cell life-span, and viral generation time *Science* **271** 1582–6

Perelson A S and Weisbuch G 1997 Immunology for physicists *Rev. Mod. Phys.* **69** 1219–68

Perra N 2021 Non-pharmaceutical interventions during the COVID-19 pandemic: a review *Phys. Rep.* **913** 1–52

Poletto C, Meloni S, Colizza V, Moreno Y and Vespignani A 2013 Host mobility drives pathogen competition in spatially structured populations *PLoS Comput. Biol.* **9** e1003169

Poliero T, Mancini L, Caldwell D G and Ortiz J 2019 Enhancing backsupport exoskeleton versatility based on human activity recognition *2019 Wearable Robotics Association Conf. (WearRAcon)* (IEEE) pp 86–91

Pons J L 2008 *Wearable Robots: Biomechatronic Exoskeletons* (New York: Wiley)

Popp J (ed) 2014 *Ex-vivo and In-vivo Optical Molecular Pathology* (Weinheim: Wiley-VCH)

Popp J and Bauer M (ed) 2015 *Modern techniques for Pathogen Detection* (Weinheim: Wiley-VCH)

Popp J and Strehle M (ed) 2006 *Biophotonics* (Weinheim: Wiley-VCH)

Popp J, Tuchin V V, Chiou A and Heinemann S H (ed) 2011 *Handbook of Biophotonics* **vol 1–3** (Weinheim: Wiley-VCH)

Priplata A A, Niemi J B, Harry J D, Lipsitz L A and Collins J J 2003 Vibrating insoles and balance control in elderly people *Lancet* **362** 1123–4

Prvu Bettger J and Resnik L J 2020 Tele-rehabilitation in the age of COVID-19: an opportunity for learning health system research *Phys. Ther.* **100** 1913–6

Pullano G *et al* 2020 Underdetection of COVID-19 cases in France threatens epidemic control *Nature* **590** 1–9

Ramalingam B, Yin J, Rajesh Elara M, Tamilselvam Y K, Mohan Rayguru M, Muthugala M and Félix Gómez B 2020 A human support robot for the cleaning and maintenance of door handles using a deep-learning framework *Sensors* **20** 3543

Richardson C A, Glynn N W, Ferrucci L G and Mackey D C 2015 Walking energetics, fatigability, and fatigue in older adults: the study of energy and aging pilot *J. Gerontol. Ser. A: Biomed. Sci. Med. Sci.* **70** 487–94

Robinson C 1995 Rehabilitation engineering, science and technology ed J Bronzino *Biomedical Engineering Handbook* (Boca Raton, FL: CRC Press) pp 2045–54

Rossi L 2010 Superconductivity: its role, its success and its setbacks in the Large Hadron Collider of CERN *Supercond. Sci. Technol.* **23** 034001

Rossi L *et al* 2022a Preliminary study of 4 T superconducting dipole for a light rotating gantry for ion-therapy *IEEE-Trans. Appl. Supercon.* **32** 4400506

Rossi L *et al* 2022b A European collaboration to investigate superconducting magnets for next generation heavy ion therapy *IEEE-Trans. Appl. Supercon.* **32** 4400207

Rossi L and Bottura L 2012 Superconducting magnets for particle accelerators *Rev. Accel. Sci. Technol.* **5** 51–89

Rossi L and Bruning O 2016 The LHC upgrade plan and technology challenges *Challenges and Goals for ACCELERATORS in the XXI Century* ed O Bruning and S Myers (Singapore: World Scientific Publisher) pp 467–97

Rossi L and Brüning O 2019 *Progress with the High Luminosity LHC Project at Cern Proceedings of IPAC2019 (Melbourne, 19–24 May)* (Jacow Publisher) pp 17–22

Rossi L and Senatore C 2021 HTS accelerator magnet and conductor development in Europe *Instruments* **5** 33

Rossi L and Todesco E 2006 Electromagnetic design of superconducting quadrupoles *Phys. Rev. Spec. Top. Accel Beams* **9** 102401 1–20

Rossi L and Todesco E 2007 Electromagnetic design of superconducting dipoles based on sector coils *Phys. Rev. Spec. Top. Accel. Beams* **10** 112401

Rossi L and Todesco E 2011 Conceptual design of the 20 T dipoles for high-energy LHC *Proc. of the High-Energy Large Hadron Collider Workshop (Malta, Oct. 2010)* ; E Todesco and F Zimmermann CERN-2011-003 (8 April 2011) pp 13–9

Saglia J A *et al* 2019 Design and development of a novel core, balance and lower limb rehabilitation robot: Hunova® *2019 IEEE 16th Int. Conf. on Rehabilitation Robotics (ICORR) (2019)* pp 417–22

Saglia J A, Tsagarakis N, Dai J S and Caldwell D G 2013 Control strategies for patient-assisted training using the ankle rehabilitation robot (ARBOT) *IEEE/ASME Trans. Mechatron.* **18** 1799–808

Satava R 2002 Surgical robotics: the early chronicles *Surg. Laparosc., Endosc. Percutan. Techn.* **12** 6–16

Schlichting I 2015 Serial femtosecond crystallography: the first five years *IUCrJ* **2** 246–55

Schlosser F *et al* 2020 COVID-19 lockdown induces disease-mitigating structural changes in mobility networks *Proc. Natl Acad. Sci.* **117** 32883–90

Sorbi M *et al* 2013 Measurements and analysis of the SIS-300 dipole prototype during the functional test at LASA *IEEE Trans. Appl. Supercond.* **23** 400205

Soto F, Wang J, Ahmed R and Demirci U 2020 Medical micro/nanorobots in precision medicine *Adv. Sci.* **7** 2002203

Spence J C H 2017 XFELs for structure and dynamics in biology *IUCrJ* **4** 322–39

Stahl B C and Coeckelbergh M 2016 Ethics of healthcare robotics: towards responsible research and innovation *Robot. Auton. Syst.* **86** 152–61

Starr T N *et al* 2020 Deep mutational scanning of SARS-CoV-2 receptor binding domain reveals constraints on folding and ACE2 binding *Cell* **182** 1295–1310.e20

Stower R 2019 The role of trust and social behaviours in children's learning from social robots *8th Int. Conf. Affect. Comput. Intell. Interact. Work Demos (Cambridge, UK, 2019)* pp 1–5

Tannert A, Grohs R, Popp J and Neugebauer U 2019 Phenotypic antibiotic susceptibility testing of pathogenic bacteria using photonic readout methods: recent achievements and impact *Appl. Microbiol. Biotechnol.* **103** 549–66

Tollestrup A and Todesco E 2008 *Review of Accelerator Science and Technology* ed A Chao and W Chou (Singapore: World Scientific) vol 11 pp 185–210

Toxiri S, Koopman A S, Lazzaroni M, Ortiz J, Power V, de Looze M P, O'Sullivan L and Caldwell D G 2018 Rationale, implementation and evaluation of assistive strategies for an active back-support exoskeleton *Front. Robot. AI* **5** 53

Toxiri S, Naf M B, Lazzaroni M, Fernandez J, Sposito M, Poliero T, Monica L, Anastasi S, Caldwell D G and Ortiz J 2019 Back-support exoskeletons for occupational use: an overview of technological advances and trends *IISE Trans. Occup. Ergon. Hum. Factors* **7** 237–49

Tsagarakis N G, Metta G, Sandini G and Vernon C D G 2007 iCub: the design and realization of an open humanoid platform for cognitive and neuroscience research *Adv. Robot* **21** 1151–75

Tsimring L S, Levine H and Kessler D A 1996 RNA virus evolution via a fitness-space model *Phys. Rev. Lett.* **76** 4440–3

Vanderborght B *et al* 2013 Variable impedance actuators: a review *Rob. Autom. Syst.* **61** 1601–14

Very High Energy 2020 Electron Radiotherapy Workshop (VHEE'2020) (https://indico.cern.ch/event/939012/)

Volz E *et al* 2021 Assessing transmissibility of SARS-CoV-2 lineage B.1.1.7 in England *Nature* 1–17

Wang S *et al* 2015 Manipulating the selection forces during affinity maturation to generate cross-reactive HIV antibodies *Cell* **160** 785–97

Wang X V and Wang L 2021 A literature survey of the robotic technologies during the COVID-19 pandemic *J. Manuf. Syst.* **60** 823–36

Wilson M N 1983 *Superconducting Magnets* (Oxford: Oxford University Press)

Wilson M N 1999 Superconductivity and accelerators: the good companions *IEEE Trans. Appl. Supercond.* **9** 111–21

World Health Organization (WHO) 2008 10 facts on health workforce crisis(http://who.int/features/factfiles/health_workforce/en/) Accessed 6 Aug 2020

Wu J T, Leung K and Leung G M 2020 Nowcasting and forecasting the potential domestic and international spread of the 2019-nCoV outbreak originating in Wuhan, China: a modelling study *Lancet* **395** 689–97

Wu Y H, Wrobel J, Cornuet M, Kerhervé H, Damnée S and Rigaud A S 2014 Acceptance of an assistive robot in older adults: a mixed-method study of human–robot interaction over a 1-month period in the Living Lab setting *Clin. Interv. Aging.* **8** 801–11

Yang G Z, Bellingham J and Dupont P *et al* 2018 The grand challenges of science robotics *Sci. Robot.* **3**

Zemmar A, Lozano A M and Nelson B J 2020 The rise of robots in surgical environments during COVID-19 *Nat. Mach. Intell.* **2** 566–72

Zhang S-Q *et al* 2018 High-throughput determination of the antigen specificities of T cell receptors in single cells *Nat. Biotechnol.* **36** 1156–9

IOP Publishing

EPS Grand Challenges
Physics for Society in the Horizon 2050

Mairi Sakellariadou, Claudia-Elisabeth Wulz, Kees van Der Beek, Felix Ritort, Bart van Tiggelen, Ralph Assmann, Giulio Cerullo, Luisa Cifarelli, Carlos Hidalgo, Felicia Barbato, Christian Beck, Christophe Rossel and Luc van Dyck

Chapter 6

Physics for the environment and sustainable development

Ankit Agarwal, Philippe Azais, Luisa Cifarelli, Jacob de Boer, Deniz Eroglu, Gérard Gebel, Carlos Hidalgo, Didier Jamet, Juergen Kurths, Florence Lefebvre-Joud, Søren Linderoth, Alberto Loarte, Norbert Marwan, Natalio Mingo, Ugur Ozturk, Simon Perraud, Robert Pitz-Paal, Stefaan Poedts, Thierry Priem, Bernd Rech, Marco Ripani, Shubham Sharma, Tuan Quoc Tran and Hermann-Josef Wagner

6.1 Introduction

Luisa Cifarelli[1] and Carlos Hidalgo[2]
[1]University of Bologna, Bologna, Italy
[2]CIEMAT, Laboratorio Nacional de Fusión, Madrid, Spain

One of the most crucial and challenging developments of recent decades has been the discovery that the environment is fragile. This discovery shows that we cannot afford to delay the implementation of actions to tackle climate change if the long-term objective is to limit the increase in temperature of the planet at an affordable cost. Although the effects of climate change on the environment are too complex to admit simple solutions, recent developments illustrate how basic science can be pulled together successfully with social awareness and political action to avert an environmental tragedy.

This section presents work done in a wide range of research areas, illustrating how humanity has the responsibility to preserve our delicate planet but also the power to affect its environment. The chapters highlight the strength of fully interdisciplinary effort among physicists, mathematicians, and chemists as well as multilateral science to address global challenges that affect societies at their core.

© 2024 The Authors.

doi:10.1088/978-0-7503-6342-6ch6   6-1   Published under license by IOP Publishing Ltd

A thorough understanding of the Earth's system is essential for the life quality of modern society. It is important to be able to define the conditions for sustainable development of humanity in order to maintain the Earth's system within habitable limits, predict critical transitions and events in the Earth's dynamics, and effectively mitigate and adapt to changes and events related to the Earth's system to prevent the disastrous consequences of natural hazards. Section 6.2 deals with Earth system analysis from a nonlinear physics perspective. It describes key concepts from nonlinear physics and shows that they enable us to treat challenging problems of Earth sciences.

Energy is the lifeblood of today's society and one of the factors that has decisively contributed to improving humanity's quality of life. Section 6.3 deals with the description of physics fields with relevance for energy technologies. It addresses the further development of energy sources, such as solar, wind, nuclear fission energy, and storing energy storage systems as well as the quest to develop nuclear fusion, since the dominance of fossil fuels must decline. It addresses the potential challenges and opportunities in the development of global energy systems, emphasising how deeply interconnected the energy and climate debates are.

The invention of the internal combustion engine radically transformed industrial and personal transport and, consequently, our social organization system. Section 6.4 deals with transport electrification for green cities. It addresses research and development to deploy technologies that enhance the performance of electric drive vehicles.

Hazardous wastes and materials are diverse, with compositions and properties that vary significantly between industries and related energy sources. Section 6.5 deals with environmental safety from a chemical perspective to address how environmental emissions and waste disposal can be managed to meet sustainable development criteria.

Finally, space weather describes the way in which the Sun, through emergence of magnetic field into its atmosphere, flares, coronal mass emissions, high-energy particles, and subsequently induced space conditions, affects human activity and technology both in space and on the ground. Section 6.6 invites us to understand and predict space weather.

6.2 Earth system analysis from a nonlinear physics perspective

Juergen Kurths[1], Ankit Agarwal[1,2], Ugur Ozturk[3,4], Shubham Sharma[3], Norbert Marwan[1,5] and Deniz Eroglu[6]

[1]PIK—Potsdam Institute for Climate Impact Research, Member of the Leibniz Association, 14473 Potsdam, Germany

[2]Department of Hydrology, Indian Institute of Technology Roorkee, 247667 Roorkee, India

[3]Helmholtz Centre Potsdam–GFZ German Research Centre for Geosciences, 14473 Potsdam, Germany

[4]Institute of Environmental Science and Geography, University of Potsdam, 14476 Potsdam, Germany

[5]Institute of Geosciences, University of Potsdam, 14476 Potsdam, Germany

[6]Faculty of Engineering and Natural Sciences, Kadir Has University, 34083 Istanbul, Turkey

A reliable understanding of the Earth's system is essential for a good quality of life for modern society. Natural hazards are the cause of most life and resource losses. The ability to define the conditions for sustainable development of humankind, to keep the Earth's system within the boundaries of habitable states, and to predict critical transitions and events in the dynamics of the Earth's system are crucial to mitigate and adapt to Earth system–related events and changes (e.g., volcanic eruptions, earthquakes, and climate change) and to avert the disastrous consequences of natural hazards. In this chapter, we discuss key concepts from nonlinear physics and show that they enable us to treat challenging problems of Earth sciences which cannot be solved by classic methods. In particular, the concepts of multiscaling, recurrence, synchronization, and complex networks have become crucial in recent decades for a substantially more profound understanding of the dynamics of earthquakes, landslides, and (paleo)climate. They can even provide a significantly improved prediction of several high-impact extreme events. Additionally, crucial open challenges in the realm of methodological nature and applications to Earth sciences are given.

6.2.1 Introduction

The invention of thermoscopes and barometers in the early 17th century enabled the study of physical parameters of climate variables, such as precipitation, temperature, and pressure. Exploring the Earth's system in detached disciplinary practices became convenient with these early instruments for limited geographic locations. The disciplinary assessment of individual Earth system components continues to help in understanding fundamental mechanisms. They have been regarded as autonomous systems in their own right and further broken down into more specialized subsystems. One standard topic is to study, for instance, precipitation concerning more prominent atmospheric modes [1]. Until recent decades, this traditional practice of studying the four major spheres of the Earth's system, that is, the atmosphere, hydrosphere, biosphere, and geosphere, continued independently. However, the Earth behaves as an integrated complex system with nonlinear interactions and feedback loops between and within them [2]. For example, the influence of significant volcanic eruptions on climate oscillations proves a vital link between the geosphere and the atmosphere [3]. The increasing availability of data and the rising concerns related to shifts in the global climate system, concomitant extremes, and natural hazards have urged the development of a more *holistic understanding of the Earth's system* in recent decades (figure 6.1).

Furthermore, it was assumed that various Earth processes are *scale-invariant*, that is, we can expect a phenomenon to occur in several scales when we observe its occurrence on only one scale [4]. Indeed, the scale invariance theory was applied to many fields, such as the frequency–size distributions of rock fragments, faults, earthquakes, volcanic eruptions, landslides, and oil fields, but not all on the Earth's systems. Nevertheless, even the lack of scale invariance means that information is stored and perceived differently at different scales, resulting from mutual interactions of intertwined subcomponents interacting over a wide range of scales. Generally, a deep understanding of these multicomponent interactions between the different subsystems of the Earth's system, including human activities, requires an interdisciplinary approach in which concepts from various fields of physics and complex systems science are vital elements [5].

Trying to understand interacting Earth systems as a giant complex system using only instrumental records is insufficient, since such measures cover only a very

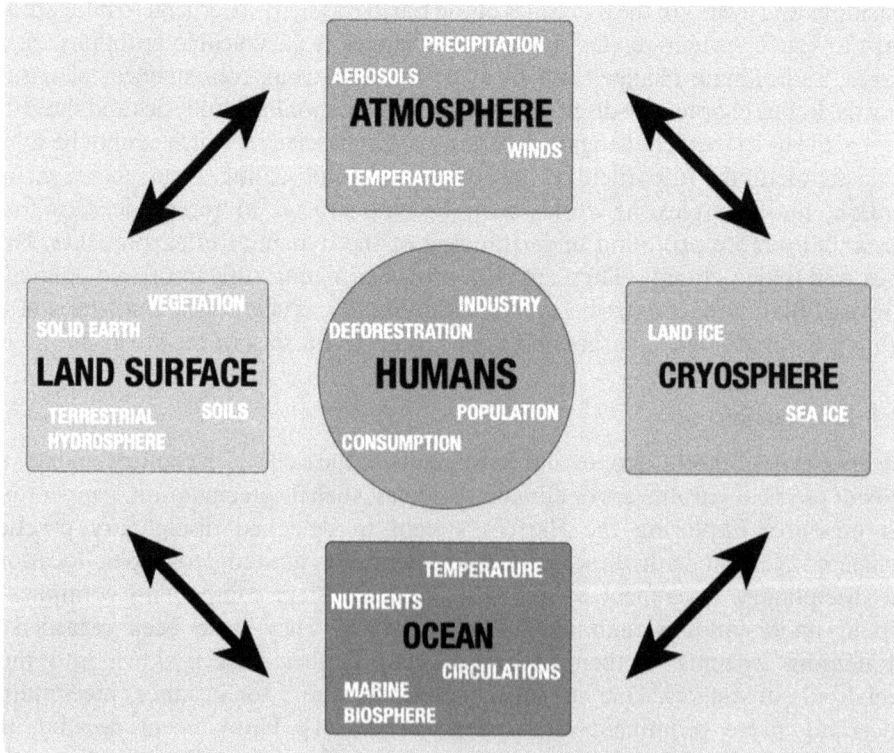

Figure 6.1. Scheme of rich connections within main components of the Earth's system.

narrow window of the planet's history. The Earth is continuously experiencing natural events such as geological and tectonic processes, climate change, and biological and chemical activities. Although the instruments to record such events were not available before the 17th century, various natural and complex formations, such as stalagmites, marine and lake sediments, and trees, have recorded such events in their structures as proxy records. Investigating these archives to reveal the hidden preceding events helps us understand the dynamics and predict the oncoming behavior of the associated natural events on the Earth. For this purpose, paleo-climatology, a field of climate science that works to understand ancient climates without direct measurement, has reached sufficient maturity to reveal significant climate periods, such as glaciations and abrupt global temperature rises, by dating and analyzing the proxies [6].

Whereas paleoclimate variations as derived from the geoscientific archives are only estimates and contain a degree of uncertainty, the significant climate periods of the driving processes such as the Milankovich cycles can be determined with high accuracy because the equations of motion for the dynamics of the Earth's orbit in space can be solved with a reasonable approximation using Hamiltonian mechanics. However, the celestial sign of objects in the solar system is, in general, a many-body system in which the planets' gravitational fields mutually influence their orbits around the Sun. Solving such a many-body problem (and even that of a three-body system) is not simple and was at the forefront of science for a long time [7]. In this spirit and in honor of the 60th birthday of the King of Sweden, Oscar II, in 1887, a prize for solving the many-body problem was announced. The French mathematician Henry Poincaré finally won this prize with his seminal work on the three-body system and discovering the chaotic nature of the planets' orbits [8]. In this work, he proved an important theorem that affects the recurring orbits of the interacting objects in a celestial system and is also a fundamental property of many complex dynamical systems: the now well-known recurrence theorem, which states that a (conservative) system recurs infinitely many times as closely as one wishes to its initial state. The property of recurrence is not only of fundamental importance in the study of dynamic systems; it is also a fundamental principle in Earth sciences at all temporal and spatial scales.

6.2.2 Nonlinear concepts

The vigorous progress in exploring nonlinear dynamics in the 1980s and 1990s opened new doors for a more appropriate analysis of complex nonlinear systems, such as lasers, the human brain, power grids, and the Earth's system [9]. Techniques for estimating fundamental characteristics of nonlinear systems, such as fractal dimension, Lyapunov exponents, Kolmogorov entropy, and Hurst exponents, were developed and applied to various disciplines [10]. However, these methods are mainly helpful in low-dimensional processes and are not appropriate for under-standing the Earth's system from data.

Other essential concepts, such as recurrence plots [11, 12], synchronization [13], wavelets [14], and complex networks [15], have been developed to explore dynamical and structural properties in high-dimensional spatiotemporal systems. They have been proven to be very promising even for the study of the Earth's system. In the following subsections, we describe basic nonlinear concepts and present some paradigmatic applications in Earth sciences.

6.2.2.1 Multiscaling

Various Earth processes are assumed to be *scale-invariant* [4]. An essential law is the size distribution of natural events, meaning that prominent events are less frequent when compared to smaller ones. Deriving an adequate size distribution of natural events would take into account the rarity and likelihood of a specific event. Hence, one major challenge of studying the occurrence, frequency, and intensity of climate-driven natural extremes and natural hazards is determining these events' spatial and temporal scaling to derive adequate risk estimates. One way to analyze the scaling of natural hazards is to use the frequency–size distribution $p(x)$ (x stands, e.g., for landslide area). For instance, $p(x)$ of landslides follows a power law probability density function in an area with arbitrary dimensions independently of their source mechanism (e.g., earthquake- or rainfall-induced):

$$p(x) = (\alpha - 1)x_{\min}^{\alpha-1}x^{-\alpha} \qquad (6.1)$$

with α the power exponent, valid for $x \geqslant x_{\min}$ [16].

Similarly, the famous Gutenberg–Richter power law [17, 18] scales the seismic activity to assess earthquake hazards for different events magnitudes m. It states that earthquake magnitudes m are distributed exponentially as

$$\log N_{m \geqslant M} = a - bM \qquad (6.2)$$

where $N_{m \geqslant M}$ is the number of earthquakes with magnitude $m \geqslant M$, a is a constant, and b is the scaling parameter. The scaling parameter b determines the relative frequency of small and large earthquakes. The estimation of b is around 1.0, with deviations up to 30% in seismically active regions [19]. A real example of this particular case is presented in section 6.2.3.1.

Information about the Earth's system's processes can be stored and perceived differently at multiple scales. The information observed at one scale often cannot be directly used as information at another. Scaling approaches address the changes at the measurement scale and play an essential role in Earth sciences by providing information at the scale of interest.

Determining scaling properties of geophysical variables provides an alternating way to obtain information about the associated process. The processes with similar statistical properties at different scales are said to be self-similar, which can be described mathematically as [20]

$$\phi(x) = \lambda^{-\beta}\phi(\lambda x) \qquad (6.3)$$

where x is the finer spatial resolution (scale), β is the scaling exponent, λ is the ratio of the large resolution, λx to the small resolution x, and ϕ is the geophysical property or variable of interest. A field is said to be spatially scaling with respect to the moment, q, if the following relationship holds [21]:

$$E\left[\left(\phi_{\lambda}\right)^{q}\right] \propto \lambda^{K(q)} E\left[\left(\phi_{1}\right)^{q}\right] \tag{6.4}$$

where $K(q)$ is the scaling exponent associated with the moment of order q. If the exponent $K(q)$ is linear with regard to q, the process has simple scaling. On the other hand, if the scaling exponents, or slopes, are a nonlinear function of q, then the process is said to be *multiscaling*. This concept of scaling and multiscaling has been used widely in many scientific fields, including hydrology and ecology. For instance, wavelet analysis can decompose high-resolution nonstationary spatial information into nonstationary fields of increasingly coarse spatial scales [22]. The wavelet and the corresponding scaling function are a function to decompose spatial information into directional components explained by the wavelet coefficients.

6.2.2.2 Recurrence analysis
The seminal work of Poincaré in 1890 [8] played a central role in the qualitative theory for nonlinear dynamics (see section 2.1). Poincaré presented a method that provides a local and global analysis of nonlinear dynamical systems by the Poincaré recurrence theorem and stability theory for fixed points and periodic orbits. This theoretical finding is compellingly confirmed by the real world, where recurrences can be observed in our daily life and across all scientific disciplines. Therefore, the investigation of recurrences has attracted much attention, and several approaches have been developed for this purpose.

Among the various methods for studying recurring processes, power spectrum analysis is one of the best known and most widely used techniques for identifying periodicities in time series [23]. Wavelet analysis reveals similar information, additionally providing the change of the detected periods over time (see section 2.1). Coming from the theory of dynamical systems and based on Poincaré's recurrence theorem, the *recurrence plot* (RP) is another fundamental approach that can be used to investigate recurring features in time series and even in spatial data [11, 12]. In a given m-dimensional phase space, two neighboring points are called recurrent if the distance between their state vectors is closer than the threshold ε. Formally, for a given trajectory x_i ($i = 1,..., N, x \in \mathbb{R}^{m}$), the recurrence matrix R is defined as

$$R_{i,j} = \begin{cases} 1, \text{ if } \left\|x_i - x_j\right\| \leqslant \varepsilon \\ 0, \text{ otherwise} \end{cases} \tag{6.5}$$

where $\|\cdot\|$ is a norm of the adopted phase space. The graphical representation of the recurrence matrix R is the RP (figure 6.2). RP of different dynamical behavior represents different particular features (figure 6.2). Such differences can be quantified with the measures of recurrence quantification analysis, such as determinism (the fraction of recurrence points that form diagonal lines in the recurrence plot), laminarity (the fraction of recurrence points that form vertical lines), and recurrence rate (the percentage of recurrence points in a recurrence plot). These measures are used to find changes in the dynamics of a process (e.g., in climate), to classify the dynamics (e.g., random, chaotic, regular) [12], or to identify interrelationships and coupling directions in coupled systems [24].

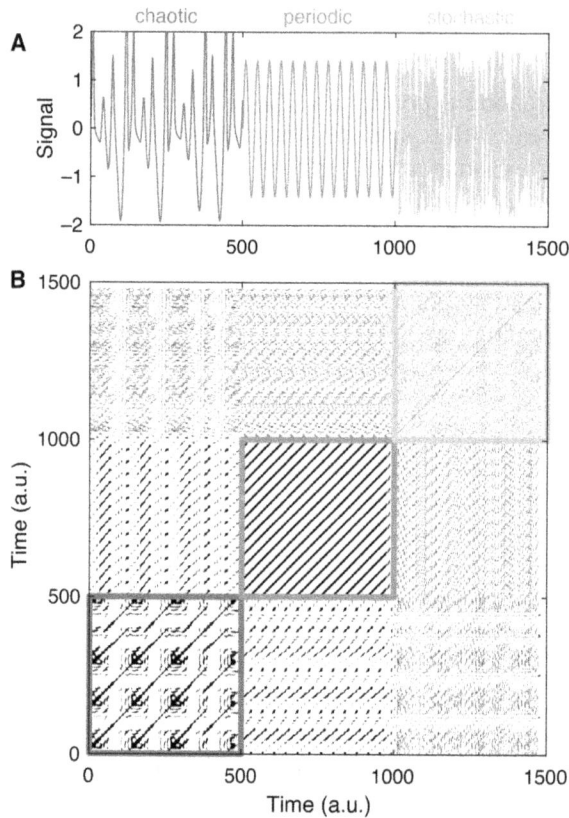

Figure 6.2. (A) Time series representing switching between different dynamical regimes, from chaotic via periodic to stochastic, each lasting 500 time steps. (B) A recurrence plot (RP) represents the recurrence of a state at a given point in time (x-axis) at another point in time (y-axis). Different dynamics cause typical recurrence patterns, which can be used to detect these changing dynamical behaviors. Continuous long diagonal lines in the RP indicate the periodic window, shorter diagonals show the chaos, and single points appear in the stochastic part.

6.2.2.3 Complex networks and event synchronization

Essential challenges in climatology are quantifying the spatial extent of climate extremes and early forecasting procedures of their dynamical behavior. Such forecasting relies predominantly on numerical models which solve physics-based coupled systems of partial differential equations. Starting with Richardson in the 1920s, it has been a long way to the first successful prediction in 1950 and eventually to today's highly sophisticated general circulation and Earth system models. Despite multiple efforts using these methods, their predictive power, especially for extreme events, can be rather limited. A primary reason for this is that in particular long-range interactions, called *teleconnections*, and their interaction with more regional interactions may not be well represented or may even be absent in such models.

Therefore, a quite different approach has been suggested: a network-based presentation of climate phenomena called *climate networks*. The main idea is to get additional information by capturing the evolving interactions of different locations, regarded as nodes, through similarity measures, such as the Pearson correlation, mutual information, or the Granger causality, from spatiotemporal observational data. An important description of such similarity of strong events is the *event synchronization* approach [25], inspired by Christiaan Huygens' detection of synchronization in the 17th century. Here, we consider the occurrence of extreme events, such as rainfall, in a synchronized manner at different locations, even faraway ones.

The final complex network is then represented by an adjacency matrix A, which encodes the links between the nodes i and j as follows:

$$A_{i,j} = \begin{cases} \text{nonzero,} & \text{if variability at node } j \text{ is similar (or synchronized) to node } i \\ 0, & \text{otherwise} \end{cases} \tag{6.6}$$

The value of the elements of A represents the weight of the link obtained from quantifying similarity (figure 6.3).

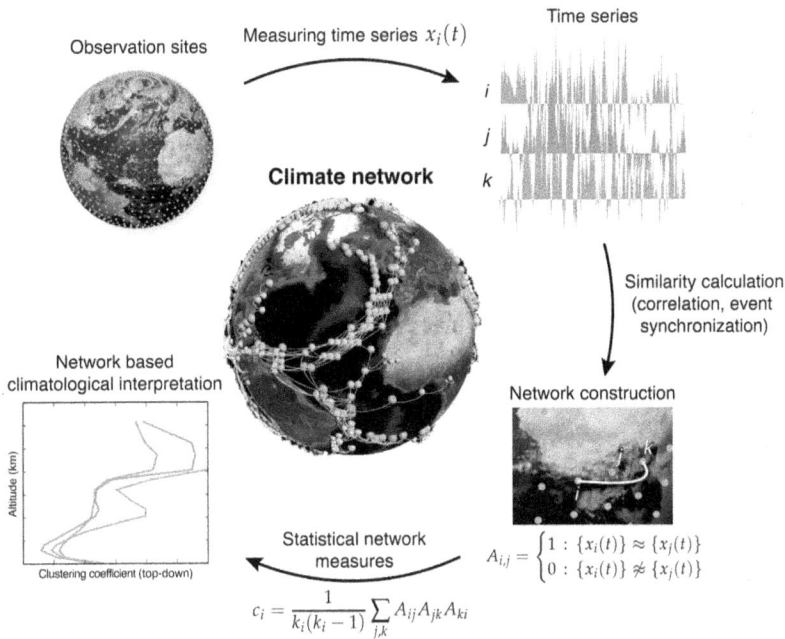

Figure 6.3. The climate network framework as a tool for prediction. Observational data of physical quantities, such as temperatures, are available at different geographical locations. These data can be used directly or via a reanalysis (numerical weather model) which assimilates and maps them onto a regular grid. Thus, a time series of the regarded physical quantity is available for each climate network node (observational site or reanalysis grid point). Cooperativity between nodes can be detected from the similarity in the evolution of these time series and translated into links connecting the corresponding nodes. The links or their strengths may change with time. These nodes and their links constitute the evolving climate network represented by the adjacency (connectivity) matrix A (equation (6.6)). The analysis of this network can enable early predictions of climate phenomena and provide insights into the physical processes of the Earth's system.

There are various generalizations of this construction, particularly to emphasize multilayer networks, which enable variables from different subsystems.

The reconstructed adjacency matrix A allows us to calculate standard network measures such as degrees, clustering coefficients, or betweenness and to identify teleconnections. It has been shown recently that climate networks provide ideal tools for exploring even large amounts of climate data to uncover spatiotemporal patterns, leading to new physical insights into the climate system [1]. Moreover, they have a strong predictive potential; that is, they enable the development of new forecasting methods. Examples of up-and-coming applications are given in sections 3.3–3.5.

6.2.3 Applications of nonlinear dynamics in the Earth's system

Vigorous progress in nonlinear science contributed to detecting, attributing, and understanding the Earth's system, reducing uncertainties, and projecting future climate changes. In this section, we discuss some significant contributions of nonlinear physics in Earth system sciences.

6.2.3.1 Earthquakes and the Gutenberg–Richter law

A proper fitting of the power law is essential to study most natural hazards, particularly earthquakes (equation (6.1)). The Gutenberg–Richter law (equation (6.2)) represents scaling in earthquakes, as power law distribution makes it scale-invariant. An example of the scale parameter b for central California for 20 years (2001–20) is illustrated in figure 6.4. The California region has a b of 1.0, which is as per the global average, meaning that central California has the same relative

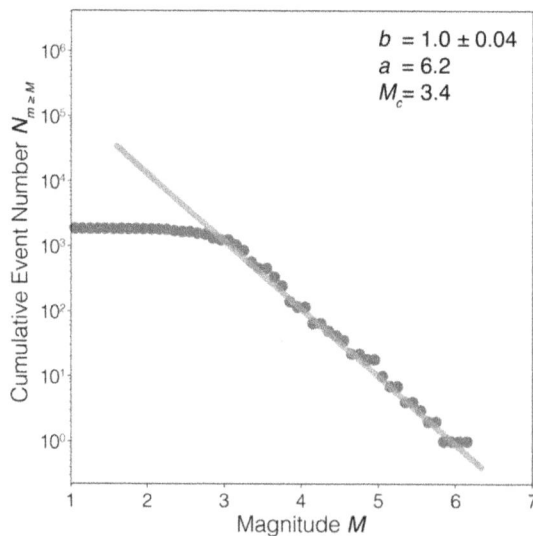

Figure 6.4. Frequency–magnitude distribution for earthquakes in central California between 2001 and 2020. The red line shows a fit to the cumulative frequency and has a slope (b-value) of 1.0. The magnitude cutoff, $M_c = 3.4$, is used for estimating the scaling parameter b.

frequency of small and large earthquakes. However, the magnitude threshold parameter M_c must be selectively applied above crossover magnitude for larger earthquakes with significant seismic moments [26]. The Gutenberg–Richter law accurately describes the shallow seismicity. However, it is not the only scaling law for all levels of earthquake events; the distribution of deeper earthquakes was observed to follow a bimodal (multiscaling) pattern [27].

It is also crucial to accurately estimate scaling parameter b (equation (6.2)) from the earthquake events to characterize the seismicity activity (see section 2.1) sensitively. There is an inverse correlation between b and the differential stress, which was revolutionary in that b can act as an indicator of stress accumulated around the fault volume [28]. This observation was used in the study done before and after the vast 2011 Tohoku-Oki earthquake with a high slip area, where an increase in b is observed as a large amount of stress was released [29]. Another use for this observation is studying the structural anomalies in the crust and identifying the volumes of magma in an active volcano. A study performed at two active volcanoes [30], Mt. St. Helens and Mt. Spurr, shows a relatively high b ($\geqslant 1.3$) due to the presence of material heterogeneity and high thermal gradient. This high b is why these volcanoes are less likely to host large earthquakes but frequent small ones. A typical intraplate b is around 0.8, making intraplate regions prone to large earthquakes over a short recurrence time. However, the scaling parameter b is not the perfect parameter to measure seismicity at all magnitude scales. The tail of the $\log(N_m > M)$ versus M relation holds for only a certain range of magnitudes. A nonlinear fit is a better approximation for smaller ($M_c \leqslant 3.4$) and larger ($M_f \gtrsim 7$) magnitudes. A reason for the deviation from the power law for earthquakes smaller than $M_c \leqslant 3.4$ (figure 6.4) is the incompleteness of catalogs. For large earthquakes, a reason is the saturation of the magnitude scale and the long recurrence time; they are missing from the catalogs because they are often too short.

A high scaling parameter b indicates a lower chance of observing significant seismicity while the frequency of small earthquakes is high. However, smaller-magnitude events are observed much less often than indicated by b due to insufficient seismic network coverage.

6.2.3.2 Recurrence plot application

Recurrence is a fundamental principle in Earth sciences at all temporal and spatial scales, from the key principle of the doctrine of uniformity, over the rock cycle, glaciation cycles, and active geysers, to alternating sediment layers, to mention only a few. One crucial phenomenon with complex recurrence patterns is climate. One of the primary drivers of climate is solar insolation, modulated by mutual variations of the Earth's orbit around the Sun and the tilt of the Earth's axis, which are responsible for seasons, changes in global temperature, and glaciations. This influence was discovered in the first half of the 20th century by investigating annually layered lake sediments [31] and considering the Earth's orbital parameters [32].

Recurrence plots (see section 2.2) provide a powerful framework to study the dynamics of the climate by their recurrence properties. As an application, the dynamics of the Cenozoic climate will be investigated by recurrences properties in a selected paleoclimate proxy record. Such studies are essential to advance our understanding of the past and will help to improve climate models to better forecast future climate change and its impacts, as well as increasing our understanding of climate dynamics.

Calcareous lake sediments, speleothems, and benthic foraminifera store environmental conditions by changing their geochemical and petrographic composition. The study of stable isotopes is an active field to derive past environmental and climatic conditions. For example, the temperature-dependent fractionation of oxygen isotopes is the key to reconstructing global seawater temperatures and ocean circulation by using planktonic and benthic foraminifera. Ongoing deep ocean drilling programs and novel quantitative methods such as clumped isotope thermometry provide new insights with improved quantification, increasing temporal resolution, and ever-smaller time uncertainties. The recently developed temperature reference curve for the Cenozoic [33] is an example with a temporal resolution of up to 2000 years and covering 66 million years. This period is crucial because it provides an analog of future greenhouse climate and how (and which) regime shifts in large-scale atmospheric and ocean circulation can be expected in a warming world [34]. The outstanding high resolution of this record allows study and comparison of recurrence properties of selected time intervals. The recurrence plot indicates the different climate regimes of hothouse, warmhouse, coolhouse, and icehouse by their very distinct recurrence pattern (figure 6.5). During the Miocene (18–14 Ma ago), the climate was in a warmer state more similar to the warmhouse than the coolhouse, visible by some recurrences linking this period to the late Eocene. The fine-scale pattern of the recurrence plot reveals more details, such as the change from the 41-ka cycles to 100-ka Milankovich cycles of glaciation during the mid-Pleistocene transition.

Recurrence analysis of climate time series indicates different dynamical regimes, such as chaotic or predictable dynamics, thus enabling detection of critical transitions between different climate periods.

6.2.3.3 Extreme rainfall teleconnections and monsoon prediction

The Indian summer monsoon is an intense rainy season lasting from June to October. The monsoon delivers more than 70% of the country's annual rainfall, which is India's primary source of freshwater. Although the rainy season happens every year, the monsoon onset and withdrawal dates vary within a month from year to year. Such variability strongly affects the life and property of more than a billion people in India, especially those living in rural areas and working in the agricultural sector, which employs 70% of the entire population. So far, only Kerala in South India receives an official monsoon forecast two weeks in advance, while the other 28 states rely on the operational weather forecast of about five days [35]. A much better forecast has been recently reached by combining two nonlinear concepts: complex climate networks and a tipping element approach.

In the first step, from rainfall data from the Asian Precipitation Highly Resolved Observational Data Integration Towards the Evaluation of Water Resources

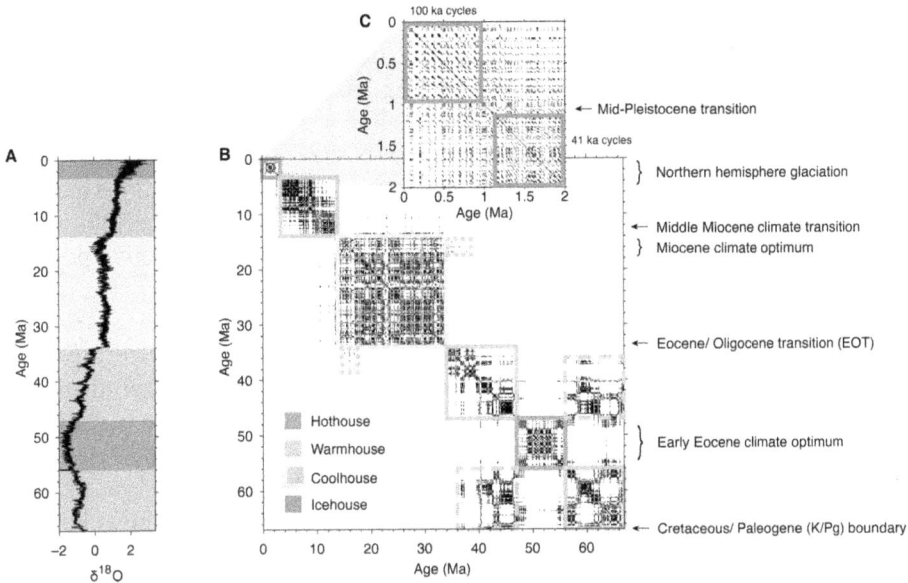

Figure 6.5. RP of a paleoclimate time series. (A) Paleoclimate variation indicated by oxygen isotope measurements from marine sediments (CENOGRID). Lower values correspond to a warmer global climate. (B) The RP indicates the different climate regimes of hothouse, warmhouse, coolhouse, and icehouse by their very distinct recurrence pattern. During the Miocene (18–14 Ma ago), the climate was in a warmer state more similar to the warmhouse than the coolhouse, visible by some recurrences linking this period to the late Eocene (marked by the dashed box). (C) The fine-scale pattern of the RP reveals more details, such as the change from the 41-ka cycles to 100-ka cycles of glaciation during the mid-Pleistocene transition.

(APHRODITE) and the high-resolution satellite product Tropical Rainfall Measurement Mission (TRMM) 3B42 dataset, complex networks were retrieved via the event synchronization technique (see section 2.3). This exploratory network-based analysis of extreme rainfall across the Indian subcontinent enabled for the first time the identification of critical geographical domains displaying far-reaching links, influencing distant grid points [36]. In particular, North Pakistan and the Eastern Ghats turn out to be crucial for the transport of precipitation across the subcontinent.

In the second step, a tipping elements approach of the measured daily mean air temperature and the relative humidity at these two sensitive regions allowed us to uncover the critical nature of the spatiotemporal transition to the monsoon. It was especially found that the temporal evolution of the daily mean air temperature and the relative humidity exhibits critical thresholds on the eve of the monsoon. A highly developed instability occurring in these regions creates the conditions necessary for spatially organized and temporally sustained monsoon rainfall.

Based on this knowledge, a scheme was developed for forecasting the upcoming monsoon onset in the central part of India 40 days in advance, thus considerably improving the time horizon of conventional forecasts. The new scheme not only has proven its worth (73% of onset predictions have been correct) in retrospect (for the years 1951–2015) but also has already been shown to be successful in predicting future monsoons five years in a row since its introduction in 2016. The methodology appears to be robust under climate change and has proven its skill also under the extreme conditions of 2016, 2018, and 2019.

Further successful applications of this network-based concept are El Niño forecasts beyond the spring barrier, predicting droughts in the central Amazon 12–18 months in advance, and forecasting extreme rainfall in the Eastern Central Andes [37].

Thus, a network-based analysis of climate data can provide predictive power for mitigating the global warming crisis and societal challenges.

6.2.3.4 Understanding landslide distributions

As explained in section 6.2.2.1, successfully fitting a global power law distribution (equation (6.1)) to landslides would help us to understand whether we lack information in hazard and risk models. However, the distribution of spatial landslides follows a power law distribution. Just as in the case of the Gutenberg–Richter law, the power exponent is valid to a minimum value (equation (6.1)) [16], and the rollover below the minimum is found in two different forms: (1) the double Pareto distribution and (2) the inverse Gamma distribution according to different studies [38, 39]. Like other universal scaling laws [40], it is expected to have a universal power exponent for the landslide events. However, a lack of data makes studying the problem impossible at a better resolution, especially at the function's tail [38, 39]. Most studies rely primarily on landslide inventories collected after a significant landslide triggering event, such as the 1994 Northridge earthquake (M_W 6.4). Landslides have also been found to exhibit temporal scaling or clustering besides spatial and geometric ones. Although some studies suggest a global power exponent $\alpha = 2.3 \pm 0.6$, the physical process is not known to implement a functional probabilistic multihazard assessment [41].

Besides the power law–based approximation models, ample practice has offered linear solutions to study natural hazards, making a nonlinear application redundant. An example is Newmark's sliding block analysis. It estimates the displacement potential of hillslopes under seismic loading (i.e., acceleration). This hypothetical displacement aims to indicate the likelihood of failure under seismic loading as a function of hillslope inclination and seismic acceleration. For example, landslides related to the 2016 Kumamoto earthquake (M_W 7.1) caused significant damage, especially to infrastructure such as highways (figure 6.6(A)). Although landslide locations correlate well with the seismic waveforms based on a physics-based ground motion model [42], the Newmark's distances highlight particularly elevated gradients in the landscape (figure 6.6(B)).

Rainfall decreases the slope stability by altering cohesion, elevating the landslide susceptibility in most cases. In some other cases, rainfall could also mobilize the superficial surface material leading to the debris flows. However, in contrast to an earthquake, rainfall is not introducing a direct force on the hillslopes to estimate rainfall impact on landslides. Hence, most of the time, statistical methods are applied to forecast rainfall-induced landslides. One standard tool is to use statistically derived rainfall intensity–duration thresholds above which landslides are

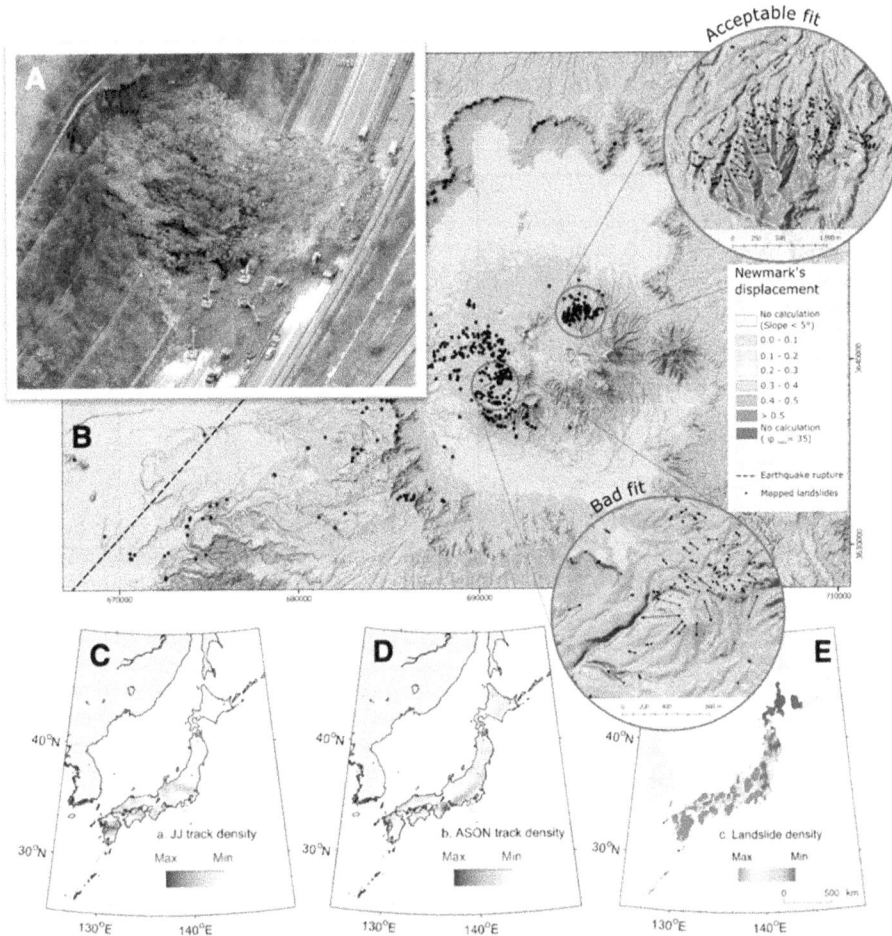

Figure 6.6. (A) Example of a cut slope failure by the Oita Expressway following the 2016 Kumamoto earthquake (M_W 7.1). The photo is taken from Dave Petley's landslide blog (https://blogs.agu.org/land-slideblog/2016/04/18/kumamoto-Earthquake-1/). (B) Newmark's displacement of the 2016 Kumamoto earthquake (M_W 7.1) in Kyushu, Japan (UTM-52). In certain regions, the elevated displacement correlates well with the mapped landslides, while in some others, it is relatively poor. The concentration of extreme precipitation streamlines during (C) June and July (JJ), and (D) August to November (ASON), normalized by cumulative above 95% extreme rainfall for the same period between 1998 and 2015 based on TRMM (Tropical Rainfall Measurement Mission) rainfall estimates. (E) Normalized rainfall-triggered spatial landslide density -weighted by log-transformed landslide volumes calculated from an inventory of 4744 events and smoothed by kernel density estimation onto a 5 × 5 km grid by [44]; white areas have no data.

triggered. The logic behind is that high-intensity rainfall triggers landslides and moderate intensity, but long-duration events would increase the landslide suscept-ibility. Therefore, several spatial classification models are developed to try to relate landslide activity to rainfall distribution.

Another notorious example is that the extreme rainfall flux over a region during tropical storms, as previously explained, might already highlight the landslide-prone regions on large spatial scales. It is possible to estimate or cluster the rainfall motion over large areas, such as countries or continents, by blending event synchronization and complex network methods (see section 2.3). These results can help track landslide activity along the path of extreme rainfall. As an application, the extreme rainfall trajectories over the Japanese archipelago were estimated using event-synchronization [43]. The density of extreme rainfall tracks aligns well with the landslide distribution (figures 6.6(C)–(E)).

The power law distribution can model landslide distributions, and by using nonlinear methods such as event-synchronization, it is possible to describe spatial landslide distributions as a function of rainfall distributions.

6.2.3.5 Multiscale sea surface temperature (SST)

Climatic systems are complex systems composed of multiple feedback loops and interactions. In such systems, the coupling between climate variables takes place at different time and spatial scales. Untangling this multiscale variability and inter-actions of a climatic process is vital, as it would improve the understanding of global climate and its variability. Hence, climate networks are constructed (section 2.3) at different time scales considering each sea surface temperature (SST) grid cell as a node, and edges are created between all pairs of nodes based on statistical relation-ships. First, SST data are decomposed at different time scales using wavelet (section 2.1), and then the Pearson correlation between all pairs of nodes is calculated at a corresponding time scale. Finally, significance-based pruning is applied to retain only highly correlated edges in the network. The network is constructed by applying a 5% link density threshold, which is well accepted for the network construction. Multiple testing was employed to avoid false links.

The network visualization of the original SST data (all scales) reveals short-range and long-range connections between various regions of the Earth. As at a finer scale, there is no significant correlation, and that is expected, since we have removed the annual cycle using anomalies. Interestingly, at 8–16 months, we observe mainly two zones with many significant correlations in the equatorial Pacific and Indian Ocean dipole, which are known to affect each other via the atmosphere (figure 6.7(A)). In the next period of

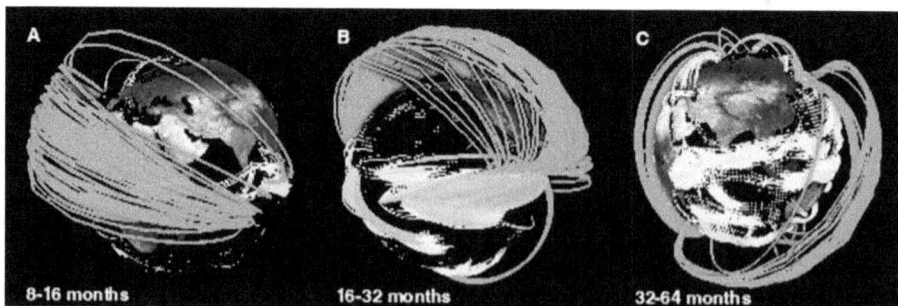

Figure 6.7. Spherical three-dimensional globe representation of the long-range teleconnections at different timescales in a sea surface temperature network. Reproduced from [45] CC BY 4.0. Edge color represents the geographical lengths.

32–64 months, these patterns become more prominent as known ENSO events act on scale up to two years (figure 6.7(B)). There is a link between SST in the Southern Ocean to ENSO events via the Southern annual mode, that is, the north–south movement of the westerly wind belt that circles Antarctica (figure 6.7(B)). The three-dimensional visualization (figure 6.7(C)) shows several links from the North Atlantic to the South Atlantic. This negative correlation likely exhibits the seesaw response due to the transport of heat from the Southern Ocean to the North Atlantic via the Atlantic meridional overturning circulation (AMOC) [46]. If the AMOC is stronger than it was before 2000 (as it has been in the period since the year 2000 compared to the years before [47]), more heat is transported towards the North, which leads to a cooling in the Southern Ocean and warming in the subpolar North Atlantic.

Multiscale analysis of climatic processes helps to uncover the time scales of interaction and feedback in the climate system that may be missed when processes are analyzed at one timescale only.

6.2.4 Outlook

We have shown that basic concepts of nonlinear physics and complex systems science have a strong potential for treating important problems in Earth systems sciences. We have argued that they complement established concepts with new possibilities to reveal entire causal chains of complex phenomena in the Earth's system, primarily to reveal new precursor processes of extreme events.

However, it is essential to emphasize that these interdisciplinary approaches are in their infancy and the subject of ongoing research. There are various open challenges in the realm of methodological nature and applications. Some of them are as follows:

- There is a growing recognition in the scientific community and more broadly that the Earth's functions have to be regarded as an interconnected complex system with properties and behaviors characteristics of the system as a whole. These include tipping points, critical thresholds, 'switch' or 'control' points, strong nonlinearities, teleconnections, chaotic elements, and uncertainties of different origins. Understanding the components of the Earth's system is important; however, that is insufficient for understanding the functioning of the Earth's system as a whole. Humans are now a significant force in the Earth's system, altering key process rates and absorbing global environ-mental changes. Human activities' environmental significance is so profound that the current geological era is called the Anthropocene [48]. Therefore, there is a strong need to develop a complex global model involving Earth system dynamics, human activities, and environmental boundaries to system-atically study the planetary boundaries and tipping points and uncover fundamental principles.
- An important task is to improve our capabilities regarding data-driven inference of governing principles to reach a deeper understanding of the connection between the microscopic dynamics of the constituents of the Earth's system and their nonlinear interactions on the one hand and the dynamics emerging from these interactions at the macroscopic level on the other hand.

- Combining traditional physics-based modeling and statistical approaches with state-of-the-art machine learning (ML) techniques is necessary to efficiently include the huge amount of available data in a model. However, we would like to emphasize that neither an ML-only nor a scientific knowledge-only approach is sufficient for complex Earth system applications. Hence, we must explore the continuum between mechanistic and ML models, where both scientific knowledge and data are integrated synergistically [49, 50]. This approach has picked up momentum just in the last few years [50] and is being pursued in Earth system science [51], climate science [52], and hydrology [53].
- A key driver of further advances is the desire to improve predictions of the behavior of complex systems and especially—for example, in the context of the ongoing global warming driven by the anthropogenic release of greenhouse gases—of the response of complex systems to time-varying external forcing.
- The study of surface processes with nonlinear tools is still not common. Combining nonlinear approaches with linear methods could advance the existing forecasting schemes, especially predicting extreme events. The European floods in summer 2021, which claimed 184 lives in Germany alone [54], are a matchless example that emphasizes that more effort has to be placed to forecast extreme incidents to prevent life loss.
- Climate-driven hazards are rarely the output of a single system. Many of those are in the form of hazard cascades. For example, extreme rainfall initiates a flash flood, high waters lead to carving the river banks and trigger landslides, and dislocated loose landslide mass mixes with the high waters and is transported downstream as a debris flow. However, most of the research that links urban interaction with climate-driven natural hazards consists of empirical studies. Only recently, the first numerical model (CHASM) has been able to describe the informal housing-related changes in the topography and link it to the occurrence rates of landslides [55]. Models such as CHASM could connect different earth systems. Blending such physics-based models and a nonlinear causation metric such as event synchronization as a preceding step could enhance our capacity to forecast extreme rainfall-driven natural hazard cascades, such as flash floods and landslides.
- The simultaneous occurrence of two or more natural extremes affects society much more strongly than their univariate counterparts do [56]. For instance, a hazard resulting from a summer that is both dry and hot is higher than that resulting from a univariate drought extreme, given that the hot, dry summer has a severe impact, such as a reduction in agricultural productivity, irretrievable loss to property and health, and damage to natural ecosystems and public infrastructure. These manifold extremes are called compound extremes or compound events [57]. The investigation of compound extremes has received less attention so far; nevertheless, it has recently gained significant momentum across the globe [56, 58–60].
- Overall transient central components of Earth systems, such as temperature and rainfall, and their effects on other processes, such as concomitant natural hazards of droughts or landslides, should be emphasized and studied using more recent comprehensive data.

6.3 Physics fields with relevance for energy technologies

6.3.1 Solar energy

Robert Pitz-Paal[1] and Bernd Rech[2]
[1]German Aerospace Center (DLR), Cologne, Germany
[2]Institut für Silizium-Photovoltaik, Berlin, Germany

6.3.1.1 The solar resource and the global potential of solar energy technologies

Solar energy is the most abundant resource of energy in the world. With approximately 23 000 TW per year reaching the world's surface, solar energy exceeds the current world energy use by approximately a factor of 3800. However, its average energy density is low (≈ 160 W m^{-2}) compared to that of other energy sources. Less than 1% of solar energy is converted to other renewable energy sources such as wind, biomass, or hydro power. Fossil and nuclear reserves together account for less than 10% of the yearly solar resource. Based on these numbers it is not surprising that sustainable energy scenarios consider solar energy to become one of or even the major energy source [61].

However, two major challenges need to be overcome: First, the low energy density requires a cost-efficient technology for the collection. The second challenge is the inhomogeneous distribution of solar radiation in space and time (see figure 6.8), which requires cost-efficient energy transport and storage technology.

Two different technical concepts have been developed successfully over the last five decades that will be discussed in this section.

Photovoltaics (details in box 6.1) converts high-energy photons into an electric current, taking advantage of Einstein's photoelectric effect in a semiconductor. Cost efficiency is achieved by mass-produced semiconductor devices that are electrically connected to collect energy over large surface areas. Forms of electrical energy storage such as batteries are required to provide electricity on rather short-term

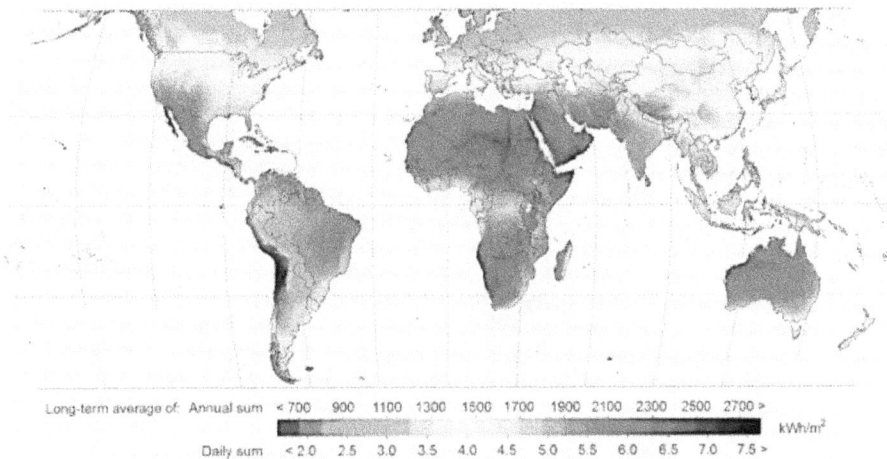

Long-term average of: Annual sum < 700   900   1100   1300   1500   1700   1900   2100   2300   2500   2700 >

kWh/m^2

Daily sum < 2.0   2.5   3.0   3.5   4.0   4.5   5.0   5.5   6.0   6.5   7.0   7.5 >

Figure 6.8. Global map of annual solar horizontal radiation [62]

demand, while the production of hydrogen via electrolysis opens the path towards long-term storage.

Box 6.1 Photovoltaics.

Solar cells directly convert light into electricity. The thermodynamic efficiency of the conversion process is determined by the surface temperature of our Sun with approximately 5800 K enabling an efficiency potential above 80%. The basic working principle of solar cells relies on a semiconductor material which is embedded between selective contacts, collecting the electron and holes which are generated in the semiconductor by the photoelectric effect (also see figure 6.9). The prerequisite is

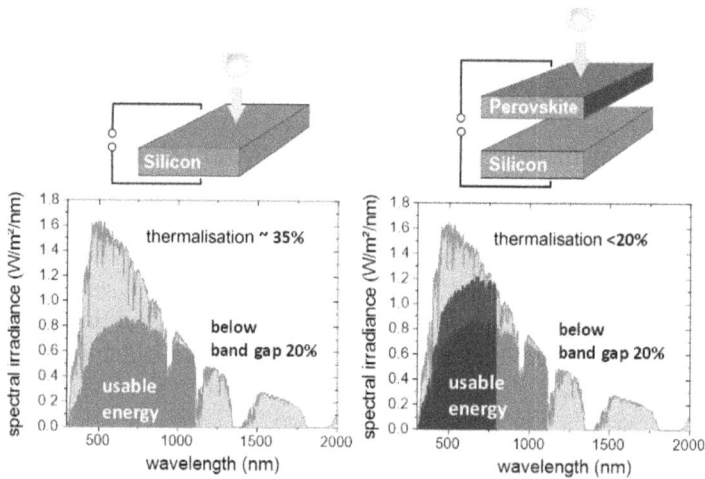

Figure 6.9. Working principle and usable energy of a single junction and a tandem junction solar cell. The solar irradiance with an energy higher than the bandgap energy E_g of the semiconductor is absorbed and converted into electrical energy. The photon energy higher than E_g is lost through thermalisation. The multijunction solar cell architecture (in this case two junctions combining silicon and a higher band gap perovskite base semiconductor) utilises the higher photon energy range (lower wavelength regions) at higher voltages, thus enhancing the overall solar conversion efficiency.

that the photon energy is larger than the band gap of the semiconductor. Note that the photo-excited electrons and holes lose their excess energy (photon energy minus band gap energy) in a very fast process called thermalization. Further loss processes are reflection losses at the front side of the solar cells and recombination losses when carriers recombine before they reach the carrier-selective contacts. Shockley and Queisser [63] derived an efficiency limit for a single junction solar cell of a few percent above 30% under one sun and 40% under highly concentrated sunlight. The theoretical limit for multijunction cells (shown as a tandem cell combining two semiconductor materials) with an infinite (or very large) number of component cells, each adapted to a specific part of the solar spectrum, is higher than 80% because virtually no excess photon energy is lost. Today, the world record cell efficiency measured under concentrated sunlight with a multijunction solar cell is 47.1%, while the benchmark for crystalline Si cells—the PV mainstream today—is 26.7% in conversion efficiency for laboratory-type small-area cells [64].

Solar thermal technology (details in box 6.2) is based on the broadband absorption of solar photons that are converted to heat. In combination with concentrating devices a high-temperature heat source can be established that can power a thermodynamic cycle to generate mechanical energy used for electricity production. In this concept, heat is stored rather than electricity to provide electricity on demand, as this can be realized at significantly lower cost. High-temperature thermal systems rely on low-cost sun-tracking mirrors spread over large surfaces that concentrate the energy on a small receiver device, where it is absorbed and converted to heat. Concentration devices can use only the direct component of the solar radiation, as diffuse radiation (i.e., photons that were scattered on the way through the Earth's atmosphere) cannot be concentrated, according to the laws of physics. This limits the cost-efficient application of this technology to sites with a high direct component, such as we find in the Earth's sunbelt.

Box 6.2 Solar thermal power.

Solar thermal power plants use mirrors to gather direct sunlight and convert it into heat. This is used to generate steam to operate a turbine, which in turn drives a generator to convert kinetic energy into electrical energy. The integrated heat storage system enables the power plant to accurately generate electricity when needed without being affected by fluctuations in solar radiation intensity throughout the day (see also figure 6.10). Additionally, the use of fossil or renewable fuels can compensate for

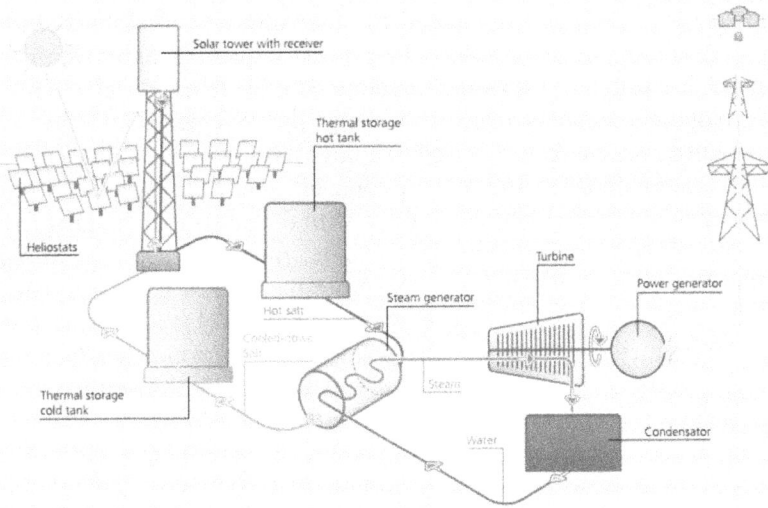

Figure 6.10. Schematic of a solar thermal power plant: a concentrator composed of heliostats, a receiver on top or a central tower, two storage tanks containing molten salt as a storage and heat transfer fluid, and a power block built of a steam generator, a turbine, and a condenser.

longer periods of low radiation. Since steam turbines can operate economically only above a certain minimum size, the rated output power of today's solar thermal power plants ranges from 50 to 200 mW. The main difference from traditional steam power plants is the solar field, which provides heat for the steam generator. In order to reach the high temperature required to produce steam, solar radiation needs to be strongly concentrated.

For this, only direct sunlight can be used. A mirror that tracks the path of the Sun focuses it on a focal point or a focal line. The higher the concentration, the higher the temperature that can be reached. According to the laws of thermodynamics [65], higher temperatures increase the efficiency of the power plant process. The higher the efficiency, the smaller the collector area required by the power plant to produce the required power output. The technical challenge in the solar field is to achieve the required optical accuracy and robustness against environmental influences such as wind and temperature fluctuations at the lowest possible cost.

As shown in figure 6.8, the solar energy distribution on the Earth's surface is quite inhomogeneous. As the output of a solar energy converter is almost proportional to its input, the cost of solar energy for both photovoltaics (PV) and concentrating solar power (CSP) is strongly related to the selected site.

6.3.1.2 State of the art and future perspectives

Photovoltaics

The worldwide PV market is dominated by solar modules consisting of individual crystalline Si wafer–based solar cells. The process sequence has proven robust and easily expandable using new process developments in research labs. Thus, evolutionary development benefits from technology improvements and the scaling of global production. Both effects have led to a huge reduction of module costs over the past decades (see figure 6.13) and solar module efficiency are typically around 20%.

Such modules can be applied on rooftops with a peak power ranging from a few kWp for residential or large commercial roofs or can be installed in solar parks with several 100 kWp or even MWp capacity. Note that the largest solar parks today have already GWp capacity. As an example, figure 6.11 shows a solar-powered facade of a laboratory building. The blue solar modules are electrically connected in larger strings, and the produced dc electrical current is transmitted by a dc/ac inverter to the electricity grid with an efficiency above 98%.

To calculate the energy produced by 1 kWp of installed PV per year, several parameters have to be considered, including real operation temperature, solar spectrum, and sunshine hours per year. The latter is the most significant contribution to the energy yield per year and is typically 1000 h under moderate climate conditions and above 2000 h under very sunny conditions. Depending on the installation site and application (rooftop, solar park, etc), the levelized costs of electricity provided by PV can range from below 3 cents/kWh up to 7 cents/kWh [66], making PV already the most cost-efficient solution for a growing number of applications. For the future additional cost reductions are expected. Finally, a short

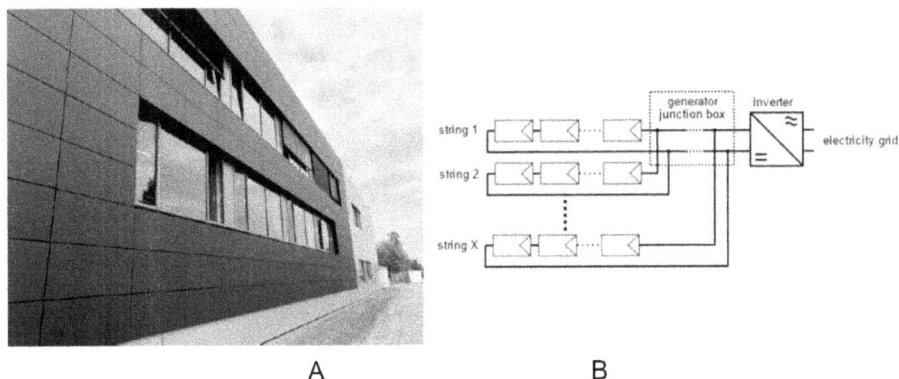

Figure 6.11. (A) Building integrated PV facade as an example of an installation. (B) The electrical layout of the installation. The solar modules are electrically in series connected in strings and bundled in the generator box, and the inverter transfers the dc electricity to the ac internal or external electrical grid.

remark on the energy payback time. With state-of-the-art technologies the amount of energy needed to produce and install a PV rooftop system is delivered back after 1–1.5 years of operation [67], while the operation time of a PV system is longer than 20 years.

PV is still a very young player in the energy sector. As has already been mentioned, significantly higher conversion efficiencies are feasible, already proven by the multijunction concept. Only one highlight example is the unprecedented rise of metal halide perovskites as a new class of PV absorber materials [68], achieving efficiencies well above 20% [64] and close to 30% in tandem configurations with Si [64, 69] within only a few years of systematic research. If the technology can be further developed to module efficiencies surpassing 30% and with low production costs, such tandem PV technologies will pave the way towards further cost reductions.

Solar thermal

In practice, two different basic principles are used to concentrate solar radiation: solar towers and parabolic troughs. In solar tower power plants (figure 6.12(A)), dual-axis tracking mirrors (called heliostats) direct solar radiation to a central receiver mounted on the tower. The heat transfer medium, usually molten salt, water/steam, or air, absorbs energy there and transfers it to the heat storage system and power plant circuit. The surface area of a single heliostat can reach 200 m². In commercial power plants, there are thousands of solar towers arranged in a semicircle or circle. Their intense radiation concentration can produce temperatures in excess of 1000 °C at the receiver. In practice, the system operates between 300 °C and 700 °C, depending on the heat transfer medium used [70, 71].

Parabolic trough power plants (figure 6.12(B)) are the most common commercial implementation variant so far. The parabolic mirror trough tracks the Sun uniaxially and focuses the light on an absorber tube running along the focal line.

A   B

Figure 6.12. (A) Photo of solar tower plant and (B) parabolic trough plant.

The absorption tube contains a special heat transfer oil. The optically selective coating on the tube absorbs visible light while suppressing heat radiation. The absorption tube is surrounded by a glass envelope tube, similar to a Thermos flask, with a vacuum between the two tubes. This further reduces heat loss. The collector is 7 m wide and 200 m long and uses a hydraulic drive to track the Sun. Today's commercial heat transfer oil allows operating temperatures up to 400 °C.

An important component of solar thermal power plants is the thermal storage system. Two-tank systems with molten salt as the storage medium are most frequently used commercially. In tower power plants, the salt is pumped from the 'cold' tank (at around 300 °C) directly to the solar receiver, where it is heated to over 500 °C and fed to the 'hot' tank. A second circuit takes hot salt as needed and feeds it to the steam generator, from where it is pumped back into the 'cold' tank. In parabolic trough power plants with thermal oil as the heat transfer medium, the salt storage tank is loaded and unloaded indirectly via heat exchangers. As heat can be stored more easily and more economically than electricity, solar thermal power plants can produce solar electricity cost-effectively even after sunset.

Solar thermal power plants are characterized by very low environmental impacts. In particular, greenhouse gas emissions over the entire life cycle are comparatively low. The land requirement roughly corresponds to that of large photovoltaic systems. In the operation of newer power plants, the use of dry cooling significantly reduces water consumption. The effects on the flora and fauna are minor, and only very small amounts of pollutants need to be safely disposed of. In addition, solar thermal power plants have a long service life of up to 40 years [72].

6.3.1.3 Current market situation and perspectives

Photovoltaics

During the past decade PV has developed from a niche technology to a pillar of the electricity supply in a several countries. Global installation reached 700 GWp in 2020 [67].

Expectations for the role of PV in the energy system have changed dramatically. Initially thought to be an economic option only for satellites or remote places, PV is now ubiquitous.

The World Energy Outlook 2020 states that 'solar PV is set for the largest growth of any renewable source. Average generation costs for solar PV have fallen 80% since 2010, and it enjoys support of one kind or another in over 130 countries' [73]. According to the stated Policies Scenario worldwide, solar PV capacity additions will reach 150 GWp per year in 2030. Considering the need to defossilize not only electricity generation, but also the huge industrial sector, transportation, and heating, the global need for PV as a source for electricity but also for 'green hydrogen' is expected to increase significantly. 'In the Sustainable Development Scenario, worldwide solar PV capacity additions reach 280 GWp/year in 2030 and ca. 320 GWp/year in 2040' [73]. The path towards a global installation of 30–70 TWp is discussed in [74] by an international group of PV experts. This scenario is based on very low-cost PV, the enhanced electrification of the energy system, and the ability and strong need to convert power to fuels and chemicals in huge quantities.

Regarding the target for photovoltaic technologies to provide power on the terawatt scale, next-generation PV devices have to provide conversion efficiencies that are as high as possible, provide long-term stability, and are embedded in a circular economy.

A surface area footprint per GWp of solar capacity is estimated to be between 5 and 15 km^2, depending on technology choice. The total required surface area would be up to 400 × 400 km for the targeted 70 TWp. Accordingly, the added value, for example, by being part of the building (BIPV) or of vehicles (VIPV), will help to identify suitable areas. In addition, the use of PV in agriculture needs to be explored and developed. Floating PV plants, using PV on lakes or at sea, will enhance the usable area on our planet.

CSP

In sun-rich countries, CSP technology is suitable to take over the role that today is predominantly filled by fossil fuel power plant, that is, to provide a flexible capacity for periods with low solar or wind input. This capability has been shown by more than 100 commercial solar thermal power plants with a total installed capacity of 6.2 gW. Worldwide, 21 gWh of thermal storage capacity are already installed in CSP power plants, which corresponds to approximately 3 gW of electrical output with an average storage capacity of 7 h [75]. This power can be made available to the power grid as required.

Today, solar power plants are already being planned as an integrated solution to combine PV and CSP power plants at one location, using thermal energy storage to ensure the requirements for security of supply in a cost-effective manner. Thereby, large shares of solar power in the energy system of sunny countries are possible [76].

A significant cost reduction of CSP electricity has been achieved along with its deployment (see figure 6.13). The cost decrease per doubling of the installed capacity (learning rate) is similar to that for PV and higher than that for wind energy; however, since the overall installed capacity is two orders of magnitude smaller than

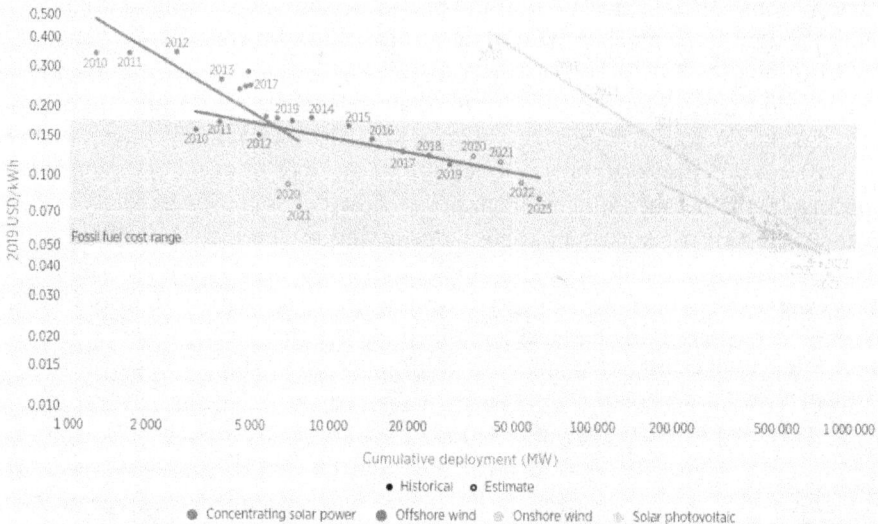

Figure 6.13. Cost reduction of renewable energy technology as a function of its global deployment. Reprinted with permission from [77] IRENA.

that for PV, the levelized cost of energy is higher. With figures approaching 7 US cents per kilowatt-hour in good locations, electricity from solar thermal power plants is already competitive with electricity generated using fossil fuels in places. PV and wind power are offered at lower costs with their integrated thermal storage systems, and solar thermal power plants are the less expensive option for a reliable power supply in times of insufficient input from sunlight and wind, reaching around 5 US cents per kilowatt-hour by 2030 with the help of technical innovations [78].

6.3.1.4 The role of solar technologies for the decarbonization of the global energy system

The threat of climate change requires a fast reduction of anthropogenic greenhouse gas emissions. As most of them are related to the burning of fossil fuels, a massive change in the existing energy system is required. Climate science provided evidence that it is necessary to limit global warming in the next century to less than 2 °C compared to the preindustrial period in order to avoid a catastrophic change of human living conditions on our planet. The cumulative CO_2 emission budget for the energy system that may be acceptable to stay below this target was estimated to be 790 GT [79]. If the emission were kept constant, this budget would be used up in less than 24 years. In the Paris Agreement, the global community set the target to limit global warming to a 1.5 °C increase, based on updated risk analysis. For this goal, less than 7 years are left at today's emission level [80].

Decarbonization of the energy sector to reach the 2 °C target is considered technically feasible if the energy sector can be decarbonized completely by renewable energy and if the share of renewable energy is strongly increased in all other sectors in part through electrification. In addition, a significant increase in energy efficiency in all sectors is required.

To spell this out for solar energy in detail is a bit difficult, as it is not yet clear how the energy will be split among the different renewable resources (in particular wind and solar) and potentially nuclear options and how big the contribution will be in other sectors, for example, through solar fuel production or solar-powered heat pumps.

According to the World Energy Outlook 2020, the 'decisions over the next decade will play a critical role in determining the pathway to 2050' [74]. In all scenarios, renewable energies will play a key role for a sustainable future, and solar energy will become a lead source of our future energy supply. The IEA Net Zero Emissions Scenario [81] predicts a growth of the solar energy supply by a factor of 22 from around 5 EJ solar energy in 2020 to 109 EJ in 2050. Solar energy is expected to provide around 20% of the global energy supply in a decarbonized world. Other scenarios propose even higher shares of solar energy technologies. Thus, it can be considered as one of the most essential technologies to combat climate change.

The installed PV capacity needs to grow from 700 GWp installed in 2020 to many TWp capacity to contribute accordingly. Also, solar thermal power combined with thermal energy storage, today below 10 GW installed capacity, is expected to scale to several 100 GW that will be applied in the sunbelt regions, in particular to supply electricity after sunset. In addition, contributions to cover industrial heat demand is expected by this technology. A possible energy mix progression towards 2050 is depicted in figure 6.14 as a scenario from the IEA [81]. It shows an increasing contribution of solar, wind, biomass, and CCS but also nuclear technologies. This scenario is one of many, with others projecting even larger contribution from solar energy. As scenarios are not predictions but self-consistent options of the future energy system under certain boundary conditions, there is considerable uncertainty

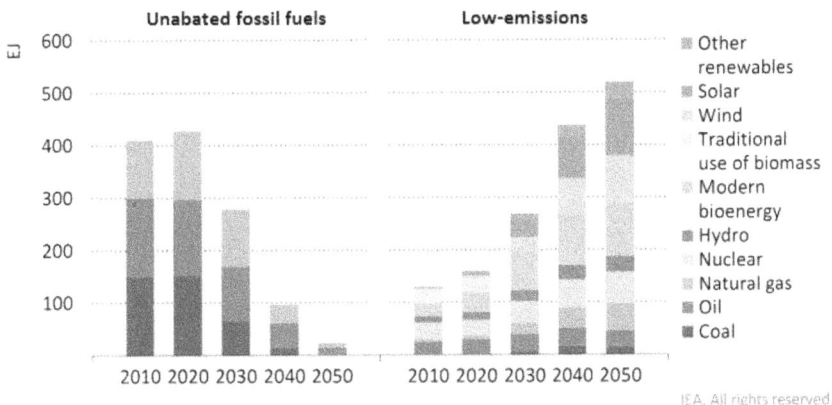

Some fossil fuels are still used in 2050 in the production of nonenergy goods, in plants equipped with CCUS and in sectors where emissions are hard to abate

Figure 6.14. Contribution of different technologies o the global energy supply based on the IEA Net-zero Emissions Scenario [IEA 2021]

about the true future energy mix. However, the inevitable contribution of solar energy is undisputed.

Finally, the following arguments support the massive deployment of solar energy over other technology options that may be considered for defossilization:

1. Today's solar energy technologies are mature and already provide electricity at a cost below that of of all other alternatives at more and more sites.
2. Physical principles allow still much higher performance.
3. Learning curves are steep but promise further cost reduction.
4. A combination of PV and solar thermal offers competitive low electricity cost during nonsunshine hour.
5. The cost of electricity storage such as battery storage is declining quickly, allowing low-cost energy storage in countries not suited for CSP.
6. Low-cost solar electricity is also required to produce green H_2 as a key requirement to decarbonize the transport and industry sector.
7. Solar energy technologies have a low environmental impact, and existing technologies do not run into material resource constraints even if scaled as discussed.
8. Production facilities are available and quickly scalable to the required magnitude if the demand is there.

We conclude that a massive exploitation of solar energy is a feasible, low-risk, cost-efficient, and thus no-regret option to achieve the defossilization the global energy system. A continuation of fossil fuel use without considering its climate impact is the only major threat to the rapid deployment of this technology.

6.3.2 Physics of energy: wind energy

Hermann-Josef Wagner[1]
[1]Ruhr-University Bochum, Bochum, Germany

6.3.2.1 General information and facts

Humankind has used wind energy for thousands of years. Even the ancient Persians utilized windmills for pumping water to higher levels. Today, as the leading industrialized countries are debating about the climate change and its dangers, wind energy plays one of the most relevant roles within the renewables. The biggest advantage of wind energy is its clean status. Wind turbines do not pollute the air and do not fill the atmosphere with dangerous gases; this is its strongest characteristic in contrast to conventional energy plants. Furthermore, wind as a natural physical process is available in many regions of the world. Thus, all states of the world need no longer be dependent on the small number of states that own the fossil fuel sources of energy. Unfortunately, because wind is a natural process, one cannot plan and manage its force, velocity, and intensity. This subordinates wind energy systems to conventional power plants, which can produce electricity with a reliable and constant performance. There also exists a location problem. Many people want clean energy, but they do not want a windmill in the neighborhood.

Wind turbines are mostly built in windy land regions and, in recent years, on open sea, so they are called *onshore* or *offshore* turbines. Wind on sea is stronger and more intense than wind on land. Because of the smooth surface of the sea, there is no friction between the wind and objects, so the turbines can make more use of the kinetic wind energy. Onshore turbines are not able to generate as much energy compared to those on sea. The location dilemma and the low acceptance by some citizens are small obstacles which need a solution in future.

There are various types of windmills. Generally, wind turbines are separated into horizontal axis converters and vertical axis converters. Apart from this the number of blades and the rotation velocity define different wind energy converters. Figure 6.15 shows these types.

6.3.2.2 Physics and technical overview

The typical wind turbine is a three-bladed wind energy converter as seen in figure 6.16. These more prevalent than the other types and can be seen in nearly every rural region. A wind turbine generally consists of blades, which are rotated by the kinetic energy of wind and have an aerodynamic profile similar to that of a jet wing. They are attached with bearings and screws to a nacelle, which contains a generator, safety and controlling instruments, and possibly a gearbox. Wind energy systems can contain a gearbox, but they do not have to. Generators with different physical characteristics (multipole equipment) make gearboxes unnecessary. Finally, the plant is held by a steel tower and its foundation.

The functioning of a wind turbine is made possible by physics, especially fluid mechanics. The profile of the blades has a specific design which is also seen in gas and steam turbines, jet wings, and helicopter rotors. The airflow around the blades

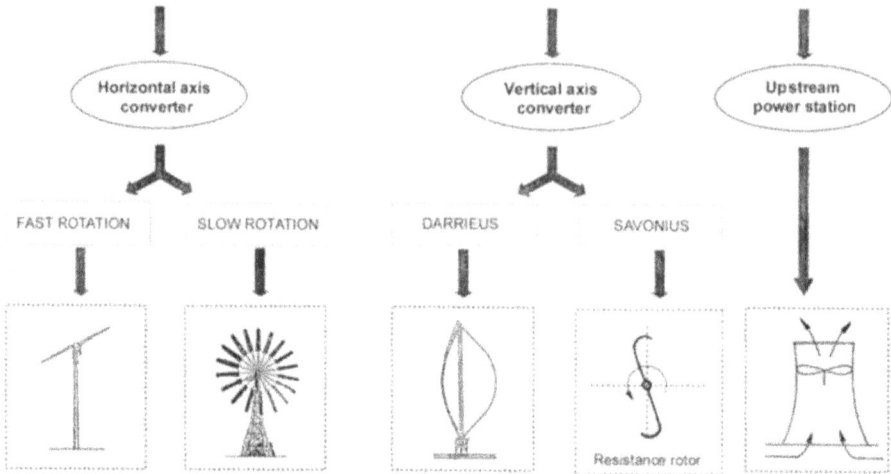

Figure 6.15. Overview of different types of wind energy converters.

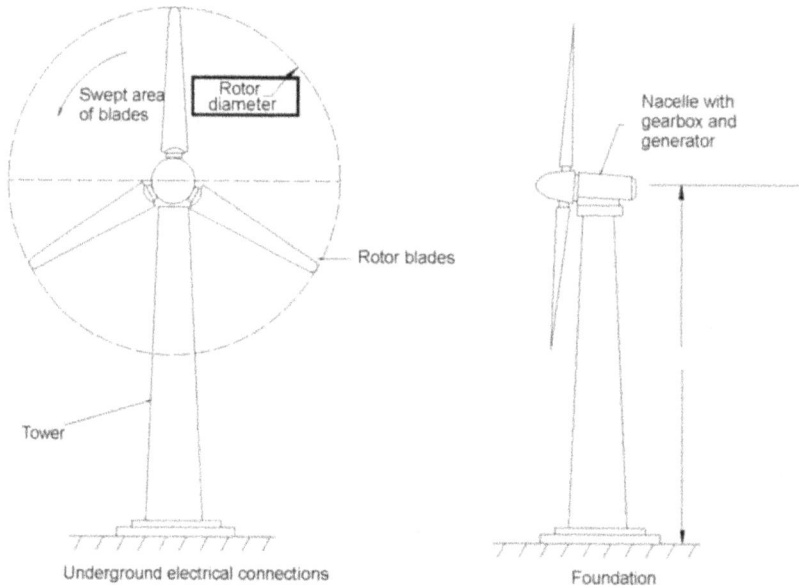

Figure 6.16. Horizontal axis three-blade wind energy converter. A mechanism enables the nacelle to turn in the direction of the wind (rotor speed: 15–25 rpm).

generates a distribution of pressure and different velocities of wind above and under the blade. While low pressure occurs above the blade, high pressure prevails underneath the blade. Based on Bernoulli's fundamental equations, velocity increases with low pressure and decreases with higher pressure. Therefore, the air under the blade flows more slowly than the air above; this results in lift force. Its tangential component lets the blades rotate (figure 6.17). The rotating blades also

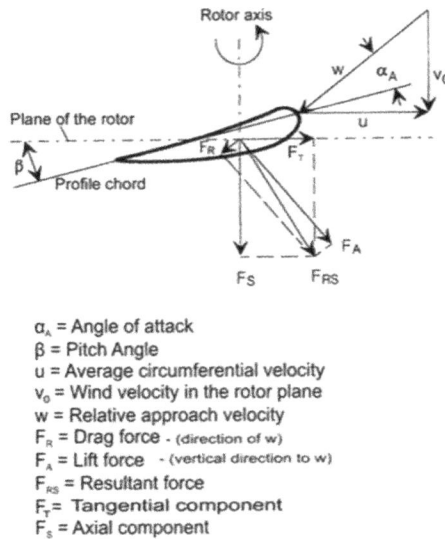

α_A = Angle of attack
β = Pitch Angle
u = Average circumferential velocity
v_0 = Wind velocity in the rotor plane
w = Relative approach velocity
F_R = Drag force - (direction of w)
F_A = Lift force - (vertical direction to w)
F_{RS} = Resultant force
F_T= Tangential component
F_s = Axial component

Figure 6.17. Velocities and forces acting on a blade. The tangential component creates the rotation force.

rotate the shaft, which is connected to a gearbox and the generator. The gear ratio changes the rotation speed and allows the generator to build up voltage.

The nacelle of a wind turbine contains a lot of different elements. As was mentioned earlier, gearboxes can be included but do not have to be. Figures 6.18 and 6.19 provide schematic overviews.

With a gearbox: Although the blades are turning slowly, the gearbox can transmit the velocity into higher levels. So the generator, which contains a few pole pairs, is seated on a second shaft and is independent from the rotor velocity. Nevertheless, a gearbox is a complicated mechanical product which needs maintenance. These types can also include a brake, which stops the rotor from turning at stormy winds. Modern plants with gearboxes do not have brakes anymore. They will be stopped by changing the blades angles.

Without a gearbox: These types contain only one shaft, so the velocities of the turning blades and the generator are equal. The loss of many rotating parts is an advantage; possible failures can be reduced. Due to a high number of pole pairs, the generator is still effectively working. Another difference to the model on figure 6.18 is the braking system. The blades can change their angle towards the wind and stop creating the lifting force, which automatically slows down the rotor. A disadvantage of this model is the big diameter of the generator, which is necessary because of the high number of pole pairs in the generator.

Possible electricity generation rates of windmills are as follows:

- Europe: 1500–2500 kWh kW^{-1} installed power
- North Sea: 3800–4500 kWh kW^{-1} installed power

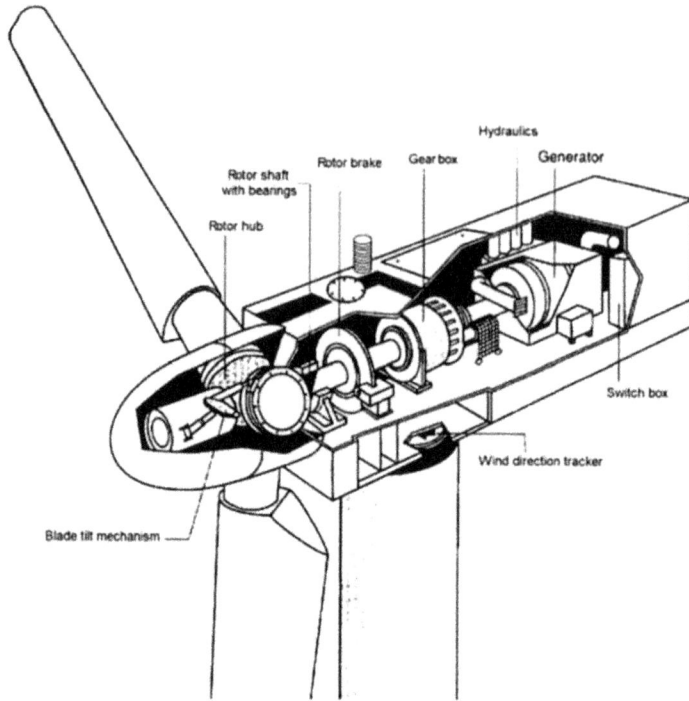

Figure 6.18. Wind turbine with a gearbox.

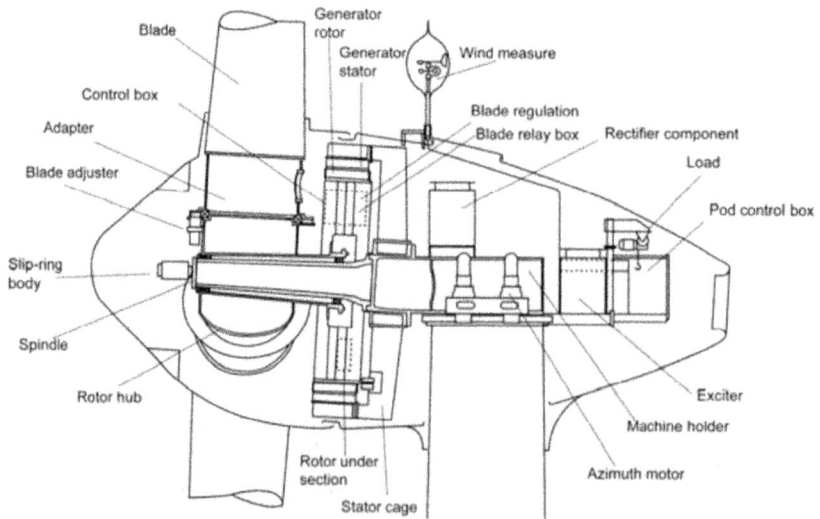

Figure 6.19. Wind turbine without gearbox and details of technical equipment (design of ENERCON Company).

6.3.2.3 The future of wind energy

Today's wind energy technology is still a young form of energy conversion, so scientific and academic research is highly necessary [82]. Because of its clean status and biggest chance to challenge the global climate dilemma, governments all over the world have been helping to expand the wind energy industry through funding. International agreements, such as the Paris Agreement of 2015, obligate all participated countries to follow some rules to keep the climate change low and reversible.

There are numerous new wind energy system variants. One of them is the so-called floating converter. The converter floats like a ship or an oil/gas production facility on the sea. It is fixed with steel cables to the seabed. The biggest advantage is its independence of sea level. Thus, these types can be installed far out in the deep sea (water depth from 50 m to hundreds of meters). This opens opportunities to install large wind parks far out in international sea areas. Of course, maintenance of the turbines and corrosion protection in open sea must be improved.

Furthermore, wind energy converters have a potential to work side by side with gas- and steam-powered plants and hydrogen technology. Unfortunately, engineers still have not found a way of cheaply saving energy created by wind. A battery which can save power on windy days and use it while doldrums could expand the possibilities with wind to higher levels.

One of the main challenges in the next years will be the recycling of old wind energy plants. Like every other mechanical product, wind converters have only a limited lifetime. Especially in the early 2000s a lot of plants were installed, and their deconstruction dates will occur soon. Rotor blades, for example, are made of specific wooden materials such as balsa wood or new materials with higher strength, such as carbon fibre for the blades. The industry still has not found a way to successful recycle these materials in an environmentally friendly way. Researchers are working intensively on this problem.

Ecobalances made for the offshore windpark Alpha Ventus were excellent, but it is necessary to improve them every ten years because of the strong changes in energy supply (due to CO_2 emission reduction activities). The payback time for Alpha Ventus is less than one year.

Important aims which must be fulfilled are:

- More acceptance by citizens living near windmills (through noise reduction, decreased impacts on the local ecology, etc),
- Environmentally friendly processes for recycling of old turbines (especially rotors),
- Connecting wind converters with energy storage modules, such as batteries, for an intermediate time with gas-fired backup power stations until cheap batteries are available.

Finally, the social effects must be calculated too. New wind converter types and their installation support local employment and investment and taxation revenues.

6.3.3 Energy storage

Søren Linderoth[1]
[1]Technical University of Denmark, Lyngby, Denmark

6.3.3.1 History of energy storage

Ever since human beings found out how to utilize energy, we have made use of energy storage. For many years, wood was the way to store energy. To store energy for longer times and make it easier to transport, wood can be converted into charcoal. This is done by pyrolysis, which is a technique in which the wood is converted into charcoal in an oxygen-deficient atmosphere. Nature has also made black coal, which was created millions of years ago. Nature has also made energy-rich gases (natural gas) and oil. Because coal, gas, and oil were created from fossils, they are termed fossil fuels. Fossil fuels were discovered in the 18th century.

When we harvest this stored fossil energy, we may refine and store it further until we need to use the energy. For example, oil is extracted from underground and refined into gasoline that is stored in big tanks; the gasoline is later transported to petrol stations for use by drivers of cars and trucks. We depend on the ability to store the energy.

Fossil fuels are what we use most today worldwide. This fossil energy has advanced the quality of life of humans tremendously. However, now we have to pay the bill. A problem is that large amounts of CO_2, which comes with the burning of the fossil fuels, have been released in a relatively few years. This has caused a significant increase of CO_2 in the atmosphere, which has the effect of warming up the Earth. The atmosphere has become warmer, the polar ice caps are melting, the water level in the oceans is rising, and temperature and weather conditions are changing rapidly all over the world. This is causing severe challenges for many societies. For that reason, significant steps are being taken to replace fossil fuels with energy from renewable sources, such as power from wind turbines and solar cells, solar heating and cooling, use of biomass (wood, straw, etc), and biogas made from waste.

6.3.3.2 Storage of renewable energy

The most efficient way to use renewable energy is to use it directly. However, power from wind and sun fluctuates, so it is not well aligned with our need for energy use. We therefore have to make ourselves independent of when the wind is blowing, and when the sun is shining; that is, we have to find ways to store the energy. Hence, energy storage continues to be a very strong need; in fact it is essential for the transition into a society free from fossil energy supplies. In a sense, we have to do something like what Nature has done for millions of years: make fuels from solar energy and to store the energy, in this case in electrical batteries, as thermal energy, or as mechanical energy such as in flywheels (kinetic energy) or as hydropower (potential energy).

Here, we will discuss some of the important ways to store energy in the future. In order to be useful, the storage technologies must be sufficiently cheap, and the energy must be available in the form we want when we want it. In the same way as energy production must be sustainable, so must energy storage be economically, environmentally, and socially sustainable.

The most important ways to store energy are as follows:

- Electrochemical energy storage
    1. in batteries
    2. by conversion into chemicals (e.g., hydrogen, ammonia, methanol)
- Thermal energy storage
- Potential energy storage, as in dams for hydropower
- Kinetic energy storage, as in flywheels

6.3.3.3 Electrochemical storage

The types of electrochemical energy storage have some common features. They rely on a membrane, also called a separator or an electrolyte, which can conduct only ions and not electrons (or holes), that is, they are solely ionic conductors and not electronic conductors. The electronic conduction must then take place through electronic conductors that can pass through the membrane, leading the electrons from one side of the membrane to the other. The electrodes on the two sides are called the anode and the cathode, depending on whether they are 'charging' or 'discharging'. The key for all is that they consist of an electrolyte (or membrane) and two electrodes. The electrolyte must in all cases conduct only ions and not electrons (figure 6.20). That is an important aspect of electrochemical devices such as batteries

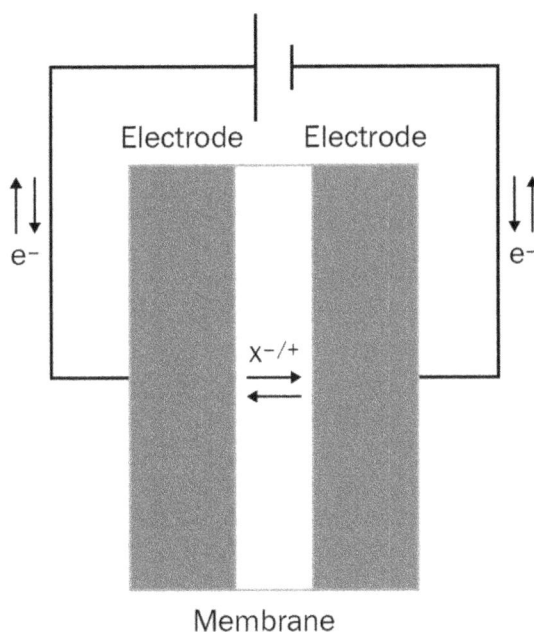

Figure 6.20. Sketch of the heart of electrochemical conversion and storage. The membrane is sandwiched between two electrodes (called anode and the cathode, depending on the direction of the current). The membrane can conduct only ions, not electrons.

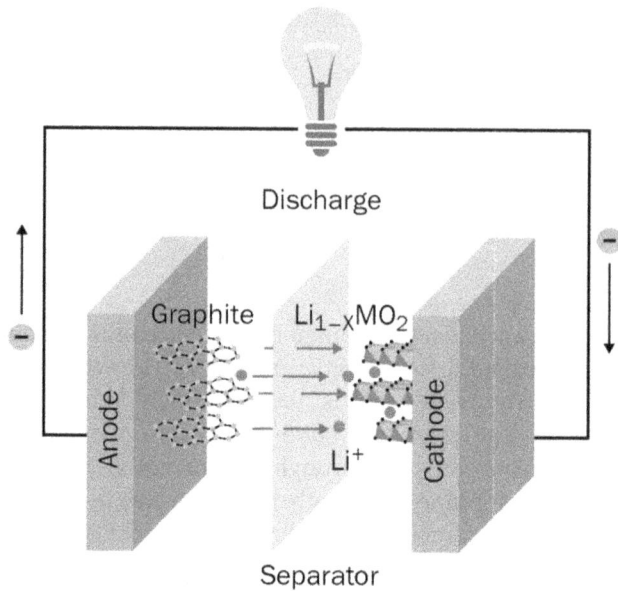

Figure 6.21. Sketch of the function of a Li-ion battery. The 'M' is Mn, Co, Ni, or a mixture of those. When the battery is charged by applying a voltage across the cell, the Li^+ ions are forced to leave the $Li_{1-x}MO_2$ phase through the separator and intercalate instead in the graphite. The separator conducts Li^+ ions. The figure shows the function in the discharging mode, where Li^+ ions intercalated in carbon are transferred through the separator to the cathode, where they are incorporated in the $Li_{1-x}MO_2$ phase (thereby decreasing the x, the deficiency of Li in the phase).

and electrolysers. Examples of ions that can conduct in the electrolyte are H^+, OH^-, O^{2-}, Li^+, and Na^+.

6.3.3.4 Electrochemical storage in batteries

Batteries are electrochemical storage units that contain the stored energy. Many types of battery are portable, like the lead-acid batteries in vehicles and the Li-ion batteries in mobile phones (figure 6.21). Lead-acid batteries are still the most produced and used battery type, especially in cars and trucks. However, the use of Li-ion batteries is increasing fast with the implementation of battery-powered vehicles.

Flow batteries are a quite different type of battery (figure 6.22). The principle is the same: the ions are stored in phases in the anode and cathode, and only ions pass through a membrane. In flow batteries the anodes and cathodes are in a liquid form. The electrodes are the current conductors, and electrons move in and out of the system during charging and discharging. The ions are not stored in the electrodes, as is the case in Li-ion batteries, but are stored in liquids, called anolytes and catholytes.

Yet another type of flow batteries, which will be able to store large amounts of energy, are called gas flow batteries. For gas flow batteries, the liquids in the traditional flow batteries are replaced by gases such as CO_2 and methane (CH_4).

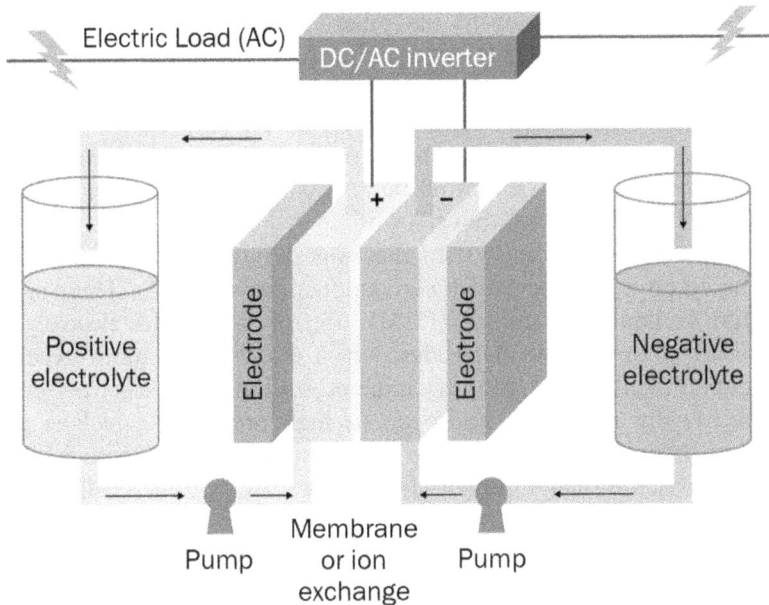

Figure 6.22. Example of the most developed type of flow batteries, termed vanadium flow batteries. Here the valencies are changing in the anolytes and catholytes by a redox cycle. The membrane is a proton-conducting membrane. Anolytes and catholytes made from organic-based materials are under development.

6.3.3.5 Electrochemical energy storage by conversion into chemicals
Electrochemical conversion of power into chemicals can be done; the simplest is hydrogen (H_2), but conversion to ammonia (NH_3), methanol (CH_3OH), methane (CH_4), and other chemicals is also possible. By being converted into chemicals, energy can be stored for a long time and can be transported in various ways and used where electricity is not easy or possible. This could be, for example, in heavy transport, such as long-range flights, or in long-range marine applications.

The core of electrochemical conversion of power is electrolysis. As in batteries, the heart of an electrolyser consists of a membrane/electrolyte/separator sandwiched between two electrodes, the anode and the cathode. The prime difference is that an electrolyser can continue to produce products (charging) as long as it fed with power and the elements for the production. In the simplest case, this is water (H_2O), and the products are hydrogen (H_2) and oxygen (O_2).

In fact, electrolysers have been used for about two centuries, primarily for the production of ammonia for agricultural use as fertilizer. Electrolysers have also been used in industry for production of chloride from NaCl. After many years of slow progress in the development of electrolysers, the new trend of move away from fossil fuels and instead use renewable energy for all energy use has revitalized greatly the need for electrolysis for energy conversion and storage.

The first generation of electrolysers was the alkaline electrolyser with potassium hydroxide and asbestos as the separator. OH^- is the moving ion that is transported from one side of the separator to the other. Asbestos has been replaced by a ceramic–polymer matrix or porous ceramic separator [83]. KOH is very corrosive, and therefore the separator material must be corrosion resistant for use with concentrated KOH. The electrodes are based on Ni [84]. In the beginning, the alkaline electrolysers operated at ambient pressure, but more recent electrolyser have been pressurized (up to about 30 bar), and lately alkaline electrolysers have been developed that operate at above 100 °C.

Another type of electrolyser system that was developed in second half of 20th century is based on a polymer which can conduct protons (H^+). These systems are called proton exchange membrane (PEM) electrolysers. PEM electrolysers need platinum and iridium for the electrodes, which makes them more expensive and therefore more difficult to bring to very large utilization. PEM electrolysers are able to electrochemically pump the hydrogen to high pressures, which is useful for storage of the hydrogen in pressurized tanks, such as is the case in the use of hydrogen in electrical vehicles using hydrogen together with fuels cells for powering the electrical motor.

A third type of electrolyser that is under development and deployment is based on a ceramic electrolyte which conducts oxygen ions (O^{2-}). The electrolyte is based on yttrium-doped zirconium; the doping makes the ceramic a good oxygen ion conductor but at elevated temperatures (700–1000 °C). The electrodes are ceramic or metallic–ceramic (cermets). This type of electrolyser is termed a solid oxide electrolyser cell (SOEC) and is so far the most efficient of types of electrolyser. Other benefits of SOEC is that it also can electrolyse CO_2 and make CO [85]. By electrolyzing simultaneously water steam and CO_2 gas, it is possible to produce a mixture of $H_2 + CO$, which is called a syngas in industry; from this, one can produce, for example, methanol and methane. Another advantage of SOEC is that the same unit can work the other way around, that is, as a fuel cell (SOFC), and thereby works similarly to a battery [86].

All types of electrolysers are being developed for upscaling and cheap production of hydrogen and other chemicals (figure 6.23).

6.3.3.6 Thermal energy storage

Thermal energy storage is typically the storage of hot water in tanks on the top of a roof or in very large reservoirs of hot water. This type of energy storage is called sensible heat storage. Another type of so-called sensible heat storage is by heating rocks to high temperatures. Here the crushed rocks are heated to around 600–800 °C by blowing high-temperature air through the system. The stored energy may then, preferably within hours or few days, be used to produce high-temperature heat for industrial use, for electricity production, or simply for heating houses. The energy may also be stored as high- or low-temperatures reservoirs in rocks [87] for higher-efficiency electrical production later on.

Molten salts are also used for heat storage. In this case the storage relies on latent heat related to the phase change between the liquid form and the solid form of the

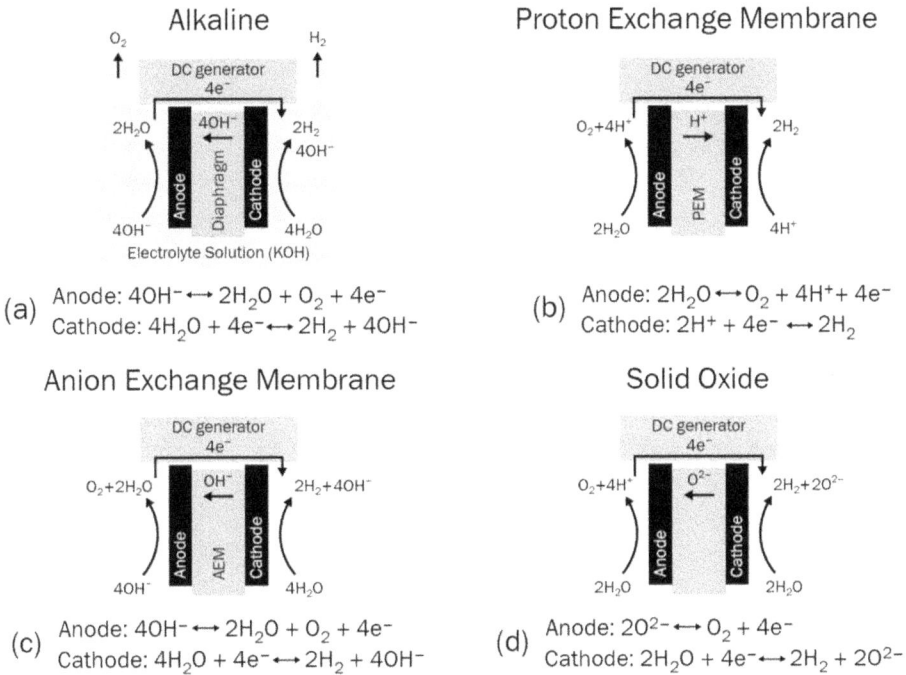

Figure 6.23. Different types of commercially available electrolysis technologies.

material. The temperature range for molten salts is typically from 200 °C to several hundred degrees Celsius. Other phase change materials [88] include some polymers that work at around room temperature and some salts dissolved in water that operates between 20 °C and 100 °C.

Another type of heat storage is thermochemical heat storage. Here heat is generated when two elements react chemically. This could be hydrogen reacting with a metal, making a metal hydride and releasing reaction heat. The system is recharged by heating and releasing the hydrogen from the hydride. Another example of thermochemical heat storage is in salts, where ammonia (NH_3) reacts with the salt structure, forming a salt–ammonia compound. Again heat is released, here at a temperature dictated by the composition of the salt.

6.3.3.7 Energy stored as potential energy
When energy is stored as potential energy, the force of gravity is utilized for the energy storage. The stored element is typically water, and when released, the potential energy converts into kinetic energy, which then drives a turbine, which produces electricity. This type of energy storage is typically used in countries with mountains, where the water is kept behind walls until the water is allowed to flow to

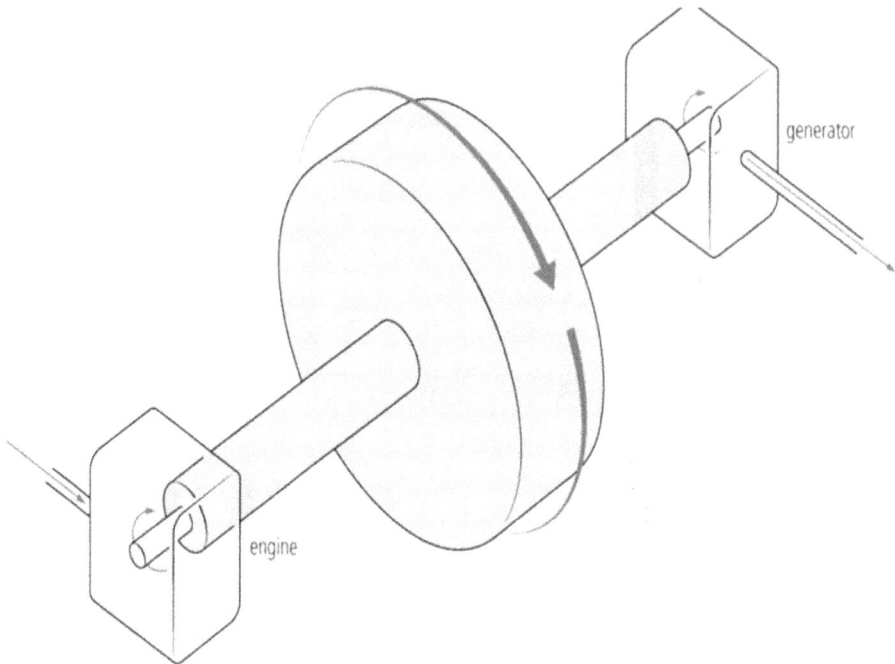

Figure 6.24. Energy can be stored as kinetic energy in a flywheel.

lower heights through turbines. The type of energy that is produced is termed hydropower.

6.3.3.8 Energy stored as kinetic energy

Energy can also be stored as kinetic energy in a flywheel. Here the amount of energy depends simply on the weight of the medium and the speed of rotation. The medium can be glass, metal, or concrete, for example. The stored energy relies on high-speed rotation, and therefore the medium must be able to withstand very high forces that comes with rotation. In order to be able to run at high speeds, the flywheel is typically in vacuum, and magnets are used to lift the system and keep it in place (figure 6.24).

6.3.3.9 Challenges and opportunities in Horizon Europe and beyond

All the energy storage technologies mentioned earlier in this section exist and are used. Therefore, it is sometimes said 'that the technologies exist'. Well, this is correct, just as electrical batteries existed 150 years ago when the first battery-powered car was on the roads and wind turbines have existed for centuries. However, neither the batteries or the wind turbines would be of any use today without the immense development of much more efficient batteries with high energy densities and power densities and the development of much more efficient wind turbines for electricity production. The present industrial revolution using mobile phones and production of cheaper renewable energy would otherwise not have been possible. This is also the case for energy storage. Energy storage has existed and been utilized for many centuries, starting primarily with the use of wood and charcoal and continuing nowadays with batteries and hydrogen. In between, we have utilized the energy storage created by Nature in the form of oil, coal, and natural gas. Because Nature has provided this energy storage, it is cheap. The great challenge now is to reinvent energy storage in many forms that enable cheap energy storage and conversion of renewable energy from solar, wind, and biomass. To make sustainable energy storage—that is the challenge we must meet.

6.3.3.9.1 Electrochemical energy storage in batteries

The most used type of battery is still lead-based batteries. They are the type of batteries used in traditional cars. These batteries are rather cheap, but the energy density per volume and weight is rather low. In addition, they contain lead, which is not good for the environment and health. On the bright side, the recycling of lead is probably around 99%, although the remaining 1% is not recycled.

To make the transport sector such as cars, buses, and trains, more able to use electrical batteries, high-energy density batteries have to be developed. In 2019 the Nobel Prize in Chemistry was given to three scientists who made it possible to develop the kind of Li-ion batteries we use in our cell phones and in battery-powered vehicles and trains today [89]. Without this breakthrough about 30 years ago, we would not be where we are today with convenient mobile phones and battery-powered vehicles. The development of today's batteries took at least 30 years, and much more has to be discovered and developed to bring the costs down, to make the batteries altogether sustainable, and to make them sufficiently safe.

Li-based batteries contains metals such as Li, Co and Ni. The mining of these elements is a problem. In general, much energy and water are used to extract the metals. Recycling of the materials in Li-ion batteries must come to a much higher level than it is today. Replacing or minimizing the Co could be one aim, and recycling in general is another important aim for Li-based batteries.

The electrolyte in Li-based batteries is a liquid polymer, which can cause fire and therefore can be a risk if used. Other Li-based batteries are based on a solid electrolyte, for example, one made of a ceramic. Such so-called solid state batteries are under development but will need much more development. Some of the challenges for this type of batteries can be dendrite formation [90] in the electrolyte, such that a short-circuit can happen, thereby destroying the battery and causing damage and accidents. Other challenges are the loss of electrical contact between the electrolyte and the electrodes and the reliance on Li metals as one electrode in current solid state batteries. This is greatly beneficial for the energy density, because the volume and the weight of the electrode are much reduced. However, Li metal reacts preferably with oxygen and can become flammable. All of these issues are likely to be solvable but need much effort.

Batteries based on Na^+ as the moving ion are also in use. One type of Na-type batteries is Na-S batteries. They operate at elevated temperatures because they use molten Na as one electrode. The electrolyte is a solid ceramic, beta-alumina (β-alumina/Al_2O_3). Beta-alumina is a good conductor of Na^+ above 250 °C but a poor conductor of electrons, thereby fulfilling the requirement for the electrolyte. Pure Na presents a hazard, because it spontaneously burns in contact with air and moisture, as is the case for Li in Li-metal batteries. Therefore, the system must be protected from water and oxidizing atmospheres. The Na reacts with S on the other electrode, creating Na_2S_4. On charging, this reverses. Na–S batteries have a high energy density and have been considered for use in vehicles. However, the high-temperature operation makes it somewhat difficult, and they are currently used for stationary applications. One challenge is that Na is quite corrosive.

Batteries based on ions such as Al^{3+}, Mg^{2+}, and Zn^+ are under development for rechargeable batteries [91]. Zn–air batteries are well known and used today as primary batteries, that is, used only once, in hearing aids. Magnesium- and aluminum-based batteries still needs much development to be of possible use.

Flow batteries based on organic materials could be very beneficial due to possible nice environmental properties. The first organic-based flow batteries have been demonstrated, but developing more stable organic materials that do not degrade too fast is one challenge. Another is to make the production of the materials cost effective, which will be a requirement for widespread use. Research projects are ongoing but will probably require several years of research and development. For example, flow batteries based on iron compounds are also under development. The materials for these batteries are easily abundant.

6.3.3.9.2 Electrochemical energy storage by conversion into chemicals

For electrochemical energy storage to be sufficiently sustainable, the cost of renewable energy must be low; therefore, it relies on making energy production from solar and wind even cheaper than it is today. There exist various types of electrolysers, whose cost must be lowered. This can happen in several ways. One is simply by upscaling, which, similarly to, for example, the rapid lowering of costs of Li-ion batteries [92] due to extensive use in battery-powered vehicles, will drive the costs down. However, the electrolysers also need to be improved, and they need to be much more sustainable. The different types of electrolysers systems have different challenge and improvements to cope with. Common challenges are to bring the total cost down for the production of the chemicals and to be able to work well under dynamic conditions, as the input from solar and wind will fluctuate.

Alkaline electrolyser systems should preferably become more efficient in the production of hydrogen. Today, much energy input is lost as heat. Also, the areal need for alkaline electrolysers should preferably be reduced, that is, the power density should be increased. This can be reached by, for example, redesigning the alkaline electrolysers by using new materials that allows the separator to become thinner, and the electrode to be more efficient. New materials and concepts to allow operation at elevated temperatures would improve the power density per volume.

PEM electrolysers would greatly benefit from lowering or replacing the need for platinum and iridium in the electrodes. The membranes of the present PEM electrolysers cannot withstand temperatures above 100 °C. Polymer membranes that can operate well and long-term at higher temperatures could bring higher efficiencies. A challenge in enabling such an increase of operating temperature could be that the very fine-grained (nanoscale) Pt and Ir agglomerate and make the electrode be less efficient and durable.

SOEC operates today at 700–1000 °C. They are very efficient, but the durability of the materials has to be improved. The ceramic materials are brittle, which limits the size of the cells. Today cells are typically around 100 cm^2. To bring down the costs, the cell areas should be larger without causing frequent failures during operation. In general, the durability during operation must be improved by optimizing the materials and the operation. The stack concepts should also be revisited and improved. The most efficient use of SOEC comes with the synergy of synthesis of chemicals, such as methanol. The interplay between these units must also be optimized and developed further.

6.3.3.9.3 Thermal energy storage

The technologies for storage of hot water in small and large quantities are relatively mature. Thermal energy storage in rocks, in phase change materials, and by thermochemical reactions have virtues for much higher energy densities and for uses other than heating of houses. However, they all need to be reduced in price, and some need to be developed further and demonstrated for up-scaling for large deployment.

The use of phase change materials, such a molten salts, for energy storage is well proven but is waiting for a breakthrough. Corrosion of molten salts is one issue; another challenge is avoiding solidification of the molten salt, which can cause severe problems due to expansion of the medium, causing cracks and faults. In general, cost is an issue.

Thermochemical energy storage has the potential for high thermal energy density, and it can be completely without energy loss, as the heat is realized only upon reaction, which can be controlled. Thermochemical units have not much been demonstrated, and there is a need for further developments to happen. The salt-ammonia system is probably the most energy-dense system. Here, one challenge can be the ammonia, which is poisonous and therefore can limit the use of this system outside industrial applications. Water or steam may replace the ammonia but with much reduced energy density. The recycling of the system must be proven for many thousands of cycles.

6.3.3.9.4 Energy stored as potential energy

The potential for the use of more potential energy is already quite limited, though it is widely used in hilly countries. Also, the impact of the environment in the hills or mountains can be bad, as the level of the water alternates greatly over the year. This causes damage to the environment. In some areas, however, there are still great possibilities for use of melting ice as a renewable resource. In Greenland more ice is melting each year as the atmosphere becomes warmer. The melting can create a flow of water, which can be used to run turbines whose power can be used for making power-to-X (hydrogen, ammonia, and more).

6.3.3.9.5 Energy stored as kinetic energy

The benefit of flywheels is the possible large power density, not the energy density. Flywheels can take up and deliver fast energy, for example, in connection with storing the electricity from the stop and start of electrical trains. Flywheels could also be used in cars and buses for storage of energy from deacceleration and used when accelerating. Flywheels have not yet found widespread use, and one aspect of this can be the cost of the systems. The cost would decrease with more use of flywheels. The high speed of the flywheel can be a risk that has to be dealt with. Improving the stability of the spinning wheel by using even better magnetic systems could be a solution.

6.3.3.10 Conclusions

New means of energy storage have to be developed for the green transition to be possible. Such energy storage technologies must come at the lowest possible costs, both in making the storage units and in their operation. Energy storage is needed both for short-term storage with high power output and long-term storage of large amounts of energy. Some types of energy storage, though not all, have been discussed here, and new technologies will appear in the future. However, it is quite certain that electrochemical energy storage in the form of batteries of various kinds is a must, as is electrochemical storage in chemicals provide by electrolysis and synthesis into chemicals and fuels, such as hydrogen, ammonia, and methanol. Thermal energy storage will also be of importance, especially in places with a need for heating or cooling.

In all cases, further research and development, demonstrations, and commercialization are needed, and we have to do this rather quickly, as the goal for many countries, regions, and companies is to be CO_2 neutral by 2045–50. This is in a short amount of time when we consider that many technologies have to be developed to a much higher level and costs have to come down drastically so that the transition from dependence on fossil fuels can be replaced by energy supply coming from renewables sources. For this we ned many good brains and much work.

6.3.4 Fusion energy development

Alberto Loarte[1]

[1]ITER Organization, Saint-Paul-lez-Durance, France

6.3.4.1 Introduction to fusion energy development and present status

The physical processes that lead to the production of energy in the Sun and stars were understood in the first third of the 20th century [93, 94]. Since then, there has been a continuous effort to make use of this understanding to develop an energy source for humankind based on the same principles. The advantages of mastering the fusion processes as an energy source are tremendous, including the wide availability of fusion fuels, no production of greenhouse effect gases, and no long-lived radioactive waste. However, the scientific and engineering challenges to realize this energy source are also considerable, as was determined in the initial research by the mid-20th century.

To overcome such challenges, researchers soon realized that concerted scientific and technical efforts were required. These were undertaken at the national and supranational levels (e.g., the European Union) with a strong international collaborative attitude from the start. The pinnacle of these collaborative efforts is the first experimental fusion reactor (ITER), now under construction in France by a consortium of international members (China, the EU, India, Japan, the Republic of Korea, the Russian Federation, and the United States). ITER's objective is to demonstrate the scientific and technological feasibility of fusion energy as a sustainable source for humankind, and experiments to address this objective will begin in the mid-2030s [95].

The most effective way to produce fusion energy on is through the reaction of two heavy forms (isotopes) of hydrogen: deuterium and tritium. Most of the deuterium was created in the early stages of the formation of the Universe and is thus widely available (typically, about 0.016% of the hydrogen in seawater is deuterium). Tritium, on the other hand, is an unstable form of hydrogen that decays radio-actively in 12.5 years and thus needs to be produced. An effective way to produce tritium for fusion energy production is by neutron irradiation of lithium, which is abundant in the Earth's crust. The nuclear reaction of deuterium (D) and tritium (T) produces helium (He) and one neutron:

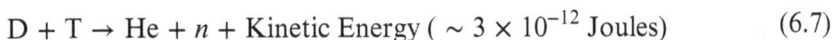

$$D + T \rightarrow He + n + \text{Kinetic Energy} \ (\sim 3 \times 10^{-12} \ \text{Joules}) \qquad (6.7)$$

Since the mass of deuterium and tritium is larger than that of helium plus the neutron, the missing mass is converted to kinetic energy of the helium nucleus and neutron according to Einstein's relativity law for conservation of mass and energy ($E = mc^2$). Because the neutron is four times lighter than the helium nuclei, it carries most of the energy produced in the reaction (75% of the total). This neutron kinetic energy can be transformed into electricity by slowing down the neutrons and heating up water or other fluids in a way similar to that used in present gas, coal, and nuclear fission power stations.

There are, however, two key differences between fusion-based reactors and present fission-based reactors that make the fusion option sustainable in the long term. The first one is that the product of the reaction is helium, which is nonradioactive. The second is that the total kinetic energy gained by the neutrons in the fusion reaction, which ultimately is transformed into electricity, is much larger than that for fission for the same mass of fuel converted into energy. In fact, the electricity consumption needs of one person for 30 years can be satisfied with the deuterium contained in the water in a bathtub and the lithium contained in the battery of a laptop computer when fused in a reactor. The widespread and long-term availability of lithium and deuterium could support the supply of fusion energy to humankind for thousands of years.

To get the deuterium and tritium nuclei to fuse, it is necessary to overcome the electrostatic repulsion force, since both nuclei are positively charged and they repel each other. When the nuclei approach sufficiently close to each other, the nuclear fusion process can effectively take place and ensures high production of fusion energy. This requires deuterium and tritium nuclei to collide at high velocities; this is realized by heating the deuterium and tritium mix to very high temperatures (hundreds of millions of kelvins) (figure 6.25); at these high temperatures the thermal agitation of deuterium and tritium ensures that a sufficient fraction of them have enough velocity for the fusion reaction to take place. In addition, the collisions between deuterium and tritium nuclei should be frequent enough for copious power (i.e., energy per unit of time) production to take place; this implies that the density of deuterium and tritium should be sufficiently high. These density and temperature requirements are met in the cores of the stars and will be met in fusion reactors; the differences are that the stars fuse protium (the lightest type of hydrogen) and not deuterium and tritium, with the Sun having a typical temperature of 10 000 000 K in its core compared to ~200 000 000 K in a fusion reactor. In these high-temperature conditions, the hydrogen gas is ionized and electrons are not bound anymore by electrostatic forces to the hydrogenic nuclei. This state of matter is called plasma; it is the most abundant state of matter in the Universe (although not on the Earth's surface) and can exist over a wide range of temperatures and densities (figure 6.26).

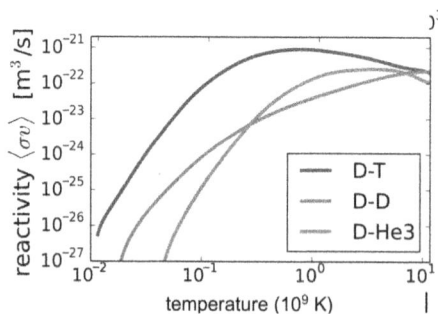

Figure 6.25. Reactivity of three fusion reactions versus temperature showing the higher reactivity of the deuterium–tritium reaction compared to the others and that it reaches its maximum value at temperatures of few hundred million kelvins. (Source: https://en.wikipedia.org/wiki/Nuclear_fusion. Credit: Dstrozzi)

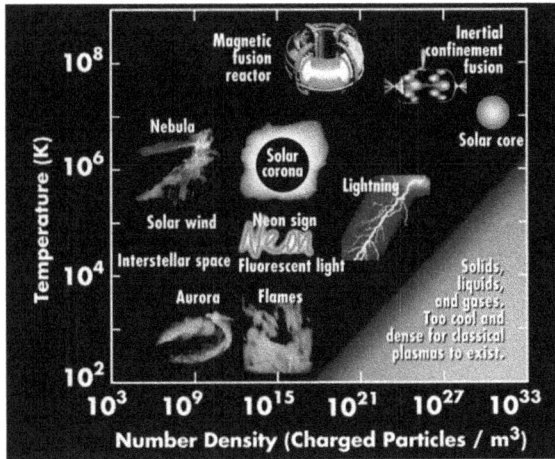

Figure 6.26. Typical temperatures and densities of plasmas on the Earth and in the Universe compared to those to be achieved in fusion reactors. (Source: https://www.cpepphysics.org/fusion.html. Credit Contemporary Physics Education Project (CPEP), used by permission.)

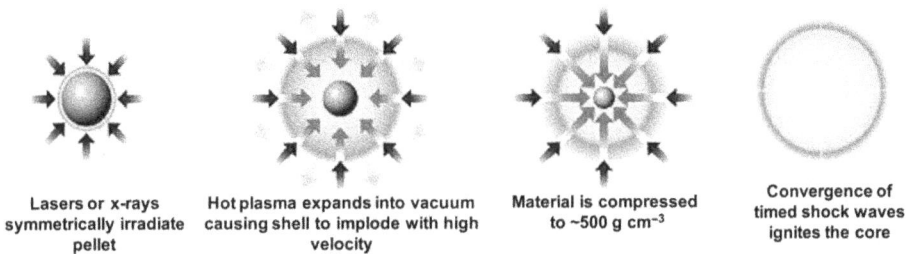

| Lasers or x-rays symmetrically irradiate pellet | Hot plasma expands into vacuum causing shell to implode with high velocity | Material is compressed to ~500 g cm^{-3} | Convergence of timed shock waves ignites the core |

Figure 6.27. Diagram of the dynamics of fusion energy production by inertial confinement by direct. (Source: https://www.lanl.gov/projects/dense-plasma-theory/background/dense-laboratory-plasmas.php. Credit Los Alamos National Laboratory.)

Such hot plasmas lose heat by conduction and convection, and they naturally expand because of their pressure in a similar way to hot air in a balloon. To sustain fusion power production in these plasmas, it is necessary to keep the plasmas hot and with sufficient pressure, since expansion and heat losses would decrease the plasma temperature and density and stop the fusion reactions. Stars achieve these goals thanks to their huge dimensions, which slow down heat losses, and to their mass, which provides the gravitational force to compensate the expansion of the plasma. To achieve the same goals as in the stars and achieve fusion power production on Earth, other physical processes are required; two approaches have been developed the so-called inertial confinement fusion and magnetic confinement fusion.

In inertial confinement fusion, a small solid deuterium–tritium spherical shell is irradiated by high-power-density light, which heats and compresses the shell to high densities and temperatures, leading to production of fusion power in short bursts (figure 6.27).

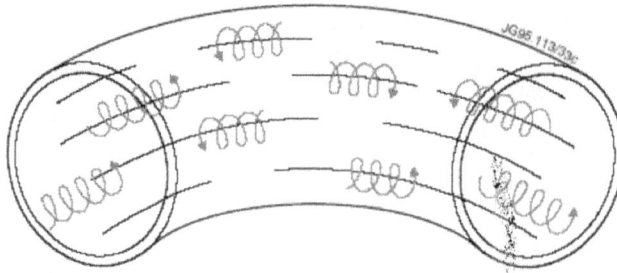

Figure 6.28. Trajectories of ionized charges particles (red) along magnetic field lines (black) in a section of a torus. (Source: https://www.euro-fusion.org. Credit EUROfusion consortium.)

Magnetic confinement fusion takes advantage of the physics processes that determine the movement of charged particles in magnetic fields. A charged particle (with charge q) with velocity $\vec{v}$ in a magnetic field $\vec{B}$ is subject to a force (so-called Lorentz force) given by

$$\vec{F} = q \ \vec{v} \times \vec{B} \qquad (6.8)$$

The Lorentz force ties the movement of particles to the lines of the magnetic fields (figure 6.28). For hot plasmas this decreases the heat losses in the direction perpendicular to the field (note that $\vec{F}$ in equation (6.8) is always perpendicular to $\vec{B}$) but has no impact in the parallel direction to the field. To avoid losses in this direction, it is necessary to use magnetic field lines that close on themselves, forming a torus (or doughnut) with a magnetic field (or toroidal field) around the axis of symmetry of the torus. This is not sufficient to ensure that the plasma is in equilibrium, and another component to the field in the short direction around the torus (or poloidal direction) must be added. The most successful magnetic field configurations for fusion development are the tokamak and the stellarator (figure 6.29). In the tokamak, the magnetic field is produced by electric currents circulating in external coils to the plasma and within the plasma itself; in the stellarator the magnetic field is produced by external coils only. Since the generation of these magnetic fields requires electric currents and thus power, which would decrease the net power production by the reactor, it is important to optimize the magnetic fields so that they provide the required thermal insulation and compression force to maintain the fusing deuterium–tritium plasma with the minimum power used for their generation. Typically, the magnetic fields to be applied to the plasma are in the multi-Tesla range (or ~100 000 times the Earth's magnetic field) and are created by coils surrounding the plasma in which large electric currents circulate, typically in the multimegaampere range (or several million times the electric currents used at home).

For both tokamak and stellarators it is necessary to heat the plasma to sufficiently high temperatures so that the fusion reaction is effective. This is done by the injection of radiofrequency waves that couple to the movement of electrons and ions in the magnetic field (shown in figure 6.27) and accelerate them (similar to the physics processes to heat food in a microwave oven) or by the injection of high-energy

A

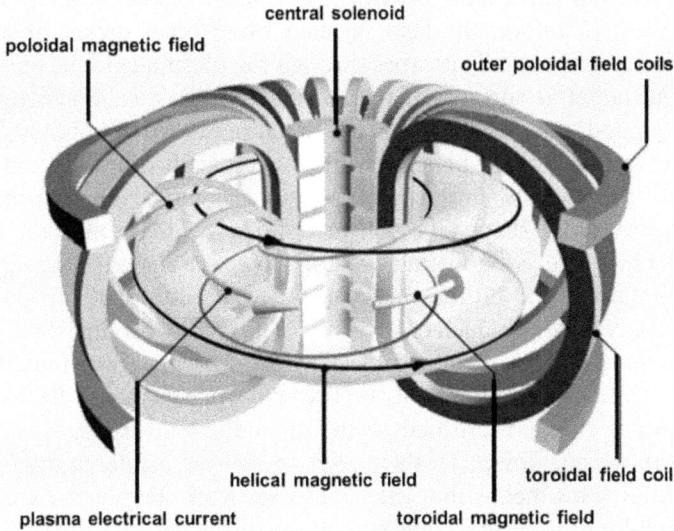

poloidal magnetic field

central solenoid

outer poloidal field coils

helical magnetic field

toroidal field coil

plasma electrical current

toroidal magnetic field

B

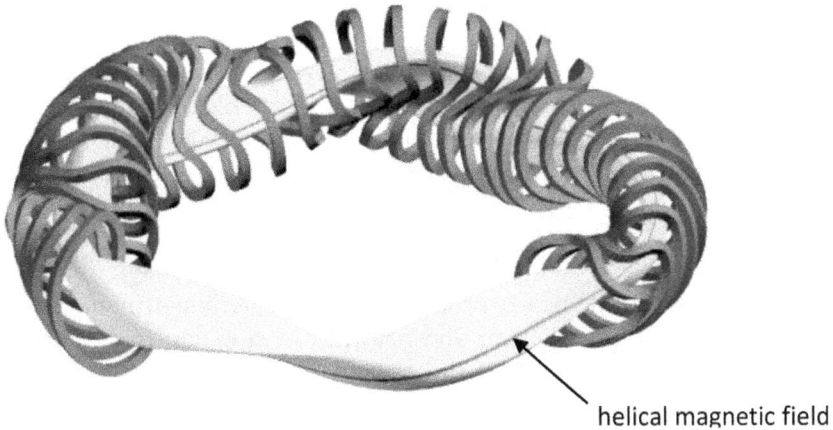

helical magnetic field

Figure 6.29. Schematic description of the coils and magnetic fields in the tokamak (a) and stellerator (b) magnetic confinement schemes and the resulting helical fields. (Source: https://www.euro-fusion.org, Credit EUROfusion consortium, and http://www.ipp.mpg.de, credit Max-Planck-Institut für Plasmaphysik, copyright IPP.)

hydrogenic neutrals that thermalize in the plasma and increase its temperature (similar to adding a bit of boiling water to cold water to get lukewarm water). Once high temperatures are achieved and the DT fusion reaction starts, the plasma starts to self-heat by the high-energy helium produced in the fusion reaction, which carry 25% of the energy produced in the DT reaction (equation (6.7)). At higher fusion reaction rates, this becomes the dominant source of plasma heating, with other heating sources external to the plasma being used to control the fusion reactivity.

In fact, the thermal insulation provided by the magnetic fields is limited by turbulent processes; therefore, the heat injected by external means and produced by plasma self-heating eventually escapes through the plasma external surface and is deposited on the material surfaces that surround the plasma. Since this thermal insulation is very good, it is possible to maintain temperatures of hundreds of million kelvins in the central part of the plasma, while it is typically a few million kelvins at the plasma external surface. In magnetic fusion reactors, the plasma itself is contained in a toroidally shaped vessel surrounded by the coils that create the magnetic fields. This vessel is clad with a wall of protective elements that are in direct contact with the plasma and ultimately receive the heat and neutrons produced by the fusion process. Neutrons, which carry 75% of the energy produced by fusion, are not electrically charged and are not subject to the Lorentz force (equation (6.8)) and thus distribute themselves uniformly on the protective elements. By contrast, the plasma particles are charged and tied to the magnetic field by the Lorentz force, which makes the plasma losses to the vessel protective elements flow along the magnetic field lines. This means that the heat losses from the plasma are deposited upon a very small fraction of the available surface of the vessel wall (typically a few percent), and this leads to very large power fluxes (flux = total heat/area). Even if only 25% of the energy produced by fusion reactions is deposited in the plasma itself by the energetic helium, the power fluxes deposited on the wall by the plasma can be 100 times larger than those from neutrons in a fusion reactor. Such concentrated plasma fluxes can cause local overheating and erosion of the wall components, which pose specific challenges, as will be discussed in the next section.

The effectiveness of fusion production is characterized by the fusion gain (Q), which measures how much power is produced by the deuterium–tritium plasma compared to that used to heat up the plasma by external means:

$$Q = \frac{P_{\text{DT fusion}}}{P_{\text{heat}}} = \frac{P_{\text{He}} + P_n}{P_{\text{heat}}} \tag{6.9}$$

where P_{He} and P_n is the power produced by the deuterium–tritium fusion reaction in the form of energetic helium and neutrons in equation (6.7) and P_{heat} is the external power required to heat the plasma. This can be directly related to the product of the plasma density (n), temperature (T), and the so-called energy confinement time (τ_E), which characterizes the time for plasma heat losses (i.e., the characteristic time in which the plasma loses its energy when heating stops). This triple product is commonly used to quantify progress in fusion energy research, since most of the present experimental facilities do not make use of deuterium–tritium plasmas but only deuterium plasmas. In these plasmas the equivalent production of fusion power by deuterium–tritium can be easily quantified by the triple product. Note that actual DT fusion power production in magnetic confinement devices has been demonstrated in only two tokamak experiments (TFTR (US) and JET (EU)), with JET having recently achieved the fusion energy production record in the magnetic confinement line. The National Ignition Facility (US) has also made significant progress recently with the achievement of the first self-sustained fusion plasma and record energy production in the inertial confinement line.

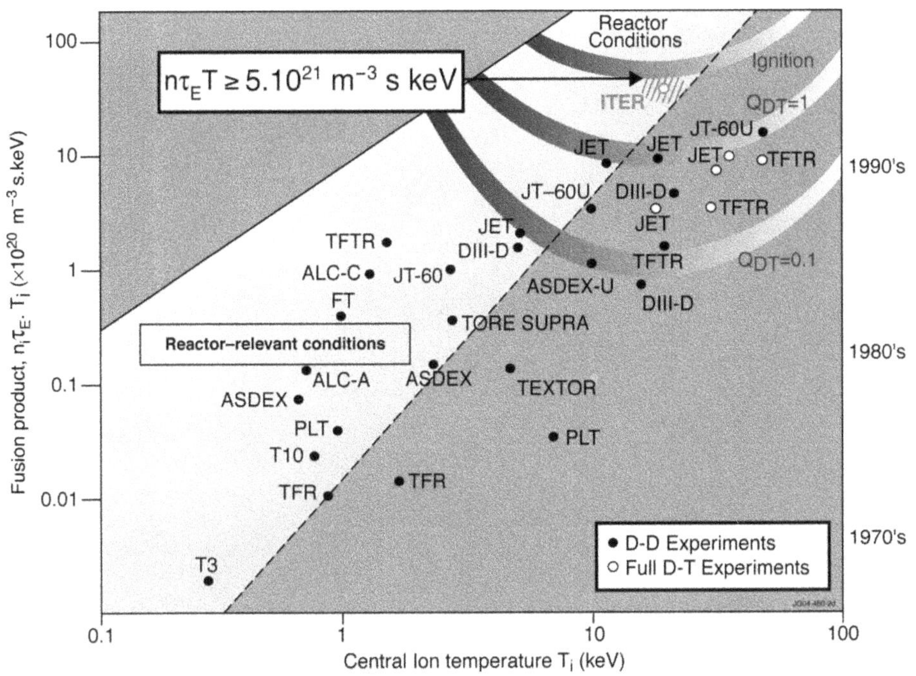

Figure 6.30. Experimentally achieved values of the fusion triple products versus temperature of the plasma hydrogenic ions in kiloelectronvolts (1 keV ≈ 12 000 000 K) in a range of tokamak devices and expected values in ITER. (Source: https://www.euro-fusion.org. Credit EUROfusion consortium.)

The challenge of magnetic confinement fusion energy development is thus the creation and maintenance of high-temperature plasmas with sufficiently high density and low heat losses (e.g., long τ_E) for the fusion reactions to occur and self-heat the plasma for long durations (minutes to hours). This has been the focus of fusion research in experimental facilities, and significant progress has been made, as summarized in figure 6.30. As shown in this figure, deuterium–tritium equivalent fusion power production in present experimental facilities has reached $Q \leqslant 1$ values, meaning that more power is required to heat the plasma up than is produced by fusion processes. Although not shown in this figure, the duration of the high fusion power production phases is restricted to a few seconds at most. Experiments have demonstrated that to achieve and sustain fusion power production over long time scales, it is necessary to understand the physics processes not only of the hot deuterium–tritium plasma but also of those at the outer colder layers of the plasma and in the interaction zone between the plasma and the wall-protective elements. Therefore, although the individual physics processes required to achieve fusion power production have been identified and thoroughly explored in present experimental facilities, their integration in a plasma dominated by self-heating from fusion reactions (i.e., the fast helium produced by deuterium–tritium fusion) remains an open challenge.

a)

b)

c)

Figure 6.31. (A) Cutaway drawing of the ITER tokamak showing the main components enclosed in the cryostat and an artistic impression of a plasma inside its vessel. (B) First portion of ITER vessel being assembled with two coils that will produced the toroidal magnetic field. (C) Aerial view of the ITER site with the buildings where the tokamak is assembled and will operate (center, in black) and the supporting ancillary plants and buildings spread over 42 hectares. (Source: https://www.iter.org. Credit ITER Organization.)

The demonstration of high Q fusion power production for extended times of minutes up to an hour is the mission of the first generation of experimental fusion reactor devices, chiefly the ITER tokamak that is presently under construction (figure 6.31). The physics basis for the design of the ITER tokamak rests solidly on the physics understanding provided by the present experimental fusion facilities and theoretical developments [96]. Specifically, ITER is designed to demonstrate fusion power production with $Q \geqslant 10$ for durations of 5–8 min and aims to demonstrate $Q \geqslant 5$ for durations of up to 1 h [95]. These goals have been considered for the determination of the physical dimensions of the plasma (with a volume of 840 m^3, i.e., ten times larger than existing fusion facilities), the magnetic fields to be applied, the electric current created in the plasma, and so on, but they also have a direct impact on technological choices for the components of the device. For example, heat dissipation by ohmic heating if the ITER coils were made of copper would be very large; therefore, superconducting materials are used. These superconducting coils have negligible electric resistance and thus are not heated up by the electric currents that circulate in them. For the superconducting coils to reach such conditions requires very low temperatures (a few degrees Kelvin) and thus strong refrigeration (by liquid helium) and very good insulation from surrounding heat sources. This is provided by installing the coils inside a cryostat (a 'Thermos flask' of very large dimensions) and in vacuum.

The size of the ITER tokamak and its ancillary systems, the complexity of the technologies required for the achievement of its goals, and the unprecedented plasma parameters that will be achieved for the first time in a self-heated plasma by deuterium–tritium fusion processes make this experimental facility unique for addressing the challenges ahead in the development of fusion as an energy source. Its international cooperation dimension is also a demonstration of the importance of ITER's goals for humankind.

ITER by itself will be a major first step in the demonstration of fusion energy as an energy source. ITER's scientific exploitation will be accompanied by other, smaller-scale devices that will address specific fusion physics issues in more detail that can be explored in ITER or will explore different approaches to the tokamak concept on which ITER is based. The knowledge acquired in ITER and these accompanying devices will be used for the construction of DEMO [97], the first demonstration fusion power plant with net electricity production from deuterium fusion reactions. In the next section, we discuss the scientific challenges that need to be faced to achieve this goal and the contributions from ITER to these challenges.

6.3.4.2 Challenges for the development of fusion energy: ITER and beyond

As was mentioned previously, the main challenges to be resolved in ITER and future magnetic confinement fusion reactors concern the simultaneous integration and control of the key physics processes that are required to achieve a plasma dominantly heated by deuterium–tritium fusion leading to high-gain fusion power production. To face these challenges, an understanding of the underlying physics processes determining heat and particle losses from the plasma, the stability of the equilibrium of forces between the plasma and the magnetic fields, and the integration of fusion plasmas with wall requirements, to cite the major key challenges, is required. Conceptually, such challenges are similar for the main magnetic confinement schemes (tokamaks and stellarator), although quantitatively they can substantially differ.

Confinement and transport in self-heated fusion plasmas

The conductive and conductive losses from the plasma are determined by turbulent transport processes driven by the variations of plasma parameters across the magnetic fields. The dominant turbulent transport processes that will take place in

ITER plasmas are well established from present experimental results and a well-developed plasma theory/modeling basis. However, the magnitude of the turbulent processes is difficult to predict quantitatively for plasmas in ITER and future fusion reactors, since they will have densities and temperatures that cannot be achieved simultaneously in present experimental devices. The precise magnitude of turbulent transport has a direct impact on the plasma parameters and thus on the self-heating of the plasma by fusion reactions. Since plasma self-heating dominates external heating in ITER and fusion reactors, unlike in plasmas of present experimental devices, a complex feedback loop among plasma parameters, self-heating, and plasma transport will be established. The detailed balance of these processes affects the efficiency in fusion power production (quantified by Q) that will be achievable in the reactor.

It is of particular importance to understand the physics of convective losses (driven by the flow of plasma ions and electrons) in comparison to conductive losses (driven by temperature gradients in the plasma) and optimize their ratio as far as possible. Reducing conductive losses increases the plasma temperature and fusion power production while reducing convective losses; this can have detrimental effects if these become too small. If plasma flows from the core to the periphery are not sufficiently intense, this can cause the accumulation in the plasma of the helium 'ash' produced by the fusion reaction and other impurities present in the plasma (resulting from interaction of the plasma with the wall-protective elements), and this can eventually stop the reaction altogether.

Convective transport also plays the essential role of replenishing the burnt deuterium and tritium in the plasma core where fusion reactions take place. If no 'fresh' deuterium and tritium are provided to the plasma core, fusion energy production will stop because of lack of fuels. Directly injecting deuterium and tritium into a hot fusion plasma is not possible; due to the high plasma temperatures of fusion plasmas (tens to hundreds of millions of kelvins), any deuterium or tritium atom entering the plasma is immediately ionized and thus tied to the magnetic field. To increase the deuterium and tritium fueling efficiency, these are injected into the plasma at high speed in the form of solid pellets, which first vaporize and then ionize. This allows increasing the penetration of the deuterium and tritium fuels up to several tens of centimeters into the plasma, which is still small compared to the typical dimensions of several meters of a fusion plasma. Therefore, detailed understanding of the convection of ionized deuterium and tritium from the plasma periphery towards the plasma core is key to achieving the replenishment of the burnt fuel; this is a rich physics area, since turbulent transport affects deuterium and tritium ions differently, due to their different masses.

Understanding the physics processes that dominate the feedback loops that will be established in self-heated plasmas and how to use external actuators (electric currents in the plasma, magnetic fields, etc) to optimize them for maximum fusion power production is a major objective of the ITER scientific program. The fusion gain of $Q \geqslant 10$, for which ITER is designed, was specifically chosen to ensure that the physics of self-heated plasmas can be investigated. For a $Q \geqslant 10$ fusion plasma, the internal plasma self-heating by fusion reactions is at least twice that provided by external heating sources. For demonstration fusion power plants with net electricity production after ITER, such as DEMO, the ratio of self-heating to external heating will increase to 5 or more.

Plasma stability control

Magnetic confinement fusion plasmas maintain equilibrium by the magnetic fields creating a force that opposes the expansion of the hot plasma, and this equilibrium may become unstable. These instabilities (so-called magnetohydrodynamic instabilities) directly affect the achievable fusion power production, since this requires high density and temperature in the reactor plasma and thus high pressures (in the range of 5–10 times the atmospheric pressure in the central plasma region). Instabilities in the plasma equilibrium can be local or global; local instabilities lead to a rearrangement of magnetic fields and plasma pressure in a toroidal annulus of the plasma and can cause a moderate decrease of fusion power production, while global instabilities cause the loss of magnetic confinement, cool-down of the plasma, and stopping of the fusion reactions. An example of such instability is the so-called edge localized mode (ELM), which affects the periphery of the plasma leading to the expulsion of plasma filaments that impact the wall-protective elements (figure 6.32(A)). If not controlled, in some cases local instabilities can grow and eventually trigger global instabilities.

The basic physics processes behind such instabilities, how to avoid them, and approaches to control them are well established. However, challenges remain for ITER and future fusion reactors because of the specificities of self-heated plasmas and because of the high plasma pressures which are required to achieve effective fusion power production. The latter implies the need to maintain the plasmas in the stable equilibrium zone but near stability boundaries. The presence of energetic helium ions (and from external heating sources) in the plasma can cause specific instabilities (so-called Alfvén instabilities) [98] that can cause the expulsion of these energetic ions from the plasma and reduce plasma self-heating and fusion power production. These need to be avoided or reduced to a sufficiently low level that the impact on fusion production is small. Operation of fusion reactors near stability boundaries implies the need to quantitatively predict these boundaries and to develop the capabilities to control the parameters of the plasma by external actuators (external heating, magnetic fields, etc) to avoid overstepping them; this is particularly complex in self-heated fusion plasmas because of the feedback loops described previously. Finally, if uncontrolled instabilities develop, it is necessary to mitigate their consequences. A prototypical example is the tokamak disruptive instability, which is a global instability in which the plasma loses its thermal energy in timescales of milliseconds and its magnetic energy in timescales of tens of milliseconds, leading to large forces being exerted and large plasma fluxes being deposited on the tokamak wall and components.

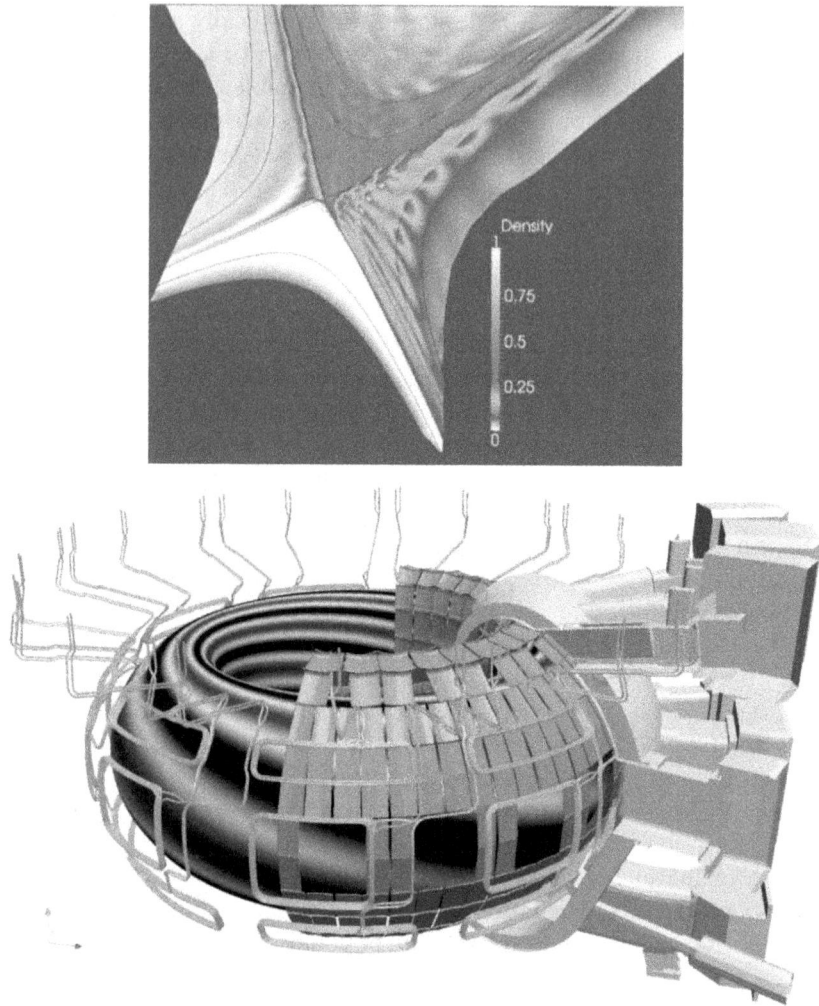

Figure 6.32. (A) Modeled plasma density contours showing the expulsion of plasma filaments from the peripheral plasma to the wall by ELMs in ITER. (Source [99].) (B) Set of 27 coils (in blue) to control ELMs in ITER. (Credit ITER Organization, courtesy of G Huijsmans.)

A major scientific objective of ITER is thus to demonstrate the avoidance and control of plasma instabilities in self-heated fusion plasmas and the mitigation of their consequences when these cannot be avoided. To achieve this goal, ITER is equipped with a set of versatile systems to create local currents in the plasma (by selectively acceleration of electrons with radiofrequency waves) to modify locally the magnetic fields at the plasma periphery (see figure 6.32(B)), and so on. Specifically for disruptions, ITER is equipped with a sophisticated system to reduce the heat fluxes and forces when disruptive instabilities cannot be avoided by the injection large amounts of hydrogen and neon in small solid pellets. This injection leads to a transient increase of the plasma density that allows the thermal and magnetic energy of the plasma to be lost as a flash of light emission (thus the need for neon) that is uniformly deposited on the tokamak walls rather than by plasma fluxes and electric currents on the tokamak wall and components.

Integration of fusion power producing plasmas with tokamak wall requirements

As discussed in the previous section the energy lost by the plasma by conduction and convection is ultimately deposited on the tokamak walls by plasma electrons and ions. Since both move along magnetic field lines, this causes concentrated heat fluxes in a very small fraction of the total wall area. For ITER and future fusion reactors these power fluxes are comparable to those at the Sun's surface ($\sim$60 MW m^{-2}), and there is no technological solution to evacuate such levels of heat for any significant length of time. For example, ITER's wall components are cooled by a high-pressure water flow and have demonstrated capabilities to extract heat fluxes up to 10 MW m^{-2} for long durations, but higher heat fluxes, when sustained over time, cause component degradation and reduction of lifetime. In addition, the impact of a high-temperature plasma causes erosion of the wall by energetic ion impact, similar to sandblasting of a surface, and this also limits its lifetime. Replacing wall components in ITER and fusion reactor is a complex operation that can take several months and thus reduces the availability of the reactor for experiments (ITER) or electricity production (DEMO and later fusion reactors). Therefore, wall component degradation and erosion need to be minimized.

The solution to this problem has been demonstrated in present experimental facilities by the injection of so-called impurities such as noble gases (e.g., neon) which transform the heat flux from the plasma into light, which deposits over a wide wall area, in a similar way to a fluorescent lamp. The intense light emission decreases the local heat fluxes on the wall and cools the peripheral plasma in contact with the tokamak wall down to temperatures of $\sim$10 000 K rather than the several million kelvins that the plasma in contact with the material would have otherwise. This large decrease of plasma temperature implies that the ions impacting the wall do so at very low energy and erosion of the wall is thus negligible. On the other hand, if the impurities for heat flux dissipation enter the core plasma from the periphery where they are injected, they can cool and dilute the deuterium–tritium plasma and decrease fusion power production. To optimize heat flux dissipation versus impurity penetration into the core plasma, specific configurations of the peripheral magnetic field have been explored in present experimental devices (so-called divertor configurations) and have been adopted for ITER and future fusion reactors. The challenge remains for ITER to demonstrate that this approach can provide the required heat flux dissipation and sufficiently 'clean' plasmas for high Q fusion power operation. An important ingredient in this challenge concerns the understanding of the physics processes that govern the transport of plasma particles (including impurities) and heat at the plasma periphery. It is unclear whether the same processes that govern these phenomena in present experiments will be at work in ITER or whether other processes will dominate peripheral plasma transport, given the large densities, temperatures, and gradients expected at the ITER plasma edge.

Therefore, for ITER to achieve its goals it is necessary to demonstrate that a core plasma, with the high temperatures and density required for deuterium–tritium fusion to take place, can coexist with the peripheral plasma conditions required to have fluxes on the wall that can be handled by present technologies in terms of both heat fluxes and erosion. This requires a detailed understanding and control of the interaction between the plasma and the wall materials, the dynamics of the injected impurities, and so on. This is a prototypical example of the core–peripheral plasma integration issues that need to be demonstrated in ITER and future fusion reactors, but there are many others issues, such as the exhaust of the helium ash and the control of power fluxes due to ELMs. In general, such integration issues become more challenging for fusion reactors beyond ITER because of the higher fusion powers required for the demonstration of electricity production. In this regard, ITER is an essential first step on the way to the solution of such integration issues, but additional steps beyond ITER will be required to demonstrate the degree of integration required for future fusion reactors.

To conclude this subsection, we note that the challenges that we have discussed focus on plasma and plasma–material interaction physics, but these are not the only issues that need to be faced for fusion energy development. For example, it is necessary to understand the behaviour of materials to be used for the construction of fusion reactor vessels and in-vessel components under the high-energy neutron fluxes produced by deuterium–tritium plasmas and to demonstrate *in situ* tritium production (from lithium) to the level required for self-sufficiency in a reactor. ITER will provide a first insight into such challenges and an initial demonstration of tritium production, but the knowledge required for future fusion reactors will need to be gained through future dedicated experimental facilities (see [100, 101] for a description of these issues and challenges).

6.3.4.3 Conclusions
Fusion energy is a very attractive option to supply the energy needed by humankind in the next centuries. The fuels required are abundant, and the fusion reaction produces no long-lived radioactive residues. The process of research and development of fusion energy has been a long one that started in the middle of the 20th century and has now come to fruition. The next few decades will be crucial to demonstrate the scientific and technical viability of fusion as an energy source by integrating the acquired knowledge in physics optimization (e.g., energy and particle confinement, impurity control, power exhaust) and engineering optimization (e.g., plasma-facing components, tritium technologies, heating and current drive systems, plasma diagnostic systems). A key step in this demonstration will be the scientific exploitation of the ITER experimental reactor in which deuterium–tritium fusion plasmas, producing more energy by fusion reactions than that required to heat them, will be demonstrated for the first time. This key step will be accompanied by research in other specialized experimental facilities to address additional challenges

(e.g., neutron resistant materials, advanced power exhaust schemes), which need to be addressed for the design and construction of a net electricity-producing fusion reactor.

The successful resolution of these integration challenges for fusion energy development requires further understanding and control of the physics processes that govern the behaviour of fusion plasmas. As an example, such detailed understanding is essential for maximizing power production in fusion reactors while ensuring an acceptable lifetime of the wall-protective elements whose power-handling capabilities are determined by materials properties and cooling technologies limits. This advance in physics understanding will be the focus of experimental research in the next decades for the new generation of experimental fusion devices that have recently come into operation or are presently under construction. Among them, ITER will provide unique contributions for self-heated deuterium–tritium reactor-scale plasmas, which are essential for the follow-up step of a net electricity-producing fusion reactor such as DEMO. This experimental research will be accompanied by theory and modeling developments, taking advantage of the capabilities provided by advanced supercomputers, which are necessary to understand the details of the physics processes at play in experiments and for their extrapolation. Recent record results of fusion energy production, using both inertial and magnetic confinement, are the clearest demonstration of the potential for fusion energy to deliver a safe and sustainable low-carbon energy source. These advancements bring us one step closer in the quest of fusion energy for our society.

6.3.5 Fission energy: general overview

Marco Ripani[1]
[1]National Institute for Nuclear Physics, Genova, Italy

There are three basic physical phenomena behind nuclear energy from fission. First of all, some specific heavy nuclei, called *fissile*, can easily split into two lighter fragments when they absorb a neutron[1]. Second, when such nuclei split, more neutrons with relatively high energy of motion are emitted, which can in turn produce other fissions, in a so-called *chain reaction*. Finally, thanks to Einstein's mass–energy equivalence, the total mass of the two fragments and the few neutrons emitted is less than the initial mass of the fissile nucleus; the missing mass has been transformed to energy. Each fission reaction releases about 3.2×10^{-11} J. By comparison, a chemical reaction releases energy of the order of a few 10^{-19} J.

In nature, the element uranium is found in two different forms, or isotopes[2], namely, U-235 and U-238, in the proportion of 0.7% for the first and 99.3% for the second. U-235 is fissile and can undergo fission with high probability when exposed to slow neutrons. Nuclei such as U-238 are called *fissionable*, that is, they can give rise to a fission reaction, but only for fast neutrons above a certain energy of motion. Nuclei such as U-238 are also called *fertile*, because when hit by a neutron, they start a series of nuclear reactions that lead to the formation of new elements. In the case of U-238, irradiation with neutrons leads to the formation of plutonium isotopes, among which the most relevant is Pu-239, which is fissile. Therefore, starting from a mixture of U-235 and U-238, not only fission of U-235 occurs, but also production, or *breeding* of new fuel, in terms of the mixture of plutonium nuclei containing Pu-239. Most reactors using mixed U-235/U-238 fuel typically need 3%–5% U-235, so a procedure called uranium *enrichment* is needed to fabricate reactor fuel. Other reactors work by using natural uranium without enrichment. It is worth mentioning that thorium, an abundant element in nature, can in principle be used to produce (or breed) a different type of fuel. Indeed, in neutron absorption by thorium the final product is U-233, yet another fissile nucleus. Use of thorium as a breeder of U-233 is studied internationally as a possible way to increase the available stock of fissile fuel.

The whole process of mining, extraction of the uranium mineral from the ore, purification, fuel fabrication through enrichment, irradiation, and final storage or disposal is called the *fuel cycle*. In the *open* or *once-through* fuel cycle the spent fuel is put in temporary storage waiting for final disposal. In the *recycling* (sometimes called *closed*) fuel cycle the spent fuel is reprocessed to extract the plutonium, which is then used to fabricate new fuel, typically in the form of mixed oxide of uranium and plutonium.

---

[1] The neutron is one of the two components of atomic nuclei (protons, which are positively charged, and neutrons, which have no electric charge).

[2] Isotopes are nuclei of the same chemical element that contain the same number of protons but different numbers of neutrons.

The fragments emerging from fission are typically radioactive, so the fuel extracted from the reactor, *irradiated* or *spent* fuel, is a highly radioactive material that needs to be protected and shielded. While most of the radioactive fragments have relatively short half-lives[3], the longest being of the order of a few tens of years, a small group of fragments has very long half-lives of the order of 10^5 years, the so-called long lived fission products. The aforementioned plutonium production process is one example of a process leading to the appearance of *transuranics* in the fuel, that is, chemical elements beyond uranium that do not exist in nature. Many transuranics have long half-lives, from a few hundred years to a few hundred thousand years (e.g. Pu-239 has a half-life of 24 000 years). Reprocessing the irradiated fuel gives the possibility of recycling the Pu to fabricate new fuel, but whenever reprocessing is not applied or is stopped, the resulting spent fuel is considered to be *nuclear waste* or *radioactive waste* which needs to be properly stored and eventually disposed of. Indeed, such radioactive nuclei can be dangerous for the environment and for human health (due to direct exposure, ingestion, or inhalation). Additional radioactive waste is produced in the reactor by neutrons interacting with the surrounding materials, producing *activated* materials with varying degrees of radioactivity. The International Atomic Energy Agency has come up with a classification scheme that is used as a basis by each country to define categories of waste [102].

In terms of fuel consumption, in a 1-GWe reactor[4] at 80% load factor (the actual operational time) the annual consumption of fissile material is about 900 kg: in volume of pure metallic heavy elements, this would be a cube with sides of about 35 cm. If we consider an actual oxide fuel enriched in U-235 to be 3.5%, the annual fuel consumption will be of the order of 30 tonnes. A comparison between the annual consumption of different fuels is shown in table 6.1[5].

All nuclear reactors today (so-called Generation II, III, and III+) are based on fuels containing U-235 and/or Pu-239. With a few exceptions (to be discussed later in this section), they contain light materials such as water, *heavy water*[6] or graphite to slow down the fast fission neutrons and increase the fission probability[7]. Water or

Table 6.1. Comparison between different annual fuel consumption rates, for a 1-GWe plant at 80% load factor.

Nuclear fuel (tonnes)	Natural gas (cubic meters)	Coal (tonnes)
30	1.4 billion	2.1 million

---

[3] Radioactivity declines with time following an exponential law. The time after which the amount of radioactivity has halved is called the half-life.
[4] A reactor producing about 3 GWth of thermal power converted to about 1 GW of electric power at 33% conversion efficiency.
[5] Assuming a 50% conversion efficiency for gas and 40% for coal.
[6] In heavy water, instead of the ordinary H_2O molecule, hydrogen is replaced by its heavier partner deuterium.
[7] These materials are called *moderators*.

gas is used to cool the fuel and transport the heat produced by the fission reactions to the equipment, producing electricity. In *fast reactors*, the materials crossed by the neutrons are chosen to keep the neutrons fast and energetic. This can be accomplished if the core contains mostly heavy materials. Therefore, for core cooling, the choice will be a liquid metal such as sodium, lead, or a lead–bismuth mixture; alternatively, the coolant should be a low-density material such as helium gas. On the other hand, the fission probability is smaller for fast neutrons, which means that a much higher percentage of fissile material, of the order of 20%, is needed to sustain the chain reaction. Only two examples of fast reactors connected to the electricity grid exist today (to be discussed later in the section), while many others are prototypes or research reactors, and a few more are planned for the future.

6.3.5.1 Worldwide figures

Worldwide electricity consumption increased annually by 2.9% on average in 2007–20 [103], regardless of progress in transmission or utilization efficiency, while a slight decrease occurred in 2020 due to the COVID-19 pandemic. Nuclear energy from fission continues to have an important share in the electricity mix. After recovering from the significant post-Fukushima drop due to several shutdowns and stress tests, production returned to pre-Fukushima levels, apart from COVID-affected 2020, with 2% more electricity supplied in 2019 with respect to the average of 2007–10 (see figure 6.33). However, while total electricity production increased by 31% in the same years, the worldwide percentage of nuclear electricity went from about 14% to around 10% [103, 104] (see figure 6.34). As of July 2021, there were 443 reactors in operation in the world, for a total net installed capacity of about 393 GWe, and 51 under construction, for a total capacity of about 53.9 GWe [105].

6.3.5.2 Cost of electricity

One of the parameters used to compare the economics of different electricity technologies, independent of their very different characteristics, is the levelized cost of electricity (LCOE). It is calculated by dividing costs of a plant over its whole

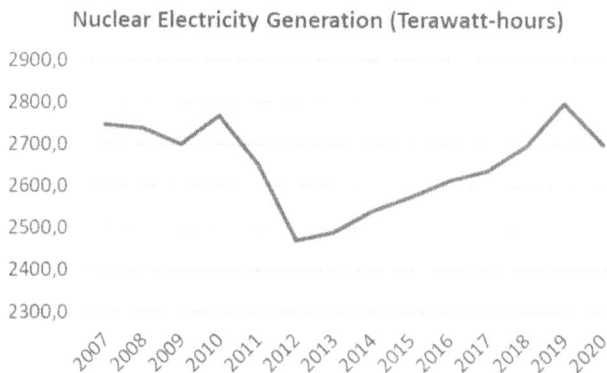

Figure 6.33. World nuclear electricity generation from 2007 to 2020. Data from [103].

Electricity generation by fuel

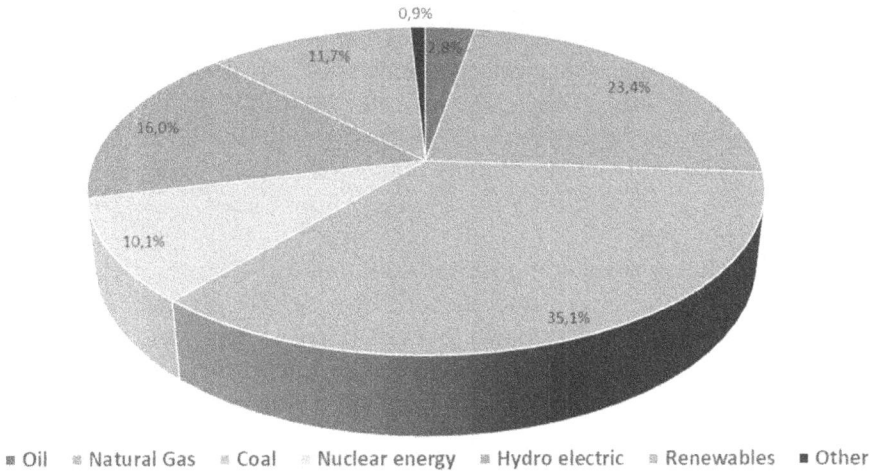

Figure 6.34. World electricity generation by fuel (2020). Data from [103].

lifetime by energy production and is typically given in USD/MWh. In the IEA/NEA/OECD study of 2020 [106], at a 7% discount rate, nuclear energy's LCOE appears to be definitely lower than that of coal and combined cycle gas turbines with carbon capture and storage, as well as lower than that of most renewables, with the exception of utility-scale photovoltaic, onshore wind, and hydropower. When considering lifetime extensions of existing nuclear plants, the nuclear LCOE is definitely lower than any other source, be it coal or renewables, with only a marginal superposition with utility-scale photovoltaics.

A similar study was performed by the Intergovernmental Panel on Climate Change (IPCC) [107] that assumed a 10% discount rate scenario, two scenarios for the number of operational hours of the plants, and two scenarios for the cost of carbon emissions for the CO_2 emitting technologies. This study, too, shows that nuclear energy features an interesting LCOE, especially when carbon emission costs are factored in, and is cheaper than many renewable or frontier technologies. Therefore, the nuclear LCOE appears to be competitive with that of other sources. This is particularly true in low discount rate scenarios, given that nuclear energy is a capital-intensive energy source, requiring large investments at the construction stage.

6.3.5.3 Carbon emissions
Although nuclear power plants do not emit CO_2 (nor other greenhouse gases or fine particulates) during electricity production, CO_2 can be emitted along the whole plant lifetime, from fuel mining and extraction to final decommissioning. A detailed study comparing emissions among electricity production sectors in terms of CO_2 per unit of produced energy has been performed by the IPCC [107]. Their report clearly

shows that the small amount of CO_2 produced as a result of the whole life cycle is much lower than that from fossil fuel–fired power plants, is comparable with that of many renewables, and is somewhat lower than that of solar photovoltaic. A similar conclusion is found in [108] for the case of the twice-through cycle adopted in France, in which the nuclear spent fuel undergoes one cycle of reprocessing to fabricate new fuel from uranium and plutonium, with 5 g of CO_2 equivalent production per kilowatt-hour of nuclear electricity produced, the smallest value in a group including coal, oil and gas, photovoltaic, hydropower, and wind.

6.3.5.4 Resources
World annual uranium consumption amounted to 62 825 tU (tonnes of uranium) in 2017, and assuming it to be constant, identified recoverable resources should suffice for over 130 years [109]. It is estimated that additional undiscovered and unconventional resources could suffice for well over 400 years, but this would require significant efforts in securing all resources for an effective use. Moving to advanced technology reactors and recycling fuel could increase the long-term availability of nuclear energy from hundreds to thousands of years. Thorium, which is more abundant than uranium in the Earth's crust, is another potential source of alternative fuel if used to breed U-233, a fissile nucleus that does not exist in nature. Even when relying only on identified resources, assuming an increase in nuclear energy production as in the IEA NZE scenario and considering only current reactor technologies, there should be enough uranium supply for the rest of the century. This implies that nuclear energy from fission can have the important role of a bridging technology until nuclear fusion becomes commercially available. However, research on fuel recycling and innovative reactor technologies should continue to be pursued to have the possibility to extend the availability of nuclear power from fission well beyond the end of the century. This is also important because both fuel recycling and the introduction of innovative reactors, in particular fast reactors, would help to reduce the amount of radioactive waste to be disposed of.

6.3.5.5 Accidents and safety
A lot of public concern has been raised by nuclear accidents, most notably the ones at Chernobyl in 1986 and Fukushima in 2011. Obviously, the use of nuclear power must be appropriately regulated and managed with maximum care, given the potential impact of accidents entailing release of radioactive substances into the environment. Even though no industrial activity (and no human activity in general) can be made totally risk-free (indeed, several accidents have happened in the fossil fuel sector, including oil spills in the sea, fuel truck and train carriage explosions, and fires), the goal of nuclear plant designers is to incorporate as many safety measures as possible to prevent any anomaly from escalating to a significant nuclear accident, following the principle of defence in depth [110]. Safety is an aspect of utmost importance in nuclear power production. The core of a nuclear reactor produces radiation both while in operation and after shutdown, because of the high content of radioactive materials in the fuel and because of the neutron irradiation of structural materials, which produces radioactive nuclei starting from stable ones. Lessons

learned from past accidents have been implemented in the design of plants, with particular regard to engineered safety measures. Attention to safety is present from the design stage all the way through to operation to shutdown and decommissioning of the plants, and particular care is devoted to the analysis of possible accidental scenarios. A fundamental principle is the *defence in depth* [110], which means that a series of barriers, starting from the solid fuel form itself up to the reactor containment building, are put in place to contain radioactivity as much as possible at all times and in all instances. Moreover, research specifically focused on the safety of nuclear plants is being conducted worldwide to pin down and address residual weaknesses, also taking advantage of lessons learned from the past plant operational history. Obviously, in this respect, the role of independent regulatory authorities is crucial in overseeing all stages of a plant lifetime, from design to operation to decommissioning and dismantling, and corresponding radioactive waste management. Clearly, scientific and technical aspects are the main practical objectives of a safety assessment. However, human and organizational factors are also deemed to be of high importance, as, for example, human errors may still have an impact even in a system which is designed and operated according to the state of the art. Therefore, what is called culture for safety means that technical achievements and established engineering standards are a must, but a continuous process of self- and reciprocal assessment on human and organizational aspects is also necessary, involving all actors, from industry to regulators, including the stakeholders for whose benefit nuclear energy is ultimately utilized [111].

6.3.5.6 Radioactive waste

The decay time of fission products is the reason why disposal of spent fuel and waste requires very long term storage, for which one possible solution envisaged is storage deep underground in the so-called geological repositories. According to the IAEA categorization [102], depending on quantity and decay time, nuclear waste is classified into Ex-emptExempt waste (EW), Very short lived waste (VSLW), Very low level waste (VLLW), Low level waste (LLW), Intermediate level waste (ILW), and High level waste (HLW). VLLW is suitable for disposal in near surface facilities of the landfill type with limited regulatory control. LLW requires robust isolation and containment for periods of up to a few hundred years and is disposed of in engineered near-surface facilities. ILW requires a greater degree of containment and isolation, which can be provided by disposal at greater depths, of the order of tens of metres to a few hundred metres. For HLW, disposal in deep stable geological formations, usually several hundred metres or more below the surface, is the generally recognized option for disposal. In the case of HLW, the internal heat generated by radioactive decay may have to be taken into account. Near-surface disposal facilities are safely in operation since many years in several countries around the world [112]. The IAEA estimated that, since the start of the nuclear power era in 1954 to the end of 2013, a total of about 370 000 t HM (tonnes of heavy metal) of spent fuel was discharged from all nuclear power plants worldwide (excluding India and Pakistan), of which about one third was reprocessed to produce a second round of energy production with recycled fuel [113]. This total amount of discharged spent fuel would correspond to a few meters over the surface of a standard soccer field. International recommendations (e.g. [114]) have guided national policies, strategies and programs for the management of spent fuel and radioactive waste. Surface or near-surface facilities are a reality in several countries and demonstrated that safe disposal of the corresponding classes of waste can be easily realized. Continuing research on the optimal choices of sites and the prediction of the time evolution of the waste packages underground is performed at international level. In particular, Euratom has promoted the advancement of research on several aspects of decommissioning technologies (e.g. SHARE Euratom project [115]), as well as on aspects and issues arising in interim storage, pre-disposal and final disposal, where in the latter point the scientific aspects of deep geological disposal are of special interest. Alternatively to geological disposal, it has been proposed to burn at least part of the nuclear waste with accelerator-driven systems, based on a subcritical reactor core [116]. A subcritical core is one where the fission chain reaction does not occur spontaneously in a steady state but rather requires an external neutron source to work, such as that produced by an accelerated beam of protons impinging on a thick target. Thanks to the subcriticality, there would be the possibility to mix a certain quantity of minor actinides (MA) into the fuel, which make it possible to incinerate them by fission. This is called transmutation of the MA into fission products, which can reduce the radiotoxicity of the final materials in the fuel in the long term, such that the confinement time would be reduced by orders of magnitude, from hundreds of thousands of year to a few hundred years. In order to set up such an incineration cycle, both partitioning and transmutation are needed. In the partitioning stage, plutonium and MA are separated from the spent fuel; then in the transmutation stage they are irradiated in a special core to be incinerated [117]. Supporting a robust research program is necessary because radioactive waste management programs present special challenges in their planning and execution. Indeed, besides involving science and technology, they also involve programming, regulatory, public participation, and stakeholder involvement aspects [118]. While engineering aspects are generally well understood, the necessary scientific topics of site characterisation, process modeling, safety assessment, and so on can evolve over time and require the capability of adapting the program planning to new findings [118].

6.3.5.7 Nuclear power in future scenarios

There are several projections about the future development of electricity and energy supply sources. Nowadays, several scenarios consider how to develop a low-carbon economy [119]. Typically, nuclear energy plays a role in all projections, even when a general trend towards phasing out nuclear power is conservatively assumed. Indeed, nuclear power plants are capable of delivering a large quantity of electric power (typically a power of the order of 1 GWe with high availability), providing a solid contribution to baseload electricity production. Like hydropower, nuclear power requires much less material resources for the same electricity output, due to high power density and capacity factor. As a result, nuclear power plants have relatively low emission levels per unit of energy produced [107, 108]. Nuclear plants provide dispatchable electricity and can be regulated according to the system needs, although rapid load following is not yet a common feature of these types of plants. Moreover, they have the potential to offer grid services needed to ensure continuous balance between load and generation [119].

The IPCC report on limiting global warming within 1.5 °C [120] discusses four illustrative pathways by which different measures are adopted across energy production, industry, and agriculture to mitigate the temperature increase on the planet. In all four pathways, nuclear energy is expected to increase in 2030 with respect to 2010 by amounts that vary between 60% and 100%. Variations from 2010 to 2050 are even higher, between 100% and 500%. Even though there is a large variation among the many different models considered, 50% of the models indicate an increase in nuclear energy from 2010 to 2030 between 44% and 102%[8], while 50% of the models indicate an increase in nuclear energy from 2010 to 2050 between 91% and 190%.

Among the most recent studies is the IEA Roadmap to Net Zero [121]. This study considers three scenarios: the Stated Policies Scenario (STEPS), based on well-established policies and up-to-date information about energy, climate plans, and related policies based on explicit laws or regulatory actions, where the latter also come from commitments following the Paris Agreement and can cover economic measures as well; the Announced Pledges Case (APC) scenario, based on timely and full realization of all national net-zero emissions pledges; finally, the Net-Zero Emissions by 2050 scenario (NZE) discusses what is actually needed, and when, to achieve net-zero CO_2 emissions by 2050 worldwide in energy production and industry. In both STEPS and APC, nuclear power's role in the total energy supply will increase from now to 2050. In STEPS, nuclear energy is projected to grow by 15% between 2020 and 2030 and still more until 2050, while in APC the nuclear growth is projected to be around 25% by 2030, so nuclear energy would maintain its 10% share of electricity production and would maintain a similar trend until 2050. In NZE, the role of nuclear power is essential, and its contribution to the total energy supply is projected to increase by nearly 100% between 2020 and 2050.

6.3.5.8 Obsolescence of the reactor fleet and economics

Most reactors in operation today were built between the 1970s and the 1980s [122], and the typical lifetime of a plant ranges between 40 and 60 years. As a consequence of that and of the fact that some countries decided to phase out nuclear power, many plants are already being decommissioned or will be in the next couple of decades worldwide. On the other hand, in recent US and Western Europe experience, construction costs of nuclear power capacity have increased from about 2000 USD/KWh in 2005 to between about 7000 and 8000 USD/KWh [122]. Moreover, for

---

[8] This is the interquartile range of the models, meaning that models predicting variations below 44%—even negative ones where nuclear energy decreases—represent 25% of the total, while another 25% predict variations above 102%.

example, in the US, nuclear power is experiencing a strong market pressure due to low wholesale electricity prices, driven by low gas prices and rising renewable power capacity (see, for instance, [119]), yet another economical aspect that is affecting the ability of the utilities to remain profitable and stay on the market producing nuclear power.

6.3.5.9 *Innovative reactors*

In the effort to improve the safety, security, and efficiency of nuclear plants, new concepts of reactors have been developed with goals including the minimization of the production of MAs, a better and more efficient use of the fuel, a better thermodynamic efficiency, the possible production of hydrogen at high temperatures and finally improved safety features to minimize the risk of accidents. All these various concepts are considered within the so-called Generation IV reactors, which are the subject of an international initiative [123] and are one of the pillars of the European Sustainable Nuclear Energy Technology Platform [124]. In particular, fast reactors use liquid metals or gas as coolants, resulting in more energetic neutrons traveling within the reactor core. In such reactors, it becomes possible to partly burn not only fissile elements such as U-235 and Pu-239, but also fissionable elements such as U-238 and some transuranics, for which fission occurs only above a certain minimal neutron energy. Typically, the fuel has to be richer in U-235 or Pu-239 content, but on the other hand, to some extent U-238 becomes a fuel as well, which means that uranium resources are used more efficiently and will last longer. Another important feature of fast reactors is that the ratio between fission and production of transuranics is higher, so relatively lesser amounts of transuranics are formed. At the same time, thanks to the more energetic neutrons, in a fast reactor the transuranics that are produced can be partly incinerated in the reactor itself by fission, resulting in less long-lived radioactive waste. Besides several prototypes based on different technologies around the world, two sodium-cooled commercial reactors have been in operation for several years in Russia [125].

Recently, there has been growing interest in the concept of small modular reactors (SMR), delivering up to about 300 MW of electric power, with several ongoing projects around the world [126, 127]. Besides advantages in terms of improved safety features and ease of deployment due to high standardization, SMR are getting serious attention for district heating, where a major impact on CO_2 emissions can be expected.

6.4 Towards green cities: the role of transport electrification

Natalio Mingo[1], Gérard Gebel[1], Philippe Azais[2], Thierry Priem[2], Tuan Quoc Tran[3], Didier Jamet[1], Florence Lefebvre-Joud[1] and Simon Perraud[1]

[1]Université Grenoble-Alpes, CEA, Liten, Grenoble, France
[2]Université Grenoble-Alpes, CEA, DPE, Grenoble, France
[3]Université Grenoble-Alpes, CEA, Liten, INES campus, Le Bourget du Lac, France

6.4.1 Transport, climate change, and air pollution

Global warming has reached a point where there is widespread concern about the large impacts that are being observed on ecosystems, human health, food security, and water supplies [128]. CO_2 constitutes the main source of anthropogenic global warming, representing around 70% of greenhouse gas emissions [129]. Global CO_2 emissions are rising steadily [130], being about 60% higher in 2018 (33.5 Gt[CO_2]) than in 1990 (20.5 Gt[CO_2]) when the first IPCC report was edited. The transport sector, which relies heavily on fossil fuels (mainly oil), accounts for one fourth of the global CO_2 emissions.

Apart from CO_2 emissions, transport has also a major impact on air pollution and hence on human health [131]. The combustion of fossil fuels for transport applications leads to the emission of a range of air pollutants, such as particulate matter (PM), nitrogen oxides (NO_x), and carbon monoxide (CO), which have a significant impact on human health, reducing life expectancy and increasing medical costs. The European Environment Agency estimates that in 2018, long-term exposure to particulate matter with a diameter of 2.5 μm or less ($PM_{2.5}$) led to around 379 000 premature deaths in the EU-28 [132].

When comparing various cities in countries having similar GDPs [133], the example of Hong Kong stands out. The city has a rate of energy consumption of 80 kWh/person-day, significantly smaller than that of cities in France (150 kWh/person-day) or in the US (>200 kWh/person-day). Being densely populated, Hong Kong has an efficient public transportation, compact housing, and almost nonexistent private car ownership; these are presumably part of the explanation. This suggests that there is a lot that infrastructure and lifestyle, in particular transport, can help with when it comes to achieving greener cities.

6.4.2 How to reduce urban transport emissions?

In the EU, significant efforts have led to energy efficiency improvements in the transport sector. For example, in the case of road transport, the average CO_2 emissions from new passenger cars [134], measured in g[CO_2]/km, decreased by 30% between 2000 and 2018. However, transport demand continues to grow, and the sector remains the only one that has not been able to reduce its CO_2 emissions. In the EU, transport emissions have increased by around 20% compared to 1990 levels, while the EU total emissions have decreased by around 20% in the same period.

In order to lower the CO_2 and air pollutant emissions of urban transport sector, several routes can be considered in parallel, including the following:

1. Replacement of motorized transport by nonmotorized transport modes, such as bicycles (generating no emissions).
2. Replacement of individual transport by public transport modes, such as trains, trams, subways, and buses (generating much lower emissions, on a passenger and kilometer basis, than passenger cars).
3. Development of modular mobility with multimodal transportation networks, mixing public and individual transport modes with digital management of fleets and fluxes.
4. Replacement of internal combustion engine vehicles by electric vehicles.

Here, we focus on the last of these routes, that is, the electrification of transport using batteries or hydrogen fuel cells. The main advantage of electric vehicles (EV), such as battery electric vehicles (BEV) or hydrogen fuel cell electric vehicles (FCEV), compared to vehicles with internal combustion engine (ICE), is that EVs have much lower local emissions (since there is no combustion of fossil fuel), and therefore they are very good solutions to reduce air pollution, notably in urban areas. Furthermore, if the electricity used to recharge the battery of the BEV (or to synthesize the hydrogen by water electrolysis for the FCEV) is made from low-carbon primary sources (nuclear or renewable energy), then the electric vehicle has a low life-cycle carbon footprint compared to fossil fuel–powered vehicles, hence contributing to significantly reducing CO_2 emissions and fighting against climate change.

6.4.3 Transport electrification: state of the art

6.4.3.1 Electric vehicles and their powertrains
The electrification of vehicles has a long history. At the beginning of 20th century, there were more EVs in New York than ICE vehicles. However, the rapid progress of the latter, an attractive range of vehicles, and the development of the gas station network quickly overtook EVs. Since the 1990s, the emergence of efficient electrochemical storage solutions, the threat of fuel dependency, and increasing concerns about pollution have brought renewed interest in partial or complete electrification of vehicles' powertrain.

In the case of hybrid electric vehicles (HEV) (figures 6.35 and 6.36), the high efficiency of the electric motor (>90%) and its very high torque even at low speed are advantageously exploited to reduce emissions in the torque/engine speed range where the combustion engine has poor efficiency. Even if the only fuel remains fossil, it is used more efficiently. HEV are currently undergoing strong commercial development and should rapidly replace the NiMH battery-based 'full hybrid' solutions that have been widely marketed since 1997 (especially by Toyota).

A more complex solution, the plug-in hybrid electric vehicle (PHEV), generally a parallel hybrid type, aims at coupling two solutions requiring two energy sources: electricity from the grid and a fossil fuel. This solution makes it possible to travel a significant number of kilometers in all-electric mode (typically 50 km), which

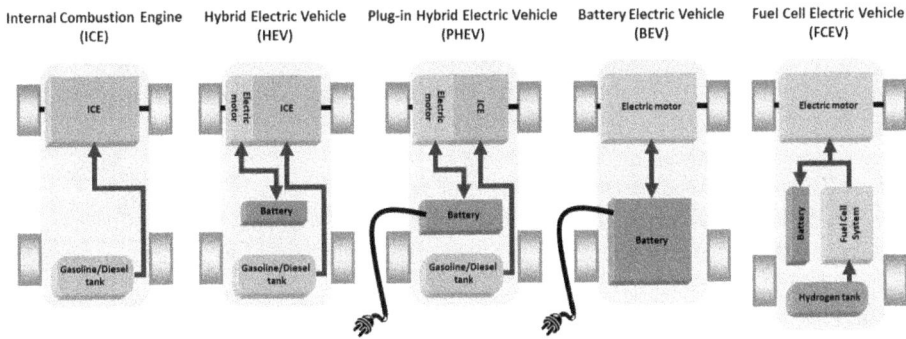

Figure 6.35. Simplified layouts of different powertrain configurations. (This figure is from the CEA HPC website (http://www-hpc.cea.fr). All reproduction rights are reserved.)

E-powertrain type		Micro	Mild / HEV	Full	PHEV	BEV
Functions	Start Lighting Ignition	X	X	X	X	
	Start & Stop	X	X	X	X	
	Boardnet	X	X	X	X	X
	Brake energy regeneration		X	X	X	X
	Boost		X	X	X	X
	Motor downsizing		X	X	X	
	ZEV		(a few 100m)	a few km	40 to 120km	> 150km
Main characteristics	System voltage	12V	48V	<300V	200 to 400V	300 to 800V
	System Power	< 5kW	10 to 30kW	40 to 165kW	> 30 kW	> 50 kW
	Usable energy/cycle	< 5Wh	0.3 to 1.3kWh	< 400Wh	5 to 18kWh	> 20kWh
	Installed energy	<2kWh	1 to 3kWh	< 2kWh	6 to 20 kWh	> 22kWh
	CO$_2$ decrease ratio	<5%	< 15%	< 20%	> 35%	100%
	Electrochemical Energy Storage System	EDLC, Pb Acid and / or Li-ion	Li-ion	NiMH	Li-ion	Li-ion

Figure 6.36. Main characteristics and functions of electrified powertrain types in passenger vehicles. Micro: micro-hybrid vehicle; Mild/HEV: Mild-hybrid vehicle/hybrid electric vehicle; Full: Full hybrid vehicle; PHEV: Plug-in electric vehicle; BEV: Battery electric vehicle. ZEV: Zero emission vehicle range. (This figure is from the CEA HPC website (http://www-hpc.cea.fr). All reproduction rights are reserved.)

generally covers urban daily use. For longer trips, long-distance driving is possible using the ICE powertrain.

Finally, it is possible to electrify the powertrain completely, as in BEVs, totally getting rid of fossil fuels on board the vehicle. The major drawbacks of this solution are the charging time and the considerable mass of the Li-ion battery that is required for long journeys. Typical energy consumption of such a BEV is 10 kWh/100 km/ton in a Worldwide Harmonized Light Vehicles Test Procedure (WLTP) cycle [135].

In the case of FCEVs, a fuel cell is used to produce electricity for powering the electric motor. Hydrogen, which feeds the fuel cell, is stored in a high-pressure tank (usually 350 bar) containing a few kilograms of hydrogen[9]. In this configuration, the range of the FCEV is directly correlated to the size of the hydrogen tank. A battery is usually coupled to the fuel cell to manage power demands. This solution is therefore very close to a full-hybrid solution.

6.4.3.2 Brief introduction to batteries and fuel cells

In an electrochemical generator, an exothermal chemical reaction is converted into two separated electrochemical reactions involving the same ion shuttle and generating electrons.

In a Li-ion battery the ion shuttle is Li^+. A Li-ion battery cell consists of two electrodes (a cathode, that is, positive electrode, and an anode, that is, negative electrode) separated by a porous separator (to avoid short-circuiting between the electrodes) impregnated with an electrolyte (one or more organic solvents with a lithium salt dissolved in them). The electrolyte is the Li ion carrier between the two electrodes. The name 'lithium-ion' comes from the fact that Li ions are transported between the host electrodes duringthe charge and discharge cycles (figure 6.37).

The nominal voltage of a Li-ion battery cell is generally around 3.7 V. Therefore, it is necessary to combine several cells in series to increase the battery voltage and in

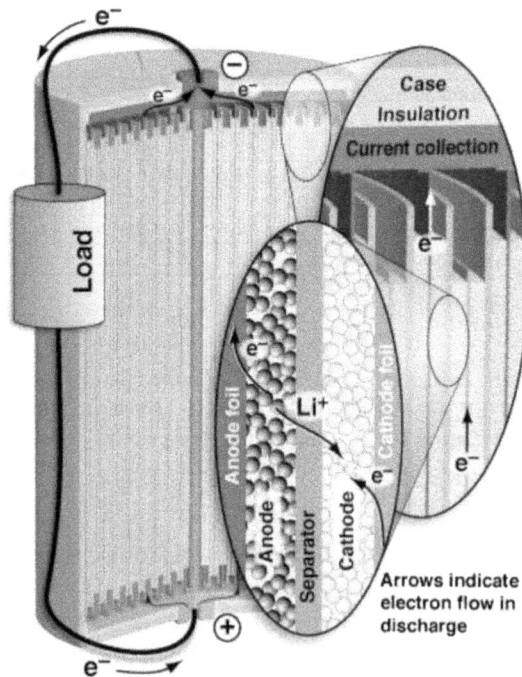

Figure 6.37. Principles of a lithium-ion battery cell. Reprinted from [136], copyright (2010), with permission from Elsevier.

---

[9] Approximately 1 to 1.5 kg of hydrogen are needed to travel 100 km.

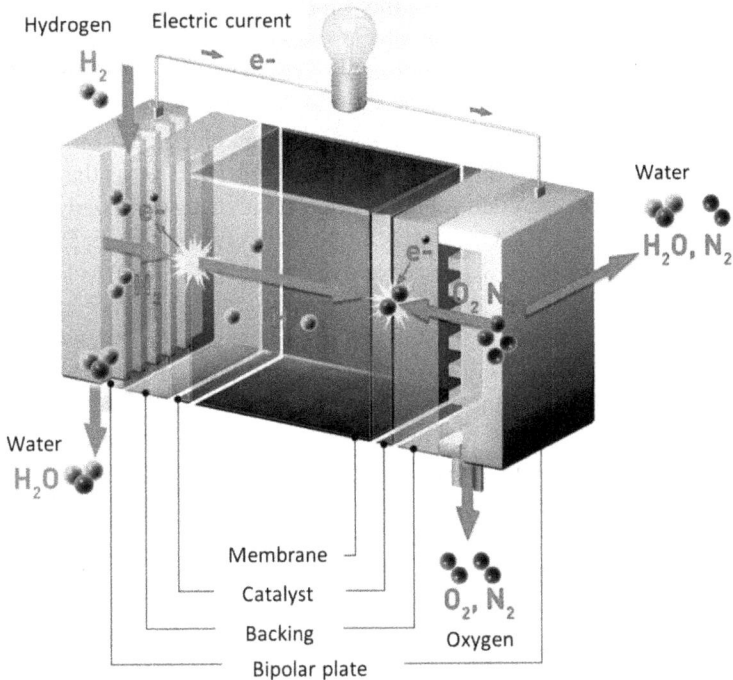

Figure 6.38. Principles of a PEMFC.

parallel to increase its capacity. In order to operate the battery in a safe and durable way, additional elements for thermal management, mechanical strength, safety devices, sensors, electronics, and so on are included in the battery pack.

In a proton-exchange membrane fuel cell (PEMFC), the ion shuttle is a proton H^+ (figure 6.38). Electricity is generated by separating the exothermal chemical reaction between hydrogen and oxygen into hydrogen oxidation at the anode and oxygen reduction at the cathode. Protons, produced on platinum-based catalyst at the anode, migrate through a proton exchange membrane (generally based on NAFION™ based polymer) to react with oxygen at the cathode and generate water vapor. Nominal voltage of a PEMFC individual cell ranges around 0.6 V; it is therefore mandatory to stack several cells to get required voltage, while current is directly adjusted with cells total active surface.

6.4.3.3 Current status of electric vehicles deployment

The electrification of passenger vehicles has been growing rapidly over the last 10 years. However, if the impact of transport electrification in urban areas is immediate, its global impact for reducing greenhouse gases is directly conditioned by the carbon footprint of the electricity that is used to recharge batteries.

At the end of 2020, the global vehicle fleet consisted of more than 1.2 billion vehicles, of which about 10.1 million were plug-in EVs (BEVs and PHEVs), with about 40% of the vehicles located in China. In the same year, BEV and PHEV sales reached 3.24 million vehicles, or 4.2% of new vehicle sales. The market shares of BEVs and PHEVs vary greatly from country to country. This is mainly due to very different financial incentive strategies, with some countries making a stronger choice for BEVs (France, South Korea, China) while others are pushing electrification in the broadest sense (PHEVs and BEVs), thus directing buyers towards PHEVs, as the latter are more suitable for all uses (UK, Sweden). The European car fleet consisted of 275 million passenger cars (EU28 + EFTA), of which about 1.9 million were EVs (BEVs and PHEVs). Of the 13.7 million new passenger cars sold in 2020, nearly 1.4 million were EVs (BEVs and PHEVs), or 10.2% of sales. It is worth noting the very high penetration of BEVs in Norway, representing 55.9% of new vehicle sales in 2019, compared with 5.3% of sales in the Netherlands, 2.8% in France, and 2.0% in Germany [137].

Commercial low-carbon solutions are available today for both light-duty and heavy-duty vehicles. However, while massive electrification of transport is becoming a reality, it does not solve the problems inherent in current transport modes such as the excessive number of vehicles (expected to remain so for a long time) and the limited energy efficiency in heavy traffic.

Public transport on dedicated tracks, such as tramways, trains, or subways, allows solutions which are largely decarbonised, as they are easily connected to electrical networks. Public transport solutions on open roads, such as buses, coaches, or trolleys, with on-board stored energy are already widely deployed (electric and/or fuel cell buses and coaches), and many demonstrators have been created in the last 10 years. At the end of 2019, there were about 500 000 electric buses (battery electric buses and plug-in hybrid electric bus) in operation worldwide. In 2018, more than 92 000 new electric buses were registered [138], mostly in China [139], compared to 104 000 in 2017. Plug-in hybrid electric buses account for 98% of new electric bus registrations. Outside China, about 900 electric buses were registered in 2018, mainly in Europe. China accounts for 98% of the global electric bus market. The city of Shenzhen had already completed the full electrification of its approximately 16 000 buses in 2017. European cities have only about 4000 electric buses altogether, and US cities have fewer than 400.

The World Resource Institute analysed the efforts of municipalities at different stages of electric bus adoption [140]. The main barriers to deployment were identified together with solution to overcome them:

1. Cities must first upgrade their electrical grid and develop a charging infrastructure. One of the difficulties in Shenzhen was the time it took to build a charging infrastructure capable of powering over 16 000 electric buses. For a range of about 200 km a bus needs to charge 250 kWh. In total, the fleet consumes over 4000 MWh per day.

2. The cost for acquiring a fleet of electric buses is often cited as the main obstacle. The price of a new electric bus ranges from US$300 000 to US$900 000 per unit. In the US, the average cost of an electric bus is US$750 000, compared to US$435 000 for a conventional diesel bus.

FCEV deployment appears to be significantly slower: About 4650 hydrogen buses were in circulation in the world by the end of 2020 [141], mostly in China with almost 4430 buses, in Europe with about 150 buses, and a little in the US with some 60 buses. This situation can be explained by several factors. BEV prices have decreased faster than expected, thanks to massive production. At the same time, numerous BEVs offer ranges of hundreds of kilometers (more than 600 km per charge for Tesla Model 3 or Xpeng P7). Finally, the deployment of fast charging stations (125 kW or more) allows an additional recharge of more than 200 km in less than half an hour. The differentiating factor in favor of FCEVs is their capability of withstanding very intensive daily use or daily long travels. These conditions can justify the deployment of expensive infrastructures (hydrogen pump, on-site hydrogen storage, etc).

6.4.3.4 Management of EV charging in smart cities

If not managed properly, the connection of a large number of EVs to the electrical grid can cause several technical problems for power systems, such as increasing peak demand and the risk of congestion and overload. However, EVs may also offer new opportunities; for example, smart charging and vehicle-to-grid can be used to support the grid by providing several local and global power- and energy-based services (frequency regulation, peak load sharing, congestion management, reduction of energy curtailment, increase of variable renewable energy source penetration, etc) [142].

Several studies have been performed recently to precisely assess the impact of EV integration into the electrical distribution networks and validate the opportunities offered by EVs. For example, a simulation tool based on a probabilistic three-phase load flow program has been developed and combined with EV usage scenarios simulated by using Monte Carlo techniques [143], by taking into account random variables of EV charging such as the battery state of charge, the starting and stop time of charging, and the EV's location. By using this tool, technical and economic impacts of EV integration on the distribution network were assessed. Studies of the potential opportunities of ancillary services provided by EV were also carried out for low voltage rural and urban networks.

Several strategies for the optimal management and real-time control of EVs in smart cities have been developed (figure 6.39), in order to do the following:

1. Minimize the cost of EV charging (economic criterion)
2. Maximize the use of solar energy to charge EVs (environmental criterion)
3. Reduce the peak load and the power exchange with the main grid (local flexibility)
4. Maximize the contribution of EVs to ancillary services (voltage and frequency control).

6.4.3.5 Major limitations to overcome

Battery deployment is having great success. However, further progress is needed, especially regarding performance, durability, robustness in all climate conditions,

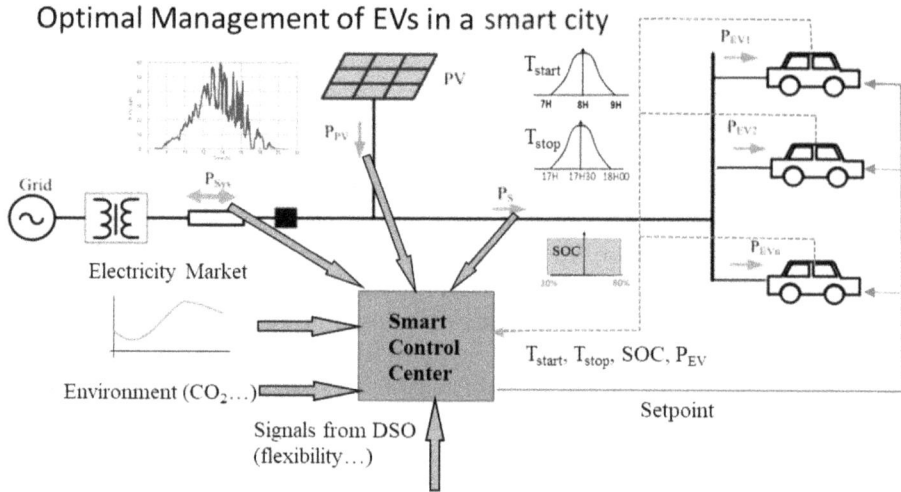

Figure 6.39. Optimal management of EV charging stations in a smart city. DSO: Distribution system operation means a natural or legal person responsible for operating, ensuring the maintenance of, and, if necessary, developing the distribution system in a given area and, where applicable, its interconnections with other systems and for ensuring the long-term ability of the system to meet reasonable demands for the distribution of electricity. SOC: State of charge is the level of charge of an electric battery relative to its capacity.

and safety during operation. In addition, high-performance batteries currently integrate critical materials such as cobalt, which will need to be reduced or substituted in the near future to ensure a large-scale deployment of electrified transport modes. Similarly, hydrogen fuel cells still do not have sufficient lifetime and robustness to be cost competitive, and the amount of critical raw materials such as platinum group metals should be reduced. Hereafter, we describe how physicists can contribute to overcoming those bottlenecks.

6.4.4 Challenges and opportunities for physicists: the role of modeling, simulation, and characterization

Physicists have a crucial role to play in the coming decades to push forward the science and technology of electrochemical power sources such as batteries and hydrogen fuel cells, which are key building blocks for transport electrification and green cities. Improving the performance, durability, and safety of batteries and fuel cells cannot be achieved by relying only on experimental work with conventional trial-and-error methods. It will be necessary to gain a deeper understanding of the complex multiphysics and multiscale phenomena occurring in batteries and fuel cells. This can be achieved by making extensive use of computational simulation in combination with advanced characterization techniques. In the remainder of this section, we describe how physicists can use can use modeling, simulation, and characterization to bring a major contribution to the field of batteries and fuel cells.

In the 250 years that have elapsed since the industrial revolution, scientific discovery has led to a quantitative understanding of matter solidly grounded on the atomistic concept. Ludwig Boltzmann originated a new scientific age by setting the bases of statistical mechanics. At a time when many physicists did not believe in atoms as real entities, Boltzmann's originality was to combine the atomistic hypothesis, with the laws of mechanics that were known at his time plus a new ingredient: statistical reasoning and the math that goes with it. Two of his equations, namely, the definition of entropy and the transport equation, constitute the foundations of equilibrium and nonequilibrium statistical physics. A few decades ago, computers began complementing this approach. Intensive computer modeling and simulation are increasingly able to predict observable macroscopic properties of materials accurately, starting from the atomistic concept and the laws of statistical mechanics.

Concerning batteries or fuel cells, the computational physicist has a lot to contribute to the materials aspects. One way is by finding new materials, screening them for desired properties and for replacing critical materials. This constitutes the goal of high-throughput (HT) approaches, which trade off accuracy for number of compounds and may investigate collections containing more than 1000 compounds [144–149]. Therefore, HT benefits greatly from the aid of machine learning methods like those used in regression analysis and pattern recognition.

However, real materials are more complex than the ideal single crystal. They contain defects and microstructural features that strongly affect their properties, which HT cannot easily handle at present. These investigations of single materials are the realm of multiscale approaches, which encompass the atomic description but extend to length and time scales several orders of magnitude above it. They play a role at least as important as that of HT, allowing us to understand the experiment when none of our expectations is matched by what we measure. Here the physicist starts from the atomic picture but gets to the observable macroscopic properties by means of intermediate equations or models of the intermediate scales. In these investigations too, machine learning methods can be a big help. They help to surmount the computational limitations inherent to *ab initio* techniques, which are often too slow to sample statistically significant numbers of atomic configurations, large system sizes, or long molecular dynamics simulations [150–153].

Both equilibrium and nonequilibrium computational physics are necessary for research into battery and fuel cell materials. Equilibrium modeling includes the prediction of phase diagrams. Since recently, finite temperature phase stability can in many cases, be predicted without adjustable parameters. These are important, for example, in understanding the different phases formed in the cathodes and anode as Li insertion takes place [154]; to estimate vacancy formation probability, which affects whether Li will be able to diffuse in solid battery cathodes; or to evaluate the amount of heat that a thermochemical material can store.

Nonequilibrium materials modeling is also essential to energy storage. On the one hand, there are the linear response conductivities, which are all interrelated via

Onsager's relations [155] to give the transport properties of the material. The electric and ionic conductivities determine how batteries charge and discharge, which is essential for use in small vehicles. In addition, the thermal conductivity is essential to battery safety and plays a fundamental role in the thermal runaway mechanism, which often leads to catastrophic ignition of a battery pack [156]. The dynamic insertion of Li in battery active materials is a complex multiphase process that requires the description of coupled dynamic phase transitions. Their description can benefit from atomic-scale models and simulations to determine both the energies of interactions and the local mobility of Li. In order to be used in larger-scale models, these complex transport mechanisms must be upscaled in the form of simpler and yet relevant laws.

An even more involved aspect of nonequilibrium physics is the prediction of reaction kinetics. Phenomena such as the oxidation of solids contain a lot of complex physics, involving phase transformations, vacancy and interstitial formation, atomic diffusion, and elastic and plastic deformations. Predictive modeling of reaction kinetics thus requires linking the atomic and microstructural descriptions. Not just the macroscopic properties of the material, but also its aging and degradation, depend on the combined physics at these different levels. A promising avenue is to combine *ab initio* and phase field methods. Reaction kinetics is also important for the safety of batteries to understand both the stability of materials and the energy released; the dynamics of this release during their rapid degradation can eventually lead to thermal runaway. An important challenge is the prediction of the chemical composition of the rapid degradation products depending on the composition of the electrodes: active materials, electrolyte, and possible degradation products due to aging.

If the material scale and the atomic scale are of primary importance because, as in any other domain of physics, all the macroscopic properties of interest depend on the atomic scale (local thermodynamic and transport properties), the upper scales are at least as important. Indeed, a battery or a fuel cell is not made of a homogeneous crystal: an electrode is a complex microstructure of heterogeneous multimaterials, a cell is made of a multilayer of electrodes and other components (separator, current collectors, bipolar plates, etc), a module or stack is made of several cells, and so on. Each of these scales presents specific heterogeneities and couplings and represents specific modeling and simulation challenges that must not be underestimated to develop a truly predictive approach to the development of battery and PEMFC technologies. Beyond the development of models at the different intermediate scales of interest, an important challenge is the link between these scales. An important scientific challenge corresponds to the upscaling process: At a given scale, the laws that account for mechanisms known at a lower scale must be determined. The importance of this approach can be illustrated in the durability challenge: To increase the lifetime of batteries and fuel cells requires their management at the system scale such that degradation mechanisms at the microstructure or even atomic scale are not triggered or at least are triggered in only a limited way. Thus, to develop the optimized management strategies, it is necessary to understand and quantify how macroscopic parameters (current, temperature, etc) influence these local phenomena and vice versa.

Because the physicochemical processes at stake are very complex, they cannot be determined *ab initio*. Therefore, they must primarily be determined from experimental observations. Moreover, the models must be fed by quantitative parameters and validated quantitatively. This threefold necessity (observe, characterize, and validate) requires the development of specific experimental characterizations.

Initially, experimental characterizations of fuel cell and battery materials were achieved by electrochemists with the aim to evaluate and improve material performance and ultimately to increase the total system power density and thus the vehicle autonomy. However, the results obtained with pure materials tested separately were often nonrepresentative of their behavior in real systems. Physicists entered the game either implementing high-resolution *ex situ* characterization tools such as transmission electron microscopy [157], x-ray photon spectroscopy (XPS) [158], or time-of-flight secondary ion mass spectroscopy. More recently, they developed *in situ* and *operando* experiments performed on neutron and x-ray sources for studying full systems. Such techniques have led to a better understanding of the transport and degradation mechanisms but require in the case of *operando* observations the development of specific cells and environments adapted to the experimental constraints.

Lithium insertion in cathodes of Li-ion batteries has been studied *operando* by x-ray [159, 160] and neutron diffraction [161] for around 10 years. Recently, using a microfocus beam, it has even been possible to follow the different steps of lithium disinsertion with a micrometer spatial resolution along the thickness of a graphite anode similar to those used in the very large majority of commercial batteries [162]. Specific electrochemical cells are still being developed to study *in situ* and *operando* the lithium insertion mechanisms using different and complementary characterization techniques [163]. XPS could also be used to characterize *operando* the formation of solid electrochemical interphase resulting from the interaction between electrolyte and electrodes [164].

Regarding fuel cells, *operando* characterization was developed, on the one hand, to study electrode reaction kinetics and specifically catalyst efficiency when decreasing platinum loading using EXAFS and, on the other hand, to study cell water management. The water distribution within an operating fuel cell was first characterized globally using neutron imaging [165] and specifically within the electrolyte membranes by small-angle neutron scattering [166]. In both cases, recent significant advances in spatial resolution have permitted researchers to get information on the water content in each electrode as a function of the operating conditions [167].

These results can be advantageously combined with modeling approaches to understand the different mechanisms that are highly coupled and multiscale. The next step will be dedicated to in-depth understanding of the aging mechanisms, which are paramount in evaluating the cost and the sustainability of an energy storage or conversion technology.

6.4.5 Conclusion

In this section we have described how transport electrification in urban areas can contribute to fight climate change and air pollution. We have also presented the role that physicists can take in the coming decades to drive forward the science and technology of batteries and hydrogen fuel cells and hence to make electric vehicles more cost effective, safe, and sustainable.

Although we have focused in this section on urban transport, long-distance transport should not be forgotten. Reaching net-zero greenhouse gas emissions by 2050, as required by the Paris Agreement, means decarbonizing long-distance transport modes such as air transport and shipping. Direct electrification with either batteries or hydrogen fuel cells cannot fully meet the needs of those types of transport modes. Energy carriers with higher volumetric or gravimetric densities are required, which means that low-carbon solutions such as sustainable biofuels (i.e., biofuels made from nonfood biomass) or synthetic fuels (i.e., carbon-based or nitrogen-based fuels, obtained by reacting hydrogen with CO_2 or N_2) [168, 169] should be developed and deployed.

6.5 Environmental safety

Jacob de Boer[1]
[1]Vrije Universiteit Amsterdam, De Boelelaan 1085, 1081 HV Amsterdam, The Netherlands

6.5.1 Introduction

When we speak about *environmental safety*, it is good to look first at the definition of the term *environment*. The environment can be defined as 'everything that is around us'. So let's take the place where you live, such as your home. What is around you? Nearby, you may see furniture, carpets, curtains, and electronic equipment, such as computers, mobile phones, and a television. You may see paint and dust, and there is (indoor) air around us. There may be other humans and pets. There may also be unwanted creatures such as insects. And of course, since 2020 we have become painfully aware of the viruses around us. Farther away, there is the outdoor air, and depending on where you live, there may be river or lake water or the sea. There could be meadows and a forest or dunes and a beach. There may also be industry nearby with wastewater and steam and smoke from their stacks. Even farther away are the upper layers of the atmosphere and the oceans. The various environmental compartments also interact, for example, through rain, wind, and temperature that can affect our homes and gardens and ourselves. In other words, the environment is actually relatively complex.

Our safety in our environment is dependent on various factors. There are the aforementioned influences of the environment itself, including biological effects of undesired pests, viruses, flooding, and so on. Even a physical effect such as noise, for example, from a nearby factory or wind turbine, could be felt. There can be also influences of a chemical nature. These can be for example a high nutrient load from wastewater that enters a river, exhaust from industries, contaminants present in our drinking water, house dust, or chemicals that migrate out of mobile phones and computers. In this section, environmental pollution will be discussed with an emphasis on its chemical nature, followed by a discussion of physical, chemical, and political measures to protect ourselves.

Have people ever lived without environmental pollution? If we go back to the Middle Ages or before, there was always personal waste: urine and feces. There were also clothes that people had worn until they were worn off, and remainders of food such as peels. People normally dumped personal waste outside their homes or outside their small villages. Two characteristics of this waste meant that it did not create an environmental problem: The waste was purely natural, and there was a lot of space because areas were never densely populated. When cities grew, environmental problems also grew. Cities began to smell due to putrefaction. This caused illnesses because bacteria started to grow and spread around. In Amsterdam, the drinking water, which initially came straight from the canals and later from the nearby river Vecht, was so bad that people decided to drink only beer (with 1%–2% alcohol) to survive. For the beer, water was used from the nearby lake Haarlemmermeer. In 1845, Amsterdam decided to install an entire new drinking

Figure 6.40. The Dutch dunes provide an excellent natural filter to produce clean drinking water.

water system with natural filtration in the Dutch dunes; it opened on 12 December 1853. This system now produces 70 000 000 m^3 of pure drinking water per year [170] (figure 6.40).

The example of the Amsterdam's drinking water shows that the initial environmental problems were of a biological nature. The industrial revolution that took place in the 1800s profoundly changed this situation. Suddenly, it was possible to synthesize chemicals for all sort of purposes. Those chemicals were very useful, such as medicines, dyes, fertilizers, and eventually plastics. Unfortunately, waste streams were created as well. These Included wastewater containing intermediates from production, polluted vapours that were released into the air, and the products themselves which did not degrade after use. Cars were produced, which produced exhaust, and trains and airplanes with comparable or more exhaust. Ships using bunker oil started to pollute harbors and ocean air. The Haber–Bosch process was invented in 1909 and created an almost endless source of fertilizer (and explosives) [171]. Due to the use of these artificial fertilizers—more than 100 000 000 metric tons annually—more food could be produced, and it was possible to feed many more people, which stimulated the growth of the world population. However, pumping massive amounts of synthetic nitrogen products into the food chain also created massive algal blooms in waterways and the proliferation of ocean dead zones. The concerns about synthetic nitrogenous fertilizer now have a greater immediacy than ever [171]. Meanwhile, the combination of a world population of more than seven billion people and a continuously growing production of chemicals have created environmental pollution at a global scale.

Unfortunately, problems with environmental safety are not the only ones caused by the strong population growth. There are serious problems with food and water scarcity. Energy production may be one of the greatest threats at the moment. If we are not able to make much more use of alternative energy resources such as solar and wind energy or nuclear energy, global warming will continue to a level that causes serious effects on many people, through flooding, droughts, hurricanes, or the temperature itself. Clearly, the problems with the COVID-19 virus (and other viruses such as other SARS viruses and MERS viruses) are also related to the dense world population. These problems may be more threatening for humankind than chemical pollution. However, the combination of the exponential growth of the world population and the industrial revolution has resulted in unprecedented levels of chemical pollution. This section will focus in particular on this aspect.

6.5.2 Current developments in production and restriction of chemicals

Between 1965 and 2006, global production of human-made chemicals increased exponentially. In 1965, 0.2 million substances were registered by Chemical Abstracts Service, and 88 million were registered in 2006 [172]. This exponential increase continues every day, with a staggering 10 million new compounds being produced each year. This is more than 1000 chemicals every hour, or 17 per minute [173]. Certainly, not all of the new chemicals find their way to commercial production and use. However, the growth in the chemical production is exponential. The United Nations (UN) Global Chemicals Outlook II predicts a doubling of the total global volume of chemicals used in the period 2020–30 [174]. There are currently no predictions for the period after 2030. According to US Environmental Protection Agency (EPA), there are approximately 85 000 chemicals in commerce in the US only. This number is, however, debated. Because of many duplications in that list, the real number would be around 35 000. Given that the number of evaluations taking place by the US EPA [175] and by the EU REACH program (Registration, Evaluation, Authorisation and Administration of Chemicals) are no more than a few hundred, it is clear that control is falling behind. To find our way in this chemical labyrinth, the most relevant categories of environmental pollutants will be discussed. This selection of chemicals is made based on their harmful properties in combination with their production volumes. The latter is important because, according to the old but still very relevant adage of Paracelsus, the dose determines the effect of a certain compound. Therefore, chemicals produced in low volumes will not be very important as regards environmental risk, except in local situations.

6.5.2.1 Persistent organic pollutants
To determine whether properties are harmful, the PBT concept is one of the most used tools. PBT stands for persistent, bioaccumulating, and toxic. A compound with this combination of properties would be one of the first to be categorized as undesirable. The PBT concept also underlines the definition of POPs: persistent organic pollutants. This definition was used by the UN in the Stockholm

Table 6.2. Current decision criteria for persistence, bioaccumulation, and toxicity of chemicals [176].

Compartment	Persistence half-life (d)	Bioaccumulation factor	Environmental toxicity, NOEC (mg L^{-1})	Human toxicology
In air				
In water	>40^a, >60^b, >60^c			
In soil	>120, 180^c			
In sediment	>120^a, >180^b, >180^c			
Biota/water		>2000, >5000^d	< 0.01^a, <0.01^b	
Humans				CRM or chronic toxicity

Notes: [a] Freshwater.
[b] Marine water.
[c] Very persistent.
[d] Very bioaccumulative.
CRM: carcinogenic (category 1 or 2), mutagenic (category 1 or 2), or toxic for reproduction (category 1, 2, or 3).
NOEC: no-observed-effect concentration.

Convention, which was set up to ban all POPs worldwide. The criteria for persistence, bioaccumulation, and toxicity have developed through the years. Especially for toxicity, criteria may change if new study results become available. In general, toxicity criteria will therefore become more strict. Table 6.2 shows the current decision criteria [176].

Given the very wide acceptance of the UN Stockholm Convention, it is fair to say that there is a global agreement on banning POPs. The initial list of POPs contained 12 substances. Unfortunately, the list had grown since the Convention entered into force in 2004, instead of shrinking. New discoveries and better toxicological insights have led to adding more substances to this Convention. At the moment there are 35 substances listed as POPs. Although it is 17 years since the Stockholm Convention came into force, no substances have been cleared. There are still many illegal dump sites with DDT or other pesticides, many old transformers containing polychlorinated biphenyls (PCBs), and so on. On the positive side, production of almost all POPs has stopped, although, in particular on the more recently added POPs, several countries have asked for and received exemptions, often for economic reasons. Because so many countries are involved, the Stockholm Convention is a rather bureaucratic organization in which progress is slow. The main problem is countries' capacity for the destruction of chemicals. The initial POP list contained PCBs, dioxins, and a number of organochlorine pesticides such as DDT, dieldrin, heptachlor, mirex, and toxaphene. The list was extended with endosulfan, several brominated flame retardants, perfluorinated alkyl substances (PFAS), and a few others. All POPs are halogenated substances. Whereas initially chlorinated compounds were considered, it appeared later that brominated and fluorinated compounds also behave as POPs. Especially, the carbon-fluorine bond is extremely strong. The persistence of the PFAS is therefore very high. An article in the *Washington Post* coined the term 'forever chemicals' because once produced, they will be present in nature for very long times, leaving a problematic heritage for future generations [177].

Positive aspects of the Stockholm Convention include creation of awareness and production of environmental data and of course the termination of production across the entire world. Global collaboration is essential to reduce the presence of these substances. Once they are released in the environment, they are distributed worldwide, mainly due to the so-called grasshopper effect [178]. They evaporate, precipitate again with rain or snow, evaporate again, precipitate, and so on. The direction is away from the equator towards the Arctic and Antarctica. Because production was mainly in the northern hemisphere, the Arctic is a very vulnerable area for POPs. This long-range transport is the fourth pillar of the Convention. Potential for long range transport is considered in the following circumstances:

1. When a chemical can be found in concerning concentrations at locations distant from the sources of its release.
2. When monitoring data show that long-range environmental transport of a chemical may have occurred via air, water, or migratory species.
3. When environmental fate properties and/or model results demonstrate that the chemical has a potential for long-range environmental transport through air, water, or migratory species. For a chemical that migrates significantly through the air, its half-life in air should be greater than two days [179].

The POPs in the Stockholm Convention are listed in three annexes, with different requirements [179]. For the initial 12 POPs (in annex A), all production and new uses should have ended in 2009, but exemptions were made for POPs that were added later to the list, such as hexabromocyclododecane, pentachlorophenol, chlorinated paraffins, lindane, endosulfan, and polybrominated diphenyl ethers. For the two POPs listed in annex B, DDT and perfluorooctane sulfonate (PFOS), no safe and affordable alternatives are yet available; therefore, they can still be produced or used for certain purposes. Notification to the Secretariat of production or use of chemicals under annex B is required. For DDT, the only acceptable purpose is disease vector control in some specific countries. For PFOS, research on alternatives is ongoing, but at this time, the production of Teflon cannot be achieved without fluorinated intermediates (figure 6.41). For use in firefighting foam and for coatings on outdoor wear, alternatives have been proposed [180, 181]. PFAS is a large group of chemicals, consisting of more than 5000 varieties with new ones being regularly added. This often leads to so-called regrettable substitution, which means that the alternative is not much better or even worse than the chemical that was replaced. One example is the substitution of perfluorinated octanoic acid (PFOA or C8) by Gen X (figure 6.42). In Gen X an oxygen atom is present which makes the compounds much less bioaccumulative. The half-life in the human body is

Figure 6.41. PFAS, often present in pizza boxes and other fast-food packaging.

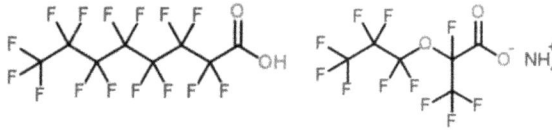

Figure 6.42. PFOA (left), used as intermediary in the production of Teflon, was replaced by Gen X (right), an example of regrettable substitution.

around 24 h instead of approximately 4–6 years for PFOA. However, Gen X dissolves much better in water, making it a much more mobile chemical. When preparing drinking water from surface water contaminated with PFOA, drinking water companies have great difficulties removing Gen X, so it arrives in drinking water, causing a daily uptake by consumers [182].

The lesson from this example is that a ban makes sense only when it includes the entire group of PFAS. Initiatives to realize this are currently being taken in Europe. Such a ban may include exemptions for essential use [183]. This concept addresses the use of chemicals for which no alternatives are yet available but which offer essential benefits, for example in medical applications. In the same way, within the group of halogenated flame retardants, one flame retardant is being replaced by another without making progress from an environmental point of view. The only solution is a ban on an entire group of compounds (figure 6.43) [184].

Chemicals in annex C are formed and released unintentionally during the production of other compounds or in thermal processes, such as during the incineration of waste. In order to prevent formation of these POPs, countries need

Figure 6.43. House dust contains high amounts of toxic flame retardants.

to develop release inventories, apply the best available techniques and promote the best environmental practices, and report progress every 5 years. Compounds in annex C comprise polychlorinated dibenzo-*p*-dioxins, polychlorinated dibenzofurans, and pentachlorobenzene. Some chemicals, such as polychlorinated naphthalenes, hexachlorobenzene, and PCBs are in both annex A and annex C [179].

6.5.2.2 *Endocrine disruptors*

Endocrine-active compounds or endocrine disruptors (EDCs) are chemicals that disturb the hormonal process in animals and humans. EDCs are distinguished in synthetic (S-) EDCs and natural (N-) EDCs [185]. The group has a high diversity and runs across other categories of chemicals, such as the aforementioned POPs. Natural hormones such as estradiol and progesterone are also included. The function of the endocrine system is strictly regulated, involving the hypothalamic–pituitary–gonad axis. The endocrine system can be modulated in two basic ways. The first is by agonists or antagonists of the respective estrogen and androgen receptors. Compounds that mimic, to some extent, the behavior of hormones can disturb the hormonal balance. A debate is going on about which compounds should be considered as EDC as which should not and also whether the N-EDCs may be having a much stronger effect than the S-EDCs [186]. Isoflavones from soy-based food and coumestans in certain vegetables are examples of N-EDCs that contribute substantially to daily human consumption. On the other hand, reports suggest an association between S-EDCs and obesity, for example, for parabens, phthalates, and bis-phenol-A. Another diverse group of bioactive chemicals receiving comparatively little attention as potential environmental pollutants includes pharmaceuticals and personal care products, both human and veterinary, including prescription drugs, diagnostic agents, sunscreen agents, fragrances, and many others. These compounds and their bioactive metabolites can be continually introduced to the aquatic environment as complex mixtures via a number of routes but primarily by both untreated and treated sewage. Many also show EDC activity [187].

6.5.2.3 Pesticides

Some chlorinated pesticides, such as DDT, dieldrin, toxaphene, chlordanes, and lindane, and their metabolites have been categorized under Stockholm Convention as POPs that should be phased out [179]. There are, however, numerous categories of pesticides—euphemistically called plant protection products—that have been developed to replace the chlorinated pesticides or for specific purposes. The enormous growth of the world population and the associated demand for food has greatly stimulated the use of pesticides. Without the use of pesticides, it is impossible to maintain even the current high food production for the entire world. That does not take away from the fact that many pesticides are toxic for the environment and humans. There are painful stories of casualties in Africa due to improper use of pesticides. This is often related to the lack of knowledge about the application of pesticides, personal protection, and proper storage.

Apart from direct intoxication, many pesticides have chronic effects. Debates are taking places on the admission of certain frequently used pesticides such as glyphosphate. In 2017, the European Commission reapproved the use of glyphosphate, the world's most widely used active ingredient in herbicides, in spite of a heavy debate [188]. These authors called for a new system of categorization of pesticides in which regulatory decision making takes into account not only the technical evidence on safety but also the societal context. Another group of heavily debated pesticides are the neonicotinoides, which are suspected to cause detrimental effects on pollinators such as honey bees. Thirty-five percent of the world's food crop production depends on pollinators [189], so although pesticides are essential for the world food production, this type of pesticide may eventually be counterproductive. In addition to the example mentioned, there is a very large collection of pesticides that are used worldwide, from easily degradable to persistent, and from relatively nontoxic to toxic. New pesticides are regularly introduced to the market, and food control agencies have a hard time continuously checking for residues.

6.5.2.4 Nutrients and fertilizers

Nutrients are essential for growth of plants, animals, and humans. However, nutrients are also excreted by animals and humans, which, when population density is high, can create environmental problems. On the other hand, to meet the global demand for food, natural nutrients fall short, so fertilizers are being added to agricultural soil to enhance food production. The enormous pressure created by the current large world population and the heavy use of fertilizers seem to have pushed the nutrient system out of balance [171]. Too large an amount of nitrates and ammonium on land may cause damage to forests. Too large an amount of phosphates and nitrates in water causes eutrophication and possibly undesired plankton blooms in our seas. In Western countries, sewage treatment systems have been built that remove most nitrates from wastewater. Experimental systems may even remove phosphates. However, in most developing countries, the situation is completely different. Sanitation is sometimes completely unavailable. Nitrous and sulfuric vapours from industries can cause acid rain and air pollution with serious risks for human health. The World Health Organization estimates that air pollution causes the annual premature death of two million people worldwide [190]. To put this into perspective, this is twofold higher than the current annual number of COVID-19 victims.

6.5.2.5 Polycyclic aromatic hydrocarbons

Polycyclic aromatic hydrocarbons (PAHs) are carcinogenic compounds that occur naturally. They may also be produced unintentionally during combustion of fossil fuels. In spite of their natural character, they can cause serious environmental pollution. Traffic, especially cars and airplanes, and heavy industry are the main producers of PAHs that are being formed during combustion processes in engines. PAHs are semipersistent, having the ability to stay for a very long time in environmental compartments such as soil and sediments. In organisms, PAHs are subject to metabolism processes, which does not mean that they are less harmful. Benz(a)pyrene in particular is highly carcinogenic. PAHs are also easily generated in open fireplaces and during barbecues, making houses and gardens sometimes a critical environment. PAHs occur in rubber, and through abrasion from car tires, they reach the environment near roads. The offshore industry and steel factories are other sources of PAHs. Also, they occur in crumb rubber, which is still used on football pitches. In addition to PAHs, methylated, oxygenated and hydroxyl PAHs also occur [191].

6.5.2.6 Synthetic dyes

The accelerating textile industry requires high volumes of dyes. The dye manufacturing industry represents a relatively small part of the overall chemical industries. The worldwide production of dyes is nearly 800 000 tons per year [192]. About 10%–15% of synthetic dyes are lost during different processes of textile industry. Textile factories, such as for example in India, struggle to get rid of the dye waste from their factories in an environment-friendly way. There are more than 10 000 dyes used in textile manufacturing alone, nearly 70% being azo dyes which are complex in structure and synthetic in nature. Textile industries produce large amounts of liquid waste. These textile effluents contain organic and inorganic compounds [193]. During the dyeing processes, not all dyes that are applied to the fabrics are fixed on them and there is always a portion of these dyes that remains unfixed to the fabrics and gets washed out. These unfixed dyes are found in high concentrations in textile effluents. Effluents are rich in dyes and chemicals, some of which are nonbiodegradable and carcinogenic and pose a major threat to health and the environment. Several primary, secondary and tertiary treatment processes like flocculation, trickling filters and electrodialysis have been used to treat these effluents. However, these treatments are not found effective for the removal of all dyes and chemicals used. The usage of cotton has been increasing constantly throughout the past century. Cotton fibers are mainly dyed using azo dyes, which are one of the largest groups of synthetic colorants used in the industry [194]. Azo dyes are difficult to degrade by the current conventional treatment processes. Advanced oxidation processes (AOPs) can be used to treat dyes. The main advantage of AOPs over the other treatment processes is its pronounced destructive nature, which results in the mineralization of organic contaminants present in wastewater [193].

6.5.2.7 Greenhouse gases and CFCs

At the moment, global warming, caused by greenhouse gases mainly, is obviously, one of the greatest threats for humans. Although three gases are always mentioned as the most important greenhouse gases, carbon dioxide, methane and nitrous oxide, water vapor is the most important actor in causing greenhouse effects, twice as powerful as carbon dioxide [195]. In addition, there are a few other gases causing global warming to a minor extent, such as volatile halogenated compounds (see below). Global warming is a rather complex process, not only related to industrial activity. Carbon dioxide, methane, nitrous oxide and water vapor are being produced by both natural and anthropogenic sources. Carbon dioxide emissions are for example large on a human scale but small compared to natural fluxes due to photosynthesis, and uptake and release into and from oceans. Methane is more related to anthropogenic sources such as mining, petroleum industry, coal combustion, biomass burning and domestic sewage treatment, than to natural sources. Nitrous oxide is mainly coming from natural sources but use of fertilizers, biomass burning and fossil fuel combustion also contribute.

Chlorofluorocarbons (CFCs) are stable, synthetic, halogenated alkanes, developed in the early 1930s as safe alternatives to ammonia and sulphur dioxide in refrigeration. Production of CFC-12 (dichlorodifluoromethane, CF_2Clz) began in 1931 followed by CFC-11 (trichlorofluoromethane, $CFCl_3$) in 1936 [196]. Many other CFC compounds have since been produced, most notably CFC-113 (trichlorotrifluoroethane, $C_2F_3Cl_3$). CFCs are nonflammable, noncorrosive, nonexplosive, very low in toxicity, and have physical properties conducive to a wide range of industrial and refrigerant applications. Many of these CFCs also have a, relatively small, effect on global warming. They are, however, more known from causing damage to the ozone layer, allowing harmful solar UV radiation reach the Earth's surface. The Montreal Protocol on Substances that Deplete the Ozone Layer was adopted in 1987. This protocol is often described as a unique example of a global agreement that successfully averted an environmental crisis [197]. It is still the only UN treaty ratified by all 197 member states. The first paper on ozone layer depletion by Molina and Rowland [198] was published in Nature in 1974. In 1995, Molina and Rowland, together with Crutzen, who focused on the effect of nitrous oxides on the ozone layer [199], received the Nobel prize for their discovery of these effect on the ozone layer and their efforts leading to the Montreal protocol. An overview of how basic research led to a worldwide agreed protocol for control of the harmful CFC's was given in the Nobel lecture by Rowland [200]. Scientists report the level of UV radiation could have quadrupled since the 1980s, whereas now UV radiation has stayed constant during the last 20 years [201]. Nevertheless, attention remains needed. Currently, in the atmosphere volumes of hydrofluorocarbons (HFCs), which have limited effect on the ozone layer, but are potent greenhouse gases, are growing rapidly, with an average rate of 1.6 ppt (parts per trillion) per year between 2012 and 2016 [202]. In 2016 the phase-down of production and consumption of HFCs was added to the Montreal protocol in the Kigali Amendment [203].

6.5.2.8 Trace metals

Although of natural origin, trace metals were one of the first compound classes to be considered as environmental pollutants. In the Minamata incident, waste from a factory producing mercury sulfate as catalyser was released into Minamata Bay (Japan), where methylmercury accumulated in fish. Fish-eating cats were the first victims, followed by birds falling from the sky and dying dogs. Fishermen caught fish with two or three heads. Finally, more than 1700 people died, including children [204]. Mercury binds to proteins that support the nervous system. Mining and steel industries initially caused local and later regional and global elevated levels of mercury in particular, but also of other trace metals such as copper, arsenic, cadmium, and lead. Nowadays, trace metals are still threatening the environment, but globally there is a lot more attention for this type of pollution. The Minamata Convention of the United Nations monitors mercury flows worldwide. Many countries have monitoring programs in which trace metals are being analyzed in water, sediments, and fish. One of the bigger remaining problems is the high level of arsenic, often of natural origin, that pollutes drinking water in many areas in India and Pakistan and some other developing countries. A shift in the use of metals is seen towards use of rare earth elements (REE), which are used in magnets and military systems. Recycling of REE would be a possible solution, but so far, no reliable and sustainable techniques are available [205]. These REE are often mined from areas that are highly vulnerable. Cobalt is another metal that indirectly threatens the environment. Cobalt is used in batteries of mobile phones, and stocks of cobalt are available in areas in Congo where the highly threatened mountain gorillas have their habitat [206]. DR Congo is a the top producer of cobalt with 40% of world production and one-third of world reserves.

6.5.2.9 Radionuclides

In the human perception the risk of radionuclides is always considered very high. This may be related to the two atomic bombs that were dropped on Hiroshima and Nagasaki in Japan and the nuclear tests during the Cold War. Also, serious accidents such as those at the nuclear power plants at Chernobyl and Fukushima, which have contaminated parts of the world, have contributed to the general fear. Indeed radionuclides can be health threatening. Some of them have very long half-lives, which can cause a risk for many generations. On the other hand, aside from these accidents, the level of pollution from nuclear reactors is extremely low. It does not threaten environmental or human lives and is often far below natural background radiation as experienced in airplanes or high in the mountains. Benefits of radionuclides and radiation should not be forgotten. Many radionuclides are essential in the treatment of cancer. X-rays are pivotal in detecting fractures and many other health-threatening impairments. Medical radiation exposures are, therefore, often life-saving, although at the same time they constitute the majority of exposures in terms of number of persons exposed and higher dose rates. In the environment, a somewhat higher risk may be found in the production of phosphate fertilizers, which often leads to pollution with the alpha-radioactive lead-210 and polonium-210 [207, 208]. These compounds contribute to the occurrence of cancer due to smoking, because tobacco plants accumulate these compounds from the fertilizers.

6.5.2.10 Plastics

Although pollution of beaches has been well known since the 1960s, scientists have ignored this topic for many years. Possibly it was seen as a problem that was so easy to avoid through human behavior that no one bothered to approach it in a scientific way. It was not until 1980 that reports started to appear in the literature on the risks of macroplastics and microplastics [209]. Large plastics field floating around in gyros in the Pacific Ocean were reported [210]. Beach monitoring of litter was started, and several initiatives involving plastic monitoring and suggestions for cleaning were observed. It appeared that microplastics with a size range of 0.1–5 mm were a large part of the problem. Microplastics can be formed, though slowly, from weathering of larger pieces of macroplastic. Spills in harbor of commodities for plastic production is another source. Microplastics are also present in many products such as toothpaste and cosmetics, and reach the rivers and sea because they pass through sewage treatment systems.

Macroplastics and microplastics are of serious concern for the environment [211] (figure 6.44). Macroplastics cause risks for turtles, which can be strangled, and for birds that consume plastics and stop feeling hungry, which eventually causes starvation because they do not take in their normal diet anymore. Microplastics can shade light and, therefore, affect the growth of algae and benthic organisms. They enter the gastrointestinal tracts of mussels, other shellfish species, and of fish. Some scientists claim that microplastics are also harmful for humans, causing inflammations. More research is needed on this subject to provide evidence. The analysis of microplastics is difficult because there is not only the chemical component, that is, the various types of the plastics, but also a physical component that can cause damage to organisms. Microplastics have many different shapes, and fibers, for example, from clothes, are included in this group. In addition, chemical contaminants in water can adsorb to the plastics and cause an additional risk for organisms. Nanoplastics are also being studied, and some scientists are worried about harmful effects because, due to their smaller size, they could more easily enter cells or organisms. This research is, however, still in its infancy.

Figure 6.44. Dumpsite in nature, Kerala, India, including a lot of plastic waste.

6.5.2.11 Fine particles, NO_x and SO_x

Although estimates vary, there is no doubt about the high mortality caused by air pollution. According to Landrigan [212], polluted air was responsible in 2015 for 6.4 million deaths worldwide: 2.8 million from household air pollution and 4.2 million from ambient air pollution. Lelieveld *et al* [213] estimated the number of death related to air pollution at 3.3 million per year, mainly caused by exposure to fine particles, smaller than 2.5 mm (PM2.5). Lelieveld *et al* [213] mention that agricultural sources are the second-largest contributor to global mortality because releases of fine particles and ammonia from livestock and fertilizers lead to atmospheric formation of ammonium nitrate and sulfate particles. Agricultural sources are leading sources of mortality in the eastern US, Russia, Turkey, Korea, Japan, and Europe, contributing to more than 40% of deaths in many European countries. Coal-fired power plants and steel industries are other dominant sources of air pollution. Apart from mercury (see section 6.5.2.8), they produce high levels of fine particles and significant volumes of NO_x and SO_x vapours. Fine particles from the steel industry pose a double risk, as they have absorbed many different metals (figure 6.45). Traffic is especially causing high levels of fine particles and NO_x exhaust. Some densely populated cities in countries such as China and India have permanently too high PM2.5 levels (figure 6.46). Bronchitis, asthma, strokes, cardiovascular diseases, and lung cancer are examples of diseases caused by air pollution, but it has also been suggested that air pollution is an important although not yet quantified risk factor for neurodevelopmental disorders in children and neurodegenerative diseases in adults [212]. Lelieveld *et al* [213] project a doubling of mortality from air pollution by 2050 on the basis of projected rates of increase in pollution and population levels. This projection should obviously sound alarm bells for public health agencies around the world.

Figure 6.45. Steel production releases lots of toxic fumes, fine particles and CO_2.

Figure 6.46. The extremely densely populated city of Jaipur (India).

6.5.3 Challenges and opportunities towards 2050

6.5.3.1 Challenges: How to cope with such a load of chemicals?

As mentioned earlier in this section, the UN Global Chemicals Outlook II expects a 6.6 trillion Euro/year market for chemicals in 2030, double the amount in 2020. That is obviously not the end of the growth. Will our world be able to cope with so many chemicals? There is no doubt about the benefits of chemicals for humankind. However, the same report predicts that hazardous chemicals will continue to be released in large quantities. International treaties and voluntary instruments may reduce the risks of some chemicals and wastes, but progress has been uneven, and implementation gaps will remain [174]. It is, therefore, obvious that more is needed than just hoping for the better. Small improvements in the current situation will not help. Similar to the increased consciousness about global warming, there is a need for a world plan for chemical pollution. Today most lactating women in the world still have DDT in their breast milk. Most people in the northern hemisphere have PFAS levels in their blood close to or over the safe levels advised by EFSA. Each year 1.6 million individuals are affected by air pollution. If we do not want to spoil this world for future generations, our ways to deal with chemicals need some fundamental changes. There are solutions that involve using better techniques, investing in innovation, and changing our policies.

6.5.3.2 Better products

To produce high-quality, useful products, many chemical industries produce substantial amounts of waste. Waste often leads to local, regional, or even global pollution. However, some industries make chemical compounds, which are chemical waste themselves as soon as they are produced. Getting rid of those substances often requires a very special type of treatment, such as burning at very high temperatures. A lot would therefore be gained if we could refrain from producing persistent, bioaccumulating, and toxic compounds. Some frameworks have been erected to realize this goal. In Europe the REACH program should ensure that such dangerous compounds are not made any longer. In the US, the EPA is trying to reach the same goal with the Toxic Substance Control Act. Unfortunately, these programs cannot keep pace with the introduction of new chemicals [174]. The only solution is to test chemicals *before* they are produced and used. The pharmaceuticals industry needs to prove the safety of drugs before they can be produced for the market. It has been a big mistake not to test bulk chemicals before production can start. That has caused the aforementioned situations with DDT in human milk and PFAS in human blood, which are basically experiments on a global scale. The chemical industry will always protest and claim that costs will be much too high. A level playing field is therefore needed globally to realize this. Until a proper system of testing has been installed, the precautionary principle needs to be applied [214]. This system prescribes that if a certain chemical substance, based on its structure and on modeling outcomes, may likely cause damage to the environment and/or to human health, it should not be produced and used. Industries may claim that alternatives are not available. However, we heard that argument many times before, such as when PCBs had to be phased out. Alternatives were found. The same argument was used when some brominated flame retardants were phased out. Again there were alternatives. The European research project ENFIRO showed that within 3 years a series of environmentally friendly flame retardants could be selected and produced for major applications [215]. Some industries do produce environmentally friendly and environmentally unfriendly compounds at the same time, showing their good will. Shareholders then press for keep selling the unfriendly alternatives, as profit is often better for those substances. Which chemical industry will stand up and decide to produce only safe chemicals?

Apart from safe alternative chemicals, there is also mileage in thinking about the purpose of using specific chemicals. Flame retardants may be necessary to delay ignition of plastics that are heated, such as in computers and television sets. But do they need to be used in paint on the walls of houses or in flags? Are we not able to find alternatives such as glass or wood to the many plastic furniture items we have in our homes? There are many ways to reduce the use of flame retardants without elevating the risk of fire. The concept of essential use [183] may help us also in the world of flame retardants to decide which ones are really necessary and which ones we can do without.

6.5.3.3 Better processes

The environment has a self-cleaning capacity. The system is able to transform and metabolize chemical compounds in a rather effective way. UV light, bacteria, temperature, and so on will cause chemicals to be converted into water and carbon dioxide. This wonderful process works fine as long as the system is not overwhelmed and as long as no chemicals that are insensitive to UV light, bacterial degradation, and so on are being discharged. It is amazing to see that during the last half century we have entirely ignored the self-cleaning capacity of the environmental system. Certainly locally and also regionally but even globally we have released so many persistent and toxic compounds into the environment that some of them will remain there for generations to come. Some pharmaceuticals in fish farming, for example, have been released in such high quantities that a phenomenon called antibiotic resistance is taking place, that is, the entire environment has been made sterile. Due to their extreme persistence the PFAS have been called *forever chemicals*, because we do not know when they will ever disappear from our systems. Microplastics and macroplastics will float around (and sink to the sea floor) for ages to come, simply because the self-cleaning capacity of the environment is close to zero for the large polymeric molecules.

The solution is actually rather simple: Go back to respecting the self-cleaning capacity of our world. That means basically, taking care of our waste, as we should do already as citizens. Where citizens are supposed to collect their waste and offer it to the garbage collection service, industry should act in a similar way. Just don't throw the waste into the environment, either into the air or into the water. Factories need to destroy their waste on their premises or send it to a service that can handle it properly. A big Teflon plant in the Netherlands has discharged tens of thousands of kilograms of PFAS into the rivers of the Netherlands. This was not illegal; the authorities had provided the plant operator with permits to do so. When we discovered that PFAS could be found in fruit in gardens around the plant [182], the owners claimed that production would be impossible without such discharge. However, after more pressure, when building companies discovered they could not build anymore because most locations exceed the maximum tolerance levels for PFAS, suddenly there appeared to be options, such as by installing carbon filters and replacing them with clean ones from time to time and sending the polluted filters to be burned at high temperature. Better filter installations, sewage treatment systems, absorption materials, and so on will help to solve these issues. This is not a matter for industry alone. Even in 2021, in the Netherlands a permit was given to a company to discharge 12 000 kg yr^{-1} of microplastics into the river Meuse, even though everyone is aware of the risks of microplastics and efforts are being made to clean microplastics from rivers and seas. It looks as if public services are mutually disconnected. When Gen X was introduced as an alternative to PFOA for producing Teflon and a permit for discharge was provided to the industry, the drinking water service situated near and downstream from the Teflon factory was not informed about this new compound. Consequently, for years they tried hard to identify the unknown peaks they observed with their instruments. One simple phone call would have been enough to avoid all these efforts. Apart from that, Gen X appeared to be a typical regrettable substitute (figure 6.42).

It is unrealistic to strive for a world without plastic or steel. Even pesticides are needed to ensure sufficient food production for the large global population. However, we need technical innovation, including better filters, clever adsorption procedures, use of safer and cleaner materials, less polluting energy systems such as using hydrogen as fuel for cars and airplanes, and solar energy to reduce air pollution. Physical and chemical designing should be much more focused on environmental and human safety, rather than on making profit.

6.5.3.4 International developments

In 2016, the UN put forth sustainable development goals (SDGs), comprising 17 goals and 169 targets [216]. Less known is that these SDGs also contain 42 targets that focus on means of implementation. As regards technology, there is a focus on transferring technologies from developed to less developed countries. This would certainly assist in developing environmentally friendly production systems and increase the safety of farmers. Although the SDGs have been criticized, for instance, for a lack of interlinkages and interdependencies among goals, the SDGs do give strong guidelines on a global scale [217]. It will be up to authorities and governments to translate these SDGs into laws and international agreements.

The EU has announced substantial action with its Green Deal on a climate-neutral continent. In the wake of the Green Deal, other plans have been introduced, such as the EU Chemicals Strategy for Sustainability, presented on 14 October 2020; the Zero Pollution Ambition; the European Industry Strategy; and the EU Hydrogen Strategy. These ambitious plans strive for a strong reduction of pollution and the phase-out of persistent chemical groups such as PFAS and emphasize the need for technical innovation.

In the US, the CSS Strategic Research Action Plan (StRAP) outlines the EPA's research on the development of innovative science to support safe and sustainable selection, design, and use of chemicals and materials required to promote human and environmental health and sustainability [218]. The StRAP will, among other things, provide a chemical safety informatics infrastructure to support decision makers, introduce high-throughput hazard and exposure approaches to fulfill data needs, and consider sensitive populations and life stages in chemical safety evaluations. StRAP will build a broader understanding of biology, chemical toxicity, and exposure while providing more rapid, cost-effective approaches that protect human health and valued ecological resources and services.

The most influential country as regards environmental chemical safety is China. According to the UN Global Chemicals Outlook II, in 2030 China will have a 49% share of the chemicals market [174]. Few global actions on environmental safety will therefore succeed without the participation of China. The EU and China have developed the EU–China 2020 Strategic Agenda for Cooperation [92]. This document contains one paragraph (V.9) on the environment, which involves cooperating on tackling air, water, and soil pollution and chemical pollution; sustainable waste management and resource efficiency within consumption and production; and environmental pollution emergency action. At this stage, it is somewhat unclear how powerful this document will be. Apart from common concern about environmental issues there are obvious competitive issues, such as shortages of rare earth metals and other chemicals which can be found in one region and not in the other. On 8 January 2019, China published the draft of its new Regulation on the Environmental Risk Assessment and Control of Chemical Substances (Consultative Draft) [219]. The new regulation intends to govern the environmental risk assessment and control of any chemical substance. It is the first chemical control legislation proposed by China which covers both existing and new chemical substances. Given also the Chinese development activities in other countries, particularly in Africa, the Chinese influence on global environmental safety in the coming decades should never be underestimated.

6.5.4 Conclusions and outlook

Within the present decade our densely populated world will see a vast expansion of chemicals that will serve many individuals but at the same threaten our environmental safety and our own health. Given the very short notice of this development, immediate action is required. The EU has set an ambitious example with a set of plans including the Chemical Strategy for Sustainability. Comparable action plans in the US and China and collaboration to meet the UN SDGs will be essential. Environmental safety is very much related to the use and policies of dealing with chemicals. However, technical innovation is required to handle these chemicals in a safe way, produce new and safer alternatives, safely collect and handle waste, and produce energy without negative effects on the environment. This is a huge challenge. These changes will not be easy. They will need to overcome strong opposition by powerful vested interests [212]. Much will depend on success in achieving a circular economy. This will be needed not only to save the environment but also to prevent depletion of valuable resources.

6.6 Understanding and predicting space weather

Stefaan Poedts[1,2]
[1]KU Leuven, Leuven, Belgium
[2]Institute of Physics, University of Maria Curie-Skłodowska, Lublin, Poland

6.6.1 General overview

In the last few decades, space-based and ground-based solar observations have disclosed that the atmosphere of the Sun is very inhomogeneous, with structures on a wide range of length scales and also very dynamic with activities on a large variety of temporal scales. Some of these solar dynamic events are so energetic that they affect the Earth at a distance of 150 million kilometres. The Sun emits a continuous outflow of solar particles in all directions, called the solar wind. This wind contains transients and interacts with our magnetosphere and the magnetospheres of other planets in the solar system. The whole set of complex effects of the radiation and the plasma stream from the Sun on the Earth and our magnetosphere, our technological systems, our climate, and humankind determines most of the so-called space weather[10]. The broader term 'heliophysics' is used for the study of the interconnectedness of the influence domain of the Sun, that is, the entire solar–heliospheric–planetary system[11].

The activity of the Sun and Sun–Earth interactions are the main drivers of the space weather. It causes substantial socioeconomic damage to human infrastructure both in space and on the Earth. This includes both direct effects on specific industry sectors, such as electric power, spacecraft, and aviation, and indirect effects on dependent infrastructure and services, such as positioning and navigation systems, electric power grids, telecommunication, and oil and gas pipelines. As human society becomes ever more dependent on technological infrastructure that fully or partially relies on the space surrounding us, the impact of space weather events becomes increasingly important.

Over recent years, programs and infrastructure have been built up by national and international agencies and corporations to predict solar activity and to forecast its potential impact on the Earth and its space vicinity. The main science questions are as follows:

- What conditions in the magnetic field emerging into the solar atmosphere lead to eruptive events, such as flares and coronal mass ejections?
- What mechanism or mechanisms cause the sudden acceleration of particles during a solar flare and the gradual solar energetic particle events in coronal mass ejection (CME) shocks?

---

[10] Definition used in the US National Space Weather Program Strategic Plan [220]: 'The term "Space weather" refers to the conditions on the Sun and in the solar wind, magnetosphere, ionosphere, and thermosphere that can influence the performance and reliability of space-borne and ground-based technological systems and can endanger human life or health.'

[11] The heliosphere is the region surrounding the Sun and the solar system that is filled with the solar magnetic field and the protons and electrons of the solar wind.

- How can we estimate the direction and strength of the particle and radiative fluxes released by these events?
- How is magnetic flux transported from the solar interior to the heliosphere?
- Is it possible to fully understand and accurately predict solar magnetic storms and quantitatively forecast the impact they will have on the Earth?
- How can human society best mitigate effects of extreme space weather events?

6.6.2 Socioeconomic importance

Our Sun is a very dynamic star that influences the Earth in many ways. The continuously changing solar wind as well as eruptions on the solar surface significantly alter the conditions in the whole solar system and more specifically the near-Earth environment (the magnetosphere, ionosphere, and thermosphere). We call these space environmental conditions 'space weather,' and its effects are noticeable in our magnetosphere, the atmosphere, and even at the Earth's surface (see figure 6.47). Because human society is becoming ever more dependent on advanced technologies and infrastructure, space weather phenomena influence our daily lives increasingly by affecting both our space-borne and ground-based technological systems [221–223]. Energetic particles that are ejected from the Sun in the heliosphere or accelerated by shock waves in the solar wind can induce magnetic fields on the Earth, which in turn can drive large currents through power networks that can interfere with the network operation, damage transformers, and

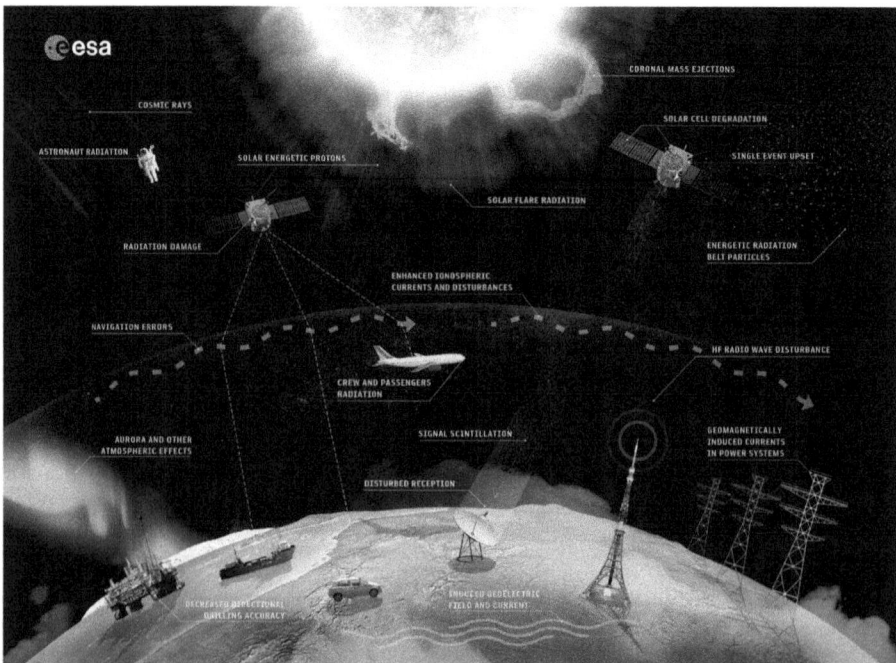

Figure 6.47. Space weather effects. (Source: ESA SSA—Space Weather Segment https://www.esa.int/ESA_Multimedia/Images/2018/01/Space_weather_effects.)

cause electric power loss and damage to oil pipelines. Such high-energy solar particles can also damage satellites and spacecraft in orbit, for instance, by bit swapping in electronics, single-event latch-up (destruction), and wear to solar panels. Astronauts could be killed during extravehicle activities by protons of more than 30 MeV, since these can penetrate spacesuits. Hard particle energy spectra can contain large fluxes of hundreds of MeV–GeV type superenergetic particles, which can reach low Earth orbit satellites and even penetrate the safest areas of spacecraft. The radiation can even endanger airplane crews and frequent airline passengers. Heat expansion of the Earth's thermosphere caused by solar storms can perturb or even change spacecraft orbits and thus disturb and even disrupt telecommunication and navigation systems. Because of this, over the past decade, space weather has been included in national risk assessment plans.

One of the key drivers of space weather are solar eruptions. These include both solar flares and CMEs (see figures 6.48 and 6.49. From a socioeconomic perspective, the CMEs are the main drivers of adverse conditions in the inner heliosphere (e.g., Richardson *et al* [224]). CMEs are large magnetised clouds, containing up to $10^{13} - 10^{16}$ g of plasma, that are ejected into interplanetary space at velocities of typically 450 km s^{-1} (1 620 000 km h^{-1}). CMEs launched in our direction typically arrive at the Earth in about 2–3 days. However, so-called extreme events involve velocities up to 3000 km s^{-1} and more (more than 10 million km h^{-1}) and can reach the Earth in as little as 12 h. While some of the less eruptive CMEs may have minor to no effects on the Earth, the strongest and fastest of them can create severe geomagnetic storms. One of the most severe space weather storms ever reported happened in September 1859 (called the Carrington event), when auroras were spotted at very low latitudes and disruptions of telegraphs were reported.

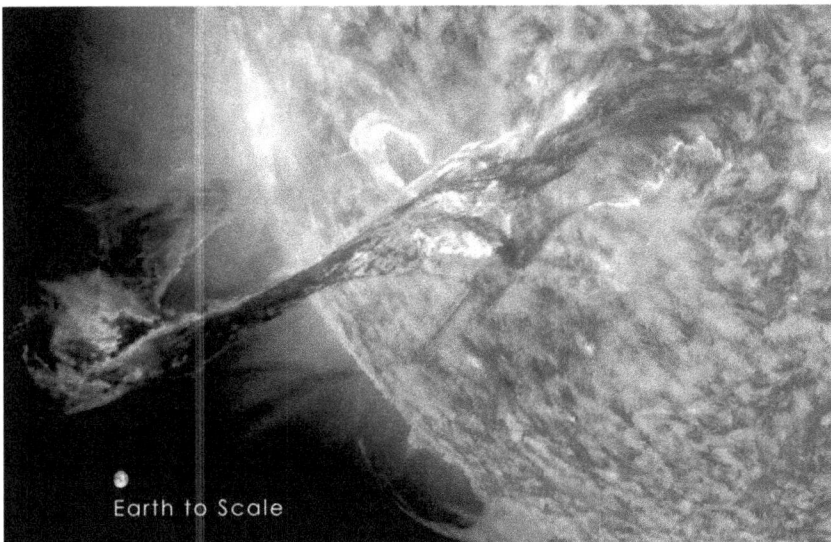

Figure 6.48. Eruption of a solar flare on August 31, 2021. The Earth is projected to scale on the image. (Source: NASA Goddard Space Flight Center.)

Figure 6.49. Coronagraph image of a CME. The red disk in the centre is the part of the instrument that covers the solar disk (the white circle indicates the solar disk). The false-colour image was taken from the Solar and Heliospheric Observatory (SoHO) spacecraft on December 2, 2002. (Source: SoHO/ESA/NASA.)

The damage was relatively modest because there were no satellites or electric power grids back then. A space weather effects report of the National Research Council of the National Academies in 2008 [225] estimated that the advent of an extreme event, such as the Carrington event, would cost our current society between 1000 billion and 2000 billion (i.e., 1–2 million million) USD. Moreover, it would take 4–10 years to repair all the damage. This is an order of magnitude worse than the harm caused by Hurricane Katrina, which caused 'only' 153 billion USD of destruction [226]. More recently, Eastwood *et al* [227] estimated that the total economic loss from an extreme event would vary from 0.5\$tn to 2.7\$tn (tn = trillion = 1000 billion, i.e., 10^{12}) based on calculations taking into account the disruption to the global supply chain. Cannon [228] claims that extreme events like the Carrington event occur only about once in 250 years with a confidence level of 95% and once in 50 years with a confidence level of 50%. However, in 2012, Riley [229] estimated that there is a 12% probability of such a superstorm occurring within the next decade. In fact, on 23 July of that year (2012) an unusually large and fast CME event occurred, but it missed the Earth, thanks to the rotation of the Sun. It would have hit us if it had occurred 9 days earlier. This most severe storm in 150 years did hit the Stereo-A spacecraft, though. Luckily, such extreme events are rare, but the more common 'normal' space weather events also have a considerable socioeconomic impact. For instance, Schrijver *et al* [230] deduced from insurance claims that geomagnetically induced currents (GICs) may cause losses of the order of 5–10 billion USD per year to the US power grid.

It is not only CMEs that are affecting space weather. Fast solar wind streams, originating from coronal holes, seen as dark areas in extreme ultraviolet and soft x-ray images of the solar corona, can also cause significant socioeconomic damage. One example of this is the failure of the Anik E2 spacecraft, caused by a fast solar wind stream that swept past the Earth in January 1994. The total damage is estimated to have cost the Telesat satellite communication company around 50–70 million USD in recovery costs and lost business. The recovery of its operational status took 6 months, during which thousands of people lost their television and data services. Satellite blackouts and temporary disruptions are frequent. Howard *et al* [231] determined that about 4500 spacecraft anomalies have been reported over the 25-year period 1974–99 (and currently there are more than 1000 satellites in orbit). The effects of solar storms, caused by CME impact or interaction with high-speed solar wind streams, in the Earth's ionosphere can distort radio waves used for communication or corrupt GPS signals, with significant impact on our current society. Furthermore, space weather events can increase the radiation dose at the altitudes of flying aircraft and, in general, the exposition to radiation of in-orbit astronauts. Therefore, in addition to limiting the socioeconomic impact of space weather events, their mitigation also implies protection of human life at large.

6.6.3 Current state of the art

Given the enormous socioeconomic impact involved, reliable predictions of space weather and its effects on technological systems, human life, and health are extremely important. Genuine predictions and forecasts require a deeper insight into the underlying space weather physics, the mechanisms behind the different phenomena and their effects. Observations are much needed but sometimes limited and/or difficult to interpret, for example, due to projection effects or complexity. Moreover, some things cannot be observed, such as the important coronal magnetic field. Numerical simulation models can provide complementary information, facts that cannot be observed directly, such as the magnetic field topology, the internal density structure of a CME, and the local velocity in and around a CME.

Thus, when trying to assess the daily space weather and its possible dangers, two main types of data are used: observational data obtained from spacecraft and synthetic data obtained from running numerical simulation models. One can compare the situation with the weather conditions on the Earth, except that the amount of observational data to work with is much more limited. There are only a few space weather satellites gathering *in situ* data, compared to the thousands of weather stations in Europe alone [232]. Moreover, one of the main drawbacks of the current observations is that the physical parameters are measured in the line of sight of the spacecraft, which are positioned in only a few different locations in the heliosphere, and until recently they were all in the equatorial plane. This gives rise to projection effects, for example, on the limb of the solar disk or problems to determine the three-dimensional structure of the ejected CME. On top of that, the *in situ* observations that are available are so close to the Earth that they do not give us enough lead time to adapt our satellite and Earth-based systems to an incoming space weather storm. It is for these reasons that we resort to both empirical and physics-based models when it comes to forecasting space weather.

6.6.4 Achievements

Solar drivers of space weather. The Sun emits a continuous stream of charged particles, known as the solar wind. It is inhomogeneous and bimodal, containing slow wind and high-speed solar wind streams. As a result, the interaction of the solar wind with our magnetosphere is not steady and causes the magnetic field of the Earth to vary in time, which in turn drives geomagnetic activity. In fact, the Sun is the main source of space weather events on the Earth. In particular, the main solar drivers are solar flares, CMEs, and solar energetic particle (SEP) events. Solar flares are enormous explosions that are observed as bright areas on the Sun in x-rays and are also visible in optical wavelengths and cause bursts of noise in radio wavelengths. Flares release magnetic energy previously stored the solar atmosphere. The electromagnetic emission produced during flares travels at the speed of light and reaches the Earth after about 8 min. CMEs are violent outbursts of plasma from the Sun's outer atmosphere (the corona). They propagate at thousands of kilometres per second and typically carry a billion tons of material into the heliosphere. When sampled *in situ* by a spacecraft, they are termed Interplanetary CMEs (ICMEs). Apart from the plasma (mostly protons and electrons), CMEs also contain powerful magnetic fields which turn out to play a key role when CMEs interact with our magnetosphere. Unlike solar flares, CMEs are not particularly bright, and they are much slower and typically need 2–4 days to travel to the Earth. The fastest CMEs, however, can reach the Earth in less than a day. In SEP events, large numbers of high-energy charged particles, predominantly protons and electrons, are released. They can be caused by magnetic reconnection-driven processes during solar flares resulting in impulsive SEPs, but also at CME-driven shock waves that produce large gradual SEP events.

Modeling of the solar wind, CMEs, and SEPs. A lot of data are available nowadays, including full-Sun observations from below the solar surface into the heliosphere at multiple wavelengths and far-side and stereoscopic imaging by the instruments aboard the STEREO, SDO, and SoHO satellites. The growing open-access event archival databases facilitate their use for data-driven and physics-based simulation models of the variable solar wind, solar flares, and CMEs. These enable us to explore and quantify the conditions leading to solar storms and their initial development with increasing detail. Also within the heliosphere we now have observations of propagating perturbations traveling from the Sun to the Earth and beyond, including continuous (long-term) *in situ* monitoring of the solar wind and SEP properties at the L1 point, about one million miles upwind from Earth along the Sun–Earth line, and multiscale measurements in the near-Earth solar wind. Such observations provide vital information on the internal structure of CMEs and the CME-driven shock fronts and help us to understand the dynamics of our magnetosphere in response to the variable solar wind conditions and CME impacts.

These observations are crucial for mathematically modeling the phenomena. The data are not only key to validation of the models; state-of-the-art models are data-driven, that is, they use the available data as input, for instance as boundary or initial conditions. Examples of such data-driven, operational solar wind and CME propagation and evolution models are ENLIL [233] and EUHFORIA [234] (see figures 6.50 and 6.51). Both these models use magnetograms of the solar photosphere as boundary conditions for the solar coronal model and various other observational data to determine the launch parameters of the CMEs, such as the initial velocity, spread angle, location, and magnetic flux. These models are made available to the wider community via the NASA Community Coordinated Modeling Center (CCMC, http://ccmc.gsfc.nasa.gov) or the ESA Virtual Space

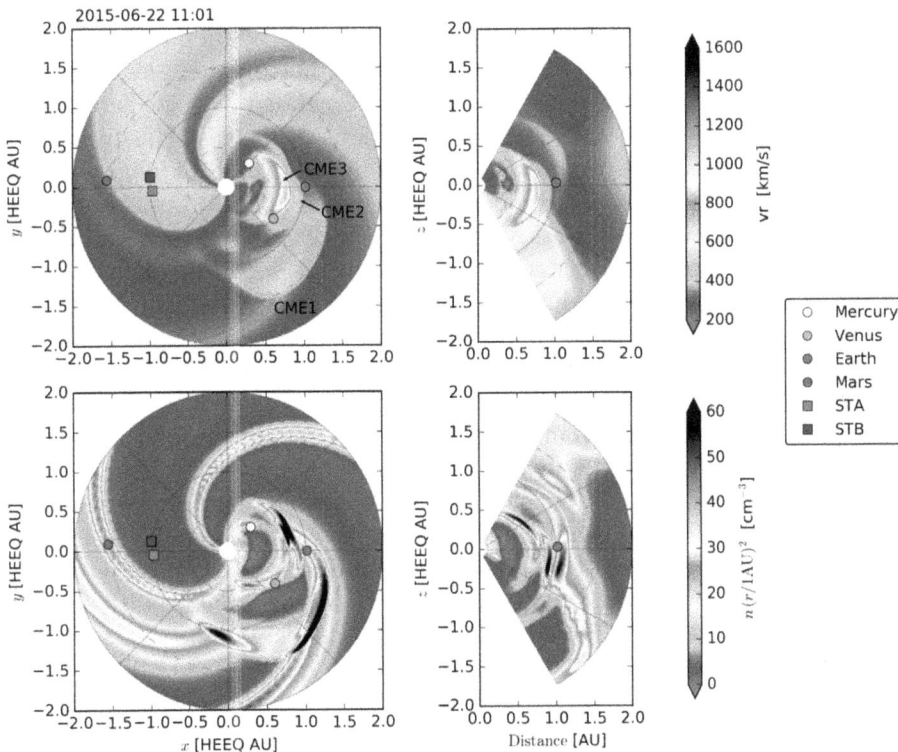

Figure 6.50. Snapshot of an EUHFORIA simulation of an event in June 2015. The snapshot was taken at 03:03 UT on June 21, 2015. The top row shows the radial speed, while the bottom row shows the scaled number density. The left panels depict the solution in the heliographic equatorial plane, while the right panels show the meridional plane that includes the Earth, indicated by the blue dot. Reproduced from [234] CC BY 4.0.

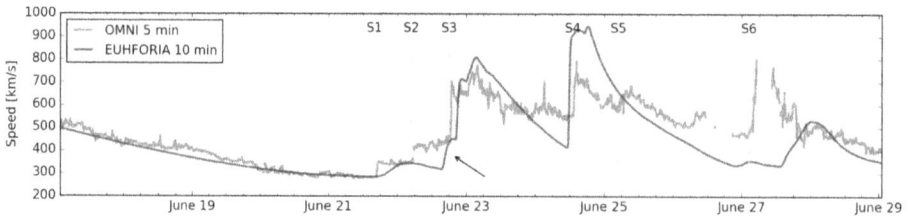

Figure 6.51. Radial speed at the position of the Earth as a function of time. The blue curve shows the data from the EUHFORIA simulation also shown in figure 6.50, saved at 10 min cadence, while the red curve shows OMNI 5 min data. (OMNI is an hourly resolution multisource data set of the near-Earth solar wind's magnetic field and plasma parameters.) Reproduced from [234] CC BY 4.0.

Weather Modelling Centre (VSWMC) [235]. They are used on a daily basis to forecast the arrival of the CIRs and CME shocks and magnetic clouds at the Earth (and other planets) with relative success. They are also continuously adjusted and improved, as they are of course not perfect. Their weaknesses include the use of very simplified (semiempirical) background solar wind models as well as the use of oversimplified CME models that do not take the structure of the magnetic field within the CME itself into account or do so only marginally. They also do not describe the CME onset self-consistently nor its early propagation as the CMEs are introduced only at 0.1 au[12]. These models also do not provide any information about the SEP emission and transport properties generated by solar flares and the CME leading shock fronts and offer no predictions of their geoeffectiveness, the impact they have on the Earth, because they are not coupled with magnetospheric/ionospheric models or with effects models. However, numerical simulation models for particle acceleration at shocks do exist. The Coronal Shock Acceleration simulation model [236], for instance, traces energetic ions and self-consistently determined power spectra of Alfvénic fluctuations that scatter them in prescribed large-scale fields upstream of a shock (including global heliospheric field configurations). Some numerical simplifications made in describing the wave–particle interactions allow the model to be run over global time scales. The SOLar Particle Acceleration in Coronal Shocks model is more recent and uses a physically accurate description of the microphysics involved. However, it is presently limited to local simulation volumes around the shock [237]. The test particle Monte Carlo simulation model called DownStream Propagation with Magnetohydrodynamics models the downstream side of the shock, and the Shock-and-Particle model [238] solves a focused transport equation assuming a constant solar wind flow with a Parker spiral magnetic field. The novel PARADISE model [239] simulates energetic particle distributions in the solar wind provided by EUHFORIA [234] using a quasilinear approach to capture the interaction between solar wind turbulence and energetic particles. An illustration of the possibilities is shown in figure 6.52.

Geoeffects. The variable solar wind conditions causes changes in our magnetic shield, the magnetosphere, the area of space around the Earth that is controlled by

---

[12] 1 astronomical unit is 149 597 871 kilometres, more or less the average Sun–Earth distance.

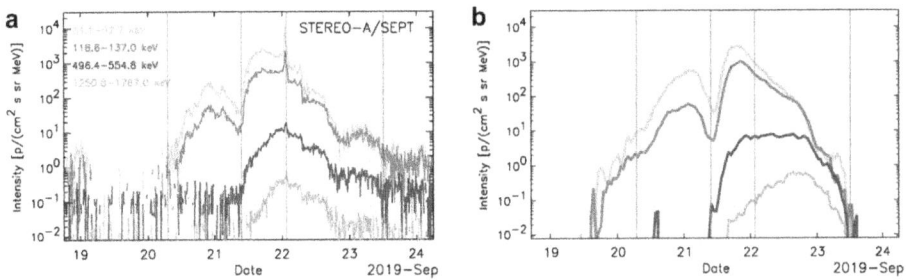

Figure 6.52. PARADISE result: Observed (left) and simulated (right) omnidirectional ion intensities at STEREO-A due to a solar wind stream interaction region (SIR). Vertical lines indicate the onset time of the SIR event (20 September 09:00 UT), the stream interface (21 September 09:30 UT), the developing a reverse shock (22 September 01:35 UT), and the stop time of the SIR event (23 September 12:00 UT). Reproduced from [240], copyright IOP Publishing Ltd. All rights reserved.

its magnetic field. The Earth's magnetosphere is blasted by the solar wind, compressing its sunward side to a distance of only 6–10 times the Earth's radius and creating the bow shock, a supersonic shock wave on the day side. This bow shock heats and slows down the majority of the solar wind particles and causes them to detour around the Earth in the magnetosheath. The night-side magnetosphere is extended by the solar wind to possibly 1000 times the Earth's radius, known as the magnetotail. The geomagnetic response to transients in the solar wind (CIRs, CMEs, magnetic clouds, etc) is expressed in terms of indices such as Kp, Dst, and magnetopause stand-off distance, and these are often used as metric in forecasts of space weather impact on the Earth.

The Kp index is a global geomagnetic activity index based on 3-hourly measurements of magnetometers spread over the whole planet. It consists of 10 values ranging from 0 to 9, where a Kp of 0 indicates quiet conditions, and a Kp of 9 means extreme storm conditions. The disturbance storm time (Dst) index gives information about the strength of the ring current around the Earth caused by solar protons and electrons. The magnetic field induced by this ring current is directly opposite the Earth's magnetic field. Hence, when the difference between solar electrons and protons becomes higher, the Earth's magnetic field becomes weaker. During solar storms, the Dst value becomes negative, meaning that the Earth's magnetic field is weakened.

The Earth's magnetopause separates the solar wind from the Earth's magnetosphere. Its location changes due to different solar wind conditions. Its location is determined by the balance between the pressure of the magnetic field of the Earth and the dynamic pressure of the solar wind. When the solar wind pressure increases, the magnetopause moves inward, and the reverse occurs when it decreases.

Modeling of geoeffects. The observational coverage of the ionosphere–thermosphere–mesosphere domain is growing, which enables near-real-time maps of the total electron content and critical frequency maps for radio communication. Our insight in the contribution of the plasmasphere to the ionospheric total electron content is improving, and so is our understanding of subauroral effects and the troposphere–ionosphere coupling as well as the storage and release of solar wind momentum in the thermosphere.

Global magnetosphere models aim to simulate the detailed spatial and temporal response of the magnetosphere–ionosphere system under solar wind forcing. GUMICS-4 [241] is a coupled magnetosphere–ionosphere simulation model including a global magnetohydrodynamics (MHD) magnetosphere model and an electro-static ionosphere model. It is capable of predicting the magnetopause distance, interplanetary magnetic field penetration in the magnetotail, ionospheric field-aligned current pattern, and so on. At Imperial College London, a model is being developed based on the global MHD code Gorgon, a three-dimensional (3D) resistive MHD code originally developed to simulate laboratory plasmas [242]. Other models focus on specific parts of the magnetospheric system, such as the 3D model of the plasmasphere coupled with the ionosphere developed by Pierrard *et al* [243] and the Salammbô code, the 3D electron and proton radiation belt model that gives the instantaneous particle flux inside the radiation belts, taking into account wave–particle interactions as well as atmosphere–particle and plasmasphere–particle interactions and even radial diffusion [244]. But we still do not fully understand the acceleration mechanisms of energetic electrons and protons within the radiation belts, and quantitative prediction models are needed.

Simple empirical models and more complicated neural network models exist to derive the geoeffectiveness parameters. They are usually based on L1 data, which means they provide forecasts of the immediately expected parameter values, so-called nowcasts. The current state-of-the-art in Europe, in terms of geomagnetic activity at ground level and in GIC estimation in power grids, is also real-time monitoring. In the UK the MAGIC (Monitoring and Analysis of GIC) power system tool, developed by UKRI-BGS in association with National Grid, represents this state of the art. But real-time limits the reaction time and mitigation options available to the power industry.

6.6.5 Most urgent needs

As was mentioned earlier, there exist some basic operational models for forecasting space weather and its effects. However, many of them are (semi-)empirical and need to be upgraded to physics-based models in order to improve the forecasts, and predict the parameters a few days ahead to enable mitigation.

The operational solar wind models are 3D and data-driven (based on magneto-grams of the solar surface, the photosphere), and they are still very much simplified in order to speed up the numerical simulations. The mentioned operational models ENLIL and EUHFORIA both are based on a semiempirical coronal model and provide a steady background solar wind corotating with the Sun starting only at 0.1 au, that is, beyond the 'sonic' point where the solar wind becomes superfast, faster than the characteristic speeds in a magnetised plasma, so the boundary conditions to implement at the inlet boundary (0.1 au) are simple. We need an operational full-MHD coronal model, containing the most important physical effects, such as radiation, thermal conduction, and heating/acceleration. Such models exists, but they require way too much CPU power and wall clock time to be used in operational settings.

We also need more advanced 'flux rope' models for CMEs that are capable of predicting the evolution of the internal magnetic structure of the ICMEs and, in particular, the component of the magnetic field perpendicular to the equatorial plane, as this is key for the severity of the impact of the CMEs on the Earth. The internal magnetic structure of the CMEs evolves due to its interaction with the ambient solar wind, which is also magnetized. This yields deformations and erosion of both the plasma and the magnetic flux in the magnetic cloud. These processes need to be understood very well in order to be able to predict the magnetic structure of the ICME upon impact. In fact, this leads to a more fundamental problem and challenge: We need to better understand the magnetic reconnection process because this generic plasma physics phenomenon is playing a key role in the onset of flares and CMEs (forming the flux rope and cutting it loose from the Sun), the already mentioned deformation and erosion of ICMEs, the interaction of CMEs with the magnetospheric bow shock, and so on. A full quantitative understanding of magnetic reconnection is thus absolutely necessary to understand most of the space weather phenomena. This requires the development of combined full-scale fluid and kinetic models for it.

Another major, related modeling challenge is to understand why certain solar events insert SEPs into the heliosphere and how such SEPs propagate through the heliosphere. We mentioned some advanced models for SEPs before, but these still contain ad hoc terms to take into account the turbulence before the shock, the particle acceleration in the shock, and cross-field diffusion. In order to model impulsive SEP events properly, one would need a self-consistent CME launch model. Also genuine gradual SEP event modeling should include the early evolution phase of the CMEs in the low corona. Such models exist to some extent, but again, they require way too much CPU time to be useful in an operational environment.

In short, we need to take into account more physics in the space weather models. This is because we need real quantitative predictions, preferably a few days ahead of time so that mitigation is possible. Many current space weather models are empirical or semiempirical and facilitate only nowcasts. We urgently need to develop more physics-based models that enable realistic simulations to the extent that they can predict events a few days ahead so that mitigation is possible. Of course, such models are data-driven, and we thus also need more and continuous *in situ* and remote sensing data.

6.6.6 Coupling different simulation models

Because of the huge temporal and spatial extent of space weather phenomena, it is almost impossible to model the entire phenomena from the start (e.g. before eruption) to the end (e.g. a possible space weather storm arriving at the Earth). For this reason, over the past decade, scientists have developed a large variety of models for each of the parts of the larger puzzle. Ranging from close to the Sun at the solar surface, we can model from the preeruption to the propagation of solar wind and CMEs through the solar system all the way to the Earth, where we can model the magnetosphere and even effects on the power grid. Most of these models solve a set of mathematical equations, taking into account the physical and/or chemical processes behind the phenomena of interest. Some of the model inputs are based on observations, for example, the CME structure, but they contain both observational and fitting errors. Numerical simulations based on these models require high-performance computer clusters. But after appropriate validation, some of these models have a predictive value, so they can be used for forecasting, for instance, the arrival of a CME shock at the Earth or the radiation to be expected at the location of a satellite, enabling, in some cases, mitigation of its destructive effects. As each model tries to gain deeper insight into the physical mechanisms that are at hand, they usually focus on a specific area, for example, the solar surface, the Earth's magnetosphere, and so on. However, the output of one numerical model can be used as input for other models, thus creating a chain of models [235]. Thus, we are able to model the propagation of the solar wind and CMEs from the solar surface all the way to the Earth and even specific effects on the Earth, such as power grid effects. The different complicated numerical models usually have a rather compli-cated installation procedure, and operating these models requires experience. Moreover, the efficient use of these models often requires a considerable amount of computer power (CPU time) and computing memory. Such model runs also produce an enormous amount of data as output, which needs to be stored, interpreted, analysed, and visualised. A preliminary example of the possibilities that the coupling of models creates is given in figure 6.53 where the current nowcasts (a few hours ahead) are compared with forecasts (3 days ahead) obtained by using synthetic (simulation) data at L1 instead of measurements.

Because of the need for this chain of models, scientists started to develop integrated space weather model frameworks and infrastructures in which these models can easily be run and also be coupled. A good example is the CCMC (see [247]). This is a multiagency (NASA and the National Science Foundation) initiative that 'enables, supports and performs the research and development for next-generation space science and space weather models'[245]. Another good example of such an integrated framework is the University of Michigan Center for Space Environment Modeling (CSEM) (see [246]). CSEM also develops physics-based, high-performance computer models and applies them to predict the space weather and its effects. These frameworks provide a kind of standard environment for the computer models and at the same time serve as model and data repositories. They thus enable model simulation runs and some even facilitate the coupling of different submodels that are integrated in the framework. This way the framework is able to support space weather forecasters and researchers and even space science education. Also the VSWMC developed by ESA provides a common framework for running space weather models and easily couple them together to create more complex model chains [235].

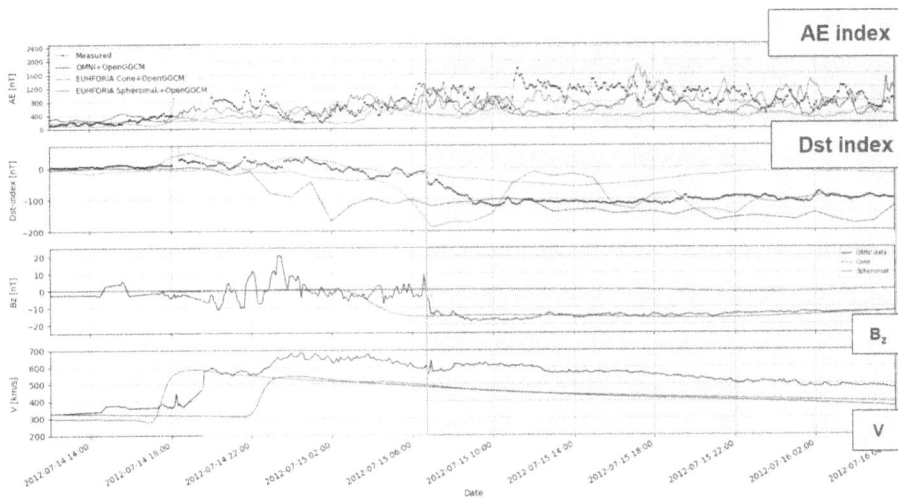

Figure 6.53. Preliminary result from coupling EUHFORIA to the magnetospheric model OpenGGCM. Reprinted by permission from Springer Nature, [247], copyright (2008). The standard nowcast, using measurements at L1, are compared with forecasts (3 days ahead) using synthetic L1 data from EUHFORIA. When a spheromak (magnetized) CME model is used, the forecast can stand the comparison with the nowcast. From top to bottom: comparison of AE index measurements to nowcast (OMNI-OpenGGCM) and forecasts based on EUHFORIA with a cone CME and a spheromak CMEs; the same for the Dst index; the measured B_z component compared to the predictions; and the same comparison for the radial velocity (credit: Anwesha Maharana).

6.6.7 Challenges

6.6.7.1 Scientific challenges

The goal in space weather–related scientific research is to develop robust methods and models that enable the prediction of space weather events and their geoeffects with high accuracy. The aim is to do better than nowcasts and provide the predictions at least 12 h in advance, preferably even 2–3 days in advance, in order to enable to mitigate the impact on human infrastructure and society as well as to protect human health in flight or in orbit.

The global space weather road map for 2015–25 [222] commissioned by COSPAR and the International Living With a Star program, recognises the progress that has been made regarding both ground-based and space-based observations of the Sun–Earth system. However, a deeper understanding of the complex space weather events involving a variety of physical and chemical effects on many different length scales requires a multidisciplinary and internationally coordinated approach. The roadmap advocates physics-based and data-driven models and prioritises the scientific focus areas and research infrastructure needed to deepen our under-standing of space weather and how it affects our daily life. The identified highest-priority research areas include the quantification of the active-region magnetic structure, which is necessary to model and predict nascent coronal mass ejections; the quantification of the solar wind–magnetosphere–ionosphere coupling dynamics;

knowledge of the global coronal field required to drive solar wind models; and understanding and modeling of the radiation belts and solar energetic particles. Moreover, these recommendations are expanded into pathways with focus questions and urgent actions, for instance, to maintain crucial infrastructure and expand it where necessary. The focus should be on observation-based modeling throughout the Sun–Earth system, which should yield prediction models of the internal magnetic structure of coronal mass ejections that enable forecasts more than 12 h ahead of their impact at the Earth. Also the geoeffectiveness of the CME impacts needs to be better understood and how they induce intense geomagnetically induced currents that affect our power and transport infrastructures and radiation storms that cause aging and malfunctions of space-based and sometimes even ground-based assets. The combined effect of radiation and current flow results in ionospheric disturbances that affect navigation and telecommunication systems, cause satellite drag, and lead to ageing of satellites.

The more recent European Space Weather Assessment and Consolidation Committee (ESWACC) report [248] was commissioned by the European Space Science Committee of the European Science Foundation. It first discusses the ongoing European space weather activities and problems and then gives recommendations for better consolidated efforts. It identifies six areas where European-wide coordination is required. These include 'Enabling critical science to improve our scientific understanding of space weather', the 'Development *and coupling* of advanced models by applying a system-science approach which utilises *physics-based* modeling', the 'Assessment of risks at National, Regional and European levels', the 'Consolidation of European User Requirements', 'Support to R2O (and O2R)', and 'Define and implement an operational network for future space weather observations'.

It is extremely difficult to correctly predict the component of the magnetic field of a ICME upon impact at the Earth's bow shock, especially when the CME hits the Earth not fully but gives it only a glancing blow [249, 250]. The fact that the Earth is about 150 million kilometres away from the Sun makes it such that a minor error in the initial position parameters may result in the CME 'missing' the Earth. To mitigate such errors on the input parameters, so-called ensemble simulations are performed where a few dozen of combinations of input parameters are considered (based on the uncertainty of the observations), yielding a prediction windows for, for example, the shock arrival time, that is, a window in which 75% of the ensemble CME shocks arrived.

Extreme space weather events are a particular challenge, as they occur only every 100 or 200 years and moreover do not follow the solar cycle, that is, they can occur anytime, including during solar minimum [251]. Moreover, it has been shown that CME-CME interactions can turn two average CMEs into a very geoeffective combination under the right circumstances [250]. Hence, the potential geoeffectiveness of such CME–CME interactions needs to be investigated more, as they occur very frequently.

6.6.7.2 The data avalanche challenge

The solar and space environment (and the broader heliophysics) research community is experiencing a rapid and radical transformation with an enormous increases in data rates and archive volumes. There are urgent needs for theoretical support for the interpretation of the intricate, nonlinear systems that couple the Sun's deep interior all the way to its atmosphere and further out to the planetary climate systems. In the subdiscipline of solar physics, for example, the currently operative Solar Dynamics Observatory (launched in 2010) sends down over 2 terabytes of data per day. This constitutes a hundredfold increase over the TRACE mission and a thousandfold increase of the SOHO/EIT instrument, both of which are just over 25 years old. This flood of data cannot be routinely inspected and analysed by the scientific community in the traditional ways of personal inspection and detailed analysis of much of the data. Instead, the instrument teams are guiding the community towards automated feature recognition, pattern finding, and knowledge base–supported guidance to preselected subsets of the full data archive that are deemed to be the most helpful as we seek to further our understanding of solar activity and the space weather phenomena that it drives.

On the theoretical/numerical front, we are increasingly aware of the fact that the solar interior, the solar atmosphere, and the heliosphere form a nonlinear coupled system with multiple complex feedback pathways. The interpretation of the observational data has moved past the intuitive or elemental model phase and urgently requires advanced numerical models that are able to simulate the intricate coupled processes in the system.

6.6.7.3 The research coordination challenge

Developing such advanced physics-based numerical simulation models for the intricate 'big problems' in solar physics and in heliophysics in general requires a different way to manage, perform, and fund scientific research. The current practice resulted in small research groups with unpredictably fluctuating financial resources and few long-term, fully supported scientists, postdoctoral researchers, and PhD students. These have revealed the potential value of such models, but we have now reached the phase where larger, collaborative, and more coordinated efforts are required. The obvious analogy is the similar experience in climate research with the development of general circulation models (GCMs). In the 1990s, the GCMs became so complex, as they were incorporating ever more coupled processes, that individual research groups were forced to collaborate on large community-based models to make progress. The basis of a similar effort in solar and space weather physics is currently being created by the ESA VSWMC. However, the VSWMC will integrate and couple only currently existing models, while these also need to be advanced to a larger scale that enables us to address the big problems of solar and space weather physics. This is a very important key competence, from both strategic (independence, military, etc) and socioeconomic points of view, of the EU research community that needs to be developed and maintained.

The needed coordinated development of complex, efficient, high-resolution, and high-fidelity numerical codes is very different from, for example, having an environment in which existing codes are available to the community (such as at the CCMC in the US and ESA's VSWMC) or in which codes are strung together to cross disciplinary interfaces (such as at the Center for Space Environment Modeling in the US), and something on a substantially larger scale than the International Solar-Terrestrial Physics Geospace Model [252] is needed. In order to develop powerful 'missions to the virtual reality' we will need to treat them as space missions, starting with science definition teams, followed by competitive selection of projects, led from a concept–study report phase into a detailed construction phase with well-defined interfaces between elements and subsequently operated by a staff of experts who aid a user community in operating the codes and in archiving, retrieving, and interpreting the results.

6.6.7.4 *The evolving space weather landscape*
The ESWACC report [248] also mentions other issues requiring attention, such as the need for 'continuous elaboration of the analysis including assessment of space weather risks on European infrastructure and understanding of the user needs' and a better coordination of the 'scattered' EU space weather funding ('±60 M€, over the last 10 years') and the preoperational space weather services currently developed by ESA in the optional Space Situational Awareness Programme. Recently, a bottom-up initiative arose within the European space weather community. A group of researchers started to organise itself in a reaction to the changing space weather landscape in Europe and the concern to sustain and further develop the research and user community and the space weather assets (the journal, the yearly European Space Weather Week conference, the medals) and thus to somehow continue to develop the successful efforts made thus far [253]. This resulted in the European Space Weather and Space Climate Association (E-SWAN), an international non-profit association established in 2022 (https://eswan.eu/).

6.6.8 Opportunities

6.6.8.1 *Increasing high-performance computing power*
High-performance computing is a game-changing technology in this field (and in many other fields as well). As a matter of fact, with the rapid increase in computational power, a new opportunity is arising: Numerical experiments and simulations now provide access to a virtual world that has the potential of being as revealing about the world around us as direct observations. This is particularly true for the many intrinsically nonlinear processes where numerical experiments can reveal instabilities and bifurcations to guide us to the correct parameter range that applies to the heliophysical processes under study. A few such examples are already found among the most-cited papers in solar and space weather physics[13] (apart from the mission and instrument papers): a model for CME initiation [254], the study of radiative transfer in near-surface convection (also for abundance determinations) [255], and a study of particle acceleration at the Sun and in the heliosphere [256].

---

[13] Determined from a search of refereed publications with ADS, containing any of the following terms in the abstract: 'solar atmosphere', 'solar corona', 'solar photosphere', 'solar chromosphere', 'solar transition region', 'solar region', 'solar flare', 'space weather', or 'coronal mass ejection'.

6.6.8.2 Funding programs

ESA started preparing for space weather in 2001 [257], but the Space Weather (SWE) Portal was opened only in 2012. It has been expanded ever since, and the first SWE Service elements transited to operations in 2020 in the Space Situational Awareness (SSA) program. ESA's Space Weather Network is currently managed in a preoperational framework and is coordinated by the Space Weather Coordination Centre, located in Brussels, Belgium. It organises a large network consisting of more than 40 teams from organisations across Europe who are spread over five Expert Service Centres. These are collaborating to provide tailored products and services for Space Weather Network customers. In the meantime, ESA continuous to create opportunities via different technology programs and the SSA program.

ESA's new Space Safety Programme (S2P) replaces and continues the SSA program and adds space mission elements to it. It will test the Space Weather Service Network's operational capacity to support ESA's space operations and mission design. This includes nowcasting and forecasting all aspects of space weather in the heliosphere but also in the magnetosphere, ionosphere, and atmosphere. The S2P also includes missions such as the Lagrange mission, the first ever operational space weather mission and the first mission to L5, that is, outside of the Earth–Sun line. The mission is aimed at space weather monitoring and has been developed in close collaboration with NOAA and NASA; its launch is foreseen in 2027. In fact, the Lagrange mission envisions launching two spacecraft at Lagrangian points L1 and L5, which will enable scientists to research solar flares and CMEs and improve forecasts of the solar wind at 1 au. S2P has other components, such as the in-orbit servicing/removal mission (ADRIOS), which will accomplish the first ever removal of a piece of space debris; the HERA mission, which will aim to develop planetary defence technologies (to be launched in 2024); and the CREAM mission, involving a variety of activities aimed at automatic collisions avoiding.

The role of space weather has also become ever more important for EU/EC research programs. The results of the many different research programs are available on CORDIS, the Community Research and Development Information Service [258]. In June 2021, a search on 'Space Weather' in CORDIS yielded 369 hits in the Horizon 2020 program for all collections together. Limiting the selection to projects yielded only 50 hits. The first two 'space weather' projects were funded already in Framework Programs 4 (FP4). In both FP5 and FP6 there were six 'space weather' projects funded (and 14 hits over all collections in FP6), and in FP7, 82 projects were funded (with 215 'space weather' hits over all collections).

Horizon Europe, the EU's key funding program for research and innovation, has a budget of €95.5 billion for the period 2021–27. However, as of June 2021 there was only one forthcoming 'Space Weather' call: the HORIZON Research and Innovation Action HORIZON-CL4–2022-SPACE-01–62[14]. Clearly, other, more general calls on AI and ML, for instance, also include opportunities for space weather research.

---

[14] Searching for 'space weather' on https://ec.europa.eu/info/funding-tenders/opportunities/portal/screen/opportunities/topic-search

It is good to notice that the advice from the mentioned ESWACC report [248] on the ESA–EU coordination is being implemented. As a matter of fact, the last Horizon 2020 space weather calls insisted on using existing EU and ESA infrastructures as much as possible.

Of course, additional funding for space weather activities is provided by individual European states. However, this funding source is very fragmented and localised, often focusing on the local space industry. As was mentioned earlier, the complexity of space weather requires coordinated scientific research and harmonised provision of services with the ESA–EU coordination and the ESA Space Weather Service Network as good examples, respectively.

Acknowledgments

UO has been supported by research focus point Earth and Environmental Systems of the University of Potsdam. UO and AA acknowledge Co-PREPARE of DAAD (DIP Project No. 57553291). SS acknowledges support from the DFG research training group 'Natural Hazards and Risks in a Changing World' (Grant No. GRK 2043/1). DE acknowledges TÜBİTAK (Grant No. 118C236). We thank Georg Veh for helping with visualization.

References

[1] Boers N, Goswami B, Rheinwalt A, Bookhagen B, Hoskins B and Kurths J 2019 Complex networks reveal global pattern of extreme-rainfall teleconnections *Nature* **566** 373–7

[2] Steffen W, Sanderson R A, Tyson P D, Jäger J, Matson P A, Moore B, Oldfield F, Richardson K, Schellnhuber H-J and Turner B L 2006 *Global Change and the Earth System: A Planet Under Pressure* (Berlin: Springer Science & Business Media)

[3] Mann M E, Steinman B A, Brouillette D J and Miller S K 2021 Multidecadal climate oscillations during the past millennium driven by volcanic forcing *Science* **371** 1014–9

[4] Ge Y, Jin Y, Stein A, Chen Y, Wang J, Wang J, Cheng Q, Bai H, Liu M and Atkinson P M 2019 Principles and methods of scaling geospatial Earth science data *Earth Sci. Rev.* **197** 102897

[5] Fan J, Meng J, Ludescher J, Chen X, Ashkenazy Y, Kurths J, Havlin S and Schellnhuber H J 2020 Statistical physics approaches to the complex Earth system *Phys. Rep.* **896** 1–84

[6] Bradley R S 2015 *Paleoclimatology: Reconstructing Climates of the Quaternary* (New York: Academic)

[7] Muller R A and MacDonald G J 2002 *Ice Ages and Astronomical Causes: Data, Spectral Analysis and Mechanisms* (Berlin: Springer Science & Business Media)

[8] Poincaré H 1890 Sur le problème des trois corps et les équations de la dynamique *Acta Math.* **13** A3–270

[9] Nicolis G and Nicolis C 2012 *Foundations of Complex Systems: Emergence, Information and Prediction* (Singapore: World Scientific)

[10] Kantz H and Schreiber T 2004 *Nonlinear Time Series Analysis* (Cambridge: Cambridge University Press) vol 7

[11] Eckmann J 1987 Recurrence plots of dynamical systems *Europhys. Lett.* **5** 973–7

[12] Marwan N, Romano M C, Thiel M and Kurths J 2007 Recurrence plots for the analysis of complex systems *Phys. Rep.* **438** 237–329

[13] Pikovskij A, Rosenblum M and Kurths J 2007 *Synchronization: A Universal Concept in Nonlinear Sciences* (Cambridge: Cambridge University Press)

[14] Torrence C and Compo G P 1998 A practical guide to wavelet analysis *Bull. Am. Meteorol. Soc.* **79** 61–78

[15] Newman M E J 2010 *Networks: An Introduction* (Oxford: Oxford University Press)

[16] Jafarimanesh A, Mignan A and Danciu L 2018 Origin of the power-law exponent in the landslide frequency-size distribution *Nat. Hazards Earth Syst. Sci. Discuss.* 1–28

[17] Gutenberg B and Richter C F 1944 Frequency of earthquakes in California *Bull. Seismol. Soc. Am.* **34** 185–8

[18] Gutenberg B and Richter C F 1956 Earthquake magnitude, intensity, energy, and acceleration: (second paper) *Bull. Seismol. Soc. Am.* **46** 105–45

[19] Frohlich C and Davis S D 1993 Teleseismic b values; or, much ado about 1.0 *J. Geophys. Res.: Solid Earth* **98** 631–44

[20] Blöschl G 1996 *Scale and Scaling in Hydrology* (Techn. Univ., Inst. f. Hydraulik, Gewässerkunde u. Wasserwirtschaft)

[21] Peters-Lidard C D, Pan F and Wood E F 2001 A re-examination of modeled and measured soil moisture spatial variability and its implications for land surface modeling *Adv. Water Res.* **24** 1069–83

[22] Kumar P and Foufoula-Georgiou E 1993 A multicomponent decomposition of spatial rainfall fields: 1. Segregation of large- and small-scale features using wavelet transforms *Water Resour. Res.* **29** 2515–32

[23] Schuster A 1898 On the investigation of hidden periodicities with application to a supposed 26 day period of meteorological phenomena *J. Geograph. Res.* **3** 13–41

[24] Eroglu D, McRobie F H, Ozken I, Stemler T, Wyrwoll K-H, Breitenbach S F, Marwan N and Kurths J 2016 See–saw relationship of the Holocene East Asian–Australian summer monsoon *Nat. Commun.* **7** 1–7

[25] Malik N, Marwan N and Kurths J 2010 Spatial structures and directionalities in Monsoonal precipitation over South Asia *Nonlinear Processes Geophys.* **17** 371–81

[26] Pacheco J F, Scholz C H and Sykes L R 1992 Changes in frequency–size relationship from small to large earthquakes *Nature* **355** 71–3

[27] Zhan Z 2017 Gutenberg–Richter law for deep earthquakes revisited: a dual-mechanism hypothesis *Earth Planet. Sci. Lett.* **461** 1–7

[28] Scholz C H 1968 The frequency-magnitude relation of microfracturing in rock and its relation to earthquakes *Bull. Seismol. Soc. Am.* **58** 399–415

[29] Tormann T, Enescu B, Woessner J and Wiemer S 2015 Randomness of megathrust earthquakes implied by rapid stress recovery after the Japan earthquake *Nat. Geosci.* **8** 152–8

[30] Wiemer S and McNutt S R 1997 Variations in the frequency-magnitude distribution with depth in two volcanic areas: Mount St. Helens, Washington, and Mt. Spurr, Alaska *Geophys. Res. Lett.* **24** 189–92

[31] De Geer G 1926 On the solar curve: as dating the ice age, the New York Moraine, and Niagara Falls through the Swedish Timescale *Geogr. Ann.* **8** 253–83

[32] Berger A L 1978 Long-term variations of caloric insolation resulting from the Earth's orbital elements *Quat. Res.* **9** 139–67

[33] Westerhold T, Marwan N, Drury A J, Liebrand D, Agnini C, Anagnostou E, Barnet J S, Bohaty S M, De Vleeschouwer D and Florindo F 2020 An astronomically dated record of Earth's climate and its predictability over the last 66 million years *Science* **369** 1383–7

[34] Burke K D, Williams J W, Chandler M A, Haywood A M, Lunt D J and Otto-Bliesner B L 2018 Pliocene and Eocene provide best analogs for near-future climates *Proc. Natl. Acad. Sci.* **115** 13288–93

[35] Pai D S and Nair R M 2009 Summer monsoon onset over Kerala: new definition and prediction *J. Earth Syst. Sci.* **118** 123–35

[36] Stolbova V, Surovyatkina E, Bookhagen B and Kurths J 2016 Tipping elements of the Indian monsoon: prediction of onset and withdrawal: tipping elements of monsoon *Geophys. Res. Lett.* **43** 3982–90

[37] Ludescher J *et al* 2021 Network-based forecasting of climate phenomena *Proc. Natl Acad. Sci.* **118** e1922872118

[38] Stark C P and Hovius N 2001 The characterization of landslide size distributions *Geophys. Res. Lett.* **28** 1091–4

[39] Malamud B D, Turcotte D L, Guzzetti F and Reichenbach P 2004 Landslides, earthquakes, and erosion *Earth Planet. Sci. Lett.* **229** 45–59

[40] West G B 2017 *Scale: The Universal Laws of Growth, Innovation, Sustainability, and the Pace of Life in Organisms, Cities, Economies, and Companies* (London: Penguin)

[41] Liu Z, Nadim F, Garcia-Aristizabal A, Mignan A, Fleming K and Luna B Q 2015 A three-level framework for multi-risk assessment *Georisk: Assess. Manag. Risk Eng. Syst. Geohazards* **9** 59–74

[42] von Specht S, Ozturk U, Veh G, Cotton F and Korup O 2019 Effects of finite source rupture on landslide triggering: the 2016 MW 7.1 Kumamoto earthquake *Solid Earth* **10** 463–86

[43] Ozturk U, Marwan N, Korup O, Saito H, Agarwal A, Grossman M J, Zaiki M and Kurths J 2018 Complex networks for tracking extreme rainfall during typhoons *Chaos* **28** 075301

[44] Saito H, Korup O, Uchida T, Hayashi S and Oguchi T 2014 Rainfall conditions, typhoon frequency, and contemporary landslide erosion in Japan *Geology* **42** 999–1002

[45] Agarwal A, Caesar L, Marwan N, Maheswaran R, Merz B and Kurths J 2019 Network-based identification and characterization of teleconnections on different scales *Sci. Rep.* **9** 8808

[46] Rahmstorf S 2002 Ocean circulation and climate during the past 120,000 years *Nature* **419** 207–14

[47] Caesar L, McCarthy G D, Thornalley D J R, Cahill N and Rahmstorf S 2021 Current Atlantic meridional overturning circulation weakest in last millennium *Nat. Geosci.* **14** 118–20

[48] Crutzen P and Stoermer E 2000 International geosphere biosphere programme (IGBP) *Newsletter* 41

[49] Rai R and Sahu C K 2020 Driven by data or derived through physics? A review of hybrid physics guided machine learning techniques with cyber-physical system (cps) focus *IEEE Access* **8** 71050–73

[50] Karpatne A, Atluri G, Faghmous J H, Steinbach M, Banerjee A, Ganguly A, Shekhar S, Samatova N and Kumar V 2017 Theory-guided data science: a new paradigm for scientific discovery from data *IEEE Trans. Knowl. Data Eng.* **29** 2318–31

[51] Reichstein M, Camps-Valls G, Stevens B, Jung M, Denzler J and Carvalhais N 2019 Deep learning and process understanding for data-driven Earth system science *Nature* **566** 195–204

[52] Faghmous J H and Kumar V 2014 A big data guide to understanding climate change: the case for theory-guided data science *Big Data* **2** 155–63

[53] Xu T and Valocchi A J 2015 Data-driven methods to improve baseflow prediction of a regional groundwater model *Comput. Geosci.* **85** 124–36

[54] Dietz M *et al* 2022 More than heavy rain turning into fast-flowing water – a landscape perspective on the 2021 Eifel floods *Nat. Hazards Earth Syst. Sci.* **22** 1845–56

[55] Bozzolan E, Holcombe E, Pianosi F and Wagener T 2020 Including informal housing in slope stability analysis–an application to a data-scarce location in the humid tropics *Nat. Hazards Earth Syst. Sci.* **20** 3161–77

[56] Hao Z, Singh V P and Hao F 2018 Compound extremes in hydroclimatology: a review *Water* **10** 718

[57] Zscheischler J, Martius O, Westra S, Bevacqua E, Raymond C, Horton R M, van den Hurk B, AghaKouchak A, Jézéquel A and Mahecha M D 2020 A typology of compound weather and climate events *Nat. Rev. Earth Environ.* **1** 333–47

[58] Raymond C, Horton R M, Zscheischler J, Martius O, AghaKouchak A, Balch J, Bowen S G, Camargo S J, Hess J and Kornhuber K 2020 Understanding and managing connected extreme events *Nat. Clim. Change* **10** 611–21

[59] Hao Z and Singh V P 2020 Compound events under global warming: a dependence perspective *J. Hydrol. Eng.* **25** 03120001

[60] Boers N, Kurths J and Marwan N 2021 Complex systems approaches for Earth system data analysis *J. Phys.: Complex.* **2** 011001

[61] Perez R and Perez M 2015 A fundamental look at energy reserves for the planet *Int. Energy Agency SHC Program. Sol. Updat* (https://iea-shc.org/data/sites/1/publications/2015-11-A-Fundamental-Look-at-Supply-Side-Energy-Reserves-for-the-Planet.pdf)

[62] Solarmap 2021 (https://google.com/search?q=global+solar+energy+distribution&client=firefox-b-e&source=lnms&tbm=isch&sa=X&ved=2ahUKEwjjjYnAzrnyAhWICewKHTdvAnsQ_AUoAnoECAIQBA&biw=1264&bih=573#imgrc=3jhhX0D-D8vxUM)

[63] Shockley W and Queisser H J 1961 Detailed balance limit of efficiency of p-n junction solar cells *J. Appl. Phys.* **32** 510

[64] NREL *Best Research-Cell Efficiency Chart* (https://nrel.gov/pv/cell-efficiency.html) Accessed 13 September 2021.

[65] Stieglitz R and Heinzel V 2012 *Thermische Solarenergie—Grundlagen, Technologie, Anwendungen* (Heidelberg: Springer)

[66] International Technology Roadmap for Photovoltaic (ITRPV) 2020 *Results* 12th edn (https://itrpv.vdma.org/)

[67] Philipps S and Warmuth W 2021 *Photovoltaics Report* (https://ise.fraunhofer.de/en/publications/studies/photovoltaics-report.html) Accessed 27 July 2021

[68] Lee M *et al* 2012 Efficient hybrid solar cells based on meso-superstructured organometal halide perovskites *Science*

[69] Al-Ashouri A *et al* 2020 Monolithic perovskite/silicon tandem solar cell with >29% efficiency by enhanced hole extraction *Science*

[70] Pitz-Paal R 2020 *Concentrating Solar Power in Future Energy* (Amsterdam: Elsevier) 3rd edn pp 413–30

[71] DLR 2021 *German Aerspace Center: Solar Thermal Power Plants, Heat Electricity and Fuels Form Concentrating Solar Power* (https://dlr.de/sf/en/DownloadCount.aspx?raid=533626&docid=13013&rn=adfa4881-0ef9-4df1-b131-7bd11bfb8758)

[72] EASAC 2011 *Concentrating Solar Power: Its Potential Contribution to a Sustainable Energy Future* EASAC Policy Report 16 (European Academies Science Advisory Council EASAC) 2011, Halle (Saale) (https://easac.eu/publications/details/concentrating-solar-power-its-potential-contribution-to-asustainable-energy-future/) Accessed 05 February 2021

[73] IEA 2020 *World Energy Outlook 2020* (Paris: IEA) https://iea.org/reports/world-energy-outlook-2020

[74] Heagel N *et al* 2019 *Terawatt-Scale Photovoltaics: Transform Global Energy* (United States: N.p.) (https://doi.org/10.1126/science.aaw1845)

[75] REN21 2020 *Renewables 2020—Global Status Report* UN Environment Programme, REN21 Secretariat, Paris (https://ren21.net/wp-content/uploads/2019/05/gsr_2020_full_report_en.pdf) Accessed 18 December 2020

[76] Denholm P and Hand M 2011 Grid flexibility and storage required to achieve very high penetration of variable renewable electricity *Energy Policy* (Amsterdam: Elsevier) vol 39 pp 1817–30

[77] IRENA 2020 *Renewable Power Generation Costs in 2019* (Abu Dhabi: International Renewable Energy Agency) https://irena.org/publications/2020/Jun/Renewable-Power-Costs-in-2019)

[78] Solar Energy Technologies Office, U. S. Department of Energy 2017 *The Sunshot 2030 Goals: 3¢ per Kilowatt Hour for PV and 5¢ per Killowatt Hour for Dispatchable CSP. The Sunshot Goals, DOE/EE-1501* (Office of Energy Efficiency & Renewable Energy) (https://energy.gov/sites/prod/files/2020/09/f79/SunShot%202030%20White%20Paper.pdf)

[79] IRENA 2018 *Global Energy Transformation: A Roadmap to 2050* (https://irena.org/publications/2018/Apr/Global-Energy-Transition-A-Roadmap-to-2050)

[80] MCC 2021 *Mercator Research Institute on Global Commons and Climate Change: Remaining Carbon Budget* (https://mcc-berlin.net/forschung/co2-budget.html)

[81] IEA 2021 *Net Zero by 2050, IEA Flagship Report* (https://iea.blob.core.windows.net/assets/beceb956-0dcf-4d73–89fe-1310e3046d68/NetZeroby2050-ARoadmapfortheGlobalEnergy Sector_CORR.pdf)

[82] Wagner H-J 2018 *Introduction to Wind Energy Systems: Basics, Technology and Operation* (Berlin: Springer) 3rd edn

[83] Vermeiren P, Adriansens W, Moreels J P and Leysen R 1998 The composite Zirfon separator for alkaline water electrolysis *Hydrogen Power: Theoretical and Engineering solutions* (Dordrecht: Springer)

[84] Hall D E 1981 Electrodes for alkaline water electrolysis *J. Electrochem. Soc.* **128** 740

[85] The Engineer 2017 Hot rocks offer solution to grid-scale energy storage *The Engineer* Nov 2017

[86] Santhanam S, Heddrich M P, Riedel M and Friedrich K A 2017 Theoretical and experimental study of reversible solid oxide cell (r-SOC) systems for energy storage *Energy* **141** 202–14

[87] Phase change materials for thermal energy storage | Climate Technology Centre & Network | Tue, 11/08/2016 (ctc-n.org).

[88] advanced-chemistryprize2019.pdf (https://nobelprize.org).

[89] How to prevent short-circuiting in next-gen lithium batteries | MIT News | Massachusetts Institute of Technology

[90] Battery prices have fallen 88 percent over the last decade | Ars Technica

[91] Xu Wangwang and Wang Ying 2019 Recent progress on zinc-ion rechargeable batteries *Nano-Micro Lett.* **11** 90

[92] https://ec.europa.eu/environment/international_issues/relations_china_en.htm

[93] Eddington A 1920 The internal constitution of the stars *Nature* **106** 14

[94] Bethe H A 1939 Energy production in stars *Phys. Rev.* **55** 434

[95] Bigot B 2019 Progress toward ITER's first plasma *Nucl. Fusion* **59** 112001

[96] Progress in ITER Physics Basis 2007 Overview and summary *Nucl. Fusion* **47** S1–S404

[97] Federici G *et al* 2019 Overview of the DEMO staged design approach in Europe *Nucl. Fusion* **59** 066013

[98] Alfvén H 1942 Existence of electromagnetic-hydrodynamic waves *Nature* **150** 405

[99] Huijsmans G T A and Loarte A 2013 Non-linear MHD simulation of ELM energy deposition *Nucl. Fusion* **53** 123023

[100] Giancarli L M *et al* 2020 Overview of recent ITER TBM Program activities *Fusion Eng. Design* **158** 111674

[101] Ibarra A *et al* 2018 The IFMIF-DONES project: preliminary engineering design *Nucl. Fusion* **58** 105002

[102] Classification of Radioactive Waste 2004 *No. GSG-1, General Safety Guide* IAEA.

[103] BP Statistical Review of World Energy 2021 (www.bp.com/en/global/corporate/energy-economics/statistical-review-of-world-energy.html)

[104] Key World Energy Statistics 2020 International Energy Agency (www.iea.org/reports/key-world-energy-statistics-2020)

[105] IAEA Power Reactor Information System (PRIS) 2021 (pris.iaea.org/PRIS/home.aspx) *IAEA*

[106] Projected Costs of Generating Electricity 2020 (www.iea.org/reports/projected-costs-of-generating-electricity-2020)© *OECD/IEA and OECD/NEA 2020*

[107] Climate Change 2014 *Mitigation of Climate Change—Working Group III Contribution to the Fifth Assessment Report of the Intergovernmental Panel on Climate Change (IPCC)* Cambridge University Press (www.ipcc.ch/report/ar5/wg3/)

[108] Poinssot C *et al* 2014 Assessment of the environmental footprint of nuclear energy systems. Comparison between closed and open fuel cycles *Energy* **69** 199–211

[109] Uranium 2018 *Resources, Production and Demand* (OECD) (www.oecd-nea.org/jcms/pl5080/) *NEA No. 7413*, 2018

[110] Implementation of Defence in Depth 2016 *at Nuclear Power Plants—Lessons Learnt from the Fukushima Daiichi Accident* (OECD) (www.oecd-nea.org/jcms/pl14950)

[111] Human and Organizational 2016 *Aspects of Assuring Nuclear Safety* (IAEA).

[112] Storage and Disposal of Radioactive Waste *2016–2021 World Nuclear Association, Registered in England and Wales, Number 01215741*

[113] Status and Trends in Spent 2018 *Fuel and Radioactive Waste Management* IAEA (www.iaea.org/publications/11173/status-and-trends-in-spent-fuel-and-radioactive-waste-management)

[114] Disposal of Radioactive 2011 *Waste—Specific Safety Requirements No. SSR-5* IAEA (www.iaea.org/publications/8420/disposal-of-radioactive-waste)

[115] StakeHolder-based Analysis 2019–2022 of Research for Decommissioning, share-h2020.eu *Research Funded by Euratom, Grant Agreement ID: 847626*

[116] Nifenecker H, Meplan O and David S 2001 *Accelerator Driven Subcritical Reactors* (Bristol: Institute of Physics Publishing)

[117] Aït Abderrahim H *et al* 2020 Partitioning and transmutation contribution of MYRRHA to an EU strategy for HLW management and main achievements of MYRRHA related FP7 and H2020 projects: MYRTE, MARISA, MAXSIMA, SEARCH, MAX, FREYA, ARCAS *EPJ Nucl. Sci. Technol.* **6** 33

[118] Strategic Research Agenda of EURAD 2019–2024 European Joint Programme on Radioactive Waste Management (www.ejp-eurad.eu/sites/default/files/2020-01/2._eurad_sra.pdf)*Research Funded by Euratom, Grant Agreement ID: 847593*

[119] Climate Change and Nuclear Power 2020 (www.iaea.org/publications/14725/climate-change-and-nuclear-power-2020)

[120] Global Warming 2019 *of 1.5 °C—An IPCC Special Report on the Impacts of Global Warming of 1.5 °C above Pre-Industrial Levels and Related Global Greenhouse Gas Emission Pathways, in the Context of Strengthening the Global Response to the Threat of Climate Change, Sustainable Development, and Efforts to Eradicate Poverty* (www.ipcc.ch/sr15/). Intergovernmental Panel on Climate Change.

[121] Net Zero by 2050 2021 *A Roadmap for the Global Energy Sector* (www.iea.org/reports/net-zero-by-2050)

[122] Nuclear Power in a Clean Energy System 2019 International Energy Agency (webstore.iea.org/nuclear-power-in-a-clean-energy-system)

[123] Generation IV International Forum^{SM} (www.gen-4.org)

[124] Sustainable Nuclear Energy Technology Platform (snetp.eu/)

[125] Pakhomov I 2018 BN-600 and BN-800 operating experience *Gen IV International Forum^{SM}* (www.gen4.org/gif/upload/docs/application/pdf/2019-01/gifiv_webinar_pakhomov_19_dec_2018_final.pdf).

[126] Advances in Small Modular 2020 Reactor Technology Developments—A Supplement to: IAEA Advanced Reactors Information System (ARIS) 2020 edn (aris.iaea.org/sites/Publications.html)

[127] Small Modular Reactors 2021 *Challenges and Opportunities* OECD (www.oecd-nea.org/jcms/pl) (57979/small-modular-reactors-challenges-and-opportunities), NEA No.7560

[128] IPCC 2018 Summary for policymakers *Global Warming of 1.5 °C. An IPCC Special Report on the Impacts of Global Warming of 1.5 °C above Pre-Industrial Levels and Related Global Greenhouse Gas Emission Pathways, in the Context of Strengthening the Global Response to the Threat of Climate Change, Sustainable Development, and Efforts to Eradicate Poverty* ed V Masson-Delmotte *et al* (https://ipcc.ch/sr15/chapter/spm/)

[129] United Nations Environment Programme 2019 *Emissions Gap Report 2019* (Nairobi: UNEP)https://unep.org/resources/emissions-gap-report-2019

[130] www.iea.org/data-and-statistics

[131] Hoffmann B 2019 *Air pollution in cities: urban and transport planning determinants and health in cities Integrating Human Health into Urban and Transport Planning* ed M Nieuwenhuijsen and H Khreis (Cham: Springer)

[132] European Environment Agency 2020 *Air Quality in Europe* https://op.europa.eu/en/publication-detail/-/publication/447035cd-344e-11eb-b27b-01aa75ed71a1/language-en

[133] MacKay D J C 2009 *Sustainable Energy—Without the Hot Air* p 231 (http://withouthotair.com/)

[134] European Environment Agency 2020 Monitoring CO2 Emissions From Passenger Cars and Vans in 2018 https://op.europa.eu/en/publication-detail/-/publication/6bac010d-dc45-11ea-adf7-01aa75ed71a1/language-en

[135] WLTP 2017 *Worldwild Harmonised Light Vehicle Test Procedure*

[136] Colclasure A M and Kee R J 2010 Thermodynamically consistent modeling of elementary electrochemistry in lithium-ion batteries *Electrochim. Acta* **55** 8960–73

[137] IEA 2020 *Global EV Outlook 2020* (Paris: IEA) https://iea.org/reports/global-ev-outlook-2020

[138] Bloomberg New Energy Finance 2018 *Electric Buses in Cities Driving Towards Cleaner Air and Lower CO_2* (https://about.bnef.com/blog/electric-buses-cities-driving-towards-cleaner-air-lower-co2/)

[139] https://iea.org/data-and-statistics/charts/new-electric-bus-registrations-in-china-2015–2019

[140] https://wri.org/insights/how-did-shenzhen-china-build-worlds-largest-electric-bus-fleet

[141] https://sustainable-bus.com/fuel-cell-bus/fuel-cell-bus-hydrogen/

[142] IRENA 2019 *Innovation Outlook: Smart Charging for Electric Vehicles* (Abu Dhabi: International Renewable Energy Agency) https://irena.org/publications/2019/May/Innovation-Outlook-Smart-Charging

[143] Tran-Quoc T and Le Pivert X 2016 Stochastic approach to assess impacts of electric vehicles on the distribution network *IEEE PES 3rd Innovative Smart Grid Technologies EUROPE (ISGT Europe)(October 2012) (Berlin, Germany)*
Tuan Tran Q and Van Linh Nguyen L V Integration of electric vehicles into industrial network: impact assessment and solutions *IEEE/PES, General Meeting(17–21 July) (Boston, MA)*

[144] Hautier G 2019 Finding the needle in the haystack: materials discovery and design through computational ab initio high-throughput screening *Comput. Mater. Sci.* **163** 108–16

[145] Curtarolo S, Hart G L W, Nardelli M B, Mingo N, Sanvito S and Levy O 2013 *The high-throughput highway to computational materials design Nat. Mater.* **12** 191–201

[146] Saal J E, Kirklin S, Aykol M, Meredig B and Wolverton C 2013 *Materials design and discovery with high-throughput density functional theory: the open quantum materials database (OQMD) JOM* **65** 1501–9

[147] Cheng L *et al* 2015 Accelerating electrolyte discovery for energy storage with high-throughput screening *J. Phys. Chem. Lett.* **6** 283–91

[148] Aykol M *et al* 2016 High-throughput computational design of cathode coatings for Li-ion batteries *Nat. Commun.* **7** 13779

[149] Hautier G *et al* 2011 Phosphates as lithium-ion battery cathodes: an evaluation based on high-throughput ab initio calculations *Chem. Mater.* **23** 3495–508

[150] Wallace S K, Bochkarev A S, van Roekeghem A, Carrasco J, Shapeev A and Mingo N 2021 Free energy of $(CoxMn_{1-x})_3O_4$ mixed phases from machine-learning-enhanced ab initio calculations *Phys. Rev. Mater.* **5** 035402

[151] Wallace S K, van Roekeghem A, Bochkarev A S, Carrasco J, Shapeev A and Mingo N 2021 Modeling the high-temperature phase coexistence region of mixed transition metal oxides from ab initio calculations *Phys. Rev. Res.* **3** 013139

[152] Gubaev K, Podryabinkin E V, Hart G L W and Shapeev A V 2019 Accelerating high-throughput searches for new alloys with active learning of interatomic potentials *Comput. Mater. Sci.* **156** 148–56

[153] Novikov I S, Gubaev K, Podryabinkin E V and Shapeev A V 2021 The MLIP package: moment tensor potentials with MPI and active learning *Mach. Learn.: Sci. Technol.* **2** 025002

[154] Chandesris M, Caliste D, Jamet D and Pochet P 2019 Thermodynamics and related kinetics of staging in intercalation compounds *J. Phys. Chem.* C **123** 23711–20

[155] Onsager L 1931 Reciprocal relations in irreversible processes I *Phys. Rev.* **37** 405–26

[156] Feng X, Ouyang M, Liu X, Lu L, Xia Y and He X 2018 Thermal runaway mechanism of lithium ion battery for electric vehicles: a review *Energy Storage Mater.* **10** 246–67

[157] Lopez-Haro M, Guetaz L, Printemps T, Morin A, Escribano S, Jouneau P H, Bayle-Guillemaud P, Chandezon F and Gebel G 2014 Three-dimensional analysis of Nafion layers in fuel cell electrodes *Nat. Commun.* **5** 5229

[158] Dupre N, Moreau P, De Vito E, Quazuguel L, Boniface M, Bordes A, Rudisch C, Bayle-Guillemaud P and Guyomard D 2016 Multiprobe study of the solid electrolyte interphase on silicon-based electrodes in full-cell configuration *Chem. Mater.* **28** 2557–72

[159] Leriche J B *et al* 2010 An electrochemical cell for operando study of lithium batteries using synchrotron radiation *J. Electrochem. Soc.* **157** A606–10

[160] Tonin G, Vaugham G B M, Bouchet R, Allouin F, Di Michiel M and Barchasz C Operando investigation of the lithium/sulfur battery system by coupled x-ray absorption tomography and x-ray diffraction computed tomography *J. Power Sources* **468** 228287

[161] Bianchini M, Leriche J B, Laborier J L, Gendrin L, Suard E, Croguennec L and Masquelier C 2013 A new null matrix electrochemical cell for rietveld refinements of in situ or operando neutron powder diffraction data *J. Electrochem. Soc.* **160** A2176–83

[162] Tardif S, Dufour N, Colin J F, Gebel G, Burghammer M, Johannes A, Lyonnard S and Chandesris M 2021 Combining operando x-ray experiments and modelling to understand the heterogeneous lithiation of graphite electrodes *J. Mater. Chem.* A **9** 4281–90

[163] Tripathi A M, Wei-Nien Su W N and Hwang B J 2018 In situ analytical techniques for battery interface analysis *Chem. Rev. Soc.* **47** 736–851

[164] Benayad A, Morales-Ugarte J E, Santini C C, Bouchet R and Operando X P S 2021 A novel approach for probing the lithium/electrolyte interphase dynamic evolution *J. Phys. Chem.* A **125** 1069–81

[165] Satija R, Jacobson D L, Arif M and Werner S A 2004 In situ neutron imaging technique for evaluation of water management systems in operating PEM fuel cells *J. Power Sources* **129** 238–45

[166] Xu F, Diat O, Gebel G and Morin A 2007 Determination of transverse water concentration profile through MEA in a fuel cell using neutron scattering *J. Electrochem. Soc.* **154** B1389–98

[167] Martinez N, Porcar L, Escribano S, Micoud F, Rosini S, Tengattini A, Atkins D, Gebel G, Lyonnard S and Morin A 2019 Combined operando high resolution SANS and neutron imaging reveals in situ local water distribution in an operating fuel cell *ACS Appl. Energy Mater.* **2** 8425–33

[168] IEA 2019 *The Future of Hydrogen* (Paris: IEA) https://iea.org/reports/the-future-of-hydrogen

[169] The Royal Society 2019 *Sustainable Synthetic Carbon Based Fuels for Transport* (https://royalsociety.org/topics-policy/projects/low-carbon-energy-programme/sustainable-synthetic-carbon-based-fuels-for-transport/)

[170] Geelen L H W T, Kamps P T W J and Olsthoorn T N 2017 From overexploitation to sustainable use, an overview of 160 years of water extraction in the Amsterdam dunes, the Netherlands *J. Coast Conserv.* **21** 657–68

[171] Paull J 2009 A century of synthetic fertilizer: 1909–2009 *Element.—J. Bio-Dynam. Tasmania* **94** 16–20

[172] Binetti R, Costamagna F M and Marcello I 2008 Exponential growth of new chemicals and evolution of information relevant to risk control *Ann. Ist Super Sanità* **44** 13–5

[173] Daley J 2017 Science is falling woefully behind in testing new chemicals *Smithsonian Magazine*

[174] United Nations Environment Assembly of the United Nations Environment Programme (UNEP) 2019 *Global Chemicals Outlook II: Summary for Policymakers* (Geneva, Switzerland: UNEP/EA.4/21, UNEP)

[175] Card M L, Gomez-Alvarez V, Lee W-H, Lynch D G, Orentas N J, Lee M T, Wong E M and Boethling R S 2017 History of EPI Suite™ and future perspectives on chemical property estimation in US Toxic Substances Control Act new chemical risk assessments *Environ. Sci.: Processes Impacts* **19** 203–12

[176] Matthies M, Solomon K, Vighi M, Gilman A and Tarazona J V 2016 The origin and evolution of assessment criteria for persistent, bioaccumulative and toxic (PBT) chemicals and persistent organic pollutants (POPs) *Environ. Sci.: Processes Impacts* **18** 1114–28

[177] Allen J G These toxic chemicals are everywhere—even in your body. And they won't ever go away, 'Opinions' article published in the *Washington Post* 02 January 2018 (https://washingtonpost.com/opinions/these-toxic-chemicals-are-everywhere-and-they-wont-ever-go-away/2018/01/02/82e7e48a-e4ee-11e7-a65d-1ac0fd7f097e_story.html) (Accessed 15 March 2021)

[178] Gouin T, Mackay D, Jones K C, Harner T and Meijer S N 2004 Evidence for the 'grasshopper' effect and fractionation during long-range atmospheric transport of organic contaminants *Environ. Pollut.* **128** 139–48

[179] Fiedler H, Kallenborn R, de Boer J and Sydnes L K 2019 The Stockholm convention: a tool for the global regulation of persistent organic pollutants *Chem. Int.* **41** 4–11

[180] Sontake A R and Wagh S M 2014 The phase-out of perfluorooctane sulfonate (PFOS) and the global future of aqueous film forming foam (AFFF), innovations in fire fighting foam *Chem. Eng. Sci.* **2** 11–4

[181] Holmquist P H, Schellenberger S, van der Veen I, Peters G M, Leonards P E G and Cousins I T 2016 Properties, performance and associated hazards of state-of-the-artdurable water repellent (DWR) chemistry for textile finishing *Environ. Intern.* **91** 251–24

[182] Brandsma S H, Koekkoek J C, van Velzen M J M and de Boer J 2019 The PFOA alternative substitute GenX detected in grass and leaves near a fluoropolymer manufactory plant in the Netherlands *Chemosphere* **220** 493–500

[183] Cousins I T *et al* 2019 The concept of essential use for determining when uses of PFASs can be phased out *Environ. Sci.: Processes Impacts* **21** 1803–15

[184] de Boer J and Stapleton H M 2019 Toward fire safety without chemical risk *Science* **364** 231–2

[185] Autrup H *et al* 2020 Human exposure to synthetic endocrine disrupting chemicals (S-EDCs) is generally negligible as compared to natural compounds with higher or comparable endocrine activity. How to evaluate the risk of the S-EDCs? *Arch. Toxicol.* **94** 2549–57

[186] Zoeller R T *et al* 2014 A path forward in the debate over health impacts of endocrine disrupting chemicals *Environ. Health* **13** 118

[187] Daughton C G and Ternes T A 1999 Pharmaceuticals and personal care products in the environment: agents of subtle change? *Environ. Health Persp.* **107** 906–10

[188] Legler J and van Straalen N 2018 Decision-making in a storm of discontent *Science* **360** 958–60

[189] Blaquière T, Smagghe G, van Gestel C A M and Mommaerts V 2012 Neonicotinoids in bees: a review on concentrations, side-effects and risk assessment *Ecotoxicol* **21** 973–92

[190] Walters R 2010 Toxic atmospheres air pollution, trade and the politics of regulation *Crit. Crim.* **18** 307–23

[191] Skoczynska E 2021 Development and appliucation of comprehensive chemical analytical; methods for the analysis of polyaromatic compounds *PhD Thesis* Vrije Universiteit, Amsterdam, The Netherlands p 190

[192] Hassaan M A and El Nemr A 2017 Health and environmental impacts of dyes: mini review *Am. J. Environ. Sci. Eng.* **1** 64–7

[193] Kdasi A, Idris A, Saed K and Guan C T 2004 Treatment of textile wastewater by advanced oxidation processes: a review *Global Nest. Int.* J **6** 222–30

[194] Mohan V, Rao C and Karthikeyan J 2002 Adsorptive removal of direct azo dye from aqueous phase onto coal based sorbents: a kinetic and mechanistic study *J. Hazard. Mater.* **90** 189–204

[195] Wallington T J, Srinivasan J, Nielsen O J and Highwood E J 2009 Greenhouse gases and global warming *Environmental and Ecological Chemistry* ed A Sablic (Oxford: Eolss Publ. Co. Ltd) vol 1

[196] Plummer L N and Busenberg E 2000 Chlorofluorocarbons *Environmental Tracers in Subsurface Hydrology* ed P G Cook and A L Herczeg (Boston, MA: Springer)

[197] Birmpili T 2018 Montreal Protocol at 30: the governance structure, the evolution, and the Kigali Amendment *C. R. Geosci.* **350** 425–31

[198] Molina M J and Rowland F S 1974 Stratospheric sink for chlorofluoromethanes: chlorine atom-catalysed destruction of ozone *Nature* **249** 810–2

[199] Crutzen P J 1970 The influence of nitrogen oxides on the atmospheric ozone content *R. Meteorol. Suc. Quart. J.* **96** 320–5

[200] Rowland F S 1996 Stratospheric ozone Depletion by chlorofluorocarbons (Nobel lecture) *Angew. Chem. Int. Ed. Engl.* **35** 1786–98

[201] McKenzie R, Bernhard G and Liley B *et al* 2019 Success of Montreal Protocol demonstrated by comparing high-quality UV measurements with 'World Avoided' calculations from two chemistry-climate models *Sci. Rep.* **9** 12332

[202] Montzka S A, Velders G J M, Krummel P B, Mühle J, Orkin V L, Park S, Shah N, H and Walter-Terrinoni 2018 Hydrofluorocarbons *Scientific Assessment of Ozone Depletion, Ch. 2: Global Ozone Research and Monitoring Project-Report No. 58* (Geneva: Chemical Sciences Laboratory)

[203] Flerlage H, Velders G J M and de Boer. J 1992 A review of bottom-up and top-down emission estimates of hydrofluorocarbon(HFCs) in different parts of the worlds *Chemosphere* **283** 131208

[204] Fujika M and Tajima S The pollution of Minamata Bay by mercury *Water Sci. Technol.* **25** 133–40

[205] de Boer M A and Lammertsma K 2013 Scarcity of reare earth elements *ChemSusChem* **6** 2045–55 2013

[206] Boekhout van Solinge T 2008 Crime, conflicts and ecology in Africa *Global Harms: Ecological Crime and Speciesism* ed T Sollund (Nova Science Publ.) pp 13–35

[207] Aoun M, El Samrani A G, Lartiges B S, Kazpard V and Saad Z 2010 Releases of phosphate fertilizer industry in the surrounding environment: Investigation on heavy metals and polonium-210 in soil *J. Environ. Sci.* **22** 1387–97

[208] Zagà V, Lygidakis C, Chaouachi K and Gattavecchia E 2011 Polonium and lung cancer *J. Oncol.* **2011** 860103

[209] Day R H 1980 The occurrence and characteristics of plastic pollution in Alaska's marine birds *M.S. Thesis* University of Alaska. Fairbanks, AK, USA p 111

[210] Moore C J, Moore S L, KLeecaster M and Weisberg S B 2001 A comparison of plastic and Plankton in the North Pacific Central Gyre *Mar. Pollut. Bull.* **42** 1297–300

[211] Law K L 2017 Plastics in the environment *Ann. Rev. Mar. Sci.* **9** 205–29

[212] Landrigan P 2017 Air pollution and health *Lancet* **2** E4–5

[213] Lelieveld J, Evans J S, Fnais M, Giannadaki D and Pozzer A 2015 The contribution of outdoor air pollution sources to premature mortality on a global scale *Nature* **525** 367–71

[214] Kriebel D, Tickner J, Epstein P, Lemons J, Levins R, Loechler E L, Quinn M, Rudel R, Schettler T and Stoto M 2001 The precautionary principle in environmental science *Environ. Health Perspect.* **109** 871–6

[215] Hendriks H S, Meijer M, Muilwijk M, Van den Berg M and Westerink R H S 2014 A comparison of the *in vitro* cyto- and neurotoxicity of brominated and halogen-free flame retardants: prioritization in search for safe(r) alternatives *Arch. Toxicol.* **88** 857–69

[216] Hák T, Janoušková S and Moldan B 2016 Sustainable development goals: a need for relevant indicators *Ecol. Indic.* **60** 565–73

[217] Stafford-Smith M, Griggs D, Gaffney O, Ullah F, Reyers B, Kanie N, Stigson B, Shrivastava P, Leach M and O'Connell D 2017 Integration: the key to implementing the sustainable development goals *Sustain. Sci.* **12** 911–9

[218] EPA 2019 Chemical safety for sustainability national research program strategic research action plan 2019–2022 *Office of Research and Development Chemical Safety for Sustainability, No. CSS FY2019-FY2022 StRAP* (Washington DC: EPA)

[219] https://www.mondaq.com/china/chemicals/775388/china-publishes-draft-regulation-on-the-environmental-risk-assessment-and-control-of-chemical-substances-the-real-china-reach

[220] Robinson R M and Behnke R A 2001 *The U.S. National Space Weather Program: A retrospective* (Washington, DC: American Geophysical Union Geophysical Monograph Series) 125 10

[221] Hapgood M A 2011 Towards a scientific understanding of the risk from extreme space weather *Adv. Space Res.* **47** 2059–72

[222] Schrijver C J *et al* 2015 Understanding space weather to shield society: a global road map for 2015–2025 commissioned by COSPAR and ILWS *Adv. Space Res.* **55** 2745–807

[223] Green L and Baker D 2015 Coronal mass ejections: a driver of severe space weather *Weather* **70** 31–5

[224] Richardson I G, Cliver E W and Cane H V 2001 Sources of geomagnetic storms for solar minimum and maximum conditions during 1972–2000 *Geophys. Res. Lett.* **28** 2569–72

[225] National Academies Press 2008 *Severe Space Weather Events–Understanding Societal and Economic Impacts* (National Academies Press)

[226] Kreutzer R 2009 Space weather *Global Risk Dialogue* (Berlin: Springer) pp 12–7

[227] Eastwood J P, Biffis E, Hapgood M A, Green L, Bisi M M, Bentley R D, Wicks R, McKinnell L A, Gibbs M and Burnett C 2017 The economic impact of space weather: where do we stand *Risk Anal.* **37** 206–18

[228] Cannon P S 2013 Extreme space weather—a report published by the UK Royal Academy of Engineering *Space Weather* **11** 138–9

[229] Pete R 2012 On the probability of occurrence of extreme space weather events *Space Weather* **10** 02012

[230] Schrijver C J, Dobbins R, Murtagh W and Petrinec S M 2014 Assessing the impact of space weather on the electric power grid based on insurance claims for industrial electrical equipment *Space Weather* **12** 487–98

[231] Howard J W Jr and Hardage D M 1999 Spacecraft environments interactive: space radiation and its effects on electronic system *Technical Report, NASAs* https://ntrs.nasa.gov/citations/19990116210

[232] https://euweather.eu/stations_stats.php?en

[233] Odstrcil D 2003 Modeling 3-D solar wind structure *Adv. Space Res.* **32** 497–506

[234] Pomoell J and Poedts S 2018 EUHFORIA: European heliospheric forecasting information asset *J. Space Weather Space Clim.* **8** A35

[235] Poedts S *et al* 2020 The virtual space weather modelling centre *J. Space Weather Space Clim.* **10** 14

[236] Vainio R and Laitinen T 2007 Monte Carlo simulations of coronal diffusive shock acceleration in self-generated turbulence *Astrophys. J.* **658** 622–30

[237] Afanasiev A, Battarbee M and Vainio R 2015 Self-consistent Monte Carlo simulations of proton acceleration in coronal shocks: effect of anisotropic pitch-angle scattering of particles *Astron. Astrophys.* **584** A81

[238] Pomoell J, Aran A, Jacobs C, Rodrguez-Gasén R, Poedts S and B S 2015 Modelling large solar proton events with the shock-and-particle model. Extraction of the characteristics of the MHD shock front at the cobpoint *J. Space Weather Space Clim.* **5** A12

[239] Wijsen N, Aran A, Sanahuja B, Pomoell J and Poedts S 2020 The effect of drifts on the decay phase of SEP events *Astron. Astrophys.* **634** A82

[240] Wijsen N, Samara E, Aran À, Lario D, Pomoell J and Poedts S 2021 A self-consistent simulation of proton acceleration and transport near a high-speed solar wind stream *Astrophys. J. Lett.* **908** L26

[241] Lakka A, Pulkkinen T I, Dimmock A P, Kilpua E, Ala-Lahti M, Honkonen I, Palmroth M and Raukunen O 2019 GUMICS-4 analysis of interplanetary coronal mass ejection impact on Earth during low and typical Mach number solar winds *Ann. Geophys.* **37** 561–79

[242] Eggington J W B, Eastwood J P, Mejnertsen L, Desai R T and Chittenden J P 2020 Dipole tilt effect on magnetopause reconnection and the steady-state magnetosphere-ionosphere system: global MHD simulations *J. Geophys. Res. (Space Phys.)* **125** e27510

[243] Pierrard V and Voiculescu M 2011 The 3D model of the plasmasphere coupled to the ionosphere *Geophys. Res. Lett.* **38** L12104

[244] Bourdarie S A and Maget V F 2012 Electron radiation belt data assimilation with an ensemble Kalman filter relying on the Salammbô code *Ann. Geophys.* **30** 929–43

[245] cf http://ccmc.gsfc.nasa.gov/

[246] https://clasp.engin.umich.edu/research/theory-computational-methods/center-for-space-environment-modeling/

[247] Raeder J, Larson D, Li W, Kepko E L and Fuller-Rowell T 2008 OpenGGCM simulations for the THEMIS mission *Space Sci. Rev.* **141** 535–55

[248] Opgenoorth H J *et al* 2019 Assessment and recommendations for a consolidated European approach to space weather—as part of a global space weather effort *J. Space Weather Space Clim.* **9** A37

[249] Scolini C, Rodriguez L, Mierla M, Pomoell J and Poedts S 2019 Observation-based modelling of magnetised coronal mass ejections with EUHFORIA *Astron. Astrophys.* **626** A122

[250] Scolini C *et al* 2020 CME-CME interactions as sources of CME geoeffectiveness: the formation of the complex ejecta and intense geomagnetic storm in 2017 Early September *Astrophys. J. Suppl.* **247** 21

[251] Kilpua E K J, Olspert N, Grigorievskiy A, Käpylä M J, Tanskanen E I, Miyahara H, Kataoka R, Pelt J and Liu Y D 2015 Statistical study of strong and extreme geomagnetic disturbances and solar cycle characteristics *Astrophys. J.* **806** 272

[252] Horwitz J L, Gallagher D L and Peterson W K 1998 *Geospace Mass and Energy Flow: Results from the International Solar-Terrestrial Physics Program* (Washington, DC: American Geophysical Union Geophysical Monograph Series) p 104

[253] Lilensten J *et al* 2021 Quo vadis, European Space Weather community? *J. Space Weather Space Clim.* **11** 26

[254] Antiochos S K, DeVore C R and Klimchuk J A 1999 A model for solar coronal mass ejections *Astrophys. J.* **510** 485–93

[255] Lodders K 2003 Solar system abundances and condensation temperatures of the elements *Astrophys. J.* **591** 1220–47

[256] Reames D V 1999 Particle acceleration at the Sun and in the heliosphere *Space Sci. Rev.* **90** 413–91

[257] Daly E J 2001 *ESA Space Weather Activities* ed I A Daglis (Hazards: Space Storms and Space Weather) p 459

[258] https://cordis.europa.eu/

IOP Publishing

EPS Grand Challenges
Physics for Society in the Horizon 2050

Mairi Sakellariadou, Claudia-Elisabeth Wulz, Kees van Der Beek, Felix Ritort, Bart van Tiggelen, Ralph Assmann, Giulio Cerullo, Luisa Cifarelli, Carlos Hidalgo, Felicia Barbato, Christian Beck, Christophe Rossel and Luc van Dyck

Chapter 7

Physics for secure and efficient societies

Felicia Barbato, Marc Barthelemy, Christian Beck, Jacob D Biamonte, J Ignacio Cirac, Daniel Malz, Antigone Marino, Zeki Can Seskir and Javier Ventura-Traveset

7.1 Introduction

Christian Beck[1] and Felicia Barbato[2]
[1]Queen Mary University of London, London, UK
[2]Gran Sasso Science Institute, L'Aquila, Italy

As we have seen in this book, physics research contains two very different aspects. There is the fundamental research driven by curiosity, with the ultimate aim of understanding very small interacting systems, very large interacting systems, and their complex behaviour on intermediate scales. There is also the applied side, where physics is applied to develop new technologies, new analysis methods, and new concepts and insights that are useful for society. Ultimately, much of what physics for interacting systems encounters is nonlinear, high-dimensional, and complex, and the final goal is to apply novel physical insights to real-world systems, providing useful applications that are helpful for society in general. The topic of this chapter, physics for secure and efficient societies, is very broad and has many different aspects, and this chapter in no way makes an attempt to treat it in full. Rather, we have selected a few topics that we find particularly interesting, with the emphasis of looking into the future—perhaps looking towards the year 2050 or towards a similar time scale.

One important aspect for the future of society is the further development of information technologies; proper communication and information processing and enhanced computer development are absolutely essential. Our world today is dominated by computers in their various shapes and sizes, from small to big, from personal to institutional, from local to worldwide. Science has made immense

doi:10.1088/978-0-7503-6342-6ch7

© 2024 The Authors.
Published under license by IOP Publishing Ltd

progress by implementing modern machine learning technologies and artificial intelligence, so there are some natural questions: Where is computing going? What is the next generation of computers made of? And what is the next generation of algorithms? Still in its infancy today, quantum computing may hold the key for outstanding novel computational developments of the future. Some problems are so complex that they cannot be solved with present conventional computers but require something that is orders of magnitudes faster, or they need algorithms and novel approaches that are very different from what is currently used in mainstream simulations.

The section of this chapter by Daniel Malz and J Ignacio Cirac summarizes the most important principles of quantum computing, exploiting quantum superpositions, and entanglement for the purpose of future quantum computers. The aim is to solve certain problems much faster than is possible with conventional computers. The section by Zeki Can Seskir and Jacob Biamonte looks at the historical development of quantum computing research and in particular quantum algorithms and new types of machine learning models, which are expected to be very relevant in the future.

How do we actually get the data that we feed into our physical models to make accurate predictions for the future, using the best computers and analytical techniques available? The problem is nontrivial, as bad data yield biased and unprecise predictions. Sensor technology has made immense progress recently. The convergence of multiple technologies, real-time analytics, machine learning, ubiquitous computing, and embedded systems gave birth to the Internet of Things (IoT), and the automation and control of industrial processes can be seen as the fourth industrial revolution, also known as the Industrial Internet of Things (IIoT). In her section, Antigone Marino describes the historical basic steps of sensor developments, arriving then at the future challenges set by Europe's climate change strategy for 2050. In this framework, smart sensors are fundamental for monitoring all services related to the automation of processes as regards to waste reduction, clean water, and environmental control and to improve the quality of life in the workplace.

The use of smart sensors is also open to the space sector, which is gaining more and more importance and is going to enter its golden age driven by the longstanding dream of humankind, the exploration of space, with many interesting new perspectives and applications for the benefit of humans on the horizon. In his section on the space sector, Javier Ventura-Traveset reviews the current status of next future missions and explores the prospects of the space sector beyond 2030–35. Many intriguing topics are covered, starting with more gnoseological problems such as space science, going through futuristic scenarios about human and robotic exploration of space, and finally touching on more practical issues such as understanding the climate change trend, its sources, its dynamics, and the major anthropogenic impacts. Javier Ventura-Traveset makes it very clear how space exploration has had, and will have even more in the future, huge impacts on our society from both economic and social points of view and how future society can benefit from this emerging sector.

Finally, another problem of utmost relevance for future societies is understanding the complexity that underlies the real-world systems that surround us and the daily aspects of our lives. Here, statistical physics, in its modern form, has a lot to say. One particular example is the science of cities. A very large part of the world population these days live in cities, but how do cities actually function, how do they evolve, and how can we improve their day-to-day structures and life quality in a sustainable and environmentally friendly way? Cities are spatially extended complex systems, and statistical physics, in its modern formulation, can be applied. In the historical Boltzmann formulation, particles are replaced by agents (companies, vehicles, people, sustainable energy sources, etc), interactions are replaced by communications (mobile phones, e-mail, Twitter, etc), phase transitions correspond to an abrupt change of relevant observables (opinions, prices, behavioural patterns, etc), and so on. In his section, Marc Barthelemy provides a state-of-the-art overview of city modeling, city growth aspects, traffic congestion, and much more, using the tools of statistical physics and complex network theory.

Overall, the example topics treated in this chapter show that often there is initially fundamental basic physical science, which then feeds into more advanced applied models relevant for future development. For example, starting from quantum physics, we proceed to modern methods and algorithms of quantum computing; starting from classical equilibrium and nonequilbrium statistical physics, we proceed to a modern science of cities; and so on. Better predictions and better models can be made if we have access to better data obtained with more powerful sensors by better satellite navigation methods, and so on. Let's hope that in 2050, when a reader looks back into this book, most of the world's population will be living in a clean, peaceful, and sustainable environment where physics helped a lot to attain this stable state.

7.2 Second quantum revolution: quantum computing and cybersecurity

7.2.1 The second quantum revolution: quantum computing and information

Daniel Malz[1] and J Ignacio Cirac[2]

[1]Department of Mathematical Sciences, University of Copenhagen, Universitetsparken 5, 2100 Copenhagen, Denmark

[2]Max Planck Institute of Quantum Optics, Garching, Germany

Even though quantum physics is now more than a hundred years old, it still offers new challenges, insights, and applications. One particularly exciting possibility is to use quantum physics to process and transmit information in more efficient and secure ways. In this section we briefly summarize the principles of quantum computing (QC), which exploits quantum superpositions and entanglement to solve certain particular problems much faster than any classical device. We also review some algorithms and applications and the experimental state of the art, and we conclude with an outlook towards more powerful devices and their potential impact.

7.2.1.1 Introduction

Quantum mechanics explains all microscopic phenomena around us. As well as being of fundamental interest, it is crucial to understand materials, chemistry, and molecular biology and thus much of modern science and technology. In the past decades, it has become increasingly clear that the properties of quantum mechanics, such as superposition and entanglement, allow one to process information in a fundamentally different way, which may revolutionize communication and computation. This has led to the emergence of a new multidisciplinary field, quantum information science, which combines mathematics, computer science, and quantum physics. The ambitious key objective of this field is the construction of a quantum computer—a device that manipulates information according to the laws of quantum mechanics. Originally proposed in a seminal talk by Richard Feynman [1], quantum computers are now known to be capable of solving problems that that are too complex for classical computers. There are significant challenges that need to be overcome on the way to a fully functional quantum computer, but the journey there will provide scientists with novel insights and change our understanding of Nature.

In the second half of the 20th century, the computer revolutionized virtually all areas of society and industry. This development was enabled by electrical engineering, based on the modern understanding of classical electrodynamics and classical information theory, but also other concepts building on quantum mechanics, such as the physics of semiconductors and transistors. It is tempting to draw an analogy to the current development. Scientists around the world are making great strides towards the construction of a quantum computer, based on other aspects of quantum mechanics and quantum information theory, enabled by a technological leap in atomic physics, optics, and solid-state materials. While the potential of quantum computers is vast, there are still relatively few good algorithms available today, and building one appears very challenging. As a result, it is very difficult to gauge the impact that quantum computers will have on society, even in the short run. Here we will nevertheless try to paint a rough picture of what awaits us.

A common misconception is that quantum computers are superior to conventional computers and that it is only a matter of time until your smartphone's processor will run quantum algorithms. In fact, tasks that classical computers are good at will very likely always run faster on classical processing devices. Instead, there are many important problems for which the best classical algorithms require exponential time in the input size, which means that adding merely one bit to the input causes the runtime of the algorithm to be multiplied by a constant factor greater than 1. Such problems quickly become infeasible to solve, even for moderate input sizes. There exist very important problems with this property, including the simulation of quantum systems such as molecules and materials, as well as optimization problems, for example, to schedule trains, route traffic, and distribute resources optimally. Scientists have invented quantum algorithms that solve a subset of these problems in polynomial time and thus remain feasible for large inputs. Which problems belong to which complexity class is subject to change as new algorithms are invented, but very likely there will always remain some that can be efficiently solved only with a quantum computer. At the risk of oversimplification, this means that there are important problems that cannot be solved at all on classical computers but become accessible with a quantum computer. For the interested reader, these fascinating concepts are explained in greater detail in the excellent texts by Nielsen and Chuang [2] and Aaronson [3].

We are now at a very exciting point in the development of quantum computers and quantum technologies (QT) in general. During the last decades, leading technology companies have announced programmes to develop quantum computers. This has boosted the pace of engineering efforts significantly. In 2019 the first experiment was performed in which a quantum processor completed a task that, according to our current knowledge, would take many years to complete on even the best supercomputers [4]. Since then, these results have been reproduced with more qubits and on other platforms [5, 6].

Naturally, these developments have attracted the attention of the media, politicians, and society as a whole. One should be careful, however, to distinguish the current generation of quantum processors from fully fault-tolerant and scalable quantum computers in order not to raise unrealistic expectations that can only be disappointed. Quantum systems are very susceptible to noise, which causes errors to accumulate during a computation. At the same time, these systems require an exceptional level of control, which makes them difficult to scale to large sizes. As a result, current devices, also referred to as noisy intermediate-scale quantum (NISQ) devices [7], can run only short computations before their output becomes completely scrambled. The previously mentioned algorithms with exponential speedup require large computers that are capable of correcting errors that arise during the computation, which places very stringent requirements on their properties. This is challenging to achieve and may take decades. Yet, as is often argued, there are important applications of NISQ devices, such as analog simulation, and even before fault tolerance becomes a reality, there is much to be learned and gained from this research program as a whole [7, 8].

In this section, we give a very brief and basic introduction to quantum information theory, review some of the algorithms and applications of quantum computers, and discuss the most advanced platforms in which they are being implemented. We also give an outlook, trying to balance the substantial and very real promise that quantum computers hold with the equally real and formidable challenges that have to be met on the way.

As this field is changing continuously and new algorithms, platforms, and protocols, appear almost every day, we provide just few basic references here so that the interested reader has access to more detailed material. We have often given references not to original papers, but rather to reviews where those can be easily found. For a comprehensive, textbook-style account of the field, we point the reader to the book by Nielsen and Chuang [2], which is perhaps the most important reference text for the field. Another deeply insightful text in a somewhat lighter style is by Aaronson [3]. For technical reviews, which contain many further references, we point to those on quantum algorithms [9], quantum algorithms for chemistry [10], quantum machine learning [11], and specific platforms [12–14].

7.2.1.2 Quantum information theory basics
We briefly introduce the principles of quantum information theory, in which information is stored on quantum bits (qubits), computation corresponds to unitary operations, and the computational advantage relies on quantum superposition and entanglement.

7.2.1.2.1 Qubits
In quantum mechanics, energy is quantized into bits called excitations. As applied to light, which is quantized in terms of photons, this means that a system may contain zero, one, or ten of them, but not half a photon. The state of the system can thus be expressed in terms of those discrete states. Restricting our attention to just two states, which we label $|0\rangle$, $|1\rangle$, we can define a quantum bit (a qubit): a quantum system with two states. Note that the states do not have to be distinguished by photon number; they might correspond to any other degree of freedom as well, such as, for example, spin or collective excitations in a circuit. In a classical computer, this is the end of the story: Computation is done by manipulating many bits, each of which can assume either state 0 or 1 (see figure 7.1). Quantum mechanics, however, permits superpositions. The general state of a qubit is described by the following vector:

$$|\Psi\rangle = c_0 |0\rangle + c_1 |1\rangle, \tag{7.1}$$

where c_0 and c_1 are complex numbers that obey $|c_0|^2 + |c_1|^2 = 0$ and $|0\rangle$ and $|1\rangle$ form the basis of the vector space. It is important to note that a qubit really can be a superposition in the sense that it is distributed over both basis states. This should be contrasted with the state of a coin after it has been flipped but before the result has been recorded. The coin may be either in 0 or in 1; the qubit is in both. The reason for this subtle difference relies on the basic principles of quantum physics. For

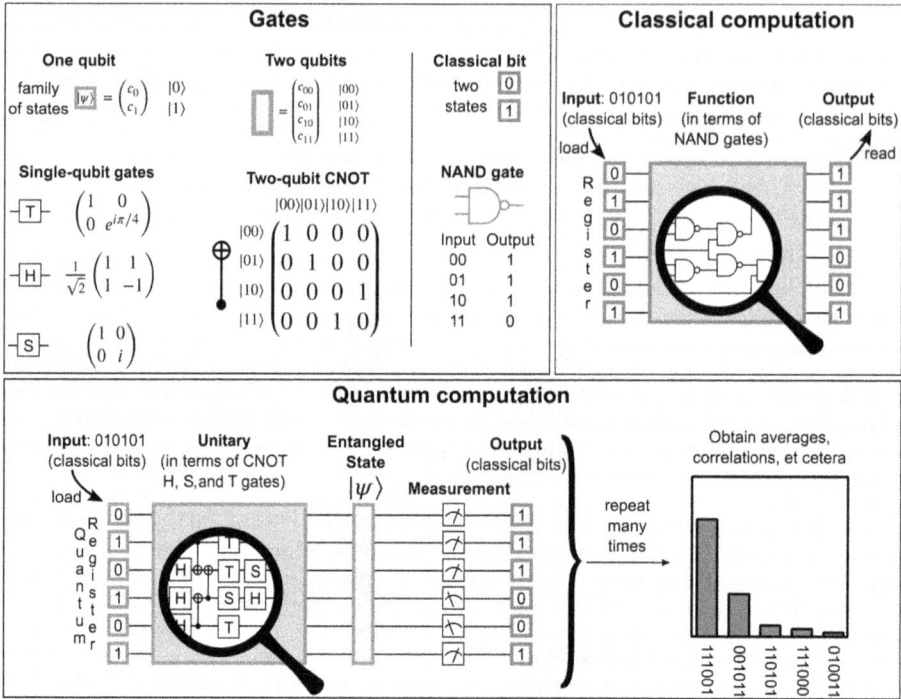

Figure 7.1. Comparison between quantum and classical information processing. Top left: A universal set of quantum gates is (in this case) composed of one two-qubit gate (CNOT) and three single-qubit gates (H, S, T). Classically, the NAND gate (and a copy operation) is universal. Top right: Classical computation can be understood as computable functions acting on an input, provided as a string of classical bits. Bottom: In quantum computation, the input is (usually) again classical data, which are loaded into a quantum register made out of qubits. A quantum algorithm is a unitary operation acting on the qubits, expressed in terms of a quantum circuit. The outcome of a computation is an entangled state. The experiment is repeated many times, which allows one to sample from the probability distribution generated by the state.

example, the state cannot be measured directly. Even if we know exactly which state the qubit is in, if we measure it, the measurement device will record either 0 or 1 at random, with probabilities $|c_0|^2$ and $|c_1|^2$. But this is different from the lack of knowledge present in the coin example. Indeed, knowing that the system is in the state equation (7.1), we could apply an (invertible, deterministic) operation that turns the system state into $|0\rangle$ without needing to perform a measurement whether the qubit was in $|0\rangle$ or $|1\rangle$. The fact that randomness comes out of deterministic operations and thus is an intrinsic property of Nature initially led to a lot of debate, but since then an incredible amount of effort has gone into making sure that this description is correct. Nowadays, such true randomness finds important application, for example, in cryptography (see section 7.2.1.8.4).

7.2.1.2.2 Entangled qubits
If we take two qubits, their possible states are $|00\rangle$, $|01\rangle$, $|10\rangle$, $|11\rangle$. For a classical computer, this is the end of the story. Quantum mechanics again permits the system

to be in a superposition but now encompassing both particles. One example is the Bell state:

$$|\Psi^+\rangle = (|01\rangle + |10\rangle)/\sqrt{2}. \tag{7.2}$$

This state has the fascinating property that when the constituent qubits are considered individually, there is an equal chance to find each in 0 or 1, just as a coin could come up heads or tails with equal probability. When the state of both qubits is measured, however, these random outcomes are perfectly (anti-)correlated and come up either as 0–1 or 1–0. This correlation is intrinsic in the state and has nothing to do with an interaction between them. The qubits could reside at opposite ends of the Universe and they would still exhibit this property. We call this astonishing property *entanglement*. Einstein, Podolsky, and Rosen famously questioned the reality of this 'spooky action at a distance' [15], but experiments have since confirmed this property, thereby ruling out local realistic theories, which are incompatible with quantum physics [16, 17].

Entanglement has a further important consequence: Writing down the state of N qubits requires specifying 2^N parameters, unlike the state of N bits, which is contained in a string of N 0s and 1s. Specifically, an N-qubit quantum state can be written as

$$|\Psi^+\rangle = \sum_{i_1, i_2, \ldots, i_N = 0}^{1} C_{i_1, i_2, \ldots, i_N} |i_1, i_2, \ldots, i_N\rangle, \tag{7.3}$$

where the sum on the right-hand side runs over all possible configurations, which are sometimes referred to as computational states, that is, all numbers in binary from 0 to $2^N - 1$. In the worst case we thus have to keep track of 2^N numbers, which is the reason why simulating quantum systems on classical computers is difficult. A rough estimate shows that even with only 300 qubits the number of states ($2^{300} \approx 10^{90}$) exceeds the number of baryons in the observable universe. Quantum computing aims to exploit this complexity.

7.2.1.3 Classical versus quantum computing

7.2.1.3.1 Models of computation

There are many different models of computation, each describing a way in which information can be processed. Those models that are universal, or Turing complete, are all equivalent in the sense that the class of functions whose output they can compute coincides. In this sense, quantum and classical computers are equivalent. However, quantum computers disprove the common belief that all universal models of computation have the same complexity classes. Specifically, quantum computers can solve certain problems in polynomial time that classical computers cannot.

One universal model of computation is the circuit model, where the computation (the evaluation of a function on the input) is represented by a sequence of logic gates acting on the input, as illustrated in figure 7.1. Turing completeness implies that any computable function can be represented in this way. Classically, one can show that any computation can be decomposed into a series of NAND gates (and a copy operation), which is a gate that takes two inputs and returns 1 except when both inputs are 1, in which case it returns 0 (see figure 7.1 for the truth table).

The standard model of quantum computation is also gate based and is shown in figure 7.1. The operations performed by a quantum computer on its input (a quantum state) are so-called unitary transformations (or *unitaries* for short). One can show that all unitaries can be decomposed into a quantum circuit comprising only single-qubit unitaries and one type of two-qubit gate, the CNOT gate [18], which constitutes the analog of universality for a quantum computer. Solovay and Kitaev have proven that the continuous set of single-qubit gates can be efficiently decomposed into just few gates drawn from a universal gate set. There are many such universal sets; a commonly employed one is shown in figure 7.1 and comprises the H, T, and S gates. Computational universality also exists in the quantum realm, in that different models of quantum computation are equivalent and possess the same complexity classes. In practice, this means that one can get clever in choosing the allowed operations to reduce errors or overhead, but this will not affect the computational power in a fundamental way. The key point is that the classical and quantum complexity classes are distinct, which in the circuit language means that there are quantum circuits that are exponentially shorter than the best available classical circuits tackling the same problem.

7.2.1.3.2 Accessing the outcome of the computation

The outcome of applying a given quantum circuit is a N-qubit state such as the one specified in equation (7.3). It is important to note that we cannot access the full N-body state directly. Instead, we typically measure each qubit, determining whether it is in 0 or 1. This measurement necessarily destroys the state and yields at most N (classical) bits of information. The measurement outcome is a randomly drawn bit string out of those that make up the state (cf equation (7.3)), where the string $(i_1, i_2, ..., i_N)$ occurs with probability $|C_{i_1, i_2, ..., i_N}|^2$. Repeating the same experiment many times thus yields not a single result but a probability distribution. The goal of a quantum algorithm is that this probability distribution peaks at the desired solution, such that it can be extracted in a small (polynomial in N) number of measurements. In a simple case, we might have a way to check whether a solution is correct and would like it to come up with reasonable probability during the experiment. In this case, repeating the algorithm a sufficient number of times solves the posed problem.

7.2.1.4 Quantum algorithms

The principles of quantum mechanics allow one to design algorithms that work in fundamentally different ways and thereby can be executed more efficiently. Efficiency is judged via the time and memory requirements of an algorithm given its input size, which is referred to as its complexity. The most important distinction that is made is whether algorithms run in polynomial time or in exponential time, that is, whether for large inputs N the time or memory scales as N^k or as $\exp(kN)$ for some fixed k. The most useful and important quantum algorithms offer exponential speedup, which means that their complexity behaves like the logarithm of the complexity of the best classical algorithm. The nature of exponential scaling means that problems for which only exponential time algorithms are available are in practice unsolvable for moderately large inputs. In this sense, quantum computers can solve problems classical computers cannot.

A simple academic example, introduced by David Deutsch in 1985 [19], illustrates how quantum mechanics can help solve a problem. The problem he considered is that we are provided with a quantum system (an oracle) that implements an unknown function h: $\{0, 1\} \rightarrow \{0, 1\}$. If we send the oracle a two-qubit state $|x, y\rangle$, it returns the output two-qubit state $|x, y \oplus h(x)\rangle$, where the function h can either (1) do nothing, $h(y) = y$, (2) flip the qubit, $h(y) = \bar{y}$, (3) always return 0, $h(y) = 0$, or (4) always return 1, $h(y) = 1$. We are asked to determine whether h is balanced, as in (1) or (2), or constant, as in (3) or (4). While classically we would have to check the output on two different inputs and thus need two function calls, in quantum mechanics this can be done using just one function call. The key is to prepare the first qubit in the superposition $|+\rangle = (|0\rangle + |1\rangle)\sqrt{2}$ and the second qubit in the superposition $|-\rangle = (|0\rangle - |1\rangle)/\sqrt{2}$. One can then show that the output of the machine is $|--\rangle$ if h is balanced and $|+-\rangle$ if h in constant. These two states can be distinguished in a single measurement, which means that only one function call is required, in contrast to the classical case.

While the preceding algorithm was an academic toy problem, there are known useful algorithms, which, however, are substantially more complex. We list some of the best-known in the next subsection. This shows that there are problems for which quantum computers are required. Nevertheless, we would like to draw the analogy with the development of classical computers, in which most of the important algorithms in use today were invented only after classical computers were available to a broad community. Similarly, we expect the class of useful quantum algorithms to grow significantly as this technology becomes available.

7.2.1.4.1 List of algorithms

Two key distinguishing properties of quantum algorithms should be kept in mind, as they determine the practical use of the algorithms. One is whether they use an oracle or not. An oracle is a supposed black box with arbitrary computing power that implements a certain function or algorithm, as in the preceding example. This is an important tool in theoretical quantum information research, as it is typically easier to prove statements about the complexity of such algorithms, but they usually have no real-world applications, as implementing the oracle requires too many resources. The second is whether the quantum speedup is polynomial or exponential. Whether or not a polynomial speedup is useful will have to be decided on a case-by-case basis. The reason is that since quantum processors come with significant overhead, it may happen that the crossover beyond which a quantum computer actually has a smaller runtime occurs for very large inputs and runtimes of years or more, which makes them impractical despite the theoretical speedup. In contrast, exponential speedup means that even modestly sized quantum computers can perform calculations beyond the ability of any classical computer in existence. There are several algorithms that offer exponential speedup, including algorithms to calculate the discrete logarithm, solve Pell's equation, evaluate Gauss sums, determine abelian hidden subgroups for a large variety of groups, or determine Jones' polynomials [9].

We also include heuristic algorithms in the list. These algorithms do not have a provable advantage, but if classical computing is any guide, heuristic algorithms are often very successful, even beyond the scope of their intended use. Prominent examples include simulated annealing [20], which is a very successful algorithm in a great variety of hard optimization problems, and machine learning, which has revolutionized many fields even beyond typical applications in computer science, the lack of theoretical guarantees notwithstanding. By their very nature, heuristic algorithms are difficult to develop if one cannot try them out. Therefore, it is reasonable to expect that the best and most impactful algorithms in this class will be found only once quantum computers have arrived.

- **Deutsch–Josza algorithm** (Deutsch and Josza [21]). This is an oracle-based algorithm similar to the one described in the previous section, but with large inputs. While the quantum algorithm requires only one query, the classical algorithm requires exponentially many. If a small probability of returning the wrong result is permitted, the classical algorithm requires only a constant number of queries. This was one of the first algorithms with provable quantum speedup.

- **Simon's algorithm** (Simon [22]). Simon's algorithm tackles a similarly academic problem, namely, to find the string s given the black box function (oracle) that obeys $f(x) = f(y)$ if and only if $x \otimes y \in \{0^N, s\}$, where N is the length of the bit string. While in the Deutsch–Josza algorithm, quantum algorithms only provide an advantage if the algorithm must always return the correct result, Simon's algorithm provides an exponential speedup even if we allow a small (<0.5) constant probability that the algorithm fails.

- **Shor's algorithm to factor large numbers and to find discrete logarithms** (Shor [23]). This algorithm was the first quantum algorithm with exponential speedup and a practical (oracle-free) application. It was inspired by Simon's algorithm and uses what is now known as the quantum Fourier transform. The technique can be adapted to solve all abelian hidden subgroup problems. Relatedly, the quantum phase estimation algorithm (Kitaev [24]) employs the quantum Fourier transform to find the eigenvalue of a unitary operator U that corresponds to a given eigenstate $|\Psi\rangle$.

- **Grover search** (Grover [25]). This is a generic algorithm to do an unstructured search. Given a desired output, it finds the correct input to a black box function (an oracle) among N choices using $O\sqrt{N}$ function calls, whereas classical algorithms require $O(N)$ calls. The central element of Grover search is amplitude amplification, which gives a quadratic speedup for generic optimization problems [26].

- **Adiabatic algorithms** [27]. This is a heuristic algorithm based on physical insight which often works well but for which no convergence guarantees exist. Many problems (notably optimization problems and many-body problems in physics and chemistry) can be encoded by specifying a Hamiltonian H, which is a Hermitian operator that determines the system properties and its time evolution, and asking what its ground state (eigenstate with the lowest eigenvalue) is. The adiabatic theorem states that if a system starts in the ground state, and its Hamiltonian is changed very slowly, the system will remain in the ground state as long as the ground state remains unique; similar to how one should avoid jerky movements in an egg race or when carrying a full glass of water. The adiabatic algorithm proceeds by preparing the (known) ground state of a different operator H', applying the time evolution, and slowly changing the Hamiltonian to H. It fails if the energy difference between lowest and second-lowest energy eigenstate becomes too small.

- **Variational algorithm.** This is another heuristic algorithm that aims to find the ground state by defining a quantum circuit whose gates depend on some parameters and optimizing the energy (or any other cost function) with respect to the parameters. It may fail if it gets stuck in a local minimum or if the ground state is not inside the variational class.

7.2.1.5 Key applications
We now outline some known applications using the previously introduced algorithms.

- **Many-body quantum simulation.** A clear and natural use case for quantum computers is modeling quantum systems. There is a vast array of applications ranging from solid-state materials (as used in classical computers, batteries, and electric or magnetic materials) to chemistry (modeling reactions and catalysis) and high-energy physics (e.g., phases of lattice gauge theories). Numerical simulation is a key tool employed by a large fraction of researchers in these fields, but the computational complexity of simulating quantum systems has stymied progress. Thus, a quantum computer has the potential to enable breakthroughs in these fields with important indirect applications.
- **Prime factorization.** Many cryptographic protocols are based on the (presumed) hardness of prime factorization, including the ubiquitous Rivest–Shamir–Adleman (RSA) public key algorithm. While this could be regarded an antiapplication, as breaking cryptographic schemes is undesirable, it is nevertheless very important. Variations of Shor's algorithm can break other cryptographic schemes based on the hidden subgroup problem [9]. An alternative are postquantum cryptographic schemes that use functions that cannot be inverted even by a quantum computer (at least not with known quantum algorithms) or quantum cryptography, which makes it physically impossible to eavesdrop.
- **Polynomial speedup for optimization problems.** There are several optimization problems where a polynomial speedup can be shown. Finding the optimal solution to a problem (e.g., finding the fastest route) can typically be connected to some optimization problems and attacked with amplitude amplification, which gives a quadratic speedup. Other speedups exist for a

variety of problems [9]. Other potential (heuristic) algorithms may be found by mapping the optimization to finding the ground state of a Hamiltonian, which can sometimes be solved with adiabatic algorithms or quantum annealing.

- **Solving linear and nonlinear systems of equations.** A number of algorithms have been developed to solve linear and nonlinear (differential) equations with exponential speedup. These algorithms are typically formulated in terms of an oracle or access to a quantum random-access memory but may offer an advantage in some instances.
- **Inspiration for classical algorithms.** The quantum point of view has already proven to be very fruitful to understand classical problems and will continue to be important, regardless of the success of quantum computers. For instance, there are examples of 'quantum-inspired' classical algorithms that compete with or beat previously known best classical algorithms to solve certain problems, such as the Netflix recommendation problem, in which first a quantum algorithm was found that gave exponential speedup and later a classical algorithm was found that achieved the same [28].

7.2.1.6 Error correction and fault tolerance

A major challenge on the road to QC is that in a real-world device, errors can always occur. Given some error rate p per qubit per unit time, the probability that no error occurs in a single qubit ($N = 1$) and in one unit of time ($T = 1$) is $1 - p$. Thus, the probability for a large computation to succeed tends to zero exponentially fast, $(1 - p)^{NT} \rightarrow 0$, which means that we need exponentially many tries to get it right, which eliminates all speedups.

While classical error correction can be achieved simply by a repetition code (if '0' is encoded as '000', a single bit flip error can be corrected), in quantum mechanics this is much harder, since (1) the state of a qubit is continuous (see equation (7.1)), (2) quantum information cannot be copied [29], and (3) (related to (2)) a measurement of a qubit to detect whether an error that occurred alters the state uncontrollably and destroys encoded information. This problem was solved by Shor [30] and Steane [31], who independently invented codes in which a qubit is represented as the entangled state of many qubits in a way that allows one to correct all single-qubit errors.

This by itself does not suffice to show that computation can succeed, since, for example, two errors might happen at once. Worse still, the correction of the error might itself introduce further errors. For classical computers, von Neumann proved a threshold theorem showing that errors can be arbitrarily suppressed. Fault-tolerant quantum computation, introduced by Shor [32], is a scheme that removes the errors during the computation, which succeeds if the individual gate errors lie below a certain threshold. These results ensure that quantum computers are scalable in principle.

There are two main practical implications of these results:

1. There exists an error threshold of the order of 10^{-2}. The device must allow one to perform quantum gates with errors occurring with a probability below this threshold.

2. There is a large overhead. It may require thousands of physical qubits to represent a logical qubits, and logical operations may take many orders of magnitude longer to complete than the underlying physical operations.

These conditions pose substantial challenges, such that a practical fault-tolerant quantum computer is likely still decades away, as we discuss in section 7.2.1.8.

7.2.1.7 Implementations

7.2.1.7.1 Criteria

In the past two decades, several different platforms for quantum computation have been explored, and there exist working prototypes for a number of them. While these prototypes do not achieve fault tolerance, they can still be used for useful computations and simulations.

In 2000, DiVincenzo summarized the criteria that devices ought to fulfil to be viable candidates for quantum computers [33]. They are as follows (bold words correspond to the steps outlined in figure 7.1):

1. *A scalable physical system with well-characterized qubits.*
   This criterion arises from using qubits as the fundamental unit of a quantum computer. Scalable means that increasing the number of qubits should come at a moderate cost only.
2. *The ability to initialize the qubits in a simple fiducial state.*
   This is required to be able to load data into the computer (**input**).
3. *Long coherence times, much longer than the gate operation time.*
   If this is not fulfilled, one cannot apply a **unitary** of sufficient complexity, or else the **entangled state** is randomized.
4. *A universal set of quantum gates.*
   Without a universal set, one cannot implement the desired **unitary**. In figure 7.1, this was taken to be the T, H, S, and CNOT gates, but other sets can be chosen.
5. *A qubit-specific measurement capability.*
   This is needed to obtain **output** from the device.
   More recently, it has been understood that this list has to be supplemented with further conditions to ensure error correction.
6. *Scalability.*
   The error for qubit gates must lie below the error threshold for the used code, and the gate error must be independent of system size.
7. *Parallelization.*
   The number of gates that can be applied in parallel must grow linearly with system size. If this is not the case, the time taken between two gates acting on the same qubit grows with system size, which means that the intrinsic error rate increases.

7.2.1.7.2 *Leading platforms*

The requirements for quantum computers are steep, but several platforms are under active investigation. It is unclear which one of the following will be the platform of choice for eventual fault-tolerant quantum computers. It may even be that a new platform will be invented, in which the complications associated with fault tolerance can be overcome more naturally. This calls for parallel investigation and an open mind. We show sketches for the main platforms in figure 7.2 and briefly describe them. We provide references to only some of the theoretical proposals, leaving out many groundbreaking experimental results, which can be found in [36] or in more specialized reviews [12–14].

    (a) **Trapped ions** [37]. Ions (black) are levitated in vacuum and tightly confined along two dimensions. In the third dimension they naturally form a string, stabilized by electrostatic repulsion (yellow). The ions are typically chosen to be an element with naturally two valence electrons, such that after ionization one remains. The qubit is formed by two internal states of the ion that are chosen as to have a large coherence time, such as dipole-forbidden excited (clock) states or hyperfine levels. Interaction occurs via the coupled motion of the ions and is controlled via applied lasers (red). This approach features very long coherence times and extremely high gate fidelities. Readout occurs in the transverse direction. A number of groups and companies have pursued this approach, and several tens of qubits have been fully controlled in this setup already.

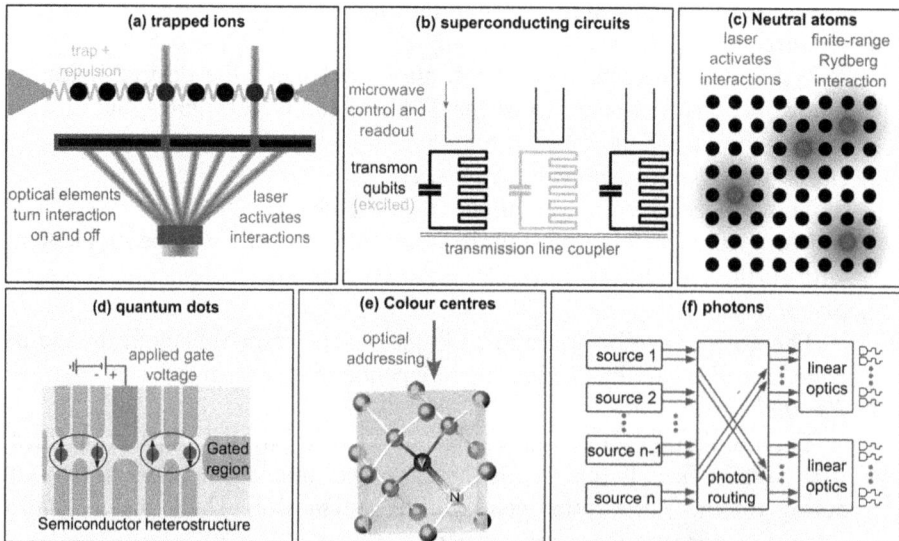

Figure 7.2. Platforms for quantum computation. Explanation in main text. Part (e) is from the NIST public domain [34]. Part (f) is inspired by reference [35].

(b) **Superconducting circuits** [38]. Nanofabricated circuits of a superconducting material are placed on a substrate. The qubit is defined as the presence or absence of a single excitation in a nonlinear oscillator. The qubits are driven with microwave tones supplied via transmission lines, which can enact both single-qubit gates and two-qubit gates; the latter use the interaction mediated through a shared resonator.

Superconducting chips are being researched by several major companies and many groups and are among the largest devices produced so far, with qubit numbers approaching the low hundreds.

(c) **Neutral atoms** [39]. Individual atoms are trapped in a lattice defined through either back-reflected laser beams or optical tweezers (not shown). A qubit is defined through the electronic ground state and another long-term stable electronic state. Interactions can (for example) be mediated by driving a transition to a Rydberg state on both atoms using a laser (shown in red). The interaction is present as long as the corresponding atoms lie within the Rydberg blockade radius (blue) of each other. Readout is done optically in the transverse direction. Hundreds of atoms have been trapped in such devices, but full control has not yet been achieved.

(d) **Quantum dots** [40]. A layered semiconductor structure is used to confine electrons to move in two dimensions (light gray). Additional gates (dark gray) are used to expel electrons from selected regions so as to define islands (those with red dots) that can host an electron. An electron is a spin-1/2 particle with two spin states that are used as the logical 0 and 1 of the qubit. Interactions can be controlled by changing the gate voltage between qubits, which brings electrons closer together or moves them further apart, depending on their spin. Being based on semiconductors, this approach can benefit from the large semiconductor industry, but progress has not moved beyond a few qubits yet.

(e) **Colour centres** [41]. Point defects (substituted atoms) in crystals such as diamond yield localized electrons in very clean environments that consequently are very stable. This makes them promising candidates for qubits, but fabrication of arrays of colour centres and control of their interactions are challenging and are only now starting to become practical.

(f) **Photons, linear optics, and detectors** [42]. Photons are mainly investigated as carriers of quantum information, as they can travel long distances before being absorbed or perturbed. Qubits can be defined through the presence or absence of photons or through their polarization. The main challenge in using them to perform computations are two-photon gates, as they hardly interact with each other. One approach uses single-photon detectors to perform gates probabilistically. This has been successfully used to entangle tens of photons.

7.2.1.7.3 Other important platforms

There are a number of platforms for which some important ingredients are unclear or have not been demonstrated experimentally as yet but that nevertheless look very promising.

(a) **Nuclear magnetic resonance quantum computer** [43]. An NMR quantum computer builds on manipulating the spin states of nuclei in molecules. It is generally understood that this platform scales poorly. Alternative proposals, based on solid-state nuclear spins, may remove this roadblock [44].

(b) **Topological quantum computer** [45]. This is a fundamentally different proposal, in which quantum gates are implemented by braiding anyonic quasiparticles in two-dimensional materials with topological order. The topological nature of the operations imply resistance to local noise and imperfections, which may make fault-tolerance much easier to achieve. Experimentally, this platform is still in its infancy, and no clear candidate material that hosts appropriate excitations has been realized.

(c) **Quantum annealers** [46]. Quantum annealing is a heuristic procedure to reach the ground state of a Hamiltonian (and thereby solve an encoded problem) by slowly lowering the temperature. These systems are not universal quantum computers but may still offer a quantum advantage for certain problems.

(d) **Measurement-based quantum computation** [47]. This is an alternative way to perform quantum computation that proceeds by (i) preparing an entangled state of many qubits and (ii) measuring the qubits sequentially in appropriate bases. It can be shown that by choosing a state with appropriate entanglement, universal quantum computation can be performed in this way. This is a very appealing model for photonic platforms.

7.2.1.8 Present situation and outlook

For several years now, the basic ingredients of quantum processors have been experimentally demonstrated in most of the platforms introduced in section 7.2.1.7. That is, several qubits have been prepared in the 0 state, manipulated with single- and two-qubit gates, and then measured. None of these operations is perfect, but each of them can be performed with errors of around 1% (or even smaller). This has also made it possible to perform small proof-of-principle quantum computations with a few qubits, which illustrate how the algorithms operate.

In some systems (most notably trapped ions, Rydberg atoms, and superconducting circuits), prototypes of several tens of qubits have been built. It is estimated that, at 30 qubits and above, it is virtually impossible to simulate the operation of a quantum computer with existing classical computers, so we are reaching a situation where the former can outperform the latter. However, there are several reasons for tempering optimism, at least as far as the range of applications of quantum computers is concerned. On the one hand, since the errors accumulate during the whole process of initialisation, execution, and measurement, the success probability becomes exponentially small in the number of elementary operations performed. For example, even with errors at the 1% level or below, a quantum algorithm requiring 500 operations will almost certainly give a wrong result. Moreover, even if errors are reduced further, many quantum algorithms that have been developed over the years do not work correctly in the presence of errors. On the other hand, there are powerful classical algorithms for classical computers that can simulate some quantum computer operations efficiently. For example, if each qubit participates in only two (or fewer) quantum logic gates, then tensor network-based methods can be used to predict the outcome with a classical device. This sets a high bar for quantum computers to clear.

It is very difficult to use current prototypes or those to be developed in the near future to gain a real computational advantage over existing computers. Therefore, it seems that the only way to exploit computers in the future is to try to correct the errors and scale up these prototypes. This may take several years or even decades, since, as we shall see later in the chapter, it requires solving a number of major technological challenges. However, there is the possibility that these prototypes, which still produce errors, could be useful for some concrete tasks. In what follows, we will briefly describe those possibilities, as well as the road ahead for creating scalable quantum computers.

7.2.1.8.1 *Noisy intermediate-scale quantum (NISQ) devices*

To date, only prototypes exist in the platforms listed above. They contain several tens (or, potentially, hundreds or thousands) of qubits and operate without fault-tolerant error correction. In a recent breakthrough, researchers at Google ran an algorithm on a 53-qubit device that would take years for a classical computer to complete [4]. This and other results have opened the door to the hope that such devices will have some application in the near future.

Google's experiment, later replicated and partially improved by a team at the University of Science and Technology of China [6], consists of the following. With the $N = 53$ qubits arranged in a square geometry and initialized in 0, one defines a quantum circuit comprising logic gates between neighbouring qubits. The gates are chosen at random at the beginning of the calculation (and are then kept the same). As a result of the random circuit, measurement outcomes at the end of the whole computation are close to random (but not quite), such that the probability P_x for each outcome $x = (x_1, ..., x_N)$ (with $x_i = 0, 1$) is very close to $1/2^N$ for each of the 2^N possible outcomes. In the blue panel shown in figure 7.1, this would produce an essentially flat histogram of outcomes. We can then specify a computational task, namely, to sample bit strings according to this probability distribution, with a small prescribed error, chosen to be about 10^{-3} with respect to the probability. This means that the simulated (quantum or classical) probability distribution $\tilde{P}_x$ should obey $\sum_x |P_x - \tilde{P}_x| \leqslant 10^{-3}$. The quantum algorithm that performs this sampling is trivial, since it consists of executing the quantum logic gates and then measuring, a procedure that is repeated to obtain several samples (keeping the same logic gates). However, there is no known classical algorithm that can perform this task efficiently, even if it is allowed to have the mentioned error. Note that even though each $P_x \approx 1/2^N$, replacing P with the uniform distribution is not sufficiently precise, as small deviations (e.g., by a factor of 2 or 1/2) accumulate to yield a sizeable error. In the experiment, the error per logic gate was approximately 0.3%, which allowed the task to be performed successfully. An important point is how one can verify that the NISQ device executed the task correctly, since it is not possible to verify the result with a classical computer (and, moreover, it would be necessary to obtain an exponential number of samples to be able to see that they correspond to the correct result). In the experiment, verification was done using careful analysis and extrapolation of results obtained with a smaller number of qubits (or a subset of logic gates) for which the result could be predicted with a classical computer.

The problem solved by Google is purely academic, chosen to demonstrate so-called quantum supremacy (a better term might be *quantum advantage*), although with no obvious application. It nevertheless demonstrates that NISQ devices are already capable of performing tasks that are not possible with standard computers. The question naturally arises whether there are real applications for which these devices provide a quantum advantage. This question is being studied intensively in many universities and research centres and in industry. Among the most studied candidate problems are process optimization problems. These are usually combinatorial problems, where one has to decide which configuration optimizes a variable. A standard example is the traveling salesman problem, in which, given a set of geographical points, the objective is to find the path through all of them that minimizes its length. There are many other similar problems with wide applications in industry and beyond that are difficult to solve with classical computers and that could potentially be accelerated with NISQ devices.

However, there are at least two reasons to be cautious. First, long quantum circuits lead to large errors, which on the one hand may mean that one needs to repeat the simulation an unworkable number of times, leading to a large overhead. Second, the presence of errors means that we should compare these to classical algorithms that obtain the result with a certain margin of error, which are often substantially faster. We note here that even when devices become fault tolerant, the large overhead of error correction may imply that for problems with a modest polynomial speedup, quantum advantage (i.e., the crossover between classical and quantum runtime) is achieved only for very large problem sizes and at runtimes of years, which would render them impractical. Nevertheless, it is worth insisting that it is also not clear that NISQ devices *cannot* accelerate optimization. For this reason, the current research efforts are very important. Moreover, as was mentioned earlier, research into quantum algorithms may inspire novel, more efficient classical algorithms, as has happened before.

Another area in which NISQ devices may give rise to an advantage is in the area of artificial intelligence (AI) and, more specifically, machine learning. This advantage can come from several directions. On the one hand, quantum states express certain probability distributions very efficiently. In fact, this is precisely what Google's experiment demonstrates: If a process we want to learn were given as a probability distribution obtained from a quantum circuit, we likely would not be able to sample from it efficiently with a classical computer. Machine learning may offer problems that require this kind of probability distribution. On the other hand, an advantage may arise if the training process is more efficient on a quantum computer. Since training can be considered an optimization process, the arguments from the previous paragraph apply.

Other applications concern the solution of equations of many variables (linear, nonlinear, or differential). In this case, there are several algorithms that require an oracle or quantum random access memory (i.e., where it is assumed that there is a process that performs a subroutine whose execution time is ignored). Ignoring the runtime of the subroutine means that these algorithms are generally not useful in practice (they serve to illustrate the power of quantum computers in an abstract way), but there may exist situations in which an advantage remains even if the subroutine is taken into account. How to adapt these algorithms to NISQ devices and account for errors is also an area of active research.

In addition to the problems mentioned previously, it is to be expected that in the near future, new applications of NISQ devices will be discovered that are difficult to imagine today. The reason is that these devices have just been created, and hence algorithmic research is still in its infancy. This leads to considerable optimism and motivates further research. To mention one concrete example, it has recently been discovered that a QC device that can demonstrably execute quantum circuits can be used to generate certified random numbers. That is, we can ensure that these numbers are generated on the spot and are intrinsically random (i.e., neither precalculated nor from a quasirandom classical algorithm). This is something for which there is no classical analog and provides an unexpected new application for quantum computers.

7.2.1.8.2 *Quantum simulation*

The most promising application of quantum computers, both current (NISQ) and future (fault-tolerant) devices, is the simulation of complex quantum systems. Complex quantum systems appear in various incarnations in physics and chemistry and consist of a set of objects interacting with each other according to the laws of quantum physics. The objects can be atoms, electrons, spins, photons, or any other particle (elementary or not). In the case of physics, quantum many-body problems naturally appear in condensed matter physics, high-energy physics, or atomic physics, to name a few. Examples include the electrical or magnetic properties of materials at low temperatures, quantum electrodynamics, quantum chromodynamics, and strongly interacting ultracold atoms. In chemistry, they appear when studying the geometrical structure of molecules, their electronic or spin properties, their dynamics, or their reactions. On classical computers, the solution of the associated problems requires resources (memory or computing time) that grow exponentially with the number of objects (atoms, electrons, etc) or the volume of space in which they are located (cf equation (7.3) and subsequent discussion). Because of this, we usually have to use approximate methods or restrict to a few particles.

A quantum computer can circumvent this problem, as it can naturally implement quantum states. To simulate a many-body quantum system, a quantum computer must simply prepare the desired state Ψ (e.g., the ground state or the state resulting from time evolution) in N qubits and subsequently can obtain the required physical properties by making the appropriate measurements. The device naturally circumvents the memory problem, needing N qubits instead of 2^N bits. To simulate quantum dynamics on a quantum computer, one can execute the quantum logic gates that correspond to the evolution of the system, which requires a number of steps that scales polynomially with the simulation time, the number of qubits, and the inverse of the prescribed error tolerance. If one wishes to determine the ground state or equilibrium properties of a system, there are several different algorithms that perform this task. Although in the most general case, the execution time scales exponentially with N even for a quantum computer, in certain cases this time can be shortened using adiabatic and variational algorithms (see section 7.2.1.4). Either way, it is clear that quantum computers are ideally suited for quantum simulation.

Quantum simulation is also a very interesting prospect for NISQ devices. The reason is that, even with the presence of errors, it is still conceivable that useful results can be obtained in simulation problems. For example, to determine the phase of a material (e.g., is a magnet ferromagnetic or paramagnetic?), it is not necessary to obtain absolute precision in the computation of its physical properties, so even in the presence of sufficiently small errors, a quantum computer could solve problems that would be difficult to address with a classical one. An even more attractive option is that of analog quantum computation. In that case, the idea is to build a laboratory experiment in which the interactions between the qubits (or other particles) can be tuned to some degree. The goal is to engineer this system such that it emulates a certain real-world system. Apart from the obvious advantage that building one of these computers can be easier, because less precise control is needed, an analog computer is most appropriate for simulating the dynamics of quantum many-body systems, obviating the need to discretize time and avoiding the decomposition into logic gates and, therefore, errors. Platforms based on neutral atoms (in either Rydberg or hyperfine levels) are, together with trapped ions, the most advanced.

7.2.1.8.3 Fault tolerance and beyond

In the long term, the goal of QC is the development of scalable computers, that is, computers that are error-free and whose qubit number can be increased arbitrarily with reasonable effort. This is possible only with fault-tolerant quantum error correction, which, as was mentioned earlier, requires that both the initialisation error of each qubit and the error per logic gate (and, if possible, in measurement) are sufficiently small and do not grow as the system scales up. In addition, it is necessary to be able to apply the logic gates in parallel. This is an extraordinary challenge for all current platforms. First, according to current error-correction codes, many physical qubits are needed to encode a single logical qubit. This multiplies the number of required qubits by a large constant factor, which is several thousands for errors of the order of 0.3%. This also means that many more logic gates are required for the dynamic correction of errors as they occur. It is estimated that for many applications of quantum computers, quantum advantage will require the use of hundreds of thousands if not millions of physical qubits. Engineering such large devices constitutes a formidable technological challenge and is therefore unlikely to happen in the next decade. Nevertheless, it is well worth the effort. Advancement in the field must be made not only by further developing the existing platforms, but also by looking for new ones as well as developing more efficient error correction schemes that are adapted to the specific weakness of each implementation.

How will quantum computers will affect us in the next few years? As we stated in the introduction, this question is exceedingly difficult to answer, as it heavily depends on future insights and technological breakthroughs, which, by definition, are not available to us now. We have gone to great lengths to emphasize the challenges connected to scaling up quantum computers. However, even if building fault-tolerant quantum computers takes a long time, NISQ and analog quantum computers will still be of great use in the mean time.

In the long run, the question is not whether a scalable quantum computer will be built but rather when. What seems clear, looking back at the history of classical computation, is that the most impactful applications of quantum computers are yet to be discovered. In the end, the joint research effort, uniting fundamental and applied research, academy and industry, theoreticians and experimentalists, will surely bring new and exciting avenues for research and development.

7.2.1.8.4 *Other applications*

In this section we concentrated on QC, as it is the quantum technology that has raised the most interest not only in the scientific world, but also in industry and society. However, quantum science gives rise to other applications as surprising as or more surprising than this first one. We conclude by mentioning some of them.

Quantum communication consists of sending messages encoded in quantum superpositions. This way, two objectives can be achieved. First, information can be encrypted in such a way that even a quantum computer cannot decrypt it. This is in contrast to many of the traditional forms of encryption we currently use, which can be attacked by quantum computers. This new form of encryption has given rise to what is known as quantum cryptography and, more specifically, quantum key distribution. Quantum cryptography is now a reality and is even being exploited commercially. The transmission is done through photons sent through optical fibres. At the moment, given that we do not have quantum computers that can attack traditional systems and that it requires specific hardware, it is not very active. Moreover, current cryptographic systems are not completely secure, as they would require hardware conditions (such as high-efficiency photon detectors) that cannot be easily incorporated into them. Furthermore, absorption of photons in the fibres limited such secure connections to a few kilometres. Two different ways of extending these ranges are currently being investigated: through quantum repeaters and on by routing them through satellites. Although there are very promising results, further developments are still needed to achieve these goals. We should mention that alternative strategies exist to make cryptographic schemes robust against attack from quantum computers. To do so, one has to replace the current algorithm, which uses the fact that finding primes is easy but prime decomposition is difficult, by another algorithm based on a problem that is believed to be hard even for a quantum computer. One large class of such schemes are based on learning with errors. Postquantum schemes come with the caveat that one typically cannot prove that the chosen problem is hard for a quantum computer and thus may fail suddenly when such an algorithm is discovered, which may not be made public. Quantum cryptography does not have such asterisks and may thus be safer in the long run. Second, quantum communication can also be more efficient. Tasks such as voting or agreeing on a date can be done more efficiently. Similar to speedups in QC, sending qubits in specific entangled states can lead to vastly reduced resource requirements in terms of the number of photons that need to be exchanged to solve a remote problem. In addition to these applications, there are a variety of other applications including unforgible quantum money or quantum tokens.

Quantum randomness is an intrinsic property of quantum systems (see section 7.2.2) which cannot, even in principle, be predicted. Using quantum randomness thus eliminates one potential attack on protocols relying on randomness, such as encryption.

Another field in which quantum physics offers advantages is that of *sensors and metrology*. It has long been known that atomic clocks base their accuracy on the possibility of creating quantum superpositions and that the use of entangled states can further increase that accuracy. Also, these quantum properties can help build better gravitometers, accelerometers, or other sensors in general. There is a lot of research and development activity in this field, and these sensors are likely to have important applications in other fields, such as medicine.

7.2.2 Milestones of research activity in quantum computing

Zeki Can Seskir[1] and Jacob Biamonte[2]
[1]Karlsruhe Institute of Technology—ITAS Karlstraße 11, 76133 Karlsruhe, Germany
[2]Skolkovo Institute of Science and Technology Bolshoy Boulevard 30, bld. 1, Moscow 121205, Russian Federation

We argue that QC underwent an inflection point circa 2017, when long-promised funding materialised, which prompted public and private investments around the world. Techniques from machine learning suddenly influenced central aspects of the field. On one hand, machine learning was used to emulate quantum systems. On the other hand, quantum algorithms became viewed as a new type of machine learning model (creating the new model of variational quantum computation). Here, we sketch some milestones which have lead to this inflection point. We argue that the next inflection point will occur around when practical problems are first solved by quantum computers. We anticipate that by 2050 this will have become commonplace, while the world will still be adjusting to the possibilities brought by quantum computers.

7.2.2.1 General overview

What is quantum computing?
Today's computers which we know and love—whether smartphones or the mainframes behind the Internet—are all built from billions of transistors. (A transistor is an electrically controlled switch which is ultimately the power behind any electronics.) While transistors utilize quantum mechanical effects (such as tunneling, in which an electron can both penetrate and bounce off an energy barrier concurrently), the composite operation of today's computers is purely deterministic or classical. By classical, we mean classical mechanics, which is exactly the physics (also known as mechanics) that we would anticipate day to day in our lives. Quantum computers are not intended to always replace classical computers. Instead, they are expected to be a different tool to complement certain types of calculations. We will elaborate shortly.

The term *quantum mechanics* (in German, *quantenmechanik*) dates to 1925 in work by Born and Jordan [48] and comprises the physics governing atomic systems. Quantum mechanics contains principles and rules that appear to contradict the classical mechanics we are so intuitively familiar with. Such counterintuitive phenomena provide new possibilities to store and manipulate (quantum) information. This is exactly what a quantum computer should do. The information must be stored and processed in the matter. You can think of a quantum computer as providing new mechanisms to store, process, and generally manipulate data. Indeed, the ultimate limitation of computational power is given by quantum physics.

How did quantum computing begin?

QC dates back at least to 1979, when the young physicist Paul Benioff [49] at Argonne National Laboratory in the US proposed a quantum mechanical model of computation. Richard Feynman [50] and independently Yuri Manin [51] suggested that a quantum computer had the potential to simulate physical processes that a classical computer could not. Such ideas were further formulated and developed in the work of Oxford's David Deutsch [52], who formulated a quantum Turing machine and applied a sort of anthropic principle to the plausible computations allowed by the laws of physics. What we now call the Church–Turing–Deutsch principle asserts that a universal (quantum) computing device can simulate any physical process. (This hypothesis does not give an algorithm but just provides an assertion that such an algorithm exists.) Yet even the most elementary quantum systems appear impossible to fully emulate using classical computers, whereas quantum computers would readily emulate other quantum systems [50, 63].

Early insights into the computational power of quantum computers were based on the assertion that it is impossible to develop an efficient classical algorithm able to accurately emulate quantum systems [50, 53]. While this claim has not been formally proven, ample empirical evidence supports it. Yet there are many computational problems where the required computational resources are better understood than simulating physics. These problems arise in the form of, for example, the theory of numbers, groups, or properties of graphs. For decades a small number of researchers worked to understand if one might expect an exponential quantum speedup for problems long studied in computer science. The early algorithms, such as Deutsch–Jozsa [54] and Simon's [55] solved elegantly contrived problems, making prima facie practical merit difficult to envision.

A more practical breakthrough occurred in 1994 when Peter Shor proposed an efficient quantum algorithm for factoring integers. This would open the door to decrypt RSA-secured communications [56]. Shor's algorithm, if executed on a quantum computer, would require exponentially less time than the best-known classical factoring algorithms. Other seminal early findings include Grover's 1996 quantum algorithm [57], which can search through N items in a database in $\sqrt{N}$ steps. It turns out that Grover's algorithm is provably optimal: It is not possible for a quantum computer to search N unstructured elements any faster than $\sqrt{N}$ steps [58], which could be significant in practice. Can these and other quantum algorithms (see table 7.1) be realized experimentally [59]?

7.2.2.2 *Where is quantum computing heading?*

As of today, billions of dollars of public and private investment are being spent to build quantum computers [68]. In October 2019, Google, in partnership with NASA, performed a quantum sampling task that appears infeasible on any classical computer [69]. Late in 2020, Chinese scientists [70] reported results of like significance. At the core of these developments is the quantum mechanical bit: the *qubit*, as first coined by Benjamin Schumacher [71], who asserts that the term arose

Table 7.1. Expected quantum complexity. Theory predicts that quantum computers have the potential to rapidly execute several important algorithmic tasks. Originally developed for the gate model, variational counterparts to Grover's search [61], optimisation (QAOA [62]), linear [63] (and nonlinear [64]) systems, and quantum simulation (VQE, first proposed in [65] and first demonstrated in [66]) have been developed. Factorization and discrete log can readily be mapped to Ising optimisation problems, yet the scaling remains unclear. Polynomial time variational quantum factoring might instead be accomplished by means of the circuit to variational algorithm mapping [67].

Problem class	Quantum complexity
Factorization	Polynomial [56]
Discrete log	Polynomial [56]
Complete search	$\mathcal{O}(\sqrt{N})$ [57]
Sparse linear systems	Polynomial [60]
Quantum simulation	Polynomial [53]

Table 7.2. A brief comparison of traditional bits and quantum bits (qubits).

The quantum bit or qubit.

A traditional computer operates using bits, 0 and 1. Each classical register or classical memory must be in a single classical state, which is represented by a string of bits (e.g., 0011 or 0100000111). Quantum computers allow quantum registers and memories to be in multiple states concurrently. For this reason, quantum computers are often described as enabling parallel processing.

out of a 1996 conversation with William Wootters (for qubit, see table 7.2). Simply, a qubit is a two-state (binary) quantum system.

There are a variety of physical approaches to creating qubits. For example, a single photon of light can represent two quantum states given by vertical and horizontal polarization. Common qubit realisations include superconducting electronics [69], trapped ions [72], and various optical realisations [73]. Qubits are the building blocks of fully programmable (e.g., universal) quantum computers. The universal quantum Turing machine's abstraction from physics makes it harder to realize; consequently, this model has fallen from interest [49].

Qubits should be isolated from their surroundings yet also be made to interact. Quantum algorithms are described by quantum circuits; such circuits depict actions on and interactions between qubits by quantum gates and encompass today's *de facto* universal model of quantum computation. In practice, design imperfections and random noise can not be avoided, meaning that the ideal qubit can never exist. Such noise processes serve to restrict quantum circuit depth. Over the last several decades researchers have developed a powerful theory of quantum error correction, proving that qubits need not be perfect to realise high-depth quantum circuits [74]. Indeed, quantum error correction proves that a small amount of noise can be tolerated: called the error tolerance threshold [75]. Hence, according to the laws of physics, nothing fundamentally prevents humans from building a quantum processor capable of executing high-depth quantum circuits.

A universal quantum computer is assumed to be error tolerant through error correction [76, 77]. Throughout the history of quantum computation, several universal models of quantum computation have been developed and shown to be computationally equivalent to the *de facto* quantum circuit model [74]. This includes adiabatic quantum computation [78, 79] both discrete and continuous quantum walks [80, 81], measurement-based quantum computation [82] as well as one of the authors installment proving universality of the variational model [67].

Assuming this idealized (universal) setting, several miraculous quantum algorithms have been developed which would offer an advantage over the best-known classical algorithms. Lower bounding the computational resources required in quantum (and classical) algorithms has, however, proven extremely difficult. How might we rule out the existence of a better algorithm when the computational power of the class of possible algorithms is not fully understood?

For example, regarding recent quantum adversarial advantage demonstrations [68, 70], who is to say that a classical algorithm will not one day be discovered which can replicate the reported sampling task(s)? This does not imply that such assumptions are not without formal footing. Elegant methods exist to compare the power of a classical computer to the power of a quantum computer [83]. It does, however, make the timeline for a practically meaningful quantum computation difficult to predict. So with all of the dramatic progress, what might we expect from NISQ era quantum processors [84]?

7.2.2.3 *Design constraints*

Design constraints limit manufacturing qubits. Working with current design constraints means that we must find ways to utilize imperfect qubits. The question is whether we can still build meaningful systems with imperfect devices.

Despite outperforming classical computers at an adversarial advantage [69, 70], such a demonstration was tailored to favor quantum processors and has unclear practical applications. Even with Moore's law failing [85], traditional computing resources are ever improving and increasingly accessible. Moreover, the von Neumann architecture has a first-mover advantage: The entire tool chain, the compilers, algorithms, and so on that are in use today are tailored for this architecture. To instead utilize quantum systems as a computational paradigm represents a dramatic change in how problems must be decomposed, encoded, and compiled and in how we think about computing. Perhaps gleaning lessons from sampling [69, 70], we must learn to look towards problems that are more amenable to quantum processors, with desirable criteria such as the following:

1. Problems which bootstrap physical properties of a quantum processor to reduce implementation overhead(s).

Table 7.3. The memory scaling argument asserts that even the worlds largest computer cannot store into its memory any but the simplest quantum states.

The memory scaling argument asserts that quantum states exist which cannot be stored using even the largest classical memories. Early arguments for the quantum computing advantage considered an ideal state of interacting qubits, requiring about $2^{n+1} \cdot 16$ bytes of information to store assuming 32-bit precision. This reaches 80 terabytes (TB) at just less than 43 qubits and 2.2 petabytes (PB) at just under 47, for example, the world's largest memory of the supercomputer Trinity. Hence, applications with $\geqslant 47$ qubits might already outperform classical computers at certain tasks. While this argument did not account for noise and approximation/compression schemes to reduce required memory, similar arguments are considered valid lines of reasoning today.

2. Systems and/or models of computation which offer some inherent tolerance to noise and/or systematic faults.
3. Problems which utilize the ability of quantum systems to efficiently represent quantum states of matter, called the *memory scaling argument* (see table 7.3).

These constraints have led us to what is now called the variational model of quantum computation. In the absence of error correction, NISQ era quantum computation is focused on quantum circuits that are short enough and with gate fidelity high enough that these short quantum circuits can be executed without quantum error correction, as in the recent quantum supremacy experiments [69, 70]. Herein lies the heart of the variational model: by adjusting parameters in an otherwise fixed quantum circuit, low-depth noisy quantum circuits are pushed to their ultimate use case.

NISQ circuits typically bootstrap experimentally desirable regularities inline with criterion 1: The gate sequence itself is fixed, while the gate angles can be varied. A classical computer will adjust parameters of a circuit. Measurements will be used to calculate a cost function, and the process will be iterate (see table 7.4 for some examples).

The prospects of the variational model are limited by the computational overhead of outer-loop optimisation. This requires significant classical computing resources. Variational model proposes some alternatives to this. (For a further comparison of these models, see table 7.5.)

7.2.2.4 How did quantum computing develop prior to 2017?
A turning point in the development of quantum computation appears around 2017. At this point, several long-promised large funding programs began, such as the European Quantum Flagship and the American National Quantum Initiative Act (which happened around the world and was in the billions of US dollars). Most national investments appear to keep a country competitive in technological development. There are many initiatives around the world, adding up to more than

Table 7.4. Hamiltonian complexity micro zoo. Anticipated computational resources to determine ground state energy and calculate energy relative to a state. Restricted Ising denotes problems known to be in P. An asterisk denotes expected and not formally proven conjectures. Electronic structure problem instances have constant maximum size and so are assumed to be in BQP, whereas the ZZXX model admits a QMA-complete ground state energy decision problem.

Problem Hamiltonian	Finding ground energy (classical/ quantum)	Calculating state energy (classical/ quantum)
1-Local Hamiltonian	Polynomial	Polynomial
2-Local ising	*Exp	Polynomial
Electronic structure	*Exp	*Exp/polynomial
ZZXX model	*Exp	*Exp/*polynomial

Table 7.5. Comparison of standard gate model quantum computation versus variational quantum computation. Variational quantum computation trains short quantum circuits to reach their maximum use case yet requires significant classical coprocessing to train these quantum circuits.

Variational	Traditional
• Agnostic to systematic errors • Tightly connects hardware with software to overcome hardware constraints • Optimizes short depth circuits for optimal use • Emulates Hamiltonians by local measurements • Outer loop optimization can require significant classical computing resources • Coherence time and error rates limit circuit depth	• Intuitive and familiar, textbook quantum algorithms adhere to the circuit model • Theoretical analysis, including complexity, has largely been proven possible • Impossible to execute all but the shortest circuits (smallest examples) with current hardware • Ignores hardware constraints and susceptible to both systematic and random errors

20 billion USD committed in public funding. Many private companies also invested dramatically around this time.

Meanwhile, quantum computation was merged with machine learning in two different ways (see [86]). First, quantum circuits can be trained variationally. In other words, quantum circuits can be viewed as machine learning models. Second, machine learning can be applied to a host of problems faced in building and emulating quantum systems. These two facts encouraged the tech industry to participate in QC research and development. (In fact, a new model of computation was developed and proven to be universal [67].)

While those working in the field might readily agree that things have rapidly developed since around 2017, putting data behind this claim is the focus of this

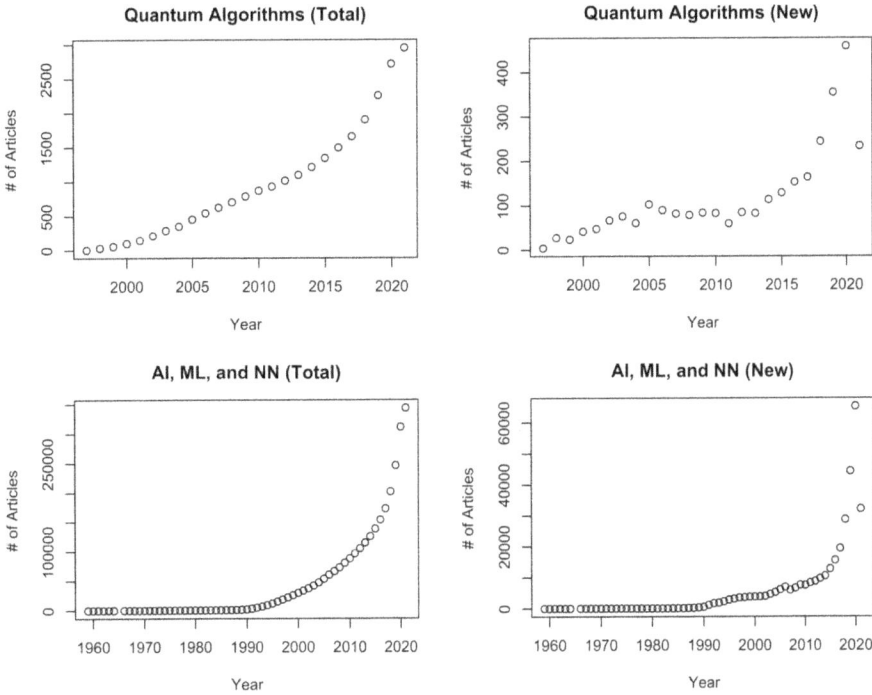

Figure 7.3. The numbers of new and total published articles in the quantum algorithm and artificial intelligence fields.

section. Hence, we will quantitatively describe the levels of activity before 2017, especially focusing on the field of quantum algorithms.

We have utilized two data sources for this section, one for academic publications and one for patent publications: Web of Science owned by Clarivate Analytics PLC and the Cipher platform owned by Aistemos Ltd. However, our readers can use the same queries[1] to reproduce our results. To generate our queries, we have used the previous ones created by previous publications in the literature [87, 88], and personal expertise.

To build our study, we first consider figure 7.3. We see in the upper left-hand panel the growth of quantum algorithms articles. We see what appears to be an increase between 2015 and 2020. By *total*, we mean the entire sum up until that

---

[1] For quantum software and algorithms: (('quantum machine learning') OR ('qml' AND 'quantum') OR ('quantum approximate optimization') OR ('vqe' AND 'quantum') OR ('variational quantum eigen*') OR ('quantum algorithm*') OR ('quantum software') OR ('quantum Machine Learning') OR ('Classical-quantum Hybrid Algorithm*') OR ('quantum PCA') OR ('quantum SVM') OR ('variational Circuit*' AND 'quantum') OR ('quantum Anneal*' AND 'algorithm*') OR ('quantum Enhanced Kernel Method*') OR ('quantum Deep Learning') OR ('quantum Matrix Inversion') OR ('quantum embed*') OR ('quantum neural') OR ('quantum perceptron') OR ('quantum tensor network*')) For artificial intelligence: (('machine learning') OR ('artificial intelligence') OR ('neural network*')) Date: 11 June 11 2021.

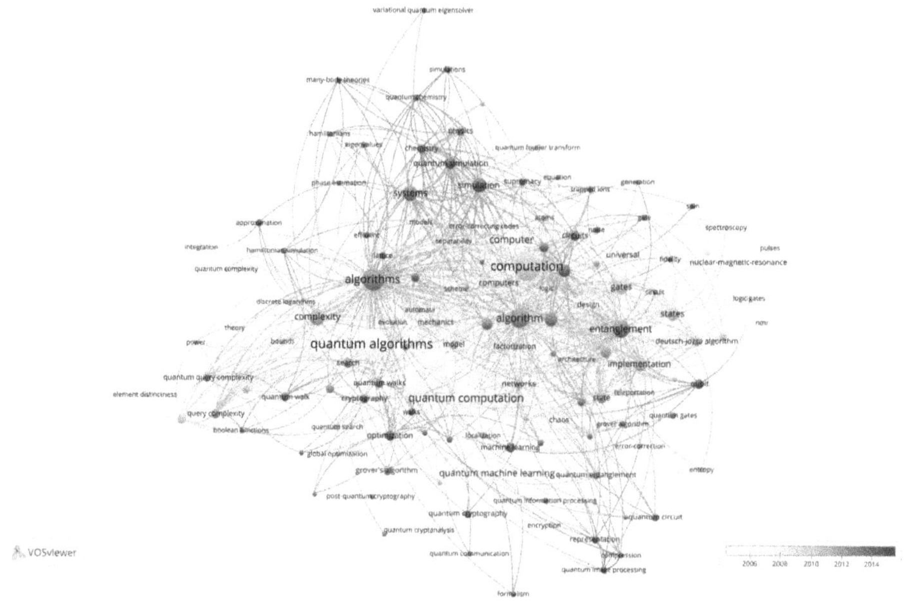

Figure 7.4. Overlay visualization of author keywords used in academic articles in time.

point. By *new*, we mean the number of articles in a given year. In the left top panel of figure 7.3, we again see a jump between 2015 and 2020.

Additionally, we ran the publication dataset through a software tool for constructing and visualizing bibliometric networks (VOSViewer) [89] to run author keyword clustering, which resulted in figure 7.4. In this figure, each node represents an author keyword, the size of the node is correlated with how many times the keyword appears in the dataset, connections between nodes represent co-occurrences of those keywords, and the color scale represents the average year of the keyword in the literature.

Here, one can notice the following point. Keywords such as *quantum algorithms*, *quantum computation*, and *quantum computer* are mainly the keywords utilized by older literature. In recent years they have been replaced with more field specific terms sch as *quantum chemistry, variational quantum eigensolver,* and *quantum image processing*. This indicates an evolution of the literature into partially distinct lines of research, which are more developed topics in terms of maturity, compared to earlier keywords utilized in the literature.

We ran a similar analysis for the collaboration between countries (figure 7.5). The country-level data are associated with the affiliations of authors, and this map represents only the academic literature created by cross-country collaboration. This visual reveals that, although countries such as England, Germany, China, and the US are located in the centre of the collaboration network, a considerable number of new countries joined this network in recent years (represented by dark purple in the figure).

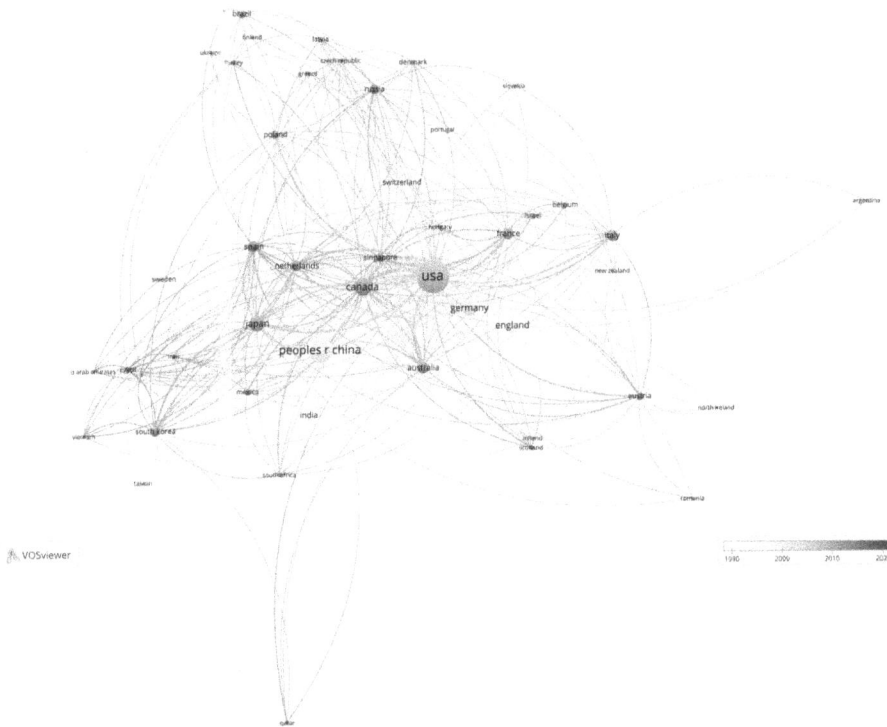

Figure 7.5. Overlay visualization of countries of origin of academic articles in time.

In terms of numbers, the patent literature reveals results similar to those of academic publications. We see a rise in the early 2000s in the top left panel in figure 7.6. We again see a jump before 2020. That sharp jump is the turning point we are discussing. The new and total numbers reflect this. We see similar jumps in the AI fields at a much larger scale (bottom two panels). When we compare the trends in figure 7.6 with those in figure 7.3, it is clear that scientific interest in both fields has been steadier than commercial interest until around 2016–17, and then both show a strongly upward trend.

This exploration of the academic and patent landscapes reveals several important insights into the current state of the field. First of all, comparison between AI and QC in terms of commercialization is clearly an overstatement in terms of patents. Using the queries given, we were able to find 122 609 patent families in the AI field versus 144 in quantum algorithms. Similarly, there were 343 808 articles in AI and 2951 in quantum algorithms. These represent differences of two to three orders of magnitude in publications and patents between these fields. In this sense, quantum algorithms can be related to the early 1990s when compared to AI, which might provide some insight into how the algorithms and QC might evolve in the coming decades.

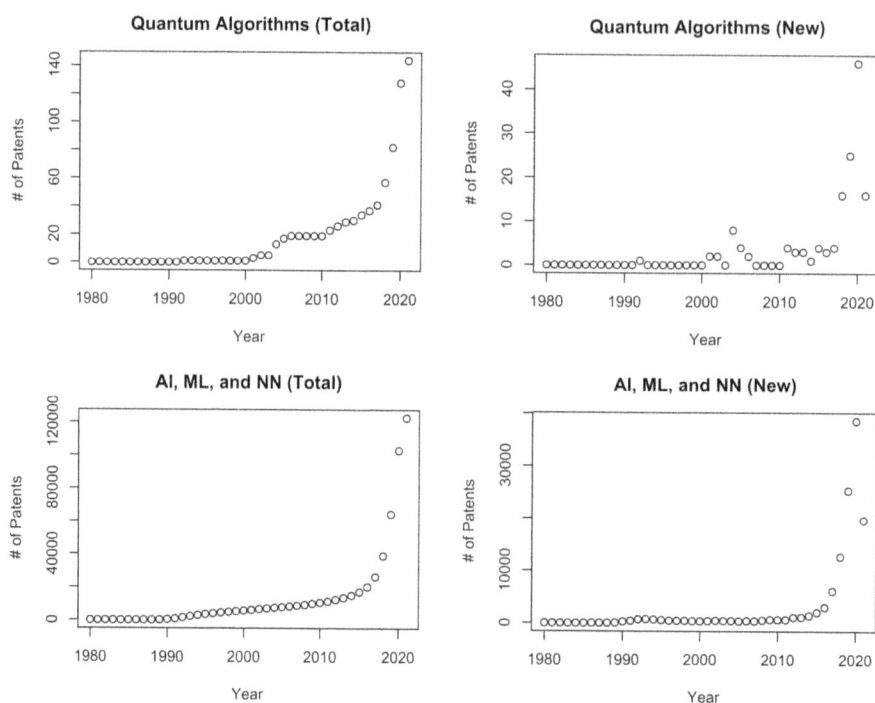

Figure 7.6. The numbers of new and total patents in the quantum algorithm and artificial intelligence fields.

Finally, to see how this activity has been translating in the entrepreneurial realm, we gathered a list of companies[2] in QT and identified the startups related to QC in them to compare them in figure 7.7. It should be noted that other fields under QT (such as quantum sensing and quantum communication/cryptography) have also been gaining popularity in the recent years. From the figure, we can see that startups in QC are a relatively new phenomenon compared to companies in QT, but both have been rapidly increasing in number during the last five years.

In summary, some historical differences and similarities between these fields can be seen in figures 7.3 and 7.6. One clear difference is of the scales, as there are orders of magnitude between the fields. The second difference is that since early 1990 there has been a steady increase in the total number of patents obtained in the field of AI compared to the almost zero activity in the field of quantum algorithms except a brief and short-lived interest in the mid-2000s. One clear similarity is the sudden spike in the late 2010s, especially after 2016–17, which is also evident in figure 7.7. This can be attributed to long-promised funding materialising, which prompted public and private investments around the world. The origins of some of

---

[2] This list was collected by us manually and contained 439 companies as of June 2021. We used open access resources such as Crunchbase, LinkedIn, The Quantum Insider, and Quantum Computing Report.

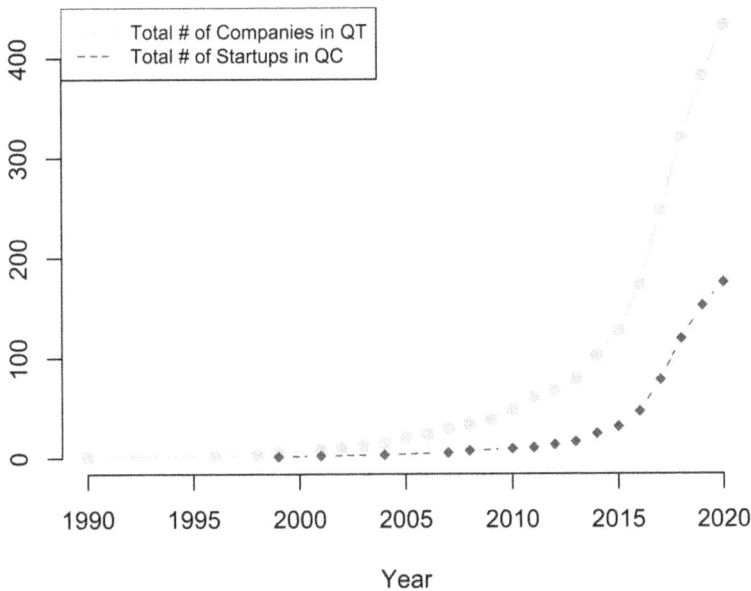

Figure 7.7. Total number of companies operating in quantum technologies and startups in quantum computing in years.

the public funding schemes can be attributed to fear of missing out for countries with existing scientific investments in the field (figure 7.5). Some can be explained by public demonstrations of IBM's and Google's superconducting quantum processors, signaling to the public (and investors) that QC is coming into the realm of calculable risk from Knightian uncertainty [90]. Regardless of the specifics, it is clear that both the number of new academic articles and new patents per year and the number of startups in QC have increased significantly, starting from 2016 to 2017.

This inflection point has not gone unnoticed in the ethical, legal, and social aspects field as well, and one of the first special issues on QT titled *The Societal Impact of the Emerging Quantum Technologies* was published in the journal *Ethics and Information Technology*[3]. Topics such as access to QT [91], the impact of QC on the future of scientific computing [92], responsible research and innovation in QT [93], and the potential impact of quantum computers on society [94] were discussed in the special issue. Since 2017, there has been growing interest in the societal impact of QT (and QC in particular). As of 2021, many researchers and commercial actors in the field have been calling the community to action regarding quantum ethics, which is a clear indication that there is a strong belief in the community that this technology will play a huge role in our future and should be developed ethically to avoid any undesired consequences while taking this quantum leap.

---

[3] ISSN 1388-1957 (print) 1572-8439 (web).

7.2.2.5 What is the next turning point in the development of quantum computing?
It is hard to predict technology. We can assume that everything works and imagine
best-case scenarios. We can assume that nothing works and imagine a sort of
quantum winter. In reality, it is likely the case that the changes ahead are impossible
to envision today.

QC violates no known laws of physics. It, therefore, is thought of as perhaps the
world's most challenging engineering problem. So when will we engineer such
devices? Currently, we have seen progress so dramatic that it would be impossible
for those working in the field to have imagined even five years ago. Perhaps this
means that the state of QC even five years from today is difficult to predict. We still
believe that five years from now, quantum computers will be in the early stages of
development and still lack error correction and other essential features to realize all
their potential.

Having said that, we do think that a quantum future is inevitable. This is the
natural progression of technology. This is probably the ultimate limit of computers,
perhaps until we can harness new powers in the cosmos. Humankind's trajectory is
set on a quantum course. Companies such as Google, IBM, and the like are in this
for research and long-term prospects.

When will we see another inflection point? It's hard to tell. The saying goes that
knowledge begets knowledge, so development always seems to go increasingly fast.
But the next jump might have to wait until practical problems of commercial value
are sufficiently solved. This should take place perhaps around 2050. We do imagine
that by then, this technology will already have changed the world in ways we cannot
predict now.

7.3 Sensors and their applications

Antigone Marino[1]
[1]Institute of Applied Sciences and Intelligent Systems, National Research Council
(CNR), Naples, Italy

The etymology of the word *sensor* already explains its meaning in a simple way. The sensor is a device that senses. The human body is one of the most incredible devices that nature has created; it is composed of various sensors that allow us to manage the five senses, namely, sight, smell, touch, taste, and hearing. Scientific research and the development of technology have allowed the human race to develop a huge number of sensors, capable of 'feeling' where humans do not have the right sensors to do so. Sensors are among the technologies that will enable us to address many of the global challenges of the 21st century and beyond. Let's think, just to give a few examples, of food conservation, monitoring of transport, air, water, and health. The first sensors are more than a century old. However, smart sensors, with integrated Information and communications technology (ICT) capabilities, have been around for only a few decades. The 20th century saw the birth of a wide range of sensors, but it will be in the 21st century that their application, driven by the union of sensing and ICT, will affect many aspects of our life.

Due to a scholastic legacy, we are used to talking about one industrial revolution, which refers to the industrialization process of turning society from agricultural, artisanal, and commercial into a modern industrial system characterized by the generalized use of machines powered by mechanical energy and the use of new energy sources. The industrial revolution has never stopped. Since it started, there have already been four industrial revolutions. The first mainly concerned the textile and metallurgical sectors. The second industrial revolution is conventionally said to have started in 1870 with the introduction of electricity, chemicals, and fuels. Since 1970, the massive introduction of electronics, telecommunications, and information technology into industry identifies the third industrial revolution. Finally, the convergence of multiple technologies, real-time analytics, machine learning, ubiquitous computing, and embedded systems gave birth to the IoT. In parallel, the automation and control of industrial processes led to the fourth industrial revolution, also known as the Industrial Internet of Things (IIoT) (figure 7.8).

The history of sensors has gone hand in hand with that of industrial revolutions. Similar to the four phases of industrial development, we have four phases of the evolution of sensors (figure 7.9). In 1844 the French physicist Lucien Vidie invented the barograph, a device to monitor pressure, a recording aneroid barometer. This was indeed a sensor, but not a modern one, as it was a purely mechanical indicator not equipped with any electronics. The thermostat patented by Warren Seymour Johnson in 1883 [95] is considered by some to be the first modern sensor. The introduction of integrated circuits, capable of detecting a specific physical parameter and converting it into an electrical signal, led at the beginning of the 20th century to the birth of 'Sensors 2.0'. Since 1970, electronic devices have been used to measure physical quantities such as temperature, pressure, or loudness and convert them into electronic

Figure 7.8. The Industrial Internet of Things has led to the development of Industry 4.0. Augmented reality and smart sensors will allow greater control of production processes.

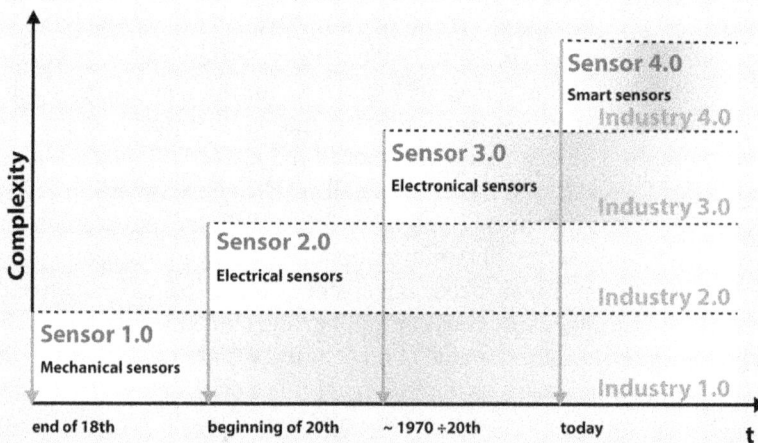

Figure 7.9. From the end of the 18th century up to today, scientific research has allowed the development of four sensor technologies, each one strongly connected to one of the four industrial revolutions: Sensors 1.0, intended as mechanical sensors, without electrical output; Sensors 2.0, thus electrical sensors; Sensors 3.0 or electronical sensor; and finally Sensors 4.0, also known as smart sensors.

signals, leading to the family of 'Sensors 3.0'. Finally, thanks to smart sensors, everything is getting clever, and the data collected during the manufacturing process are used to improve the quality of the product itself. We are now in the fourth industrial revolution thanks to 'Sensors 4.0' [96]. The importance of sensors is evident, since the industrial evolution is strongly connected to sensors and instrumentation.

In this section, we will look at how far scientific research has gone in the field of sensors and what the growth prospects are for the future.

7.3.1 Sensor characteristics

The operation process of a sensor is simple: The sensor detects a physical, chemical, or biological quantity (e.g., bacteria, proteins, waves, movement, or chemical agents). Then the measurement is processed by a transducer, which converts it into an output signal, often an electrical one (figure 7.10).

Each sensor has its own characteristics which outline its performance. These are categorized as systematic, statistical, or dynamic. Systematic characteristics are those which can be exactly quantified by mathematical means. Statistical characteristics are those which cannot be exactly quantified. Dynamic characteristics are those who describe the ways in which an element responds to sudden input changes [97].

The sensor range is a static characteristic that describes both the minimum and maximum values of the input or output [98]. The full-scale input, called span, describes the maximum and minimum input values that can be applied to a sensor without causing an unacceptable level of inaccuracy. It is also called the dynamic range. If we speak of the sensor output, it is the algebraic difference between the output signals measured at maximum and minimum input stimulus. Finally, the operating voltage range describes the minimum and maximum input voltages that can be used to operate a sensor.

The sensor transfer function describes the relationship between the measurand and the electrical output signal. If it is time independent, we can write it as $S = F(x)$, where x is the measured quantity and S is the electrical signal produced by the sensor. This function can be a very complex one. The simplest case is the one of a linear transfer function, $S = A + Bx$, where A is the sensor offset and B is the sensor slope. The sensor offset is the output value of the sensor when the input is zero. The slope of a linear transfer function is the sensor's sensitivity, which we will define shortly. Many sensors do not have a linear response; rather, it is approximated to be

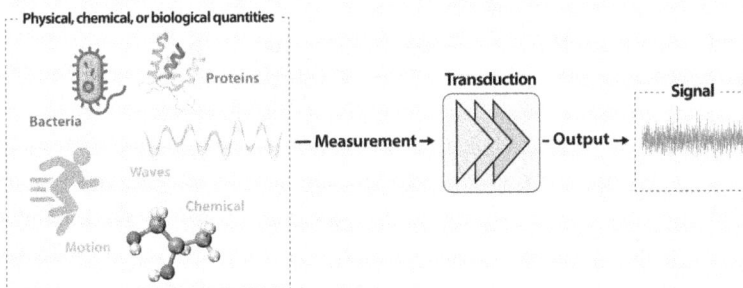

Figure 7.10. A sensor is a device that detects a physical quantity and converts its measurements into a human-readable output, such as an electric signal.

linear within certain limits. As you can imagine, having a linear or linearized function offers many advantages, such as being able to calculate the quantity we are analyzing from the output, to predict the output based on the analyzed quantity value, or even to easily obtain the offset and sensitivity. In modern sensors, such as smart sensors, the presence of the digitization of the output makes its linearization considerably easier. In the case of nonlinear transfer functions, nonlinearity is defined as the difference between the real line and the ideal straight one. As the nonlinearity can vary along the input–output graph, it is customary to indicate as characteristic of the sensor the maximum nonlinearity. Typically, this value is expressed in percentage of the span.

The sensitivity S of the sensor is the ratio between the variation of the output quantity ΔS and that of the input Δx which determined it:

$$S = \frac{\Delta S}{\Delta x} \tag{7.4}$$

A sensor is very sensitive when a small variation of the input quantity corresponds to a large variation of the output quantity. If the sensor transfer function is linear, the sensitivity will be constant over all of the sensor range. Otherwise, it will change. The range of values for which a sensor does not respond is called the dead band. In this range the sensitivity is zero. Dead band is also usually expressed as a percentage of the span. The sensor saturation point is the input value at which no more changes in the output can occur.

A sensor should be capable of following the changes of the input parameter regardless the direction in which the change is made. However, this does not always happen. The output of a sensor may be different for a given input, depending on whether the input is increasing or decreasing. The hysteresis is the measure of this property, it quantifies the presence of a 'memory' effect of the sensor whose output, with the same measurand value, could be affected by the previous operating condition. Like nonlinearity, hysteresis varies along the input–output plot; thus, maximum hysteresis is used to describe the characteristic. This value is usually expressed as a percentage of the sensor span.

The sensor resolution is the smallest change that can be detected in the quantity that is being measured. It is one of the features that most affects the cost of a sensor. Precisely for this reason it is important to understand, based on the application to be implemented, what resolution is needed. A sensor with a low resolution could cause the fail of detecting the signal, while a resolution that is too high could be unnecessarily expensive.

The sensor accuracy represents the maximum error between the real and ideal output signals. It is the sensor's ability to provide an output close to the real value of the analyzed quantity. It can be quantified as a percentage relative error using the following equation:

$$\text{Percentage Relative Error} = \frac{\text{Measured Value} - \text{True Value}}{\text{True Value}} \tag{7.5}$$

The concept of precision refers to the degree of reproducibility of a measurement. In other words, if exactly the same value were measured a number of times, an ideal sensor would output exactly the same value every time. But real sensors output a range of values distributed in some manner relative to the actual correct value. As precision relates to the reproducibility of a measure, it can be quantified as percentage standard deviation using the following equation:

$$\text{Percentage Standard Deviation} = \frac{\text{Standard Deviation}}{\text{Mean}} \times 100 \qquad (7.6)$$

Precision is often confused with accuracy. Figure 7.11 illustrates the key difference. If we repeat a measurement five times, the distribution of our data will change depending on the sensor accuracy and precision [99]. Obviously, the best sensor is the one that has greater accuracy and precision.

Like any measuring device, sensors are subject to measurement error, that is, the difference between the measured value and the true one. The errors can be either systematic or random. Systematic error always affects measurements by the same amount or the same proportion, provided that a reading is taken the same way each time. Systematic error is predictable. Random error causes one measurement to differ slightly from the next. It comes from unpredictable changes. Random errors cannot be eliminated, while most systematic errors may be reduced with compensation methods, such as feedback, filtering, and calibration [98]. Systematic errors result from a variety of factors: interfering inputs, modifying inputs, changes in

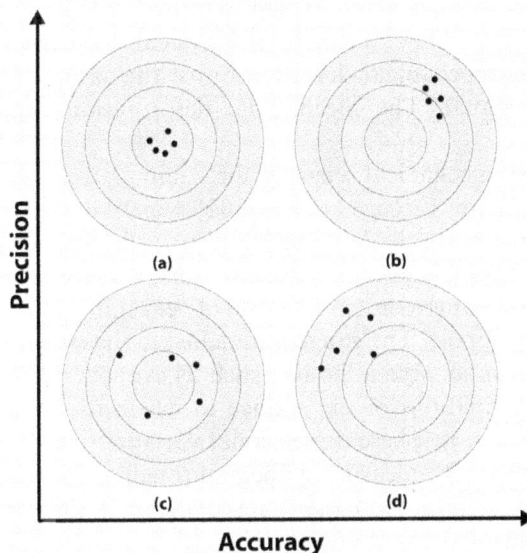

Figure 7.11. Repeatability and accuracy. If we repeat a measurement five times, the distribution of our data will change depending on the sensor accuracy and precision: (a) good accuracy and precision; (b) low accuracy and good precision; (c) good accuracy and low precision; (d) low accuracy and precision.

chemical structure or mechanical stresses, interference, signal attenuation or loss, and even human factors. Random error, also called noise, is a signal component that carries no information. The quality of a signal is expressed quantitatively as the signal-to-noise ratio, which is the ratio of the true signal amplitude to the standard deviation of the noise.

Sensors' dynamic characteristics are those that are time-dependent. These are relevant when the sensor inputs are not constant over time. The most common dynamic characteristics are response time and dynamic linearity. Sensors do not change output state immediately when an input parameter change occurs. Rather, the output will change to the new state over a period of time, called the response time. This can be defined as the time required for a sensor output to change from its previous state to a final settled value within a tolerance band of the correct new value. The tolerance band is defined based on the sensor type, the sensor application, or the preferences of the sensor designer. The dynamic linearity of the sensor is a measure of its ability to follow rapid changes in the input parameter.

7.3.2 General overview of sensors

If we stop for a moment, and observe the world around us, we will immediately realize that in our daily life we use much more sensors than we imagine. It is not trivial to classify them, or to review them all. Conventionally, they can be classified according to the measurement principle they use.

7.3.2.1 Mechanical sensors

Mechanical sensors form a class of sensors that are sensitive to changes in mechanical properties. They measure a physical quantity resulting from a stimulus that causes mechanical deformation of the sample [100] and translate it into an electric signal. The mechanical deformation can be realized with different stimuli.

The most common mechanical sensor is the strain gauge, invented by Edward E Simmons [101] and Arthur C Ruge in 1938 [102], which is used to measure strain on an object. Strain gauges are the key sensing element in a variety of sensors types, including pressure sensors, load cells, torque sensors, and position sensors. The most common type of strain gauge consists of a grid of very thin metal wire rigidly applied to a plastic material support. The gauge is attached to the object whose strain gauge is to be measured by a suitable adhesive, such as cyanoacrylate. The strain gauge wire follows the deformations of the surface to which it is glued, elongating and shortening together with it. These dimensional variations cause fluctuations of the electric wire resistance. By measuring them through a Wheatstone bridge, it is possible to trace the deformation that caused them. A key problem with strain measurements is that of thermal effects. The electric current flowing through the strain gauge causes heating by the Joule effect. Temperature compensation is required to address the problem, using a Wheatstone bridge connected to two strain gauges: the one for measurement on one side and another equal strain gauge on a piece of the same material not subjected to any stress but exposed to the same temperature. Due to the characteristics of the Wheatstone bridge, the deformations due to temperature produce the same variation in resistance in modulus but with a different sign so as to cancel each other out. It is not always possible to know in advance the direction in which the deformation of the material will occur. In this case, it becomes necessary to use a system of strain gauges with axes oriented in different directions.

Piezoelectric sensors are a particular type of strain gauges [103]. They differ slightly in terms of physical operating principle. In fact, in piezoelectric materials the resistance change is due to resistivity, while in strain gauges the resistance varies almost exclusively because of the length and the section of the conductor that makes up the sensor. Thus, piezoelectric strain gauges have even greater sensitivity to temperature variation.

Sensors based on both mechanical and electrical variations are known as micro electromechanical systems (MEMS) [98]. These devices have been recognized as one of the promising technologies of the 21st century, capable of revolutionizing both the industrial world and that of consumer products. They are three-dimensional, miniaturized, mechanical and electrical structures, typically ranging from 1 to 100 μm, that are manufactured using standard semiconductor manufacturing techniques. MEMS sensors are widely used in the car industry and, since the early 1990s, to realize accelerometers in airbag systems, electronic stability programs, and antilock braking systems. The availability of inexpensive, ultracompact, low-power multiaxis MEMS sensors has led to rapid growth of their use in customer electronics devices. MEMS can be found in smartphones, tablets, game console controllers, portable gaming devices, digital cameras, and camcorders. They have also found application in the healthcare domain in devices such as blood pressure monitors, pacemakers, ventilators, and respirators. Two of the most important and widely used forms of MEMS are accelerometers and gyroscopes [104].

An accelerometer is a sensor capable of measuring acceleration. The use of the accelerometer has increased considerably in recent years since, alongside traditional applications in the scientific and aerospace fields, it has been adopted in numerous civil fields such as the automotive and consumer electronics industries. In most accelerometers, the operating principle is the same: It is based on detecting the inertia of a mass when subjected to acceleration. The mass is suspended from an elastic element, while some type of sensor detects its displacement with respect to the fixed structure of the device. In the presence of an acceleration the mass, which has its own inertia, moves from its rest position in proportion to the detected acceleration. The sensor transforms this displacement into an electrical signal.

MEMS are also used in gyroscopes, sensors that are able to measure the angular rate of rotation of one or more axes. As they have no rotating parts, they can be easily integrated into even very small devices. They use vibrating mechanical elements to detect rotation based on the transfer of energy between two modes of vibration caused by Coriolis acceleration.

7.3.2.2 Optical sensors

Optical sensors work by detecting light, from ultraviolet to infrared (IR). They can be designed in various ways in order to acquire the signal [105]: They can measure the light intensity change related to light emission or absorption by a quantity of interest, they can measure phase changes occurring in light beams due to interaction or interference effects, or they can measure a simple interruption of a light source. Let us now look at the most common types of optical sensors.

Photodetector sensors measure the sample photoconductivity, or the material change of conductivity when illuminated. There are many types of photodetectors, such as active pixel sensors, charge-coupled devices (CCD), light-dependent resistors, photodiodes, and phototransistors [106].

IR sensors measure and detect IR radiation in their surrounding environment [107]. There are two types of IR sensors: active and passive. Active IR sensors both emit and detect IR radiation. They are realized with a light-emitting diode (LED) and a receiver (photoelectric cells, photodiodes, or phototransistors). When an object comes close to the sensor, the IR light from the LED reflects off the object and is detected by the receiver. Active IR sensors act as proximity sensors, and they are commonly used in obstacle detection systems. Passive IR sensors detect only IR radiation. Inside, there is one or more pyroelectric materials that generate energy when exposed to heat. They are commonly used in motion-based detection, such as in-home security systems. When a moving object that generates IR radiation enters the sensing range of the detector, the difference in IR levels is measured.

Optical fiber sensors use an optical fiber as the sensing element. They are widely used in construction, such as bridge monitoring [108], and aircraft security [109], as they allow monitoring of different physical quantities. Strain can be measured as it changes the geometric properties of the fiber and so the refraction of the light passing through it; in the same way, a temperature change can be detected as it causes fiber strain; pressure sensing can be realized with an intensity sensor [110] or an interferometric sensor [111]; humidity sensing can be achieved in several ways, such as luminescent systems with fluorescent dyes that are humidity sensitive or reflective thin film-coated fibers which change their refractive index, resulting in a shift in resonance frequency [105].

Finally, interferometric sensors measure changes in a propagating light beam, such as path length or wavelength along the path of propagation [112].

Optical sensors are widespread, due to the many advantages they show: high sensitivity, high integration, suitability for remote sensing, wide dynamic range, wide range of chemical and physical parameters detection, and many others. However, they can be costly and susceptible to physical damage.

7.3.2.3 Semiconductor sensors

Semiconductor sensors owe their popularity to their low cost, high integrability, and long life. They are mostly used for gas monitoring, but they also allow the detection of temperature or optical physical quantities, as in CCDs [98].

In gas sensors, the operation of a semiconductor sensor is determined by the variation in conductivity of a semiconductor element caused by the chemical absorption of the gases in contact with the porous surface of the semiconductor, electrically heated to a predetermined temperature. The temperature of the sensitive element (depending on the type of gas to be detected) is a determining parameter for the sensitivity and selectivity of the sector.

Semiconductor temperature sensors commonly use a bandgap element which measures variations in the forward voltage of a diode to determine temperature. They are designed with materials showing a strong thermal dependence. To achieve reasonable accuracy, these are calibrated at a single temperature point. Therefore, the highest accuracy is achieved at the calibration point, and accuracy then deteriorates for higher or lower temperatures. For higher accuracy across a wide temperature range, additional calibration points or advanced signal processing techniques can be employed.

Semiconductor magnetic field sensors exploit the galvanomagnetic effects due to the Lorentz force on charge carriers. When the sensor is placed inside a magnetic field, the voltage difference of the semiconductor depends on the intensity of the magnetic field applied perpendicular to the direction of the current flow. Electrons moving through the magnetic field are subjected to the Lorentz force at right angles to the direction of motion and the direction of the field. The Hall voltage, which is generated in response to the Lorentz force on the electrons, is directly proportional to the strength of the magnetic field passing through the semiconductor material. The output voltage is often relatively small and requires amplification of the signal. Integrated semiconductor magnetic field sensors are manufactured using integrated circuit technologies [113].

The most common optical semiconductor sensor is the photodiode, a photodetector that converts light into either current or voltage. The photodiode is a diode that works as an optical sensor by exploiting the photovoltaic effect. It is able to recognize a certain wavelength of the incident electromagnetic wave and to transform this event into an electrical signal of current by applying an appropriate electric potential to its ends. It is therefore a transducer from an optical signal to an electrical signal. Another form of photodetector is the phototransistor, which is essentially a bipolar transistor with a transparent window that allows light to hit the base–collector junction. Phototransistors have the advantage of being more sensitive than photodiodes. However, they have a slower response time.

7.3.2.4 Electrochemical sensor

Electrochemical sensors are devices that give information about the composition of a system in real time by coupling a chemically selective layer to an electrochemical transducer. In this way, the chemical energy of the selective interaction between the chemical species and the sensor is transduced into an analytically useful signal. Due to the simplicity of the procedures and instrumentation required, they are the largest and the oldest group of chemical sensors. They attract great interest nowadays because they are easy to miniaturize and integrate into automatic systems without compromising analytical characteristics.

Different families of electrochemical sensors can be recognized depending on the electrical magnitude used for transduction of the recognition event [114]:

potentiometric (change of membrane potential) [115], conductometric (change of conductance), impedimetric (change of impedance), voltammetric (change of current for an electrochemical reaction with the applied voltage), and amperometric (change of current for an electrochemical reaction with time at a fixed applied potential) [116].

Electrochemical sensors have a number of advantages, including low power consumption, high sensitivity, good accuracy, and resistance to surface-poisoning effects. However, their sensitivity, selectivity, and stability are highly influenced by environmental conditions, particularly temperature.

7.3.2.5 Biosensors

A biosensor is a device consisting of a biologically active sensitive element and an electronic part. The operating principle is simple: The biological element interacts with the substrate to be analyzed, and a transduction system converts the biochemical response into an electrical signal. A common example of a biosensor is the glucometer used by diabetics to measure the concentration of glucose in the blood. The sensitive element of this biosensor is an enzyme, glucose oxidase, which converts glucose into gluconic acid.

The analyte is a substance that we intend to measure, such as glucose, cardiac biomarkers, or tumor biomarkers. The bioreceptor is the sensitive element, it can be an enzyme (catalytic biosensor), an antibody (affinity biosensor), DNA/RNA, or an aptameter. The transducer can be electrochemical (potentiometric, amperometric), optical, electromechanical, mechanical, or even acoustic [117].

The main feature of a biosensor is the specificity, which is guaranteed by the use of biological receptors, which by their intrinsic nature are specific to particular analytes. Specificity is the ability to react only with a certain analyte and not with others that may be present in the measurement environment. In other words, the other analytes present in the measurement environment are influencing quantities with a negligible effect. The main characteristics of the biosensors are their high sensitivity, measurement speed and cost-effectiveness.

7.3.3 Sensors challenges

In 2015 the United Nations General Assembly set up a collection of 17 interlinked global goals designed to be a 'blueprint to achieve a better and more sustainable future for all'. These are called Sustainable Development Goals (SDGs). They are included in a UN Resolution called the 2030 Agenda [118], as they are intended to be achieved by the year 2030. Among the SDGs are Good Health and Well-being, Clean Water and Sanitation, Affordable and Clean Energy, Industry Innovation and Infrastructure, Sustainable Cities and Communities, Responsible Consumption and Production, Climate Action, and Life Below Water, Life On Land.

In 2020, the European Commission (EC) signed The European Green Deal [119], a set of policy initiatives with the overarching aim of making Europe climate neutral in 2050. The plan is to review each existing law on its climate merits and introduce new legislation on the circular economy, building renovation, biodiversity, farming, and innovation. The climate change strategy is focused on a promise to make Europe a net-zero emitter of greenhouse gases by 2050 and to demonstrate that economies will develop without increasing resource usage. Also, the Green Deal has measures to ensure that nations that are reliant on fossil fuels are not left behind in the transition to renewable energy.

These community actions, at both global and European levels, require the support of scientific research in order to achieve the set goals. In the last century, scientific research has allowed the development of many types of sensors. Basical,ly more than one type of sensor is available to measure a quantity of interest. Of course, each sensor option has its advantages and disadvantages. These need to be weighed against the context in which the sensor will be used in order to determine which sensor technology is the most suitable one.

Let's examine some of the main challenges that our society is facing and how sensors can be one of the necessary tools to address them.

7.3.3.1 Environmental and earth sensing

Intensive agriculture, growing urbanization, industrialization, the increasing demand for energy, and climate change are putting planet Earth at risk. Environmental monitoring is therefore essential to reveal its state, to assess the progress that has been made to achieve certain environmental objectives, and to help detect new environmental problems.

The Joint Research Centre (JRC) is the EU's science and knowledge service which employs scientists to carry out research in order to provide independent scientific advice and support to EU policy. The results are of fundamental importance to environmental management in general, as the drafting and prioritisation of environmental policies is based on the findings of environmental monitoring. The JRC's work supports many actions, such as Copernicus (the European Earth Observation Programme), the Water Framework Directive, the Marine Strategy Framework Directive, EU Food Security Policy, European Climate Policy, EU Strategy for Sustainable Development, the Directive on Ambient Air Quality, and the Clean Air For Europe programme.

The protection of the human race and the Earth's ecosystem is the highest priority. Scientific research is trying to create scalable sensors capable of detecting pollutants, even at very low concentrations and on a widespread basis, quickly and sensitively, While many environmental monitoring solutions are already available, what sensors are being asked for is to work *in situ* and online, allowing real-time decision making. Many analyzes, such as those for bacterial contamination, still require *in situ* sampling with laboratory analyzes, which also entails a slowdown due to bureaucracy. Sensors that operate *in situ* are the new challenge for environmental monitoring, organized in a distributed network that can have an ever greater geographical distribution.

Power plants, agriculture, industrial manufacturing, and vehicle emissions are all sources of air pollutants such as sulfur dioxide, carbon monoxide, nitrogen dioxide, and benzene. Air pollution is a problem in both the developed world and the developing world. Energy production linked to fossil fuels is the main source of air

Figure 7.12. Agriculture will make use of smart sensors to remotely control the parameters that regulate crop growth, such as humidity and temperature.

pollution. The second source is urbanization and the consequent increase in vehicular emissions.

Air monitoring stations are now equipped with multiple sensors so that not only one gas can be analyzed, but also particulates, hydrocarbons, and metals, according to local regulatory requirements for air quality [120]. Air monitoring requires very sophisticated and therefore expensive instrumentation, such as spectroscopic analysis. This explains why air monitoring has a low distribution density in many countries. In the future, we will see the production of low-cost and smart sensors, which may require the most widespread and real-time air analysis (figure 7.12).

Several companies are studying the possibility of creating devices designed to transform smartphones into an environmental monitoring centre capable of recording levels of environmental humidity, carbon monoxide, fine dust, light intensity, pressure, and many other factors [121]. The spread of these integrated sensors would be a revolution in the field of air monitoring, as each citizen would constitute a data collection unit.

Regarding air monitoring, the JRC supports many actions, such as the Climate Service of Copernicus, the European Earth Observation Program relating to climate change, which among its purposes has not only monitoring, but also ensuring compliance with international standards. The JRC also maintains a worldwide database, the Emissions Database for Global Atmospheric Research (EDGAR), which allows monitoring and verification of emissions around the world, providing the information needed to determine appropriate policies.

The growing need for clean water for drinking and industrial use requires water quality monitoring. Similar to air quality, strict standards are set by national bodies and geopolitical bodies, resulting in the need for reliable sensor technologies that can monitor different water quality parameters with the required sensitivity. Again, the market requires sensors that provide real-time readings to ensure that any abnormal changes in water quality have the least impact on human health or manufacturing operations [122].

There are three main categories of interest: physical (turbidity, temperature, conductivity), chemical (pH, dissolved oxygen, concentration of metals, nitrates, organic substances), and biological (biological oxygen demand, bacterial content). Scientific research is developing smart water networks and predictive models [123]. Sensors implemented directly in the water distribution network will contribute to the identification of leaks through pressure drops, thus allowing minimization of water losses. The development of predictive models for water quality monitoring, based on the fusion of water quality data, environmental sensors, and metrology sensors, will allow to predict potential changes in future water quality.

The JRC's work on water monitoring covers the monitoring of water quality and assessment of the impact of pollutants and chemicals, the monitoring of water and marine ecosystems, the provision of early warnings and risk management, the monitoring of floods and droughts, and the monitoring of water quantity in Europe and worldwide.

7.3.3.2 Global navigation satellite systems

The development of satellite technologies has made possible one of the most fascinating and promising projects in scientific research. The Global Navigation Satellite System (GNSS) is a constellation of satellites providing signals from space that transmit positioning and timing data to GNSS receivers [124]. The receivers then use the data to determine location. This network provides global coverage. Examples of GNSS include Europe's Galileo and the USA's NAVSTAR Global Positioning System (figure 7.13).

The satellite transmits a signal that contains the position of the satellite and the time of transmission of the signal itself, obtained from an atomic clock in order to

Figure 7.13. Satellites equipped with various sensors, organized in a sophisticated network, can transmit information about the planet and space in real time. This is the concept behind the Global Navigation Satellite System.

maintain synchronization with the other satellites in the constellation. The receiver compares the transmission time with that measured by its own internal clock, thus obtaining the time it takes for the signal to arrive from the satellite. Several measurements can be made simultaneously with different satellites, thus obtaining the positioning in real time. Each distance measurement, regardless of the system used, identifies a sphere that has a satellite as its centre; positioning is obtained from the intersection of these spheres.

GNSS systems have many applications: navigation, both with portable receivers, for example for trekking, and with devices integrated between the controls of means of transport, such as cars, trucks, ships, and aircraft; time synchronization in electronic devices; monitoring; search and rescue service; geophysics; topographic applications; machine automation applications (automatic driving of machines) for earthmoving and agriculture; wild animal tracking devices; and satellite alarm.

GNSS systems are also being studied to face a possible Kessler effect in the future. This is a scenario proposed by NASA scientist Donald J Kessler in 1978 [125]. It is a theoretical scenario in which the density of objects in low Earth orbit, due to space pollution, is high enough that collisions between objects could cause a waterfall where each collision generates space debris, which increases the likelihood of further collisions. The direct consequence of this scenario is that the increasing amount of waste in orbit would make space exploration and even the use of artificial satellites impossible for many generations. A science-fiction version of Kessler's syndrome is depicted in the 2013 movie *Gravity*, directed by Alfonso Cuarón. He imagines that the reckless downing of a spy satellite with a missile causes a chain reaction that ends in the destruction of a space shuttle and the death of its crew, the Hubble Space Telescope, the International Space Station (ISS), and the Chinese space station. The film focuses on the fate of the protagonist and does not analyze the consequences for future space travel.

The problem of space waste is very difficult to solve directly, since the small size and high speeds that characterize most of the waste make their recovery and disposal practically impossible. However, GNSS systems allow the monitoring of space waste and help institutions to eventually run emergency procedures.

7.3.3.3 Security sensors

Physical security and safety have always been critical for the welfare of individuals, families, businesses, and societies. In the past, castles were surrounded by water ditches and cities by thick walls to defend themselves from invasions.

The IoT has enabled growth in the residential and commercial security sectors, thanks to sensors that power these solutions. IR security sensors, active or passive, utilize IR light to detect motion in order to trigger an alarm. Photoelectric sensors are now more common in spy films than in real life, but you can find them at work in specific security settings. Photoelectric sensors would be helpful in, for example, an environment that contains a space that humans or objects may not enter. Also, photoelectric sensors use invisible IR light but can travel significantly farther than

IR sensors. Photoelectric beams establish an invisible barrier that, when broken, triggers a security notification. Microwave sensors are also used in order alarm systems. Like active IR sensors, microwave sensors emit and receive a signal to detect an object in motion. These sensors are generally much more sensitive than IR sensors. Moreover, they can sense motion through nonmetal materials such as wood, plastic, and drywall. One of the most recent technologies, which will see a wide diffusion in the coming years, is that of tomographic motion detection sensors. Motion tomography technology does not require a direct line of sight to trigger a safety alert. It uses a mesh network of radio emitters and receivers to detect any movement within the mesh network. This sensor technology works by detecting interruptions in signals between emitters and receivers, which it interprets as motion. The tomographic motion technology has been around for only about a decade, but it is a promising technology for the high-security commercial and industrial sectors.

Safety is a very important factor in various contexts, such as airports, transport control, detection of weapons or hazardous materials, drug detection, and even nuclear safety. In this type of application, real-time detection is required in order to be able to suddenly face dangerous situations. When very large control areas must be covered, it is necessary to use networks of smart sensors, which allow a capillary control of the situation. Sensors are an important part of the technology behind modern security systems. As processing power and software capabilities continue to increase, smart sensors will also keep pace in the security market.

7.3.3.4 Industrial IoT

Collecting and analyzing all data coming from production sensors is for the industrial market a new key to competitiveness. It means being able to analyze all the process variables, such as energy consumption, temperatures and pressures, speed, and all the measurable physical quantities within the production cycle. The IIoT is able to connect every process line, every production phase, and every single machine thanks to different types of sensors positioned at sensitive and critical points [126]. Thanks to the data generated by the objects and transmitted to the system, it is possible to obtain an accurate monitoring of the plant, the quality of the system, energy efficiency, and timely feedback criticalities. The management and integrated analysis of data coming from sensors allow successful management of real-time monitoring, alerting, energy management, quality prediction, and predictive maintenance.

7.3.4 Conclusions

Over the last decade, the term *smart sensor* has become increasingly widespread. This is the result of the development of new technologies capable of implementing communication and data processing, thanks to increasingly complex digital systems. A smart sensor is an intelligent sensor, which is a sensor that is not only capable of detecting electrical, physical, or chemical quantities, but also capable of reprocessing the information that is collected and then transmitting it in the form of an external digital signal. A smart sensor includes an interface designed for communication and an analog-to-digital converter. These elements are essential for proper functioning.

IoT and IIoT seek to allow individuals to interact and connect with electronic tools and sensor networks. The IoT is considered a very important basis for monitoring all services related to the automation of processes as regards the sphere of waste reduction and environmental control and finally to expand the quality of life in the workplace. Conceived to be able to increase efficiency, smart sensors are designed to ensure four main features: Advanced sensitivity allows the various anomalies present during operation to be detected in a minimum time; smart tasks, on the other hand, have the function of processing data directly from the sensor, making data transmission fast and efficient; efficient communication allows a bidirectional data exchange between the sensor and the control unit; finally, diagnostics have the function of recognizing anomalies present in the system and at the same time carrying out preventive maintenance that bases interventions only on real needs and requirements. In this way, maintenance costs are halved. Errors can be identified very simply, thanks to the display modes that guarantee minimum resolution times.

In science fiction, the future of humanity has often been imagined with robots, and they will surely soon be part of our daily life, which is partially already happening. Obviously, cinema loves fiction, especially when it comes to a subject such as science. But even before robots, what has changed our lives are sensors, and in the next few years their use will become widespread in our lives. Almost all of us have occasionally checked the ambient temperature on a smartphone or used a device to check how many steps we have taken in a day. Cars available on the current market are equipped with sensors that allow us to check tire pressure without getting out of the vehicle. Medical devices allow us to monitor the oxygen level of the blood or blood pressure without having to go to a doctor. The use of sensors in medicine will allow human beings to do more prevention and to use telemedicine where health systems do not allow widespread coverage of the territory. That is why improving public understanding and perceptions of sensor science, through education and communication, is fundamental.

The seed technologies are now being developed for a long-term vision that includes smart sensors as self-monitoring, self-correcting and repairing, and self-modifying or evolving not unlike sentient beings. The ability for a system to see (photonic technology), feel (physical measurements), smell (electronic noses), hear (ultrasonics), think and communicate (smart electronics and wireless), and move (sensors integrated with actuators) is progressing rapidly and suggests an exciting future for sensors.

7.4 The space sector: current and future prospects

Javier Ventura-Traveset[1]
[1]European Space Agency, Centre Spatial de Toulouse, 18, Avenue Edouard Belin, 31401 Toulouse, France

In this section, we review the current status and future prospects of several key domains within the space sector and their associated technologies. These include space science (a golden age), human and robotic exploration, climate change and earth monitoring, and satellite navigation. We complete the section with a discussion of the problems and perspectives of space debris mitigation, the prospects of planetary defence, and the ongoing NewSpace revolution.

7.4.1 The space sector today: a quick balance

The space sector today is a mature and diversified sector without which the modern world, as we conceive it today, simply could not exist. As of 2022, and according to data from the UN Office for Outer Space Affairs, 10 countries (Europe grouped as one) have the capability to put satellites into orbit independently, and about 85 countries have (or have had) their own national satellites [127]. Currently, over 9000 operational satellites orbit our planet, and the expectation is that tens of thousands of new satellites will join them in just a few years, including the ongoing and planned megaconstellations for broadband Internet services, a revolution in the space telecommunications sector.

In 2020, the direct economic impact of the space sector was estimated at around $366 billion [128], having almost doubled in a period of 10 years. Interestingly, only around 25% of that corresponds to institutional expenditure, incurred by government agencies. The remaining 75% comes from the private sector; this is a fundamental paradigm shift experienced by the space sector in the last 30 years, with a major increasing trend, as we will discuss later.

Taken in a wider scope, space economy can be defined as 'the full range of activities and the use of resources that create and provide value and benefits to human beings in the course of exploring, understanding, managing and utilising space' [129]. If we consider this global definition, the figures are much higher and actually difficult to compute, given the high penetration of space activities in our society. For example, it is estimated that more than 10% of the gross domestic product (GDP) of the EU currently depends on the availability of satellite services, a figure that is far higher than the sector's direct turnover, and that a space blackout would imply a loss of between 500 000 and 1 000 000 jobs on the European continent [130].

A detailed analysis of the space economy also reveals an aspect of particularly high interest in this sector. Indeed, if we examine the overall space-related economic volume, we can conclude that the actual expenditure in manufacturing and launching satellites accounts for about only 6% of the total business generated by this sector. This actually means that investment in space infrastructures provokes a major multiplication effect in the global space economy, about 15 on average. This is the case, for example, for the largest European space institutional systems, such as Galileo, Copernicus, and the meteorological Meteosat and Metop satellites, whose actual impact on the overall economy versus their development cost is even greater.

Beyond these remarkable commercial and social impacts, the contribution of space to the growth in our scientific knowledge has also been (and will continue to be) extraordinary. The data from our space science missions have contributed to remarkable breakthroughs during the last 50 years, notably in the fields of fundamental physics, solar system science, astrometry, astronomy, and astrophysics, contributing towards a much better understanding of our universe and of the fundamental physical laws governing it. Linked to space exploration, significant research advances have been attained in fields such as space medicine, space life sciences, biotechnology, space material science, microgravity fluid physics, radiation, space weather, aerodynamics, space geosciences, and so on. Moreover, linked to earth science research, we could mention the advances in geodesy, geophysics, volcanology, geochemistry, meteorology, and oceanography; the understanding of earth magnetic field dynamics, the lithosphere, and the Earth's interior; hydrology; biodiversity; and certainly, the understanding of trends in climate change trends, its sources, dynamics, and the major anthropogenic impacts.

In the following sections, we will briefly review the current status and future prospects of some of the main space sector domains, covering the topics of space science, human and robotic exploration, climate change and earth monitoring, and satellite navigation. We will complete this chapter by discussing the problematic and future prospects of space debris mitigation, the challenges of planetary defence, and the ongoing revolution in space commercialisation, often referred to as NewSpace.

7.4.2 Space science: a golden age

7.4.2.1 Current prospects (up to 2035)
We live in extraordinary times in space science with a plethora of novel and ambitious missions already planned for this decade, anticipating a scientific knowledge revolution. Three research fields are the focus of particular attention today:

- The observation of our universe through gravitational waves and multi-messenger astronomy.
- The quest for biological activity (life) beyond the Earth.
- The understanding of the dark universe: dark matter and dark energy.

Gravitational waves: a new window to observe our universe
After several decades of effort, the Laser Interferometry Gravitational Wave Observatory detected the first gravitational waves on 14 September 2015 (event GW150914), the ripples in the space–time fabric resulting from a binary black hole merger at about 410 Mpcs (about 1.3 billion light-years away). On that historic day, astronomy, as we perceived it up until then, changed forever. As is often stated, we could now 'add sound to the film of our universe', it having previously consisted only of images, the result of observation in the electromagnetic band. Just two years later,

on 17 August 2017, LIGO and Virgo detectors made another significant discovery when they detected, for the first time ever, the merging of two neutron stars, triggering a kilonova explosion. This event was subsequently observed by numerous telescopes worldwide, generating a total of 84 scientific papers in just one day. Since then, the global gravitational wave detectors network has identified over 80 mergers of black holes, two potential mergers of neutron stars, and a handful of events believed to involve black holes merging with neutron stars. Today, the quest for gravitational waves, is further intensified by the the LIGO–Virgo–KAGRA (LVK) collaboration, with enhanced instruments and various improvements in place, and will be further boosted by the future ESA's Laser Interferometer Space Antenna (LISA) Mission, recently confirmed and planned for launch in 2035. The LISA Mission, consisting of a constellation of three spacecraft, forming an extremely accurate equilateral triangle in space trailing Earth in its orbit around the Sun, will become the first scientific mission ever to detect and study gravitational waves from space. LISA will complement Earth-based gravitational wave detectors by enabling the detection of gravitational waves between 0.1 Hz and 0.1 MHz, a frequency sensitivity which should allow the capture of gravitational waves associated with events as extraordinary as the merger of supermassive black holes. LISA observations will be complemented by other ESA's planned large mission currently under study, NewAthena, scheduled for launch around 2037, and aimed at becoming the largest x-ray observatory ever built: an unprecedented temporal synergy of complementary observers from both Earth and space (please refer to section 2.6 for a detailed discussion).

The dark universe

The European Space Agency (ESA) Planck mission (put into orbit in 2009) has contributed to the most detailed knowledge to date of the cosmic microwave background (CMB) radiation, the radiation left over from the Big Bang, when the Universe was born, nearly 13.8 billion years ago. The CMB shows tiny fluctuations in temperature, anisotropies that correspond to regions that had a slightly different

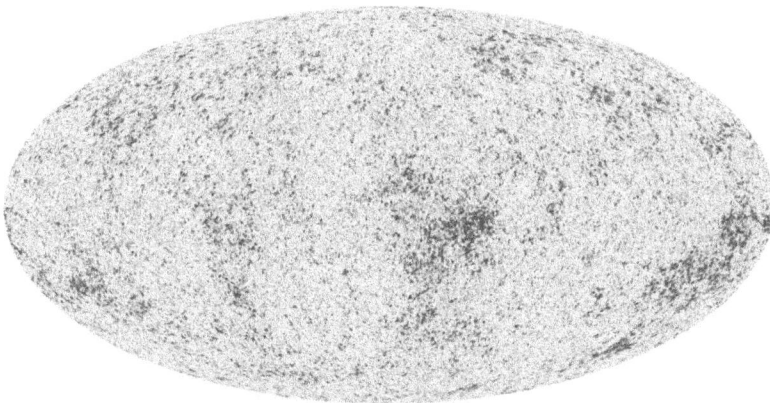

Figure 7.14. Cosmic microwave background image obtained by ESA's Planck telescope. (Credits: ESA and the Planck Collaboration.)

density in the initial moments of the history of the Universe (figure 7.14). Planck has allowed us to refine our knowledge of age, expansion, and history and to extract the most refined values yet of the Universe's ingredients [131].

Thanks to data from the Planck telescope, we now know that conventional matter, of which stars, galaxies, and all baryonic matter are made, constitutes only about 5% of the total mass–energy density of the Universe. Dark matter, which until now could be detected only indirectly through its gravitational interaction with galaxies or clusters of galaxies, constitutes 27%, according to our current estimates, thanks to Planck. On the other hand, dark energy, the mysterious force that we believe to be responsible for the expansion of the Universe, represents, according to our cosmological calculations, 68% [131]. In conclusion, in simple terms, we do not know the composition of 95% of our universe. Space (e.g. through the observations of the recently launched James Webb Space telescope and ESA's Euclid telescope or future planned NASA's Nancy Grace Roman Space Telescope or ESA's NewAthena missions) will provide an unprecedented contribution towards advancing our understanding of this matter over the next 10–15 years (please refer to section 2.5.1 for a detailed discussion).

Alongside gravitational wave observations and the understanding of the dark universe, an essential third pillar emerges strongly in space science in this decade: the search for life in the Universe.

The search for life beyond the Earth
The search for biological activity outside our planet focuses today on three main research axes: the possibilities of finding traces of life on the planet Mars, the possibility that life may have appeared in the subsurface oceans of the icy moons of Jupiter or Saturn, and the search for traces of biological activity outside the Solar System, on exoplanets orbiting other stars. The sequence of missions on-going or planned for this decade makes up the largest technological crew of seekers of life beyond the Earth ever, focusing upon those three objectives (please refer to section 4.2 for a detailed discussion).

7.4.2.2 Beyond 2035: space science priority themes
ESA has recently embarked upon the exercise of defining the scientific priorities for the 2030–50 period, a process known as 'Voyage 2050'.

Following a well-established and efficiently proven methodology, the first step consisted of a general call for ideas from the scientific community. This call, launched in 2019, generated over 100 different proposals or white papers [132], which were grouped according to specific scientific topics. Next, a defined Senior Scientific Committee was tasked with recommending the top-priority scientific topics for future ESA medium and large class missions, recommendations that were formally adopted by ESA's Science Programme Committee in June 2021 [133].

As a result, three identified top-priority scientific topics for future large missions have been proposed [133]:

- Research on giant planet moons, notably researching their potential habitability and capitalising on past Cassini-Huygens and future JUICE (Jupiter Icy Moons Explorer) missions. As noted by the science programme committee, these missions could include landers or drones for *in situ* research.

- Temperate exoplanets, proposing the detailed characterisation in the mid-IR domain of temperate exoplanet atmospheres in order to assess their habitability conditions.
- Development of new physical probes to study the early universe, which could include improving gravitational wave detection capabilities and expanding their frequency ranges, or the development of high-precision spectroscopy technologies to improve the measurement precision of the cosmic microwave background, capitalising on future LISA and past Planck missions, respectively.

Concerning medium-class missions, themes across all domains of space science have been proposed, including solar system research, fundamental physics, astrometry, astrophysics, and new astronomic observatories.

Following the same approach that was implemented for Cosmic Vision, this selection of topics was complemented by the identification of the technologies that would be needed to enable those missions. These include key technologies such as advanced x-ray interferometry for the analysis of astronomic compact objects, the development of more efficient and innovative power sources for solar system exploration, and improvement of the technologies associated with the storage of cryogenic samples for future sample return missions in the Solar System.

The identification of scientific topic priorities and the development of the associated technologies will allow ESA to launch individual calls for mission proposals during the coming years, which will detail, in turn, the new space science agenda for Europe in the 2030–50 period.

7.4.3 A new era for space exploration

7.4.3.1 Current prospects (up to 2035)
Since 19 December 1972, when the astronauts of the Apollo 17 mission returned to the Earth, no human being has left low-Earth orbit again. This is going to change during the current decade, and we may also witness a major revolution in space exploration during the next three decades, as we will discuss here.

The space exploration agenda for the next 10–15 years has been defined around four complementary elements:

1. To continue exploiting the ISS until at least 2030, the nominal date currently planned for its conclusion, maximising cooperation with private companies and facilitating the development of new business models.
2. To build a cislunar orbital station, the Gateway, which is the result of international collaboration by the main agencies of the world and is accessed through the Orion spacecraft of NASA and ESA.
3. To conduct robotic and manned lunar surface missions and the emergence of a lunar economy.
4. To perform a round-trip mission to Mars, the Mars sample return, bringing back samples of rocks from the red planet to the Earth for detailed scientific analysis.

These four phases are discussed briefly in the next part of this section.

The evolution of the ISS

The ISS is probably the most complex space project in history. A total of 15 countries have contributed to its construction, with the collaboration of the main agencies in the world. The ISS station has been inhabited without interruption since November 2000. The outcome of these 20 years of exploitation of the ISS is very positive, with important scientific advances in human physiology, molecular biology, biotechnology, materials science, fluid physics, combustion field, Earth observation, and fundamental physics.

The current proposal is to extend the useful life of the ISS until 2030, with a transition plan that enables its financial sustainability thanks to the additional contribution of private funds. Among the options that NASA is considering is the possibility of including new infrastructures, which are developed and operated privately. The ESA shares this vision and has already begun partnership activities in this regard. A good example of this are the IceCubes and Bartolomeo initiatives for experimentation marketed in microgravity conditions within the Columbus pressurised module or on the outside of it in outer space conditions, respectively (figure 7.15).

Returning to the cislunar orbit

The evolution of the ISS and its transition towards a more commercial model should allow the main space agencies to focus on new challenges beyond the Earth's orbit, with the Moon now being the natural intermediate step before a manned mission to Mars could be contemplated.

The lunar exploration roadmap for this decade is essentially defined, based on an open architecture around two main elements:

Figure 7.15. Europe's Columbus space laboratory. Image taken by ESA astronaut Luca Parmitano during his spacewalk on 9 July 2013. (Credit: ESA.)

1. A crew transport vehicle, the Orion spacecraft, capable of carrying a crew of up to six astronauts.
2. The development of a lunar station in retrograde cislunar orbit, the Gateway station, because of international collaboration, led by NASA.

The first mission of the Orion spacecraft, Artemis-1, was launched on November 16th, 2022 and had a duration of 25.5 days (figure 7.16). Passing as close as 130 km from the lunar surface, the spacecraft utilized the Moon's gravity to propel it into lunar orbit, and subsequently adjusted its trajectory to return to Earth. The first lunar flyby occurred on November 21st, with ESA's Service Module (ESM) activating its main engine to maneuver Orion behind and around the Moon. Ten days after liftoff, Orion entered the Moon's orbit on November 25th, when the ESM fired its main engine. Following a flight of 2.3 million km, the Orion capsule splashed down in the Pacific Ocean west of Baja California on December 11th, 2022. [134].

The Artemis-2 mission, an already manned spacecraft, completing a slightly different flight path, will follow the Artemis-1 mission. Artemis-2 crew members will reach a distance of 70 000 km beyond the Moon, the longest distance from the Earth that a human has ever traveled, before completing a lunar flyby and returning to the Earth. Artemis-2 is today scheduled to launch not earlier than September 2025, while Artemis III, aiming to land the first astronauts near the lunar South Pole, is today scheduled for September 2026. Artemis IV, the inaugural mission to the Gateway lunar space station, is today scheduled for 2028. The Artemis programme will establish the foundation for long-term scientific exploration at the Moon and will boost the emergence of a new lunar economy.

Figure 7.16. Orion crew transport vehicle (credits: ESA–D Ducros).

Figure 7.17. Future Gateway cislunar station. (Credits: Thales Alenia Space/Briot.)

The second element in the lunar exploration strategy, the Gateway space station, is conceived as a multimodular station, with six modules (figure 7.17). With a mass of about 40 tonnes, the Gateway station is planned to include a service module, a communications module, a connection module, a spacewalk access hatch, a habitable module, and an operations station to command the future robotic arm of the station or future lunar rovers. Europe, through the ESA, will contribute to the Gateway station with the ESPRIT (European System Providing Refuelling Infrastructure and Telecommunications) modules and the I-HAB (International Habitational module), the module that will constitute the main habitat for astronauts when they visit the Gateway station.

Astronauts are expected to inhabit the Gateway station for up to a total of 90 days consecutively. Gateway will open up enormous possibilities for future space exploration and will enable advanced technological testing and scientific advances. Gateway will also allow the remote control of robotic missions on the lunar surface and facilitate access to future manned missions. Carrying out such missions will represent the future of international collaboration and the cooperation of private industries.

Together with the contribution to the Orion spacecraft and the Gateway station, two additional axes currently form the basis of the European contribution to lunar exploration during the next 10 years:

- The Argonaut EL3 (European Large Logistic Lander) logistics vehicle, being developed as a versatile means of accessing the Moon. The ESA Argonaut programme will provide a series of recurrent landers dedicated to delivering cargo and infrastructure to support scientific operations on the lunar surface, including rovers and power stations. The first mission, Argonaut-1, is projected to land on the Moon around 2031, with subsequent Argonaut landers planned for launch approximately every two years thereafter. Argonaut, together with the Moonlight service (referenced below), is poised to serve as a significant European contribution to NASA's Artemis lunar exploration program, providing capabilities for cargo delivery and scientific endeavours.

- The Moonlight/LCNS system, consisting of a miniconstellation around the Moon to provide a permanent satellite navigation and communication service to future lunar exploration missions [134]. Moonlight is expected to be able to offer its initial operational services in 2027–28 and its final operational capability around 2030. The Moonlight infrastructure will support both current and future generations of institutional and commercial lunar explorers, presenting a unique opportunity for European industry to assume a prominent role in the future Lunar Economy. The significance of the Moonlight program is enormous, potentially becoming the first ever extraterrestrial human infrastructure that provides commercial communication and navigation services: an extraordinary paradigm shift in the field of human exploration.

These technologies will undoubtedly contribute to the sustainable development of future recurrent robotic and manned missions to our satellite. They will, in turn, lay the foundations for the development of the necessary technologies for future manned missions to Mars, including protection against space radiation for human beings, advanced life support systems, and the possibility of exploiting lunar resources for stable human settlements. We can anticipate that this will also bring many benefits to our planet and will open up new commercial and business opportunities.

Mars sample return mission

Despite the large number of scientific missions to Mars to date, none has yet made the return trip to the Earth. Before we may consider a manned Mars return mission, it seems natural to implement that intermediate step with a robotic mission. That is the goal of the future Mars sample return mission: to make that journey to Mars and back, bringing samples from Mars to the Earth for in-depth scientific analysis. This mission, in addition to its extraordinary scientific value, should allow us to develop several of the technologies required to plan future manned missions to the Martian surface (figure 7.18).

But bringing Martian samples to the Earth is not an easy task. The current concept identifies the need to perform a minimum of three missions from the Earth and the launch of a rocket from the Martian surface, something that has never been done before.

Due to its nature, this mission also faces some key challenges concerning the environmental and biological protection of the Martian samples and Mars. For example, the perfect preservation of Martian samples is an essential element, minimising any possible organic contamination or chemical alteration of them. At the same time, all spacecraft destined for the Martian surface must incorporate unique planetary protection requirements [136].

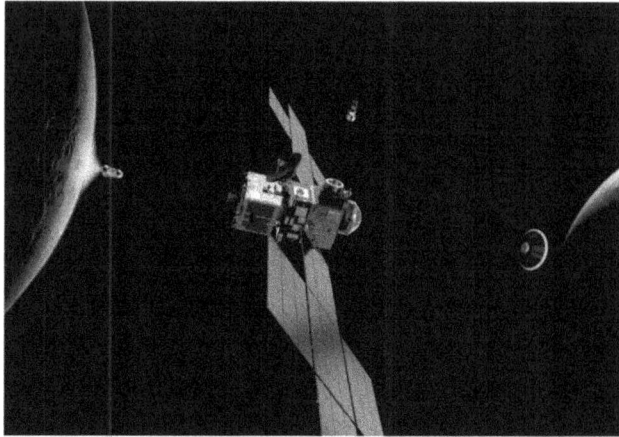

Figure 7.18. Mars sample return mission. (Credit: ESA/ATG Medialab.)

7.4.3.2 Space exploration beyond 2035

Towards a sustainable human lunar presence and research infrastructure
The idea of a permanent lunar presence on our Earth's satellite was strongly supported by ESA's former Director General, Jam Woerner, through a concept he named *Moonlight village* [137], intended to be developed through private and public cooperation, involving space and nonspace partners. This may become a reality in the next decade (figure 7.19).

During the next 10–15 years, dozens of missions, both institutional and commercial, are already envisaged. These will imply the development of much more accessible regular trips to our Earth's natural satellite and the setting up of a long-term infrastructure facilitating fundamental services such as communication and navigation services. It is also expected that these missions will enable the development of the necessary technologies to support a long-term human presence, such as the construction of radiation-shielding structures built with bulk regolith, regular water generation for life support and rocket propellant production, thermal and electrochemical processes to convert lunar regolith into oxygen and water, the consolidation of 3D printer technologies to build structures from local lunar materials, and even inflatable habitats and the development of regenerative closed-loop systems, such as those being currently studied at the ESA Melissa research infrastructure [138].

On the scientific front, the radio-quiet lunar far side, the so-called shielded zone of the Moon (SZM), may enable unique astronomical investigation, through the deployment of dedicated radio telescope infrastructures, for which specific frequency protection recommendations already exist [139]. China has initiated this research path by setting up some radio astronomy instruments as part of its recent Chang'e-4 lander [140], and several ambitious projects are being conceived today for the next decade.

Figure 7.19. Concept of a semiinflatable lunar habitat near the lunar south pole, designed by Skidmore, Owings and Merrill. (Credit: SOM.)

Future Mars human exploration: key technology needs

The experience that will be acquired in the next 10–15 years with Moon exploration, the Mars sample return, and associated robotic Mars missions should allow us to plan a crew mission to Mars in the approximate timeframe of 2035–45.

Some of the technologies needed for this human mission to Mars have been identified as part of the Global Exploration Roadmap [141]. This document is the result of the cooperative effort of 14 space agencies, with European contributions from ESA, CNES, ASI, and the UK Space Agency. Some of the identified key technologies are the following:

- Radiation protection technologies, including protecting constructions built with *in situ* regoliths and radiation-shielding technologies for human protection to sustain a long-term human presence.
- Advances in microgravity countermeasures, including the development of compact devices to limit microgravity disorders and possible medical countermeasures.
- Enhanced reliability life support systems, eliminating the dependence on Earth supply logistics and increasing systems autonomy, failure detection capabilities, and in-flight repair capabilities.
- Advances in human surface suits and EVA mobility, aimed at providing extended thermal, radiation, and vital life support.
- New propulsion technologies, based on the exploitation of *in situ* liquid oxygen–methane propellant production.
- Dust mitigation technologies to support both long-term lunar and Mars missions.
- Advances in autonomous systems enabling crew operations without the need for Earth-based support.
- Exploitation of *in situ* resources for life support, including O_2/CH_4 generation from the atmosphere and LOX/LH2 generation from soil and the

exploitation of many other chemicals and minerals to support a sustained long-term human presence.

- Advances in solar arrays and the development of fission power for surface missions' energy autonomy.

As space history has already demonstrated on multiple occasions, this focus and colossal planned technological effort towards achieving an enhanced human exploration capability in the Solar System will also bring extraordinary benefits to our life on the Earth, such as new technological capabilities with applications on the Earth, improved generation and efficiency in the use of the Earth's resources, and major advances in medical and life support technologies.

Towards a new space industry based on lunar and asteroid mining

It is known that the Moon and other celestial bodies may contain materials of great value for future space exploration or for use on the Earth. In the case of the Moon, the presence of ice in the polar regions could enable future exploration missions to produce oxygen, drinking water, or lunar rocket fuel. The Moon also contains minerals such as titanium, iron, and aluminium and an abundant presence of helium-3 from the solar wind.

Beyond the Moon, serious consideration is now also being given to the possible future mining exploitation of asteroids. Some asteroids may contain gold, silver, platinum, nickel, or cobalt, which some analysts believe could generate an extraordinary commercial opportunity if access and exploitation costs are lowered.

7.4.4 Space and climate change

The evolution of our planet and climate change are arguably recognised today as the most important global challenges. This is clearly reflected in the 2021 Annual Global Risk Report of the World Economic Forum, identifying extreme weather events, the inability to mitigate climate change trends, and anthropogenic effects on the environment as the three most important global risks [142].

The global monitoring of our planet and the continuous assessment of the effectiveness of future mitigation measures therefore becomes vital. Already today, Earth observation (EO) satellites represent more than a third of all operational satellites orbiting the Earth [143]. In the case of Europe, for example, about 30% of ESA's overall annual budget in 2024 is devoted to EO missions [144], including meteorological and scientific missions and the Sentinel satellites of the EU Copernicus Programme.

This proliferation of EO missions is essential for global, continuous, and long-term data relating to the so-called essential climate variables [145], defined through the Global Climate Observing System, jointly sponsored by the World Meteorological Organisation, the UNESCO Intergovernmental Oceanographic Commission, the UN Environment programme, and the International Science Council. The monitoring of these essential climate variables is key to understanding the overall health status of our planet, diagnosing and predicting its short-, mid-, and long-term evolution and to supporting government's decisions for its mitigation and

Figure 7.20. Approximately 50% of essential climate variables require satellite monitoring for global coverage. (Credit: ESA.)

risk management. In the case of Europe, this continuous monitoring will also be key in supporting the EU objective to be climate-neutral by 2050 [146], regularly assessing the current status, trends, and efficiency of the measures implemented, fully adhering to the European Green Deal strategy [147].

A total of 54 essential climate variables have been defined, grouped into three main categories: atmospheric, oceanic, and terrestrial [145]. Of these, approximately 50% require satellite measurements for their global assessment (figure 7.20).

This global and continuous observation is precisely one of the key objectives of the EU's Copernicus programme, which is currently providing the largest volume of Earth observation data worldwide. The space component of Copernicus includes the ESA-developed Sentinel satellites (the core of the programme) and other national and international contributions. All of these satellites provide essential data for the Copernicus Services, addressing challenges such as urbanisation, food security, rising sea levels, diminishing polar ice, natural disasters, and, of course, climate change [148].

There are currently eight Sentinel satellites in orbit and multiple national contributions (figure 7.21). To these we must add an extensive portfolio of more than 30 European Earth observation missions, which have already been defined for this decade, including the recently selected 7–12 Sentinel Mission families [149]:

- Sentinel 7 (Copernicus Anthropogenic Carbon Dioxide Monitoring, CO2M)
- Setinel 8 (Land Surface Temperature Monitoring, LTSM)
- Sentinel 9 (Copernicus Polar Ice and Snow Topography Altimeter, CRISTAL)
- Sentinel 10 (Copernicus Hyperspectral Imaging Mission for the Environment, CHIME)

Figure 7.21. The future Sentinel 7, Copernicus Anthropogenic CO2 emissions monitoring mission. (Credit: ESA.)

- Sentinel 11 (Copernicus Imaging Microwave Radiometer, CIMR)
- Sentinel 12 (Radar Observing System for Europe - L-Band, ROSE-L).

These satellites are equipped with advanced instruments and sensors, including sea, land, and surface radiometers; altimeters; multiple-band spectrometers; synthetic aperture radars; photometers; multispectral and thermal cameras; trace gas detectors; atmospheric aerosols; very precise orbit determination devices; and so on. Their combined use enables a continuous and fine global analysis of our planet in all of its dimensions, namely, the atmosphere, hydrosphere, cryosphere, biosphere, and lithosphere, allowing us to distinguish between natural events and anthropogenic effects, which are the consequence of human activity.

The longer time period that these satellites will permit will enable the analysis of decadal trends. This is the purpose of ESA's Climate Change Initiative [150], which is aimed at generating accurate and long-term satellite-derived data to characterise the long-term evolution of the Earth's system. Further, these technologies will enable us to address issues closely linked to societal concerns such as the availability of food, water, energy, resources, public health, continuous climate monitoring, and the evaluation of the effectiveness of related policies to confront those.

7.4.4.1 Moving towards a scientific understanding of Earth dynamics
The Copernicus programme is currently the most ambitious Earth observation programme worldwide and a source of pride for Europe. Along with continuous monitoring, it is essential so as to have a deep scientific understanding of the critical variables of the Earth's system and to develop and test new technologies for potential use in future Sentinel satellites. This is precisely the objective of the ESA's Earth Explorers Programme: to contribute towards a significant improvement in our fundamental knowledge of the five main Earth science disciplines: atmosphere,

cryosphere, land surface, ocean, and solid Earth. As of June 2023, this programme was composed of 12 Earth explorers, among which five were put in orbit between 2009 and 2018:

- GOCE: ESA's gravity mission (launched in 2009; ended in 2013).
- SMOS: ESA's water mission (launched in 2009; mission extended until 2025).
- CryoSat: ESA's ice mission (launched in 2010; mission extended until 2025).
- Swarm: ESA's magnetic field mission (launched in 2013; mission extended until 2025.).
- Aeolus: ESA's wind mission (launched in 2018; ended in 2023).

Seven others are scheduled for launch during the next 10–15 years:

- EarthCARE: ESA's cloud and aerosol mission, planned for launch in 2024.
- Biomass: ESA's forest mission, planned for launch in 2024.
- FLEX: ESA's photosynthesis mission, expected to be lunched in 2025.
- FORUM: ESA's planet radiation budget mission.
- Harmony: a constellation of two SAR (Synthetic Aperture Radar) equipped satellites that will orbit in formation with one of the Copernicus Sentinel-1 satellites to address key scientific questions related to ocean, ice and land dynamics.
- Earth Explorers 11 and 12: selection process ongoing.

7.4.4.2 Moving towards an accurate Earth digital twin

The continuous provision of Earth observation satellite data and an improvement in the understanding of the scientific principles governing the Earth's dynamic natural processes can contribute to the development, during this decade, of what has been named the Digital Twin Earth: a reliable digital replica of our planet which accurately mimics the Earth's behaviour and enables us to anticipate climate change evolution for the coming decades (figure 7.22).

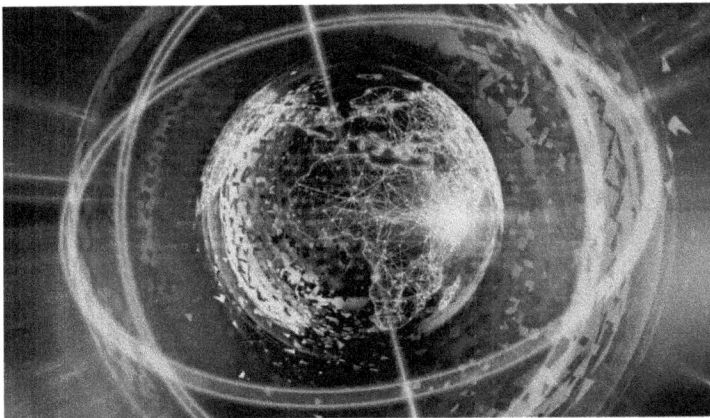

Figure 7.22. A digital twin Earth should allow us to reliably monitor and forecast the effects of natural and anthropogenic events on our planet. (Credit: ESA.)

A digital twin Earth should allow the monitoring and forecasting of natural and anthropogenic effects on our planet. The use of machine learning and AI technologies may prove to be necessary in order to obtain an accurate digital representation of our planet, notably for extreme weather events and accurate numerical forecasting models. This model could be extremely helpful in performing detailed simulations of the interaction between Earth's interconnected systems and human behaviour, supporting the field of sustainable development.

7.4.4.3 ESA and NASA strategic partnership for Earth monitoring
In July 2021, ESA and NASA formalised a strategic partnership for Earth science and climate change [151]. Through this alliance, the two agencies will join forces in Earth science observation satellites, research, and applications to improve the monitoring of the Earth and its environment. This strategic agreement may also set the standard for future international collaboration, encouraging all key space stakeholders to actively share relevant satellite data in fighting against the extraordinary challenge of climate change.

7.4.5 Satellite navigation

7.4.5.1 Current prospects (up to 2030)
Knowing where we are and our time reference are two inherent measures of our human existence, linked to most of our daily activities. Not surprisingly then, satellite navigation systems have become essential in our societies, and this technology is considered by many as the 'fifth utility', along with water, electricity, gas, and telecommunications. All sectors of the economy benefit from it today: transport, energy, tourism, agriculture, fishery, livestock, civil engineering, telecommunications, the financial sector, and so on; around 40 000 different applications have been identified so far.

Because of this extraordinary economic and strategic importance, each relevant space nation in the world has or will soon have its own global or regional satellite navigation system. Altogether, over 130 operational navigation satellites are currently in orbit, of which over 100 correspond to the four GNSS constellations (figure 7.23).

Current estimates indicate that around 10% of the European Union's GDP depends, to a greater or lesser extent, on the availability of satellite services. This economic dependence was central to Europe's decision to implement its own navigation systems, namely, EGNOS and Galileo, which are described next.

The European Geostationary Navigation Overlay Service
The European Geostationary Navigation Overlay Service (EGNOS) is Europe's regional satellite-based augmentation system. It is used to improve the performance of GPS (today) and of Galileo (in the near future) for safety of users. EGNOS provides an augmentation message to GPS L1 frequency by providing corrections and integrity information to GPS, and the ionosphere delays affecting the users. The EGNOS system began its open service in 2009 and was certified for civil aviation in 2011. The next generation of EGNOS, EGNOS V3, planned to be operational

Operational* GNSS Satellites
MEO only

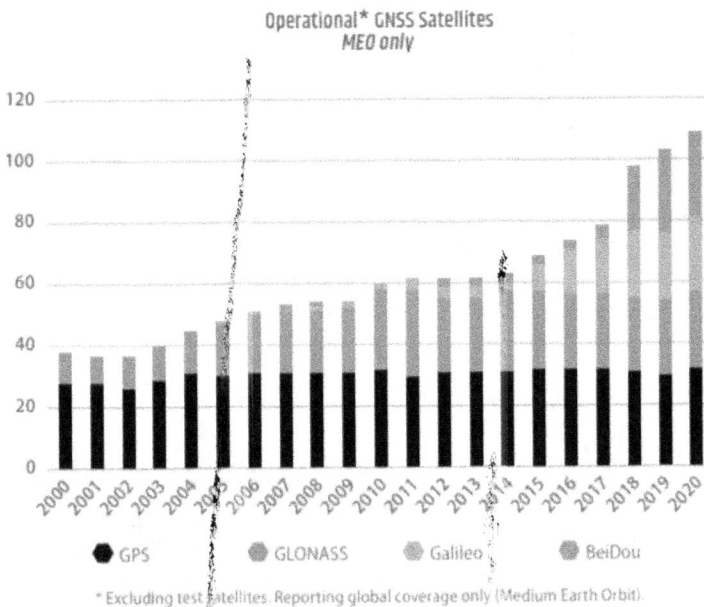

Figure 7.23. Evolution in the number of operational MEO satellite navigation satellites.

around 2028, will augment both **GPS** and Galileo constellations across their two operational frequencies: L1 and L5 bands, becoming the first multi-constellation and multi-frequency satellite based augmentation service.

The European Galileo system

At the end of the 1990s, in full development of the EGNOS system and thanks to the knowledge that had being acquired by the European industry in these technologies, the interest in designing a global satellite navigation system under European control arose. Several system studies were then launched, culminating in a system of 30 satellites at an altitude of 23 222 km and distributed between three orbital planes. This system, initially known as GNSS-2, was officially renamed the Galileo system in 1999. Since 15 December 2016, Galileo has officially been an operational system.

Today, Galileo is Europe's own GNSS, providing a highly accurate, guaranteed global positioning service under civilian control. By offering dual frequencies as standard, Galileo is able to deliver real-time positioning accuracy down to the metre range (figure 7.24).

Galileo features several technology and system differentiators, which allow it to reach higher levels of precision than other GNSS systems. First and foremost, Galileo satellites now include four clocks based on two different clock technologies: two rubidium atomic frequency standard (RAFS) clocks and two passive hydrogen maser (PHM) clocks. In the PHM clocks, molecules of hydrogen are dissociated into atomic hydrogen, entering a resonance cavity by passing through a collimator and a magnetic state selector, which means that only those atoms at the desired energy level enter a resonant cavity, where they tend to return to their 'fundamental' energy state, emitting a microwave frequency with very high stability. Galileo PHM clocks

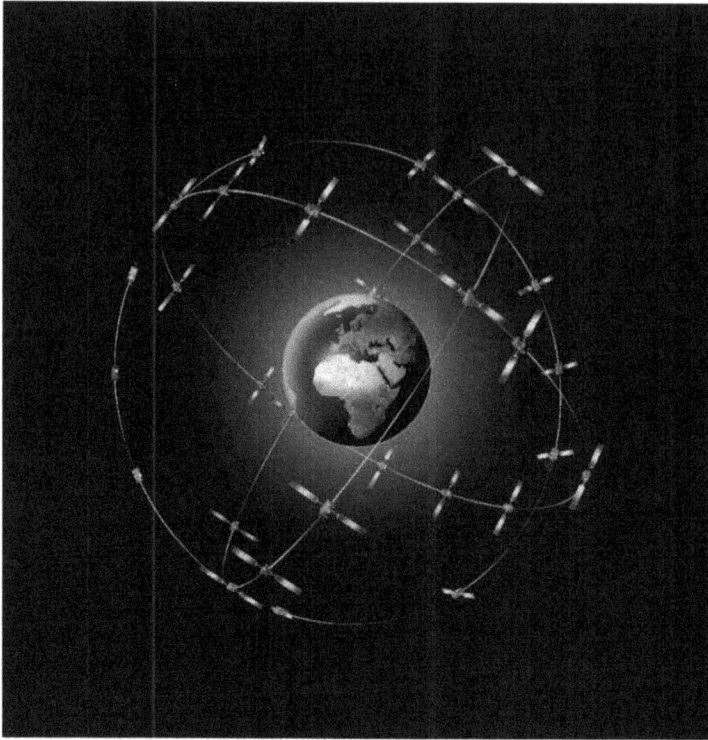

Figure 7.24. The Galileo system constellation. (Credit: ESA.)

are about 1 order of magnitude more stable that the traditional Rb clocks, with stabilities in the range of 1 s in 3 million years. Thanks to a large and well-distributed set of Galileo sensor stations and very accurate modeling of the Galileo affecting nongravitational forces, the Galileo orbits are also computed in real time with very high accuracy, on the order of a few decimetres. Furthermore, the Galileo selected navigation signal modulation, based on what is known as the binary offset carrier family, provides more power at high frequencies away from the centre frequency, thereby providing better ranging performances and a more robust tracking capability in multipath conditions than standard binary phase-shift keying signals [152]. Finally, Galileo satellites are also equipped with on-board laser retroreflectors, which allow periodic high calibration of the Galileo orbits with ground-based laser ranging stations and enhance its scientific utilisation potential.

With 28 satellites in orbit as of December 2023, Galileo allows Europe to have an autonomous and global system, making the continent one of the world's leading players in this field. Galileo currently offers a service of the highest quality, and all the world's leading mobile handset manufacturers include it (figure 7.25).

Beyond its open global service, Galileo also provides a robust and encrypted service for government applications (PRS, or public regulated service) and is currently also the only civilian GNSS including an operational search and rescue service, integrated as part of the COSPAR-SARSAT service [153].

Figure 7.25. Ariane 5 (VA244) launch with Galileo satellites 23, 24, 25, and 26 on board, July 2018. (Credit: ESA.)

Galileo provides a free of charge High Accuracy Service (HAS) by offering high accuracy Precise Point Positioning (PPP) corrections through the Galileo signal (E6-B) and Internet. Galileo HAS offers today real-time improved user positioning perform-ances down to a decimetre level [154]. This new service becomes a major differentiator for Galileo and will prove to be an enormous facilitator in the development of new applications and services that require high precision. Good examples include enhanced drone navigation, augmented reality, autonomous cars, and vehicle-to-everything services, in which connected vehicles will communicate wirelessly with other vehicles and infrastructure to avoid collisions.

Another important differentiator of Galileo today is the provision of signal authentication via its Open Service Navigation Message Authentication (OSNMA) service of free access. This service provides users with assurance that the Galileo navigation message they receive originates from the system itself and has not been altered. [155]. This makes its positioning service robust against possible malicious attacks that could emit signals similar to Galileo signals with the aim of causing errors in the receivers.

Galileo second generation

The European Union and ESA are currently deeply engaged in the development of the second generation of the Galileo system, known as G2G. Galileo G2G will incorporate various new technologies, including the integration of ion propulsion in future satellites, establishment of intersatellite links between Galileo satellites, implementation of several signal enhancements for quicker acquisition and reduced energy consump-tion of receivers, and a heightened level of digitalization of the navigation signal generator payloads to enhance their flexibility and reconfigure their on-board capacity. These advancements will result in new service capabilities, enhancements to the existing service, a more robust and secure service, and reduced operational and maintenance costs of the system. It is envisaged that the first satellites of this Galileo second generation will be deployed into orbit around 2025, with the G2G constellation expected to achieve its initial operational capability before 2030.

7.4.5.2 Satellite navigation: technology prospects beyond 2030

Beyond the development of new generation Galileo and GPS infrastructures, Galileo second generation and GPS III, respectively, which should be in full operation by the end of this decade, there are several technology trends, currently in the research field, which may be operational by the next decade. Some of them could include the following:

- The comprehensive implementation of a specialized low-Earth orbit position, navigation, and time system, referred to as LEO-PNT, presently undergoing thorough evaluation. The envisioned LEO-PNT constellation will facilitate a multilayered 'system of systems' navigation paradigm, wherein signals from medium-Earth orbit are complemented by those from low-Earth orbit (LEO) satellites positioned at altitudes below 2000 km—along with additional inputs from terrestrial PNT systems and user-based sensors.
- The regular operational use in future GNSS satellites of space optical clocks, based on the use of lasers with a frequency stabilised relative to an atomic transition, which could provide clock stabilities several orders of magnitude better than those of today, with a major impact on the final system performance. For example, technologies based on the Doppler-free spectroscopy of molecular iodine are under development [156].
- The complete fusion of GNSS and mobile telecommunications infrastructures in a seamless way, including the full integration of indoor and outdoor navigation services in a unique device. This, together with the full development of sensor fusion technologies with GNSS signals and the application of AI technologies, should make it possible to reach real-time accuracies at a millimetre range and global level.
- The full integration of QT, AI, and cybersecurity technologies as part of future GNSS systems enhancing security and system performances.
- The modernisation of regional augmentation systems conceived to extend the integrity services beyond aviation to all transportation services and for centimetre accuracies.
- Improvements in the Earth International Terrestrial Reference Frame references accuracy, reaching submillimetre level precision.
- The extension of advanced PNT technologies to the field of exploration, starting with the Moon (the Moonlight concept already having been planned for this decade [135]), including the deployment of local lunar differential systems, and then extended to Mars beyond 2030.
- The deployment of GNSS-like nodes in strategic space locations within the solar system, such as the Lagrangian points of the Sun, aiming to enhance deep space Positioning, Navigation, and Timing (PNT) capabilities.
- The potential extension of PNT services to deep oceans by exploiting the use of neutrino beams for navigation and time dissemination [157].

7.4.6 The problematic of space debris

Since the beginning of the space age, according to 2023 data at the ESA, it is estimated that a total of about 17000 satellites have been launched, of which about

11500 are still in orbit; of that number, about 9000 remain active [158]. In addition, the number of events that have produced ruptures, explosions, collisions, or fragmentations in orbit are estimated at more than 640. As a result, the current balance is a total mass in orbit exceeding 11 500 tonnes, with an estimated count of about 36 500 objects measuring 10 centimetres or more, approximately 1 million objects ranging between 1 and 10 centimetres, and roughly 130 million objects sized between 1 millimetre and 1 centimetre [158]. Of all these objects, currently, we are monitoring approximately 48,000, primarily through the US catalogue of the Space Surveillance Network [159].

Nowadays, the problem is further aggravated in the LEO orbit due to the dramatic increase in the number of small satellites launched into near-Earth orbit during the last 10 years, owing to the emergence of cubesats and large satellite constellations, as illustrated in figure 7.26.

As a result, the risk of collision with functioning satellites is increasing, notably in low-Earth orbit, where a large number of Earth observation and climate change monitoring satellites operate. Unfortunately, we have already seen some real examples.

On 23 August 2016, after about two years of operations, we observed a sudden loss of power on our Copernicus Sentinel Satellite 1A. Our investigations revealed that this loss of power was preceded by a slight change in the attitude and orbit of our satellite [160]. All this made us suspect a possible collision with an object in orbit. Our analyses confirmed this, and we concluded that the collision was caused by an object of only 1 cm in length and 0.2 g of mass, which had collided at a relative speed of about 11 km s^{-1} with one of our solar panels (figure 7.27).

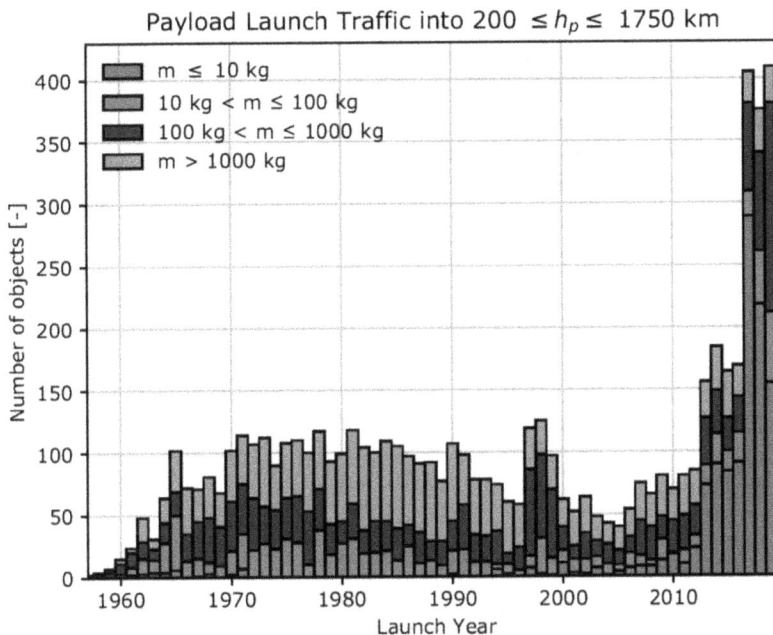

Figure 7.26. Number of satellites launched below 1750-km orbit and their associated mass [56]. (Credit: ESA.)

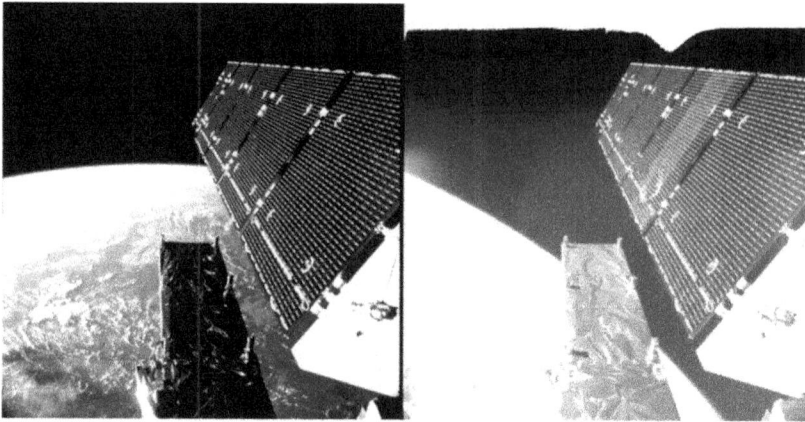

Figure 7.27. One of the sentinel 1A satellite's solar panels before and after a collision with an object of space debris in orbit. (Credit: ESA.)

Therefore, we clearly see that the problem of space debris can have an impact on the availability of our critical satellite infrastructures.

The data are very worrying and require a clear sustainability policy. According to our ESA estimates, an uncontrolled long-term evolution of the space debris environment could lead to a cascading effect, rendering the orbits between 900 and 1400 km unusable for operational services. This effect, known as the Kessler syndrome [161], has been demonstrated through simulations in which fragments currently in orbit collide over time with larger objects, resulting in more fragments, causing more and more collisions. A cascading effect would have catastrophic consequences for satellites in low orbit, which could be rendered useless, and certainly for our planet (figure 7.28).

International cooperation between the different agencies and governments is, thus, essential. An important step in this direction is the effort invested by the Inter-Agency Space Debris Coordination Committee (IADC), an international governmental forum that was specifically established for coordination on this issue. Through the IADC, a specific code of conduct has been defined, and different axes of action are proposed to mitigate this problem [162].

Along with these operational, legal, and regulatory measures, it is essential to analyse strategies for reducing the number of the most dangerous passive objects that remain in orbit. In this regard, it is of interest to highlight the ESA Clean Space initiative, which, among its many proposals, includes the first demonstration mission for the active elimination of space debris: the ClearSpace 1 mission, currently planned for launch around 2026 [163].

7.4.7 The prospects of planetary defence

We can all remember the impact, on 15 February 2013, of a meteorite striking the Earth in the Russian region of Chelyabinsk, leaving more than 1500 people injured and affecting six cities in the region. The destruction, according to the best estimates,

Figure 7.28. Artist's impression of space debris in low orbit. (Credit: ESA.)

was the result of the impact of a meteorite just 20 m in diameter that had not been monitored in advance [164].

Although the frequency of large asteroid impacts is not very high, the consequences may be of catastrophic dimensions. To deal with this risk in an efficient way, we need to develop technologies for our planetary defence, which we consider to be within reach over the coming decades.

The first obvious step is to be able to achieve accurate monitoring of near-earth objects, which are likely to involve a risk of collision with our planet [165]. The second planetary defence security objective, as defined by ESA, which is undoubtedly much more ambitious, would be to have the technological capacity to be able to deflect asteroids up to 1 km in diameter if they were identified more than 2 years in advance. This is certainly a great technological challenge, but it could be achieved in the next 30 years.

In preparation for this, the ESA is currently implementing the Hera mission. Hera is the European component of a full-scale planetary defence mission, complementing NASA's Double Asteroid Redirection Test (DART) mission in its analysis of the Dydimos dual asteroid system, composed of two asteroids measuring 780 m and 160 m in diameter (figure 7.29).

NASA's DART spacecraft successfully executed a kinetic impact on the smaller asteroid, known as Dimorphos, on September 26, 2022, shortening its 11-hour 55-minute orbit around its parent asteroid Didymos by approximately 33 minutes and altering and remodelling its entire shape. This marked the world's first planetary defence technology demonstration. ESA's Hera mission, scheduled for launch during 2024 and estimating to reach the asteroid system in 2026, will conduct a thorough post-impact survey, becoming Europe's flagship Planetary Defender.

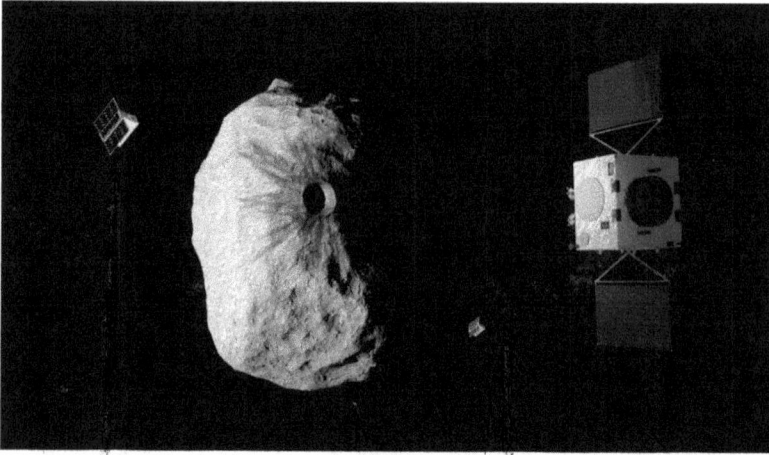

Figure 7.29. ESA's future Hera mission, aimed at contributing to the development of some of the basic technologies for future planetary defence. (Credit: ESA.)

The ultimate goal of this joint mission is to validate the basic technologies necessary for the development of a reliable planetary defence strategy that could be implemented in the forthcoming decades.

7.4.8 The prospects of NewSpace

Around 75% of the world's space economy is today of a private nature, with a turnover of approximately 270 billion euros (data from 2021), the downstream space business being responsible for more than 95% of that figure [166]. The current dynamic of the space industry is enormous, having doubled its turnover over the last 10 years. But this is just the beginning. Several organisations have conducted prospective studies on the future scale of the space economy, projecting figures ranging from 1 trillion to 2.7 trillion USD for the period spanning 2040–2045 [167].

An important part of this revolution is linked to the advances in digitalisation technologies and to the miniaturisation of components, which, together with an extraordinary reduction of the launch costs, makes the space sector accessible to newcomer industries: a new paradigm often referred to using the term *NewSpace*. This extraordinary transformation of the space industry is enormous, having attracted over 14.8 billion euros in private investment between 2000 and 2018 [168].

An integral aspect of this NewSpace is the emergence today of megaconstellations, comprising networks of hundreds, thousands, or even tens of thousands of small satellites. Private initiatives like SpaceX's Starlink and Amazon's upcoming Project Kuiper now offer broadband internet services, facilitating streaming, video calls, online gaming, and more. This has resulted in a completely new paradigm and revolutionized the commercial space communications sector, with a trend of a clear steady growth.

Another sector with great dynamism and from which we can expect another revolution over the coming decades is linked to space tourism.

7.4.9 Summary

The space sector is today a mature and diversified sector, present in all facets of our society and with enormous growth potential. Over 9000 operational satellites orbit our planet today and tens of thousands of new satellites are scheduled to join them in just a few years. Multiple satellite services are now simply essential for the smooth operation of our society and the global economy. In the European Union, for example, it is estimated that more than 10% of its GDP currently relies on the availability of satellite services, a figure far exceeding the sector's direct turnover. Today's space sector dynamic is enormous, and according to some recent forecasts, the space economy could grow by a factor of 10 in 2045, reaching a potential business of about 2.7 trillion dollars [167].

In this section, we have reviewed the status and future prospects of some of the main space sector domains and their associated technologies, covering the topics of space science, human and robotic exploration, climate change and earth monitoring, and satellite navigation.

Multimessenger astronomy, gravitational wave detection from space telescopes, the understanding of our dark universe, the evidence of life beyond Earth, a lunar and Martian sustainable human presence, and regular asteroid mining exploitation are just some of the examples explained here, illustrating the extraordinary prospects of space science and space exploration which could be within reach in the coming decades.

In the applications field, we explained the growing importance of earth observation in the monitoring and mitigation of climate change, our most urgent global challenge today. Scheduled space missions will allow a continuous, global, and complete monitoring of the 54 essential climate variables, providing us with a more complete understanding of climate change trends, their sources, their dynamics, and the major anthropogenic impacts. These precious data, together with a complete scientific understanding of the critical variables of the Earth system, thanks to future scientific satellite missions, and the development of a reliable digital twin Earth will provide us with the tools and knowledge to respond efficiently to our most urgent challenges in terms of food security, urbanisation, rising sea levels, diminishing polar ice, natural disasters, extreme weather events, and, of course, climate change mitigation.

We have also discussed here the current and future prospects of satellite navigations, with the planned development of second- and third-generation global navigation satellite systems, which could allow us to reach millimetre position accuracies in real time and at a global level in the coming decades. We have identified here also some of the future PNT technologies under development, which could allow us to contemplate the extension of navigation services to deep oceans or to the near Solar System, with dedicated infrastructures around the Moon and Mars, already planned today.

But the growth of the space sector also implies the need to take effective and urgent measures against space debris. Major improvements in space debris monitoring, international cooperation, the development of effective legal and regulatory measures to avoid new debris, and new technologies for the active reduction of the most dangerous passive objects or automatic collision warnings on-board our future satellite, are some of the expected advances for these decades, which were briefly discussed here.

Protecting our planet also implies protection against possible extreme space weather events, which could endanger many of our space and critical digital Earth infrastructures, and the development of active monitoring and planetary defence technologies to protect us against near-earth objects, which could involve a risk of collision with our planet. As we have discussed here, current technology plans and some pioneering demonstration missions already undertaken in this decade could lead us to the availability of reliable early warnings in the case of dangerous asteroids measuring over 40 m in diameter and to the capacity to deflect asteroids of up to 1 km in diameter, if identified more than two years in advance.

We concluded this section by addressing some of the extraordinary prospects linked to the NewSpace paradigm and the major transformation of the space sector that this implies.

7.5 Large-scale complex sociotechnical systems and their interactions

Marc Barthelemy[1]
[1]Université Paris Saclay, Gif-sur-Yvette, France

7.5.1 Complexity and modeling

7.5.1.1 Cities are complex systems

Cities are certainly among the most complex systems built by humans. They are made of a large number of different constituents, such as individuals, various institutions, and private companies, that interact in various ways at different time and spatial scales. All the ingredients are present for giving rise to emergent collective behaviours and consequently to difficulties in understanding and modeling these systems. In particular, there is the issue of unpredictability; varying some parameters will not cause any change to the system, while varying others leads to dramatically different behaviours and outcomes. This is one of the difficulties of modeling complex systems: A naive modeling can completely fail, and it is not by adding various parameters and mechanisms that we can increase the realism of the model and the validity of its predictions.

An important thing to notice is that thanks to the variety of sensors and other devices such as mobile phones, GPS, or RFIDs, the amount of available data about urban systems increased dramatically this last decade. For example, origin-destination matrices, a fundamental ingredient that tells us where people are living and where they are working, was essentially obtained by surveys during censuses. The data were then under the form of locations and estimate of trip duration and so on. Now, thanks to the large variety of new data sources, we have access to the trips of millions of individuals at an unprecedented resolution. More generally, we have data at all scales (see table 7.6) from the minute (or less) with mobile phones and GPS to longer time scales such as months or years with socioeconomic data (e.g., taxes or real estate transactions). Even longer timescales with the digitalization of historical maps are now accessible; we can study the evolution of the road network over centuries [169]. Despite this recent availability, there is still a lot of do in terms

Table 7.6. Data sources according to their typical timescale and some phenomena occurring on these timescales.

Timescale	Data sources	Phenomena
Minutes to days	GPS, mobile phones RFID	Spatial structure mobility, urban activity
Month to years	Surveys, censuses	Social organization housing market
Decades to centuries	Historical documents and maps	Urban growth, self-organization impact of planning

of data accessibility. Some countries are better than others and the European initiative EuroStat can certainly be improved. Ideally, cities should open their own data platform that would certainly trigger the interest of many scientists.

This flow of data about many different aspects of cities allowed scientists to test their theory. While urban economics [170] mainly developed as a mathematical field with few connections to reality, we can now construct a model and test its predictions against empirical observations. This increasing availability of pervasive data opened the exciting possibility of constructing a 'new science of cities'. A new problem that we have to solve is then to extract useful information from these huge datasets and construct theoretical models for explaining empirical observations. In particular, a common difficulty shared by many complex systems such as cities is the existence of a large number of agents, acting through a variety of processes that occur over a wide range of time and spatial scales. It is then necessary to disentangle these processes and to single out the few that govern the dynamics of cities. Albeit difficult, the hierarchization of processes is of prime importance, and a failure to do so leads to models which are either too complex to give any real insight into the phenomenon or too simple to provide a satisfactory framework which can be built upon. In this context, statistical physics might bring convenient tools for both the empirical analysis and modeling, with the possibility of characterizing emergent macroscopic phenomena in terms of relevant parameters describing the basic elements of the system.

7.5.1.2 How can we model a complex system?
As became obvious in interdisciplinary studies, the notions of a model and modeling are not unique and well defined, and this ambiguity is even bigger in complex systems studies. In statistics a model is what we call in physics a fit; it is a mathematical representation of observed data. The question is then what is the best fit, but there are no mechanistic interpretations here. In contrast, in physics a model is a simplified version of the reality that is supposed to capture the essence of the phenomenon by omitting some details that we hope will be irrelevant or play the role of noise. This model usually allows a mathematical study of the system, and in some cases we need to study it numerically. The loop between the model and comparing its prediction with experiments or empirical data is what worked very well for physics these last centuries. The virtue of these models lie essentially in their simplicity and their parsimonious use of parameters. The main function of such models is to explain what happens in the system and to help us understand its functioning and identifying critical parameters.

Numerical approaches modified a bit this type of approach, especially nowadays when CPU power is large enough that implementing many details of a system is not really a problem anymore. In this sort of approach, the system contains many parameters and mechanisms, and it is tempting to simulate it directly by implementing the behaviour of its constituents (the agents) and their interactions. We usually refer to this sort of simulations as agent-based ones, and in the context of cities, the extreme version of this is called a 'digital twin,' which is supposed to model all aspects of a city. For complex systems, we immediately face several problems with this type of approach. First, these numerical models are difficult to validate, and their sensitivity when varying their parameters can be very large and difficult to assess. Second, these models are usually very specific to a particular system (a region or a city) and sometimes cannot be easily transferred to another case. Related to this, these models act usually as black boxes and do not help us to understand what is really going on in these systems. In order to get around these criticisms, scientists use these models as a tool, not for forecasting and predicting, but to explore various scenarios and the impact of different measures or strategies. We can thus obtain projections that can be helpful for policymakers, for example. We saw this, for example, in the case of the COVID-19 pandemic. The simulations are, however, not the ground truth here, and this must be clearly stated, in particular when presented to policymakers or stakeholders. Digital twins that are exact replica (if that is even possible) of some systems and their behaviour of course diverges in general quickly from the real systems. Even if smart cities–related analyses seem to rely on this type of approach, it is unclear at this point how they will surpass other tools such as agent-based ones or more physical approaches.

Finally, there is machine learning that might revolutionize how we do science. At this point, these approaches are still limited, but we can expect to see their importance growing quickly, especially for real-world applications. So far, these algorithms that encode the complexity of data in a very large number of parameters act as black boxes and produces an output that is difficult to challenge and to understand. However, we can envision a possible future where machine learning techniques will become a tool among others and will assist researchers to make progress in our understanding. This sort of AI-assisted theoretical research could be the path to many important discoveries.

Going back to cities, an important aspect is interdisciplinarity. If we want to construct solid, scientific foundations of urban systems [171, 172], we need to produce an effort that is necessarily interdisciplinary, which is not always easy (see [173]); we have to build on early studies in urbanism to discuss morphological patterns and their evolution and on quantitative geography and spatial economics to describe the behaviour of individuals, the impact of different transportation modes, and the effect of economic variables (e.g., income, the rental market). In the following subsection, we will illustrate different aspects of this type of approach that combines ingredients coming from different fields and statistical physics using various examples ranging from the temporal evolution of cities (population and area), to congestion and CO_2 emission modeling.

7.5.2 Results and challenges in urban systems

There are obviously many aspects in cities. We will focus here on problems that can be tackled by quantitative methods. In addition, we will discuss problems that possess some duality; that is, they have a theoretical component, but also a practical one. This dual aspect is what usually make complex systems interesting, as they are rooted in reality but challenge our theoretical understanding. In each case, we will also mention interesting questions and challenges for future studies.

7.5.2.1 Searching for the equation of cities

Apparently, the simplest question about cities concerns the time evolution of urban populations. This question appeared more than a century ago with Auerbach [174], a German physicist who took some interest in the statistics of cities. Auerbach remarked that if the population of German cities are sorted in decreasing order $P_1 > P_2 \cdots > P_N$ (so that the rank $r = 1$ corresponds to the largest city), then the product $r \times P_r$ is approximately constant, implying that the population is inversely proportional to its rank (Auerbach, ahead of Zipf, also discussed that this type of relation could go well beyond cities and could be applied to other systems). This result was generalized by Zipf [175], who constructed a graphical representation of this, now known as rank–size plots. Zipf then plotted the population P_r versus its rank r and confirmed Auerbach's result for many countries. This is now called Zipf's law:

$$P_r \sim \frac{1}{r} \qquad (7.7)$$

More generally (see, e.g., figure 7.30), we don't observed a strictly inverse proportional relation but something of the form $P_r \sim 1/r^\nu$, where the exponent ν is in general close to 1 (empirically, we also observe that this law is more accurate for smaller cities (see, e.g., figure 7.30).

This statement is equivalent to saying that the population is distributed according to the power law:

$$\rho(P) \sim \frac{1}{P^{1+1/\nu}} \qquad (7.8)$$

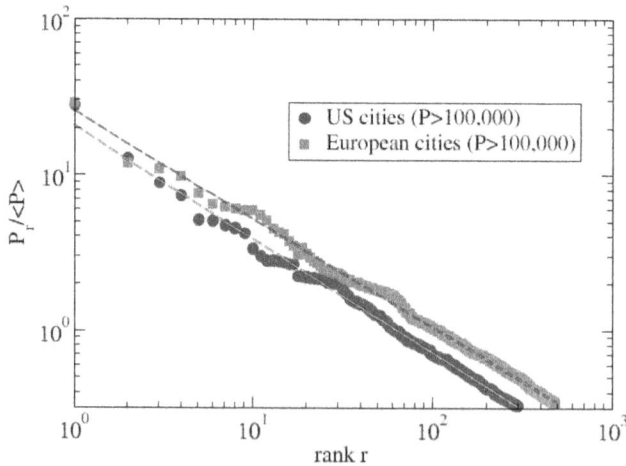

Figure 7.30. Zipf's law for European and US urban areas with population larger than 100 000 inhabitants (data from Eurostat for Europe and the S Census Bureau for the US). We normalized here the values by the average, and the plot is shown in loglog. The dashed lines represent power law fits (here the exponents are close to 0.70 in both cases).

The exponent ν is close to 1, which implies that the population distribution behaves as $\rho(P) \sim 1/P^2$, indicating a very large heterogeneity of urban systems. Quantitatively, the most fundamental problem here is to understand the hierarchical organization of city populations and the statistical occurrence of megacities described by such a distribution. On one hand, many authors have tried to explain this fact theoretically; on the other, the always growing availability of data showed that the exponent ν is not universal and can display large variations [176].

In addition, the dynamics of cities is not a smooth one; throughout history, we have observed many rises and falls of cities. A theoretical model must then be able to explain not only Zipf's law and its variation, but also this turbulent dynamics of cities. This problem was only recently answered with an approach that combined data and statistical physics tools [177]. More precisely, a stochastic equation for modeling the population growth in cities was constructed from an empirical analysis of recent datasets (for Canada, France, the UK, and the US). This equation reveals how rare but large interurban migratory shocks dominate population growth, and it predicts a complex shape for the distribution of city populations and shows that, owing to finite-time effects, Zipf's law does not hold in general, implying a more complex organization of cities. It also predicts the existence of multiple temporal variations in the city hierarchy, in agreement with observations [177, 178]. This result underlines the importance of rare events in the evolution of complex systems and, at a more practical level, in urban planning.

This long-standing problem about urban population could naively be thought as being the simplest one, but in fact it took almost a century for a more precise understanding. This problem is, however, probably much less involved than another crucial issue that mixes space and population and represents an important challenge for studies in this field: urban sprawl. In other words, how does a city evolve in space?

It is now generally accepted [178] that urban sprawl causes the loss of farmland, threatens biodiversity, and affects local climate. Despite these negative effects, urban land area increased by $\sim$60 000 km^2 from 1970 to 2000, with China, India, and Africa having the highest urban expansion rate. Although urban sprawl can be considered a local issue, its impact in terms of biodiversity and vegetation loss is at a global scale [180]. A theoretical understanding of urban growth could be extremely helpful for proposing mitigation strategies for these problems. Growth in GDP per capita and population seem to be the important drivers for the observed urban land expansion, but much of the observed variation in urban expansion is not captured by either population, GDP or other variables [179]. This clearly shows the limits of pure econometric studies, and a more mechanistic approach, inspired by physics studies could be of great help here. The spatial evolution of a city is, however, certainly a difficult problem. For example, we show in figure 7.31 the spatial representation for the city of London (UK) for different dates from 1800 to 1978 showing how the built area spreads in space.

For a physicist, the image in figure 7.31 triggers many questions. In particular, the natural question to ask is: What is the growth process (diffusion or faster?) and can we write the corresponding growth equation? There are of course many quantities

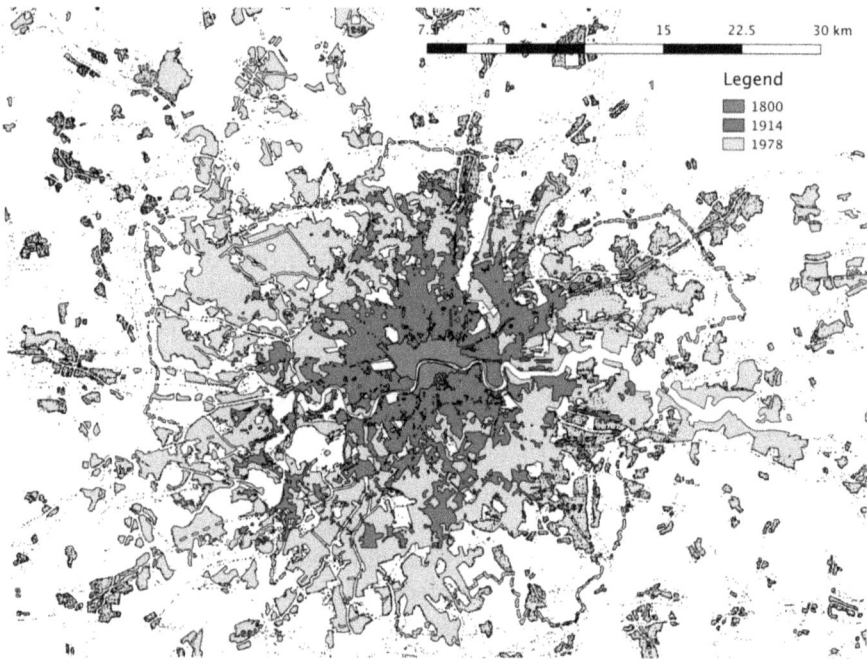

Figure 7.31. Spatial evolution of the built-up area of London for three different dates: 1800, 1914, and 1978. Data from [181].

that we can measure, but the first natural one is the surface area and how it varies with time or population. Answering this question and understanding this behaviour would represent important progress for the science of cities. Even further progress would be to describe the evolution of the frontier of the city and to find the corresponding equation. We can probably expect the effect of many factors, such as the presence of the road system, which evolves also in time and space. In addition, the city is not evolving in free space, and we expect population density and the transportation network to coevolve with the built area. This last remark actually points to the problem of the coupling between transportation network and population density: Better transportation attracts more individuals, and a large number of individuals pushes decision makers to construct better transportation. The coevolution of these quantities certainly represented an important ingredient for understanding urban sprawl. As we see with these short discussions, finding the equation that governs urban sprawl is certainly an extraordinary challenge but represents also a fantastic crossroad between the statistical physics of surface growth and the science of cities.

7.5.2.2 Congestion, CO_2 emission, and energy use
The evolution of urban sprawl triggered an increase use of cars, leading in turn to increased congestion, longer travel times, and more CO_2 emissions. More generally, understanding mobility patterns in urban areas [182] has become paramount in

reducing transport-related greenhouse gas emissions and crucial for proposing efficient environmental policies. In a seminal paper, Newman and Kenworthy correlated transport-related quantities (e.g., gasoline consumption) with a determinant spatial criterion: urban density [183]. Higher population density areas were shown to have reduced gasoline consumption per capita and thus reduced gas emissions. Their result had a significant impact on urban theories over the last decades and has become a paradigm of spatial economics [184]. This study is, however, purely empirical and has no theoretical foundation, which casts some doubts about the importance of density as the sole determinant of gasoline consumption and other car-dependent quantities. A way to test this theoretical assumption is to construct a simplified model that is validated by empirical observations and that eventually could help us to understand the effect of population density on car traffic. In order to illustrate this type of approach, we combine economic and transport ingredients into a statistical physics approach and construct a generic model [185] that predicts for different cities the share of car drivers, the CO_2 emitted by cars, and the average commuting time.

According to the classical urban economics model of Fujita and Ogawa [186], individuals choose job and dwelling places that maximize their net income after deduction of rent and commuting costs. More precisely, an agent will choose to live in x and work at location y such that the quantity

$$Z(x, y) = W(y) - C_R(x) - G(x, y) \qquad (7.9)$$

is maximum. The quantity $W(y)$ is the typical wage earned at location y, $C_R(x)$ is the rent cost at x, and $G(x, y)$ is the generalized transportation cost to go from x to y. In order to simplify the discussion, we assume here that employment is located at a unique centre $y = 0$ and that wages and rent costs are of the same order for all individuals. (In fact, most large cities have many different activity centres [187, 188], but this dramatically changes our argument here.) We also assume that the residence location x is given and random; residence choice is obviously a complex problem, and replacing a complex quantity by a random one is a typical assumption made in the statistical physics of complex systems. Within these assumptions, we obtain a simplified Fujita–Ogawa model where we focus on the mode choice; individuals have already a home and work at the central business district located at 0, and the problem is about choosing a transportation mode. Within all these assumptions, the maximization of $Z(x, 0)$ implies the minimization of the transport cost $G(x, 0)$:

$$\max Z(x, 0) \Rightarrow \min G(x, 0) \qquad (7.10)$$

where $G(x, 0)$ is the generalized transportation cost from home located at x and the office (located at 0). In other words, individuals will choose in this model (see figure 7.32) the transportation mode that minimizes their commuting costs to go the office. In order to discuss commuting costs, we assume that a proportion p of the population has access (meaning having to walk less than 1 km) to the subway, whereas a share $1 - p$ of the population has no choice but to commute by car. (We assume that all individuals can drive a car if needed.) Even if an individual has access

(a)   (b)

CBD   CBD

subway
station

x d_0 x

Probability 1-p: Probability p:
no subway station   subway station

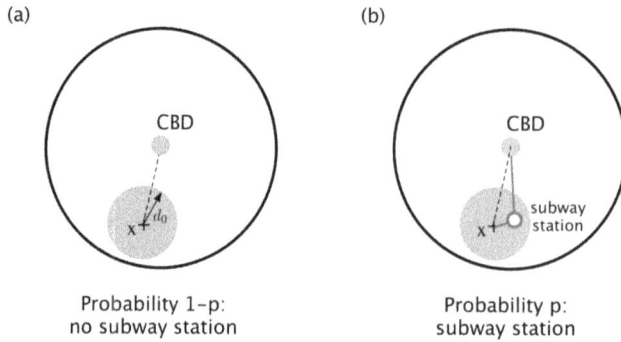

Figure 7.32. Sketch of the model. (a) For a given agent located at a random location x, there is no subway station located at a distance less than d_0 with probability $1 - p$ (in the data used, $d_0 = 1$ km). In this case, the journey to work located at the central business district (CBD) is made by car (dashed line). (b) With probability p there is a subway station in the neighbourhood of x, and the agent has to compare the cost G_{car} of car (dashed line) and the cost G_{subway} of subway (the trip is depicted by the red line) in order to choose the less costly transportation mode to go to the CBD.

to the subway, that does not necessarily mean that it is the mode chosen, and the individual needs to compare the costs G_{car} and G_{subway} in order to choose the least costly transportation mode.

The generalized cost of a mode can be written under the form $G = C_f + V \Delta t$, where C_f is the financial cost (per year, for example) of the mode, while Δt is the time needed for the trip under consideration, and V, which is in units of money per time, is called in this field the 'value of time'. It characterizes the propensity of an individual to choose a rapid transportation mode over a cheaper one. We can even include congestion effects [189] in the trip duration Δt. We can then write an expression for both modes, car and subway, that depends on the distance to go to the centre and different parameters, such as the value of time, the average velocity of car and subway, and the cost of a car. (For the subway, we neglect its monetary cost, which is small in comparison with that for cars. For details see [185].) We can then study the statistics of the minimum cost, and we find the following result [185]. For the 25 megacities in the world that are considered in this study, the subway is always more advantageous than the car. Public transportation is so economical (compared to cars) that people living near rapid transit stations are highly likely to ride them. Thus, traffic does not appear to be a determinant parameter in individual mobility choices, as it concerns mostly individuals who have no choice but to drive and who suffer from onerous commuting costs and unavoidable time-consuming trips as traffic increases. The consequence is then a simple relation between the car traffic T and population P of the form

$$\frac{T}{P} \simeq 1 - p \tag{7.11}$$

which is a nontrivial consequence of rapid transit cheapness and individual choices of mobility. We compare the empirical car modal share $\frac{T}{P}$ for these cities to this

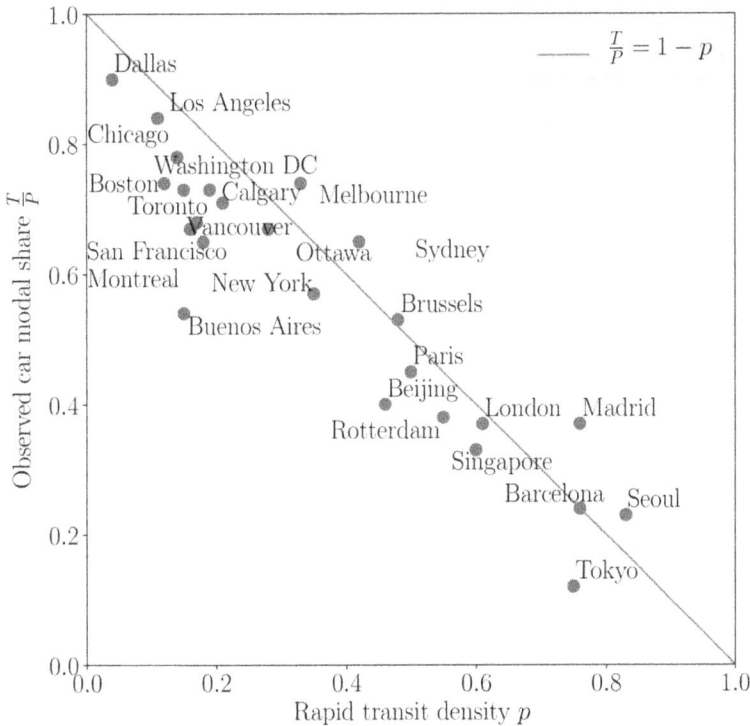

Figure 7.33. Comparison between the observed car modal share T/P and the share of population p living near rapid transit stations (less than 1 km) for 25 metropolitan areas in the world. The red line is the prediction of the model ($R^2 = 0.69$). Given the absence of any tunable parameter, the agreement is satisfactory, and discrepancies are probably mostly due to the existence of other modes of transport (walking or cycling), lower car ownership rates, a higher cost of mass rapid transit, and so on. Figure from [185] CC BY 4.0.

prediction on figure 7.33 and observe, considering the simplicity of the model, very good agreement and a relevant linear trend, highlighting the efficiency of public transportation in reducing traffic.

In particular, most of the European cities are well described by this prediction, though we observe a few deviations. These discrepancies probably have their origin in the existence of other modes of commuting, lower car ownership rates (e.g., in Buenos Aires), lower road capacities, higher cost of subway and mass rapid transit, or a high degree of polycentrism.

This model is also able to provide a prediction for the CO_2 emitted by cars. We will use the simplest assumption, where these emissions are proportional to the total time spent on roads, and we then obtain [185]

$$\frac{Q_{CO_2}}{P} \propto \sqrt{A}(1 - p)(1 + \tau) \tag{7.12}$$

where P is the population, τ is the average delay due to congestion and is empirically accessible from various databases (e.g., the TomTom database [190]), and A is the surface area of the urban system. We note that Q_{CO_2} is the product of three main terms: the typical size $\sqrt{A}$ of the city × the fraction $1 - p$ of car drivers × the congestion effects, which correspond indeed to the intuitive expectation about the main ingredients governing car traffic. We compare this prediction equation (7.6) to disaggregated values of urban CO_2 emissions and observe good agreement [185]. We observe some outliers, such as Buenos Aires, which has a very small car ownership rate and thus lower than expected CO_2 emissions, and New York, which appears to be one of the largest transport CO_2 emitters in the world [191].

This result obtained with a simple model and validated with empirical data illustrates the role played by public transport and traffic in modulating transport-related CO_2 emissions. Most important, we identify urban sprawl ($\sqrt{A}$) here as a major criterion for transport emissions. We note that if we introduce the average population density $\rho = \frac{P}{A}$, we can rewrite equation (7.6) approximately as $\frac{Q_{CO_2}}{P} \propto \rho^{-1/2}$, since $\sqrt{P}$ is a slowly varying function within the scope of large urban areas. We understand here how Newman and Kenworthy [183] could have obtained their result by assuming the density to be the control parameter. However, even if fitting data with a function of ρ is possible, the analysis presented here shows that it is qualitatively wrong; the area size A and the public transport density p seem to be the true parameters controlling car-related quantities such as CO_2 emissions. Mitigating the traffic is therefore not obtained by increasing the density but by reducing the area size and improving the public transport density. Increasing the population in a fixed area would increase the emission of CO_2 (due to an increase of traffic congestion, leading to an increase of τ), in contrast with the naive Newman–Kenworthy assumption where increasing the density leads to a decrease of CO_2 emissions.

The density is therefore not that pivotal, but different factors affect CO_2 emissions. In order to reduce these emissions, this model suggests increasing public transport access either by increasing the density around subway stations or by increasing the density of public transport (in contrast with the conclusions of an econometric study in the US [192]) or reducing the urban area size (impossible in most contexts). Increasing the cost of car use seems actually unable to lower car traffic in the absence of alternative transportation means. This model obviously ignores many parameters, but the main point is to fill a gap in our understanding of traffic in urban areas by using a parsimonious model with the smallest number of parameters and the largest number of predictions in agreement with data. Given the simplicity of the model, we cannot expect perfect agreement, but we may be able to capture correctly all the trends observed empirically and to identify correctly the critical factors for car traffic.

This parsimonious and generic model for the car traffic illustrates how a combination of statistical physics, economic ingredients, and empirical validation can lead to a robust understanding of systems as complex as cities. The next step would then be to understand other crucial quantities, such as the total carbon footprint of a city or the energy use of a city. Empirical studies such as [193] usually

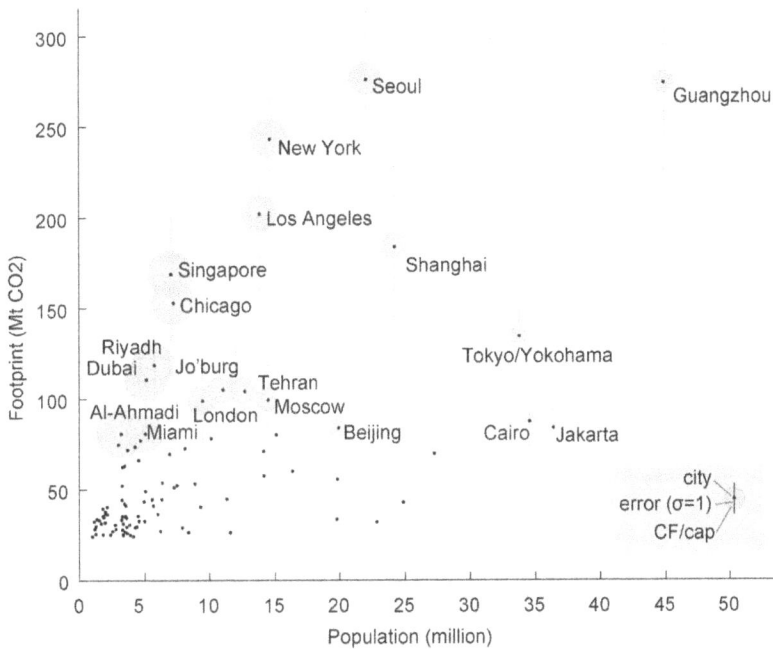

Figure 7.34. Carbon footprint of an urban area versus its population (the named cities are the top 20). The size of the disk corresponds to the carbon footprint per capita. The vertical lines correspond to one standard deviation in the carbon footprint. Figure reproduced from [193]. Copyright The Author(s). Published by IOP Publishing Ltd. CC BY 3.0.

plot the carbon footprint of an urban area (in millions of tons of CO_2) versus population (see figure 7.34).

This figure shows clearly that population is not the only determinant of the carbon footprint. The question is then how we can model this aspect of a city, and the physicist's approach could then help in identifying and understanding what are the critical parameters and their quantitative effects.

Another crucial aspect of cities is their energy use. If the current urban expansion trend continues, energy use will increase more than threefold at the horizon of 2050 [184], leading inevitably to many environmental and economic problems. Modeling this problem could help in devising urban planning and transport policies in order to limit this future increase in urban energy use. An empirical analysis over worldwide cities showed that economic activity, transport costs, geographic factors, and urban form explain 37% of urban direct energy use and 88% of urban transport energy use. As for the carbon footprint, we can illustrate the problem by plotting the energy use versus the GDP per capita (see figure 7.35).

We observe that on average, energy use increases with the economic activity measured by the GDP per capita, followed by a plateau. However, we observe large fluctuations around this average behaviour, showing that very likely the GDP per capita is not the only important factor here. In particular, the authors of [184] observed that higher population density is certainly a positive factor (and,

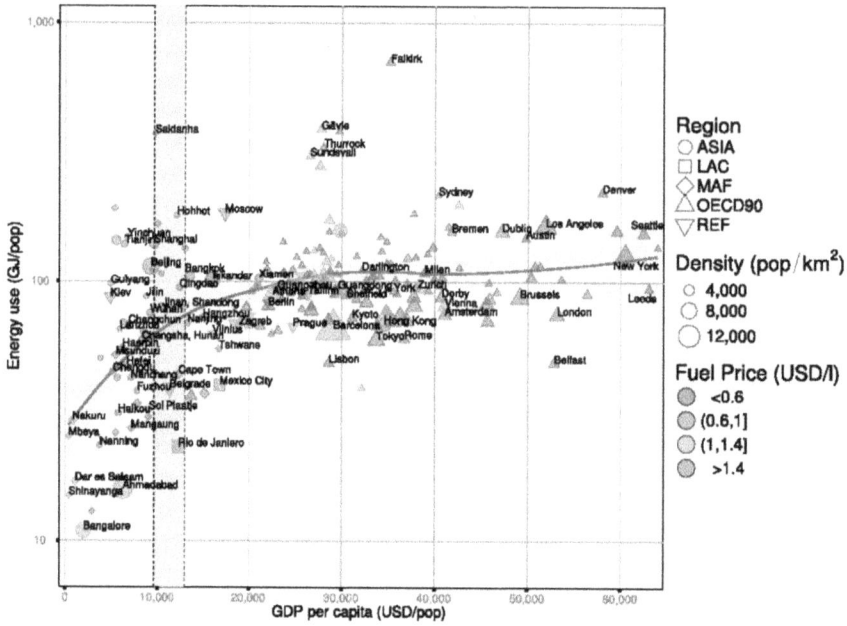

Figure 7.35. Energy use versus economic activity measured by the GDP per capita. Figure reproduced from [184], with permission from PNAS (2015), where more details can be found.

consequently, that developing countries with emerging infrastructures, urban form, and transport planning could avoid larger energy use by increasing local density). Here again, these results challenge our theoretical understanding, and a simple model, in addition to improve our fundamental understanding, could have important practical implications and effects on policies and planning.

7.5.2.3 A transition in urban traffic

The study of traffic and congestion is an old topic and has inspired physicists for a long time. The first works on the subject dealt with traffic on highways, applying various methods coming from fluid mechanics [194] or cellular automata [195]. Concepts such as the carrying capacity of a highway, the formation and progression of congestion on a highway, or the impact of multilane roads with traffic traveling at different speed on the capacity are well understood and described in the literature [196–198]. Different studies [199–201] have conducted experiments on one-lane roads, confirming that the switch from free flow to congestion is brutal and analogous to a phase transition. In [202, 203], models are proposed for the formation of congestion on a regular two-dimensional lattice, displaying a first-order phase transition between free flow and a jammed state.

In contrast, in more complex networks of roads, the mechanisms of congestion spread remain poorly understood. Urban road networks are strongly connected systems, and scaling from single-lane roads to a nonregular two-dimensional network is not an easy task. Agent-based models simulate the flow of cars on the network based on the 'microscopic' interaction between cars on the network and between cars and the network itself (e.g., traffic lights). This type of simulations requires many assumptions and parameters, and the simulations are difficult to validate. More recent studies have analyzed the urban traffic at a macroscopic level, focusing on the description of the emergence of large-scale congestion as a phase transition. Indeed, it is known that the jamming of a node of the network which happens when the demand exceeds its capacity can have a macroscopic impact, as shown by Echenique *et al* [204] in the context of Internet protocols. However, the exact response of the network depends strongly on the network itself and on the way the information propagates on the network. Understanding the origin and the propagation of congestion on urban road networks is therefore the goal of most recent studies on the subject. At a theoretical level, studies in [205, 206] have proposed arguments to understand the emergence of congestion on networks. In [207, 208], it has been proposed that reaction–diffusion equations are relevant for traffic congestion, where the congestion spreads from a congested link to its neighbors, and Saberi *et al* [209] proposed a model inspired by epidemiological studies to describe the evolution of the number of congested links, regardless of the actual structure of the network.

At a more empirical level, a lot of attention has been given to the study of congestion in the framework of percolation [210], where the main idea is to look at the structure of the set of links at a certain level of congestion (which depends then on the hour of the day). For each date, the authors of [210] divide the network into functional and congested roads, based on a velocity threshold v^*. This leads to clusters of roads functioning with a speed $v > v^*$, separated by congested roads with speeds $v < v^*$. By varying the threshold v^*, they observe a percolation transition (which depends on the hour of the day) with a breakdown of a giant functional cluster into several small clusters. The value of the percolation threshold v_c^* can then be seen as a measure of the state of the network at that date, as it measures effectively the maximal velocity at which one can travel over the main part of the network (described by the giant component).

Generally speaking, a typical marker of a phase transition in classical statistical physics is the divergence of the correlation length close to the critical value of the control parameter (see, e.g., [211]). If there is some 'jamming' transition in urban systems, we should then observe this type of behaviour. The goal is to find an intrinsic marker of a phase transition in the data and to show that congestion in a complex road network during rush hours can be viewed as a jamming transition, going beyond one-dimensional cases. We therefore analyze the correlation between the delays on all roads of the network and identify the correlation length and its variation.

In order to do this, we use hourly traffic data for the city of Paris (France) and study the correlation function of delays [212]. More precisely, we compare roads based on the relative delay experience by users, and for each year and each hour we measure the delay $d_i(t)$ experienced on link i. We also introduce the quantity $T_i(t)$, which denotes the average travel time for that year and hour on this link. The relative delay is then given by $\tau_i(t) = d_i(t)/T_i(t)$ and indicates the importance of congestion on this link. We consider the correlation function between links i and j of this quantity and measure it for the year y and hour t

$$C_{i,j}(y, t) = \langle \tau_i(t)\tau_j(t) \rangle - \langle \tau_i(t) \rangle \langle \tau_j(t) \rangle \tag{7.13}$$

The averages (denoted by the brackets $\langle \cdot \rangle$) are performed over all working days of a given year y and at a given hour t. From this definition, we construct a distance-dependent correlation function by sorting the pairs i, j according to their distance r. After having averaged over all links, we then find in the dataset that the correlation function displays a typical behaviour of the form

$$C(r, y, t) = \frac{1}{r^\eta} \exp\left(-\frac{r}{\xi(y, t)}\right) \tag{7.14}$$

where r is the distance between two roads and where $\xi(y, t)$ is the correlation length, depending on the hour of the day t and calculated for each year y (in this study, distances, including the correlation length, are converted in the time needed to travel them). This correlation function depends *a priori* on the year and the hour of the day, and its analysis enabled us to extract the correlation length and to examine its variation shown in figure 7.36.

We observe that during night hours, the correlation length is very small, typically of the order of the time needed to travel from a link to two or three links further. Delays appearing on a link during the night propagate to the vicinity of this link only. In sharp contrast, we observe during daytime hours a dramatic increase in the correlation length to values close to 1500 s, and during rush hours, the correlation length displays a peak with values well above 1500 s, which corresponds to a trip whose order of magnitude is the size of the system. Having $\xi \approx 1500$ s indicates a correlation at the scale of the whole system. This divergence is the sign of a transition

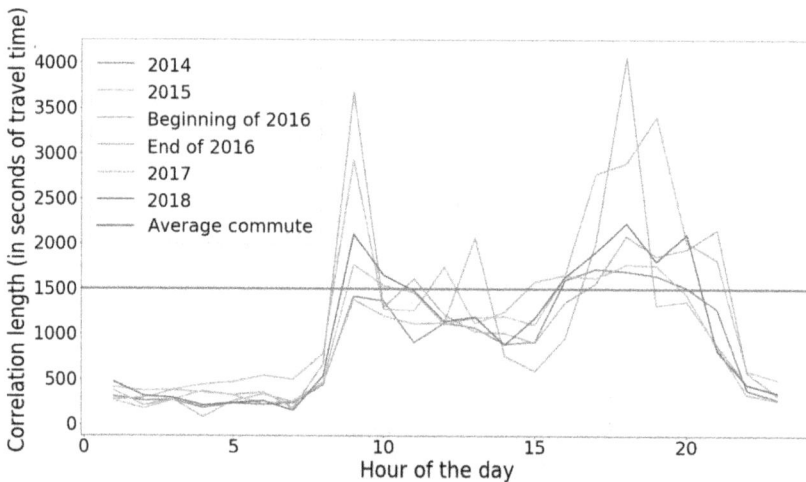

Figure 7.36. Correlation length ξ for the delays on the links of the network, as a function of time and of the average occupancy rate. Figure taken from [212] CC BY 4.0.

that occurs during rush hours with congestion expanding over the whole network. This result therefore suggests the existence of a transition between a state where congestion is localized to another state where congestion occurs over the whole network during rush hours.

This result is important empirical evidence for the existence of a jamming transition in a complex network of roads. However, this points to our lack of understanding of traffic in urban networks and opens an important research direction where the challenge is to construct a consistent theory that allows us to understand this transition but also the macroscopic behaviour of this system and how macroscopic parameters (e.g., the total traffic, the total capacity) are related to each other.

7.5.3 Discussion and perspectives

There is obviously a large number of results and of open and interesting problems about cities, and we could not discuss them all here. Among those, the evolution of infrastructures, the spatial structure of cities, and mobility patterns have made a lot of progress, thanks to new data sources such as mobile phones [213, 214] (see also [182] for a review about mobility). Important studies have also examined processes that take place in cities, such as epidemic modeling in urban areas (see, e.g., [215] and references therein) or resilience in front of flooding or other natural disasters (see, e.g., [216, 217] and references therein). Another particularly important result was obtained in [218], stating that connections between different networks increases the fragility of this multilayered network system. This is particularly relevant for cities where many networks—transportation, energy, water, and so on—are interdependent. Finally, on a larger scale, cities are not isolated but belong to a network of cities (see, e.g., [219] and references therein), and a consistent theoretical approach to cities should integrate this aspect.

We saw in this short review that the increased availability of data has already led to many results and has strengthened our knowledge of the functioning of these complex systems. There are, however, many challenges left. We saw some of them here, but more generally, the most important challenge is probably how to study a complex system such as a city in order to get reliable projections and scenarios. An important part concerns the data availability that could certainly be improved in many countries. So far, data is scattered all over the Internet, and a European or even global initiative such as a common data platform could accelerate research, facilitate reproducibility, and promote international collaborations. From a more theoretical point of view, there are challenges of very different natures. Some are purely mathematical, such as the optimal shape of a transportation network, for example [220], and some others are closer to applications, such as the traffic in urban networks. In all cases, the analysis of data is a crucial component, and we will probably assist in the future in finding hybrid approaches that combine more standard tools coming from applied mathematics or statistical physics with new algorithms, such as machine learning. It seems that we have now all the ingredients —data and tools—for constructing a robust science of cities where fundamental understanding will be able to help urban planners mitigating the large variety of issues posed by large cities.

Acknowledgments

JB acknowledges support from the project Leading Research Center on Quantum Computing (Agreement No. 014/20).

References

[1] Feynman R P 1982 Simulating physics with computers *Int. J. Theor. Phys.* **21** 467
[2] Nielsen M A and Chuang I L 2011 *Quantum Computation and Quantum Information: 10th Anniversary Edition* 10th edn (New York: Cambridge University Press)
[3] Aaronson S 2013 *Quantum Computing Since Democritus* (New York: Cambridge University Press)
[4] Arute F *et al* 2019 Quantum supremacy using a programmable superconducting processor *Nature* **574** 505
[5] Zhong H-S *et al* 2020 Quantum computational advantage using photons *Science* **370** 1460
[6] Wu Y *et al* 2021 Strong quantum computational advantage using a superconducting quantum processor *Phys. Rev. Lett.* **127** 180501
[7] Preskill J 2018 Quantum computing in the NISQ era and beyond *Quantum* **2** 79
[8] Bharti K *et al* 2021 Noisy intermediate-scale quantum (NISQ) algorithms *Rev. Mod. Phys.* **94** 015004
[9] Montanaro A 2016 Quantum algorithms: an overview *npj Quantum Inf.* **2** 15023
[10] Bauer B, Bravyi S, Motta M and Chan G K-L 2020 Quantum algorithms for quantum chemistry and quantum materials science *Chem. Rev.* **120** 12685
[11] Biamonte J, Wittek P, Pancotti N, Rebentrost P, Wiebe N and Lloyd S 2017 Quantum machine learning *Nature* **549** 195
[12] Blais A, Girvin S M and Oliver W D 2020 Quantum information processing and quantum optics with circuit quantum electrodynamics *Nat. Phys.* **16** 247
[13] Blatt R and Roos C F 2012 Quantum simulations with trapped ions *Nat. Phys.* **8** 277
[14] Browaeys A and Lahaye T 2020 Many-body physics with individually controlled Rydberg atoms *Nat. Phys.* **16** 132
[15] Einstein A, Podolsky B and Rosen N 1935 Can quantum-mechanical description of physical reality be considered complete? *Phys. Rev.* **47** 777
[16] Bell J S 1964 On the Einstein Podolsky Rosen paradox *Phys. Phys. Fiz.* **1** 195
[17] Aspect A, Grangier P and Roger G 1982 Experimental realization of Einstein–Podolsky–Rosen–Bohm Gedanken experiment: a new violation of Bell's inequalities *Phys. Rev. Lett.* **49** 91
[18] DiVincenzo D P 1995 Two-bit gates are universal for quantum computation *Phys. Rev. A* **51** 1015
[19] Deutsch D 1985 Quantum theory, the Church–Turing principle and the universal quantum computer *Proc. R. Soc. A* **400** 97
[20] Kirkpatrick S, Gelatt C D and Vecchi M P 1983 Optimization by simulated annealing *Science* **220** 671
[21] Deutsch D and Jozsa R 1992 Rapid solution of problems by quantum computation *Proc. R. Soc. A* **439** 553

[22] Simon D 1994 *Proc. 35th Annu. Symp. Found. Comput. Sci.* (IEEE Comput. Soc. Press) pp 116–23

[23] Shor P 1994 *Proc. 35th Annu. Symp. Found. Comput. Sci.* (IEEE Comput. Soc. Press) pp 124–34

[24] Kitaev A Y 1995 Quantum measurements and the Abelian Stabilizer Problem ArXiv: quant-ph/9511026

[25] Grover L K 1996 *Proc. 28th Annual ACM Symp. Theory Comput.—STOC'96* (New York: ACM Press) pp 212–9

[26] Brassard G, Hoyer P, Mosca M and Tapp A 2000 Quantum amplitude amplification and estimation ArXiv: 0005055 [quant-ph]

[27] Das A and Chakrabarti B K 2008 Colloquium: quantum annealing and analog quantum computation *Rev. Mod. Phys.* **80** 1061

[28] Tang E 2019 *Proc. 51st Annu. ACM SIGACT Symp. Theory Comput* (New York: ACM) pp 217–28

[29] Wootters W K and Zurek W H 1982 A single quantum cannot be cloned *Nature* **299** 802

[30] Shor P W 1995 Scheme for reducing decoherence in quantum computer memory *Phys. Rev. A* **52** R2493

[31] Steane A M 1996 Error correcting codes in quantum theory *Phys. Rev. Lett.* **77** 793

[32] Shor P W 1996 Fault-tolerant quantum computation ArXiv: quant-ph/9605011

[33] DiVincenzo D P 2000 The physical implementation of quantum computation *Fortschr. Phys.* **48** 771

[34] McMichael R D https://nist.gov/programs-projects/diamond-nv-center-magnetometry (Accessed 19 July 2021)

[35] Wang J *et al* 2018 Multidimensional quantum entanglement with large-scale integrated optics *Science* **360** 285

[36] Ladd T D, Jelezko F, Laflamme R, Nakamura Y, Monroe C and O'Brien J L 2010 Quantum computers *Nature* **464** 45

[37] Cirac J I and Zoller P 1995 Quantum computations with cold trapped ions *Phys. Rev. Lett.* **74** 4091

[38] Shnirman A, Schoen G and Hermon Z 1997 Quantum manipulations of small Josephson junctions *Phys. Rev. Lett.* **79** 2371

[39] Jaksch D, Cirac J I, Zoller P, Rolston S L, Coˆte´ R and Lukin M D 2000 Fast quantum gates for neutral atoms *Phys. Rev. Lett.* **85** 2208

[40] Loss D and DiVincenzo D P 1998 Quantum computation with quantum dots *Phys. Rev. A* **57** 120

[41] Nizovtsev A P 2005 A quantum computer based on NV centers in diamond: optically detected nutations of single electron and nuclear spins *Opt. Spectrosc.* **99** 233

[42] Knill E, Laflamme R and Milburn G J 2001 A scheme for efficient quantum computation with linear optics *Nature* **409** 46

[43] Gershenfeld N A and Chuang I L 1997 Bulk spin-resonance quantum computation *Science* **275** 350

[44] Kane B E 1998 A silicon-based nuclear spin quantum computer *Nature* **393** 133

[45] Kitaev A 2003 Fault-tolerant quantum computation by anyons *Ann. Phys. (N. Y).* **303** 2

[46] Brooke J 1999 Quantum annealing of a disordered magnet *Science* **284** 779

[47] Raussendorf R and Briegel H J 2001 A one-way quantum computer *Phys. Rev. Lett.* **86** 5188

[48] Born M and Jordan P 1925 Zur quantenmechanik *Z. Phys.* **34** 858–88
[49] Benioff P 1980 The computer as a physical system: a microscopic quantum mechanical hamiltonian model of computers as represented by turing machines *J. Stat. Phys.* **22** 563–91
[50] Feynman R P 1985 Quantum mechanical computers *Optics News* **11** 11–20
[51] Yuri M 1980 *Vychislimoe i nevychislimoe [Computable and Uncomputable]* (original in Russian: Soviet Radio)
[52] Deutsch D 1985 Quantum theory, the Church—turing principle and the universal quantum computer *Proc. Royal Soc. Lond. Ser. A, Math. Phys. Sci.* **400** 97–117
[53] Lloyd S 1996 Universal quantum simulators *Science* **273** 1073–8
[54] Deutsch D and Jozsa R 1992 Rapid solution of problems by quantum computation *Proc. Royal Soc. Lond. Ser. A, Math. Phys. Sci.* **439** 553–8
[55] Simon D R 1997 On the power of quantum computation *SIAM J. Comput.* **26** 1474–83
[56] Shor P W 1997 Polynomial-time algorithms for prime factorization and discrete logarithms on a quantum computer *SIAM J. Comput.* **26** 1484–509
[57] Grover L K 1996 A fast quantum mechanical algorithm for database search Proc. of the 28th Annual ACM Symp. on Theory of Computing pp 212–9
[58] Bennett C H, Bernstein E, Brassard G and Vazirani U 1997 Strengths and weaknesses of quantum computing *SIAM J. Comput.* **26** 1510–23
[59] Aaronson S 2015 Read the fine print *Nat. Phys.* **11** 291–3
[60] Harrow A W, Hassidim A and Lloyd S 2009 Quantum algorithm for linear systems of equations *Phys. Rev. Lett.* **103** 150502
[61] Morales M E S, Tlyachev T and Biamonte J 2018 Variational learning of grover's quantum search algorithm *Phys. Rev. A* **98** 062333
[62] Farhi E, Goldstone J and Gutmann S 2014 A quantum approximate optimization algorithm arXiv:1411.4028
[63] Bravo-Prieto C, LaRose R, Cerezo M, Subasi Y, Cincio L and Coles P J 2020 Variational quantum linear solver arXiv:1909.05820
[64] Lubasch M, Joo J, Moinier P, Kiffner M and Jaksch D 2020 Variational quantum algorithms for nonlinear problems *Phys. Rev. A* **101** 010301
[65] Yung M-H, Casanova J, Mezzacapo A, McClean J, Lamata L, Aspuru-Guzik A and Solano E 2014 From transistor to trapped-ion computers for quantum chemistry *Sci. Rep.* **4** 3589
[66] Peruzzo A, McClean J, Shadbolt P, Yung M-H, Zhou X-Q, Love P J, Aspuru-Guzik A and O'brien J L 2014 A variational eigenvalue solver on a photonic quantum processor *Nat. Commun.* **5** 4213
[67] Biamonte J 2021 Universal variational quantum computation *Phys. Rev. A* **103** L030401
[68] Biamonte J D, Dorozhkin P and Zacharov I 2019 Keep quantum computing global and open *Nature* **573** 190–1
[69] Arute F *et al* 2019 Quantum supremacy using a programmable superconducting processor *Nature* **574** 505–10
[70] Zhong H-S *et al* 2020 Quantum computational advantage using photons *Science* **370** 1460–3
[71] Schumacher B 1995 Quantum coding *Phys. Rev. A* **51** 2738–47
[72] Bruzewicz C D, Chiaverini J, McConnell R and Sage J M 2019 Trapped-ion quantum computing: Progress and challenges *Appl. Phys. Rev.* **6** 021314

[73] Kok P, Munro W J, Nemoto K, Ralph T C, Dowling J P and Milburn G J 2007 Linear optical quantum computing with photonic qubits *Rev. Mod. Phys.* **79** 135–74

[74] Nielsen M A and Chuang I L 2009 *Quantum Computation and Quantum Information* (Cambridge University Press)

[75] Shor P W 1996 Fault-tolerant quantum computation Proc. of 37th Conf. on Foundations of Computer Science (IEEE Comput. Soc. Press)

[76] Shor P W 1995 Scheme for reducing decoherence in quantum computer memory *Phys. Rev.* A **52** R2493–6

[77] Georgescu I 2020 25 years of quantum error correction *Nat. Rev. Phys.* **2** 519 519

[78] Farhi E, Goldstone J, Gutmann S, Lapan J, Lundgren A and Preda D 2001 A quantum adiabatic evolution algorithm applied to random instances of an np-complete problem *Science* **292** 472–5

[79] Aharonov D, van Dam W, Kempe J, Landau Z, Lloyd S and Regev O 2008 Adiabatic quantum computation is equivalent to standard quantum computation *SIAM Rev.* **50** 755–87

[80] Childs A M 2009 Universal computation by quantum walk *Phys. Rev. Lett.* **102** 180501

[81] Lovett N B, Cooper S, Everitt M, Trevers M and Kendon V 2010 Universal quantum computation using the discrete-time quantum walk *Phys. Rev.* A **81** 042330

[82] Van den Nest M, Miyake A, Dür W and Briegel H J 2006 Universal resources for measurement-based quantum computation *Phys. Rev. Lett.* **97** 150504

[83] Harrow A W and Montanaro A 2017 Quantum computational supremacy *Nature* **549** 203–9

[84] Preskill J 2018 Quantum computing in the nisq era and beyond *Quantum* **2** 79

[85] Waldrop M M 2016 The chips are down for Moore's law *Nature* **530** 144–7

[86] Biamonte J, Wittek P, Pancotti N, Rebentrost P, Wiebe N and Lloyd S 2017 Quantum machine learning *Nature* **549** 195–202

[87] Pande M and Mulay P 2020 Bibliometric survey of quantum machine learning *Sci. Technol. Libr.* **39** 369–82

[88] Seskir Z C and Aydinoglu A U 2021 The landscape of academic literature in quantum technologies *Int. J. Quantum Inform.* **19** 2150012

[89] van Eck N J and Waltman L 2009 Software survey: Vosviewer, a computer program for bibliometric mapping *Scientometrics* **84** 523–38

[90] Watkins G P and Knight F H 1922 Knight's risk, uncertainty and profit *Quart. J. Econ.* **36** 682

[91] DiVincenzo D P 2017 Scientists and citizens: getting to quantum technologies *Ethics Inform. Technol.* **19** 247–51

[92] Möller M and Vuik C 2017 On the impact of quantum computing technology on future developments in high-performance scientific computing *Ethics Inform. Technol.* **19** 253–69

[93] Coenen C and Grunwald A 2017 Responsible research and innovation (rri) in quantum technology *Ethics Inform. Technol.* **19** 277–94

[94] de Wolf R 2017 The potential impact of quantum computers on society *Ethics Inform. Technol.* **19** 271–6

[95] Johnson W S 1883 Electric tele-thermoscop *US Patent* 281,884

[96] Schütze A, Helwig N and Schneider T 2018 Sensors 4.0—smart sensors and measurement technology enable industry 4.0 *J. Sensors Sensor Syst.* **7** 359–71

[97] Bentley J P 1995 *Principles of Measurement Systems* (New York: Longman Publishing Group) 3rd edn

[98] Wilson J S 2005 *Sensor Technology Handbook* (Amsterdam: Elsevier)

[99] Taylor J R 1996 *An Introduction to Error Analysis: The Study of Uncertainties in Physical Measurements* (Mill Valley, CA: University Science Books)

[100] Fink J K 2012 *Polymeric Sensors and Actuators* (New York: Wiley)

[101] Simmons E E 1940 Material testing apparatus *US Patent* 2,350,972

[102] Ruge A C 1939 Strain gauge *US Patent* 2,350,972

[103] Wasley R J, Hoge K G and Cast J C 1969 Combined strain gauge—quartz crystal instrumented hopkinson split bar *Rev. Sci. Instrum.* **40** 889–94

[104] Barbour N and Schmidt G 2001 Inertial sensor technology trends *IEEE Sens. J.* **1** 332–9

[105] Narayanaswamy R and Wolfbeis O S 2004 *Optical Sensors* (Berlin: Springer)

[106] Santos J and Farahi F 2014 *Handbook of Optical Sensors* (Boca Raton, FL: CRC Press)

[107] Jain Y K, Alex T K and Kalakrishnan B 1980 Ultimate ir horizon sensor *IEEE Trans. Aerospace Electron. Syst.* **16** 233–8

[108] Casas J R and Cruz P J S 2003 Fiber optic sensors for bridge monitoring *J. Bridge Eng.* **8** 362–73

[109] García I, Zubia J, Durana G, Aldabaldetreku G, Illarramendi M A and Villatoro J 2015 Optical fiber sensors for aircraft structural health monitoring *Sensors (Switzerland)* **15** 15494–519

[110] Udd E and Spillman W B 2011 *Fiber Optic Sensors: An Introduction for Engineers and Scientists* 2nd edn (New York: Wiley)

[111] Lee B H, Kim Y H, Park K S, Eom J B, Kim M J, Rho B S and Choi H Y 2012 Interferometric fiber optic sensors *Sensors* **12** 2467–86

[112] Baldini F and Mignani A G 2002 Optical-fiber medical sensors *MRS Bull.* **27** 383–7

[113] Baltes H P and Popovic R S 1986 Integrated semiconductor magnetic field sensors *Proc. IEEE* **74** 1107–32

[114] Janata J 2009 *Principles of Chemical Sensors* (New York: Springer US)

[115] Bănică F-G 2012 *Chemical Sensors and Biosensors: Fundamentals and Applications* (New York: Wiley)

[116] Baron R and Saffell J 2017 Amperometric gas sensors as a low cost emerging technology platform for air quality monitoring applications: a review *ACS Sens.* **2** 1553–66

[117] Thévenot D R, Toth K, Durst R A and Wilson G S 2001 Electrochemical biosensors: Recommended definitions and classification *Biosens. Bioelectron.* **16** 121–31

[118] United Nations 2015 *Transforming Our World: The 2030 Agenda for Sustainable Development* (New York: United Nations)

[119] A European green deal–striving to be the first climate neutral continent 2021 (https://ec.europa.eu/info/strategy/priorities-2019–2024/european-green-deal_en)

[120] Ho C K, Robinson A, Miller D R and Davis M J 2005 Overview of sensors and needs for environmental monitoring *Sensors* **5** 4–37

[121] Lewis A and Edwards P 2016 Validate personal air-pollution sensors *Nature* **535** 29–31

[122] Barabde M and Danve S 2015 Real time water quality monitoring system *Int. J. Innov. Res. Comput. Commun. Eng.* **3** 5064–9

[123] Pasika S and Gandla S T 2020 Smart water quality monitoring system with cost-effective using iot *Heliyon* **6** E04096

[124] Groves P D 2015 Principles of gnss, inertial, and multisensor integrated navigation systems *IEEE Aerosp. Electron. Syst. Mag.* **30** 26–7

[125] Kessler D J, Johnson N L, Liou J-C and Matney M 2010 The kessler syndrome: Implications to future space operations *Adv. Astron. Sci.* **137** 61

[126] Wang Q, Zhu X, Ni Y, Gu L and Zhu H 2020 Blockchain for the IoT and industrial IoT: a review *Internet of Things* **10** 100081

[127] UN Online Index of Objects Launched into Outer Space (https://unoosa.org/oosa/osoindex/search-ng.jspx?lf_id=)

[128] 2020 Global Space Economy at a Glance (https://brycetech.com/reports)

[129] Measuring the Ecomomic Impact of the Space Sector, OECD Background Paper for the G20 Space Economy Leaders' Meeting, October 7, 2020

[130] Dependence of the European Economy on Space Infrastructures, Potential Impacts of Space Assets Loss, Directorate-General for Internal Market, Industry, Entrepreneurship and SMEs, European Commission

[131] Akrami Y *et al* 2020 Planck 2018 results: overview and the cosmological legacy of Planck *Astron. Astrophys.* **641** A1

[132] Voyage 2050: Long-Term Planning of the ESA Science Programme. White Papers. (https://cosmos.esa.int/web/voyage-2050/white-papers)

[133] Voyage 2050 Sets Sail: ESA Chooses Future, ESA—Voyage 2050 Sets Sail: ESA Chooses Future Science Mission Themes

[134] Artemis-1 NASA website (https://www.nasa.gov/mission/artemis-i/)

[135] Moonlight: Connecting Earth with the Moon (https://esa.int/Applications/Telecommunications_Integrated_Applications/Lunar_satellites)

[136] Allen C *et al* 2012 Challenges of a Mars simple return mission from the samples' perspective —contamination, preservation, and planetary protection *Concepts and Approaches for Mars Exploration* June 12–14, 2012, Houston, Texas

[137] Moon Village, A Vision for Global Cooperation and Space 4.0 (https://esa.int/About_Us/Ministerial_Council_2016/Moon_Village)

[138] *Melissa, Micro Ecological Life Support System Alternative* (https://melissafoundation.org/page/melissa-pilot-plant)

[139] Protection of Frequencies for Radioastronomical Measurements in the Shielded Zone of the Moon, RECOMMENDATION ITU-R RA.479–5, ITU Radiocommunication Assembly

[140] Barhels M 2019 *Radio Telescope Unfurls 3 Antennas Beyond the Far Side of the Moon* 2 December (https://space.com/radio-telescope-beyond-moon-far-side-antennas-deploy.html)

[141] Global Exploration Roadmap, Supplement Augsut 2020, Lunar Surface Exploration Scenario Update, ISECG, August 2020 (https://globalspaceexploration.org/wp-content/uploads/2020/08/GER_2020_supplement.pdf)

[142] *The Global Risk Report 2021* 16th edn (http://www3.weforum.org/docs/WEF_The_Global_Risks_Report_2021.pdf)

[143] *UCS (Union of Concerned Scientists) Satellite Database*, Jan 2019 (https://ucsusa.org/nuclear-weapons/space-weapons/satellite-database?_ga=2.21168895.576778038.1554626468-1092692463.1554216706#.XDZDs2l7ksc)

[144] *ESA Budget 2024* (https://www.esa.int/ESA_Multimedia/Images/2024/01/ESA_budget_by_domain_2024)

[145] *Essential Climate Variables, World Metereological Organisation* (https://public.wmo.int/en/programmes/global-climate-observing-system/essential-climate-variables)

[146] *EU Climate Strategies & Targets, 2050 Long-Term Strategy* (https://ec.europa.eu/clima/policies/strategies/2050_en)

[147] *The European Green Deal, Communication from the Commission to the European Parliament, The Euroepan Council, The Council, The European Economica and Social Committee and the Committee of the Regions* (https://eur-lex.europa.eu/legal-content/EN/TXT/?qid=1596443911913&uri=CELEX:52019DC0640#document2)

[148] *A Brief Outlook on Future Copernicus Missions* (https://sentinels.copernicus.eu/web/sentinel/missions/copernicus-expansion-missions)

[149] *Contracts Awarded for Development of Six New Copernicus Missions* (https://esa.int/Applications/Observing_the_Earth/Copernicus/Contracts_awarded_for_development_of_six_new_Copernicus_missions)

[150] *ESA Earth Explorers: Understanding Our Planet* (https://esa.int/Applications/Observing_the_Earth/Future_EO/Earth_Explorers/About_Earth_Explorers2)

[151] *ESA and NASA Join Forces To Understand Climate Change* (https://esa.int/Applications/Observing_the_Earth/ESA_and_NASA_join_forces_to_understand_climate_change)

[152] Avila-Rodriguez J-A *et al* 2006 The MBOC modulation: the final touch to the galileo frequency and signal plan *Proc. of the Int. Technical Meeting of the Institute of Navigation (Fort Worth, TX, 25–28 September)* ION-GNSS 2006

[153] *Galileo's Contribution to COSPAS-SARSAT Programme* (https://ec.europa.eu/growth/sectors/space/galileo/sar_en)

[154] Galileo High-Accuracy Services 2020 *Information Note* GSA (https://gsc-europa.eu/sites/default/files/sites/all/files/Galileo_HAS_Info_Note.pdf)

[155] *Assuring Authentication for all, European Global Navigation Satellite Systems Agency, GSA* (https://gsc-europa.eu/news/assuring-authentication-for-all)

[156] Schuldt T *et al* 2021 Optical clock technologies for global navigation satellite systems *GPS Solutions* **25** 83

[157] Navigation system based on neutrino detection, US. Patent, US 8,849,565 B1, 2014

[158] *The Current State of Space Debris* (https://esa.int/Safety_Security/Space_Debris/The_current_state_of_space_debris)

[159] Space Surveillance Network (SSN), U.S. Army, Navy and Air Force (http://au.af.mil/au/awc/awcgate/usspc-fs/space.htm)

[160] Krag H *et al* 2017 A 1 cm space debris impact onto the Sentinel-1A Solar Array *Acta Astronaut.* **137** 434–43

[161] Kessler D *et al* 1978 Collision frequency of artificial satellites: the creation of a debris belt *J. Geophys. Res.* **83** 2637–46

[162] *IADC Space Debris Mitigation Guidelines*, IADC-02-01 Revision 1 September 2007

[163] *ESA Purchases World-First Debris Removal Mission from Start-up* (https://esa.int/Safety_Security/ESA_purchases_world-first_debris_removal_mission_from_start-up)

[164] *Planetary Defence* (www.esa.int/Safety_Security/Hera/Planetary_defence)

[165] *ESA Sfatey & Security Missions: Plans for the Future* (www.esa.int/Safety_Security/Plans_for_the_future)

[166] Euroconsult December 2019 *The Space Economy Report 2019*

[167] Crane K W *et al* 2020 *Measuring the Space Economy: Estimating the Value of Economic Activities in and for Space* (Science &Technology Policy Institute)

[168] *The Future of the European Space Sector How to Leverage Europe's Technological Leadership and Boost Investments for Space Ventures* (European Investment Bank, 2019)

[169] Geohistorical data website https://geohistoricaldata.org

[170] Fujita M, Krugman P R and Venables A J 2001 *The Spatial Economy: Cities, Regions, and International Trade* (Cambridge, MA: MIT Press)

[171] Barthelemy M 2016 *The Structure and Dynamics of Cities* (Cambridge: Cambridge University Press)

[172] Batty M 2013 *The New Science of Cities* (Cambridge, MA: MIT Press)

[173] O'Sullivan D and Manson S M 2015 Do physicists have geography envy? and what can geographers learn from it? *Ann. Assoc. Am. Geogr.* **105** 704–22

[174] Auerbach F 1913 Das gesetz der bevölkerungskonzentration *Petermanns Geogr. Mitt.* **59** 74–6

[175] Zipf G K 1949 *Human Behavior and the Principle of Least Effort* (Reading, MA: Addison-Wesley)

[176] Cottineau C 2017 Metazipf. a dynamic meta-analysis of city size distributions *PLoS One* **12** e0183919

[177] Verbavatz V and Barthelemy M 2020 The growth equation of cities *Nature* **587** 397–401

[178] Batty M 2006 Rank clocks *Nature* **444** 592–6

[179] Seto K C, Fragkias M, Güneralp B and Reilly M K 2011 A meta-analysis of global urban land expansion *PLoS One* **6** e23777

[180] Seto K C, Güneralp B and Hutyra L R 2012 Global forecasts of urban expansion to 2030 and direct impacts on biodiversity and carbon pools *Proc. Natl Acad. Sci.* **109** 16083–8

[181] Atlas of urban expansion 2024 http://atlasofurbanexpansion.org

[182] Barbosa H, Barthelemy M, Ghoshal G, James C R, Lenormand M, Louail T, Menezes R, Ramasco J J, Simini F and Tomasini M 2018 Human mobility: models and applications *Phys. Rep.* **734** 1 74

[183] Newman P W G and Kenworthy J R 1989 Gasoline consumption and cities: a comparison of us cities with a global survey *J. Am. Plan. Assoc.* **55** 24–37

[184] Creutzig F, Baiocchi G, Bierkandt R, Pichler P-P and Seto K C 2015 Global typology of urban energy use and potentials for an urbanization mitigation wedge *Proc. Natl Acad. Sci.* **112** 6283–8

[185] Verbavatz V and Barthelemy M 2019 Critical factors for mitigating car traffic in cities *PLoS One* **14** e0219559

[186] Fujita M and Ogawa H 1982 Multiple equilibria and structural transition of non-monocentric urban configurations *Reg. Sci. Urban Econ.* **12** 161–96

[187] Louf R and Barthelemy M 2013 Modeling the polycentric transition of cities *Phys. Rev. Lett.* **111** 198702

[188] Louf R and Barthelemy M 2014 How congestion shapes cities: from mobility patterns to scaling *Sci. Rep.* **4** 5561

[189] Branston D 1976 Link capacity functions: a review *Transp. Res.* **10** 223–36

[190] TomTom International BV (2008–2017) TomTom Traffic Index

[191] OECD 2016 The OECD Metropolitan Areas Database visualized through the Metropolitan eXplorer

[192] Duranton G and Turner M A 2011 The fundamental law of road congestion: evidence from us cities *Am. Econ. Rev.* **101** 2616–52

[193] Moran D, Kanemoto K, Jiborn M, Wood R, Többen J and Seto K C 2018 Carbon footprints of 13000 cities *Environ. Res. Lett.* **13** 064041

[194] Lighthill M J and Whitham G B 1955 On kinematic waves ii. a theory of traffic flow on long crowded roads *Proc. Royal Soc. Lond. Ser. A. Math. Phys. Sci.* **229** 317–45

[195] Nagel K and Schreckenberg M 1992 A cellular automaton model for freeway traffic *J. Phys. I* **2** 2221–9

[196] Blandin S, Argote J, Bayen A M and Work D B 2013 Phase transition model of non-stationary traffic flow: definition, properties and solution method *Transp. Res. B* **52** 55

[197] Gartner N H, Messer C J and Rathi A 2002 Traffic flow theory—a state-of-the-art report: revised monograph on traffic flow theory https://rosap.ntl.bts.gov/view/dot/35775

[198] Nagel K and Paczuski M 1995 Emergent traffic jams *Phys. Rev. E* **51** 2909

[199] Sugiyama Y, Fukui M, Kikuchi M, Hasebe K, Nakayama A, Nishinari K, Tadaki S-ichi and Yukawa S 2008 Traffic jams without bottlenecks—experimental evidence for the physical mechanism of the formation of a jam *New J. Phys.* **10** 033001

[200] Tadaki S-ichi, Kikuchi M, Fukui M, Nakayama A, Nishinari K, Shibata A, Sugiyama Y, Yosida T and Yukawa S 2013 Phase transition in traffic jam experiment on a circuit *New J. Phys.* **15** 103034

[201] Tadaki S-ichi, Kikuchi M, Nakayama A, Shibata A, Sugiyama Y and Yukawa S 2016 Characterizing and distinguishing free and jammed traffic flows from the distribution and correlation of experimental speed data *New J. Phys.* **18** 083022

[202] Biham O, Middleton A A and Levine D 1992 Self-organization and a dynamical transition in traffic-flow models *Phys. Rev. A* **46** R6124

[203] Cuesta J A, Martnez F C, Molera J M and Sánchez A 1993 Phase transitions in two-dimensional traffic-flow models *Phys. Rev. E* **48** R4175

[204] Echenique P, Gómez-Gardenes J and Moreno Y 2005 Dynamics of jamming transitions in complex networks *EPL (Europhys. Lett.)* **71** 325

[205] Carmona H A, de Noronha A W T, Moreira A A, Araújo N A M and Andrade J S 2020 Cracking urban mobility *Phys. Rev. Res.* **2** 043132

[206] Lampo A, Borge-Holthoefer J, Gómez S and Solé-Ribalta A 2021 Multiple abrupt phase transitions in urban transport congestion *Phys. Rev. Res.* **3** 013267

[207] Bellocchi L and Geroliminis N 2020 Unraveling reaction-diffusion-like dynamics in urban congestion propagation: Insights from a large-scale road network *Sci. Rep.* **10** 1–11

[208] Jiang Y, Kang R, Li D, Guo S and Havlin S 2017 Spatio-temporal propagation of traffic jams in urban traffic networks *arXiv preprint* arXiv:1705.08269

[209] Saberi M, Hamedmoghadam H, Ashfaq M, Hosseini S A, Gu Z, Shafiei S, Nair D J, Dixit V, Gardner L and Waller S T *et al* 2020 A simple contagion process describes spreading of traffic jams in urban networks *Nat. Commun.* **11** 1–9

[210] Li D, Fu B, Wang Y, Lu G, Berezin Y, Stanley H E and Havlin S 2015 Percolation transition in dynamical traffic network with evolving critical bottlenecks *Proc. Natl Acad. Sci.* **112** 669–72

[211] Kadanoff L P 2000 *Statistical Physics: Statics, Dynamics and Renormalization* (Singapore: World Scientific)

[212] Taillanter E and Barthelemy M 2021 Empirical evidence for a jamming transition in urban traffic *J. R. Soc. Interface* **18** 20210391

[213] Louail T, Lenormand M, Cantu Ros O G, Picornell M, Herranz R, Frias-Martinez E, Ramasco J J and Barthelemy M 2014 From mobile phone data to the spatial structure of cities *Sci. Rep.* **4** 1–12

[214] Ratti C, Frenchman D, Pulselli R M and Williams S 2006 Mobile landscapes: using location data from cell phones for urban analysis *Environ. Plan. B: Plan. Design* **33** 727–48

[215] Chang S, Pierson E, Koh P W, Gerardin J, Redbird B, Grusky D and Leskovec J 2021 Mobility network models of Covid-19 explain inequities and inform reopening *Nature* **589** 82–7

[216] Ganin A A, Kitsak M, Marchese D, Keisler J M, Seager T and Linkov I 2017 Resilience and efficiency in transportation networks *Sci. Adv.* **3** e1701079

[217] Wang W, Yang S, Stanley H E and Gao J 2019 Local floods induce large-scale abrupt failures of road networks *Nat. Commun.* **10** 1–11

[218] Buldyrev S V, Parshani R, Paul G, Stanley H E and Havlin S 2010 Catastrophic cascade of failures in interdependent networks *Nature* **464** 1025–8

[219] Sanders L, Pumain D, Mathian H, Guérin-Pace F and Bura S 1997 Simpop: a multiagent system for the study of urbanism *Environ. Plan. B: Plan. Design* **24** 287–305

[220] Aldous D and Barthelemy M 2019 Optimal geometry of transportation networks *Phys. Rev. E* **99** 052303

IOP Publishing

EPS Grand Challenges
Physics for Society in the Horizon 2050

Mairi Sakellariadou, Claudia-Elisabeth Wulz, Kees van Der Beek, Felix Ritort, Bart van Tiggelen, Ralph Assmann, Giulio Cerullo, Luisa Cifarelli, Carlos Hidalgo, Felicia Barbato, Christian Beck, Christophe Rossel and Luc van Dyck

Chapter 8

Science for society

Tobias Beuchert, Alan Cayless, Frédéric Darbellay, Richard Dawid, Bengt Gustafsson, Sally Jordan, Philip Macnaghten, Eilish McLoughlin, Christophe Rossel, Pedro Russo, Luc van Dyck, François Weiss and Ulrich von Weizsäcker

8.1 Introduction

Christophe Rossel[1] and Luc van Dyck[2]

[1]IBM Research Europe - Zurich, Switzerland
[2]Euro-Argo ERIC, Plouzané, France

During the second part of the 20th century, the social contract between science and society was merely a tacit agreement foreseeing that public money would finance the research that would sustain technology development and innovation and enhance the socioeconomic well-being of our society. The spheres of science, politics, and society were largely separate.

In the last 25 years, this model has been broadly questioned. Blurred ethical standards and catastrophes in addition to the dissemination of 'fake news' have repeatedly undermined some people's faith in science. Innovation has not always been driven by the common good or the needs and expectations of the citizens. Most important, there has been increasing awareness that the world is facing drastic new challenges, from climate change and food security to migrations and energy supplies, which will determine its future.

doi:10.1088/978-0-7503-6342-6ch8 © 2024 The Authors.
Published under license by IOP Publishing Ltd

Against this background, a new normal is arising. It involves the interplay of all sciences, including natural, social, and human sciences, without which societal challenges cannot be solved. Education and training must be rethought to foster interdisciplinarity and transdisciplinarity. A democratic governance of science and innovation will, while protecting the inspiration and creativity that drives research, facilitate the participation of all stakeholders in developing choices and processes. Citizens will have greater expectations regarding communication and accountability from scientists at a time when the Internet revolution and social media make it possible for all to access, understand, and share knowledge and scientific data. At the dawn of the open science era, we are seeing the benefits of information technology and artificial intelligence (AI) in consolidating and speeding up the research and innovation process. Finally, we must have trust between citizens and science, conditioned by aspects of research such as ethics, integrity, and transparency.

A global goal is to generate the new knowledge that will help us to better understand and address the major challenges of our time and facilitate the transfer and integration of scientific findings into politics and society. But science has its own limits, whether theoretical, experimental, ethical, or philosophical.

All these issues that will determine the future of scientific research—and *ipso facto* of humankind—are addressed and discussed in this chapter by a panel of prominent contributors. The chapter is divided into five main sections: *Education and research in an interdisciplinary environment, Science with and for the citizens, Open communication and responsible citizens, Science and ethics,* and *Limits of science.*

8.2 Education and research in an interdisciplinary environment

8.2.1 The future of physics education

Eilish McLoughlin[1]

[1]Dublin City University, Dublin, Ireland

8.2.1.1 Introduction
Physics for Society at the Horizon of 2050 will focus on addressing societal grand challenges and enabling individuals and societies to prosper in a globally connected society. The Physics Education Division of the European Physics Society has identified physics education as vital to ensuring active citizenship in democratic societies, as well as for supplying and training a wide range of scientists and engineers. This mission is echoed in the Organisation for Economic Cooperation and Development (OECD) Learning Compass 2030, an evolving learning frame-work that sets out an aspirational vision for the future of education [1]:

> *How can we prepare students for jobs that have not yet been created, to tackle societal challenges that we cannot yet imagine, and to use technologies that have not yet been invented? How can we equip them to thrive in an interconnected world where they need to understand and appreciate different perspectives and world views, interact respectfully with others, and take responsible action toward sustainability and collective well-being?*

The learning compass (figure 8.1) offers a vision of the types of interconnected competencies that students will need to succeed in 2030, namely, knowledge, skills, attitudes, and values. The concept of student agency—defined as the capacity to set a goal, reflect, and act responsibly to effect change—is central to this framework. This concept is rooted in the principle that when students are agents in their own learning, they are more likely to have 'learned how to learn', an invaluable skill that they can and will use throughout their lives.

It is widely recognised that individuals who have studied physics are eminently suited to roles in a wide range of jobs, industries, and organizations; therefore, their physics education needs to support the development of interconnected competencies across the domains of knowledge, skills, attitudes, and values. Over the past decade,

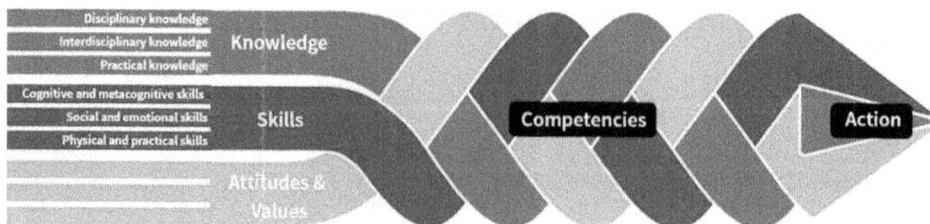

Figure 8.1. Organisation for Economic Cooperation and Development learning compass 2030 [1], reproduced courtsey OECD (OECD, 2019).

physics curricula in schools, colleges, and universities have adopted learning goals that involves developing student's scientific abilities, skills, and competencies alongside physics-specific knowledge. It is less common, however, for physics programmes to explicitly consider knowledge and skills associated with the application of physics in interdisciplinary contexts and in the wide variety of career settings in which many graduates find themselves [2]. Crosscutting, interdisciplinary connections are becoming important features of future-generation physics curricula and define how physics should be taught collaboratively along with other STEM disciplines. Recent studies highlight that an integrated approach to STEM education can be effective in supporting students to develop a range of transversal competencies such as problem solving, innovation and creativity, communication, critical thinking, meta-cognitive skills, collaboration, self-regulation, and disciplinary competencies [3]. Indeed, individuals' long-term professional success is often attributed to their having developed transferrable skills that can be applied in diverse career directions.

8.2.1.2 Pedagogical practices in physics education

Despite the frequent use of a variety of representations, such as graphs, symbols, diagrams, and text, by physics teachers, educators, and researchers, the notion of using the pedagogical functions of multiple representations to support teaching and learning is still a gap in physics education. Studies have shown that when students use representations in multiple formats during the learning process, their conceptual understandings of physics concepts as well as their problem-solving skills are enhanced [4].

While psychologists and educational scientists seem to converge on the notion that student involvement is critical to successful learning, there is much debate around how students should be facilitated in the learning process. Bao and Koenig [5] report as follows:

> Based on a century of education research, consensus has settled on a fundamental mechanism of teaching and learning, which suggests that knowledge is developed within a learner through constructive processes and that team-based guided scientific inquiry is an effective method for promoting deep learning of content knowledge as well as developing high-end cognitive abilities.

In disciplines such as physics, inquiry-based learning (IBL) has been promoted as being a more effective pedagogical approach compared to more expository instructional approaches—as long as learners are supported adequately. John Dewey developed the model of learning known as 'the pattern of inquiry' and was the first to use the word inquiry in education: 'Scientific principles and laws do not lie on the surface of nature. They are hidden, and must be wrested from nature by an active and elaborate technique of inquiry' [6]. Distinct levels of inquiry instruction have been described in terms of the level of student guidance, for example, structured, guided, and open inquiry. Research conducted on the effectiveness of IBL pedagogy emphasises that guidance is pivotal to successful IBL at all levels; learners who are given some kind of guidance act more skilfully during the task, are more successful in obtaining topical information from their investigational practices, and score higher on tests of learning outcomes administered after the inquiry [7]. These benefits are largely independent of the specificity of the guidance. Even though performance success tends to increase more when more specific guidance is available, learning activities and learning outcomes improve equally with specific and nonspecific types of guidance. Interestingly, the effectiveness of guidance is shown to apply equally to children, teenagers, and adolescents; this finding offers physics educators scope to design effective IBL learning opportunities at all educational levels.

8.2.1.3 Strategies for assessment of learning in physics education

While there is growing agreement on the competencies individuals should possess as well as the pedagogies to develop them, there is still much debate about what strategies are effective in assessing these competencies. There are multiple forms of assessment that vary depending on the purpose they are intended to serve. Common forms of assessment include the assessment of student performance through high-stakes exams or through collection of information about students' achievement with the intent to assign a grade, usually at the end of instruction. These are examples of summative assessment in the sense that their purpose is to provide evidence of learning after instruction. Formative assessment, by constrast, is centred on (1) the collection of useful assessment information through appropriate means, (2) the meaningful interpretation of the assessment information, and (3) the process of acting on the interpretation of the assessment information to enhance teaching and learning while they are still in progress [8].

The use of formative assessment has received much recognition as a powerful means of enhancing students' learning in physics. In particular, the use of self-assessment and peer assessment is considered essential in achieving honest collaboration, critical thinking, and respect between learners. Feedback is an indispensable component of formative assessment and is considered pivotal to its potential effectiveness. It is important to emphasise that feedback needs to be formative in that it should extend beyond merely indicating performance (success/failure) to explicitly seeking to influence students' subsequent actions in a manner that should facilitate learning. Interest in digital formative assessment has grown rapidly in the past two decades for several reasons, including the provision of feedback in a timelier manner, the assessment of hard-to-measure constructs and processes that were previously inaccessible, the inclusion of new item types capable of providing more nuanced information about learning, automation of the feedback process, access for students with disabilities, and greater opportunities for student collaboration.

However, a critical barrier to the use of effective strategies for assessing student learning in physics across range of competencies (knowledge, skills, attitudes, and values) is teachers and students' engagement and competence with assessment and feedback practices. There is an urgent need to provide tailored professional learning opportunities and resources for teachers of physics as well as explicit support and guidance for students on how to productively engage in assessment. If physics education is going to contribute to a connected and complex society, educators and researchers needs to collect and disseminate more evidence about what are effective teaching and learning processes in physics. This will require further development and sharing of valid and reliable assessment strategies and tools for assessment of a range of 21st century competencies.

8.2.1.4 Equity and inclusion in physics education

There is increased emphasis on the need for schools and education systems to provide equal learning opportunities to all students. Equity does not mean that all students must obtain equal outcomes but rather that, provided with the same opportunities, differences in students' outcomes are not driven by individual factors such as gender, race, socioeconomic status, immigration background, or disabilities [9]. For example, the socioeconomic status of students' families is widely recognised as a reliable predictor of student academic performance and, indirectly, success in life. There is a growing body of evidence showing that gender stereotypes—both implicit and explicit—affect engagement and professional interactions of women in physicas and affect their careers [10]. Recent studies have focused on how a student's sense of belonging in physics can be adversely affected by intersections of these factors, such as interpersonal relationships, perceived competence, personal interest, and science identity.

Statistics compiled by the American Physical Society indicate that women typically represent only 20% of undergraduate and postgraduate physics student cohorts [11]. Given the historic and continued underrepresentation of women in physics, it is important to understand the role that secondary-level physics education plays in attracting young women to physics degrees and careers. An examination of women's experiences in high school physics education has identified three key barriers to their participation: students' perceptions of the field and what type of person practices physics, their personal experiences of learning physics, and their experiences with gender and physics identity. Ultimately, these barriers to young women's self-efficacy, competence, performance, and recognition in physics inhibit their developing a physics identity and often result in their lack of participation in further physics studies [12].

What is being done to address the gender gap in physics? The American Physical Society has partnered in a US national movement [13] to promote physics identity development and empower high school teachers to recruit women to pursue physics degrees in higher education. Initial studies from this intervention report significant changes in high school students' physics identity, that is, their recognition beliefs (feeling recognized by others as a physics person) and beliefs in a future physics career. In the UK, the Institute of Physics (IOP) has reported that there has been very little change in the proportion of girls studying physics at the upper second level over the past 30 years [14]. To tackle this issue, the IOP has implemented programmes to improve gender balance in England, Scotland, Ireland, and Wales through collaboration with schools and education–research partnerships. These programmes have facilitated professional learning opportunities for teachers to develop their knowledge and awareness of physics careers, unconscious bias, and

Table 8.1. Tackling the misconceptions that affect gender balance in physics [15].

- There is more variance within groups of boys and within groups of girls than there is between boys and girls. Gender differences are learned, not innate.
- One group should not be preferentially treated compared to any other group.
- Unconscious bias and normalisation of stereotypes means that there are often unspoken barriers.
- One-off activities or interventions do not have a lasting impact. They need to be part of a wider strategy.
- Role models can have a positive impact but usually only where there is an ongoing relationship.
- A teacher's gender does not have a large influence on subject choice. The majority of students respond to good teaching, irrespective of whether the teacher is male or female.
- Attempts to make a subject more appealing by reinforcing a stereotype are unlikely to be effective.

inclusive teaching approaches. Through these programmes, schools are supported to address equity and inclusion in physics education by tackling common misconceptions (as presented in table 8.1). The findings of these programmes emphasise the importance of raising awareness in schools of the inherent barriers that learners face in accessing opportunities.

Some of these barriers to participation in physics may be alleviated through recent technological developments that provide digital ecologies and create respective ecologies in education. Such ecologies can enhance teaching, learning, and assessment by offering innovative and inclusive learning opportunities for students that would otherwise be difficult to achieve and can result in physics education becoming more equitable and inclusive.

- Physics curricula

  Over the last three decades, physics education research has provided substantial evidence of the impact of innovative teaching, learning, and assessment strategies on the development of physics understanding and competencies. However, the uptake of physics in schools, colleges, and universities continues to be less popular than that of other natural sciences, and physics suffers from a stigma of being a science that is very mathematical, abstract, and complicated. Physics is a core pillar of all natural sciences; it is about understanding the basic laws of nature and explains how the world around us and within us functions. This fundamental understanding of physics is often not appreciated by students, and physics curricula in schools need to evolve to address this issue. The use of investigative and inquiry-based learning approaches is strongly encouraged for introducing physics concepts, starting with our youngest learners. Such approaches can support learners to 'learn by doing' and develop their knowledge and understanding of physics through real-world examples. In this way, physics education will play an important role in underpinning the development of STEM knowledge and competencies required in later curricula.

While efforts to update physics curricula in schools has gained momentum over recent decades, in many cases, contemporary and modern physics topics are introduced only by using a reductionist approach at the end of a curriculum in topics such as special relativity, particle physics, quantum physics, and liquid crystals. The particle physics community believes that exposure to particle physics and its technological applications increases the interest of students in physics and can change their perceptions of the role of physics and physicists in today's society [16]. Many physics educators emphasise the urgent need for collaborative research to design coherent learning paths and scaffold student's conceptual development in modern physics topics [17]. The inclusion of contemporary and modern physics topics in school curricula offers an opportunity to provide authentic and engaging learning experiences for students and ultimately change their perceptions of the role of physics and physicists in society.

- Supporting physics teachers

It is broadly recognised that the quality of an education system is highly dependent on (1) getting the right people to become teachers, (2) developing them into effective educators, and (3) ensuring that the system is able to deliver the best possible education for every child. However, the lack of qualified teachers of physics in secondary-level schools is a matter of international concern.

A new report from the American Physical Society prioritises actions to be taken to ensure that the US continues to be a global leader in science, technology, and innovation [18]. These include policies to 'address the urgent shortage in qualified STEM teachers, so that aspiring STEM-professionals have the opportunity to join the workforce of the industries of the future and eliminate hostile workplaces and pathways to all that want to contribute to innovation and a better society'.

Over the past two decades, the OECD Directorate for Education and Skills, together with its member and partner countries, has collected significant volumes of data on teachers, school leaders, and students that allow educators and policymakers to learn from the policies and practices that are being applied in other countries. The 2021 OECD report presents combined data collected from both the 2015 Programme for International Student Assessment (PISA) and 2018 Teaching and Learning International Survey (TALIS) [9]. This report highlights that since PISA was carried out in 2015, expenditure on schooling has climbed steadily. However in 2018, PISA showed that students' performance scores in reading, mathematics and science in the Western world have flat-lined. In particular, this report highlights the following:

Teachers and schools make an important difference to how a student performs and feels. More specifically, it is the time teachers spend actually teaching in class, not disciplining or taking care of administrative work, and the hours they spend marking and correcting work, and going over this feedback with their students that links to how well students do academically, and how motivated and optimistic they are about their learning and prospects.

The report highlights the presence of differential effects across subjects, as teachers' satisfaction with their work environment seems to be more closely related to student performance in science than in reading and mathematics. While this may be explained by the fact that, unlike reading, students mainly acquire their knowledge in science at school, it also recognises that requirements such as a well-equipped school lab are critical for science teachers to supporting student learning in science [9]. A critical review of the role of laboratory work in physics teaching and learning presented by Sokołowska and Michelini [19] highlights the essential role of laboratory work in supporting and extending student learning in physics, that is, conceptual under-standing, creativity, metacognition, modeling, and problem-solving skills. Fostering collaboration between researchers and teachers to design interactive learning environ-ments, such as augmented or virtual reality laboratories, can also provide unique opportunities to deepen student's understanding and engagement with both funda-mental and contemporary topics in physics. However, the widespread adoption of these approaches in classroom practice is highly dependent on teachers' competence and confidence in designing appropriate learning opportunities. Appropriate and sustained professional learning opportunities need to be provided for teachers to extend and deepen their own content knowledge for teaching physics. A variety of models have been proposed for physics teacher education. Recent strategies advocate supporting teachers collaborating as part of a professional learning community to carry out practitioner inquiry on their own practice.

8.2.1.5 Conclusion

In conclusion, several actions are needed to refocus the vision for physics education in Europe so as to make valuable contributions to Physics for Society at the Horizon of 2050. These changes need to be embedded in all education levels, from early childhood to higher education, and careful attention needs to be given to supporting learners across educational transitions. First, physics curricula and pedagogies need to be revised and focus on the development of competencies across the domains of knowledge, skills, attitudes, and values. Second, strategies for achieving equity and inclusion in physics education need to be adopted in classroom practice. Third, interdisciplinary collaboration between physics educators, researchers, and policy-makers is critical to attracting and supporting future generations of both physics students and physics teachers.

8.2.2 The challenges of physics education in the digital era

Alan Cayless[1] and Sally Jordan[1]
[1]School of Physical Sciences, The Open University, Milton Keynes, UK

8.2.2.1 Introduction

In this section, we consider the impact on physics education of the rapid development of digital technology, looking at challenges and opportunities in teaching, assessment, and experimental work.

Technological progress has changed the types of employment for which we are preparing students as well as their expectations and aspirations. Scientific equipment depends increasingly on electronics and computerisation, and students need skills that are appropriate for a data-rich world. Scientific investigation is an ideal platform for developing transferable skills such as IT literacy, programming, scientific communication, and collaborative working, all highly valued by employers.

Education has increased its use of digital technology, and the global response to the COVID-19 pandemic has accelerated the pace of change with lasting effects. Physics educators should be prepared to understand and respond to new technologies as they emerge.

8.2.2.2 Digital teaching and learning

While distance-learning institutions such as the UK Open University (OU) have used online tuition for some time, many other providers moved to online teaching during 2020, thanks to the widespread availability and adoption of online videoconferencing. Many students are now familiar and confident with online collaboration tools in other settings, leading to a better acceptance of this technology in educational context.

However, many challenges remain. Computer equipment and the Internet are not accessible to all students and educators, or equipment might be obsolete with bandwidth and reliability issues. Digital technology can enhance learning capabilities for some students but create barriers for others. We anticipate that these difficulties will reduce towards 2050 as stable platforms with high-bandwidth connectivity become more widely available. AI will certainly solve accessibility issues, for example, by enabling accurate real-time captioning.

Students and educators frequently have concerns regarding feedback and interactivity in an online setting. However, videos, interactive activities, and quizzes as well as text chat in videoconferencing platforms enable students to participate in discussions with more confidence.

We advocate a blended approach, mixing the strength of online delivery with more conventional methods. Using printed materials in conjunction with onscreen study can reduce the cognitive load on learners, making learning more efficient and leading to improved outcomes [20]. The study process is increasingly taking place on a wide range of digital devices, including tablets and e-readers, and improvements in these technologies may reduce the gap between online and paper-based reading and learning [21].

Students' perceptions and expectations of online study are important, as they can affect their performance on science modules. Onscreen study skills can be better developed if students are exposed early to online materials and activities. A sense of community and involvement with peers can be built through online tutorials and discussion forums and through collaborative working in group projects and remote experiments [22]. This is especially important in an entirely distance-learning environment, where students are unlikely to have access to traditional laboratories and classrooms.

Online teaching has considerable benefits. Analytics can be used to gather statistics on students' interaction with online content, allowing us to better understand their progress and to personalise educational materials and activities. By eliminating the need for travel, international students and students who are housebound or in remote locations can participate equally, without the financial and environmental cost of traveling. Thoms and Girwidz [23] describe a virtual remote laboratory that combines the benefits of remote experiments, interactive screen experiments, and simulations via live internet connections or offline. The Global Hands-on Universe project [24], which includes the Galileo Teacher Training Programme, is an example of an international initiative making online science investigation accessible in less developed countries. It provides tools, software, and teacher training materials free of charge to locations with fewer resources and encourages participation in international scientific projects.

8.2.2.3 Digital assessment

Digital assessment usually involves multiple-choice quizzes. These bring the advantages of apparent objectivity, reduced marking time, especially for large classes, and rapid feedback. Multiple-choice quizzes can be used to trigger deep learning, for example by way of peer instruction [25]. However, such quizzes pose a number of challenges for physics educators. They have limited scope in assessing written or mathematical exercises, raising the question of whether correct answers indicate genuine understanding or guesswork.

Technological and pedagogical developments have extended the potential of computer-marked assessment beyond multiple-choice questions. Systems using computer algebra to mark and give feedback on algebraic responses and systems that mark free-text written answers are readily available [26]. Technology can be used to enhance the assessment of student learning in many ways, including online submission of assignments, feedback delivered by audio or video, and e-portfolios for assessment.

8.2.2.4 Experimental work

Traditional labs with students working onsite and interacting directly with experimental equipment are increasingly replaced by online remote experiments, as is done in many real-life scientific investigations. Indeed, today many researchers can work and acquire data on a particle accelerator or a telescope remotely. Thus, many of the skills developed by students on remote experiments are directly relevant in modern research environments.

An early example of a physical experiment in progress is a spectroscopy interactive screen experiment (ISE) [27], where students use real experimental data based on hundreds of photographs of a physical experiment. ISEs were designed to supplement other methods for teaching experimental physics and to help people with disability, illness, or other employment or caring responsibilities. ISEs also provide additional practice for students to complement conventional experiments.

8.2.2.5 Remote experiments in physics
In contrast to ISEs, true remote experiments involve real equipment operated via an Internet connection. Video and software feedback enable live interaction with real-time control of experimental equipment and collection of data.

Two experiments offered to undergraduate students at the UK OU illustrate the current state of the art in remote control. The first is a traditional Compton scattering experiment; the second is an infrared (IR) spectroscopy experiment involving up-to-date space science technology.

8.2.2.5.1 Compton scattering
In the online Compton scattering experiment (figure 8.2), x-rays from a tube are scattered by electrons in a Perspex acrylic target and captured by a solid-state detector on a rotatable arm. Spectra are analysed to confirm the change in x-ray energy versus angle. In doing this, students obtain the experimental evidence that photons carry a momentum of $p = h/\lambda$, thus confirming a fundamental relationship in quantum mechanics.

8.2.2.5.2 Planetary atmosphere gas cell
The hardware for the newer planetary atmosphere gas cell experiment employs a system of valves allowing analysis of different gas mixtures representing possible planetary atmospheres. The thermal valve, originally developed for the Rosetta exploration mission, controls the pressure of the gas cell where the infrared spectroscopy takes place. Students gain experience with components from actual space missions and learn experimental design and control techniques (figure 8.3).

8.2.2.6 Benefits and drawbacks of remote experiments
The advantage of remote experiments is that they offer access to more advanced and ambitious equipment and can be operated outside regular laboratory or teaching hours. UK OU students have also access to professional optical and radio telescopes and even to a Mars rover facility where teams can control a custom-built rover in a recreated Martian landscape, following protocols modeled on NASA mission operations.

Remote experiments enable students to work with hazardous materials and environments, such as radioactive sources, x-rays, and compressed gases. Health, safety, and risk assessment are built into planning and teaching materials. Thus

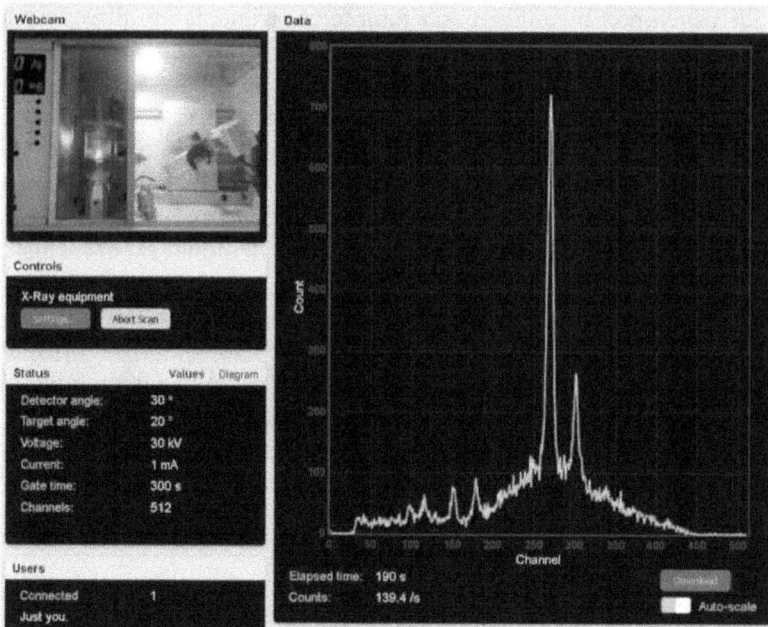

Figure 8.2. Compton scattering remote experiment. Computer interface with webcam view, control and status panels with setting parameters, and data display.

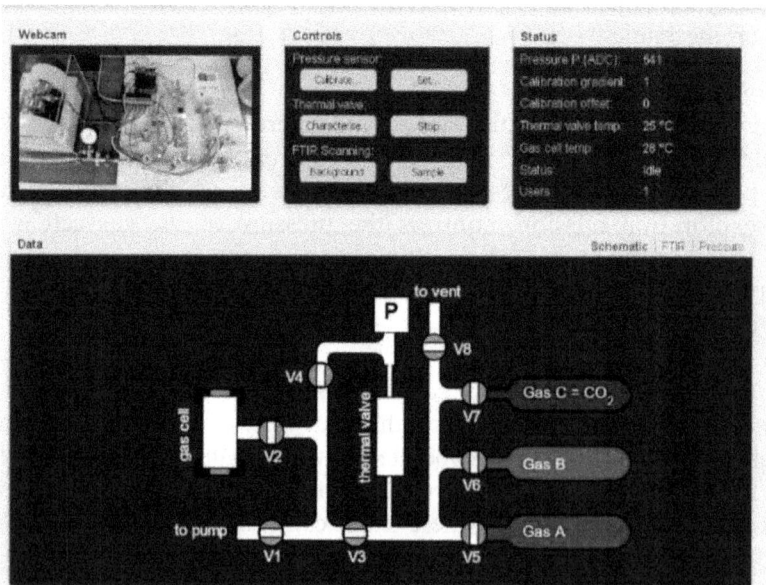

Figure 8.3. Control interface for IR spectroscopy of various gas mixtures. Computer interface with webcam view, control and status displays, and interactive representation of the experimental setup. Spectra are downloaded for analysis as in the Compton experiment.

students are educated in experimenting within safe limits, considering the safety of the personnel operating the actual equipment.

One drawback or concern relates to the effectiveness of remote training experiments regarding the development of procedural skills as compared to direct hands-on manipulation.

Other benefits and drawbacks are summarised in [28], where the authors conclude that well-designed online laboratories can be as effective and motivating as traditional ones if experiments are involving, interactive, and engaging. Dintsios *et al* [29] reported an increased acceptance of remote experiments among secondary students when the students are directly involved in their operation instead of having them demonstrated.

8.2.2.7 Opportunities and further developments to the Horizon of 2050

It is inevitable that science will continue to evolve with wider use of remotely operated equipment, large-scale collaborations, and more computational power for big data handling. This will drive new developments in experimental physics education, involving the latest technologies and fostering skills in future professional researchers. A major challenge will be to create online experiments that are as lifelike and immersive as possible. Stereoscopic views and virtual reality headsets could offer a form of telepresence, involving direct manipulation of equipment with kinesthetic feedback. Augmented reality will be used to overlay readouts and data directly onto live views of the experiment and to operate by virtual touch or gestures. AI may be used both in the design of experiments and in data analysis. The current rate of progress suggests that at least some of these technologies will become available within the next few years. Other unexpected developments will enable us to meet new challenges as 2050 approaches.

Recent events have highlighted the strengths of remote experiments in the most dramatic fashion. During the COVID-19 pandemic, with laboratories closed and students learning online, the remote experiments described previously continued to operate, enabling students to train under lockdown conditions. As described in [30], the crisis accelerated the uptake and acceptance of remote experiments among mainstream research facilities as well as in education, with potentially lasting effects.

8.2.3 The interdisciplinary challenge: the why, the what, the where, the who, and the how

Frédéric Darbellay[1]
[1]University of Geneva, Geneva, Switzerland

Physics, like a large majority of other disciplines, is a limited area of teaching and research in the academic and social fields. It is itself fragmented into specialised subdisciplines representing many potential silos in the organisation of scientific knowledge, know-how, and interpersonal skills. At the same time, this partitioning between physics as a discipline in its own right, as well as between its constituent subdisciplines, is very relative insofar as the complex field of physics participates in scientific progress by cross-fertilisation with other disciplines and/or between the subdisciplines that compose it. The general trend in physics as elsewhere is to increase interdisciplinary work by interaction between disciplines [31]. These interdisciplinary openings promote scientific success and the expression of the talents of researchers who venture beyond disciplinary boundaries by making room for randomness, serendipity, and creativity in the research process [32, 33].

Why are teaching and researching done from an interdisciplinary perspective? Among the various reasons which motivate teachers and researchers to embark on the interdisciplinary path, there is agreement on the identification of four major drivers [34]: (1) the inherent complexity of physical, natural, and social phenomena; (2) the desire to explore theoretical questions and/or practical problems which are not reducible to a single disciplinary point of view; (3) the need to solve problems; and (4) the impact of the power of new technologies. Between science, technology, and society, the plural scientific field of physics and its multiple areas of application replay and combine these drivers which guide interdisciplinary work. This desire for interdisciplinary collaboration—in which physics participates with full rights and obligations—for scientific progress and the need to solve in the short, medium, and long terms the urgent problems of society (health, climate, financial, social crises, etc) does not go without disciplinary resistance, epistemological controversies, and institutional obstacles. The creation of an efficient and sustainable interdisciplinary environment should be based on a constructive dialog between disciplines and interdisciplines.

What is interdisciplinarity and in what network of concepts does it make sense? The definitions of disciplinary, multidisciplinary, interdisciplinary, and transdisciplinary approaches express this productive tension between disciplines and their progressive decompartmentalisation (by degree) in a collaborative and integrative dynamic [35–37] (figure 8.4). If disciplinarity allows the deepening of specialized knowledge in a strictly delimited field of study, it can also be juxtaposed with other disciplinary perspectives to analyze and understand a theoretical or practical problem through different facets. This openness to multifaceted disciplinary pluralism is likely to be reconfigured in a more interdisciplinary dynamic which aims to overcome the juxtaposition of heterogeneous points of view to create interactions and links between (*inter-*, which is at the interface) the disciplines in order to

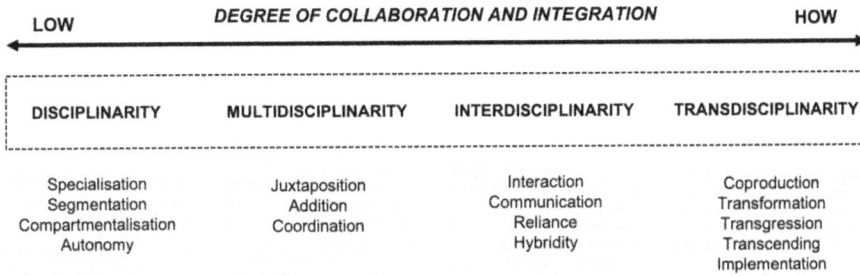

Figure 8.4. Degree of collaboration and integration between disciplines.

understand an object of study in its complexity and hybridity. Transdisciplinarity aims at the coproduction of knowledge and practical solutions between academic teachers and/or researchers and stakeholders by directly involving them from the start in the research process.

Interdisciplinarity in the broad and canonical sense is therefore:

> a mode of research by teams or individuals that integrates information, data, techniques, tools, perspectives, concepts, and/or theories from two or more disciplines or bodies of specialized knowledge to advance fundamental understanding or to solve problems whose solutions are beyond the scope of a single discipline or area of research practice [34].

Integration is at the heart of the interdisciplinary process which aims to bring into coherence and synthesize two or more disciplines to advance the global understanding of a theoretical or practical question and to solve complex problems. Integrative practices are also at work in transdisciplinarity, which integrates extrascientific actors in teaching and research, they transcend, transgress, and transform the borders between the academic world and the real world [38].

In what context (the 'where' of the interdisciplinary challenge) can these integrative interdisciplinary and transdisciplinary practices take place in a productive, efficient, and sustainable way? It goes without saying that universities—historically and still today structured in terms of faculties, disciplines, and subdisciplines—seem *a priori* to be places that are not very conducive to interdisciplinary work. The fact remains that academia is opening up more and more to the decompartmentalisation of disciplinary tribes and their intellectual territorialism. Positive institutional governance evolution is seen in the creation of structures dedicated to interdisciplinary and transdisciplinary collaboration which are located at the interface between several faculties and disciplines; the provision of appropriate financial, technological, and human resources; and the recognition and increased promotion of interdisciplinary networking. This new organizational culture [39] aims to develop not only at the research level, but also at the complementary levels of basic education (bachelor and master degrees) and doctoral training.

The establishment of interdisciplinary places—of third places for boundary work at the interface between disciplines and between academic and socioprofessional spheres—promotes teamwork [40] and increases the capacity to create new and original knowledge in numerous fields of research, from science and engineering to social sciences, humanities, and the arts [41]. The added value of interdisciplinary collaboration, at long distance between disciplines and not strictly between very close disciplines, can be seen in atypical combinations of knowledge from different disciplinary horizons, as shown among others by bibliometric research in citation impact term [42]. Scientific advances are thus made on a complementary double axis, aiming to balance an extension through innovation and creativity by atypical combinations of ideas, theories, and methods with pursuing the deepening of thoughts that are more in line with disciplinary paradigms. This is necessary for their recombination in new interdisciplinary fields of research.

The concrete realization of interdisciplinarity certainly requires knowing what it is, why we do it, and in what context, but it also requires teachers and researchers (the 'who' of the interdisiciplinary challenge) open to this approach and capable of developing and applying it. A spirit of openness to other disciplinary languages makes it possible to develop transversal skills, skills for dialog, and interdisciplinary and transdisciplinary communication, which facilitate the connection and integration of knowledge from at least two disciplines. The interdisciplinary teacher or researcher would therefore be the one who demonstrates in-depth knowledge in a given specialty while being able to open up and connect to a universe of knowledge extending beyond their disciplinary origin. Between disciplinary depth/verticality and transdisciplinary breadth/horizontality, the figure of the interdisciplinarian can be metaphorically embodied in a 'T-shaped person' (figure 8.5) [43]. The vertical bar of the T represents the depth of expertise in a given field, while the horizontal bar indicates the ability to collaborate with other disciplines, to import tools or other methods into one's own field of concepts, and, conversely, to export one's own tools or methods to other scientific fields. These new academic and professional profiles are being developed and gradually recognized in universities. They represent a new generation [44, 45] of teachers, researchers, and practitioners who should perhaps be able to share common values of empathy, tolerance, benevolence, and respect between disciplines, forming a common ground on which to coconstruct interdisciplinary and transdisciplinary work.

As in any scientific approach, there is no definitive recipe on the 'how' of carrying out interdisciplinary work with certainty and success. There are, however, handbooks which offer avenues for reflection and present tools, methods, and good practices (see, e.g., [46–48]), without forgetting that very often we learn from our productive mistakes [49].

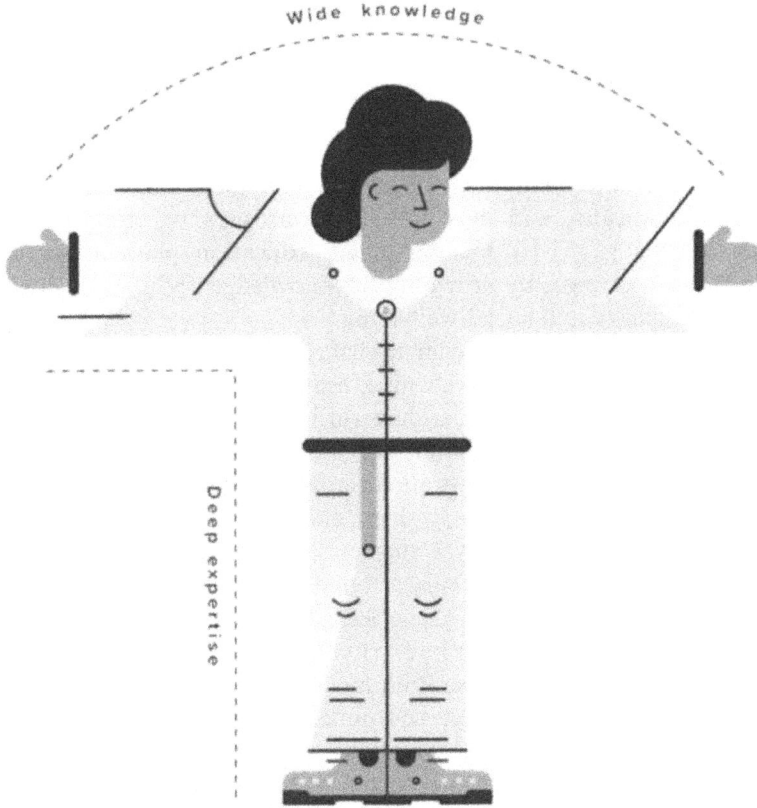

Figure 8.5. T-shaped person illustrating the relationship between wide and deep knowledge. Source: Joonas Jansson. https://dribbble.com/shots/3787357-T-Shaped-People.)

8.2.4 Education and training across sectoral borders

Francois Weiss[1]
[1]LMGP Grenoble INP, Grenoble, France

The world, confronted with societal and technological challenges, is changing rapidly and dramatically, and it is moving towards an uncertain future. In an increasingly knowledge-driven society, higher education plays a major role in responding to these changes by providing the knowledge and the educated citizens necessary for prosperity and social well-being.

The societal challenges, including climate change, environmental protection, sustainable resources, energy, demographic changes, health, security, and economic competitiveness, are intertwined and complex, as shown in figure 8.6. They call for paradigm shifts and disruptive innovation. Solutions to these challenges will require convergent thinking across disciplines, creativity, and cooperation across sectoral and geographic borders.

At the education level, the goal is to shape graduates and professionals who have deep expertise in science capped by a substantial breadth of perspective, excellent collaboration and communication skills, a sound understanding of their impact, and a good sense of responsibility towards society. That means more multidisciplinary and experiential learning as well as interactive problem solving inside and outside the classroom or laboratory. Cooperation has long existed in research, and major breakthroughs have emerged often at the frontier between disciplines. The challenge now is to foster a real dialog among science, industry, policymakers, and society, which has too often been neglected before now. Universities must adapt to fulfil these new missions at the interface of education, research, and innovation and to increase the efficiency and adequacy of the system.

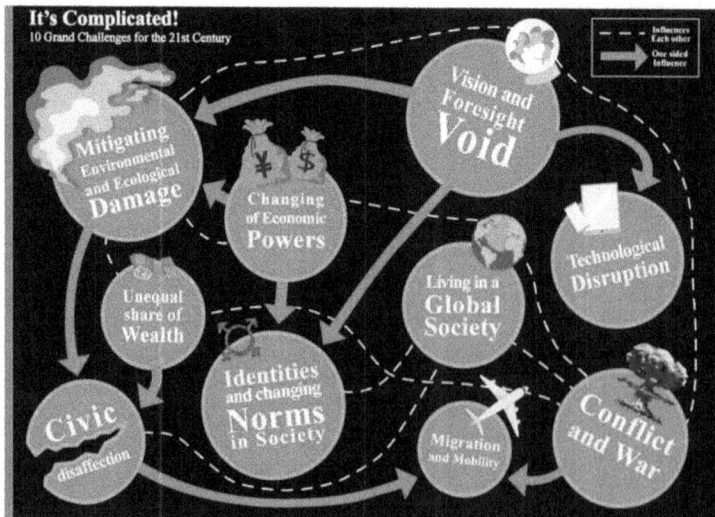

Figure 8.6. The 21st century challenges are interrelated. (Source: Ten grand challenges for the 21st century, 21st century Lab, University of Lincoln, UK, http://21stcenturylab.lincoln.ac.uk/ten-grand-challenges/.)

A possible approach in the short term is to build on solutions developed in the past 20 years, notably in the successive Framework Programmes for Research and Innovation of the European Union. For instance, EMMI, the European Multifunctional Materials Institute, is an association of academic and industry partners created on the bases of the European FP6 Network of Excellence FAME dedicated to Functional Advanced Materials and Engineering of Hybrids and Ceramics[1]. One of the objectives of EMMI is to serve as a European platform to create and conduct projects in research and education in the field of multifunctional materials science. The FAME+ Master Programme is the flagship developed within EMMI[2]. The FAME+ consortium is built with seven leading European universities in materials sciences and endorsed by seven industrial partners, seven European research and technology organisations, and 13 worldwide academic partners. Its main objective is the education of graduates with masters degrees in materials science while developing advanced skills and awareness to societal and industrial needs (see figure 8.7). Over the last 10 years, more than 200 graduates from all over the world have been awarded with the FAME+ masters degrees.

Another potential approach is exemplified by the Knowledge and Innovation Communities (KIC) catalysed by the European Institute of Technology and Innovation (EIT)[3]. Addressing economic sectors with a large potential for

---

1. Address Global Challenges such as Energy, Environment, Health, and Security by fostering:

- Creative thinking
- Effective use of global resources
- Increased cooperation across regions, sectors, and academic disciplines
- A cadre of well-trained global leaders
- Assembling of interdisciplinary and international teams of researchers to develop collaborative projects based on creative approaches

2. Build Global Leadership capabilities for young scientists and engineers to solve the world's most pressing problems by developing specific skills and competencies such as:

- Collaborating across sectors and disciplines
- Building global teams with complementary strengths
- Understanding the broader implications of their research
- Communicating their research to diverse stakeholders
- Understanding policy, manufacturing, and technology management
- Knowledge of international R&D infrastructures
- Interdisciplinary and intercultural communication skills

Figure 8.7. Dual Mission of the FAME+ Master Programme. (Source: https://www.fame-master.eu/brochures/.)

---

[1] https://www.emmi-materials.cnrs.fr/
[2] https://www.fame-master.eu/
[3] https://eit.europa.eu/

innovation and a real impact on society, the current eight KICs are partnerships that bring together businesses, research centres, and universities. Each KIC addresses a specific challenge, such as climate, food, digital, health, energy, manufacturing, raw materials and urban mobility. The main objectives are to develop innovative products and services in every imaginable area in order to create new companies and train a new generation of entrepreneurs. The training part is very important, and the EIT is strongly engaged in upgrading curricula or establishing new ones in emerging fields as well as in providing better access to research and industrial infrastructures. The creation of a forum for stakeholders (students and experts) to exchange information on education, research opportunities, and career developments is also part of the programme. Good training is very relevant for speeding up the process of technology development and the transfer of new discoveries or innovations into the marketplace.

In the KICs, each partner finances its own primary missions while all stakeholders finance together the value-added activities (mobility, skills, competencies, etc) that are developed in the frame of joint education, research, or innovation programmes. As a body of the EU, the EIT finances the KICs up to 25% of their overall resources over their typical lifetime of 15–20 years. They must be sufficiently agile and creative to become self-sustainable, generating the necessary income to finance the investments that are put in the creation of new education and research programmes. The creation of KICs should be a dynamic process, new ones being added in the future according to pressing societal and/or technological needs.

As successful as they are, the two examples above also show the limitations of the approach: These education, research, and innovation ecosystems work only at the programme and/or departmental level. The challenge of the future is to transform universities and other institutions of higher education at an institutional level in order for them to fulfil their social contract with society and adapt to a rapidly changing world [50, 51].

8.2.4.1 Conclusions

To prepare for the society of the future and improve the integration of education and research across sectoral borders, further efforts must be made today. Scientists and engineers need to become more flexible and adaptable across disciplines to be better prepared to solve societal challenges. New modes of cooperation among universities, research institutes, and industry must be developed to create a fertile ground for innovation and better address the urgent needs of society. A new paradigm in education will emerge, involving all stakeholders and relying on best practices to support decision making and new policies. Stronger computational capabilities based on AI, machine learning, and quantum computing will dominate our world, requiring more specialized and well-trained professionals across all different disciplines. Solving the actual and future societal challenges is a complex enterprise involving many interrelated parameters, as shown in figure 8.6. However, scientific and technological progress should be aligned with solid principles of governance, such as ethics, transparency, openness, participation, accountability, effectiveness, and coherence, just to name a few.

8.3 Science with and for the citizens

Philip Macnaghten[1]
[1]Knowledge, Technology and Innovation group, Department of Social Science, Wageningen University and Research, Wageningen, The Netherlands

Innovations that fail to take into account how people use them, copyright Katerina Kamprani.

If the physical sciences are to be successful in addressing some of the biggest problems facing humankind, it is imperative that they understand how to engage with society. This proposition is more challenging that it might appear, as there are various, and at times competing, models of how such engagement should take place. In this chapter I discuss four paradigmatic ways of governing the relationship of science and technology with society, situating each in a historical context. Starting with the ubiquitous linear model of innovation, I locate its origins and provenance and discuss how it came to be replaced, at least in part, through a 'grand challenges' paradigm of science policy and funding. I then describe how this paradigm in turn has been subjected to rigorous analytical critique by a coproduction model of science and society and how this model, in part, is being put into practice through a framework of responsible research and innovation (RRI). I conclude with reflections on how a framework of RRI can help us to navigate the grand challenges affecting the lives of citizens in the coming decades.

8.3.1 The linear model of innovation

World War II and its immediate aftermath signaled a critical moment in the unfolding relationship between science, society, and the state, especially in the United States. The Manhattan Project, involving the coordination of infrastructure and personnel in the development and production of the US nuclear programme, demonstrated the utility of science in public policy, in this case its role in helping to win the war through the detonation of two atomic bombs in Japan. In November

1944, President Roosevelt commissioned Vannevar Bush, who had had a formative role in administrating wartime military research and development through heading the US Office of Scientific Research and Development, to produce a report laying out the contributions of science to the war effort and their wider implications for future governmental funding of science. What emerged in July 1945 was the Bush report, *Science—The Endless Frontier* [52], which became the hallmark of US policy in science and technology, and the blueprint and justification for many decades of increased funding in US science.

The Bush report is associated with the linear model of innovation, postulating that the knowledge creation and application process starts with basic research, which then leads to applied research and development, culminating with production and diffusion, and associated societal benefit. Even if this sequential linkage may have been added, partially and imperfectly reflected in Bush's report [53], nevertheless it developed an iconic status as the origin and source of a dominant science policy narrative in which pure curiosity-driven science was seen as both opposed to and superior to applied science, effectively operating as the seed from which applied research grows, the economy grows, and society prospers [54]. As Sheila Jasanoff [55] has argued, the metaphor that gripped the policy imagination was the pipeline: 'With technological innovation commanding huge rewards in the marketplace, market considerations were deemed sufficient to drive science through the pipeline of research and development into commercialisation' [55]. This logic was given further impetus by the diffusion of innovation literature, notably in E M Rogers' classic text [56], which again adopted a linear and determinist model of science-based innovation diffusing into society with beneficial consequences.

Central to the post–World War II science policy narrative was the concept of the social contract, namely, that in exchange for the provision of funds, scientists, with sufficient autonomy and minimal interference, would provide authoritative and practical knowledge that would be turned into development and commercialisation. The linear model understands science and policy as two separate spheres and activities. The responsibility of scientists is first and foremost to conduct good science, typically seen as guaranteed by scientists and scientific institutions upholding and promoting the norms of communalism, universalism, disinterestness, and organised scepticism [57]. The ideal of science was represented as 'the Republic of Science' [58], separate from society, and as a privileged site of knowledge production. The cardinal responsibility of science, according to this model, is primarily to safeguard the integrity and autonomy of science, not least through practices of peer review as the mechanism that guarantees the authority of science in making authoritative claims to truth, thus ensuring its separation from the sphere of policy and politics.

This division of powers served the interests of both actors. For scientists, it meant a steady and often growing income stream as well as considerable autonomy; for politicians and policymakers, it provided a narrative that enabled them to claim that their policies were grounded in hard, objective evidence and not in subjective values or ideology. This division was also written into institutional arrangements for science policy. The Haldane Principle, for instance, by which the decision-making powers about what and how to spend research funds should be made by researchers rather than politicians, was written into national science funding bodies in the UK as far back as 1918, operating especially following World War II as a powerful narrative for self-regulation and for safeguarding the autonomy of science.

So far, we have described the linear model of science and technology, the assumptions that underpin its governance, including its optimistic and deterministic view of the relationship between pure science and social progress. However, as the 20th century progressed, this model came increasingly to be under strain as providing robust governance in the face of real-world harms that derived from scientific and technological innovation. Whereas the traditional notion of responsibility in science was that of safeguarding scientific integrity, responsibility in scientific governance came to include responsibility for impacts that were later found to be harmful to human health or the environment. The initial governance response was to acknowledge that science and technology could generate harms but that these could be evaluated in advance and within the bounds of scientific rationality through practices of risk assessment. Following a report from the US National Research Council [58], systematising the process of risk assessment for government agencies through the adoption of a formalised analytical framework, a rigorous and linear scheme was promoted and disseminated in which each step was based on available scientific evidence and in advance of the development of policy options. Risk assessment was thus a response to the problems of the linear model but was still very much within the linear model's framing and worldview.

Notwithstanding the efficacy of risk assessment to mitigate the harms associated with science and technology, notably in relation to chemicals and instances of pollution, it did little to anticipate or mitigate a number of high-profile technology disasters that took place throughout the latter half of the 20th century. Those disasters demonstrated that science and technology can produce large-scale effects that evaded the technical calculus of science-based risk assessment [59]. High-profile disasters included the accident at the Three Mile Island nuclear power plant in the US in 1979; the toxic gas disaster at the Union Carbide pesticide plant in Bhopal, India, in 1984; the disaster at the Chernobyl nuclear power plant in Ukraine in 1986; the controversy over bovine spongiform encephalopathy (BSE), also known as 'mad cow disease,' in the UK and Europe throughout the late 1980s and 1990s; and the controversy over genetically modified food and crops in the 1990s and 2000s first in Europe and then across much of the Global South. The nuclear issue was a focal point throughout the 1970s and 1980s as a result of wider concerns about technological modernity, manifested in large social movements that mobilised against the potential of science-led innovation to produce cumulative unknown and potentially cataclysmic risks. This was theorised most famously by the sociologist Ulrich Beck. Through his notion that modernity had entered a new phase, dubbed *the risk society*, science and technology were seen as having produced a new set of global risks that were unlimited in time and space, manufactured, potentially irreversible, incalculable, uninsurable, difficult or impossible to attribute, and dependent on expert systems and institutions for their governance. In this view, society operates as an experiment in determining outcomes [60].

The saga of BSE in the UK and Europe illustrates one such risk that was woefully and inadequately governed by a reliance on formal processes of science-based risk assessment, where the political controversy derived from the inadequate handling of a new disease in cattle under conditions of scientific uncertainty and ignorance, in the context of Britain's laissez-faire political culture. In this case, despite reassurances from government ministers, who claimed to be innocently following scientific advice that a transmission across the species barrier would be highly unlikely, a deadly degenerative brain disease spread from cattle to humans, escalating to such proportions as to threaten the cohesion of the EU [61].

More generally, risk assessment as a formal mechanism of scientific governance came under sustained criticism [62]. First, it embodies a tacit presumption in favour of change in assuming that innovations should be accepted in the absence of demonstrable harm. Second, it prioritises short-term safety considerations over long-term, cumulative, and systemic impacts, including those on the environment and quality of life. Third, it prioritises *a priori* assumptions of economic benefits with limited space for public deliberation of those benefits and their effects on society. Fourth, it restricts the range of expertise that is considered to be scientific expertise, typically from a restricted set of disciplines, with limited scope to access the knowledge of ordinary citizens. Fifth, it ignores the values and deep-seated cultural presuppositions that underpin how risks are framed, including the legitimacy of alternative framings.

8.3.2 The grand challenge model of science for society

While the linear model has been criticised for failing to account for the risks associated with late modernity, the model has also come under sustained criticism as offering an inadequate account of how the innovation system is structured and for what ends. Throughout the latter part of the 20th century, science and innovation became increasingly integrated and intertwined. The knowledge production system moved from the rarefied sphere of elite universities, government institutes, and industry labs into new sites and places that now included think tanks, interdisciplinary research centres, spinoff companies, and consultancies. Knowledge itself became less disciplinary based and more bound by context and practical application. Traditional forms of quality control based on peer-based systems became expanded to include new voices and actors, adding additional criteria related to the societal and economic impact of research. Variously framed using new intellectual concepts that included 'mode 2 knowledge' [63], 'post normal science' [64], 'strategic science' [65], and the 'triple helix' [66], a new model of knowledge production emerged in which science came to be represented as the production of socially robust or relevant knowledge, alongside, and often in conflict with, its traditional representation as knowledge for its own sake. Interestingly, in a later book, some of the same authors contextualised this transformation to accounts of societal change, particularly the Risk Society and the Knowledge Society, where 'society now speaks back to science' [67, 68].

One institutional response to critiques of the linear model has been the development of initiatives aimed at ensuring that science priorities and agenda-setting processes respond to the key societal challenges of today and tomorrow. The grand challenge approach to science funding best illustrates this approach. Historical examples of grand challenges range from the prize offered by the British Parliament for the calculation of longitude in 1714 to President Kennedy's challenge in the 1960s of landing a man on the Moon and returning him safely to the Earth. However, it was in the 2000s that the concept developed into a central organising trope in science policy, propelled *inter alia* by the Gates Foundation as a way of mobilising the international community of scientists to work towards predefined global goals [69]. In European science policy, the Lund Declaration in 2009 was a critical moment, which emphasised that European science and technology must seek sustainable solutions in areas such as global warming, energy, water and food, ageing societies, public health, pandemics, and security.

More generally, the concept has been embedded across a wide array of funding initiatives. Most recently, the European Commission (EC) instituted the Framework 8 Horizon 2020 programme, in which €80 billion of funding was to be available over the 7 years from 2014 to 2020, as a challenge-based approach that reflected both the policy priorities of the EU and the public concerns of European citizens. It was legitimated as responding to normative targets enshrined in treaty agreements; these included goals regarding health and well-being, food security, energy, climate change, inclusive societies, and security. In other words, it was based on the assumption that science does not necessarily, when left to its own self-regulating logic and processes, respond to the challenges that we, as a society, collectively face. Some degree of steering or shaping on the part of science policy institutions is needed to ensure alignment. It is thus embedded in a discourse about the goals, outcomes, and ends of research.

Over the last decade, the grand challenge concept has become deeply embedded in science policy institutions as a central and organising concept that appeals to national and international funding bodies, philanthropic trusts, public and private think tanks, and universities alike. It operates not only as an organising device for research calls but also as a way of organising research in research-conducting organisations, notably universities. For example, my university, Wageningen University, configures its core mission and responsibility in strategic documents as producing 'science for impact', principally through responding to global societal challenges of food security and a healthy living environment [70].

The grand challenge concept is clearly aligned with the 'impact' agenda, in which researchers increasingly must demonstrate impact in research-funding applications and evaluation exercises. These concepts help to reconfigure the social contract for science such that, at least in part, the responsibility of science is to respond to the world's most pressing societal problems, while the responsibility of science policy institutions is to ensure that the best minds are working on the world's most pressing problems [69]. Perhaps not surprisingly, these initiatives have proved to be controversial within the scientific community, as, for example, was shown by backlash from the scientific community to an initiative from one of the UK research councils, the Engineering and Physical Science Research Council (EPSRC), to prioritise its funding for grants, studentships, and fellowships according to national importance criteria [71].

More recently, scholars in the field of science and technology have added a further analytical layer [72]. They analysed the concept of the grand challenge scientist and the ways in which this has replaced the concept of the scientist that was prevalent in the linear model of innovation. Since at least Vannevar Bush's report *The Endless Frontier* (1944) [52], the dominant figure of the scientist was that of a lone individual discovering the frontiers of knowledge through pioneering or frontier research at the rock face of knowledge. However, while the ideal type of this kind of scientist was one who practised 'the risk taking behavior of rugged competitive individualists pioneering into the unknown' [72], the grand challenge concept configured a different kind of scientist. The grand challenge scientific endeavour still remains competitive but now has become collective, even sports-like, in the ways in which teams are presented as fighting to achieve a significant long-term goal, the accomplishment of which will have significant societal impacts. This tends to favour the organisation of science in highly interdisciplinary and collaborative units, as has become the case in systems biology and synthetic biology. Yet even though grand challenges are attempts to respond to society and to the public interest, the choice and framing of the challenges themselves have tended to remain those that have been chosen top-down by funding organisations [73] and in ways that often lend themselves to 'silver bullet' technological solutions [69]. Nevertheless, the grand challenge concept can be seen as part of an attempt to establish a new social contract for the public funding of science and as an important counterweight to the other dynamic that has affected the autonomy of science: the relentless influence of economic drivers that has come to dominate research policy agendas [74].

8.3.3 The coproduction model of science and society

If the grand challenge science policy model seeks to reconfigure the social contract of science such that its core value lies not with the pursuit of pure knowledge but in providing solutions to the world's most pressing problems, the coproduction model and approach seek to reconfigure the social contract in another direction. While the linear model views science as the motor of societal progress and the grand challenge model views science as the provider of solutions for society, the model of coproduction views the spheres of science and social order as mutually constitutive of each other.

Developed by Sheila Jasanoff and colleagues and building on decades of scholarship in science and technology studies (STS), the coproduction concept criticises the idea of science as producing incontrovertible fact. AAccording to Jasanoff and Simmet, 'Facts that are designed to persuade publics are coproduced along with the forms of politics that people desire and practice' [75]. This takes place in deciding which facts to focus on, in identifying whose interests the facts are used to support, and in observing that public facts are achievements, or what Jasanoff and Simmet call 'precious collective commodities, arrived at … through painstaking deliberation on values and slow sifting of alternative interpretations based on relevant observations and arguments' [75].

There are three broad implications that derive from this approach. First, if the authority and durability of public facts depend not on their status as indelible truths, but on the virtues and values that have been built into the ethos of science over time, it follows that we need to give special attention precisely to these virtues and to how they have been cultivated over time by institutional practice as an important constituent of democratic governance. As Jasanoff and Simmet claim, 'building strong truth regimes requires equal attention to the building of institutions and norms' [75].

Second, if science and social order are coproduced, then it becomes incumbent on the research enterprise to examine precisely the relationship in practice between scientific knowledge production and social order as evinced in particular sites. Variously studying in depth the operation of scientific advisory bodies, technical risk assessments, public inquiries, legal processes, and public controversies, STS scholars have identified both the values out of which science is conducted, including the interests it serves, and the ways in which these configurations can, over time, contribute to the formation of new meanings of life, citizenship, and politics, or what more generally can be dubbed 'social ordering' (see, among many others, [76–80]).

Third, if it is acknowledged that science and social order are coproduced, even if unwittingly through forms of practice, the question arises as to what values underpin the scientific knowledge production system and the extent to which these align with broader societal values. Indeed, to what extent have the values and priorities that are tacitly embedded in scientific innovation been subjected to democratic negotiation and reflection? Perhaps more worryingly, to what extent do dominant scientific values reflect those of incumbent interests that may be, perhaps unwittingly, closing possibilities for different scientific pathways linked to alternative visions of the social good [81, 82]. Responding to these questions, a line of research has emerged since the late 1990s, particularly prevalent in northern parts of Europe, aimed at early-stage public and societal participation in technoscientific processes as a means of fostering democratic processes in the development of, approach to, and use of science and technology. Such initiatives, funded both by national funding bodies and by international bodies, such as the EC, are typically aimed at improving relations between science and society and restoring legitimacy [83]. In practice, they have been developed for reasons that include the belief that they will help restore public trust in science, avoid future controversy, lead to socially robust innovation policy, and render scientific culture and praxis more socially accountable and reflexive [84, 85]. Initiatives aimed at public engagement in science have become a mainstay in the development of potentially controversial technology, notably in the new genetics, and have even been institutionally embedded into the machinery of government in initiatives that include the UK Sciencewise programme dialogs on science and technology [86]. In academia, they have contributed to institutional initiatives that include Harvard University's Science and Democracy Network and to the subdiscipline of public engagement studies [87].

8.3.4 A framework of responsible research and innovation

RRI concept represents the most recent attempt to bridge the science and society divide in science policy. Promoted actively by the EC as a cross-cutting issue in its

Horizon 2020 funding scheme (2014–20), and embedded in its subprogramme titled Science with and for Society, RRI emerged as a concept that was designed both to address European (grand) societal challenges and to 'make science more attractive, raise the appetite of society for innovation, and open up research and innovation activities; allowing all societal actors to work together during the whole research and innovation process in order to better align both the process and its outcomes with the values, needs and expectations of European society' [88]. To some extent RRI has been an umbrella term and is operationalised through projects aimed at developing progress in traditional domains of EC activity, nominally in the so-called five keys of gender, ethics, open science, education to science, and the engagement of citizens and civil society in research and innovation activities (Rip 2016). In this interpretation, RRI is a continuation of initiatives aimed at bringing society into EU research policy, starting with its Framework 6 programme (2002–6) titled Science and Society and its follow-on Framework 7 programme (2007–13) titled Science in Society. It has been identified as a top-down construct, introduced by policymakers and not by the research field itself [89], standing 'far from the real identity work of scientists' [72].

Another articulation of the RRI concept is also available. Alongside colleagues Richard Owen and Jack Stilgoe, I have been involved in developing a framework of responsible innovation for the UK research councils. Our intention at the time was to develop a framework out of at least three decades of research in STS, building on the coproduction model as articulated earlier in this section. Our starting point drew on the observation that from the mid-20th century onwards, as the power of science and technology to produce both benefit and harm became clearer, it had become apparent that debates concerning responsibility in science need to be broadened to extend both to their collective and to their external impacts on society. This follows directly from the coproduction model as articulated previously.

Responsibility in science governance has historically been concerned with the 'products' of science and innovation, particularly impacts that are later found to be unacceptable or harmful to society or the environment. Recognition of the limitations of governance by market choice has led to the progressive introduction of post hoc and often risk-based regulation, such as in the regulation of chemicals, nuclear power, and genetically modified organisms. This has created a well-established division of labour in which science-based regulation, framed as accountability or liability, determines the limits or boundaries of innovation and the articulation of socially desirable objectives—or what Rene von Schomberg describes as the 'right impacts' of science and innovation—is delegated to the market [90]. For example, with genetically modified foods, the regulatory framework is concerned with an assessment of potential risks to human health and the environment rather than with whether this is the model of agriculture we collectively desire.

This consequentialist and risk-based framing of responsibility is limited, because the past and present do not provide a reasonable guide to the future and because such a framework has little to offer to the social shaping of science towards socially desired futures [91, 92]. With innovation, we face a dilemma of control [93] in that we lack the evidence that can be used to govern technologies before pathologies of path dependency, technological lock-in, entrenchment, and closure set in.

Dissatisfaction with a governance framework that is dependent on risk-based regulation and with the market as the core mediator has moved attention away from accountability, liability, and evidence towards more future-oriented dimensions of responsibility—encapsulated by concepts of care and responsiveness—that offer greater potential for reflection on uncertainties, purposes, and values and for the cocreation of responsible futures.

Such a move is challenging for at least three reasons: first, because there exist few rules or guidelines to define how science and technology should be governed in relation to forward-looking and socially desirable objectives [94]; second, because the implications of science and technology are commonly a product of complex and coupled systems of innovation that rarely can be attributed to the characteristics of individual scientists [60]; and third, because of a still-pervasive division of labour in which scientists are held responsible for the integrity of scientific knowledge and in which society is held responsible for future impacts [95].

It is this broad context that guided our attempt to develop a framework of responsible innovation for the UK research councils [96, 97]. Building on insights and an emerging literature largely drawn from STS, we started by offering a broad definition of responsible innovation, derived from the prospective notion of responsibility described previously:

> Responsible innovation means taking care of the future through collective
> stewardship of science and innovation in the present [96].

Our framework originates from a set of questions that public groups typically ask of scientists. Based on a meta-analysis of cross-cutting public concerns articulated in UK government-sponsored public dialogs on science and technology, we identified five broad thematic concerns that structured public responses. These were concerns with the purposes of emerging technology, with the trustworthiness of those involved, with whether people feel a sense of inclusion and agency, with the speed and direction of innovation, and with equity, that is, whether the technology would produce fair distribution of social benefit [86]. This typology, which appears to be broadly reflective of public concerns across a decade or so of research and across diverse domains of emerging technology (including our own, [98–102]), can be seen as a general approximation of the factors that mediate concern and that surface in fairly predictable ways when people discuss the social and ethical aspects of an emerging technology. If we take these questions to represent aspects of societal concern about research and innovation, responsible innovation can be seen as a way of embedding deliberation about these issues within the innovation process. From this typology we derived four dimensions of responsible innovation—anticipation, inclusion, reflexivity, and responsiveness—that provide a framework for raising, discussing, and responding to such questions. The dimensions are important characteristics of a more responsible vision of innovation, which can, we argue, be heuristically helpful for decision making on how to shape science and technology in line with societal values.

Anticipation is our first dimension. Anticipation prompts researchers and organisations to develop capacities to ask 'what if...?' questions, to consider contingency, what is known, what is likely, and what are possible and plausible impacts. Inclusion is the second dimension, associated with the historical decline in the authority of expert, top-down policymaking and the deliberative inclusion of new voices in the governance of science and technology. Reflexivity, the third dimension, is defined, at the level of institutional practice, as holding a mirror up to one's own activities, commitments, and assumptions; being aware of the limits of knowledge; and being mindful that a particular framing of an issue may not be universally held. Responsiveness is the fourth dimension, requiring science policy institutions to develop capacities to focus questioning on the three other dimensions and to change shape or direction in response to them. This demands openness and leadership within policy cultures of science and innovation such that social agency in technological decision making is empowered.

To summarise, our framework for responsible innovation starts with a prospective model of responsibility, works through four dimensions, and makes explicit the need to connect with cultures and practices of science and innovation. Since its inception our framework has been put to use by researchers, research funders, and research organisations. Indeed, since we developed the framework in 2012, one of the UK research councils, the EPSRC, has made an explicit policy commitment to it [103, 104]. Starting in 2013, using the alternative 'anticipate-reflect-engage-act' (AREA) formulation [105], EPSRC has developed policies that set out its commitment to developing and promoting responsible innovation and its expectations both for the researchers it funds and for its research organisations.

8.3.5 Challenges and opportunities

In this section I have discussed four paradigmatic ways of governing the relationship of science and technology with society. I began with the linear model, in which science is represented as the motor of prosperity and social progress and in which the social contract for science is configured as that of the state and industry providing funds for science in exchange for reliable knowledge and assurances of self-governed integrity. I then explored the dynamics and features which contributed to a new social contract for science in which the organisation and governance of science became explicitly oriented towards the avoidance of harms and the meeting of predefined societal goals and so-called grand challenges. A coproduction model of science and society was subsequently introduced to provide a better understanding of how science and social order are mutually constitutive and of the implications of such an approach for science and democratic governance. Finally, I set out a framework of responsible innovation as an integrated model of aligning science with and for society.

These four models should not be seen as wholly distinct or unrelated. Typically, they operate in concert, sometimes harmoniously, other times less so, in any governance process. Nevertheless, the broad move beyond the linear model of science and society must be applauded, both because science devoid of societal shaping is clearly poorly equipped to respond to the societal challenges we collectively face and because the premises that underpin the linear model, such as the fact–value distinction, are clearly poorly aligned with contemporary intellectual debate.

The world of 2050 is likely to be very different from that of today. Advances in science and technology are likely to present all kinds of challenges—as well as opportunities—at various scales and temporalities. Advances in digital technologies, AI, machine learning, nanotechnology, robotics, gene editing, synthetic biology, and quantum mechanics are undoubtedly going to have transformative impacts on everyday life, for good and ill. Not only will we need such advances to help tackle some of the profound challenges of today and tomorrow, ranging from climate change to global pandemics, food security, and healthy soils and oceans, but we need to engage citizens in partnership with science in the coproduction of solutions. A framework of RRI offers opportunities, tools, and possibilities to make science and its governance more responsive to the grand challenges of the 21st century by helping to ensure that the formulation of responses is aligned to the question as to what kind of society we want to be [106].

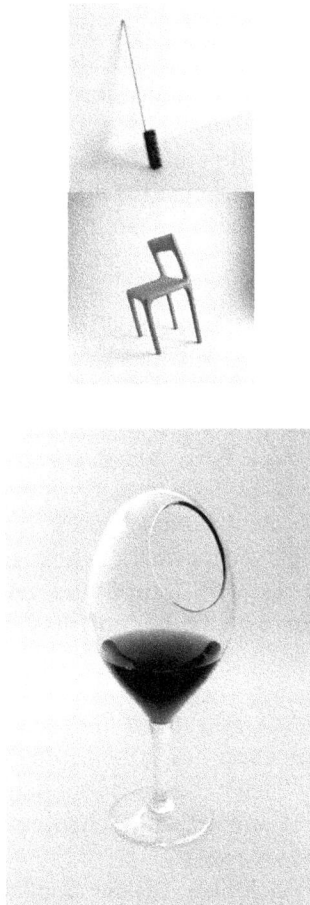

More useless innovations, copyright Katerina Kamprani.

8.4 Open communication and responsible citizens

Tobias Beuchert[1] and Pedro Russo[2,3]
[1]German Aerospace Center (DLR), Earth Observation Center (IMF-DAS), Oberpfaffenhofen, 82234 Weßling, Germany
[2]Department of Science Communication & Society and Leiden Observatory, Leiden University, Leiden, The Netherlands
[3]Ciência Viva, National Agency for Scientific and Tecnological Culture, Portugal

8.4.1 Open communication: public engagement with science

We live in challenging times, and the future lies in our hands. How is it possible for humankind to follow its basic need to expand horizons and gather knowledge while maintaining responsibility for a sustainable future? We believe that one of the key elements of responsible research is not only keeping best practices, but also making the past achievements of humankind accessible to future generations.

'Scientists keep an open line of communication with the public[4]' [107]; this call for action from the editor of *Nature Medicine* in October 2020 came at a time when our society was struggling with a global pandemic, experiencing fights for social justice, and suffering from the climate crisis. Scientists need to openly communicate with the public and engage fellow citizens with research activities. Public engagement with science is no longer a 'nice-to-have' activity but a 'must-have' one.

'Science communication', 'public engagement', and 'education and public outreach' are blanket terms covering any related topics, from goals to methodology. These terms describe the many ways in which the scientific community can share its research activities and their benefits with society. B Lewenstein [108] categorizes public engagement in two main aspects: as a learning activity and as a public participation in science (table 8.2). Both have in common that scientists and science educators reach out to individuals and society at large with various programs. Engagement is a two-way communication process, involving listening and interaction for mutual benefit. Public engagement is also an essential tool to build and strengthen public support for research. Indeed, the trend for evidence-based public policy increasingly relies on access to a wide variety of specialists, many based in universities or research facilities.

8.4.1.1 Developments in public engagement

Historically, science communication has gone through three main phases: science literacy (also known as the deficit model), public understanding of science, and, recently, public engagement of science and technology. These phases have moved forward because of several policy reports.

In 1985, the Royal Society identified a systematic lack of interest and literacy in science. The result can be imagined as a trench gaping between the science domain

---

[4]We use the term public as an umbrella term for non-university audiences, including fellow citizens, education stakeholders and policymakers.

Table 8.2. Overview of the main categories of public engagement initiatives based on [109].

Category	Characteristics	Examples
Developing an interest in science	Experience excitement, interest, and motivation to learn about science	• Exhibits (e.g.: CERN's Universe of Particles [110]) • Media: TV news, newspapers, magazines, etc • Social media
Understanding (some) science	• Understand concepts, explanations, arguments, models, and facts related to science • Manipulate, test, explore, predict, question, observe, and make sense of science	• Public talks (e.g. Physics Matters Lecture Series [111]) • Documentaries (e.g., BBC's 'The Secrets of Quantum Physics' [112]) • Popular science books and magazines (e.g., Stephen Hawking's seminal book *A Brief History of Time* [113]) • Workshops and hands-on exhibitions • Public websites
Using scientific reasoning and reflecting on science	Reflect on science as a way of knowing; on processes, concepts, and institutions of science; and on their own process of learning about phenomena	• Community and dialog initiatives (e.g., Quantum Delta Living Lab [114])
Participating in the science enterprise	Public participates in scientific activities and learning practices with others, using scientific language and tools Public identifies as people who know about, use, and sometimes contribute to science	• Citizen science projects (e.g., Steelpan Vibrations [115])

and society (figure 8.8). Science communication ought to be key to bridge that trench. The initial top-down or one-way strategy for communicating to the public, however, has resulted in 'very little improvement in adult scientific literacy' [116]. Improvement was expected when the UK's House of Lords Science and Society report spearheaded the public engagement model. Recently, science has progressively advanced into various areas of society [117]. As Schäfer notes, while scientists and journalists had attempted to increase citizens' science literacy by popularizing science, several events of global reach made the public more aware of the societal impact of science. A two-way street as communication strategy turned out to be better (see figure 8.8). A direct dialog with the public is essential not only for

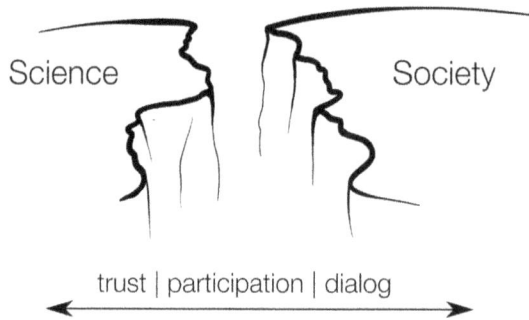

Figure 8.8. Illustration of the trench between science and society and the latest efforts to bridge that trench with a two-way communication strategy.

democratic participation in research [118], but also to facilitate more trust and confidence in science [119]. In addition, new trends in crowd-sourcing have emerged in science: notably, citizen science projects with crowd-based data collection and analysis or crowd-funding platforms [120].

Despite these efforts in pushing for more public engagement, the Public Attitudes to Science report of 2019 [121] documented an evident gap between citizens and science. While 51% of respondents felt too little exposed to science, 69% of them agreed that 'scientists should listen more to what ordinary people think'. This gap has been starting to shrink, thanks also to the increasing significance of science communication. The report moreover states that citizens have developed more positive attitudes towards science and more trust in science. They feel better informed, and science seems more accessible (see section 8.4.1.2). It is, however, worrisome that only a minority of respondents considered science important and useful in everyday life.

8.4.1.2 Open science and public engagement

Public engagement is an endeavor that takes many forms, ranging from education programmes to citize -science projects and science festivals. All of these help researchers to disseminate the societal benefits of their work while keeping abreast of public concerns and expectations. Public engagement activities provide a platform for researchers to discuss their projects and objectives with the wider public. For optimal benefits these actions must be practical, innovative, research-based, and educational, feeding each other with ideas, opportunities for research studies, and even financial resources [122].

Public engagement helps to maximize the flow of knowledge and cooperation between research communities and society, giving researchers the potential to create an impact through learning and innovation [123]. Strategic investment in public engagement helps to maximize this potential by focusing attention and support on how research enriches the lives of people. It also contributes to social inclusion and social responsibility and allows researchers to better respond to local and global social issues [124] with appropriate effective support to people [125]. Building trust and mutual understanding is critical to a healthy higher education and research system [126, 127], especially at a time when deference to authority and professional expertise is decreasing.

The Internet has fostered many citizen science projects in which the public gets involved in data collection, analysis, or reporting. The low-level access to the scientific process is one key advantage of such projects, and the large collaborations that are generated allow widespread research, leading to discoveries that single scientists could hardly achieve on their own [128].

When it comes to higher education and science communication, a study by the Wellcome Trust [129] shows a positive trend with regard to researchers but with clear caveats. For example, is the science communication activity always as effective as it could be? Examples stated in the report are low frequencies of engagement and a clear bias in target audiences. According to [130], there has been a lack of a coherent push for science communication in the academic system. There are also discussions in the scientific community about whether or not public engagement can harm the researcher's scientific career—also known as the Carl Sagan effect [131, 132]—alongside negative opinions. On the other hand, it is emphasized that communicating with the public can boost academic performances [131, 132]. However, new policies are needed to promote public engagement.

The good news is that science communication is today an established subject at many universities with an increasing number of options in the various curricula up to the PhD level. Its interdisciplinary and multidisciplinary nature is a 'sign of the subject's vitality, but it is also a condition of its vulnerability' [133]. Indeed, it is challenging for science communication to become a recognized teaching field when it is positioned between and across faculties and disciplines.

We believe that the academic institutions need to provide long-term support to retain the necessary skills, experience, and resources to facilitate communication efforts. This should not add to the existing pressure for publishing and being competitive but should be part of the job profile of responsible scientists [134]. We argue that to avoid a trench to form between scientists and communicators, research institutions should not rely on institutionalized science communication.

We do, however, endorse open and public spaces where innovative science communication can take place. Such 'Idea Colliders' [135] have the chance to supersede traditional science museums and promote critical scientific thinking and decision making. These spaces can trigger debates in an interdisciplinary setting involving scientists and citizens from diverse disciplines, including those from the cultural and political sectors. Design is one example of a largely interdisciplinary interface, able to facilitate communication between science and society and thus to contribute to novel and transformative narratives [136].

In terms of science communication research, as an increasingly academic discipline, theoretical and applied research in science communication attracts scientists from various backgrounds, such as the history of science, media studies, psychology, and sociology. In-depth knowledge in science communication is essential for the other core areas of public engagement activity described previously. 'Science is not finished until it is communicated,' said UK Chief scientist Sir Mark

Walport in 2013. These words reflect the urgent need for the science enterprise to fully commit to public engagement and place societal impact at the core of its research.

8.4.1.3 Responsible research and innovation

RRI is a policy principle [137] based on societal aspects of research, such as public engagement, research integrity, and ethics. It focuses on aligning research and innovation processes and outcomes with the values and needs of the greater society. It is the responsibility of individual researchers and innovators to consider these principles in their daily work and to anticipate the societal consequences of their practice. The six main dimensions of RRI are research engagement with society, gender equality, open access, science education, ethics, and governance in research and innovation. This policy further demands the following transdisciplinary approaches to modern research:

- Collaboration of societal actors (researchers, citizens, policymakers, business, third sector organisations, etc) during the whole research and innovation process so to better align with societal values and expectations.
- Enabling easier access to scientific results, including open science and public engagement.
- Uptake of gender equality and ethics in the whole process. Scientific curiosity should try to answer not only the question 'can we' (scientific or technical feasibility) but also the question 'should we' (societal acceptance).
- Formal and informal science education.

Current research policies such as RRI demand wider participation of researchers in societal issues. For example, the OECD [138] and the EU [139] have published specific societal impact frameworks for research communities and facilities. Research facilities such as ESO [140] and CERN [141] have also published reports on their societal impact. However, in general, there is still a mismatch between policy expectations and relevant scientific engagement.

8.4.1.4 Future challenges

We find ourselves in challenging times. Modern challenges are complex, broad, global, and deeply rooted in societal dimensions. The climate crisis and the COVID-19 pandemic are just symptoms but with connected causes. Our roles and responsibilities as scientists are changing at the same speed as that at which the environment changes —an environment that we have described and tried to understand for decades. The knowledge we gained in that process should help lead to fast progress in preserving our environment (or nature) but can no longer remain the exclusive property of specialized research fields. Open communication among scientists, across disciplines, and with society (public engagement) is crucial on our way towards the Horizon of 2050. While these communication efforts are key to meeting global challenges, their success depends on a well-functioning science–society relationship, which is a difficult task in a world that is rather unstable politically, socially, and economically.

8.4.1.5 Global environmental challenges

Global challenges and risks have developed together with globalization. The 2018 Global Compact on Refugees, the IPCC 6th assessment report [142], and the Global Risks Report (2020) demonstrate that it is imperative to prepare for a sustainable future. Ironically, this seems the greatest challenge to society itself [126]. Sciences comprise disciplines that seek to describe nature [143] with an understanding and analysis of complex phenomena concerning the bigger picture [144]. Science therefore helps in developing the knowledge to assess global challenges and risks. It requires also large-scale collaborative efforts to encompass the complexity of global and interconnected phenomena such as the climate crisis. The status of expertise has changed considerably from being exclusive to being broadly available and often challenged by society itself [145]. The climate crisis, for example, is no longer of interest only for scientific studies. As it implies huge risks for humanity, it puts extra pressure on all disciplines to provide the public with reliable knowledge [126] and guidelines over the upcoming decades. Scientific evidence concludes that the changing world climate will affect our everyday life in the future [142] and that swift action is required [146]. Meanwhile, citizens depend on accurate risk assessment (Global Risks Report, World Economic Forum, 2020 [147]), as evidenced by the COVID-19 pandemic and the climate crisis appearing as correlated symptoms [148–150]. Interdisciplinary research is key in tackling these grand challenges to society.

8.4.1.6 Society's need for public engagement in the face of global challenges

Considering these extensive and complex challenges that society will face in the upcoming decades, the public demands dedicated support by scientists, science communicators, and science journalists beyond solely increasing their scientific knowledge. Engaging the public with the process and method of science promotes a more comprehensive understanding, improves the relationship between science and society [127], helps citizens to appreciate the complexity of current and future challenges, and builds trust in scientists [126, 145].

The situation is still worrisome. Although 64% of Americans are at least 'somewhat worried' about global warming, only 22% actually understand how strong the level of scientific consensus is on global warming, according to a 2021 survey by the Yale Program on Climate Change Communication. The gap between science and the understanding of science by the public is fueling a lack of trust in science, feelings of uncertainty, and social inertia [151]. Worse, it gives sceptics and deniers a lot of momentum. As physicists, we have the privilege of understanding and being able to evaluate risks. This gives us the responsibility to share our knowledge with the citizens who are often left alone in assessing such risks and making decisions on issues such as extreme weather incidences [152, 153]. Public controversies arising over various scientific and technical issues [154] have luckily boosted numerous efforts in science communication across disciplines [155]. Now and in the coming years, we need to acknowledge that the path towards a sustainable future on our planet is a systemic transformative process involving the entire society worldwide.

Interdisciplinary research (in physics) and the emerging field of the science of science communication have developed at a fast pace in recent years (e.g., [156, 157]). They have not yet reached the speed at which our society must respond to accelerated socioeconomic and Earth systems–related trends [149] with sustainable solutions. Swift action is imperative for all of us to bridge the remaining gap between the sciences and society over the coming decades (see figure 8.11). In summary, present times demand for a culture in society that does the following:

- Appreciates [158] and trusts [126] science.
- Understands the scientific process as a probabilistic approach [159] and the concept of uncertainty in the interpretation of scientific results [156], differentiates between scientific uncertainties and low-quality or doubtful science [160], and disregards the negative connotations of the term 'uncertainty' [161].
- Makes scientifically motivated decisions.
- Seeks a dialog with society ([156] and references therein).

Moreover, the role of politics must not be forgotten, as science communication guarantees the 'democratic legitimacy of funding, governance and application of science' [162].

We encourage our colleagues in all fields of physics to appreciate the increasing challenging needs of our society, scientists included, and to engage in the exciting communication landscape.

8.4.1.7 Challenges and opportunities in modern approaches of science communication

The research community and citizens are both challenged by modern science communication [145], and tension has been growing between experts and the public in recent years. How can experts become more valued and respected? How can citizens identify true experts and factual knowledge in the complex web of the 'infodemic'? The 'overabundance of (also wrong) information' can 'undermine the public health response' and fuel conspiracy theories, pseudoscientific content, or large-scale 'controversies over scientific and technical issues' [154]. Typical examples are related to COVID-19 [163]; the misinterpretation of scientific knowledge in the documentary 'Cowspiracy' [164], in particular regarding the 'Global Warming Potential' (GWP) cited in the 6th IPCC report [142]; and the apparent but rebounding drop in CO_2 emissions induced by the measures that were initially taken against the spread of COVID-19 [165]. Some authors have investigated the complex diversity of how bloggers and contrarian bloggers discuss topics of public interest [166], such as the scientific consensus behind the climate crisis. Clearly, it requires specific training and expertise to (1) filter reliable knowledge in the face of complex and global crises, where opposing expert opinions are pervasive on the Internet [167], (2) manage the challenges of the 'mediatisation' of complex information and discussions [168–170], (3) avoid being diverted from biases (e.g., cognitive, self-confirmation, or anchoring), and (4) make scientifically driven decisions in everyday life [171].

Despite the tragic developments of the COVID-19 pandemic, it has also provided opportunities for researchers to 'rethink the role they want to play in society at large' [107]. For the first time, our society experienced the consequences of a global crisis that affects the lives of most of the world population at short time scales. The climate crisis, however, despite its evidently fatal consequences for humankind, is still largely perceived as an abstract issue given the longer time scales involved in climate variability [172, 173]. Sociologist Kate O'Brien argues that the pandemic has shown that society is in fact able to adapt and that a major global transformation in response to the climate crisis is possible [174, 175].

Science communication practices have proven capable of bridging science and society during the pandemic and hence contribute to a global transformation process [176, 177].

8.5 Science and ethics

Bengt Gustafsson[1,2]
[1]Department of Physics and Astronomy, Uppsala University, SE-751 20 Uppsala, Sweden
[2]Nordita, Hannes Alfvéns väg 12, S-10691, Stockholm, Sweden

8.5.1 Introduction

The phrase 'science and ethics' generates more than 20 million hits by the search engine Google, illustrating the significance of this topic. There are several reasons for this interest, one being the public understanding of the fundamental impact of science and of its applications. The increased pressure on scientists to demonstrate skills and productivity to build their careers and secure financing for their research may have tempted them to take shortcuts instead of strictly following some more or less well-established rules. There is, however, a desire within the scientific community to ensure high-quality results in general. This provides a solid base for future work, allows for fair evaluations of the work, and helps to ensure that money for science is used in a trustworthy way in order to secure further public support.

One overruling ethical principle, illustrated in figure 8.9, must be stated as a core in the ethos of science: Science should be a systematic attempt to observe the world

Figure 8.9. *Three monkeys* by the Swedish artist Torsten Renqvist (1924–2007), a representation of the East Asian maxim 'See no evil, hear no evil, speak no evil'. As usual, one monkey covers the ears, and one covers the mouth. The third, who traditionally covers the eyes, here cannot refrain from looking. Renqvist, with a strong interest in science, commented: 'He sees, and so he must. He is the one who has bitten of the apple.' Photography: Kristina Backe.

accurately and report its findings honestly. This requirement of truthfulness may seem obvious. Trust in the scientist and the scientific community that the new knowledge will indeed be beneficial for humanity, or at least not be harmful, is certainly widespread. Yet history and contemporary examples demonstrate that striving for truth is not obviously present or even generally accepted in the world. This is partly related to the fact that truthfulness sometimes causes problems for the individual scientist.

Important ethical dilemmas are involved in the selection of science projects and as regards the responsibility for the application of the results. Even if the ambition of the scientists is simply to provide fundamental knowledge, only they may be able to understand the possible consequences of their findings. This issue has, however, been discussed previously [178–180] and will not be treated here.

8.5.2 Ethics of the science process

We shall discuss the ethics of scientific work, noting that many of its conventions and ethical rules aim to promote truthfulness. In addition to honesty and objectivity, care and solidarity relative to colleagues are stressed. Certainly, these virtues are also significant in the society at large, but the demands are much stronger in the scientific community. These ethical requirements are set in order to guarantee the quality of the scientific results in the service of the needs to accumulate and apply knowledge.

A term for the upholding of these virtues is *scientific integrity*[5], which more specifically means avoiding fabrication and falsification of data, bias, plagiarism, and carelessness, all in order to guarantee objectivity, reproducibility and trust in results and reporting.

Here, I shall discuss some aspects of these principles and present associated areas where science plays important roles and where ethical problems arise. A discussion follows on possible actions to be taken by the science community and various authorities to uphold these principles. I shall not discuss ethical problems relating to the social life in research groups and institutions with leadership, power structures, or improper dependences. Although important, these topics are not limited to the scientific work environment.

8.5.3 Forgery

Wikipedia lists a depressingly extensive list of scientific misconduct incidents over the last two decades [181]. The list includes more than 110 entries[6]. Among those are over 70 forgeries, in which data were fabricated or systematically changed. There are papers in which identical sets of data were used to represent alleged outcomes of different and unrelated experiments. In other cases, data from certain experiments were used to represent data from experiments which were never carried out. A

---

[5] The term was introduced in *Best Practices for Ensuring Scientific Integrity and Preventing Misconduct*, OECD. 2007, https://www.oecd.org/science/inno/40188303.pdf (retrieved May 30, 2021).

[6] Another measure of scientific misconduct gives the data base of retracted papers https://retractionwatch.com/retraction-watch-database-user-guide/, although many papers have been retracted for other reasons. By October 2020 more than 24 000 retracted papers were listed.

relatively common type of forgery has been manipulation of images. These were possible to discern via careful studies by journal editors and referees. The forgeries abound in all fields, from physics and chemistry and biology to social sciences. They are found in different nations and institutions, from leading universities to small, less well-known places. An overrepresentation in medical work, including false statistics and unmotivated lethal experiments on patients, is particularly worrying [182]. Thousands of scientific papers have been retracted as the result of the disclosure of these forgeries.

One may speculate about the reason for these behaviours. Some of the cases seem truly pathological or due to personality disorders of histrionic or narcissistic personalities. Several of them seem to be demonstrations of alleged invincibility ('See what I can do, and nobody can hurt me!') or deliberate self-destruction. However, in studying these cases, another and worrying fact soon becomes clear. The cases are often suspicious in the sense that remarkable and unexpected 'results' are presented. In itself the selection of discovered cases may be heavily biased towards such results. Such cases are prone to be scrutinized, and, if falsifiable, to be disclosed by the community. Cases in which data seem more innocent or expected may have a larger chance of remaining undisclosed. Thus, fraudulent scientists who are anxious to remain undiscovered while generating extensive publication lists should be expected to avoid publishing unexpected results. One may therefore speculate that a consequence of this dishonesty may well be a retarding force on scientific progress.

How common are such less obvious falsifications? According to a metastudy by Daniele Fanelli [183], about 2% of scientists who were asked admitted that they had at least once fabricated or modified the data presented, while about 30% admitted that they had been involved in other questionable research practices. When asked whether the scientists had colleagues who had ever fabricated or modified data, the percentage increased from 2% to 14%. About 70% said that they had colleagues who had showed misconduct in other ways.

8.5.4 Plagiarism

Another type of scientific misbehaviour is plagiarism, which is taking someone else's ideas or work and presenting them as one's own. This was quite common in earlier epochs [184] and may not have hurt the progress of science much. However, the practice of stealth promotes secrecy, which may be harmful in our scientific culture which benefits so much from exchange of ideas and experiences. Today, with the contemporary stress of careers based on extensive publication lists, the temptation to plagiarize may be hard to resist.

Among the examples of plagiarism in the Wikipedia list [209] are a number of stolen publications or sections from work by graduate students or postdoctoral researchers. There are cases of papers to be refereed that were delayed or stolen by unscrupulous researchers, who subsequently published the work as their own, or stolen from other authors who have published in less well-known journals. In some cases, tens of papers were republished and misattributed to one individual. However, such plagiarisms are now easily traced via modern detection software, when the stolen work is available in public databases.

When studying the plagiarism cases more closely, one finds that in several cases, carelessness or a stressing timetable might be reasons. This may occur when reviews are to be produced for conferences or review journals, and excerpts are taken from various sources, including self-plagiarized ones, and in the next stage are 'forgotten' to be properly acknowledged[7]. A possible measure to diminish this problem would be to reduce the number of review papers, for example, replacing them by 'living' reviews, continually updated by experts.

It is noteworthy that a number of leading politicians, including some at the ministerial level, have been caught with plagiarism in their doctoral theses. Even though plagiarism seems to be a fairly common in the political culture, judging from the presence of numerous cases of stolen political speeches, it is astonishing that this culture trumps the more rigorous demands of honesty in the academic enterprise.

More difficult to trace, and presumably more common, is citation plagiarism, in which articles or other work is used with inappropriate citations. Another related form is bibliographic negligence, in which important work, preceding the present work and of great significance for it, was not referred to at all [185].

8.5.5 Reviewing science

The discovery of frauds in published science journals has cast doubts on the peer-review system. Obviously, any expectation that the system is free from fraud is unrealistic even while it provides some control on the quality of journal papers. The steadily increasing number of papers, driven by the demand for extensive publication lists in applications for jobs or resources, is also a threat to the system. There are difficulties in finding suitable referees and for those referees to find time to do a diligent job [186].

Other reviewing tasks, in nomination committees for positions, panels for grant allocation, and other forms of evaluations, also entail ethical dilemmas. A major reason for these problems is again the growing volume of publications, reflecting the growth in the number of applications, in combination with the requirement of repeated controls. Within the new public management (NPM) approach for making public services more 'businesslike', models from the private sector are introduced into the academic world with the aim to make it more efficient [187]. With such a system the number of evaluations may become excessive. This number should be kept at a minimum via longer grant periods, more tenure-track positions, and a robust allocation system of resources to research groups and departments.

8.5.6 Shadow zones: boasting

Although central in the scientific endeavour, truthfulness is not always fully respected in scientific enterprise. There is a clear tendency in applications for jobs

---

[7] Conversely, reviews are often used to prepare the introductory sections in papers without proper references. Information about the true origin of data or ideas may thus be lost in this process in several steps.

or grants and in other presentations to exaggerate the significance of the results that are obtained. This 'boasting' seems to have increased in volume and aggressiveness during the last decades, which may be a consequence of the fierce competition for jobs and money on the research market, now highly affected by the practices of NPM.

In an attempt to explore boasting, expressions of self-overestimation were searched for in several hundred applications for junior staff positions in natural and technical sciences at Uppsala University [188]. Although this study was methodologically problematic, clear differences were found depending on the origin of the applicants and their respective fields. The closer the research area was to industry or commercial applications, such as information technology, biomedicine, and technology, the more frequent were words such as 'world-leading' 'excellent', and 'successful', while such words were sparser in applications for jobs in mathematics, astronomy, physics, and fundamental chemistry. Interesting differences were also found in the degree of boasting between applications from different countries.

In a parallel study of the official web pages of different Swedish universities, very clear differences were found. The more established institutions with higher positions on international ranking lists had a lower degree of boasting, while the new ones, generally dependent on support from local regions and industries, more often presented themselves as 'excellent'. The technical universities had higher boasting indices than the classical ones with broader programmes. This again points to influences from the commercial sector being stronger when academia is closer to, or more dependent on, that sector.

These studies illustrate how the academic and commercial cultures interact and affect each other [189], including their demands for truthfulness.

Several universities now systematically train young scientists to 'sell themselves' on the international labour and grant markets. It is important to study whether the respect for accuracy and truthfulness in scientific judgements is hollowed out by that. One might hope that the next generation, trained in handling the commercial flow in media, will be able to distinguish between the language used in the market versus that used in scientific dialog.

Another type of boasting occurs when science journals overemphasize the significance of results that they present, probably exaggerated for commercial reasons and applauded by authors who seek approval and resources by financing agencies.

8.5.7 Shadow zones: the replication crisis

Reproducibility is a fundamental requirement for scientific work and demands control of experiments and observations and on reporting the methods and results. A paper with the provocative title 'Why most published research findings are false' by Ioannides [190] demonstrated the need to have very large samples with controlled bias in statistical studies in order to acquire safe demonstrations of causal relations. Similarly, Ziliak and McCloskey, in a book with another provocative title [191], examined a great number of economics papers and argued that only about one fourth of them showed reproducible results. A few years later, it was demonstrated that papers in social sciences as well as in social psychology, pharmacy, and medicine presented results that could not be reproduced, even by the scientists who had made the original study. This led to the introduction of the catch term 'The replication (or reproducibility) crisis'. A *Nature* poll of 1500 scientists in 2016 indicated that about 70% of them had not succeeded in reproducing results of at least one other study [192] ; indeed, half of them could not even reproduce one of their own reported findings. The results differed between areas; physicists and chemists showed the greatest trust in the published results, while medical and social scientists had less confidence. In this survey, the scientists were also asked about the reasons for the low reproducibility and what could be done to improve it. More than 60% responded that two reasons were particularly important: pressure to publish and selective reporting. Missing checks in the lab and too small sample sizes were also pointed out. When asked about how the reproducibility could be improved, more than 90% suggested 'more robust experiment design', 'better statistics', and 'better mentorship'. The current strong tendency in favour of open data [193] and open-source code [194] may contribute to relieving the replication crisis.

8.5.8 Shadow zones: cherry picking and the sheep–goat effect

One reason for the lack of reproducibility is no doubt more or less conscious cherry picking; that is, we tend to select the data we believe in and disregard the data that seem less probable to us. As much as this may lead to problems with reproducibility, it may also cause too much of it. Thus, the 'sheep effect' may result from cherry picking. As has been pointed out in several critical studies of the measurement of certain important quantities, 'classical values' tend to be replicated in repeated studies with a precision that is higher than realistic error estimates should permit. The background may be a consequence of the complexity of modern laboratory experiments and computer codes. The efforts to debug the equipment and the programs are cumbersome and tend to be pursued until reasonable—expected—results are produced. At that stage, the incentive to continue the debugging is considerably reduced, in particular if time is short, which is often the case. Thus, the results that are presented may have a bias towards the expected values.

However, there may be another factor of significance: the welcoming of some-what different values, which may then make it easier to publish the result, just because it seems to be more interesting than a simple replication. The study of stellar chemical abundances (a study which needs a number of steps in which judgment plays a role that is hard to automate) contains such examples. Results from different determinations 'jump' between different seemingly converged values as a function of time [195]. I named that that the 'sheep–goat effect' in 2004. A corresponding effect was also traced in molecular genetics research and dubbed the 'Proteus phenom-enon' [196].

8.5.9 Shadow zones: authorship

The principles of assigning authorship of papers are much discussed and certainly have ethical aspects. One question concerns the criteria for coauthorship. The International Committee of Medical Journal Editors (ICMJE) recommends authorship to be based on the four criteria [197]:

'(1) Substantial contributions to the conception or design of the work; or the acquisition, analysis, or interpretation of data for the work; (2) Drafting the work or revising it critically for important intellectual content; (3) Final approval of the version to be published; (4) Agreement to be accountable for all aspects of the work in ensuring that questions related to the accuracy or integrity of any part of the work are appropriately investigated and resolved. ... In addition, authors should have confidence in the integrity of the contributions of their coauthors.' These rules seem to be frequently disrespected by medical faculties in Sweden [198].

Often, the contributions are not significant in all these respects but may be specific in just one or two of them. In many cases, the contribution may be limited to some important idea, may be limited to some aspects such as the construction of equipment or computer programs used, or may already have been presented in previous papers. In other cases, the main contribution by a given author may be the writing of applications to secure sufficient funding for the research. Which of these different contributions is required, and how significant should they be, in order to earn a coauthorship?

Comparative studies, such as [199], show that in this respect, the culture in different fields, journals, institutions, even research groups at the same department may be quite different in spite of many local attempts to establish common practices. In some institutions a rather rigorous principle is established, restricting the authorship to those individuals who have taken part in the project from the beginning, have written parts of the final report, and are able to present and defend the project fully at an international symposium without drawing back from certain questions with the excuse 'I did not do that part'. In other places, there may be the habit of including anyone who is a member of a research team even if they contributed nothing to the paper in question.

Individuals who are included in the author list of an article without having contributed substantially are generally called 'honorary authors' or 'guest authors'. There may also be important contributors who are not included in the list, known as 'ghost authors'. In an extensive survey [200], the prevalence of honorary and ghost authorship in high-impact biomedical journal papers was found to be typically 20% for both categories.

The ICMJE rules admit another category, contributors, for individuals who have made important contributions to the paper but who did not meet all of the stated criteria. It seems that this option, with specification of what each contributor has done, would be worth adopting in wider circles. This would make it clear which individuals are most responsible for the paper as a whole, while giving proper credit to those who have contributed important parts but should not be held responsible for the paper in total.

Another issue for which the cultures are very different is the order of authors. As long as these conventions are understood and agreed on within the research group and its peers, the ethical problems are limited, although it may be important both for evaluators and users of the science presented to know who is most responsible for the work.

However, considerable ethical issues may lurk in these lists of names. 'Guest writers' appear in many lists. Typically, they are individuals with well-known names who are listed in order to give credibility and status to the science that is presented, as payment for various types of support, in exchange for positions in author lists of the guest's own projects, as 'consolation authors' whose own project did not go very well and so more papers were needed for the PhD thesis, or as 'decoys' which are there as encouragements to recruit young scientists to the group. Such bad manners may not mar the publications as such, but they certainly disturb the career and reward system which today is so dependent on publication statistics. In the worst case, this practice may lead to a culture wherein which scientific results and papers are produced for which nobody takes full responsibility.

One should realise that authorship of science papers may be bought for some tens of thousands of dollars without the purchaser having contributed any research at all [201]. This shows how publication of science papers is becoming another market.

8.5.10 Popular science and teaching the public

The monkey on the right in figure 8.9 is supposed to see *and* report what it sees and not only to the other monkeys on the bench but also to the beholders, to the public. For publicly financed science the obligation to communicate the ideas, results, and prospects of the research activity to the public should be obvious, at least in democratic states. An ethical problem, however, is how this is done.

All science communication, not least popularization, requires simplification. To simplify in a truthful way is an art in itself. If it is done with the aim of bringing forward a particular point, that bias must be balanced by presentation of other reasonable views, in particular if the speaker or writer, as is often the case, is trying to make a particular point or advocates a personal view. The requirement of a balanced presentation is at least as strong as what one should demand from any scientific paper. In the popularization case we cannot expect the recipients of the message to see the proper counterarguments and form their own balanced view. The contrast between our habit in science to require peer reviews of our manuscripts and project proposals while presenting our science subjectively to the public is striking.

In recent decades there has been a clear tendency among universities and other stakeholders of research to professionalize public communication by hiring public relation professionals with the ambition to give a rather glorious picture of science in general and their institution in particular. The need to attract more attention, funding, and students may be understandable, but it is unfortunate if one tries to attract these by selling the activity through exaggerated arguments. In the long run, public respect for science in general and the institution in particular may be eroded.

Figure 8.10. In a 1999 article in *National Geographic* this fossil was claimed to be a 'missing link' between terrestrial dinosaurs and birds. It was later shown to be a composite of fossils from several different species [202]. The forged 'Archaeoraptor liaoningensis' specimen is on display at the Paleozoological Museum of China. Photo: Jonathan Chen.

In popular presentations such as books, TV programmes, or exhibitions (as in figure 8.10), one rather often finds anecdotes or curios that are false but remain unchallenged for decades or even centuries. If you ask the authors about this, you may get the answer that 'that story is too good not to be told'. The balance between the desire to entertain and the need to inform correctly must not be lost. The nobility mark of science, the strong ambition to be truthful, should be dominating and visible in all our activities.

One very important aspect of teaching science to the public is to try to counteract superstition in an era of systematically spread misinformation ('fake news'), sometimes presented in a misleading scientific disguise, and as something that 'research has proven' without any sources given. It is nontrivial to foster balanced critical attitudes without eroding trust in science and truth as such. It is an obligation of scientists to contribute to this teaching in collaboration with schoolteachers and journalists, avoiding elitist attitudes as far as possible.

8.5.11 Science advice

The role of a science advisor to decision makers, whether in the public sector or industry, is delicate. Truth should always be presented as objectively as possible. A central question remains: Should the scientific advice be limited to presenting the factual circumstances while refraining from further value-based conclusions on what actions should be taken? After all, the scientist does not work from the direct mandate of the public and is not accountable to voters, unlike the politician in democratic states. This is an argument for the scientist to avoid being involved in the decision making. However, one may also argue, following Douglas [200], that scientists, being more knowledgeable in certain situations, are best equipped to foresee the moral and political implications. They should, if they meet ignorance or misunderstanding among politicians, consider giving concrete political advice.

An example was presented by the HBO TV series *Chernobyl*. Here, the physicist Legasov, beginning to suggest an evacuation of the nearby populations to minimize the radiation damage (which the politicians underestimated or did not understand), is told that his advice should be limited to 'answer direct questions about the function of the reactor ..., nothing else'[8].

The balance between loyalties to the decision makers and to the scientific community is another difficult balance for the advisor. My advice from personal experience in this respect would be that support from experienced mentors is vital.

8.5.12 Attempts to strengthen integrity and ethical awareness in science

A most obvious consequence of the forgery of scientific data and of other dishonest presentations of science, including plagiarism, is that the most qualified scientists may be hindered from getting positions or grants due to competition from less honest colleagues. A number of universities and funding agencies have introduced measures to reduce this risk. One of these, which is not very common, is to improve the reviewing process by giving more time and resources to experts involved in the evaluation and demanding more details from the applicants. For university research, much evaluation is basically in the hands of the science journals but could be complemented by a local review process at the universities, although this brings some risk of inappropriate censoring. Another measure is to install severe sanctions against cheaters, such as stopping funding of their projects, firing them from their positions or membership of committees or learned societies, retracting their papers from journals, or, if laws apply, bringing them to court. Such measures have been taken in a number of cases. It is, however, not clear that this has contributed very efficiently to the solution of the problem in view of its magnitude and its partly hidden character, as discussed earlier in this chapter.

Another countermeasure against forgery of scientific results and other bad practices is to promote the awareness within the scientific community of the need for scientific integrity. During the last decades a great number of initiatives have been taken in this direction, by international organizations, national authorities and universities. The US Office of Research Integrity and the European Science Foundation established a World Conference on Research Integrity with a first meeting in Lisbon in 2007 followed by one in Singapore in 2010, which produced a 'Statement on Research Integrity' [203], and a sequence of additional meetings about every second year, with hundreds of participants in each from the scientific community and administrative bodies, including some top officials. The themes discussed at these meetings have gradually widened, including self-regulation measures with peer reviews, replication and retraction, proper teaching of younger scientists, and systemic problems in research that undermine integrity and possibly generate misbehaviour. These problems were coupled to research funding structures

---

[8] For an interesting discussion, see Silk M S W 2019 The ethics of scientific advice: lessons from 'Chenobyl'. *The Prindle Post*, Ethics in the News from the Prindle Institute, July 26, 2019, https://www.prindlepost.org/2019/07/the-ethics-of-scientific-advice-lessons-from-chernobyl/ (May 2, 2021).

and processes, competition among researchers and institutions, and career systems. Further aspects discussed were the replication crisis, the role of media and outreach, quality control in laboratory contexts, and industrial research [204].

A number of different international organisations have set up research ethics committees that at times work on integrity issues and give relevant recommendations concerning those. The International Science Council, with a broad coverage of disciplines and countries, is advised by its Committee for the Freedom and Responsibility in Science to promote the freedom for scientists to pursue knowledge and interact but also to take responsibility to maintain high science ethical standards. Several organisations with universities as members, such as the European University Association, which has more than 800 universities as members, are important as platforms and bases for establishing ethically acceptable practices in research.

Several codes of conduct have been established through such organisations. For instance, the European federation of Academies, Allea, has established a European Code of Conduct [205], which is recognized by the EC as a reference document.

On the national level, a large number of initiatives have been taken by research agencies and universities, establishing ethical committees and procedures with the aim of preventing misconduct among their grant holders or employees. As bases for the work of the committees, rules of conduct have been produced[9]. Initially, the committees were in most cases organised as part of the respective agency or university. This is still so in many cases, for instance, for the proactive bioethic committees that have to approve research projects that involve animals or humans before grants can be allocated to them. For the reactive committees that handle reports on suspected misconduct by individuals, several countries have appointed committees that are independent from universities in order to reduce tendencies to suppress locally embarrassing cases, or even misuse the local committee to legitimate nonethical actions[10].

Another reason for appointing national committees is to ensure consistent nationwide criteria and sanctions. Formally, these committees should not be regarded as courts. Their task is to give recommendations to funding agencies and employers, which have the mandate to take action if the misconduct is not forbidden by law. If it is, the case may proceed to a regular court. However, it is not obvious what the adequate consequences should be. All sorts of reactions by the agency or employer occur, from friendly notices to firing. In practice, a committee decision that a certain behaviour is judged to be reprehensible may have a very severe impact on the future working conditions and career of the individual scientist and his or her research group: it may eventually lead to a professional ban. In view of these serious consequences, even in cases in which the sanctions are limited, the needs for possibilities and procedures for appeal have been stressed. Likewise, methods and

---

[9] One example is *Good Research Practice*, Swedish Research Council (2017), https://www.vr.se/english/analysis/reports/our-reports/2017–08–31-good-research-practice.html

[10] For examples of misusing the local ethics review committees for reputation management, see [192].

routines for rehabilitation of scientists who have been found guilty of misconduct are called for[11].

In spite of the arrangements mentioned previously, it is unclear whether scientific misconduct will be reduced. The possibilities for reporting suspected misconduct may keep the numbers of obvious fraud and plagiarism cases low, but it is uncertain whether less obvious, more cunningly constructed forgeries and citation plagiarism, 'bibliographic negligence', or authorship trading is affected at all. Courses and conferences on research ethics raise awareness in the scientific community, and exposure of misconduct offers interesting insights into human psychology, but it is uncertain whether these will really decrease the misconduct, any more than courses in criminology can be expected to reduce the crime rate. Such initiatives serve not only to make scientists aware of the integrity problems, but at least as much to make decision makers and funding agencies aware of the ambitions in the scientific community to come to terms with these problems.

8.5.13 Strengthening ethical behaviour by relaxing competition

If one aims at reducing the misdemeanours of scientists, one could try to harness one of its probable causes: the increased competition for positions and resources. For a discussion of this, see the review by Fang and Casadevall [207]. Classically, competition was regarded as beneficial to the scientific endeavour, stimulating scientists to work hard and publish swiftly [208]. This view is still common, not least among funding agencies. In recent decades, several groups have expressed their findings (or fears) that the strengthened competition, or 'hypercompetition', generates bad practices [209, 210], such as preventing others from using the methods developed, interfering with peer-review processes, poor publication habits, carelessness, and partial and unfair evaluations of the work of competitors. Fears have also been expressed that the competition scares young scientists, maybe in particular females, away from scientific careers [211].

Fang and Casadevall [207] argue that competition is not essential for good science. On the basis of neuropsychological and experimental psychology results, they suggest that in fact competition may have detrimental effects on creativity, citing the psychologist Theresa Amabile, who argues, on the basis of experimental studies of children, scientists, and technicians, that interest, enjoyment, satisfaction, and challenges in the work itself, not external motivators or pressures, are main factors in creativity. She describes a work environment that relies on external financial rewards, creates relentless deadlines, and subjects any proposals to 'time-consuming layers of evaluation, … and excruciating critiques' as counteracting creativity [212] and concludes that 'job security appears to be extremely important in fostering creativity' [213].

In an attempt to determine whether the conclusions of Fang and Casadevall are valid in a contemporary North European context, I prepared a small exploratory study with a questionnaire among staff and graduate students at physics and

---

[11] For rehabilitation of wrongdoers, see [206].

astronomy departments in the Stockholm–Uppsala region. In this I asked questions as to whether the effects of competition were mainly positive or negative in terms of quality of science, well-being of scientists, and whether competition attracted scientists and students to research or pushed them away from it. The vast majority of the respondents argued that the competition for jobs and resources was mostly negative for the well-being and recruitment of scientists, and most respondents also argued that negative effects dominated for science as such. Most positive answers came from well-established and well-supported senior professors. Noteworthy, however, was that the negative views of competition which were so dominant among the younger scientists also prevailed among the emeritus professors.

Fang and Casadevall claim, on the basis of studies of the origins of innovations in science and technology, that competition, if it prevents cooperation, is counterproductive.

From this discussion it seems natural to tentatively conclude that relaxing the competitive staging of the research career and financing systems would lower the degree of unethical practices in contemporary science. The effects of such reforms could also lead to more creative science and more collaboration within and between different research areas. Also, the recruitment of very talented but less competition-oriented young scientists would be beneficial. It thus seems possible that science at large would gain from such a relaxation. Further studies of the effects of competition may be warranted. However, there could be challenges for funding agencies where authority and power are partly based on the competition system itself.

Reforms to relax competition and enhance collaboration and creativity may include the science culture, career systems, and distribution of resources. For instance, concepts such as science as a Darwinian struggle in which only winners survive, the main focus is on priority in discoveries and innovation, and only principal investigators are celebrated may be toned down. A better balance between workforce and resources, sealing 'leaky pipelines' in the career flow, and long-term commitments for funding agencies, with grants given more to able groups and individuals than to short-term projects, could be considered [214]. An interesting question is whether the scientific community could abstain from the stimuli of fierce competition and instead embrace the joy of collaboration and discovery of our remarkable world.

Dan Larhammar and Michael Way are thanked for comments on the manuscript.

8.6 The limits of science

8.6.1 Technical, fundamental, and epistemological limits of science

Richard Dawid[1]
[1]Stockholm University, Stockholm, Sweden

In recent decades, a gap between two kinds of physical reasoning has opened up. Applied physics and phenomenological physics show all the basic characteristics of canonical 20th century science. But fundamental physics, represented by high-energy physics model building, quantum gravity, and cosmology, faces substantially new challenges that influence the nature of the scientific process. Those shifts can be expected to become even more conspicuous in the period up to 2050. Exploring their full scope will arguably be an important task for fundamental physics in upcoming decades.

The 20th century was a hugely successful period in fundamental physics. Developments from the advent of relativistic physics and quantum mechanics to advanced theories in particle physics played out based on the general expectation that physicists would find theories that could account for the collected empirical evidence in a satisfactory way and could (with some exceptions) be empirically tested within a reasonable time frame. In stark contrast to the 18th and 19th centuries' absolute trust in Newtonian mechanics, however, 20th century physicists assumed that none of the theories they were developing would be the last word on the general subject they addressed. Theories that were successful at the time would eventually be superseded by more fundamental ones. Physicists thus were highly optimistic about their future achievements but avoided declarations of finality with regard to any theory at hand.

Early 21st century physics finds itself in a very different place. This section will focus on three aspects of the new situation: the long periods of time in which influential theories remain without empirical testing, the long periods of time in which theories remain conceptually incomplete, and the issue of finality in contemporary physics.

8.6.1.1 Limits to empirical access

In high-energy physics, the last theory that has found empirical confirmation is the standard model, which was developed in the late 1960s and early 1970s. The empirical confirmation of the standard model's many predictions was completed in 2012 with the discovery of the Higgs particle[12]. Theory building from the mid-1970s onwards has advanced far beyond the standard model, however. New theories were motivated in various ways. Some characteristics of the data, though not contra-dicting the standard model, found no satisfactory explanation on its basis. More

---

[12] The only data-driven adaptation of the standard model happened in the late 1990s when the massiveness of neutrinos was empirically established. That feature could be accounted for within the standard model framework in an entirely coherent way, however.

substantially, quantum field theory, which provided the conceptual foundation for the standard model, was inconsistent with general relativity, the theory that described gravity. General characteristics of gravity imply that the strength of the gravitational interaction becomes comparable to the strength of nuclear interactions at a high energy scale (called the Planck scale). To describe physical processes at that scale, a new theory is needed that describes gravity and nuclear interactions in a consistent way.

During the last 50 years, the described lines of reasoning in conjunction with others have led to the development of influential theories beyond the standard model. Grand unified theories, supersymmetry, and supergravity introduced larger symmetries within the context of gauge field theory. String theory aims at full unification of all fundamental interactions by reaching beyond the confines of conventional gauge theories. Various conceptual approaches aim at the quantization of gravity from a general relativistic starting point. In cosmology, the theory of inflation provides an entirely new view of the very early phase of the universe.

Up to this point, none of the mentioned theories has achieved empirical confirmation of its core predictions. Nevertheless, some theories have been quite strongly endorsed by its exponents. The most striking example is string theory [215], which was presented in 1974 as a fundamental theory of all interactions based on replacing the point-like elementary particles of quantum field theory by extended one-dimensional objects, the superstrings. From the late 1980s onwards, string theory has assumed the status of a conceptual basis and anchoring place for much of fundamental physics. Exponents of string theory think that they have very strong arguments for the theory's viability even in the absence of empirical confirmation[13]. Critics of string theory, on the other hand, deny any epistemic justification for an endorsement of the theory in the absence of empirical confirmation [216].

A slightly different case is cosmic inflation [218]. The theory of cosmic inflation lies at the core of large parts of contemporary cosmological reasoning. It aims to explain some very general features of the niverse that would seem *a priori* inexplicable within the context of general relativity, and to account for characteristics of cosmological precision data. To that end, it posits an early phase of extremely fast (exponential) expansion of the universe. In contrast to string theory, quantitative implications of models of cosmic inflation can be confronted with empirical precision data. The theory is widely taken to be in very good agreement with available cosmological data. Many of the theory's exponents have a high degree of trust in the hypothesis of inflation on that basis. Critics of inflation take that trust to be unjustified, however. They emphasize the theory's lack of conceptual specificity, and they doubt the confirmatory value of seemingly supporting empirical evidence [219].

---

[13] Contact between string theory and characteristics of our world that does not reach the level of significant confirmation has been made in the research field of string phenomenology [217].

Maybe the most fundamental problem for conclusive empirical confirmation arises with respect to the multiverse hypothesis [220] [14]. Based on conceptual considerations in inflationary cosmology, the multiverse hypothesis posits that vast numbers of universes are generated in an exponentially expanding background space. We live in one of those many universes and, according to the present understanding, cannot possibly make any observations beyond the limits of our own niverse. The existence of the other universes therefore may be inferred based on theoretical considerations but can never be empirically confirmed. Exponents of the multiverse point out that there can be empirical confirmation of the multiverse theory based on the theory's predictions regarding our own universe. Critics of the multiverse argue that the in-principle lack of empirical access to core objects posited by the multiverse hypothesis nevertheless infringes on the principle of testability of scientific theories [221].

In all three described contexts, physics faces the problem how to deal with the absence of, or complications regarding, the empirical testing of fundamental physical theories. These problems can be expected to continue in upcoming decades and raise the general question as to what counts as a viable epistemic basis for seriously endorsing a scientific theory. Obviously, the scientific process will sustain its efforts to develop effective strategies for empirical testing wherever possible. Empirical confirmation will remain the ultimate and most trustworthy basis for endorsing a theory. A question that will become increasingly important, however, is how to assess the status of well-established theories in the absence of sufficient empirical confirmation. A research process where scientists often spend their entire career working on a theory without seeing that theory conclusively empirically tested raises this question with urgency. Characteristics of fundamental physics today may indeed offer a basis for epistemic commitment that reaches beyond the canonical confines of empirical confirmation. A philosophical suggestion to that end has been presented under the name nonempirical (or more specifically metaempirical) confirmation [222].

As was described previously, scientific praxis in recent decades has led many theoretical physicists towards having substantial trust in empirically unconfirmed or inconclusively tested theories, while other physicists have strongly criticized that process. The developments in physics in the upcoming decades will move this issue forward, one way or another. If further developments vindicate trust in theories such as string theory or cosmic inflation based on empirical confirmation, nonempirical arguments, or both, this will increase the willingness of theoretical physicists to strongly endorse theories even in the absence of strong or any empirical confirmation. In that case, physical reasoning in 2050 will be based on a significantly extended concept of theory confirmation. If trust in the above theories erodes for whatever reason, the focus will move back towards a more traditional understanding of theory assessment.

8.6.1.2 The chronic incompleteness of fundamental theory building
The second problem faced by fundamental physics may be even more significant from a conceptual point of view: fundamental theories in physics become

---

[14] The multiverse plays a pivotal role in the influential and much debated anthropic explanation of the fine-tuned cosmological constant.

increasingly difficult to spell out in a complete form. In some cases, such as the case of cosmic inflation, this problem is directly linked to the theory's limited empirical accessibility, which stands in the way of specifying the conceptual details of the theory in question. But in other cases, such as string theory, the core of the problem seems distinct from issues of empirical access.

String theory has been conceptually analyzed since the late 1960s and was proposed as a theory of all fundamental interactions in 1974. Fifty-five years of work on string theory, including four decades when a substantial part of the theoretical physics community has contributed to developing the theory, have not brought string theory anywhere close to completion. The theory today amounts to an enormously complex system of conjectures, elements of formal analysis, and calculations achieved based on simplifications or approximations. In conjunction, these strands of analysis provide a considerable degree of understanding of the coherence and cogency of the overall approach and the ways in which many of its aspects are intricately related to each other. Still, the theory's core remains elusive.

As described earlier, the development of the conceptual understanding of string theory has not found guidance in empirical data. Such data, if available, would obviously be very helpful for further conceptual work on the theory. However, the data would not in themselves provide the basis for solving the conceptual core problem: how to pin down the full formal structure of string theory.

The difficulties in developing a full-fledged string theory have their roots in quantum field theory, which was developed to describe interactions between highly energetic (that is very fast-moving) elementary particles starting in the 1930s. Calculations of specific quantitative empirical implications of quantum field theories can be extracted only on the basis of perturbation theory, which is a method of approximation. Perturbation theory provides highly accurate and reliable quantitative predictions of specific particle interaction processes if the interactions involved are weak (in a well-defined sense). For strong interactions, the method breaks down. String theory was initially developed as a perturbative theory along the conceptual lines of perturbative quantum field theory. Two aspects of string theory render the use of perturbative methods more problematic than in the case of quantum field theory, however. First the full theory to which perturbative string theory is an approximation is still unknown. Therefore, perturbative string theory serves not merely as an approximation scheme but also as an essential indicator of the character of the unknown theory to which it is supposed to be an approximation. Second, string theory has no (dimensionless) free parameters. Therefore, interaction strength, like all other parameters of low-energy physics, must emerge from the full dynamics of the theory. Thus, the reliability of the perturbative method cannot be a universal feature of string theory. At best, it could be extracted for a given regime from a nonperturbative analysis of the fundamental theory. Compared to conventional quantum field theory, the understanding of string theory thus is overly dependent on a perturbative perspective, and there is an urgent need to reach beyond it.

Recent decades have led to a deeper understanding of string theory that reaches beyond the perturbative perspective. A crucial role in these developments has been played by duality relations that establish weak and strong coupling regimes of seemingly entirely different types of string theory (and beyond) to be empirically equivalent [223]. No breakthrough towards a full understanding of the mechanisms that guide nonperturbative string theory has been achieved, however. The elusiveness of a full formulation of string theory at its core in this light is a conceptual problem rather than a problem of empirical access.

The problem is not confined to string theory. All approaches to quantum gravity that choose different starting points than string theory, such as loop quantum gravity or spin foam, have encountered similar problems. After decades of work on those approaches, they have not led to a complete theory. It is not even clear whether a coherent theory of gravitation in four extended spacetime dimensions can be formulated on the basis of those approaches.

The problem thus seems to arise once theory building aims at the level of universality needed to join the principles of quantum physics necessary for understanding microphysical phenomena with the principles that govern gravity. While 20th century physics has achieved a satisfactory understanding of those two realms of fundamental physics in separation, the 21st century may be expected to be devoted to bringing them into a coherent overall conceptual framework. It is exactly this context where the substantial roadblocks described above arise.

It will be the main task of fundamental physics in the upcoming decades to further attack those difficulties. But the substantial shift in the time scales for completing theories of fundamental physics may lead fundamental physics towards reevaluating its understanding of scientific progress altogether. Since the 19th century, this understanding has been based on the principle of theory succession: Theories are being developed within a reasonable time frame, to be empirically tested soon thereafter. Empirical tests could lead either to the theory's rejection or to its confirmation as a viable description of nature within a given empirical regime. Further tests would then test the theory's predictions with increasing accuracy until a disagreement between data and the theory's predictions was found. Such empirical anomalies could then lead to the development of a new theory that replaced the old one as the viable fundamental theory.

The timelines for empirical testing have been stretched to an extent that renders the canonical view of the scientific process insufficient. In this section we point out that the canonical view is drawn into question at a conceptual level as well. Scientific progress in 21st century fundamental physics may not amount to formulating complete scientific theories.

Today, this shift of perspective is merely a possibility. It may still happen that, by the year 2050, revolutionary changes in physics will have turned a full theory of quantum gravity into an imminent prospect or even into reality. If so, the current suspicion of a long-term change of theory dynamics would have turned out to be a transient impression provoked by a particularly difficult phase of theorizing.

If, however, the period up to 2050 prolonged the current step by step conceptual progress towards a better understanding of an elusive theory of quantum gravity, this would strengthen the case for acknowledging a lasting and substantial shift in the scientific dynamics of fundamental physics. The process of theory succession that characterized 19th and 20th century physics would seem to have been replaced by a different mode of the scientific process that is represented by continuous work on and an improving understanding of one theory or theoretical framework without prospects of formulating a complete theory in the foreseeable future. The completion of that theory would appear to be a remote endpoint of the process of physical conceptualization rather than an imminent goal for the individual scientists. To what extent that new dynamics amounted to a manifestation of fundamental limits to science and to what extent it should rather be viewed in terms of an altered concept of scientific progress would then emerge as a core question for the status of physics in the 21st century.

8.6.1.3 Signs of finality

A third important shift that has occurred in fundamental physics in recent decades is directly related to the issue of acknowledging a new phase of the scientific progress: Fundamental physics today is more conspicuously associated with issues of finality of theory building [224] than 20th century physics.

Throughout much of the 20th century, physics shunned any suggestion of finality in physical theory building for two reasons. First, the revolutions of special and general relativity and quantum mechanics served as omnipresent reminders that even with regard to a theory that was as dominant, successful, and long-living as Newtonian mechanics, claims of finality had been misplaced. Second, it became increasingly clear that the incoherence between the principles of quantum mechanics and general relativity would require at least one more fundamental conceptual step before arriving at a fully consistent overall under-standing of theoretical physics. All empirically successful theories in 20th century physics were therefore understood to be viable at most up to those energy scales where predictions had to account for gravity and nuclear interactions at the same time.

When attempts to develop a theory of quantum gravity took center stage in fundamental physics in the last quarter of the 20th century, this situation changed. The second reason for not considering issues of finality ceased to apply, since quantum gravity amounted to the projected conceptual step that had prevented finality claims regarding previous theories. Moreover, the appeal of the first reason was considerably weakened as well. Quantum gravity provided new reasons for taking finality claims seriously that went beyond anything that could have been said in support of the finality of Newtonian mechanics in the 19th century. The 19th century finality claims regarding Newtonian mechanics were based simply on the enormous and longstanding success of the theory in multiple contexts and the lack of evidence that suggested that it needed to be superseded. The empirical testing of processes where very high velocities, very small objects, or very high gravitational forced were involved did eventually reveal empirical inconsistencies with Newtonian mechanics that led to the revolutionary new theories that superseded it. Nothing in Newtonian physics apart from a crude meta-inductive assumption that a theory that has worked in so many contexts should work everywhere would have, in advance, spoken against the possibility of such an outcome.

General considerations on the nature of quantum gravity provide a stronger basis for a final theory claim. To understand the basic idea, one needs to remember once again how physics has changed in the 20th century. From a 19th century perspective, velocities, distances, and mass values were independent parameters. Special relativity then established that an object's mass was a form of energy (just like its kinetic energy), and quantum mechanics made it possible to view distance scales in terms of inverse energy scales. The move towards higher energy and respectively smaller distance scales thus became the one central guideline for finding new phenomena in fundamental physics. Quantum gravity now suggests that the notion of distance scales smaller than the scale where gravity becomes roughly as strong as nuclear interactions (the so-called Planck scale) may not make sense, due to a fundamental limit to information density. String theory, viewed by the majority of physicists to be the most promising approach of quantum gravity, offers a deeper understanding of this limit (in terms of a specific feature called T-duality) [225]. These arguments do not conclusively establish finality because their soundness relies on the truth of the theory or conceptual framework on whose basis they are developed. Nevertheless, they turn questions of finality into genuinely physical questions.

The issue of finality stands in a complex relation to the issue of chronic incompleteness addressed in the previous section. On the one hand, chronic incompleteness makes it more complicated to understand what could even be meant by a final theory claim. How is it possible to assert the finality of a theory whose full formulation is not in sight? At a different level, however, the final theory claims raised in quantum gravity seem in tune with the phenomenon of chronic incompleteness. Like chronic incompleteness, final theory claims may be taken to suggest that the paradigm of scientific progress that was prevalent throughout the 20th century is inadequate for characterizing the scientific process in 21st century fundamental physics. Based on the canonical paradigm of scientific progress, a final theory claim regarding a universal theory such as string theory would imply the completion of fundamental physics within the foreseeable future. Once one replaces that canonical paradigm by a principle of chronical incompleteness, however, nothing of this kind follows. In that view, the projected point in time when fundamental physics will have been completed has not come closer. What has changed is the nature of the scientific process that leads towards that point. Rather than a sequence of superseding complete theories, it would be step-by-step progress towards an improved understanding of the one universal and final theory physicists are working on already but whose completion is not in sight.

If the upcoming decades of physical research strengthen the tendencies described in this text, fundamental physics in 2050 will be a very different enterprise than fundamental physics half a century ago. Its new character will arguably change the human understanding of the nature of scientific reasoning. The three described developments are distinct but carry a coherent overall message. As long as physics deals with limited sets of phenomena, it is the physicist's task to identify those phenomena that allow for the development and empirical testing of appropriate theories within a reasonable time frame. Once physics approaches a fully universal fundamental theory, however, leaving out what seems too difficult to include stops being an option. Physics thus faces a situation where problems too difficult to solve and phenomena too remote to be empirically tested at the given point all live within the scope of the universal theory that is being developed. Achievable research goals in this new environment shift from the complete formulation and conclusive testing of the theory towards the more modest goals of solving specific problems within the overall theory, confronting the theory with empirical data to the extent possible, and assessing the theory's status based on all information available. Within this new framework, just as before, physics will pursue its old but still distant ultimate goal: to find a full and consistent description of the physical world we live in.

8.6.2 The future of humankind and behaviour

Ernst Ulrich von Weizsäcker[1]
[1]Professor, Freiburg, Germany

8.6.2.1 *Fascinating physics*

Discovering and visualizing gravitation waves accompanying the collision of two black holes have been a megaevent for astrophysics and for theoretical physicists remembering Albert Einstein's theory of general relativity, which predicted the existence of such waves [226].

A technological mega-adventure is it to copy on the Earth the energy production in the Sun's plasma by letting billions of deuterium and tritium atoms fuse into helium and a neutron. In Cadarache, 50 kilometres north of Marseille, the International Thermonuclear Experimental Reactor is under construction [227].

Understanding the solid-state and fluid-state physics of the vitreous body of the human eye is essential for ophthalmologists confronted with a patient's mechanical indentation stemming from a car accident [228].

These are three examples out of hundreds of fascinating (and often useful) discoveries or textbook results of modern physics. In some cases, such as in the physics of the eye, the usefulness is evident. In others, such as in the Tokamak fusion reactor, the usefulness for providing huge amounts of energy is still a guess. In some fields of physics, such as black holes and gravitation waves, humanity may never find practical advantages but will appreciate the scientific excellence. Moreover, the experimental setup leading to the visualization of those mysterious waves is likely to produce useful byproducts.

8.6.2.2 *Known and unknown dangers*

In 1938, Otto Hahn and Fritz Strassmann found the chemical element barium after shelling neutrons on uranium and were highly suprised. Lise Meitner, Hahn's earlier colleague, emigrated to the US, immediately explained that the uranium atoms must have been split. So far, it was just fascinating pure physics and chemistry. Otto Hahn rightly won the Nobel Prize (in chemistry) for it. But instantaneously, physicists around the world became aware of a huge potential of a new method of producing useful energy. Simultaneously, they also became aware of immense dangers resulting from atomic explosions using the uranium splitting.

This is perhaps the best-known example in history of the findings of physics leading to a nightmare. Atomic weapons soon became an immensely important factor for the politics of power.

Physics, chemistry, and, since the discovery of genetic engineering, biology have become ingredients of science fiction novels with a tendency to emphasize the dangers and rather underestimate the enormous benefits emerging from the natural sciences.

As we look towards the Horizon of 2050, it will be wise to further develop human understanding of the dangers and of methods of analysing and controlling them. We will need a political consensus that innovations in the sciences should be accompanied by technology assessments addressing potential dangers of criminal abuse or technological failures.

8.6.2.3 The limits to growth
Of course, dangers are not always caused by technological or scientific discoveries. Rather more often, dangers can arise from conventional developments. Wars, pandemics, famine, and crimes have caused disasters throughout human history. In some rare cases, disasters are looming just from the continuation of benign and highly popular activities.

Over the millennia, humans have always tried to create good lives for themselves, notably by overcoming hunger or famine and healing curable conditions and diseases. Nobody would have blamed such efforts as the causes of dangers or disasters. Creating economic growth has been the desire and plan of politicians around the world. But then, in 1972, a shocking book was published that expressed the unthinkable, namely, that continued growth might eventually lead to very unpleasant collapse events for the simple reason that the size of planet Earth was limited and was therefore in straightforward conflict with unlimited economic growth. The book was called *The Limits to Growth* [229]. It resulted from research initiated by the Club of Rome, which was founded in 1968.

What was and what is that mysterious Club of Rome? Its founders were Aurelio Peccei, an unusually gifted and successful Italian business man (Fiat, Olivetti) with a strong focus on world justice, and Alexander King, head of the OECD's science department. In 1968, they brought together some 20 like-minded people in Rome to discuss 'the predicament of mankind'. One systems scientist of the group was Professor Jay Forrester of the Massachusetts Institute of Technology, who had developed a smart computer programme, called Dynamo, allowing the estimation of future developments of several factors mutually influencing each other. Forrester offered to bring a team together for using Dynamo for tentatively deciphering the predicament of mankind.

The team, under the leadership of Dennis and Donella Meadows, did an impressive job calculating or rather estimating the predicament for food, mineral resources, industrial output, population, and pollution. Dynamo allowed the production of impressive visuals. The standard run of the programme ended up producing the graph shown in figure 8.11.

The book became an unprecedented world bestseller. It was translated into all major languages and sold more than 30 million copies. For the broad public, it was a shock. Growth, after all, was the symbol of a better life, more freedom, more mobility, and the end of hunger.

The assumed exhaustion of natural resources was the biggest shock. How could humanity survive if natural gas and oil or copper and iron would no longer be available? One group of countries, the Organization of Oil Exporting Countries (OPEC) was also shocked, but some in OPEC soon conjectured that oil scarcity also meant that oil countries held a very powerful weapon in their hands. Oil, after all, was one of the most demanded natural resources. In attempting revenge for the Yom Kippur War of October 1973, Arab oil-exporting countries pushed the price per

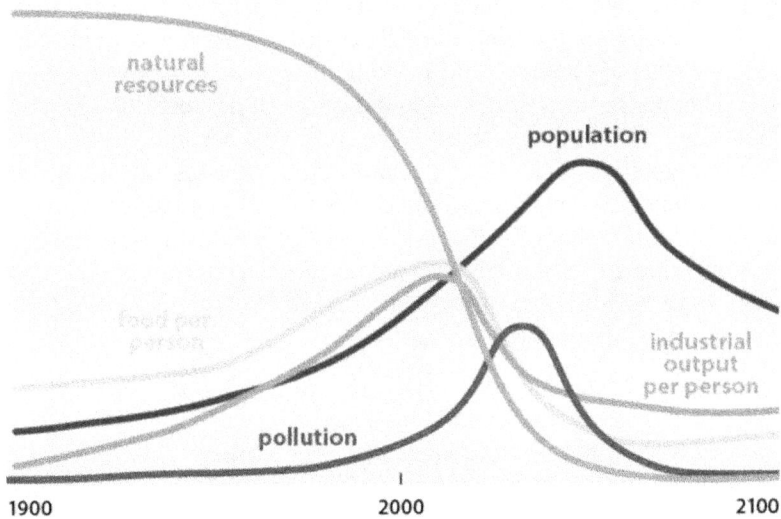

Figure 8.11. The world model brought five parameters together, mutually influencing each other. The influencing factors were empirically established from the first 70 years of the 20th century. One conspicuous prediction was the exhaustion within 100 years of natural resources (green curve). In consequence, food and industrial output per person and the human population were also sharply reduced.)Source: Meadows *et al*, p 124 [229] CC BY 4.0. The picture is styled for easy reading.)

barrel of oil from the low level of US$3 up to US$12 with the intention of punishing the countries which had supported Israel. Indeed, the price shock led to an oil crisis that caused an awful stagnation in the industrialized world.

However, one of the predictable consequences of oil scarcity and elevated oil prices was a strongly intensified worldwide oil exploration and exploitation. This turned out remarkably successfully, leading to an oil glut. The glut defeated many of the assumptions of *Limits to Growth*. Systems theory, by that time, was mature enough to realize that computer programmes must include the possibility of major changes in initial assumptions.

Nevertheless, some commentators still attempted to justify the logic of *Limits to Growth* because the basic message was so plausible. One such attempt came from Graham Turner [230], who 'updated' the *Limits to Growth* model, essentially by replacing oil scarcity with the need to reduce CO_2 emissions. It is surely correct that 40 years after the oil crisis the major concern is no longer the absence of cheap oil but global warming, caused by too much burning of fossil fuels. But in terms of the geological and mathematical correctness of the *Limits to Growth* assumptions, this is pure nonsense.

8.6.2.4 *The real scare is planetary boundaries*

Nevertheless, Graham Turner and other authors are right that the basic message of *Limits to Growth* has remained valid since the publication of the famous book. On the other hand, Johan Rockström and others managed to shift the discussion away from the simplistic Dynamo models towards more realistic concerns over specific scarcities. The first major publication of Rockström's team was published in 2009 under the title *Planetary boundaries: exploring the safe operating space for humanity* [231, 232]. The concept indicates, based on scientific research, that since the Industrial Revolution, human activity has gradually become the main driver of global environmental change. Once human activity passes certain thresholds or tipping points (defined as 'planetary boundaries'), there is a risk of 'irreversible and abrupt environmental change'. Rockström *et al* identified nine 'planetary life support systems'that are essential for human survival and attempted to quantify how far they have been pushed already.

The nine planetary boundaries are as follows:

- Stratospheric ozone depletion
- Loss of biodiversity and extinctions
- Chemical pollution and the release of novel entities
- Climate change
- Ocean acidification
- Land system change
- Freshwater consumption and the global hydrological cycle
- Nitrogen and phosphorus flows into the biosphere and oceans
- Atmospheric aerosol loading.

Surprisingly, the authors identify only two boundaries that are already in a state of high risk: genetic diversity (as part of the loss of biodiversity and extinctions) and nitrogen and phosphorous flows into the biosphere and oceans. Climate change and land system change are still located in the domain of increasing risk. This early assessment is surely up for debate. In our day and time, we would be more concerned with global warming, including its effect on weather escapades and the rise of the sea level.

8.6.2.5 *The anthropocene*

Closely related to the planetary boundaries and the safe operating space is the more descriptive definition of the 'Anthropocene'. The same Will Steffen and coauthors Paul Crutzen (Nobel Prize of Chemistry 1995) and John McNeill, coined the term [233]. They show that after the seven epochs of the Cenozoic geologic era (since some 66 million years ago), we humans have begun to massively interfere with the geological status of the planet. This gigantic human intrusion into the robust geologic and atmospheric conditions of the Earth has brought the last 'natural' epoch, the Holocene, to an end. Now, we humans are creating and dominating a new epoch. That new epoch hence should be named after its major driving force, which is humankind. In the ancient Greek language, humans are called *anthropoi*. That is the origin of the new name, Anthropocene.

Figure 8.12 shows several empirical trends of human origin (in red), and the resulting trends of physics and chemistry of the planet's condition (in green). Each of the 24 small pictures contains a thin vertical line, which marks the year 1950. It is

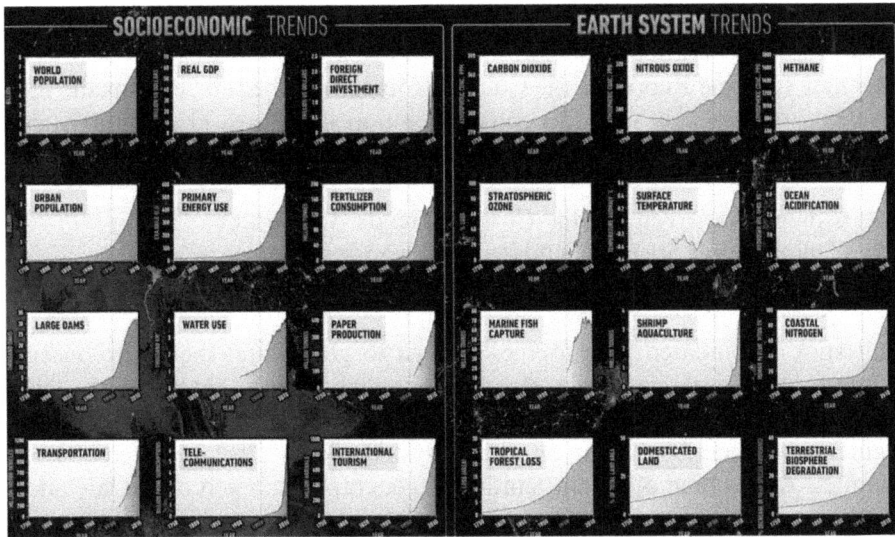

Figure 8.12. The Great acceleration of the Anthropocene. Twenty-four curves showing the dramatic changes of human population and other socioeconomic patterns (red) and Earth system changes (green). The dramatic changes occurred after 1950. (Source: Steffen *et al*, 2015 [234] with permission from Sage.)

quite surprising that thousands of years of human presence on the planet did not significantly alter the physics and chemistry of the planet, but the past 70 years have caused a biogeochemical revolution.

What is the message in all this? It says that humanity has quite recently entered a completely new historical period, in which our responsibility has begun to include the need for extremely cautious attitudes towards our planet.

It is no exaggeration to say that the fate of human society at the Horizon of 2050 will depend on the development of exactly such cautious attitudes.

8.6.2.6 A new enlightenment?

European civilizations emerged from the 'dark' Middle Ages with an enormous amount of fresh thinking in terms of logic, precise observation, and the establishment of 'natural laws' of astronomy, physics and chemistry. Also, rational and legal rules were established for the functioning of statehood and society. Moreover, European explorers and armies began to 'conquer' the rest of the planet, using technological and military advances based on scientific findings. The period in Europe from the 16th to the 18th centuries is often called the Enlightenment period. To be sure, continents outside Europe suffered considerably from the mostly arrogant and brutal European intruders. But nobody would deny that scientific understanding and philosophical clarity greatly benefited from exactly this Enlightenment.

On the other hand, the troublesome developments of the limits to growth, the population explosion, and the Anthropocene were consequences of that same Enlightenment. When we as humanity are forced to analyse and overcome our self-inflicted troubles, we can better also consider some philosophical shortcomings of the old European Enlightenment.

The new strategic Report to the Club of Rome, published in 2018 for the fiftieth anniversary of the Club, puts this consideration at its centre. The Report is called *Come On!* [235], and it contains three chapters with two different meanings of *Come On!*

1. C'mon! Don't Tell Me the Current Trends Are Sustainable!
2. C'mon! Don't Stick to Outdated Philosophies!
3. Come On! Join Us on an Exciting Journey Towards a Sustainable World.

'Don't stick to outdated philosophies' means to go beyond the anthropocentric, utilitarian, analytical, reductionist philosophy that was characteristic of the Enlightenment. We as authors discovered that some of the philosophers of the Enlightenment, including Adam Smith, are systematically misinterpreted in modern economics. At the time of Adam Smith, the geographical reach of the law and the geographical reach of the market (the 'invisible hand', according to Smith) were identical. This made for a benign balance between the market and the law. But today, markets are chiefly global, and the law remains chiefly national, if not provincial. Global financial markets, on their permanent search for maximised returns on investment, are free to identify places with the weakest laws. They actually blackmail national lawmakers to weaken laws, reducing social security costs, as a condition for the investors to invest there. Downward spirals of legal and fiscal conditions can be observed [236]. That benefits the rich and further impoverishes the poor.

We also looked at David Ricardo's description of the relative comparative advantages of different countries leading to international trade optimising cost–benefit ratios. But Ricardo at his time assumed that capital remained immobile. Under today's condition of extremely high capital mobility, we see that absolute comparative advantages determine the location of production. This gives capital markets extremely strong powers over the real economies, leading to unruly fluctuations and big losses on the part of many countries.

The neoliberal economic philosophy after massive deregulations, chiefly after the year 1990, essentially benefits the owners and speculative traders of capital and tends to create billions of losers. Also, the natural environment tends to be losing because of the enormous acceleration of all processes shown on the left side of figure 8.12.

8.6.2.7 Balance instead of dogmatism

Considering the unintended disadvantages of economic (and religious) dogmatism, we as authors felt that it was time to resurrect the philosophies of balance. We know that Asian cultures have a long tradition of balance. Mark Cartwright [237] offers a simplified definition of what is also an essential part of Confucian cosmology: 'The principle of Yin and Yang is a fundamental concept in Chinese philosophy and culture. ... This principle is that all things exist as inseparable and contradictory opposites, for example female-male, dark-light, and old-young. The two opposites attract and complement each other and, as their symbol illustrates, each side has at its core an element of the other. … Neither pole is superior to the other and, as an increase in one brings a corresponding decrease in the other, a correct balance between the two poles must be reached in order to achieve harmony.'

Somewhat related philosophies exist in the West, such as G W F Hegel's dialectic philosophy and Ken Wilber's *A Brief History of Everything.* Also Niels Bohr's complementarity has been seen by Frithjof Capra [238] as a 'door-opener' to perceiving parallels between modern physics and Eastern wisdom and religions.

We cannot here go into the fairly complicated parallels between physics and (Eastern) philosophy. Instead, I am listing examples of balances which make a lot of sense in understanding the practical usefulness of the balance concept.

We propose to seek a balance between the following:

- Humans and nature: Using remaining natural landscapes, water bodies, and minerals chiefly as resources for an ever-growing human population and the fulfilment of ever-growing consumption is not balance but destruction.
- Short term and long term: Humans appreciate quick gratification such as something to drink when thirsty. But there is a need for a counterbalance to ensure long-term, action such as policies to restabilize the Earth's climate.
- Speed and stability: Technological and cultural progress benefits from competition for temporal priority. Disruptive innovations can bring tremendous benefits. But speed by itself can be a horror for slow creatures, for most elderly humans, for babies, and for communities. The current civilizational addiction to speed is destructive to structures, habits, and cultures that have emerged under the sustainability criterion. Sustainability, after all, includes stability.
- Private and public: The discovery of the human values of individualism, private property, and protection against state intrusion has been among the most valuable achievements of the European Enlightenment. But in our times, public goods are much more endangered than private goods. The state (public) should set the rules for the market (private), not the other way round.
- Women and men: Many early cultures developed through wars during which women were chiefly entrusted with caring for the family and men for defence (or aggression). This model is outdated. Riane Eisler [239] has offered archaeological insights into cultures that thrived under partnership models and has also shown that the conventional, male-dominated 'wealth of nations' is almost a caricature of real well-being [240].
- Equity and awards for achievements: Without awards for achievement, societies can get sleepy and lose out in the competition with other societies. But there must be a publicly guaranteed system of justice and equity. Inequity, according to Wilkinson and Pickett [241], tends to be correlated with very undesirable social parameters, such as high criminality, poor education, and high rates of infant mortality.
- State and religion: It was a great achievement by the European Enlightenment to separate public from religious leadership, fully respecting religious values and communities. Religions dominating the public sector are

in high danger of destroying human rights and an independent legal system with independent high courts. On the other hand, states that are intolerant of religious communities tend to lose touch with ethical (and long-term) needs.

Many more pairs can be formulated showing that balance is essential for a high and sustainable level of culture. Exact physics does not contradict this insight. Dogmatism is always allowed in checking the validity of scientific methods, mathematical calculations, and technological applications. But arrogance of science against nonquantitative insights and goal seeking is usually counterproductive.

8.6.2.8 Conclusion

Society at the Horizon of 2050 will predictably be massively concerned with geophysical phenomena related to global warming and with biological tragedies of accelerated extinction of species. Good science will be much in demand for maintaining, enhancing, and applying measures to stop destruction and to regenerate healthy conditions for the stability of our planet. But historians will emphasize that humanity has, by a simplistic and materialistic 'pursuit of happiness', destroyed much of the earlier richness of nature.

First-class physics will maintain its high appreciation by the public under the condition of conversely appreciating the strong desire of society for long-term strategies of sustainable development and for modesty of consumption.

References

[1] OECD 2019 *The OECD Learning Compass 2030* (http://oecd.org/education/2030-project/teaching-and-learning/learning/)
[2] Phys 21: Preparing Physics Students for 21st-Century Careers 2016 *A Report by the Joint Task Force on Undergraduate Physics Programs* (American Physical Society) (https://aps.org/programs/education/undergrad/jtupp.cfm)
[3] McLoughlin E, Butler D, Kaya S and Costello E 2020 STEM education in schools: what can we learn from the research? *ATS STEM Report #1* (Ireland: Dublin City University)
[4] Munfaridah N, Avraamidou L and Goedhart M 2021 The use of multiple representations in undergraduate physics education: what do we know and where do we go from here? *EURASIA J. Math., Sci. Technol. Educ.* **17** em1934
[5] Bao L and Koenig K 2019 Physics education research for 21st century learning *Discip. Interdiscip. Sci. Educ. Res.* **1**(2)
[6] Dewey J 1938 *Logic: The Theory of Inquiry* (New York: Holt, Rinehart and Winston)
[7] Lazonder A W and Harmsen R 2016 Meta-analysis of inquiry-based learning: effects of guidance *Rev. Educ. Res.* **86** 681–718
[8] Papadouris N and Constantinou C P 2018 Formative assessment in physics teaching and learning *The Role of Laboratory Work in Improving Physics Teaching and Learning* (Nature Switzerland AG: Springer)
[9] OECD 2021 *Positive, High-achieving Students? What Schools and Teachers Can Do, TALIS* (Paris: OECD Publishing)
[10] Gonsalves A J, Danielsson A and Pettersson H 2016 Masculinities and experimental practices in physics: the view from three case studies *Phys. Rev. Phys. Educ. Res.* **12** 020120

[11] Hodapp T and Hazari Z 2015 Women in physics: why so few? American Physical Society news *The Back Page, November* (https://aps.org/publications/apsnews/201511/backpage. cfm)

[12] Hazari Z, Brewe E, Goertzen R M and Hodapp T 2017 The importance of high school physics teachers for female students' physics identity and persistence *Phys. Teach.* **55** 96–9

[13] STEP UP 4 Women (https://engage.aps.org/stepup/home)

[14] Why Not Physics? A Snapshot of Girl's Uptake at A-level 2018 *An Institute of Physics Report* (https://iop.org/sites/default/files/2018-10/why-not-physics.pdf)

[15] Gender Stereotypes and their Effect on Young People 2019 *An Institute of Physics Report* (https://iop.org/sites/default/files/2019-07/IGB-gender-stereotypes.pdf)

[16] European Strategy Group 2020 *Update on Particle Physics*

[17] Michelini M 2021 Innovation of curriculum and frontiers of fundamental physics in secondary school: research-based proposals *Fundamental Physics and Physics Education Research* (Berlin: Springer) pp 101–16

[18] Building America's STEM Workforce: Eliminating Barriers and Unlocking Advantages 2021 *A Report by the American Physical Society Office of Government Affairs* (https://aps. org/policy/analysis/upload/Building-America-STEM-workforce.pdf)

[19] 2018 *The Role of Laboratory Work in Improving Physics Teaching and Learning* ed D Sokołowska and M Michelini (Nature Switzerland AG: Springer)

[20] Chang S L and Ley K 2006 A learning strategy to compensate for cognitive overload in online learning: learner use of printed online materials *J. Interact. Online Learn.* **5** 104–17

[21] Singer L M and Alexander P A 2017 Reading on paper and digitally: what the past decades of empirical research reveal *Rev. Educ. Res.* **87** 1007–41

[22] Nicholas V 2015 Mobile learning for online practical science modules in higher education *Mobile Learning and STEM: Case Studies in Practice* ed H Crompton and J Traxler (Abingdon: Routledge) pp 234–43

[23] Thoms L-J and Girwidz R 2017 Virtual and remote experiments for radiometric and photometric measurements *Eur. J. Phys.* **38** 1–24

[24] Global Hands-on Universe 2020 *Global Hands-on Universe Projects* (https://handsonuni-verse.org/projects/) (Accessed 4 May 2021)

[25] Mazur E 1991 *Peer Instruction: A Users' Manual* (Englewood, NJ: Prentice-Hall)

[26] Jordan S 2013 E-assessment: past, present and future *New Dir. Teach. Phys. Sci.* **9** 87–106

[27] Hatherly P A, Jordan S E and Cayless A 2009 Interactive screen experiments: innovative virtual laboratories for distance learners *Eur. J. Phys.* **30** 751–62

[28] Faulconer E and Gruss A B 2018 A review to weigh the pros and cons of online, remote, and distance science laboratory experiences *Int. Rev. Res. Open Distrib. Learn.* **19** 155–68

[29] Dintsios N, Artemi S and Polatoglou H 2018 Acceptance of remote experiments in secondary students *Int. J. Online Eng.* **14** 4–18

[30] Matthews D 2021 COVID triggers remote-experiment revolution at labs across Europe *Times High. Educ.* (https://timeshighereducation.com/news/covid-triggers-remote-experi-ment-revolution-labs-across-europe) (Accessed 5 February 2021)

[31] Pan R, Sinha S, Kaski K and Jari Saramäki J 2012 The evolution of interdisciplinarity in physics research *Sci. Rep.* **2** 551

[32] Pluchino A, Burgio G, Rapisarda A, Biondo A E, Pulvirenti A, Ferro A and Giorgino T 2019 Exploring the role of interdisciplinarity in physics: success, talent and luck *PLoS One* **14** e0218793

[33] Darbellay F, Moody Z, Sedooka A and Steffen G 2014 Interdisciplinary research boosted by serendipity *Creat. Res. J.* **26** 1–10

[34] National Academy of Sciences, and National Academy of Engineering 2005 *Facilitating Interdisciplinary Research* (The National Academies Press)

[35] Klein J T 2014 Interdisciplinarity and transdisciplinarity: keyword meanings for collaboration science and translational medicine *J. Transl. Med. Epidemiol.* **2** 1024

[36] Darbellay F 2015 Rethinking inter- and transdisciplinarity: Undisciplined knowledge and the emergence of a new thought style *Futures* **65** 163–74

[37] Wernli D and Darbellay F 2016 *Interdisciplinarity and the 21st Century Research-Intensive University* Position paper (The League of European Research Universities (LERU))

[38] Klein J T 2012 Research integration, a comparative knowledge base *Case Studies in Interdisciplinary Research* ed A Repko, W Newell and R Szostak (Thousand Oaks, CA: Sage) pp 283–98

[39] Klein J T 2009 *Creating Interdisciplinary Campus Cultures: A Model for Strength and Sustainability* (New York: Wiley)

[40] Fiore S M 2008 Interdisciplinarity as teamwork: how the science of teams can inform team science *Small Group Res.* **39** 251–77

[41] Wuchty S, Jones B F and Uzzi B 2007 The increasing dominance of teams in production of knowledge *Science* **316** 1036–9

[42] Uzzi B, Mukherjee S, Stringer M and Jones B 2013 Atypical combinations and scientific impact *Science* **342** 468–72

[43] Brown R R, Deletic A and Wong T H F 2015 Interdisciplinarity: how to catalyse collaboration *Nature* **525** 315–7

[44] Stokols D 2014 Training the next generation of transdisciplinarians *Enhancing Communication et Collaboration in Interdisciplinary Research* ed M O'Rourke, S Crowley, S D Eigenbrode et and J D Wulfhorst (Thousand Oaks, CA: Sage) pp 56–81

[45] Darbellay F 2017 From monomyth to interdisciplinary creative polymathy: the researcher with a thousand faces *Special Issue: The Hero's Journey. A Tribute to Joseph Campbell and his 30th Anniversary of Death. The Journal of Genius and Eminence* 45–54

[46] Repko A and Szostak R 2016 *Interdisciplinary Research Process and Theory* (Thousand Oaks, CA: Sage)

[47] Hirsch Hadorn G, Hoffmann-Riem H, Biber-Klemm S, Grossenbacher-Mansuy W, Joye D, Pohl C, Wiesmann U and Zemp E (ed) 2008 *Handbook of Transdisciplinary Research* (Berlin: Springer)

[48] Frodeman R, Klein J T and Pacheco R C S (ed) 2017 *The Oxford Handbook of Interdisciplinarity* 2nd edn (Oxford: Oxford University Press)

[49] Fam D and O'Rourke M (ed) 2020 *Interdisciplinary and Transdisciplinary Failures* (Routledge)

[50] Duderstadt J J 2000 New roles for the 21st-century university *Issues Sci. Technol.* **16** 37–44

[51] https://21stcenturylab.lincoln.ac.uk/

[52] Bush V 1945 *Science—The Endless Frontier. A Report to the President* (Washington, DC: United States Government Printing Office)

[53] Edgerton D 2004 The linear model did not exist *The Science-Industry Nexus: History, Policy, Implications* ed K Grandin, N Worms and S Widmalm (Sagamore Beach, MA: Science History Publications) pp 31–57

[54] Godin B 2006 The linear model of innovation: the historical construction of an analytical framework *Sci., Technol. Human Values* **31** 639–67

[55] Jasanoff S 2003 Technologies of humility: citizen participation in governing science *Minerva* **41** 223–44

[56] Rogers E M 1962 *Diffusion of Innovations* (New York: Free Press of Glencoe)

[57] Merton R 1973 The normative structure of science ed N Storer *The Sociology of Science* (Chicage, IL: University of Chicago Press) pp 267–78

[58] National Research Council 1983 *Risk Assessment in the Federal Government: Managing the Process* (Washington, DC: National Academies Press)

[59] Perrow C 1984 *Normal Accidents: Living with High-Risk Technologies* (Princeton, NJ: Princeton University Press)

[60] Beck U 1992 *The Risk Society. Towards a New Modernity* (London: Sage)

[61] Macnaghten P and Urry J 1998 *Contested Natures* (London: Sage)

[62] Jasanoff S 2016 *The Ethics of Invention: Technology and the Human Future* (New York: W. W. Norton and Co.)

[63] Gibbons M, Limoges C, Nowotny H, Schwartzman S, Scott P and Trow M 1994 *The New Production of Knowledge: The Dynamics of Science and Research in Contemporary Societies* (London: Sage)

[64] Funtowicz S and Ravetz J 1993 Science for the post-normal age *Futures* **25** 739–55

[65] Irvine J and Martin B 1984 *Foresight in Science: Picking the Winners* (London: Pinter)

[66] Etzkowitz H and Leydesdorff L 2000 The dynamics of innovation: from national systems and 'Mode 2' to a triple helix of university-industry-government relations *Res. Policy* **29** 109–23

[67] Nowotny H, Scott P and Gibbons M 2001 *Re-Thinking Science. Knowledge and the Public in an Age of Uncertainty* (Cambridge: Polity Press)

[68] Hessels L and van Lente H 2008 Re-thinking new knowledge production: a literature review and a research agenda *Res. Policy* **37** 740–60

[69] Brooks S *et al* 2009 *Silver Bullets, Grand Challenges and the New Philanthropy* (Brighton: STEPS Working Paper 24, STEPS Centre)

[70] Ludwig D, Pols A and Macnaghten P 2018 *Organisational Review and Outlooks* (Wageningen University and Research)

[71] Jump P 2014 No Regrets, says outgoing EPSRC Chief David Delpy: 'Thick Skin' helped research council boss take the flak for controversial shaping capability measures *Times High. Educ.* April 17, 2014

[72] Flink T and Kaldewey D 2018 The new production of legitimacy: STI policy discourses beyond the contract metaphor *Res. Policy* **47** 14–22

[73] Calvert J 2013 Systems biology: big science and grand challenges *BioSocieties* **8** 466–79

[74] National Council on Bioethics 2012 *Emerging Biotechnologies: Technology, Choice and the Public Good* (London: Nuffield Council on Bioethics)

[75] Jasanoff S and Simmet H 2017 No funeral bells: public reason in a 'Post-Truth' world *Soc. Stud. Sci.* **47** 751–70

[76] Jasanoff S 1990 *The Fifth Branch: Science Advisers as Policymakers* (Cambridge, MA: Harvard University Press)

[77] Jasanoff S (ed) 2004 *States of Knowledge: The Co-Production of Science and the Social Order* (New York: Routledge)

[78] Miller C 2004 Climate science and the making of a global social order *States of Knowledge: The Co-Production of Science and the Social Order* ed S Jasanoff (New York: Routledge) pp 247–85

[79] Owens S 2015 *Knowledge, Policy, and Expertise: The UK Royal Commission on Environmental Pollution 1970–2011* (Oxford: Oxford University Press)

[80] Rose N 2006 *The Politics of Life Itself: Biomedicine, Power, and Subjectivity in the Twenty-First Century* (Princeton, NJ: Princeton University Press)

[81] Stirling A 2008 Opening Up' and 'Closing Down': power, participation, and pluralism in the social appraisal of technology *Sci., Technol. Human Values* **23** 262–94

[82] Stirling A 2014 *Emancipating Transformations: From Controlling 'the transition' to Culturing Plural Radical Progress* (Brighton, UK: STEPS Working paper 64, STEPS Centre)

[83] European Commission 2007 *The European Research Area: New Perspectives* (Luxembourg: Green paper 04.04.2007. Text with EEA relevance, COM161, EUR 22840 EN. Office for Official Publications of the European Communities)

[84] Irwin A 2006 The politics of talk: coming to terms with the 'new' scientific governance *Soc. Stud. Sci.* **36** 299–330

[85] Macnaghten P 2010 Researching technoscientific concerns in the making: narrative structures, public responses and emerging nanotechnologies *Environ. Plan.* A **41** 23–37

[86] Macnaghten P and Chilvers J 2014 The future of science governance: publics, policies, practices *Environ. Plan. C: Govern. Policy* **32** 530–48

[87] Chilvers J and Kearnes M (ed) 2016 *Remaking Participation: Science, Environment and Emergent Publics* (London: Routledge)

[88] European Commission 2013 *Fact Sheet: Science with and for Society in Horizon 2020*

[89] Zwart H, Landeweerd L and van Rooij A 2014 Adapt or perish? Assessing the recent shift in the european research funding arena from 'ELSA' to 'RRI' *Life Sci., Soc. Policy* **10** 11

[90] von Schomberg R 2013 A vision of responsible research and innovation *Responsible Innovation: Managing the Responsible Emergence of Science and Innovation in Society* ed R Owen, J Bessant and M Heintz (London: Wiley) pp 51–74

[91] Adam B and Groves G 2011 Futures tended: care and future-oriented responsibility *Bull. Sci., Technol. Soc.* **31** 17–27

[92] Grinbaum A and Groves C 2013 What is 'Responsible' about responsible innovation? Understanding the ethical issues *Responsible Innovation: Managing the Responsible Emergence of Science and Innovation in Society* ed R Owen, J Bessant and M Heintz (London: Wiley) pp 119–42

[93] Collingridge D 1980 *The Social Control of Technology* (Milton Keynes: Open University Press)

[94] Hajer M 2003 Policy without polity? Policy analysis and the institutional void *Policy Sci.* **36** 175–95

[95] Douglas H 2003 The moral responsibilities of scientists (tensions between autonomy and responsibility) *Am. Philos. Q.* **40** 59–68

[96] Owen R, Macnaghten P and Stilgoe J 2012 Responsible research and innovation: from science in society to science for society, with society *Sci. Public Policy* **39** 751–60

[97] Stilgoe J, Owen R and Macnaghten P 2013 Developing a framework of responsible innovation *Rese. Policy* **42** 1568–80

[98]   Grove-White R, Macnaghten P, Mayer S and Wynne B 1997 *Uncertain World: Genetically Modified Organisms, Food and Public Attitudes in Britain* (Lancaster: Centre for the Study of Environmental Change)

[99]   Macnaghten P 2004 Animals in their nature: a case study of public attitudes on animals, genetic modification and 'nature' *Sociology* **38** 533–51

[100] Macnaghten P and Szerszynski B 2013 Living the global social experiment: an analysis of public discourse on geoengineering and its implications for governance *Global Environ. Change* **23** 465–74

[101] Macnaghten P, Davies S and Kearnes M 2019 Understanding public responses to emerging technologies: a narrative approach *J. Environ. Plann. Policy* **21** 504–18

[102] Williams L, Macnaghten P, Davies R and Curtis S 2017 Framing fracking: exploring public responses to hydraulic fracturing in the UK *Public Understand. Sci.* **26** 89–104

[103] EPSRC 2013 *Framework for Responsible Innovation* (Swindon: EPSRC)

[104] Owen R 2014 The UK Engineering and Physical Sciences Research Council's commitment to a framework for responsible innovation *J. Respons. Innov.* **1** 113–7

[105] Murphy J, Parry S and Walls J 2016 The EPSRC's policy of responsible innovation from a trading zones perspective *Minerva* **54** 151–74

[106] Finkel A 2018 What kind of society do we want to be? Keynote address by Australian Government Chief Scientist *Human Rights Commission 'Human Rights and Technology' Conf. (Four Seasons Hotel, Sydney)*

[107]   Scientists, keep an open line of communication with the public *Nat. Med.* **26** 1495–5 01 10 2020

[108] Lewenstein B 2015 Informal science: a resource and online community for informal learning projects, research, and evaluation. Retrieved 14 January 2020 from http://informalscience.org/research/wiki/Public-Engagement

[109] Bell P, Lewenstein B, Shouse A W and Feder M A *et al* 2009 *Learning Science in Informal Environments: People, Places, and Pursuits* **vol 140** (Washington, DC: National Academies Press)

[110] http://visit.cern/universe-of-particles

[111] http://physicsmatters.physics.mcgill.ca/lectures/

[112] http://www.bbc.co.uk/programmes/b04v5vjz

[113] http://en.wikipedia.org/wiki/A_Brief_History_of_Time

[114] http://quantumdelta.nl/welcome-to-living-lab-quantum-and-society/

[115] http://www.zooniverse.org/projects/achmorrison/steelpan-vibrations

[116] Miller S 2001 Public understanding of science at the crossroads *Public Understand. Sci.* **10** 115–20

[117] Schäfer M S 2014 *Vom Elfenbeinturm in die Gesellschaft: Wissenschaftskommunikation im Wandel* (Zürich: Universität Zürich)

[118] Irwin A 2006 The politics of talk: coming to terms with the 'new' scientific governance *Soc. Stud. Sci.* **36** 299–320

[119] Haste 2005 *Connecting science: what we know and what we don't know about science and society* Department of Psychology, University of Bath

[120] https://www.natureindex.com/news-blog/how-to-crowdfund-your-research-science-experiment-grant

[121] https://assets.publishing.service.gov.uk/government/uploads/system/uploads/attachment data/file/905466/public-attitudes-to-science-2019.pdf

[122] NCCPE—National Coordinating Centre for Public Engagement 2012 The engaged university: a manifesto for public engagement.

[123] NCCPE—National Coordinating Centre for Public Engagement 2010 How public engagement helps to maximize the flow of knowledge and learning between HEIs and society.

[124] UUK 2010 Universities: Engaging with Local Communities, Retrieved on 15 January 2020 from https://www.universitiesuk.ac.uk/policy-and-analysis/reports/Documents/2010/universities-engaging-with-local-communities.pdf

[125] Robinson F, Zass-Ogilvie I and Hudson R 2012 *How Can Universities Support Disadvantaged Communities* (Joseph Rowntree Foundation)

[126] Oreskes N 2019 *Why Trust Science?* (Princeton, NJ: Princeton University Press)

[127] Lakomý M, Hlavová R and Machackova H 2019 Open science and the science-society relationship *Society* **56** 246–55

[128] Cavalier D, Hoffman C H P and Cooper C B 2020 *The Field Guide to Citizen Science: How You Can Contribute to Scientific Research and Make a Difference* (Timber Press Incorporated)

[129] Wellcome Trust 2006 *Science Communication*

[130] Burchell K, Franklin S and Holden K 2009 *Public culture as professional science: final report of the SCoPE project (Scientists on public engagement: from communication to deliberation?)* London School of Economics and Political Science, London

[131] Jensen P, Rouquier J-B, Kreimer P and Croissant Y 2008 Scientists who engage with society perform better academically *Sci. Public Policy* **35** 527–41

[132] Martinez-Conde S 2016 Has contemporary academia outgrown the Carl Sagan effect? *J. Neurosci.* **36** 2077–82

[133] Trench B 2012 Vital and vulnerable: science communication as a university subject *Science Communication in the World* (Berlin: Springer) pp 241–57

[134] Bubela T *et al* 2009 Science communication reconsidered *Nat. Biotechnol.* **27** 514–8

[135] Gorman M J 2020 *Idea Colliders: The Future of Science Museums* (Cambridge, MA: MIT Press)

[136] Darbellay F, Moody Z and Lubart T 2017 *Creativity, Design Thinking and Interdisciplinarity* (Berlin: Springer)

[137] https://ec.europa.eu/commission/commissioners/2019–2024/gabriel_en

[138] Reference framework for assessing the scientific and socio-economic impact of research infrastructures: https://www.oecd-ilibrary.org/science-and-technology/reference-framework-for-assessing-the-scientific-and-socio-economic-impact-of-research-infrastructures_3ffee43b-en

[139] Impact pathways: https://ri-paths-tool.eu/en/impact-pathways

[140] https://www.eso.org/public/announcements/ann21002/

[141] https://home.cern/about/what-we-do/our-impact#:~:text=CERN%20brings%20nations%20together%20and,such%20as%20health%20and%20environment

[142] IPCC 2021 *Climate Change 2021: The Physical Science Basis. Contribution of Working Group I to the Sixth Assessment Report of the Intergovernmental Panel on Climate Change* ed V Masson-Delmotte *et al* (Cambridge: Cambridge University Press)

[143] Wartofsky M W 2012 *Models: Representation and the Scientific Understanding* **vol 48** (Cham: Springer Science & Business Media)

[144] Vicsek T 2002 Complexity: the bigger picture *Nature* **418** 131–1

[145] Nichols T 2017 *The Death of Expertise: The Campaign Against Established Knowledge and Why It Matters* (Oxford: Oxford University Press)

[146] Lenton T M, Rockström J, Gaffney O, Rahmstorf S, Richardson K, Steffen W and Schellnhuber H J 2019 *Climate Tipping Points—Too Risky To Bet Against* (Nature Publishing Group)

[147] https://www.weforum.org/reports/the-global-risks-report-2020

[148] Allen T, Murray K A, Zambrana-Torrelio C, Morse S S, Rondinini C, Di Marco M, Breit N, Olival K J and Daszak P 2017 Global hotspots and correlates of emerging zoonotic diseases *Nat. Commun.* **8** 1–10

[149] Steffen W, Broadgate W, Deutsch L, Gaffney O and Ludwig C 2015 The trajectory of the Anthropocene: the great acceleration *Anthropocene Rev.* **2** 81–98

[150] Weart S 2013 Rise of interdisciplinary research on climate *Proc. Natl Acad. Sci.* **110** 3657–64

[151] Brulle R J and Norgaard K M 2019 Avoiding cultural trauma: climate change and social inertia *Environ. Polit.* **28** 886–908

[152] Swim J, Clayton S, Doherty T, Gifford R, Howard G, Reser J, Stern P and Weber E 2009 *Psychology and Global Climate Change: Addressing a Multi-Faceted Phenomenon and Set of Challenges. A Report by the American Psychological Association's Task Force on the Interface Between Psychology and Global Climate Change* (Washington, DC: American Psychological Association)

[153] Hansen J, Sato M and Ruedy R 2012 Perception of climate change *Proc. Natl Acad. Sci.* **109** E2415–23

[154] Martin B and Richards E 1995 Scientific knowledge, controversy, and public decision-making *Handbook of Science and Technology Studies* **506** 26

[155] Crow D A and Boykoff M T 2014 *Culture, Politics and Climate Change: How Information Shapes Our Common Future* (Routledge)

[156] Bowater L and Yeoman K 2012 *Science Communication: A Practical Guide for Scientists* (New York: Wiley)

[157] Sinatra R, Deville P, Szell M, Wang D and Barabási A-L 2015 A century of physics *Nat. Phys.* **11** 791–6

[158] Burns T W, O'Connor D J and Stocklmayer S M 2003 Science communication: a contemporary definition *Public Understand. Sci.* **12** 183–202

[159] http://www.oecd.org/sti/science-technology-innovation-outlook/Science-advice-COVID/

[160] Freudenburg W R, Gramling R and Davidson D J 2008 Scientific certainty argumentation methods (SCAMs): science and the politics of doubt *Sociol. Inq.* **78** 2–38

[161] E. M. A. O. National Academies of Sciences 2017 *Communicating Science Effectively: A Research Agenda* (National Academies Press)

[162] Kappel K and Holmen S J 2019 Why science communication, and does it work? A taxonomy of science communication aims and a survey of the empirical evidence *Front. Commun.* **4** 55

[163] Caulfield T 2020 Pseudoscience and COVID-19-we've had enough already *Nature (Lond.)* 27 Apr

[164] Boucher D 2016 Movie review: there's a vast cowspiracy about climate change *The Equation Blog*

[165] Le Quéré C, Jackson R B, Jones M W, Smith A J P, Abernethy S, Andrew R M, De-Gol A J, Willis D R, Shan Y and Canadell J Gothers 2020 Temporary reduction in daily global CO_2 emissions during the COVID-19 forced confinement *Nat. Clim. Change* **10** 647–53

[166] Pérez-González L 2020 Is climate science taking over the science?' A corpus-based study of competing stances on bias, dogma and expertise in the blogosphere *Human. Soc. Sci. Commun.* **7** 1–16

[167] Kreps S E and Kriner D L 2020 Model uncertainty, political contestation, and public trust in science: evidence from the COVID-19 pandemic *Sci. Adv.* **6** eabd4563

[168] Rödder S, Franzen M and Weingart P 2011 The sciences' media connection–public communication and its repercussionsvol 28 (Cham: Springer Science & Business Media)

[169] Peters H P 2013 Gap between science and media revisited: Scientists as public communicators *Proc. Natl Acad. Sci.* **110** 14102–9

[170] Lundby K 2014 *Mediatization of Communication* **vol 21** (Walter de Gruyter GmbH & Co KG)

[171] Kahneman D, Lovallo D and Sibony O 2011 Before you make that big decision *Harv. Bus. Rev.* **89** 50–60

[172] Pahl S, Sheppard S, Boomsma C and Groves C 2014 Perceptions of time in relation to climate change *Wiley Interdiscip. Rev. Clim. Change* **5** 375–88

[173] Fuentes R, Galeotti M, Lanza A and Manzano B 2020 COVID-19 and climate change: a tale of two global problems *Sustainability* **12** 8560

[174] O'Brien K 2016 *Climate Change Adaptation and Social Transformation* (International Encyclopedia of Geography: People, the Earth, Environment and Technology: People, the Earth, Environment and Technology) pp 1–8

[175] https://partner.sciencenorway.no/climate-change-covid19-crisis/the-world-with-viruses-the-corona-crisis-shows-that-rapid-change-is-possible/1666488

[176] Saitz R and Schwitzer G 2020 Communicating science in the time of a pandemic *JAMA* **324** 443–4

[177] Gross M 2020 Communicating science in a crisis *Curr. Biol.* **30** R737–9

[178] Popper K 1970 The moral responsibility of the scientist *Induction, Physics and Ethics: Proceedings and Discussions of the 1968 Saltzburg Colloquium in the Philosophy of Science* ed P Weingartner and G Zecha (D. Reidel, Publ. Company) pp 329–36

[179] Gustafsson B, Rydén L, Tibell G and Wallensteen P 1984 Focus on: the uppsala code of ethics for science *J. Peace Res.* **21** 311–6

[180] Rydén L 1990 *Etik för forskare* (in Swedish) *UHÄ FoU Report* **1990** 1

[181] https://en.wikipedia.org/wiki/List_of_scientific_misconduct_incidents (Accessed March 1, 2021)

[182] Asplund K 2021 *Fuskarna* (in Swedish), fri tanke, Stockholm

[183] Fanelli D 2009 How many scientists fabricate or falsify research? A systematic review and meta-analyis of survey data *PLoS One* **4** e:5738

[184] For example, Kennedy A 2019 A short history of academic plagiarism, https://www.quetext.com/blog/short-history-academic-plagiarism (Accessed May 29, 2021)

[185] Garfield E 2002 Demand citation vigilance *The Scientist* **16** 6

[186] Mulligan A 2004 Is the peer review in crisis? *Perspectives in Publishing* (Elsevier) (https://elsevier.com/__data/assets/pdf_file/0003/93675/PerspPubl2.pdf) (Accessed 10 May 2021)

[187] Schubert T 2011 Empirical observations on new public management to increase efficiency in public research—Boon or bane? *J. Res. Policy* **38** 1225–38

[188] Gustafsson B 2014 Om forskningsmarknaden och forskningens retorik (in Swedish) *Vetenskapsretorik: Hur vetenskapen övertygar och övertalar* ed B Lindberg (Act. Reg. Soc. Scient, Litt. Goth., Interdiciplinaria) vol 13 pp 115–32

[189] Hasselberg Y 2012 Demand or discretion? The market model applied to science and its core values and institutions *Ethics Sci. Environ. Politics* **12** 35–51
[190] Ioannidis J P A 2005 Why most published research findings are false *PLoS Med.* **2** e124
[191] Ziliak S T and McCloskey D N 2008 *The Cult of Statistical Significance: How the Standard Error Costs Us Jobs, Justice and Lives* (The University of Michigan Press)
[192] Baker M 2016 1,500 scientists lift the lid on reproducibility *Nature* **533** 452–4
[193] https://www.go-fair.org/fair-principles/ (Accessed May 16, 2021)
[194] https://opensource.org (Accessed May 16, 2021)
[195] Gustafsson B 2004 *Origin and Evolution of the Elements, Carnegie Observatories Centennial Symposia* ed A McWilliam and M Rauch (Cambridge: Cambridge University Press) p 102
[196] Ioannidis J P A and Trikalinos T A 2005 Early extreme contradictory estimates may appear in published research: the Proteus phenomenon in molecular genetics research and randomised trials . *J. Clin. Epidemiol.* **58** 543–9
[197] ICMJE: http://www.icmje.org/recommendations/browse/roles-and-responsibilities/defining-the-role-of-authors-and-contributors.html (Accessed April 29, 2021)
[198] Helgesson G, Juth N and Schneider J *et al* 2018 Misuse of coauthorship in medical theses in Sweden *J. Empir. Res. Human Res. Ethics* **13** 402–11
[199] Teixeira da Silva J A and Dobránszki J 2016 Multiple authorship in scientific manuscripts: ethical challenges, ghost and guest/gift authorship, and the cultural/disciplinary perspective *Sci. Eng. Ethics* **22** 1457–72
[200] Douglas H E 2009 *Science, Policy and the Value-Free Ideal* (University of Pittsburgh Press)
[201] Hvistendal M 2013 China's publication bazaar *Science* **342** 1035
[202] https://en.wikipedia.org/wiki/Archaeoraptor (Accessed May 19, 2021)
[203] Resnik D B and Shamoo A E 2011 The Singapore statement on research integrity *Account Res.* **18** 71–5
[204] Anderson M S 2018 Shifting perspectives on research integrity *J. Empir. Res. Human Res. Ethics* **13** 459–60
[205] *The European Code of Conduct for Research Integrity*, https://allea.org/code-of-conduct/ (Accessed May 16, 2021)
[206] Sternwedel J D 2014 Life after misconduct: promoting rehabilitation while minimizing damage , *J. Microbiol. Biol. Educ.* **15** 177–80
[207] Fang F C and Casadevall A 2015 Competitive science: is competition ruining science? *Infect. Immun.* **83** 1229–33
[208] Merton R K 1957 Priorities in scientific discovery: a chapter in the sociology of science *Am. Sociol. Rev.* **22** 635–59
[209] Anderson M S, Ronning E A and De Vries R *et al* 2007 The perverse effects of competition on scientists work and relationships *Sci. Eng. Ethics* **13** 437–61
[210] Casadevall A and Fang F C 2012 Reforming science: methodological and cultural reforms *Am. Soc. Microbiol, Infect. Immun.* **80** 891–6
[211] Adamo S A 2013 Attrition of women in the biological sciences: workload, motherhood, and other explanations revisited *BioScience* **63** 43–8
[212] Amabile T M 1998 How to kill creativity *Harv. Bus. Rev.* **76** 76–87 cited in [190]
[213] Amabile T M 1996 *Creativity in Context* (Boulder, CO: Westview Press) cited in [190]
[214] Casadevall A and Fang F C 2012 Reforming science: structural reforms *Am. Soc. Microbiol., Infect. Immun.* **80** 897–901

[215] Scherk J and Schwarz J H 1974 Dual models for nonhadrons *Nucl. Phys.* **B81** 118
Polchinski J 1998 *String Theory* **vol 2** (Cambridge: Cambridge University Press)

[216] Smolin L 2006 *The Trouble with Physics* (Houghton Mifflin)

[217] Quevedo F 2019 Is string phenomenology an oxymoron? *Why Trust a Theory?* ed R Dardashti, R Dawid and K Thebault (Cambridge: Cambridge University Press) e-print:161201589

[218] Guth A 1981 The inflationary universe: a possible solution to the horizon and flatness problems *Phys. Rev.* **D23** 347–56

[219] Ijjas A, Steinhardt P A and Loeb A 2014 Inflationary schism *Phys. Lett.* B **736** 142–6

[220] Linde A D 1986 Eternally existing self-reproducing chaotic inflationary universe *Phys. Lett.* B **175** 395

[221] Ellis G and Silk J 2014 Defend the integrity of physics *Nature* **516** 321–3

[222] Dawid R 2013 *String Theory and the Scientific Method* (Cambridge: Cambridge University Press)

[223] Witten E 1995 String theory dynamics in various dimensions, hep-th/9503124 *Nucl. Phys.* **B443** 85–126
Maldacena J 1999 The large N limits of superconformal field theories and supergravity *Int. J. Theor. Phys.* **38** 1113–33

[224] Weinberg S 1992 *Dreams of a Final Theory* (Vintage)

[225] Witten E 1996 Reflections on the fate of spacetime *Physics Meets Philosophy at the Planck Scale* ed C Callender and N Huggett (Cambridge: Cambridge University Press)

[226] Well illustrated overview by Maria Teodora Stanescu Black Holes and Gravitational Waves, 2018: https://www.mpifr-bonn.mpg.de/4171741/Maria_Stanescu.pdf

[227] https://www.iter.org

[228] Kinoshita Y *et al* 2014 Diagnosis of intraocular lesions using vitreous humor and intraocular perfusion fluid cytology: experience with 83 cases: intraocular cytology of the globe *Diagn. Cytopathol.* **43** 353–9

[229] Meadows D, Meadows D, Randers J and Behrens W 1972 *The Limits to Growth. A Report for the Club of Rome's Project on the Predicament of Mankind* (New York: Universe Books)

[230] Turner G 2008 *A comparison of Limits to Growth with thirty years of reality* (Canberra: CSIRO Working Papers)

[231] Rockström J and Steffen W *et al* 2009 Planetary boundaries: exploring the safe operating space for humanity *Ecol. Soc.* **14** 32

[232] Steffen W, Richardson K and Rockström J *et al* 2015 Planetary boundaries: guiding human development on a changing planet *Science* **347** 736–47

[233] Steffen W, Crutzen P J and McNeill J R 2007 The anthropocene: are humans now overwhelming the great forces of nature *Ambio* **36** 614–21

[234] Steffen W and Broadgate W *et al* 2015 The trajectory of the anthropocene *Anthropocene Rev.* **2** 81–98

[235] von Weizsäcker E U and Wijkman A 2018 *Come On! Capitalism, Short-terminsm, Population and the Destruction of the Planet* (New York: Springer)

[236] Rapley J 2004 *Globalization and Inequality. Neoliberalism's Downward Spiral* (Boulder, CO: Lynne Rienner Publ)

[237] Cartwright M 2012 Yin and Yang. Definition. Ancient History Encyclopedia https://www.worldhistory.org/Yin_and_Yang/

[238] Frithjof C 1975 *The Tao of Physics. An Exploration of the Parallels Between Modern Physics an Eastern Mysticism* (Boulder: Shambhala)

[239] Riane E 1987 *The Chalice and the Blade* (San Francisco, CA: Harper Collins)

[240] Riane E 2007 *The Real Wealth of Nations: Creating a Caring Economics* (San Francisco, C: Berrett-Koehler)

[241] Wilkinson R and Picket K 2009 *The Spirit Level. Why Greater Equality Makes Societies Stronger* (New York: Bloomsbury)

www.ingramcontent.com/pod-product-compliance
Lightning Source LLC
Chambersburg PA
CBHW082114210326
41599CB00031B/5768